The Gap Symmetry and Fluctuations in High-T$_c$ Superconductors

NATO ASI Series

Advanced Science Institutes Series

A series presenting the results of activities sponsored by the NATO Science Committee, which aims at the dissemination of advanced scientific and technological knowledge, with a view to strengthening links between scientific communities.

The series is published by an international board of publishers in conjunction with the NATO Scientific Affairs Division

A	**Life Sciences**	Plenum Publishing Corporation
B	**Physics**	New York and London
C	**Mathematical**	Kluwer Academic Publishers
	and Physical Sciences	Dordrecht, Boston, and London
D	**Behavioral and Social Sciences**	
E	**Applied Sciences**	
F	**Computer and Systems Sciences**	Springer-Verlag
G	**Ecological Sciences**	Berlin, Heidelberg, New York, London,
H	**Cell Biology**	Paris, Tokyo, Hong Kong, and Barcelona
I	**Global Environmental Change**	

PARTNERSHIP SUB-SERIES

1. Disarmament Technologies	Kluwer Academic Publishers
2. Environment	Springer-Verlag
3. High Technology	Kluwer Academic Publishers
4. Science and Technology Policy	Kluwer Academic Publishers
5. Computer Networking	Kluwer Academic Publishers

The Partnership Sub-Series incorporates activities undertaken in collaboration with NATO's Cooperation Partners, the countries of the CIS and Central and Eastern Europe, in Priority Areas of concern to those countries.

Recent Volumes in this Series:

Series B: Physics

The Gap Symmetry and Fluctuations in High-T$_c$ Superconductors

Edited by

Julien Bok

Ecole Supérieure de Physique et Chimie Industrielles de Paris
Paris, France

Guy Deutscher

University of Tel Aviv
Ramat Aviv, Israel

Davor Pavuna

Ecole Polytechnique Federale de Lausanne
Lausanne, Switzerland

and

Stuart A. Wolf

Naval Research Laboratory
Washington, D.C.

Plenum Press
New York and London
Published in cooperation with NATO Scientific Affairs Division

Proceedings of a NATO Advanced Study Institute on
The Gap Symmetry and Fluctuations in High-T_c Superconductors,
held September 1 – 13, 1997,
in Cargèse, France

NATO-PCO-DATA BASE

The electronic index to the NATO ASI Series provides full bibliographical references (with keywords and/or abstracts) to about 50,000 contributions from international scientists published in all sections of the NATO ASI Series. Access to the NATO-PCO-DATA BASE is possible via a CD-ROM "NATO Science and Technology Disk" with user-friendly retrieval software in English, French, and German (©WTV GmbH and DATAWARE Technologies, Inc. 1989). The CD-ROM contains the AGARD Aerospace Database.

The CD-ROM can be ordered through any member of the Board of Publishers or through NATO-PCO, Overijse, Belgium.

Library of Congress Cataloging-in-Publication Data

```
The gap symmetry and fluctuations in high-Tc superconductors / edited
   by Julien Bok ... [et al.].
       p.   cm. -- (NATO ASI series.  Series B, Physics ; v. 371)
       "Proceedings of a NATO Advanced Study Institute on the Gap
   Symmetry and Fluctuations in High-Tc Superconductors held September
   1-13, 1997 in Cargèse, France"--T.p. verso.
       Includes bibliographical references and index.
       ISBN 0-306-45934-5
       1. High temperature superconductors--Congresses.  2. Energy gap
   (Physics)--Congresses.  3. Fluctuations (Physics)--Congresses.
   4. Electron configuration--Congresses.   I. Bok, Julien, 1933-     .
   II. NATO Advanced Study Institute on the Gap Symmetry and
   Fluctuations in High-Tc Superconductors (1997 : Cargèse, France)
   III. Series.
   QC911.98.H54G38   1998
   537.6'236--dc21                                        98-37191
                                                             CIP
```

ISBN 0-306-45934-5

© 1998 Plenum Press, New York
A Division of Plenum Publishing Corporation
233 Spring Street, New York, N.Y. 10013

http://www.plenum.com

10 9 8 7 6 5 4 3 2 1

Printed in the United States of America

PREFACE

Since the discovery in 1986 of high temperature superconductors by J. G. Bednorz and K. A. Müller, a considerable progress has been made and several important scientific problems have emerged. Within this NATO Advanced Study Institute our intention was to focus mainly on the controversial topic of the symmetry of the superconducting gap and given the very short coherence length, the role of fluctuations.

The Institute on 'The Gap Symmetry and Fluctuations in High- T_c Superconductors' took place in the "Institut d'Etudes Scientifiques de Cargèse" in Corsica, France, between 1 - 13 September 1997. The 110 participants from 18 countries (yet 30 nationalities) including 23 full time lecturers, have spent two memorable weeks in this charming Mediterranean resort.

All lecturers were asked to prepare pedagogical papers to clearly present the central physical idea behind specific model or experiment. The better understanding of physics of high temperature superconductivity is certainly needed to guide the development of applications of these materials in high and weak current devices.

The chosen topics were highly controversial, so the scientific discussions were often very lively. Even now, most controversies are not quite settled as can be seen from the contributions in this Proceedings. Due to the considerable progress in preparation of better quality materials we now have good single crystals hence the high precision experiments are reproducible and reliable. Therefore it was timely to compare the results of several recent experiments with interpretations of various theoretical models. More in-depth research will ultimately give complete answers and thorough understanding of this fascinating subject. At present, this volume should provide a useful insight into our contemporary understanding of physics of high- T_c oxides and will certainly raise a few more profound questions. This was -indeed- the goal of this NATO Institute.

The Institute was funded by the Scientific Division of NATO in Bruxelles. We have also benefited from a generous support of the president of the Corsican Community, Naval Research laboratory (Washington D. C.) and the CNRS. We would like to thank all these organizations for their support and the staff of the Cargèse Institute for their professional help in the organization. Last but not least, we would like to acknowledge contributions of Mme. Suzanne Beurel, who has conscientiously executed numerous financial and secretarial tasks of this Institute.

Julien Bok, Guy Deutscher, Davor Pavuna, and Stuart Wolf
Paris, December 1997

INTRODUCTION

In a field still controversial despite a large amount of work, J. Bok and his co-organizers were right, I think, to concentrate on relatively new experimental effects, on two rather different scales: the anisotropy of superconductive gap, telling about the microscopic pair coupling, and the thermal fluctuations neat T_c, including vortex melting, related to the quasi-2d structures studied. On the theoretical side, the models exposed span the range from strong to weak electron correlations, with an unusual emphasis on the latter.

To an ex-student of N. F. Mott since the late 40's, these conflicts about electron correlations seem hardly new, dating indeed as they do from the middle of the 30's for transition and rare earth metals and compounds

A few comments might be drawn from the background.

1. Since the parallel works of N. F. Mott and L. Landau, it has been well understood that, in crystalline metallic conductors, the electron correlations did not spoil the existence of electron excitations with a definite wave vector and an inverse life time increasing from zero at the Fermi Level. The systematic observation of Fermi surfaces, starting in the 50's, came then as no great surprise.

What became clear much later was that, in cases such as the heavy fermions (non magnetic) compounds, where strong correlations only affect a limited fraction of atoms, they play less on the form of the Fermi surface than on the effective mass for excitations from the Fermi level.

Magnetic effects are an obvious sign of electron correlations, but not necessarily strong ones; strong correlations effects are more visible if they affect every atom involved in conduction. Besides Mott isolation for one electron per atomic site, electron phonon couplings leading to Kohn anomalies at 4 k_F (instead of the usual 2 k_F) or atomic pair couplings in Peierls singlet states are, for instance, observed in a number of 1d organic conductors.

2. The "strong correlation" condition is bound up with the energy required for double ionisation on an atom, compared with the independent electrons band width. This criterion separates effectively weak correlations in transitional metals from effectively strong correlations of the lanthanides.

What is now, in these domains, the situation in quasi-2d superconductive oxides? Experimental Fermi surfaces and band structure studies from electron excitations seem to point to relatively modest correlations, as in transitional metals: There is a strong analogy with band structures as compared without correlations, and the peak of two holes excitation is contained within the boundaries of the conduction band. One might object that a Mott (magnetic) insulator is observed in antiferromagnetic undoped compounds. But nothing excludes that the insulating properties below T_N are due to an antiferromagnetic gap of delocalised electrons and, above T_N, to scattering by magnetic disorder (Anderson localisation). Only a very careful and extensive study of magnetic and thermal excitations could perhaps tell the difference. This delocalised picture is also not contradictory with the occurrence of a Peierle singlet coupling in quasi-1d ladder oxide suggests as I indeed believe, that in the oxide compounds as in the organic ones, one is near borderline cases.

With a band width larger than the CuO transfer energy and comparable to O and Cu correlation energies, one can possibly treat the correlation effects as weak, with effective intra-atomic Coulomb repulsions reduced by the S matrix effect characteristic of repulsive interactions (roughly speaking, $Vo/[1+n(E_F)Vo] < Vo$ if $Vo > 0$). This is the spirit of the treatments developed here by J. Bok and by D. Pines. However, most theoreticians, following T. M. Rice and here M. Cyrot, take the opposite view, by using a "tJ" model that neglects charge fluctuations on Cu atoms, owing to correlation repulsions assumed much larger than the band width.

If weak correlations apply, an approach "à la BCS" is the most natural as long as the coherence length is definitively larger than interatomic distances, which seems to be the case in the CuO planes; correlations from simple BCS can come from strong pair coupling and from an anisotropy related to quasi-2 dimensionality and to the nature of coupling.

In the standard isotropic and weak superconducting coupling BCS approach, the details of the band structure should play a large role, as first pointed out by J. Labbé for the A15 compounds. This has been developed by J. Bok, J. Labbé and their co-workers in the high-T_c oxides, where a strong van Hove anomaly is always present near the Fermi level. Because it is the Lorentzian tail of the anomaly that gives the main contribution to T_c, the van Hove peak can be somewhat away from the Fermi level or broadened by secondary effects without decreasing T_c very much. As the center of the peak contributes little to T_c, it does matter that the BCS condition of fast electrons does not apply here, nor that the effect of the peak is somewhat reduced in a strong coupling limit.

Such a BCS approach, in the weak coupling limit, is followed by J. Bok and by D. Pines, taking explicitly or implicitly the band structure into account, with its van Hove anomaly. These authors differ in the microscopic nature of the pair coupling, phonons for J. Bok and antiferromagnetic fluctuations for D. Pines. They have both considered corrections for anisotropy, leading in both cases to anisotropic superconductive gaps; and one of the purposes of this meeting was to compare experimental data with theoretical predictions that, it must be said, do not differ very widely.

Another point of convergence/divergence refers to the AF fluctuations. Both approaches would agree that, when they are strong, they must in quasi-2d structures, produce a well marked magnetic pseudogap when the density of states should be lowered, with two peaks of density of states at the edges of the pseudogap. In LaSrCu oxides, the AF fluctuations seem to vary in period with doping as expected for a nesting condition of the Fermi surface (indicating the importance of the band structure near the Fermi level). The Fermi level should then fall always in the middle of the magnetic pseudogap: This would decrease T_c the more the AF fluctuations are marked. The maximum observed for Tc with doping would then come, in both models, from a balance between the high T_c due to an approach of the van Hove anomaly and a lower T_c due to an increase in AF fluctuations.

In oxides such as YBaCuO, the AF fluctuations seem pinned down at the wave length observed with no doping. The Fermi level should then shift with doping across the psuedogap. One could even imagine that it reaches a peak of density of states bordering the pseudogap; the effect of the pseudogap would then be to increase T_c in that range of doping. This is however very unlikely, as then one could not see where the stabilising energy of the AF fluctuations would come from. AF fluctuations should indeed be more systematically studied for doping higher then the maximum of T_c, as it is difficult to imagine a large range of doping where AF fluctuations would be weak enough to give no appreciable pseudogap, but still strong enough to provide a sizable superconductive coupling.

Jacques Friedel, Paris

CONTENTS

The Gap Symmetry and Fluctuations in High-T_c Superconductors

INTRODUCTION TO HIGH TEMPERATURE SUPERCONDUCTING OXIDES

Davor Pavuna

Department of Physics
Ecole Polytechnique Federale de Lausanne
CH - 1015 Lausanne, Switzerland

1. INTRODUCTION

1.1 Conventional superconductors

Although the main topic of this Advanced Study Institute is the symmetry of the gap and fluctuations in high temperature superconducting oxides, any pedagogical introduction has to include the conventional superconductors[1]. Namely, some of the most relevant concepts that are discussed in this Institute, like the BCS or the Landau-Fermi liquid theory[2], were developed in the 1950's in order to understand the superconducting[3] and the normal state of these conventional metals[4]. Their critical temperatures are mostly in the liquid helium range and their coherence lengths (and electron mean free paths) are rather long: in aluminium, for example, $T_c \approx 1K$ and $\xi_0 \approx 1\mu m$ (see Table 1). Therefore in this section we will briefly introduce most interesting classes of conventional superconductors. For more detailed information on properties of any of these solids the reader should consult many excellent textbooks readily available in most libraries[1-4].

Metallic elements are mostly superconductors. Their critical temperatures are typically of the order of a few Kelvin. Among metals, niobium exhibits the highest critical temperature of all the pure elements, $T_c = 9.2$ K. Noble metals, copper, silver and gold, and alkaline metals, sodium and potassium, all of which are excellent conductors of electricity at ambient temperatures, are not superconductors down to very low temperatures (if at all). Magnetic metallic elements do not exhibit superconductivity.

Hydrogen, the simplest of all elements, is in a gaseous state at normal pressures. However, there are conjectures that under tremendous pressure of ~2-3 Mbar hydrogen becomes a dense solid metallic element and superconductor with (predicted) T_c ~240 K. Helium, the next simplest of elements, is not a superconductor, but rather a superfluid below 2.2 K. The best known semiconductors, Si and Ge, become superconductors under a pressure of ~2 kbar with T_c = 7 and 5.3 K respectively. Other elements that become superconductors under pressure include P, As, Se, Y, Sb, Te, Cs, Ba, Bi, Ce and U.

Binary Alloys and Compounds. In most alloys and compounds the critical temperatures are usually somewhat higher than in elemental metals (see Table 1). Nb compounds, like Nb-Ti, are of technological interest. While the maximum current density that one can pass through the 'standard' water-cooled copper wire at 300K is about 2000

The Gap Symmetry and Fluctuations in High - T_c Superconductors
Edited by Bok *et al.*, Plenum Press, New York, 1998

1

Acm^{-2}, one can pass very high current densities of up to 10^8 Acm^{-2} in high magnetic fields of 10 Tesla at 4.2K through a wire made of NbTi without destroying superconductivity. This enables the construction of powerful magnets that provide a basis for a range of large scale applications, like energy storage or levitation trains. To find further examples of such applications see the article by Z.X. Zhao in this volume.

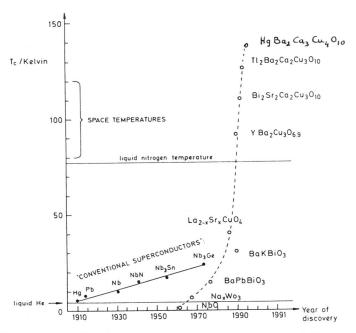

Figure 1: The evolution of critical temperatures since the discovery of superconductivity.

A-15 Intermetallic Compounds. Among the intermetallic superconductors, the most favourable group is the one based on the A_3B compound. In the cubic A-15 structure six binary compounds have T_c over 17K. The highest T_c close to 23K is obtained in Nb3Ge stabilized by traces of oxygen or aluminum; it exhibits the upper critical field of 38 T.

Chevrel Phases. In 1971, Chevrel and coworkers discovered a new class of ternary molybdenum chalcogenides of the type $M_xMo_6X_8$., where M stands for a large number of metals and rare earths (RE) and X for the chalcogens: S, Se or Te. For example, the $PbMo_6S_8$ phase exhibits the critical temperature of 'only' 15 K but with relatively high with upper critical field, B_{c2} of 60 T.

Organic superconductors. The organic superconductors are relatively novel group of materials. The first organic superconductor $[TMTSF]_2ClO_4$, where TMTSF denotes tetramethyltetraselenafulvaline, was discovered by K. Berchgaard and D. Jerome in 1980 and had a T_c of 1 K. Subsequent developments have led to higher T_c materials exhibiting a variety of novel electronic and superconducting properties. These systems were characterized by their nearly one-dimensional properties and by low carrier concentration.

Later, a new series of materials with a two dimensional character was discovered: the [BEDT-TTF]$_2$X, where BEDT-TTF denotes bis-ethylenedithio-tetrathiafulvaline. The k modification of the X = Cu[NCS]$_2$ compounds has the highest T$_c$ of 10 K.

Table 1: The critical temperature T$_c$ and upper critical fields B$_{c2}$ for selected conventional superconductors (* the thermodynamic critical field B$_c$ is given for type-I superconductors, Al and Pb):

Material	T$_c$/K	B$_{c2}$/T
Metallic Elements:		
Al	1	$\approx 10^{-2}$*
Pb	7	0.8*
Nb	9	1.9
Binary compounds:		
Nb$_3$Sn	18	24
Nb$_3$Ge	23	38
Organic phases:		
k-(BEDT-TTF)$_2$Cu(NCS)$_2$	12	20
Chevrel phases:		
PbMo$_6$S$_8$	15	60

Fullerenes. It was discovered in 1991 that a new form of solid carbon, when doped, becomes superconducting[5]. The C$_{60}$ molecule assumes the structure of the soccer ball with carbon atoms on a truncated icosahedron and is called buckminsterfullerene. The C$_{60}$ clusters crystallize and form an fcc solid. One can dope such a molecular solid with alkaline metals and observe superconductivity: T$_c$ = 18K and 30K for K$_3$C$_{60}$ and Rb$_3$C$_{60}$ respectively. By comparison, in graphite intercalated compounds of the same elements, measured T$_c$-s are only 0.5K and 0.03K for C$_8$K and C$_8$Rb, respectively.

Slow electron compounds - better known as heavy fermion systems show relatively low critical temperatures but exhibit remarkably rich (and often very anomalous) physical properties. Their properties are reviewed elsewhere in this volume by H.R. Ott.

2. HIGH-T$_c$ CUPRATES

Until 1986, most of superconducting compounds studied were metals and alloys that we briefly discussed in the first section. However, some oxide superconductors were known for decades (see Figure 1), but their transition temperatures were rather small. This was mainly due to a low number of carriers in the metallic state. Two known exceptions were LiTi$_2$O$_4$ and BaPbBiO$_3$ with critical temperatures of ~13 K. This was unusual as their densities of carriers were also very small. The breakthrough came in 1986 when Georg Bednorz and Alex Müller (IBM-Zürich), in their systematic search for new superconductors in metallic Ni- and Cu- oxides, observed an evidence for resistive superconducting transition (with an onset at ~ 30K) in a fraction of their LaBaCuO sample. This lead to the discovery of La$_{2-x}$Sr$_x$CuO$_4$ oxide with T$_c$ of ~38 K and subsequently to the widely publicized high-T_c revolution. In what follows we briefly discuss some of the characteristic physical properties of the $YBa_2Cu_3O_{7-\delta}$ and other high-T_c superconductors.

As we shall see throughthout this book the normal state of high-T_c cuprates is rather anomalous, and T_c-s are in the liquid nitrogen range: $\approx 10^2$ K, while the coherence lengths are very short, $\xi_o \approx$ few Å. While in conventional superconductors the number of Cooper pairs per coherence volume is of the order of 10^6, it is only ≈ 10 in case of high-T_c cuprates. This immediately shows the importance of critical fluctuations in high-T_c oxides that are discussed by several authors in this volume. In what follows we illustrate the characteristic properties by describing the most widely studied cuprate, $YBa_2Cu_3O_{7-\delta}$, firstly from materials scientist's and subsequently from the physicist's point of view.

2.1. Materials characteristics

i) **Highly *anisotropic, layered* structures.** Except for one material (isotropic, cubic $Ba_{1-x}K_xBiO_3$), all high-T_c superconducting oxides are layered *cuprate* perovskites. Recently discovered *non-cuprate* perovskite, Sr_2RuO_4 is important for comparisons with similar perovskites (like $La_{2-x}Sr_xCuO_4$), but it is a low-T_c ($T_c \sim 1K$) superconductor[6]. Obviously one of the important characteristics of all cuprates is the presence of CuO_2 layers which dominate most properties. If we look at the schematic structure of $YBa_2Cu_3O_6$ presented in Figure 2, we immediately notice that it is *anisotropic*: The unit cell is developed from that of a tetragonal perovskite tripled along the c-axis and it consists of a sequence of copper-oxygen *layers*.

The dimensions of the unit cell are approximately ~12 Å and ~4 Å in the c- and a- or b-axis directions respectively. An Yttrium ion in the center and barium ions above and below the copper-oxygen planes provide the vertical 'spine' of this layered cuprate structure (see Figure 2). The fact that the unit cell consists of *layers of copper oxides* is of great importance for understanding of the properties of these layered structures and is discussed in more detail in the next section.

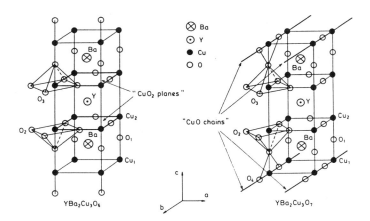

Figure 2: Structure of a) $YBa_2Cu_3O_6$ insulator (x=0); b) $YBa_2Cu_3O_{6.9}$ superconductor (x=0.9). Note the presence of oxygen ions in the 'chains' of $YBa_2Cu_3O_{6.9}$ superconductor.

ii) *Metallic* **Oxides.** The second important feature of these oxides is their (anomalous) metallic properties. While most oxides are insulating materials, HTSC oxides exhibit metallic behavior. The room temperature conductivities in the *a*- or *b*-axis direction of the cuprate crystal are of the same order of magnitude as the conductivities of some disordered metallic alloys. The conductivity is metallic mainly in the CuO_2 planes; perpendicular to these planes, the conductivity is much smaller. However, the normal state properties of these ionic metals are rather anomalous (as is amply illustrated in every article in this volume) and, as we will see, in the so-called-underdoped regime and under very high magnetic fields these superconductors exhibit phase transition directly into the insulating state ('Bobinger anomaly')[7].

iii) *Ceramic* Materials. The initially discovered materials, $La_{2-x}Sr_xCuO_4$ (from now on referred to as LSCO) and $YBa_2Cu_3O_7$ (YBCO), were synthesized by their discoverers as ceramic pellets. Such pellets look very much like pieces of ceramics used for any tea cup (apart from the difference in colour). The difference is of course that our tea cups do not superconduct while black pieces of YBCO indeed do (below ~92 K). As typical ceramics, high-T_c superconducting (HTSC) oxides also contain grains, grain boundaries, twinns, voids and other imperfections. Even some of the best thin films may consist of grains a few microns in diameter.

We emphasize that even the best single crystals of HTSC oxides often contain various defects and imperfections like oxygen vacancies, twins, impurities ... These imperfections are not only relevant to their physical properties but possibly even essential for their overall thermodynamic (meta)stability. It may well turn out that various imperfections found in HTSC crystals are intrinsic to these materials. Moreover, it is important to understand that the materials science of HTSC oxides is a non-trivial pursuit and that the understanding of phase diagrams, crystal chemistry, preparation and stability of these oxides requires an in-depth study and often hands-on experience in the laboratory. The advancement of our understanding of physics and appearance of applications depend very much on the advancements in materials research. As this introduction cannot cover more than few very basic notions on HTSC oxides, we encourage the reader to consult many more specialized books; few of more physics oriented ones are given in our references[1-4].

2.2. Characteristic physical properties

i) Superconductors with $T_c \sim 10^2$ K. This statement hardly needs much explanation, but we nevertheless emphasize the order of magnitude of T_c. Note that in A-15 compounds, the highest T_c (for Nb_3Ge) was 'only' 23 K. In general, the critical temperature corresponds to the binding energy $\sim k_B T_c$ needed to hold Cooper pairs together in the superconducting state. The fact that $T_c \sim 10^2$ K i.e. ~10 meV, as compared with <1meV in conventional superconductors, poses an appealing but profound challenge to theorists interested in the microscopic mechanism of high-T_c superconductivity.

ii) Quasi-two-dimensional doped insulators. As it stands, the schematic structure of $YBa_2Cu_3O_6$, given in Figure 2a, represents an *insulator*. It has to be doped to gradually become a metallic conductor and a superconductor below some critical temperature. The *doping* is achieved by adding additional oxygen which forms CuO 'chains'. These oxygen ions attract electrons from the CuO_2 planes which therefore become metallic.

Note that the correct formula for YBCO material is therefore: $YBa_2Cu_3O_{6+x}$, where x corresponds to partial oxygen content (see Figure 3):

For $0.0 < x < 0.4$, $YBa_2Cu_3O_{6+x}$ is an *insulator*,

for $\sim 0.4 < x < 1.0$, $YBa_2Cu_3O_{6+x}$ is a *superconductor*.

The oxygen content can be changed reversibly from 6.0 to 7.0 simply by pumping oxygen in and out of the parallel chains of CuO running along the b-axis of Figure 2. $YBa_2Cu_3O_6$ is an insulating antiferromagnet. Increasing the oxygen from $O_{6.4}$ makes the crystal metallic, nonmagnetic and superconducting: $T_c = 0 +$ first for $O_{6.64}$. General schematic phase diagram is given in Figure 3 as a function of the concentration of carriers in the CuO_2 plane. Our present understanding of the electronic phase diagram of HTSC cuprates is discussed by most authors in this volume, while the nature of insulator-metal transition is discussed by M. Cyrot .

iii) very short coherence length: $\xi \sim 1nm$. If we recall the BCS-derived formula[1-4], $\xi_0 \sim v_F/T_c$, we can immediately expect somewhat shorter coherence lengths in HTSC oxides due to their 10 times higher T_c's. However, due to the low density of carriers per unit cell the Fermi velocity in these ionic metals seems lower than in normal metals. This results in a *very short* coherence length, $\xi \sim 1nm$, which is comparable to the size of the unit cell, and it has profound consequences for the physics of HTSC oxides.

Actually, the coherence length is different for different crystallographic directions and it was experimentally found in YBa$_2$Cu$_3$O$_7$ that ξ_{ab} and ξ_c are ~15 Å and ~4 Å respectively.

Note that ξ_c is roughly equal to the interlayer distance and shorter than the corresponding unit cell length, which clearly poses some conceptual problems. These remarkably short coherence lengths dominate all material-related properties and cause a rather complex mixed state. Short coherence length also implies that HTSC oxides are *type-II superconductors* with very high upper critical fields B_{c2}.

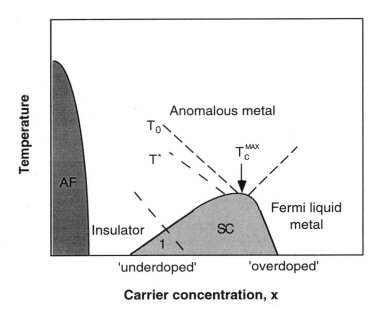

Figure 3. Schematic phase diagram of cuprate superconductors. Various authors give different name or significance to various observed lines: the 'pseudogap', T^*, 'spin-correlations', T_0, or the metallic Fermi-liquid region in the overdoped regime. Note the anomalous behaviour in the region 1 where the superconductor directly transits into the insulating state in high magnetic field[7].

3. CRYSTAL CHEMISTRY OF YBa$_2$Cu$_3$O$_{6+x}$

Crystallographers classify the structure of these oxides as of the perovskite type. The name perovskite has no special scientific meaning: it is only a label for a family of structures whose generic class is represented by SrTiO$_3$ (or 'oxygen-filled' YBa$_2$Cu$_3$O$_7$) and a derived one by K$_2$NiF$_4$ (La$_2$CuO$_4$ structure). The actual name, Perovskite, is a name of a village in Russia where over the years the crystallographers have found many oxides with similar structures (but of the non-superconducting kind).

The YBCO compound is thermodynamically stable as a tetragonal insulator. Therefore, its correct formula reads $YBa_2Cu_3O_{6+x}$ and its structure and properties depend on the exact concentration of oxygen: it becomes superconductor for $x > 0.4$. More than ~50% of all papers on HTSC materials in the past decade were devoted to studies of this compound. However, the BSCCO compounds, that can be easily cleaved (in ARPES experiments), were so far more useful in the studies of the symmetry of the gap.

If we carefully observe the unit crystallographic structure in Figure 2 we notice that the structure consists of a sequence of oxide layers perpendicular to the c-axis as follows:

- Cu-O layer which has two oxygen vacancies as compared with the 'fully oxidized' YBCO perovskite. Cu(1) site in this oxygen layer has coordination 4 and is surrounded by 4 oxygen ions. In $YBa_2Cu_3O_7$ this is the plane of CuO 'chains'.

- Ba-O layer,

- Cu-O layer in which Cu(2) has a coordination number 5 and is surrounded by 5 oxygen ions which form a polyhedra. This is the plane which we call CuO_2 plane.

- Yttrium layer which has 4 oxygen vacancies as compared with the 'fully oxidized' $YBa_2Cu_3O_7$ perovskite. The rest of the structure is symmetric with respect to Yttrium ion which can be replaced with a whole series of rare earths without losing superconducting properties (exceptions are: Pr and Yb).

Copper can be found in two different sites: Cu(1) within CuO_4 'squares' and Cu(2) within a square-based pyramid, CuO_5. The separation of Yttrium ions gives the structure its two-dimensional character.

Table 2: Most representative HTSC compounds. The index n refers to the number of CuO_2 superconducting layers within a given crystallographic structure. m refers to the number of 'chains' in the structure; m=1.5 corresponds to the case of alternating 'chains':

Compound	$T_c(K)$
$La_{2-x}M_xCuO_{4-y}$ M = Ba, Sr, Ca; x ~ 0.15, y small	38
$Nd_{2-x}Ce_xCuO_{4-y}$ (electron doped)	30
$Ba_{1-x}K_xBiO_3$ (isotropic, cubic)	30
$R_1Ba_2Cu_{2+m}O_{6+m}$ R: Y,La, Nd, Sm, Eu, Ho, Er, Tm, Lu m=1 ('123') m=1.5 ('247') m=2 ('124')	 92 95 82
$Bi_2Sr_2Ca_{n-1}Cu_nO_{2n+4}$ n = 1 ('2201') n = 2 ('2212') n = 3 ('2223')	 ~10 85 110
$Tl_2Ba_2Ca_{n-1}Cu_nO_{2n+4}$ n = 1 ('2201') n = 2 ('2212') n = 3 ('2223')	 80 100 125
$HgBa_2Ca_3Cu_4O_{10}$	133

Numerous diffraction studies indicate that most oxygen vacancies occur within planes made of CuO 'chains' rather than within the pyramids. In $YBa_2Cu_3O_6$ the chains along the b-axis are oxygen depleted and Cu(1) coordination is only 2 (only two neighboring oxygen ions). By increasing the Oxygen concentration one gradually dopes the ab-plane with charge carriers (holes) and it eventually reaches the $YBa_2Cu_3O_7$ composition in which there are no oxygen vacancies. Note that very detailed studies indicate that the maximum in T_c is reached for x~0.93 (T_c = 92 K) and that for x=1.0 the critical temperature is 'somewhat lower', $T_c \approx 90K$. There is also evidence that the best conduction channel in normal state is along chains in b-axis direction (see Figure 5).

The optimally doped compound, $YBa_2Cu_3O_{6.9}$, is usually referred to as YBCO or simply as '123'. Its 'average' structure is orthorhombic (remember 'O_6' is tetragonal) but the real material is usually full of various 'defects'. The charge-neutral formula for $YBa_2Cu_3O_7$ can be written as $YBa_2(Cu^{2+})_2(Cu^{3+})(O^{-2})_7$ or as $YBa_2(Cu^{+2})_3(O^{2-})_6(O^{-})$. The exact orthorhombic (quasi-tetragonal) cell dimensions of $YBa_2Cu_3O_7$ are: a = 3.88 Å, b = 3.84 Å, c = 11.63 Å, with a cell volume ~173 Å^3.

In $Tl_2Ba_2Ca_{n-1}Cu_nO_{2n+4}$ and the corresponding Bi-compounds, T_c increases with the number of layers of CuO_2. It has been suggested that T_c could increase further for higher n but the compound n=4 seems to have about the same T_c as n=3. The crystallographic structure shows stacking of planes; in the '2212': TlO, TlO, BaO, CuO_2, Ca, CuO_2, BaO.

We note in Table 2 that there is a series of compounds described by the formula $RBa_2Cu_3O_7$, where R represents one of the lanthanide elements that can replace Yttrium in the original '123' structure of Figure 2. This means that '123' structure exhibits superconductivity with almost any of the lanthanides. However, an exception is the $PrBa_2Cu_3O_7$ (PrBCO) compound which does not exhibit superconducting properties. This compound effectively behaves as a tunnel barrier in the temperature range where other compounds like YBCO exhibit superconductivity. PrBCO has a matching crystalline structure with YBCO and can be grown between the two $YBa_2Cu_3O_7$ superconducting layers resulting in an artificial 'in-grown' barrier, which is obviously of great potential interest for tunneling devices. Such artificial structures are of importance for tunnel junction technology or for artificial 'construction' of new oxide structures[13].

4. SIMPLE 'LEGO' MODEL FOR LAYERED OXIDES

We have mentioned in the previous section that the (super)conductivity essentially takes place within quasi-two-dimensional CuO_2 planes. In case of YBCO the unit cell can be schematically represented as a layered structure that consists of two CuO_2 planes separated by Y site. Between these bi-layers are interlayer regions which, in the case of $YBa_2Cu_3O_7$, correspond to the CuO chains (see Figure 2).

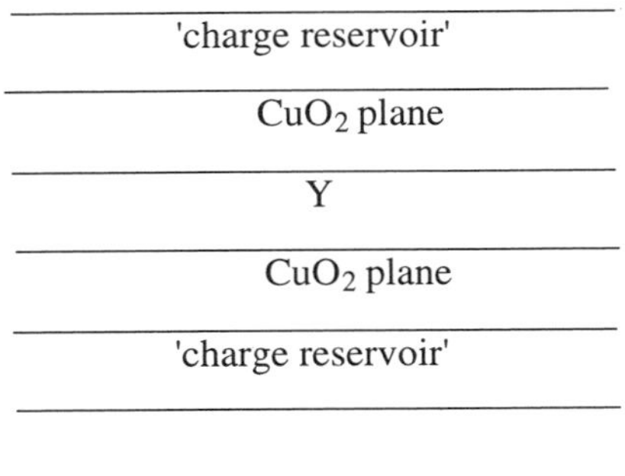

'charge reservoir'
CuO_2 plane
Y
CuO_2 plane
'charge reservoir'

Figure 4: A model unit of layered cuprate, in this case the $YBa_2Cu_3O_{7-\delta}$.

In the undoped compound, the Cu ions (2+) in this plane are in a d^9 electronic configuration and are antiferromagnetically coupled to other neighboring copper ions, and the plane is insulating. The Cu-O chains can be considered as a 'charge-reservoir' which is needed to transfer the charge into CuO_2 planes. This enables one to consider the HTSC superconductor as CuO_2 planes separated by a charge reservoir (see Figure 4)

Charge carriers are added by doping: basically by substituting divalent atoms for trivalent ones (like Sr^{2+} for La^{3+} in $La_{2-x}Sr_xCuO_4$) or by adding oxygen to $YBa_2Cu_3O_6$, which enters the compound as O^{2-} and forms CuO chains. To maintain the charge balance, electrons are removed from the copper oxide planes and the remaining holes ('missing electrons') are mobile (hence conduction) and they form the Cooper pairs below T_c (hence superconductivity).

So we can intuitively understand that adding charge carriers from the 'reservoir' into the CuO_2 planes gradually increases the conductivity within the ab-plane. While LBCO has only one 'doped unit-layer' (hence T_c of 'only' ~38 K), the $YBa_2Cu_3O_7$ has two units separated by Y site while Bi- or Tl- oxides have 2 or 3 (see Table 2). In these oxides the role of charge reservoir is evidently played by some other layer like Bi-0 rather than CuO as in YBCO. It is interesting to note that, while the conductivity of the CuO_2 planes increases by adding carriers, the superconductivity seems to increase first, reach a maximum for some 'optimal' doping, then decrease and finally vanish (in the 'overdoped regime') for about 0.3 holes per Cu (see also the phase diagram in Fig.3). There is always an optimal doping point for the the CuO_2 plane which gives the highest T_c.

5. THE NORMAL STATE

One of the unusual features of HTSC compounds is the fact that they are very close to an insulating phase. The insulator has to be slightly doped in order to obtain the superconducting state. If the doping is increased, superconductivity ultimately disappears (see Fig. 3). This rather strange behaviour poses the question concerning the physical properties above the critical temperature. These are particularly different in the underdoped regime. Only in the overdoped cuprates the normal state properties resemble those of the conventional Fermi liquid metal.

5.1. Anisotropy and Resistivity

The anisotropy of HTSC oxides is illustrated in Figure 5. where we show the temperature dependence of resistivity of the '123' compound. Electrical resistivity of the ab-plane decreases with temperature and is slightly anisotropic: the conductivity seems to be highest in the b-axis direction along the quasi-one-dimmensional CuO 'chains'.

The in plane resistivity, ρ_{ab} ~ 50 $\mu\Omega$cm at ~100 K and is proportional to the temperature for T > T_c. Actually one of the characteristics of the anomalous normal state of cuprates is the linear resistivity which in the case of LSCO extends[8] up to 1100K. Such behaviour is not readily observed in conventional metals. The departure from linearity in underdoped samples that occurs at some characteristic temperature, T^*, somewhat above the T_c, was one of the first indications of the existance of the, so called, 'pseudo-gap'. The 'pseudo-gap' concept is discussed in detail by several authors in this volume.

ρ_{ab} passes through zero in the best, optimally doped, single crystals and films, and tends to be linear at least up to room temperature. The temperature dependence along c-axis can be metallic for optimally doped crystals (x~6.93). Large anisotropy of cuprates makes an intrinsic value of ρ_c difficult to measure. Several measurements indicate that in $YBa_2Cu_3O_{6.9}$ ρ_c is ~100 times larger than ρ_{ab} at 100K. The problem of c-axis conductivity is a profound conceptual challenge to the theory[12] and is discussed at length by C. Gough elsewhere in this book.

Here we mainly discuss the properties of YBCO so it is useful to note that LSCO exhibits somewhat lower anisotropy, while Bi-, Tl- and Hg- compounds are much more anisotropic than YBCO. The measured electrical properties of Bi-, Tl- and Hg- compounds are strongly anisotropic in directions parallel and perpendicular to the layers; the ratio is of the order 10^5, i.e., roughly thousand times higher than in 'YBCO. Therefore it is unrealistic in any analysis to neglect the anisotropy of these oxides.

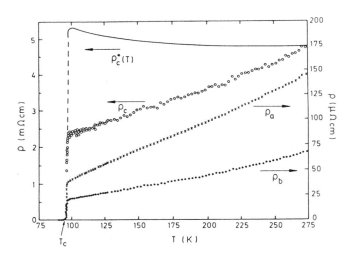

Figure 5: Anisotropy of the resistivity[1,9] of $YBa_2Cu_3O_{7-\delta}$ For c-axis resistivity both temperature dependences, $\rho_c(T)$ and $\rho^*_c(T)$, have been measured experimentally for overdoped and underdoped samples, respectively.

5.2. The Hall Number

The Hall number is defined as the inverse Hall constant $1/R_H e$ normalized to unit volume. If the conduction is due either to holes or electrons, as in simple metals, the Hall number gives an estimate of the carrier concentration per unit volume. In YBCO, with the magnetic field parallel to the c-axis and the currents in the ab-plane, the Hall number gives at 300 K a carrier concentration of $\sim 7 \times 10^{21}$ cm^{-3}. This is rather low concentration for a metal. However, if one assumes that the carriers are mostly confined to the CuO_2 planes, then the actual carrier concentration is close to that of copper.

The sign of the Hall effect of YBCO is positive. This means that the carriers are holes in the CuO_2 planes. When doping increases in $La_{2-x}Sr_xCuO_4$, superconductivity disappears and the Hall effect changes sign to negative. It has been discovered that in electron doped HTCS materials the Hall effect is negative: for example, in $Nd_{2-x}Ce_xCuO_4$. The cerium which can be replaced by thorium is 4+, i.e., gives one more electron that lanthanum or neodinium. This electron is believed to go to the CuO_2 planes which become metallic. Thus there are two classes of HTCS oxides: the "p" type where the Hall effect is positive and conductivity in the CuO_2 planes is due to holes, and the "n" type where the Hall effect is negative and conductivity is due to electrons. The properties of electron doped cuprates are discussed at length by P. Fournier et al. Due to highly anomalous normal state of cuprates some authors[12] argue that the relevant quantity that one should study in transport is the Hall angle rather than the usual Hall coefficient, as in conventional metals.

Table 3: Temperature dependences of several normal state properties, compared to behaviour expected within the Fermi liquid model. Experimentally observed dependences presented in the last column are reported for optimally doped samples[10].

Property	Fermi liquid model	HTSC oxides
Susceptibility, χ	constant	constant
Electronic specific heat, C_V(el)	$\sim T + T^3 \ln T$	$\sim T$
DC resistivity, ρ	$\sim T^2$	$\sim T$
Hall coefficient, R_H	constant	large & T-dep.
NMR relaxation rate (Y & O), $1/T_1 T$	constant	constant
NMR relaxation rate (Cu), $1/T_1 T$	constant	Non-Korringa like

5.3. Other Anomalous Normal State Properties

Most normal state properties of cuprates exhibit anomalous behaviour[10], like the NMR, for example. For brevity we do not discuss them in this introductory paper but rather encourage the reader to study other lectures in this book that all emphasize some aspects of this challenging problem of condensed matter physics.

6. THE SUPERCONDUCTING STATE

6.1. Characteristic Energy - The Gap

This important quantity, is often difficult to measure in HTCS materials and this whole Institute is dedicated to the study of the symmetry of the gap. One generally measures the gap in units of $k_B T_c$. The BCS model value is 3.5. The reported values for the HTCS cuprates vary between different experimental techniques, but seem to be higher, in the the range between 5 and 8 (or even 11 in some cases) with the characteristic energy of the order of ~50 meV. In Table 4 we present the expected behaviour of characteristic properties for two gap symmetries: the s-wave (as is the case of the conventional BCS-alike superconductors) and for the 'pure' $d_{x^2-y^2}$-wave.

In HTSC the gap is anisotropic. Excitations in the ab-plane have a different gap energy than excitations along the c-axis. Despite the difficulties in measuring the gap, one can say that the gap in the plane appears larger than along the c-axis. As this is the main topic of this book, there are several contributions that treat this problem in great detail.

6.2. Coherence Lengths

We have already seen that HTSC oxides with CuO_2 layers are built of superconducting layers separated by dielectric or weakly metallic barriers, all on an atomic scale. Moreover, we have strongly emphasized that a characteristic of these layered crystals is the short coherence length: in YBCO, $\xi_c(0) \sim 4\text{Å}$ and $\xi_{ab}(0) \sim 15\text{Å}$. $\xi_c(0)$ is practically equal to the spacing between adjacent conducting CuO_2 planes, which supports the two-dimensional (2D) model with weak coupling in the third dimension between planes (as in the simple model of Figure 4). In Table 5 we list some characteristic coherence lengths. Note that in the much more anisotropic BSSCO (and Tl- and Hg- cuprates) with T_c = 110K, the estimated $\xi_c(0)$ is ~ 1-2 Å (which is remarkably small) while $\xi_{ab}(0)$ ~13 Å.

Table 4: The expected behaviour of some characteristic properties[10] for two distinct gap symmetries: the s-wave and the d_{x2-y2} wave.

Property	s-wave	d-wave
Penetration depth, λ	~exp(-Δ/kT)	~T
Electronic specific heat, C_V(el)	~exp(-Δ/kT)	~T^2
Conductivity, σ_1	~exp(-Δ/kT)	~T^2
Coherence peak	+	+/-
Phase shift over 90°	0	π

6.3. Penetration Depths

We also list in Table 5 the other relevant length of superconductivity, the penetration depth, λ. Given a typical value of several thousand Ångströms, HTSC oxides clearly belong to extreme type-II superconductors with $\kappa = \lambda/\xi \gg 1$. The large value of λ is related to the relatively low number of carriers. Measurements of this quantity and its temperature dependence are reviewed by W. Hardy elsewhere in this volume.

Table 5: Critical temperature, penetration depth λ_i, coherence length ξ_i, and the second critical field B^i_{c2} of three HTSC oxides[1] (i = ab or c).

Cuprate	T_c(K)	λ_{ab}(Å)	λ_c(Å)	ξ_{ab}(Å)	ξ_c(Å)	B_{c2}/T	B_{c1}/T
LBCO	38	800	4500	35	7	80	15
YBCO	94	1500	7000	15	4	150	40
BSCCO	110	2500	9000	13	2	260	32

6.4. Critical Fields

As we show in Table 5, the penetration depth of YBCO, estimated from magnetization measurements, is λ_a ~1400 Å, while the coherence length is ξ_a ~15Å, estimated from critical field measurements. Thus the Ginzburg-Landau parameter, $\kappa_{ab} = \lambda_a/\xi_a$ ~50 - 100, amply satisfies the criterion $\kappa > 1$ for type-II behaviour in this compound.
Given the aforementioned crystal structures, the magnetic properties are also be expected to be highly anisotropic. Indeed, the measurements on single crystals of the LBCO and YBCO systems do show the anisotropy of the critical field.

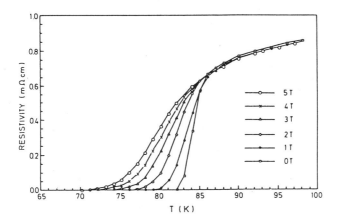

Figure 6. Broadening of the resistive transition in $YBa_2Cu_3O_7$ in the magnetic field[1].

Table 5 gives values obtained from resistivity measurements. The estimates of B_{c2} from resistivity broadening, such as shown in Figure 6, are inevitably somewhat arbitrary, since the field plotted is that for a stated fraction of the restored normal-state resistivity. It is particularly open to criticism for high-T_c superconductors, in which the resistance-temperature transition curves are especially sensitive to magnetic fields and do not behave as in conventional superconductors.

The dissipative mechanism in a rather complex mixed state in HTSC cuprates arises from activated flux line motion driven by the Lorentz force. One expects an important effect due to thermal fluctuations of the flux line lattice. Indeed, as is discussed in several papers in this Institute, the characteristic energy involved in these fluctuations is $\mu_0 H_c^2 \xi^3$ which can be of the order of $k_B T$, as B_c is of the order of ~1 Tesla and ξ is very small. These effects are often attributed to a "giant" flux creep in these layered structures.

7. CONCLUDING REMARKS

In this introductory article we have seen that the HTSC cuprates are:
- highly anisotropic, layered oxides dominated by the properties of CuO_2 planes.
- quasi-two-dimensional doped insulators with an anomalous normal state.
- superconductors with T_c ~100 K, and
- extreme type-II superconductors with a very short coherence lengths, ξ~few Å and large penetration depths, λ ~ 2000 Å.

The anomalous normal state of these layered solids and their electronic phase diagram cannot be fully understood in terms of the conventional concepts like Landau-Fermi liquid model. Moreover, given high anisotropy and very short coherence length, the superconducting state is also complex due to the presence of the pancake-type vortices. In the articles of other lecturers of this NATO Institute, as well as in some recent publications[11-13], the reader will easily find further evidence for strange behaviour and challenging physics of these fascinating superconducting solids.

Acknowledgements. This introductory article has gradually evolved from a textbook written with Michel Cyrot (ref.1) whom I gratefully acknowledge. I also thank all my colleagues at the EPFL and Julien Bok and Guy Deutscher for critical reading of the manuscript, many stimulating discussions and numerous wise suggestions.

REFERENCES

1. M. Cyrot and D. Pavuna, *Introduction to Superconductivity and High-T_c Materials*, World Scientific, London, Singapore, New Jersey 1992
2. P.G. de Gennes: *Superconductivity Of Metals And Alloys*, Benjamin, New York, 1966
   M. Tinkham: *Introduction To Superconductivity*, McGraw-Hill, New York 1996
3. J. R. Schrieffer: *Theory of Superconductivity*, Benjamin/Cunnings, 1983.
   V. L. Ginzburg and D. A. Kirzhnits (eds.): *High-Temperature Superconductivity*, Consultants Bureau, London 1982
4. R. D. Parks (ed.),*Superconductivity* vols.1 and 2, Marcel-Dekker, New York (1969), N.M. Plakida, *High Temperature Superconductivity:,* Springer Verlag, Berlin (1995)
5. R.C. Haddon et al , Nature **350**, 320 (1991)
6. Y. Maeno et al, Nature **372**, 532 (1994)
7. G. Boebinger et al, Phys. Rev. Lett. **77**, 5417 (1996)
8. M. Gurvitch and A.T. Fiory, Phys. Rev. Lett. **59** (12) 1337 (1987), S. Martin et al, Phys Rev Lett **60**, 2194 (1988)
9. T. A. Friedmann et al, Phys Rev. B**42**, 6217 (1990)
10. J. Feenstra, *Low Energy Electrodynamics of High-Tc Superconductors*, D. Sci Thesis, Univ. of Groningen (1997)
11. G. Zhao, M.B.Hunt, H. Keller, K.A. Müller, Nature **385**, 236 (1997)
12. P. W. Anderson, *'The Theory of Superconductivity in High-Tc Cuprates'* Princeton University Press (1997)
13. I. Bozovic & D. Pavuna (eds), *Oxide Superconductor Physics and Nano-engineering"* The SPIE, vols. 2058 and 2697, Bellingham, WA, USA (1996)

STATUS OF HIGH Tc

Guy Deutscher

School of Physics and Astronomy
Raymond and Beverly Sackler Faculty of Exact Sciences
Tel Aviv University, Ramat Aviv, Tel Aviv 69978, Israel

INTRODUCTION

The purpose of this chapter is to focus on the main questions that are central to our understanding of the physics of the High Tc cuprates. It is intended as a preview of the many points that are discussed in detail in the other chapters of this book, with an attempt to put them in perspective. Data and figures presented in these specific chapters are not reproduced, the reader is referred to them in the course of the discussion.

In our Advanced Institute, the normal and superconducting states were given equal importance. In this Chapter, we mostly discuss the properties of the superconducting state, the normal state general properties having been presented at the Institute by Batlogg. However, because in the case of the cuprates it is really not meaningful to discuss the superconducting state without some words on the normal state, we shall in the first section make some remarks in this respect. The other sections are devoted respectively to short coherence length effects, to the gap (values, anisotropy, symmetry),and to a more detailed discussion of the properties of weak contacts, including tunneling and Sharvin contacts.

SOME REMARKS ON THE NORMAL STATE

What is the nature of the normal state in the cuprates? Should we consider them as metals, or as doped semiconductors or insulators? Or is it a new kind of normal state, unknown until now? Depending on one's point of view on this question, one is led to very different kinds of models. If one believes that the normal state is a true metallic one (defined more exactly below), one is led to models where special properties of the

The Gap Symmetry and Fluctuations in High - T$_c$ Superconductors
Edited by Bok *et al.*, Plenum Press, New York, 1998

15

normal state, such as the linear temperature dependence of the resistivity, the anomalous Halll effect, the tunneling density of states and others, are interpreted in terms of some anomalous scattering occuring at or near special points of the Fermi surface (see the Chapters by Bok and Bouvier, and by Pines). If the normal state is not a true metallic one, other approaches, such as the one presented by Varma, are necessary.

Are there quasi-particles in the normal state?

It is of course up to the experiment to decide on the basic nature of the normal state. The general definition of a metal, as a Landau Fermi liquid, was given already a long time ago (see for instance the book of Pines and Nozieres (1)). In the case of an electron gas without interaction, the occupation number of states defined by their wave vector k, n(k), goes at T=0 from unity to zero at $k=k_F$, the Fermi wave vector. In the case of an interacting electron gas, it undergoes a smaller but *finite* discontinuity at the Fermi level. It follows that the electron-electron scattering rate varies as T^2 or as ω^2, whichever is larger.

This is in contradiction with what is observed experimentally in the cuprates, where we have ample evidence that the scattering rate varies linearly with temperature or frequency over a very large range. Yet, this observation does not prove that the cuprates are not Landau Fermi liquids, because it can be explained by some special scattering mechanisms, as mentioned above. Thus, it is difficult to decide on the existence of quasi-particles on the basis of the normal state transport properties alone. Moreover, one may argue that the normal state measurements have not in general been carried out at sufficiently low temperatures to be considered as valid tests.

Photo-emission experiments, which probe the spectral density A(k,w), provide in principle a direct test for the existence of well defined quasi-particle states at and near the Fermi level. This point is discussed in detail by Campuzzano. At first glance, his data for BiSrCaCuO samples near optimal doping do not provide much evidence for the existence of such quasi-particles. But a quantitative analysis of the data, that could place an upper bound on the value of the discontinuity, remains to be done. This would be an interesting test for the model of Pines et al. (2), who have calculated the value of the discontinuity compatible with transport data, in the framework of their model of scattering due principally to anti-ferromagnetic fluctuations at or near the $(\pi,0)$ points, the so-called "hot spots".

The situation appears to be different in overdoped samples, for which the photo-emisssion measurements give results that resemble at least qualitatively those for normal metals. One may conclude at this time that, if the cuprates are Landau Fermi liquids at all levels of doping, then the discontinuity in the occupation number at the Fermi level must be a strong function of the doping level.

Normal state tunneling in highly anisotropic materials

It is an accepted view that no information on the normal state DOS and related properties, such as the Fermi velocity and interaction parameters, can be obtained from normal state tunneling. This is because, in the expression for the tunneling current, the velocity and the DOS terms cancel each other. But in strongly anisotropic materials such as the cuprates, this statement no longer applies, and it is possible to obtain interaction parameters through *normal state tunneling* experiments. Littlewood and

Varma (3).have proposed that the inelastic scattering events, that are necessary to tunnel from a normal metal into a HTSC along the c-axis (because of the absence of coherent electronic motion in that direction), could be of intrinsic rather than extrinsic (4) origin. Within their Marginal Fermi Liquid model, the scattering rate   is proportional to the energy, or to the temperature, whichever is larger. They explain in this way the linear dependence of the tunneling conductance versus applied bias, widely reported for c-axis tunneling at frequencies above the gap. Because the slope is proportional to the coupling parameter λ, it is in principle possible to retrieve the latter from such measurements. In practice, however, this requires the independent determination of several other parameters that also enter into the tunneling conductance.

Recently, Dagan et al. (5) have shown that a determination of λ is in fact possible, by following the evolution of the tunneling conductance as a function of time, in a junction prepared in such a way that it becomes progressively thicker and less transparent (6). This produces different time dependences for the elastic and inelastic parts of the tunneling current, which allows one to eliminate most of the unknown parameters of the barrier. The point is that, as the barrier becomes thicker, the elastic part of the tunneling current goes down faster than the inelastic part. This allows one to separate the two components. One can then show that their ratio is basically a function of the ratio of λ to the Fermi energy, which is taken from photo-emission data. Another important observation of Dagan et al. is that the tunneling characteristic contains an assymmetric quadratic term, in addition to the (main) symmetric linear term (and to a small symmetric quadratic term, due to the finite barrier height). Within the Marginal Fermi Liquid theory, this quadratic term is a function of the cut-off frequency for the interaction, ω_c, which can thus also be determined from experiment. By following up the evolution of several junctions as a function of time, the authors were able to obtain a consistent picture, leading for all samples to the same values of the interaction constant and of the cut-off frequency, within the error margins. These values are $\lambda = 1$, with a 50% error margin, and $\omega_c = 0.15$ eV, with a 20% error margin. If one makes the assumption that the same interaction is at the origin of superconductivity, these numbers lead to the conclusion that High Tc is due to rather low frequency electronic excitations, with a moderately strong coupling. Spin fluctuations meet these requirements (2), although there may be other possibilities.

The normal state at low temperatures - sensitivity to impurities

Another important characterisitic of metals is that impurity scattering leads to a finite, constant value ot their electrical conductivty at low temperatures (Mathiessen rule). This finite conductivity is inversely proportional to the impurity concentration.

Recent measurements of the normal state conductivity on LaSrCuO (Ando-Bobbinger, Ref.7), obtained by applying a strong magnetic field, have shown, instead, a continuous decrease at low temperatures. This might be an indication that we are not dealing with a Landau Fermi liquid, but with a Marginal Fermi liquid in which any finite amount of impurity scattering results in an insulating normal state at low temperatures (8). Alternatively, however, this behavior might be an effect of the very strong applied fields on the magnetic state of the sample, as proposed by Pines (2). What is, in any case, missing is a study of the low temperature conductivity as a function of the impurity concentration. Such a study might shed some light on the question of the true metallicity of the cuprates.

Can superconductors be insulators in their normal state?

The weak, but continuous decrease of the normal state conductivity seen in LaSrCuO at low temperatures (7), raises the question of the transition to a superconducting state in an insulating sample. Evidently, this could not easily be explained in the framework of BCS-like theories, which assume a metallic normal state to start with.

This question is, in fact, not new. Already 15 years ago, it was shown that granular materials (somewhat improperly often called "granular metals") can be insulating in their normal state, and still undergo a transition to a coherent superconducting state (9). This behavior has been recently confirmed by Gerber et al. (10) (see Fig.1). In that case, the transition to the superconducting state by the condensation of Cooper pairs in the small metallic grains is somewhat independent of the inter-grain coupling, and the transition to a coherent superconducting state may be understood as resulting from the Josephson coupling between them. This result, obtained on a "conventional" superconductor, suggests that the superconducting transition seen in "insulating" cuprates may not be due to an exotic mechanism, but rather to some degree of granularity, related to sample inhomogeneity [intrinsic (phase separation), or extrinsic].

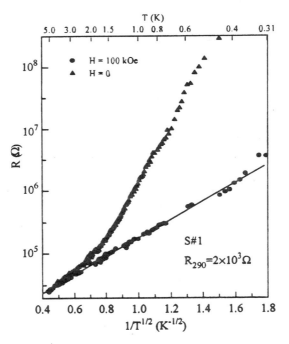

Fig.1. In sample 1, Josephson coupling is not strong enough to produce long range order: this sample is insulating in its normal state, and super-insulating when the grains become supersonductors. In sample 2, Josephson coupling is sufficiently strong to produce long range order, eventhough it is insulating in its normal state.(after Ref. 10)

SHORT COHERENCE LENGTH EFFECTS

Many properties of the superconducting state in the cuprates are also anomalous. It was noted very early on, that in the presence of an applied magnetic field, their transition is broadened rather than shifted, as is the case in the LTSC. It was also noted that their heat capacity transition was not mean field-like, i.e. did not exhibit the jump at Tc predicted by the BCS theory (11). We wish to recall here some fundamental aspects of the short coherence length, and to relate them to these two anomalous behaviors .

Some fundamental aspects of the short coherence length

According to the BCS theory, the coherence length is given by:

$$\xi = (hv_F/\pi\Delta) \tag{1}$$

where Δ is the pair potential. For a typical cuprate, such as YBCO or BSCCO, $\Delta \approx 20$ to 30 meV, and according to photoemission measurements, $v_F \approx 3.10^7$ cm/s. One then calculates $\xi \approx 20A$, i.e. only a few lattice spacings. From the same relation:

$$(\xi\ k_F)^{-1} \equiv (\Delta/E_F) \tag{2}$$

where E_F is the Fermi energy. $(\xi\ k_F)^{-1}$ is the small parameter of the BCS theory. We can estimate its value from the above relation. Both sides give a value of the order of 0.1. As a point of reference, it is of the order of 1.10^{-4} in the LTSC. The approximation of a small parameter is extremely well justified for the latter, but not for the HTSC. An immediate consequence of the border line value of this parameter in the HTSC, is that large thermodynamical fluctuations are expected in the HTSC. An easy way to see this is to notice that the number of Cooper pairs that exist within the radius of one pair is not very large:

$$N(0)\Delta\xi^3 \approx (\xi\ k_F)^2 \tag{3}$$

If we take into account the anisotropy of ξ, this number is of the order of 10 for the cuprates - there are only a few pairs within the effective radius of one pair. Therefore, fluctuation effects are large. By contrast, this number is of the order of 10^8 in the low LTSC, where:fluctuations are small, and the mean field approximation is excellent. .

The heat capacity transition and the irreversibility line as manifestations of fluctuations due to the short coherence length

Predicted as early as 1987 (12), large fluctuation effects in the heat capacity have indeed been observed in the HTSC, as discussed in detail by Junod in this volume. But there seems to be a qualitative difference between the heat capacity transition in YBCO (the least anisotropic of the HTSC), where it is highly asymmetric and resembles a mean field jump with an additional contribution coming from fluctuations -

and other, more highly anisotropic cuprates of the Bi and Hg families, where it takes the form of a nearly symmetric cusp (13). The exact nature of these transitions is still a matter of debate. It is tempting to interpret the cusp-like transition in the Bi and Hg based cuprates as the manifestation of a Bose-Einstein condensation (may be of preformed pairs?), but the last word on this point has not yet been given. We can retain the general idea that fluctuation effecs are stronger in the more anisotropic cuprates, although samples quality is evidently of utmost importance to allow meaningful comparisons.

An interesting observation is that the form of the heat capacity transition has a direct bearing on high temperature - high field - high current applications, that require strong vortex pinning. The point is that a relevant parameter for vortex pinning is the condensation energy per coherence volume, $[0.5N(0)\Delta^2\xi^3]$, divided by a thermal energy of the order of k_BT_c. When this parameter is large, there is a low probability for thermally activated motion of a vortex trapped by a defect of size ξ, the pinning is strong - and vice versa. It is easy to see that this parameter can be rewritten as $(\xi k_F)^2$, the same parameter that determines the importance of fluctuation effects. The practical conclusion is that when the heat capacity transition is quasi mean field, indicating relatively weak fluctuation effects, defects will pin vortices effectively - while in the opposite case of a transition dominated by fluctuation effects, defects will be fairly inefficient.

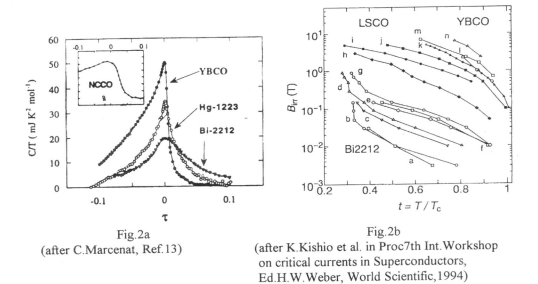

Fig.2a
(after C.Marcenat, Ref.13)

Fig.2b
(after K.Kishio et al. in Proc7th Int.Workshop on critical currents in Superconductors, Ed.H.W.Weber, World Scientific,1994)

Fig.2. The heat capacity transition (Fig.2a), and the irreversibility fields (Fig.2b) of various cuprates. YBCO has a very a-symmetric transition, resembling a BCS transition with additional contributions from critical fluctuations near Tc; it has the highest irreversibility fields. A contrario, Bi 2212 with a very symmetrical, non BCS like transition, presumably completely dominated by large fluctuations, has the lowest irreversibility fields.

This rule was proposed some time ago (14), and appears to be well obeyed in the cuprates whose heat capacity transition has been studied in detail: YBCO, with the most mean field-like transition, has the best high temperature - high field - high current performance (highest irreversibility field), and BSCCO, with the most symmetrical, non-mean field-like transition, has the poorest one (Fig.2). Considerable effort has been spent to introduce different kinds of defects to improve the performance of BSCCO at high temperature, with only modest success, and I do no think that there is much hope in that direction (except if the defects would reduce the anisotropy).

It will be very interesting to study the heat capacity transition in the Re doped mercury compound, which has been found to have a remarkably high irreversibility field (see recent results in the chapter by Zhao) - we predict that it should have a more mean field looking transition than BSCCO.

GAP VALUES , GAP ANISOTROPY AND GAP SYMMETRY

Measurements of the superconducting gap are in principle possible by a number of different methods. Historically, heat capacity, penetration depth and Far Infrared measurements have played an important role in establishing the existence of the BCS gap in the LTSC. Tunneling has been the method of choice to study the details of the superconducting density of states, including strong coupling effects through the frequency dependence of the gap. (15).

These traditional methods have not been very useful to determine gap values in the HTSC. This is due to extrinsic contributions (heat capacity), to technical difficulties (tunneling), and to the existence of intrinsic low lying excitations (penetration depth). New methods have been developed for the HTSC: photo-emission (see the Chapters by Campuzzano and Onellion) and Raman spectroscopies (Hackl and Sacuto) are applicable for the determination of large gaps, and moreover are sensitive to gap anisotropies. Recently, due to improved samples and particularly to a better surface control, there has been renewed interest in tunneling (STM) and in the study of Andreev reflections through Sharvin contacts. We shall go back to them in some detail in the last section of this Chapter.

Gap values

The large number of compounds and techniques makes a comprehensive review of gap values outside the scope of this introductory Chapter Another difficulty is that, sometimes, cited gap values involve assumptions made on the gap symmetry. For instance, in the case of tunneling results, the value of the gap is given by the maximum of the conductance if the gap is assumed to be d-wave, but to some lower value if it is assumed to be s-wave. However, some general remarks on available results can be made, and some trends outlined.

There does not appear to be any straighforward answer to the question: how large is the gap (in a given material)? Different methods of gap determination give different answers, even on high quality samples. The methods discussed in detail in our Institute included: optical conductivity, Raman scattering, photo-emission, tunneling and Andreev spectroscopies, and London penetration depth measurements.

Generally speaking, optical conductivity, Raman scattering (see the Chapters by Hackl and Sacuto), photo-emission (see the chapters by Margaritondo, Campuzzano and Onellion) and tunneling give the highest gap values, with ($2\Delta/kTc$) in the range of 6 to 12, for most cuprates, two noticeable exceptions being overdoped BSCCO 2212 (Raman, T.Stauffer et al., Ref.16; photo-emission, Ma et al.,Ref.17)) where this ratio is found to be less than 6, and NdCeCuO where it is between 4.1 and 4.9 (B.Stadlober et al.Ref.18).

The high gap values obtained by the above methods are basically temperature independent up to Tc, and increase in the underdoped regime. STM tunneling (see the Chapter by Fischer) and break junctions (19), give for BSCCO a strong coupling ratio of about 10 near optimum doping, up to 14 for an underdoped sample (Tc = 83K) and down to 6 for an overdoped one (Tc = 62K). Tunneling data on YBCO also show a strong doping dependence, varying from 20 for a strongly underdoped sample (Tc = 50K) (20) down to 5 near optimum doping (21) (this last result will be qualified later). On the other hand, gap values corresponding to weak coupling have been found by tunneling in the electron doped superconductors.

A sharp gap edge has been seen by Andreev reflections in optimally doped YBCO at 18 meV, corresponding to a strong coupling ratio of 5 (N.Hass et al., Ref.22). Low temperature penetration depth measurements, when interpreted in the framework of d-wave theory, have given a similar coupling ratio in optimally doped YBCO (23) and BSCCO (24).

Special attention has been given recently to the underdoped regime. The opening of large gaps has been observed at temperatures well above Tc. They have been interpreted as pseudo-gaps (an expression coined by Jacques Friedel) of magnetic origin, resulting from anti-ferromagnetic fluctuations, or as precursors of the superconducting gap formed below the critical temperature. The question of the actual value of the superconducting gap, in the underdoped regime, remains in discussion. As mentioned above, tunneling gives values larger than those for optimally doped samples. On the other hand, low temperature penetration depth measurements on underdoped YBCO, with a Tc of 60K, have given a linear dependence with a slope larger by a factor of about 2 than that for optimally doped samples (23), indicating a smaller gap.

Excitation gap and coherence gap

These examples are sufficient to show the problematics of the gap determination in the HTSC. Apparently equally reliable measurements give in fact, for 90K cuprates, gap values that cluster around 20 meV and 40 meV, corresponding respectively to values of ($2\Delta/k_BT_c$) of about 5 and 10. Differences become even larger in the underdoped regime.

To try to make sense of the apparently large discrepancies in the reported values of the strong coupling ratio, it is important to realise that different physical properties are sensitive to different aspects of the energy gap. For instance, tunneling and photo-emission measure the energy range where single particle states are missing (*excitation gap*), while Andreev reflections (see below the section on Sharvin contacts) measure the energy range beyond which the pair amplitude returns to zero (*coherence gap*). Other methods that measure an excitation gap are optical measurements and Raman scattering (the latter being also sensitive to coherence effects). The London penetration depth, being a property of the condensate, is a function of the coherence gap, as is the Josephson critical current.

In a conventional BCS superconductor, the excitation and the coherence gaps are identical. But let us allow for the possibility that they may be different in the HTSC. Existing data then suggest the following picture: the excitation gap (obtained for instance by photo-emission, Raman, optical conductivity, tunneling) is roughly temperature independent and a monotonously decreasing function of doping.. The coherence gap (Andreev reflections, penetration depth) is temperature dependent and follows the same doping dependence as Tc. The excitation gap is larger than the coherence gap. For both YBCO and BSCCO, near optimum doping and at low temperatures, the coherence gap is in the 20 to 25 meV range, and the excitation gap in the 35 to 40 meV range. The tunneling and Andreev spectroscopy data are discussed in detail in a later section.

There is no theory at present that predicts the co-existence of distinct excitation and coherence gaps. But it may not be unreasonable for materials that are close to being insulators. We recall the example of granular superconductor Al-Ge which, for compositions close to the metal to insulator transition, has a regime where strong localisation is already apparent above the critical temperature. The existence of two different energy gap scales in the cuprates may reflect the existence of Pines "hot" and "cold" regions of the Fermi surface. From photo-emission data (25), an excitation gap (the so-called pseudo-gap) develops already above Tc, in the hot regions; the condensate may develop mostly in the cold regions. and only below Tc. The anisotropy observed in Raman spectra may be related to the (large) excitation gap in the hot regions, and to the (smaller) coherence gap in the cold ones (26), not just to a d-wave symmetry of the gap. Note that the anisotropy in Raman spectra vanishes in the overdoped regime. Andreev reflections, on the other hand, are only sensitive to the condensate, and measure only the coherence gap. It is this smaller gap value that fits the penetration depth low temperature behavior (see the Chapter by Hardy), as it should, since the penetration depth is a property of the condensate.

Gap anisotropy

The most basic feature of the superconducting order parameter is that it has an amplitude $|\Delta|$ and a phase.φ. In a BCS superconductor, the amplitude and the phase are uniform in k space. But in general, both can vary around the Fermi surface.

The general BCS gap equation reads:

$$\Delta_k = \sum_{k'} V_{kk'}\Delta_{k'}/(\xi_{k'}^2 + \Delta_{k'}^2)^{1/2} \qquad (4)$$

where $\xi_{k'}$ is the energy of a quasi-particle in the state k', measured from the Fermi level.

In the BCS approximation, the matrix element $V_{kk'}$ is assumed to be constant around the Fermi surface, and to vanish a cut off frequency (typically, the Debye frequency). Even under this approximation, there may be a gap anisotropy due to the shape of the Fermi surface. In particular, if the Fermi level lies close to a van Hove singularity, and if in addition one assumes weak screening which makes $V_{kk'}$ decrease rapidly already for small values of (k - k'), the gap is largest near the van Hove singularities. This is discussed in detail in the Chapter by Bok and Bouvier, and leads to a gap anisotropy of the form:

$$\Delta(\varphi) = \Delta_0 + \Delta_1\cos(4\varphi) \qquad (5)$$

with Δ_0 always larger than Δ_1 (no nodes in the gap).

Sign changes of the gap occur when $V_{kk'}$ is strongly repulsive along certain directions. This is the case when the coupling mechanism is through anti-ferromagnetic fluctuations, as proposed by Pines et al. (see the Chapter by Pines), who have shown that a d $_{x2 - y2}$ symmetry of the gap is then necessary.

$$\Delta = |\Delta| \cos(2\varphi) \qquad (6)$$

We note here that the existence of strong anti-ferromagnetic fluctuations, with the spins on Cu aligned parallel on diagonal sites, is expected to lead to an extremum of an s-gap amplitude along the [110] direction, or to a node in that direction, but not to a (d+s) order parameter, for which the minimum gap is shifted from the [110] direction.

Gap symmetry

Concerning the gap absolute value, the important difference between the two above forms of $\Delta(\varphi)$, is that the first one gives a finite minimum value and the second one a node at $\varphi = 45^o$. Photo-emission measurements having a finite resolution can only eliminate the existence of a node, but strictly speaking cannot prove it. Yet, for optimally doped BSCCO, a finite minimum gap, if it exists, should be at least 20 times smaller than the maximum gap (see the Chapter by Campuzzano). Such a high anisotropy ratio is not really compatible with the spirit of the calculation of Bok and Bouvier. The feeling amongst the photo-emission community is that there are really nodes in the gap for this compound, at least at and near optimum doping. On the other hand, for strongly over-doped BSCCO, the photo-emission measurements presented by Onellion in his chapter give a finite minimum gap, at least at low temperatures. These measurements are not compatible with the existence of nodes. But they are in line with the Raman scattering data on overdoped BSCCO already quoted (16), where an almost isotropic gap fits the data (minimum and maximum strong coupling ratios respectively equal to 4.1 and 4.9).

Another case of absence of nodes amongst the hole doped cuprates is that of LSCO near optimum doping. According to Andreev reflection measurements, which are phase sensitive, there exists, at least at low temperatures, a finite minimum gap for this compound (for a more detailed discussion of Andreev reflections, see below). Finally, there is no evidence for nodes in the electron doped cuprates.

The group at IBM has examined the existence of sign reversal of the gap in a large number of hole doped cuprates by flux quantisation measurements of tri-junctions. (see the Chapter by Kirtley). Nodes have been found in YBCO, BSCCO and Tl compounds. Because these experiments are based on the use of grain boundary junctions, the doping level is not really known, but the junctions are more likely to be in the underdoped rather than in the overdoped regime. Neither LSCO, nor electron doped cuprates have been examined by this method.

Concerning the existence of nodes, there does not seem to be any specific contradiction between different experiments. There is strong experimental evidence for a gap symmetry with nodes in most of the hole doped cuprates, at least near optimum doping and in the underdoped regime. But the question of the universality of the gap symmetry, and therefore that of the mechanism for superconductivity in the cuprates, remains open at this stage, in view of the exceptions that have been found. It would really be very useful for instance to perform flux quantisation experiments on LSCO, on overdoped BSCCO (if possible), and on the electron doped cuprates.

Gap anisotropy and impurity scattering

It is important to understand the effect of impurity scattering on the gap anisotropy. It is drastically different on anisotropic s-wave and d-wave gaps.

In the case of an s-wave anisotropic gap (such as the weak anisotropy found in some pure LTSC, such as Nb), the effect of impurity scattering is to wash it out. Non-magnetic impurity scattering does not affect the average gap value (Anderson's theorem).

On the contrary, impurity scattering will destroy d-wave superconductivity, because it mixes gaps of oppposite signs. We expect superconductivity to be destroyed when the scattering rate τ^{-1} ,multiplied by Planck's constant , will be of the order of the maximum gap. This condition may be re-arranged as:

$$v_F \, \tau \cong (hv_F/\Delta) \qquad\qquad (7)$$

Expressed in this way, the condition for destruction of superconductivity in a d-wave superconductor is that the mean free path value is reduced to that of the coherence length. In other terms, d-wave superconductivity can only exist in the clean limit.

Low temperature superconductors have, as a rule, a long coherence length. For a critical temperature of 1K, the coherence length is of about 1 μm. Only extremely clean samples could be low temperature d-wave superconductors. Heavy fermion superconductors present a noticeable and interesting exception to this rule. They have in fact a short coherence length, because of the small value of the Fermi velocity, due to the heavy electron mass. Indeed, the question of the symmetry of the gap in the heavy fermions remains an interesting one (see the Chapter on heavy fermions by Ott).

The very short coherence length of the high Tc cuprates is a factor that helps d-wave superconductivity be robust in the presence of impurity scattering. One may remark that the higher the Tc, the shorter the coherence length, and the more robust d-wave superconductivity will be. One may add to this that, in the event where two mechanisms for superconductivity would co-exist, one of them leading to an s-wave and the other to a d-wave gap, impurity scattering would favor the former one.

It is impractical to probe the gap near an impurity, but it is possible to probe it at the surface of a sample. Surface scattering should act in a way similar to that of an impurity scattering, suppressing a d-wave gap while not affecting an s-wave one. In fact, in the presence of a very short coherence length, even an s-wave gap will be affected by the vicinity of a surface, but this effect is less drastic (27).. Probing the surface is thus an interesting way to probe the symmetry of the gap. This brings us to the subject of weak contacts, tunneling and Sharvin, to which the rest of this introduction is devoted. A more detailed discussion of the properties of weak contacts will be found in the Chapter by Maynard and Deutscher.

WEAK CONTACTS - SHARVIN VERSUS TUNNELING

There are some accepted views on tunneling, well established for superconducting metals and alloys, that have been frequemtly assumed to hold for the HTSC, but that in fact do not apply to them. We have already mentioned the possibility to obtain important imformation on the interaction parameters from normal state tunneling. Another interesting point is that, for unconventional gap structures including its symmetry, the study of Sharvin contacts has significant advantages as compared to tunneling. This is what we would like to underline in this section.

The nightmare - or wonderland - of High Tc tunneling

The selection criteria for a high quality low Tc junction are pretty straightforward: a low conductance at zero bias (less than 1% of the normal state conductance); and a known normal state conductance, usually pretty much bias independent. The experimental determination of the normal state conductance is absolutely necessary, since the ratio of the superconducting to normal state DOS is equal to that of the respective conductances.

When we consider High Tc junctions, these criteria look like wishful thinking. The zero bias conductance is rarely lower than 10% of the normal state one; the latter has, as we have already discussed, a significant bias dependence, and it is practically impossible to measure it at low temperatures, because of the extremely high critical fields of the cuprates. In addition, the superconducting conductance $\sigma_S(V)$ has often a complicated shape, including zero-bias anomalies. All of these have given high Tc tunneling a rather bad name. Even gap values are difficult to determine in an unambiguous fashion, not to say anything about the application of the McMillan-Rowell procedure to obtain the interaction parameters, which cannot be performed at all in any reliable way.

To give an example of the difficulties encountered in making and studying tunneling junctions, we take the example of YBCO, one of the most studied HTSC material.

Tunneling into YBCO

Normal-Superconductor junctions using a dielectric barrier (N/I/S junctions) are discussed in detail in the Chapter by Lesueur. They are typically produced by evaporating an oxidable normal metal on top of the superconductor, reliying on what is sometimes called a "natural" barrier at the surface of the latter to produce the junction. In fact, it was shown by Racah et al. (6) that, under certain circumstances, this barrier can be produced through the oxidation of the counter-electrode by oxygen diffusing out of the High Tc electrode itself. The barrier is then most probably composed of a somewhat oxygen depleted high Tc oxide, and of an oxidised counter-electrode.

This procedure for forming the dielectric barrier, already mentioned in the first section for the study of normal state tunneling into c-axis films, is particularly effective for a-axis films, because of the high diffusion rate of oxygen along the CuO planes (several orders of magnitude higher than across the planes). It was first demonstrated clearly by Racah et al.(6), who showed that oxygen diffusion can be so effective as to remove enough oxygen from a cuprate film, overlayed by an Aluminium film, to turn it into an insulator after a few days..

Due to technical problems in performing these tunneling experiments - presence of c-axis grains in the predominantly a-axis film, presence of holes and/or particulates in the film, role of these defects in the diffusion-oxidation process - the results have not yet reached the desired level of controlled reproducibility. However, Racah was able to show consistently that the tunneling characteristics of such junctions show gap like structures on two different energy scales: 20 meV, and 35 to 50 meV (28). The 20 meV structure is dominant in most junctions, but the higher energy structure is always visible, and is in fact the dominant one in the junctions that show the lowest zero-bias conductance

We show Fig.3 the results for two junctions, which we qualify as "average", and "satisfactory" in the order of decreasing values of their zero bias conductances, 50% and less than 20% of the normal state conductances, respectively. The "average" one is quite similar to STM characteristics obtained on YBCO single crystals (as reported by Fischer in his Chapter), for which one can definitely observe a double peak in the conductance..In the "satisfactory" junction, the higher energy structure is now dominant. The general trend of the data, established by Racah over tens of junctions (28), is that the higher energy structure, for these a-axis films, is growing as the ZBCP becomes smaller.

Results obtained on YBCO junctions raise many questions, and they have been variously interpreted. Most authors tend to discard the higher energy structure, which is generally not the dominant one, and to identify the 20meV one as the energy gap. Yet, for the best junctions obtained on a-axis films (junctions having the lowest zero bias conductance), it is the higher energy structure that is dominant. Let us note here that the lower and higher energy structures, if identified as superconducting gaps, correspond respectively to strong coupling ratios of 5 and 10. The last one agrees with values obtained from Raman scattering, and photoemission measurements, as already noted. But the smaller one agrees with the d-wave symmetry interpretation of London penetration depth measurements (23), and Andreev reflections studied in Sharvin contacts (see below, Fig.5)).

We tentatively conclude at this time that the lower energy structure in YBCO corresponds to the *coherence gap,* and that the higher one corresponds to the *excitation gap.* The dominant excitation gap structure seen in the best a-axis junctions (Fig.3b) shows that this gap develops along the [100] direction. The large value (50meV), obtained on an underdoped film (Tc=50K), is in agreement with photo-emission results on underdoped BSCCO (25), where a large gap is also observed along the same direction. One may however wonder why both gaps are in general seen in YBCO.

STM on BSCCO

In contrast with the YBCO results, where the main gap like structure seen by STM is at 20meV, the BSCCO characteristic shows only one structure, extremely sharp, near 40meV. This comes as a surprise: why should these two cuprates, which have the same Tc, have gaps that differ by a factor of two? Another difference is that STM observation of vortices is far more difficult in BSCCO than in YBCO (see the Chapter by Fischer). If anything, one might have expected vortices to be identified more easily in BSCCO, since its tunneling characteristic looks more ideal, judging from the lower value of the Zero Bias Conducance.

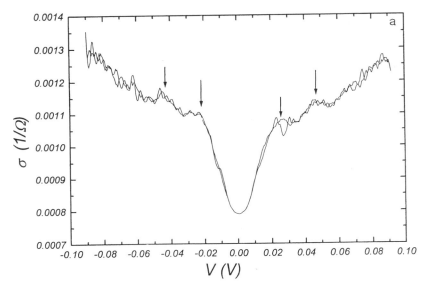

Fig.3a. Giaever tunneling into an a-axis oriented YBCO film. Note the two structures, respectively around 20 meV and 40 meV. The zero bias conductance is about 70% of the normal state value, similar to the STM data of Fig.4a.

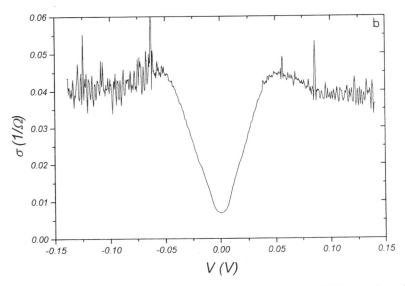

Fig.3b. Giaever tunneling into a strongly underdoped (Tc = 50K) a-axis oriented YBCO film. Note the lower zero bias conductance. Only the higher energy structure is seen, here at 50 meV, in agreement with the data on similarly underdoped Bi 2212 (Ref. 10). It is suggested that this structure may not be the superconducting gap.

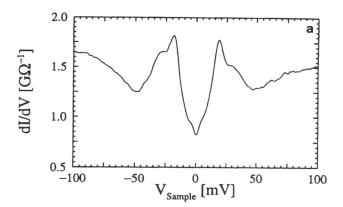

Fig.4a. STM data on an YBCO single crystal, Tc = 91K Note the main peak slightly below 20 meV (as in Fig.3b), and the additional structure around 30 meV (after I.Maggio-Aprile et al. Phys.Rev.Lett.**75**, 2754 (1995).

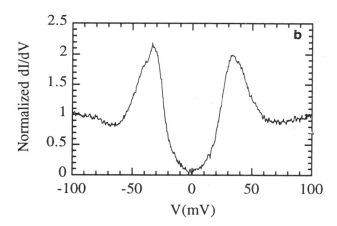

Fig.4b. STM data on a Bi 2212, Tc = 92K, crystal. The main structure is now around 35 meV, closer to the second peak in the above YBCO data, than to the first one (after Ref.30)

To try to understand these differences, it should be remembered that the STM observation of vortices is obtained by using a grey scale defined by the ratio of the conductances at the peak and at zero bias. This method is based in part upon the existence of allowed energy levels below the gap, corresponding to states localised within the core of the vortex (29). These energy levels correspond to resonances within the pair-potential well at the vortex core. They reflect the coherence of the superconducting state at energies below the coherence gap, leading to Andreev reflections from the pair potential at the walls of the vortex core (see next sub-section).

The successful STM observation of vortices in YBCO, and the unsuccessful one in BSCCO, are consistent with our assumption that the 20 meV gap structure seen in YBCO (but not in BSCCO), is the coherence gap, while the 40 meV structure is the excitation gap, dominant in BSCCO.

This interpretation implies that the STM characteristic does not actually measure the superconducting DOS. A possible reason for this may be that, as discussed in the first section on normal state properties, tunneling along the c-axis is made much easier by inelastic scattering, while the statement that c-axis STM measures the total superconducting DOS rests upon the assumption that it is entirely elastic. Now, according to Pines, the inelastic scattering rate varies enormously around the Fermi surface. It is very high at and near the "hot spots", and ordinarily small in the "cold" regions. It then seems plausible that the average made by c-axis STM around the Fermi surface is heavily weighted in favor of the "hot spots", which are precisely the regions where the excitation gap develops. The role played by inelastic tunneling should be particularly important in BSCCO, in view of its very small c-axis conductivity.

If this reasoning is correct, the superconducting STM characteristic should present some discernable differences with the widely assumed d-wave density of states. In particular, the conductance should not be a linear function of the applied bias at small bias, because this linearity is related to an equally weighted average around the Fermi surface, which we pretend does not apply here. Some recent high precision STM measurements by de Wilde et al. (30), suggest indeed that the small bias behavior is not linear. This should be carefully tested.

As a final touch in favor of our hypothesis, we note that the large gap seen by Giaever a-axis tunneling in YBCO, mentoned above, is quite similar in size to that measured by STM in BSCCO.

Sharvin-Andreev contacts

A Sharvin contact is a small area metallic contact, whose normal state conductance is essentially limited by the number of quantum channels, each having a conductance (e^2/h), and not by the presence of a barrier between the electrodes. The number of channels is basically equal to the number of atoms facing each other. The typical resistance of a submicron Sharvin contact is of the order of the Ohm. Because current densities used for Sharvin spectroscopy are high, the contact must be in the ballistic regime, where the electrodes mean free path is larger than its size, in order to avoid heating effects.

In what follows, we shall be mostly interested in the case where the interface barrier is very small. This implies that two conditions be fulfilled: that there is no dielectric barrier, and that the Fermi velocities in the two electrodes are not very different. It can be shown that a difference between the Fermi velocities is equivalent to a delta function interface barrier of height H. In units of the Fermi energy, this barrier is expressed by the parameter $Z = (2mH/h^2 k_F)$. In the absence of any dielectric barrier, it is given by:

$$Z = (v_1 - v_2)^2/4v_1 v_2 \qquad (8)$$

From photo-emission data, the Fermi velocity on the cuprates is typically 3.10^7cm/s. For a gold tip-cuprate contact, we then have for Z a rather high value, Z = 2. This is not favorable for the observation of a large fraction of Andreev reflections. Fortunately, the Fermi velocities that enter into the expression for Z, are not renormalised for mass enhancement effects (31). This explains the experimental observation of Andreev reflections in the cuprates, with a value of Z smaller than 1 (32).

If an electron is injected from the N side of an N/S contact, it can be reflected at the interface in two different ways: either as an electron (normal reflection by the barrier), or as a hole. In the second case, the hole is reflected along the trajectory of the incident electron. This is the Andreev reflection, which occurs when the energy of the incident electron, counted from the Fermi energy, is smaller than the gap. In an Andreev reflection, a charge of 2e flows across the interface per injected electron. This flow consists of the electron-hole combination on the N side, and of a Cooper pair on the S side. In the absence of any barrier (pure Sharvin contact), there are no normal reflections, only Andreev ones. Then, at applied biases smaller than the gap, the current (and therefore the conductance), is twice as large as in the normal state.

The occurence of Andreev reflections is due to the special character of the quasi-particles in the superconductor, in the energy range $(E_F - \Delta) < E < (E_F + \Delta)$. They have partly an electron like, and partly a hole like character, given by the Bogoliubov equations:

$$\varepsilon u(r) = [-(h^2/2m)\nabla^2 - E_F + U(r)]u(r) + \Delta v(r)$$

$$\varepsilon v(r) = [(h^2/2m)\nabla^2 + E_F + U(r)]v(r) + \Delta^* u(r) \qquad (9)$$

where ε is the energy of the quasi-particle counted from the Fermi level, $U(r)$ is the Hartree potential, and u and v express respectively the electron-like and the hole-like "fractions". At energies much higher than the pair potential, these equations reduce to those for usual quasi-particles, which are either electrons or holes. In the energy interval of interest, the particles have a mixed character.

Blonder, Tinkham and Klapwijk (33) have written and solved these equations for a one dimensional N/S system. The wave function on the N side can be written as the sum of three waves, corresponding to the injected electron, to the normally reflected electron, and to the Andreev reflected hole:

$$\psi(x) = \exp(ip^+x) + b\exp(-ip^+x) + a\exp(ip^-x) \qquad (10)$$

where $p^{\pm}$ is the momemtum corresponding to the energies $(E_F \pm E)$. Instead of being a pure electron (hole), the particle in N has acquired a partial hole (electron) character, because of the contact with S. On the S side, the electron-like and hole-like weights have been modified correspondingly. The Bogoliubov equations are solved in N and in S using the appropriate boundary conditions at the interface. In the simple case where there is no barrier, and the applied bias is smaller than the gap, a = 1 and b =0.

The conductance is given by:

$$(dI/dV) = C_0(1 + |a|^2 - |b|^2)$$

and in the above simple case, is equal to twice the normal state conductance C_0. As the bias is increased above the gap, the conductance goes back progressively to C_0, at the same rate as the particles return to their pure electron or pure hole states in S.

Andreev reflections, with their simple expression in the I(V) characteristics of Sharvin contacts, demonstrate beautifully one of the most fundamental properties of the condensate: the partial electron-like and hole-like character of the superconducting "particles". The coherent reflection of the hole along the trajectory of the incident electron, is a direct result of the nature of the condensate. The enhanced conductance of the contact *below* the gap: results from the structure of the superconducting particles. By contrast, the current reappears *above* the gap in a tunnel junction. A Sharvin contact measures the gap as what we have called a *coherence gap* - the energy scale above which the coherence of the superconducting state is lost. A tunnel junction measures the gap as an *excitation gap* - the energy scale below which there are no available states for quasi-particle excitations. For a BCS superconductor, the two gaps are identical, except in some special situations such as in the presence of magnetic impurities (gapless superconductivity). We have interpreted the available empirical evidence as suggesting that in the HTSC, the two gaps are distinct, possibly existing on different portions of the Fermi surface.

We show Fig.5 the conductance characteristic of a Au/YBCO contact. At zero bias, the conductance is enhanced by about 50% as compared to its high (normal state) value. Assuming that there is no dielectric at the interface, this increase corresponds from Eq.8 to Z=0.3, giving an unrenormalised Fermi velocity of 7.10^7 cm/s in YBCO, which is quite reasonable. The coherence gap is of about 20mV. As already mentioned, this value is in agreement with the value obtained by Bonn and Hardy in the d-wave interpretation of their penetration depth measurements. Incidentally, the detailed shape of the characteristic (flat below the gap), is indeed better interpreted if the gap has a d-wave symmetry than an s-wave one, taking into account that the tip is oriented along the [100]direction, and that Z=0.3

The Sharvin contact as a one-junction phase sensitive device

A Sharvin contact is sensitive to the structure of the gap in k space. This is because it probes the entire Fermi surface, contrary to a tunnel barrier which has a strong focusing effect on the injected particles. The Sharvin characteristic is sensitive to the spread of the gap around the Fermi surface and, more importantly, to possible changes of sign. This was first shown by Hu (34), and detailed calculations have been made by Tanaka et al. (35)

A detailed presentation of this topic can be found in the Chapter by Maynard and Deutscher. As an introduction, we shall limit ourselves here to a few remarks.

Let us assume that the superconductor is of the d-wave type, and consider the case where the normal metal tip is oriented along a node direction. We follow Hu and

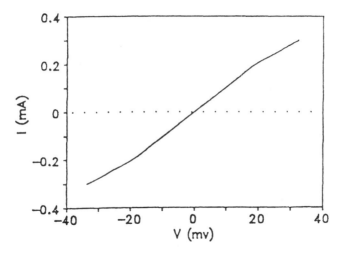

Fig.5 Characteristic of a Sharvin Au tip-YBCO contact, oriented along one of the principal axis in the (ab) plane. The gap is identified as the sharp break in the slope at 18 meV, in agreement with the tunneling (Fig.3a) and STM (Fig.4a) main structures.

model this situation with a thin normal layer in good contact with S, with an interface perpendicular to the [110] direction. Contrary to a tunnel barrier, which would simply "see" a nul gap, this contact probes gap values all around the Fermi surface. Consider a particle coming in from the N side at any finite angle with the normal to the interface. It will be Andreev reflected by a gap lobe of a certain sign, then normally reflected at the outer surface of N, then Andreev reflected by a gap lobe of the *opposite* sign, and so on. Over one cycle, the particle is Andreev reflected by two equal gaps of opposite signs - the sum of which is zero. This corresponds to a zero energy state. The argument holds for any value of the angle of incidence, and for any value of the normal metal thickness. Contrary to the finite energy bound states obtained for an s-wave symmetry gap, we have here states of zero energy, which are in fact surface bound states in the limit where the thickness of the normal layer goes to zero. These states give a maximum of conductance at zero bias, the so-called Zero Bias Conductance Peak. The argument can be generalised for all orientations of the interface, except the one that is perpendicular to the [100] direction, because in that case the gap structure is symmetrical with respect to the normal to the surface.

This method of probing the gap symmetry, where the interference between gaps of opposite signs occur at one interface, has two main advantages. It is not dependent on the existence of special grain boundary properties (Josephson grain boundaries), and it can in principle be applied at any temperature It also relies on a simple property to establish the existence, or the absence of nodes: the presence of a ZBCP.

Several materials have already been probed by this method. A ZBCP has been observed in YBCO in the [110] direction, but not in the [100] direction, as predicted for d-wave. No ZBCP is seen in NCCO, in any direction, in agreement with several other experiments which have failed to give any evidence for the presence of nodes in this material (for a review, see the Chapter by Fournier). In LSCO, no ZBCP has been seen at low temperatures, but possibly it exists above (Tc/2) (36). These materials have not yet been probed by the tri-junction method (see the Chapter by Kirtley), and may never be, because their grain boundaries do not act as Josephson junctions. Very recently, a ZBCP has been seen in a Hg cuprate (37), another material that has not yet been probed by the tri-junction method.

CONCLUSIONS

Gap values obtained with different spectroscopy methods have been reviewed. Methods that measure the single particle excitation energy give consistently higher energy gap values than methods that probe the pair amplitude energy range, such as Andreev reflections measurements. This has led us to the conjecture that in the HTSC, there are two different energy scales, which we have called excitation gap, Δ_e, and coherence gap, Δ_c. While the excitation gap increases monotonously as the doping level goes from overdoped to underdoped, the coherence gap goes through a maximum, following the same doping dependence as Tc. The excitation gap persists above Tc, while the coherence gap goes to zero at Tc. The strong coupling ratio for the excitation gap reaches values well in excess of 10, while for the coherence gap it remains in the range of 5 to 6.

This classification implies the existence of two distinct regimes: a macroscopically coherent state up to Δ_c, followed by an incoherent state up to Δ_e, with $2\Delta_e$ being the energy required to break individual Cooper pairs. To these two energy scales correspond two temperatures: Tc, below which coherence is established, and some higher temperature where the pairs are formed. These two scales become close to each other in the overdoped regime.

This description fits well with the short coherence length which, as we had understood already for some time, is responsible for strong boundary and fluctuation effects, including weak pinning and a non mean field heat capacity transition, particularly in the more anisotropic compounds.

ACKNOWLEGEMENTS

I am particularly endebted to Nicole Bontemps for her critical reading of the manuscript, and to Yoram Dagan, Elie Ferber, Amir Kohen, Ralph Krupke and Dany Racah for their continuing input during the preparation of this Cargese lecture. The hospitality of the Ecole Normale Superieure, and numerous discussions with Roland Combescot, Philippe Monod and Alain Sacuto during the preparation of this manuscript, is gratefully acknowleged, This paper owes a lot to the other Cargese lecturers, whom I would like here to thank collectively for many discussions during and after the School. This work was partially supported by the Heinrich Hertz Minerva Center for High Temperature Superconductivity, by grants from the Israel National Science Foundation and from the US Office of Naval Research, and by the Oren Family Chair for Experimental Solid State Physics.

REFERENCES

1) D.Pines and P.Nozieres, "The theory of quantum liquids", Benjamin, N. Y, 1966
2) For a review, see the Chapter by David Pines in this volume.

3) P.B.Littlewood and C.V.Varma, Pyis.Rev.**B 45**, 12636 (1992)

4) J.R.Kirtley, Phys.Rev.**B47**, 11379 (1993)

5) Y.Dagan, Mater Thesis, Tel Aviv University 1997, and Y.Dagan, A.Kohen, G.Deutscher and C.V.Varma, to be published

6) D.Racah and G.Deutscher, Physica **C 263**, 218 (1996)

7) G.S.Boebinger et al., Phys.Rev.Lett.**77**, 5417 (1996)

8) For a recent reference, see the contribuiton of C.V.Varma in this volume.

9)Y.Shapira and G.Deutscher, Phys.Rev.**B27**, 4463 (1983)

10)A.Gerber, A.Milner, G.Deutscher, M.Karpowsky and A.Gladkikh,Phy.Rev.Lett.**78**, 4277 (1997)

11) For a review, see the Chapter by A.Junod in this volume

12) G.Deutscher, in "Novel Superconductivity", Eds.S.A.Wolf and V.Z.Kresin, Plenum Pess 1987, p.293

13) C.Marcenat, R.Calemczuk and A.Carrington, in "Coherence in High Temperature Superconductors", Eds..G.Deutscher and A.Revcolevski, World Scientific 1996, p.101

14).G.Deutscher, in "Transport properties of Superconductors", Ed.R.Nicolski, World Scientific 1990, p.15

15) W.L.McMillan and J.M.Rowell, in "Superconductivity", Ed.R.D.Parks, Marcel Dekker 1969, p.561

16) T.Staufer et al., Phys.Rev.Lett **68**, 1069 (1992)

17)J.Ma et al., Science **267**, 862 (1995)

18) B.Stadlober et al., Phys.Rev.Lett.**74**, 4911 (1995)

19) N.Miyakawa et al. to appear in Phys.Rev.Lett.**80**, 157 (1998).

20) G.Deutscher and D.Racah, in "Spectroscopy Studies of Superconductors", Eds. I.Bozovic and D.van der Marel, SPIE Proceedings Vol.2696 (1996)

21) For YBCO, the strong coupling parameter value of 5 is obtained from the energy (about 20 meV) at which the major conductance peak occurs.

22) N.Hass et al., Journal of Superconductivity **5**, 191(1992)

23) D.A.Bonn et al., Phys.Rev.**B50**, 4051 (1994), and D.A.Bonn et al., in the Proceedings of the Stanford Conference on "Spectroscopy in Novel Superconductors", Stanford, May 1995.

24) T.Shibaushi, N.Katase, T.Tamegai and K.Ushinokara, unpublished

25) See for instance H.Ding et al., Nature **382**, 51 (1996)

26) An enlightening discussion with David Pines on this point is gratefully acknowledged.

27) G.Deutscher and A.K.Muller, Phys.Rev.Lett.**59**, 1745 (1987)

28) D.Racah, PhD Thesis, Tel Aviv University,1986

29) C.Caroli, P.G.de Gennes and J.Matricon, Phys.Lett.**9**, 307 (1964)

30) Y. de Wilde et al., Phys.Rev.Lett.**80,** 153 (1998); C.Renner et al., Phys.Rev.Lett.**80,** 149 (1998).

31) G.Deutscher and P.Nozieres, Phys.Rev.**B50**, 13557 (1994)

32) See for instance N.Achsaf et al., Journal of Low Temperature Physics **105**, 329(1096) for LSCO, and Ref.22 for YBCO.

33)G.E.Blonder,M.Tinkham and T.M.Klapwijk, Phys.Rev.**B 25**, 4515(1982)

34)C.R.Hu,Phys.Rev.Lett.**72**,1526(1994)

35)S.Kashiwaya et al., Phys,Rev.**B 51**, 1350 (1995)

36)A.Kohen, Master Thesis, Tel Aviv University, 1997

37)I.D'Gourno, unpublished.

SUPERCONDUCTIVITY IN CUPRATES, THE VAN HOVE SCENARIO : A REVIEW

J. Bouvier, J. Bok

Laboratoire de Physique du Solide UPR 5 CNRS
ESPCI, 10, rue Vauquelin - 75231 Paris cedex 05

INTRODUCTION

Many recent experiments of angular resolved photoemission spectroscopy (ARPES) have confirmed the existence of saddle points (van Hove singularity or v.H.s.) close to the Fermi level in five different copper oxide compounds by three different groups, in Stanford[1], in Argonne[2] and in Wisconsin[3]. These observations have been made in the following compounds : $Bi_2Sr_2CuO_6$ (Bi 2201), $Bi_2Sr_2CaCu_2O_8$ (Bi 2212), $YBa_2Cu_3O_7$ (Y123), $YBa_2Cu_4O_8$ (Y124) and $Nd_{2-x}Ce_xCuO_{4+\delta}$ (NCCO). These experiments establish a general feature : in very high T_c superconductors cuprates ($T_c \sim 90$ K) van Hove singularities are present close to the Fermi level. This is probably not purely accidental and we think that any theoretical model must take into account these experimental facts. The origin of high T_c in the cuprates is still controversial and the role of these singularities in the mechanism of high T_c superconductivity is not yet established, but we want to stress that the model of 2D itinerant electrons in presence of v.H. singularities in the band structure has already explained a certain number of experimental facts, i.e. high T_c's, anomalous isotope effect[4], marginal Fermi liquid effects[5] and the very small values of the coherence length[6]. It was also been shown that the singularity is in the middle of a wide band and that in these circumstances, the Coulomb repulsion μ is renormalized and μ is replaced by a smaller number, the effective electron-phonon coupling is $\lambda_{eff} = \lambda - \mu^*$ and remains positive[7]. We think that this fact explains the very low T_c observed in Sr_2RuO_4, where a very narrow band has been determined by ARPES[8].

We have shown by using a weakly screened electron-phonon interaction that we obtain a strong gap anisotropy[9].

We then compute the density of states (D.O.S.) of quasiparticle excitations in the superconducting state, in the frame work of this model. We also study the effect of doping, i.e. of the distance between the Fermi level E_F and the singularity E_S[10].

We apply this result to the calculation of tunneling characteristics and of the electronic specific heat C_S[10,11].

We also study the influence of doping on the screening length and on the calculation of T_c. We thus explain why the maximum T_c is not observed when $E_F - E_S = 0$.

We finally study the influence of doping, i.e. $E_F - E_S$, on the normal state properties and interpret some properties of the so-called pseudogap[12].

CALCULATION OF T_c, THE LABBE BOK FORMULA

Labbé-Bok[4] have computed the band structure for the bidimensional CuO_2 planes of the cuprates, considered as a square lattice (quadratic phase). The simplest band structure we can take for a square lattice is :

$$\xi_k = -2t\left[\cos k_x a + \cos k_y a\right]$$

where t is an interaction with nearest neighbours. This gives a square Fermi surface with saddle points, or v.H.s., at $[\,0\,,|\pi|\,]$ positions of the Brillouin zone, and a logarithmic D.O.S. with a singularity : $n(\xi) = n_1 \ln|D/(\xi - \xi_s)|$, where $D = 16t$ is the width of the singularity and ξ_s the singularity energy level. The v.H.s. corresponds to half filling. We know that is not a good representation of the high T_c cuprates because for half filling (one electron per copper site) they are antiferromagnetic insulators. We think that the Fermi level is at v.H.s. for a doping level corresponding to 20 % of holes in each CuO_2 plane or 0.40 filling of the first Brillouin zone (B.Z.). This can be achieved by taking into account the repulsive interaction between second nearest neighbours (s.n.n.) and the effect of the rhomboedric distorsion. For the repulsive interaction with s.n.n. the band structure becomes:

$$\xi_k = -2t\left[\cos k_x a + \cos k_y a\right] + 4\alpha t \cos k_x a \cos k_y a$$

where αt is an integral representing the interaction with s.n.n.. The singularity occurs for $\xi = -4\alpha t$, there is a shift towards lower energy. The Fermi surface at the v.H.s. is no longer a square but is rather diamond-shaped. More detailed calculations can be obtained in reference 6, taking also into account the rhomboedric distorsion.

The Labbé-Bok[4] formula was obtained using the following assumptions :
1- the Fermi level lies at the van Hove singularity
2- the B.C.S. approximations :
- The electron-phonon interaction is isotropic and so is the superconducting gap Δ.
- The attractive interaction V_p between electrons is non zero only in an interval of energy $\pm\hbar\omega_0$ around the Fermi level where it is constant. When this attraction is mediated by emission and absorption of phonons, ω_0 is a typical phonon frequency.
In that case, the critical temperature is given by

$$k_B T_c = 1.13 D \exp\left[-\left(\frac{1}{\lambda} + \ln^2\left(\frac{\hbar\omega_0}{D}\right) - 1.3\right)^{1/2}\right] \tag{1}$$

where $\lambda = (1/2)\,n_1 V_p$ is equivalent to the coupling constant.
A simplified version of formula (1), when $\hbar\omega_0$ is not too small compared to D, is :

$$k_B T_c = 1.13 D \exp(-1/\sqrt{\lambda})$$

The two main effects enhancing T_c are
1- the prefactor in formula (1) which is an electronic energy much larger than a typical phonon energy $\hbar\omega_0$.
2- λ is replaced by $\sqrt{\lambda}$ in formula (1) in comparaison with the BCS formula, so that in the weak coupling limit when $\lambda < 1$, the critical temperature is increased. In fact it gives too high values of T_c, we shall see later that this is due to the fact that we have neglected Coulomb repulsion between electrons. Taking this repulsion into account we shall obtain values for T_c which are very close to the observed one.

As it is however, this approach already explains many of the properties of the high T_c cuprates near optimum doping.

- The variation of T_c with doping

The highest T_c is obtained when the Fermi level is exactly at the v.H.s.. For lower or higher doping the critical temperature decreases. That is what is observed experimentally[10].

- The isotope effect

Labbé and Bok[4] showed using formula (1), that the isotope effect is strongly reduced for high T_c cuprates. Tsuei et al[13] have calculated the variation of the isotope effect with doping and shown that it explains the experimental observations.

- Marginal Fermi liquid behaviour

In a classical Fermi liquid, the lifetime broadening $1/\tau$ of an excited quasiparticle goes as ε^2. The marginal Fermi liquid situation is the case where $1/\tau$ goes as ε. Theoretically marginal behaviour has been established in two situations (a) the half-filled nearest-neighbour coupled Hubbard model on a square lattice and (b) the Fermi level lies at a v.H. singularity[13]. Experimental evidence of marginal Fermi liquid behaviour has been seen in angle resolved photoemission[14], infrared data[15] and temperature dependence of electrical resistivity[16]. Marginal Fermi liquid theory, in the frame work of v.H.s. predicts a resistivity linear with temperature T. This was observed by Kubo et al[16]. They also observe that the dependence of resistivity goes from T for high T_c material to T^2 as the system is doped away from the T_c maximum, which is consistent with our picture; in lower T_c material the Fermi level is pushed away from the singularity

INFLUENCE OF THE COULOMB REPULSION

As soon as 1959 Bogolubov et al[17] have shown that the electron-electron repulsion plays a central role in superconductivity. Assuming a constant repulsive potential $V_{kk'} = V_c$ from 0 to E_F they find that T_c is given by :

$$T_c \cong T_o \exp\left[\frac{-1}{\lambda - \mu^*}\right]$$

With $\quad \mu = N_o V_c \quad$ and $\quad \mu^* = \dfrac{\mu}{1 + \mu \ln E_F / \omega_o}$ (2)

Cohen and Anderson[18] assumed that for stability reasons μ is always greater than λ. Ginzburg[19] gave arguments that in some special circumstances μ can be smaller than λ. Nevertheless if we take $\mu \geq \lambda$, superconductivity only exists because μ^* is of the order of $\mu/3$ to $\mu/5$ for a Fermi energy of the order of $100\ \hbar\omega_o$. It is useless to reduce the width of the band W ($E_F = W/2$ for a half-filled band) because λ and μ vary simultaneously and μ^* becomes greater if E_F is reduced, thus giving a lower T_c. Superconductivity can even disappear in a very narrow band if $\lambda - \mu^*$ becomes negative.

We have shown[7] that nevertheless high T_c can be achieved in a metal containing almost free electrons (Fermi liquid) in a broad band, with a peak in the D.O.S. near the middle of the band.

Taking a D.O.S., which is a constant n_0 between energies - W/2 and W/2, (the zero of energy is at the Fermi level) and is $n(\xi) = n_1 \ln|D / \xi| + n_0$ between -D and +D we find for T_c, the following formula :

$$k_B T_c = \frac{D}{2} \exp\left[0.819 + \frac{n_0}{n_1} - \sqrt{F}\right] \qquad \text{where}$$

$$F = \left(\frac{n_0}{n_1} + 0.819\right)^2 + \left(\ln\frac{\hbar\omega_o}{D}\right)^2 - 2 - \frac{2}{n_1}\left(n_0 \ln\frac{2.28\hbar\omega_o}{D} - \frac{1}{V_p - V_c^*}\right) \qquad (3)$$

$$V_c^* = \frac{V_c}{1 + V_c \left[\frac{n_1}{2} \left(\ln \frac{D}{\hbar\omega_0} \right)^2 + n_0 \ln \frac{W}{2\hbar\omega_0} \right]}$$

We can have a few limiting cases for this formula : $n_1 = 0$: no singularity. We find the Anderson-Morel formula. $V_C = 0$ and $n_o = 0$: this gives the Labbé-Bok formula.

There are many effects enhancing T_C

$\lambda - \mu^*$ is reduced by the square root, down to $\sqrt{\lambda_1 - \mu_1}^*$ when n_1 is large enough.

As $\lambda - \mu* < 1$ the critical temperature is strongly increased because this factor appears in an exponential. The prefactor before the exponential is D, the singularity width instead of $\hbar\omega_0$. We expect $D > \hbar\omega_0$. For instance D may be of the order of 0.5 eV and $\hbar\omega_0$ about a few 10 meV ($D/\hbar\omega_0$ of the order of 5 to 10).

We have made some numerical calculations using formula (3) to illustrate the effect of Coulomb repulsion. We used two values of D : D = 0.9 eV corresponding to t = 0.25 eV and a much more smaller value D = 0.3 eV. These calculations show that the Coulomb repulsion does not kill superconductivity in the framework of the L.B. model. The general rule for high T_c in this model is to have a peak in the density of states near the middle of a broad band to renormalize the effective repulsion μ. For a narrow band, W, or D, is small, T_c decreases very rapidly as seen in figure (1). A recent case has been observed in Sr_2RuO_4 with a narrow band and T_c is small[8].

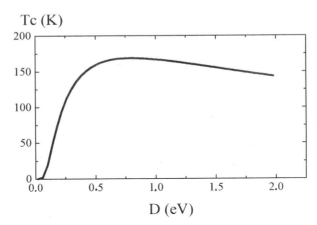

Tc (K)

D (eV)

Figure 1 : Effect of the width of the singularity D on T_c. n_0 and the total number of electrons per unit cell are maintened constant with this set of parameters. Then W = 2 eV, $n_0 = 0.3$ eV/states/Cu, $n_1 = 0.2/D$. In all these cases the calculations are made so that the total number of states of the band is one by Cu atom.

Then $n_0 W + 2 n_1 D = 1$, and $\lambda = (n_0 + n_1) Vp$. In all these cases $\hbar\omega_0 = 0.05$ eV and $\lambda = 0.5$.

GAP ANISOTROPY

Bouvier and Bok[9] have shown that using a weakly screening electron-phonon interaction, and the band structure of the CuO_2 planes four saddle points: an anisotropic superconducting gap is found.

1. Model and basic equations

We use the rigid band model, the doping is represented by a shift $D_e = E_F - E_s$ of the Fermi level. This band structure is

$$\xi_k = -2t\left[\cos k_x a + \cos k_y a\right] - D_e \qquad (4)$$

The Fermi level is taken at $\xi_k = 0$.

We use a weakly screened attractive electron-phonon interaction potential :

$$V_{kk'} = \frac{-|g_q|^2}{q^2 + q_0^2} < 0$$

where $g(q)$ is the electron phonon interaction matrix element for $\vec{q} = \vec{k'} - \vec{k}$ and q_0 is the inverse of the screening length.

We use reduced units: $X = k_x a$, $Y = k_y a$, $Q = qa$, $u = \dfrac{\xi}{2t}$, $\delta = \dfrac{D_e}{2t}$

We use the B.C.S. equation for an anisotropic gap :

$$\Delta_{\vec{k}} = \sum_{k'} \frac{V_{kk'} \Delta_{k'}}{\sqrt{\xi_{k'}^2 + \Delta_{k'}^2}} \qquad (5)$$

We compute $\Delta_{\vec{k}}$ for two values of $\vec{k}$:
$$\Delta_A \text{ for } k_x a = \pi,\ k_y a = 0 \qquad (6)$$
$$\Delta_B \text{ for } k_x a = k_y a = \frac{\pi}{2}$$

We solve equation (5) by iteration. We know from group theory considerations, that $V_{kk'}$ having a four-fold symmetry, the solution Δ_k has the same symmetry. We then may use the angle Φ between the 0 axis and the $\vec{k}$ vector as a variable and expand $\Delta(\Phi)$ in Fourrier series

$$\Delta(\Phi) = \Delta_0 + \Delta_1 \cos(4\Phi + \varphi_1) + \Delta_2 \cos(8\Phi + \varphi_2) + \ldots \qquad (7)$$

We know that $\varphi_1 = 0$, because the maximum gap is in the directions of the saddle points. We use the first two terms. The first step in the iteration is obtained by replacing Δ_k by $\Delta_{av} = \Delta_0$ in the integral of equation (5). We thus obtain, for the two computed values : $\Delta_A = \Delta_{Max} = \Delta_0 + \Delta_1$ and $\Delta_B = \Delta_{min} = \Delta_0 - \Delta_1$, the following expression :

$$\Delta_{A,B}(T) = \lambda_{eff} \int_{u_{min}}^{u_{max}} \frac{\Delta_{av}(T)}{\sqrt{u^2 + u_{av}^2(T)}} I_{(A,B)}(u) \tanh\left(\frac{\sqrt{u^2 + u_{av}^2(T)}}{k_B T / t}\right) du \qquad (8)$$

with
$$I_{A,B}(u) = \int_0^{x_o'} \frac{dx'}{\left[1 - [(\delta - u) - \cos x']^2\right]^{1/2}} \frac{(q_o a)^2}{Q_{A,B}^2 + (q_o a)^2} \qquad (9)$$

where
$$u_{min} = -\frac{\hbar\omega_c}{2t}, \quad u_{Max} = +\frac{\hbar\omega_c}{2t}, \quad u_{av}(T) = \frac{\Delta_{av}(T)}{2t}, \quad x_o' = a\cos\left(\frac{\delta - u}{2}\right)$$

ω_c is the cut off frequency. For the following part of this work we will keep the value of $\hbar\omega_c = 60$ meV for the Bi2212 compound, a characteristic experimental phonon energy. This choice respects our approximation for $V_{kk'}$.

41

- For the choice of t, the transfer integral comes from the photoemission experiments and is t = 0.2 eV as explained in reference[9].
- q_0a is adjusted, it is the Thomas Fermi approximation for small q's,
- λ_{eff} is adjusted so as to find the experimental value of Δ_{Max} and Δ_{min} and we find a reasonable value of about 0.5. λ_{eff} is the equivalent of $\lambda-\mu^*$ in the isotropic 3D, BCS model.

In fact the values of q_0a and λ_{eff} must depend of the doping level D_e. This calculation will be done later. Here $q_0a = 0.12$ and $\lambda_{eff} = 0.665$.

2. Results

In figure (2), we present the variation of the various gaps Δ_{Max}, Δ_{min} and Δ_{av} with temperature at optimum doping, i.e. for a density of holes of the order of 0.20 per CuO_2 plane, as seen before[6]. We take in that case $D_e = 0$ and we find $T_c = 91$ K and an anisotropy ratio $\alpha = \Delta_{Max}/\Delta_{min} = 4.2$ and for the ratios of $2\Delta/k_BT_c$ the following values :

$$\frac{2\Delta_{Max}}{k_BT_c} = 6. \ , \quad \frac{2\Delta_{av}}{k_BT_c} = 3.7 \ , \quad \frac{2\Delta_{min}}{k_BT_c} = 1.4$$

This may explain the various values of $2\Delta/k_BT_c$ observed in experiments. Tunneling spectroscopy gives the maximum ratio and thermodynamic properties such as $\lambda(T)$ (penetration depth) gives the minimum gap.

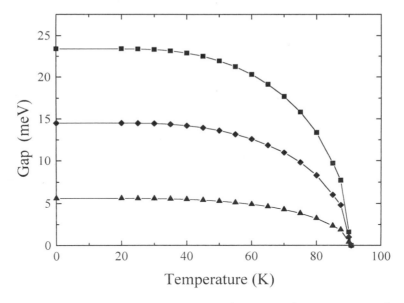

Figure 2 : Variation of the various gaps Δ_{Max}, Δ_{min} and Δ_{av} versus the temperature, at the optimum doping, i.e $D_e = E_F - E_s = 0$ in our model. With the following parameters, $t = 0.2$ eV, $\hbar\omega_c = 60$ meV, $q_0a = 0.12$, $\lambda_{eff} = 0.665$. The critical temperature found is $T_c = 90.75$ K
square symbol = Δ_{Max} diamond symbol = Δ_{av} up triangle symbol = Δ_{min}

In figure (3) we present the same results, Δ_{Max}, Δ_{min}, Δ_{av} as a function of $D_e = E_F - E_s$ (in meV).

In figure (4) we plot the variation of the anisotropy ratio $\alpha = \Delta_{Max}/\Delta_{min}$ versus D_e. In figure (5) the critical temperature T_c versus D_e and in figure (6) the various ratios $2\Delta/k_BT_c$ versus D_e.

We observe of course that T_c and the gaps decrease with D_e or dx. The agreement with experiment[20] is very good figure (7). We obtain a new and interesting result which is the decrease of the anisotropy ratio α with doping. This is confirmed by recent results on photoemission[21,22] where a maximum gap ratio $2\Delta_{Max}/k_BT_c = 7$ is observed at optimum doping with $T_c = 83$ K and $2\Delta_{Max}/k_BT_c = 3$ for an overdoped sample with $T_c = 56$ K, with a small gap $\Delta_{min} = 0$-2 meV for the both T_c, for a Bi2212 compound.

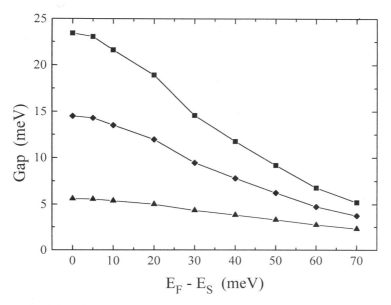

Figure 3 : Variation of the various gaps Δ_{Max}, Δ_{min}, Δ_{av} versus the doping, $D_e = E_F$-E_s, at T = 0K
square symbol = Δ_{Max} diamond symbol = Δ_{av} up triangle symbol = Δ_{min}

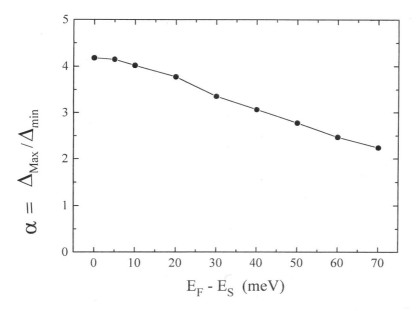

Figure 4 : Variation of the anisotropy ratio $\alpha = \Delta_{Max}/\Delta_{min}$, versus the doping, $D_e = E_F$ - E_s

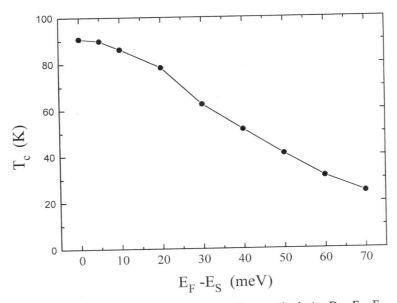

Figure 5 : Variation of the critical temperature T_c versus the doping $D_e = E_F - E_s$

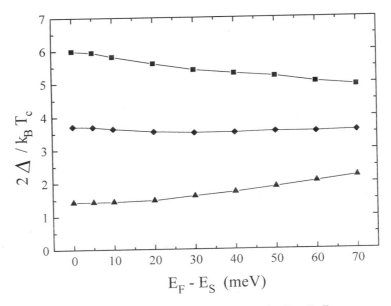

Figure 6 : Variation of the various rations $2\Delta/k_B T_c$ versus the doping $D_e = E_F - E_s$
square symbol = $2\Delta_{Max}/k_B T_c$ diamond symbol = $2\Delta_{av}/k_B T_c$ up triangle symbol = $2\Delta_{min}/k_B T_c$

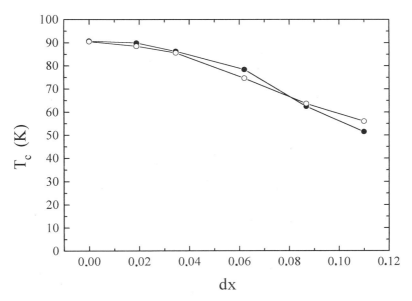

Figure 7 : Comparison of the variation of T_c versus the doping dx calculated in our model (filled circles) and the experimental results of Koïke et al ref (20) (open circles).

DENSITY OF STATES AND TUNNELING SPECTROSCOPY

We have calculated the density of states of quasiparticle excitations in the superconducting state of high T_c[10,11] cuprates using the model of anisotropic gap that we have recently developed[9,10]

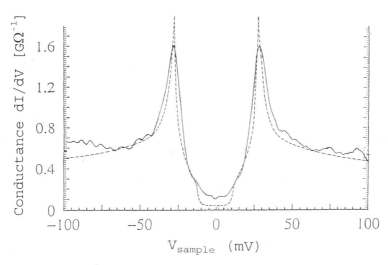

Figure 8 : The best fit of the conductance measured by tunneling spectroscopy on BSCCO, N-I-S junction, by Renner and Fischer (Fig. (10) of Ref. [23]). solid line : fitted curve with Δ_{Max} = 27 meV, Δ_{min} = 11 meV, t = 0.18 eV, Γ = 0.5 meV at T = 5 K, dashed line : experimental curve.

Here the D.O.S. is computed using the formula :

$$n(\varepsilon) = \frac{1}{2\pi^2} \frac{\partial A}{\partial \varepsilon} \qquad (10)$$

where A is the area in k space between two curves of constant energy of the quasiparticle excitation ε_k given by : $\qquad \varepsilon_k^2 = \xi_k^2 + \Delta_k^2 \qquad (11)$

where ξ_k is the band structure (eq. (4)). We use the same procedure and the same expression of Δ_k as before.

Figure (8) represents the variation of the D.O.S. as a function of ε for T = 0 K. This is similar to the experimental conductance (dI/dV versus the voltage V) of a N-I-S junction here we show the measurement made by Renner and Fisher[23] on a BSCCO sample. Δ_{Max} is located at the maximum peak and Δ_{min} at the first shoulder after the zero bias voltage, figure (9). But for different values of E_F-E_s, we see a new maximum emerging, which is a signature of the van Hove singularity and a dip between this maximum and the peak at Δ_{Max}. This dip is seen experimentally in the STM tunneling experiments of Renner et al[23], figure (10), and in photoemission measurements[24].

For the calculation of the conductance, we use the following formula

$$\frac{dI}{dV} = CN_0 \int_{-\infty}^{+\infty} N_S(\varepsilon) \left[-\frac{\partial f_{FD}}{\partial V}(\varepsilon - V) \right] d\varepsilon \qquad (12)$$

where f_{FD} is the usual Fermi-Dirac function; I and V are the current and voltage, C a constant proportional to $|T|^2$, the square of the barrier transmission, N_0 the D.O.S. of the normal metal that we assume constant, and $N_s(\varepsilon)$ the previously calculated D.O.S. in the anisotropic superconductor.

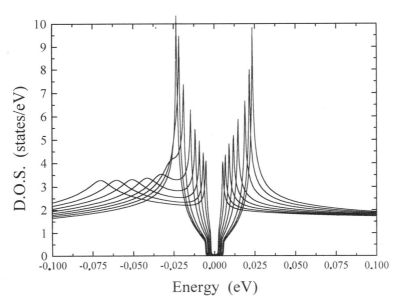

Figure 9 : Variation of the D.O.S. versus the energy ε, for T = 0 K, that is similar at a NIS junction, for different values of the doping D = E_F - E_s, i.e. 0, 10, 20, 30, 40, 60 and 70 meV with Γ = 0.1 meV and Γ' = 5 meV in the model of ref [11]

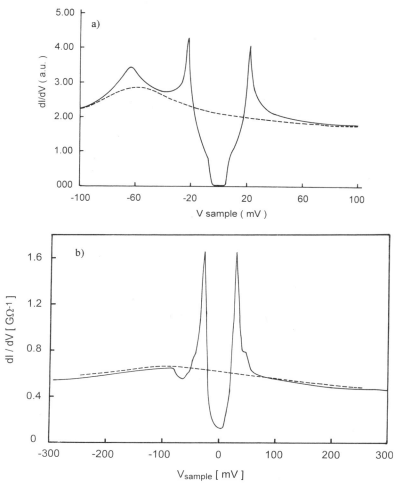

Figure 10 : (a) Curves of the conductance calculated for a N-I-S junction. Solid line: in the superconducting state at T = 5 K with Δ_{Max} = 22 meV, Δ_{min} = 6 meV, Γ = 0.1 meV, t = 0.2 eV and D_e = -60 meV, Γ' = 5 meV. Dashed line : in the normal state at T = 100 K with Δ_{Max} = Δ_{min} = 0 meV, Γ = 0.1 meV, t = 0.2 eV and D_e = -60 meV, Γ' = 5 meV. (b) For comparison we show Fig. (7) of Ref. [23]. The maximum of the normal state conductance (or D.O.S.) at negative sample bias is well reproduced.

SPECIFIC HEAT

1. Theoretical calculation

The purpose of this chapter is to evaluate the influence of the v.H.s. and the anisotropy of the gap on the specific heat calculated in the mean field B.C.S. approximation, i.e. we do not take into account the fluctuations near the critical temperature T_c. There are a great number of experiments measuring C_s. To compare our calculations to experiments, we must subtract the part due to fluctuations. These kind of adjustment have been made by various authors by using the fact that thermodynamic fluctuations are symmetric about T_c and can be easily evaluated above T_c.[25,26] Also we do not take into account the magnetic fluctuations in low temperature, nor the pair-breaking which may exist in overdoped sample. By the usual way, we obtain for C_s :

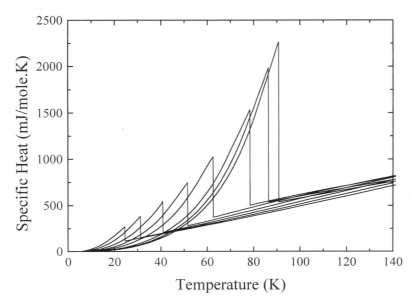

Figure 11 : The calculated specific heat versus the temperature for the different value of the doping $D_e = E_F - E_s = 0, 10, 20, 30, 40, 50, 60$ and 70 meV.

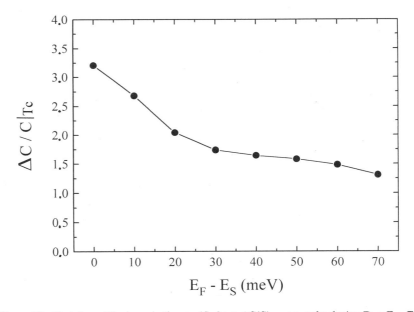

Figure 12 : Variation of the jump in the specific heat, $\Delta C/C|_{Tc}$, versus the doping $D_e = E_F - E_S$.

48

$$C_s(T) = \frac{2}{k_B T^2} \sum_k \frac{\exp(\varepsilon_k / k_B T)}{(1 + \exp(\varepsilon_k / k_B T))^2} \varepsilon_k^2 - \frac{1}{k_B T} \sum_k \frac{\exp(\varepsilon_k / k_B T)}{(1 + \exp(\varepsilon_k / k_B T))^2} \frac{\partial \Delta_k^2(T)}{\partial T} \quad (13)$$

We use the values of ε_k and Δ_k ($\Delta_{Max}(T, D_e)$ and $\Delta_{min}(T, D_e)$) given by formula (8) and (4,11) to evaluate the two integrals of (13) numerically. Near T_c we have a very good agreement between the calculated values and the following analytical formula :

$$\Delta_{Max,min} = \Delta_{Max,min}(T = 0)\, 1.7\, (1 - (T/T_c))^{1/2}$$

We see that the slopes $\partial \Delta^2 / \partial T$ do not depend on doping which simplifies the calculation of the second integral of formula (13). The results are presented in figures (11) and (12) where we plot C_s and $\Delta C/C|_{Tc}$ versus T for various doping levels D_e.

We can make the following observations :

1- The jump in specific heat varies with doping $\Delta C/C|_{Tc}$ is 3.2 for $D_e = 0$ and 1.48 for $D_e = 60$ meV compared to 1.41, the B.C.S. value for a isotropic superconductor, with a constant D.O.S., N_0 in the normal state. The high value of $\Delta C/C|_{Tc}$ is essentially due to the v.H.s when it coincides with the Fermi level and the highest value of the gap Δ_k. With doping, the v.H.s moves away from E_F and $\Delta C/C|_{Tc}$ decreases toward its B.C.S. value.

2 - There is also a difference in the specific heat C_N in the normal state. For a usual metal with a constant D.O.S. N_0, $\gamma_N = C_N/T$ is constant and proportional to N_0. Here we find $\gamma_N = a \ln(1/T) + b$ for $0 \leq D \leq 30$ meV where a and b are constant. For $D_e = 0$ this behaviour has already been predicted by Bok and Labbé in 1987[27]. The specific heat $C_N(T)$ explores a domain of width $k_B T$ around the Fermi level E_F. So for $D_e \ll k_B T_c$, the variation of γ_N above T_c is logarithmic. For $D_e > 30$ meV, at high temperature $T - T_c > D_e$, the B. L. law is observed, but for lower temperatures γ_N increases with T and passes through a maximum at T^*, following the law : T^* (meV) $= 0.25\, D_e$ (meV) or T^* (K) $= 2.9\, D_e$ (meV).

2. Comparison with experiments

Because of the difficulty to extract exactly C_s from the experimental data, we will compare only the general features to our calculation. We see that the doping has a strong influence on T_c and all the superconducting properties, so we assume that its role is to increase the density of holes in the CuO$_2$ planes. To compare our results on the effect of doping on C_s with experiments, we have chosen the family of the Tl$_2$Ba$_2$CuO$_{6+\delta}$, studied by Loram et al, fig. (9) of ref. [28], because they are overdoped samples, with only one CuO$_2$ plane. The family YBa$_2$Cu$_3$O$_{6+x}$ is underdoped for $x < 0.92$ and for $x > 0.92$ the chains become metallic and play an important role. However, recent results by Loram et al, fig. (2a) of the ref. [29] on Calcium doped YBCO, Y$_{0.8}$Ca$_{0.2}$Ba$_2$Cu$_3$O$_{7-\delta}$, which are overdoped two dimensionnal systems, show a very good agreement with our results. We notice the displacement and the decrease of the jump in specific heat C_s with doping. The jump $\Delta C/C|_{Tc} = \Delta \gamma/\gamma|_{Tc} = 1.67$ [28], and 1.60 [29] greater than the B.C.S. value 1.41 for a metal with a constant DOS. We find theoretically this increase in our model due to the logarithmic v.H.s.. The symmetrical shape of the peak of C_s, at low doping level, is due to the critical fluctuations. A subtraction of these fluctuations[25,26] gives an asymmetrical shape. For high doping levels the classical B.C.S. shape is found.

For $D_e = 0$, we find that γ_N is not constant but given by the logarithmic law[27] : $\gamma_N = a \ln(1/T) + b$. When D_e increases, the law changes, γ_N passes through a maximum for a value of T, T^*. This behaviour is clearly seen in the YBCuO$_{6+x}$ family[28]. We explain the high value $\Delta C/C|_{Tc} = 2.5$ for $x = 0.92$ in the YBCO family, and we find also the predicted variation of T^*.

Our model, neglecting magnetic fluctuations gives an Arrhenius law for C_s at low temperature with a caracteristic energy which is Δ_{min}. We see that such a law is observed in $YBaCuO_{6.92}$ and for $Tl_2Ba_2CuO_6$ at optimum doping.

EFFECT OF SCREENING ON THE GAP ANISOTROPY AND THE SPECIFIC HEAT

In the preceding parts we have taken $q_0a = 0.12$ and the effective coupling constant $\lambda_{eff} = 0.665$ in order to fit the experimental values of the gap observed by ARPES and tunneling spectroscopy. We also have stressed the importance of q_0a in the value of the anisotropy ratio $\alpha = \Delta_{Max}/\Delta_{min}$. We shall now study in more details the influence of q_0a on α and on the slope :

$$R = T_c\left(\frac{d\ln \Delta C}{dT}\right)_{T=T_c} \tag{14}$$

where $\Delta C = C_s(T) - C_s(0)$, $C_s(0)$ is computed for $\Delta_k = 0$ (normal state).
This slope R is available from many experiments for 2D superconductors.

The calculation use equation (5) where q_0a is included in $V_{kk'}$.
For this study, we adjusted our values of λ_{eff} to obtain a constant critical temperature of 90.75 K and an average gap of $\Delta_{av} = 14.50 \pm 0.15$ meV. This approximation is valid in the limit of weak screening ($q_0a < 0.2$). The results are presented in figure (13).
The specific heat is computed using formula (13). The results are presented in figures (14) and (15) for $\Delta C / C|_{Tc}$ and the slope R. The B.C.S. values for an isotropic s-wave superconductors are 1.43 for $\Delta C / C|_{Tc}$ and 2.62 for R.

Comparison with experiments :
There are no direct experiments to measure α as a function of q_0a. The photoemission experiments measure the anisotropy as a function of doping, so q_0a and E_F-E_s vary simultaneously. But there is a decrease in α when the doping is varying[21,22].

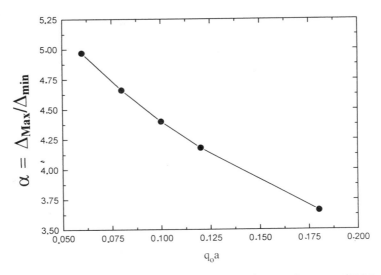

Figure 13 : The anisotropy ratio $\alpha = \Delta_{Max}/\Delta_{min}$ versus the screening parameter q_0a.

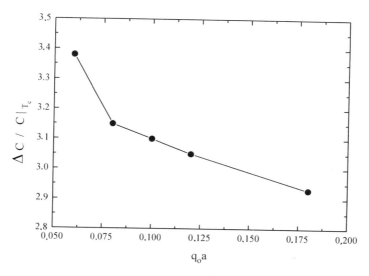

Figure 14 : The specific heat jump at T_c, $\Delta C / C\big|_{T_c}$ versus the screening parameter q_0a.

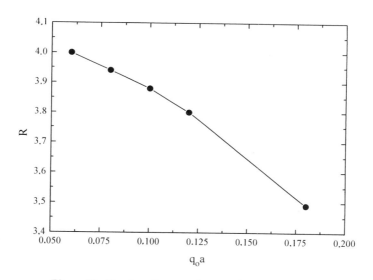

Figure 15 : The slope R versus the screening parameter q_0a.

On the other hand, there are a large number of measurements of specific heat[26,28,29,30]. The extraction of the electronic contribution to the total specific heat is rather difficult. A subtraction of the phonon part and of the influence of fluctuations near T_c is necessary. The slope R has been evaluated by Marcenat *et al*[26] for $YBa_2Cu_3O_{6.92}$. Experimental results have also been obtained by Junod *et al*[30] and Loram *et al*[28] for $Tl_2Ba_2CuO_6$ with a T_c of 85 K. For YBCO the value obtained experimentally by is $R = 6 \pm 1$ higher than our computed value. But YBCO is a special case with chains and 3D character. $Tl_2Ba_2CuO_6$ is a 2D material and the measured R is 4 ± 0.5 very close to our calculated value $3.5 < R < 4$ for reasonable values of q_0a.

In conclusion, our model explains anomalous values of $\Delta C / C|_{Tc}$ and R observed in 2D cuprates when the Fermi level is close to the van Hove singularity.

VAN HOVE SINGULARITY AND "PSEUDO-GAP"

Several experiments on photoemission, NMR and specific heat have been analyzed using a normal state pseudo-gap[31]. In fact, all what is needed to interpret these data is a density of state showing a peak above the Fermi energy. To obtain the desired D.O.S. several authors[31] introduce a pseudogap in the normal state. This seems to us rather artificial, the above authors themselves write that the physical origin of this pseudogap is not understood.

We have shown that by using a band structure of the form :

$$\xi_K = -2t\left[\cos k_x a + \cos k_y a\right] - D_e$$

where $D_e = E_F - E_s$, we may interpret the results obtained in the <u>normal metallic state</u>. We have computed the Pauli spin susceptibility[12] using the following formula :

$$\chi_p = \frac{\mu_o \mu_B}{B} \int_{-\infty}^{+\infty} n(\varepsilon)\left(f_{FD}(\varepsilon + \mu_B B) - f_{FD}(\varepsilon - \mu_B B)\right) d\varepsilon \qquad (15)$$

The results fit well the experiments. We find a characteristic temperature T* where the variation of χ_p versus T goes through a maximum. We may express D_e as a variation of doping $\delta p = p - p_0$, p_0 being the doping for which $E_F = E_s$, $p_0 = 0.20$ hole/copper atom in the CuO$_2$ plane. Figure (16) represents the various experimental points taken from figure (5) of reference [31] where the authors plot $E_g/k_B T_{cMax}$ versus p. We see that what the authors call pseudogap is exactly our $E_F - E_s$, the distance from the Fermi level to the peak in the D.O.S..

We have also computed the electronic specific heat C_s in the normal state[10] using the same D.O.S.. We find that $\gamma = C_s/T$ goes through a maximum with temperature T, at a value T* as found experimentally by Cooper and Loram[32]. In figure (17) we compare our computed T* with the experimental one (ref. [32]), the agreement is excellent.

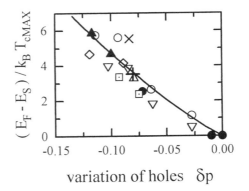

variation of holes δp

Figure 16 : $D_e = E_F - E_s$ divided by $k_B T_{cMAX}$ ($T_{cMAX} = 110$ K) versus the variation of the density of hole, calculated from the band structure of the formula (4) : solid line. The different symbols are the same as in the fig. (5) of the ref. [31]), they represent the values of the so-called normal pseudogap divided by $k_B T_{cMAX}$ ($E_g / k_B T_{cMAX}$) obtained from NMR on different compounds.
Our calculations are made with a transfer integral $t = 0.25$ eV, δp is taken as zero for $p = 0.20$.

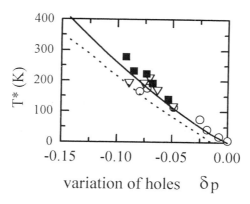

variation of holes δp

Figure 17 : The temperature, T*, where the calculated χ_p (dashed line) and the specific heat (solid line) go through a maximum, versus δp. For comparaison we show the results presented in fig. (27) of ref. [32], the symbols are the same. (solid squares : from thermoelectric power, circles : from specific heat, triangles: from NMR Knight shift data).

In conclusion, we are able to interpret the NMR and specific heat data in the normal metallic state without invoking a pseudogap, but simply by taking into account the logarithmic singularity in the D.O.S..

We explain the shift between the observed experimental optimum T_c, where p = 0.16 instead of 0.20, and the expected optimum T_c from our theory, i.e. where $D_e = 0$, by the fact that in first time in our gaps calculations we have not taking into account the variation of the 3D screening parameter q_0a in function of D_e. These calculations are in progress and show the competition between the effect of the position of the v.H.s. and the value of q_0a for getting the optimum T_c, this competition depends on the compound. When the overdoping increases, i.e. the density of free carriers increases, then q_0a increases too, and in our model this leads to a decrease in T_c. It is why for $D_e = 0$, or dx = 0.20, we have not the optimum T_c, and why the logarithmic law for χ_p is found in the overdoped range[12]. In the underdoped range in respect of the observed optimum T_c, (i.e. density of free carriers decrease), q_0a decrease too, but the Fermi level go too far away the singularity to obtain high T_c. By this way our results agree completely with the experimental observations.

CONCLUSION

Note that our model is valid only in the metallic state. It has been shown by Boebinger et al[33] that LaSrCuO for example undergoes a metal-insulator transition in the underdoped regime, fonction of temperature in the normal state. Of course our model is not valid for these very low doping levels.

We have shown the importance of the van Hove singularity in the interpretation of the physical properties of the cuprates high T_c superconductors (HTSC). The v.H.s. is essential to obtain high T_c and therefore in the coupling mechanism. In the framework of that scenario we explain a main characteristic : the anisotropic gap. V.H.s. account for several experimental features seen in conductance, specific heat, Pauli susceptibility and so on. Last the shift of Fermi level from singularity level explains the so-called pseudogap and the behaviour of HTCS versus doping.

The physic of HTSC, seeing all these convincing results, have to take into account the v.H.s..

REFERENCES

1. Z. X. Shen, W.E. Spicer, D.M. King, D.S. Dessau, B.O. Wells, *Science* **267** (1995) 343.

2. H. Ding, J. C. Campuzano, K. Gofron, C. Gu, R. Liu, B. W. Veal and G. Jennings. *Phys. Rev. B* **50** (1994) 1333.

3. Jian Ma, C. Quitmann, R.J. Kelley, P. Alméras, H. Berger, G. Margaritondoni, M. Onellion. *Phys. Rev. B* **51** (1995) 3832.

4. J. Labbé and J. Bok, *Europhys. Lett.* **3** (1987) 1225.

5. D. M. Newns, C. C. Tsuei, P. C. Pattnaik, and C. L. Kane. *Comments Cond. Mat. Physics* **15** (1992) 273.

6. J. Bok and L. Force. *Physica C* **185-189** (1991) 1449.

7. L. Force and J. Bok. *Solid Stat. Comm.* **85** (1993) 975.

8. T. Yokoya, A. Chainini, T. Takashahi, H. Katayama-Yoshida, M. Kasai and Y. Tokura, *Physica C* **263** (1996) 505.

9. J. Bouvier and J. Bok, *Physica C* **249** (1995) 117.

10. J. Bok and J. Bouvier, to be published in *Physica C*.

11. J. Bok and J. Bouvier, *Physica C* **274** (1997) 1.

12. J. Bok and J. Bouvier, "Comment on NMR evidence for a d-Wave Normal-State Pseudogap" submitted to *Phys. Rev. Lett.* (1997)
    "The van Hove scenario of high T_c superconductors : the effect of doping" published in the *proceeding of the GDR "Supraconducteurs"* n° 1063, 23-25 juin 1997

13. C.C. Tsuei, D.M. Newns, C.C. Chi et P.C. Pattnaik, *Phys. Rev. Lett.* **65** (1990) 2724. D.M. Newns, C.C. Tsuei, P.C. Pattnaik and C.L. Kane, *Comments Cond. Mat. Physics* **15** (1992) 273.

14. G.G. Olson et al, *Phys. Rev. B* **42** (1990) 381.

15. Z. Schlesinger et al, *Phys. Rev. Lett.* **67** (1991) 1657.

16. Y. Kubo et al, *Phys. Rev. B* **43** (1991) 7875.

17. N. Bogolubov, N. Tolmachev and D. Shirkov, *''A new method in the theory of superconductivity''* (1959) Cons. Bureau, NY.

18. M.L. Cohen and P.W. Anderson in *Superconductivity in d and f band Metals*, edited by D.H. Douglass (A.I.P. New York) (1972) .

19. V. Ginzburg, *Comtemporary Physics* **33** (1992) 15.

20. Y. Koike et al, *Physica C* **159** (1989) 105.

21. M. Onellion, R.J. Kelley, D.M. Poirier, C.G. Olson and C. Kendziora, preprint "Superconducting energy gap versus transition temperature in $Bi_2Sr_2CaCu_2O_{8+x}$"

22. R.J. Kelley, C. Quitmann, M. Onellion, H. Berger, P. Almeras and G. Margaritondo, *Science* **271** (1996) 1255.

23. Ch. Renner and O. Fisher, *Phys. Rev.* **B 51** (1995) 9208.

24. Y. Hwu, L. Lozzi, M. Marsi, S. La Rosa, M. Winuku, P. Davis, M. Onellion, H. Berger, F. Gozzo, F. Levy and G. Margaritondo, *Phys. Rev. Lett.* **67** (1991) 2573.

25. O. Riou, M. Charalambous, P. Gandit, J. Chaussy, P. Lejay and W.N. Hardy *"Disymmetry of critical exponents in YBCO"* preprint to be published in *LT 21 proceedings.*

26. C. Marcenat, R. Calemczuk and A. Carrington, *"Specific heat of cuprate superconductors near T_c"* published in *"Coherence in high temperature superconductors"* edited by G. Deutscher and A. Revcolevski World Scientific Publishing Company (1996).

27. J. Bok and J. Labbé, *C.R. Acad. Sci Paris* **305** (1987) 555.

28. J.W. Loram, K.A. Mirza, J.M. Wade, J.R. Cooper and W.Y. Liang, *Physica C* **235-240** (1994) 134.

29. J.W. Loram, K.A. Mirza, J.R. Cooper, J.L. Tallon. to be published in the *proceedings of M^2S-HTSC-V*, Feb 28-Mar 4, 1997, Beijing, China.

30. A. Junod, A. Beringe, D. Eckert, T. Graf and J. Müller, *Physica C* **152** (1988) 495.

31. G.V.M. Williams,J.L Tallon,E.M. Haines,R. Michalak, and R. Dupree, *Phys. Rev Lett.* **78** (1997) 721.

32. J.R. Cooper and J. W. Loram, *J. Phys. I France* **6** (1996) 2237.

33. G.S. Boebinger, Y. Ando, A. Passner, T. Kimura, M. Okuya, J. Shimoyama, K. Kishio, K. Tamasaku, N. Ichikawa and S. Uchida, *Phys. Rev. Lett.* **77** (1996).

HIGH T_C OXIDES: TWO ORDER PARAMETERS, MAGNETIC SCATTERING AND UPPER LIMIT OF T_C, NOVEL ISOTOPE EFFECTS, AND THE PHONON-PLASMON MECHANISM.

Vladimir Z. Kresin,[1] Andreas Bill,[1]
Stuart A. Wolf,[2] and Yu. N. Ovchinnikov[3]

Lawrence Berkeley Laboratory, University of California,
Berkeley, CA   94720
Naval Research Laboratory, Washington, D.C. 20375-5000
Landau Institute for Theoretical Physics,
Moscow, Russia, 11733V

INTRODUCTION

This paper is concerned with several interrelated phenomena. Let us, at first, mention some unusual properties of the overdoped state. It is well known that overdoping leads to a drastic decrease in T_c. The measurements of the critical magnetic field for Tl-and Bi-based cuprates revealed [1,2] a peculiar dependence of H_{c2} on temperature entirely different from the conventional picture [3]. In the regard to Tl-and BI-based cuprates , one should add that these materials are characterized by a strong dependence of T_c on the number of Cu-O planes in the unit cell. In addition, the isotope effect is characterized by peculiar dependence on oxygen content, Zn substitution, etc[4,5]

The present paper contains a description of the concept which allows us to explain all these phenomena.

The Gap Symmetry and Fluctuations in High - T_c Superconductors
Edited by Bok *et al.*, Plenum Press, New York, 1998

TWO ORDER PARAMETERS

Let us consider the YBaCuO compound. This compound contains a CuO quasi-one dimensional chain structure in addition to the CuO planes. The chains provide doping for the CuO planes, but, in addition, and this is particularly important for our model, they form an independent conducting subsystem. As a result, below T_c, in the superconducting state, the material displays a two-gap structure; each of the subsystems is characterized by its own energy gap. In addition, each of the gaps can be anisotropic. Let us denote by α and β the plane and chains subsystems, so that ε_α and ε_β are the corresponding energy gaps. The opportunity to observe a two-gap spectrum in the high T_c oxides was considered theoretically by us in [6,7]. The inequality $l \ll \zeta$ (l is a mean free path, ζ is the coherence length) leads to isotropization of gap in the conventional materials, as well as to the averaging into single-gap picture. The shortness of the coherence length in the high T_c oxides leads to opportunity to observe the two-gap picture.

An explicit definition of the two-gap spectrum corresponds to the presence of the two-peak structure in the superconducting density of states (there are two order parameters). The observation of such structure with the use of tunneling spectroscopy [8] provides a strong support for the picture. In the following, we will use the two-gap picture to refer to the two peaks in the density of states.

The plane (α) and chain (β) subsystems are coupled throught the charge transfer. Because of the charge transfer, the system is characterized by a single value of T_c. The double gap on the planes, $2\varepsilon_\alpha(0)$, is equal to approximately $5T_c$. The value of the smaller gap is very sensitive to the oxygen content and for x=0.4 is approximately equal to T_c. This value of the chain gap is smaller than the BCS value.

The supeconducting state in the Cu-O plane, which is a basic unit for all cuprates is caused by same intrinsic mechanisms (for our present treatment the nature of this mechanism is not essential). As for the subsystem (chains), the pairing is mainly induced by the charge transfer. There are two channels for the induced superconductivity:

1) Intrinsic proximity effect; the word "intrinsic" stresses the fact that, unlike the usual proximity effect observed in thin-film sandwich structure, we are dealing with a phenomenon occuring on the scale of unit cell. Nevertheless, the physics of the phenomenon is similar and represents the tunneling of the pair $\alpha \leftrightarrow \beta$. The process can be

described by the McMillan tunneling model [9] of the proximity effect. 2) Mediated charge transfer. More specifically, the carrier from the α subsystem radiates a phonon (or another excitation) and makes a transition to β. Another carrier absorb the phonon and also makes transition to β; as a result of this phonon exchange, these two carriers form a pair in the β subsystem.

MAGNETIC SCATTERING; PAIR-BREAKING EFFECT

As is known , the overdoped state of the cuprates is characterized by a drastic decrease in T_c upon overdoping. For example, one can observe a decrease: $T_c=90K-->T_c=14K$ for the overdoped $Tl_2Ba_2CuO_{6+x}$ compound; a similar decrease was observed for the Bi-based cuprate. We conclude that this decrease is mainly caused by the pair-breaking effect of magnetic impurities. The influence of magnetic impurities (pair-breaking, appearance of gaplessness at some value of the concentration of magnetic impurities, depression in T_c, etc.) was described in [10], see also the reviews [11]. The interaction between localized magnetic moments and Cooper pairs which are in the singlet state, destroys the pair correlation and is accompanied by spin-flip scattering (such scattering provides the conservation of the total spin).

Gapless State

Consider the case when the chain subsystem contains magnetic impurities. In the absence of the impurities the system displays the two-gap spectrum when the chains in YBCO are in the induced superconducting state. If the β-system (chains) contains magnetic impurities, then we are dealing with an unusual case of gapless superconductivity, namely at some value of the impurity concentration the energy gap becomes equal to zero, whereas the shift in T_c, unlike the usual case, is relatively small. Such a case occurs upon the depletion of oxygen. Indeed, the removal of oxygen greatly affects the chain's states. Instead of a well-developed chain structure, we have a set of broken chains with Cu atoms at the end. These Cu atom form local magnetic states Cu^{++}, similar to surface states. These magnetic moments act as strong pair-breakers in the chain band. As a result, the chain rapidly develops a gapless state. One should note that the gapless state, despite the absence of the energy gap, is still a superconducting state .

The absence of the gap leads to a power law, rather than exponential dependences of the electronic heat capacity, impedance, penetration depth, etc., but nevertheless, the material still exhibits the Meissner effect and zero resistance. Qualitatively, the gapless superconductivity can be viewed as a mixture of a "normal" component (broken pairs) and the "superconducting" component. This picture is similar to the "two-fluid" model. That's why this model provides a good description of many properties of the cuprates. Scattering by the impurities can be described by additional term in the renormalization functions Z_β (see e.g. [6]). One can calculate the critical concentration n which corresponds to an apppearance of the gapless state. It turns out that for x=0.1, that is for $Y\ Ba_2\ Cu_3\ O_{6.9}$ compound, we are dealing with the gapless state.

It is essential that, although the magnetic moments are introduced in the chain sites only, the energy gaps become equal to zero in both subsystems, chains and planes. Namely, as a consequence of the charge transfer, both densities of the states $\upsilon_\alpha(\omega)$ and $\upsilon_\beta(\omega)$ are not equal to zero up to $\omega=0$. Note also, that even the gaps are equal to zero, the density of states $\upsilon_\alpha(\omega)$ displays the peak at $\omega=\varepsilon_\alpha$, and this peak can be observed experimentally, e.g., by tunneling measurements.

Critical Field

The dependence of H_{c2} on T which is drastically different from the conventional picture, has been observed recently in the overdoped high-T_c oxides [1,2]. Conventional bulk superconductors are characterized by linear temperature dependence of H_{c2} near T_c, a quadratic behavior as T-->0 , and by negative curvature over the entire temperature region [3]. Contrary to this picture, the layered cuprates are described by linear dependence near T=0K and by positive curvature at all temperatures. In addition , the value of $H_{c2}(0)$ greatly exceeds the value that follows from the conventional theory. Similar effect has been observed in the Sm-Ce-Cu-O, La-Sr-Cu-O and, recently, in Y-Ba-Zn-Cu-O [12] systems.

The superconducting state of the Cu-O plane is greatly affected by magnetic impurities. Let us describe here a qualitative picture (a more detailed description see in [13]). Spin-flip scattering leads to a depression of the superconducting state, and this is reflected in relatively small values of T_c and H_{c2}. The magnetic impurities can be treated as independent [8], and the spin-flip scattering by the impurities provides the conservation of the total spin. However, at low temperatures (in the region T=1K), because of the correlation of the magnetic moments, the trend to ordering of the moments becomes important, and this trend frustrates the spin-flip scattering. Pairing becomes less depressed, and leads to a large increase in the value of H_{c2} and, correspondingly, to a positive curvature in H_{c2} vs T (for a more detailed discussion see [13]).

The ordering of the impurities was observed for the Sm-Ce-Cu-O compound by direct measurements of the magnetic susceptibility [12a]. The increase in the susceptibility which reflects the correlation of the moments, has been also observed in [12b]. Note also that the effect of the ordering on the curvature in $H_{c2}(T)$ has also been observed in the Ni/V system [14]. The appearance of positive curvature appears when $T_c > T_{Cu}$, T_{Cu} is the Curie temperature.

Therefore, the unusual dependence of $H_{c2}(T)$ is due to a relatively weak temperature dependence of the spin-flip relaxation time τ_s. This dependence is caused by the presence of magnetic impurities and their correlations at low temperatures. As a result, one can observe a strong deviation from the conventional theory [3].

The general equation describing a layered superconductor in the presence of magnetic impurities and an external field (at $H=H_{c2}$) has the form (see [13]):

$$\ln(2\gamma\Gamma_{cr}/\pi T) - [\psi(0.5 + (\Gamma/2\pi T)) - \psi(0.5)] = f(H_{c2},T) ; \qquad (1)$$

$$f(H_{c2},T) = (H_{c2}v^2/\Gamma) \{\Gamma^{-1} [\psi(0.5 + (\Gamma/2\pi T)) - \psi(0.5 + (\Gamma/4\pi T))] - (4\pi T)^{-1}\psi'(0.5 + (\Gamma/2\pi T))\} ;$$

Here ψ is the psi-function, $\ln\gamma = C = 0.58$ is the Euler constant, $\Gamma = \tau_s^{-1}$ is the amplitude of the spin-flip scattering, τ_s is the relaxation time; Γ_{cr} corresponds to the complete supression of superconductivity ($T_c=0$).

The theoretical approach, based on pair-breaking effect of magnetic impurities, has strong experimental support. For example, the heat capacity measurements (see e.g., review [15a]) which show low temperature Shottky anomalies, are direct experimental evidence for the presence of magnetic impurities. Magnetic impurities provide the pair-breaking effect. The μSR spectroscopy data [15b] contain direct experimental evidence for this effect.

Pair-breaking leads to a depression in T_c and, eventually, at some value of the concetration of magnetic impurities $n_M = n_{M;cr.}$, to the total supression of superconductivity. This suppression is proceeded by the appearance of the gapless state.

There is an interesting question about the nature and location of magnetic impurities. First of all, one should note that the oxygen depletion leads to formation of the Cu^{++} ions on the chains. In addition, one can observe the formation of magnetic clusters in the Cu-O planes. Moreover, the additional magnetic moments are localized on the Ba-O layer, which is adjacent the Cu-O layer for overdoped Tl- and Bi-based compounds. Localized magnetic moments can be formed on the apical oxygen site. Probably,we are dealing with the formation of the paramagnetic radical O_2^-, and this also involves the apical oxygen, as well as an additional oxygen in the Tl-O (Bi-O) layers.

The presence of magnetic moments on the apical oxygen site is a key factor which determines the large difference in the values of Tc between the 2201 and1223 (or 2223) structures. Indeed, the properties of the $Tl_2Ba_2CuO_6$ compound, and its value of Tc, in particular, are greatly affected by the two O^- apical ions. In the case of $Tl_2Ba_2Ca_2Cu_3O_{10-x}$, the pair-breaking effect is much weaker since the apical oxygens are located outside of the total set of three planes, so that the middle plane is hardly affected by the magnetic ions and the influence on the two other planes is not as strong as for the 2201 compound, where the Cu-O plane is affected on each side. As a result, the Tl-based 2223 compound has a $T_c=125K$, which is much closer to the "intrinsic" limit $T_{c;int}=155K$, see above). For compounds with more than three Cu-O planes, it is difficult to adequately dope the inner planes and T_c saturates or decreases.

Other Manifestations of the "Recovery" Effect: Josephson Tunneling, Surface Resistance.

As is well known, the temperature dependence of the amplitude of the Josephson current is described by the Ambegaokar-Baratoff (AB) equation [16]. However, if the superconductor contains magnetic impurities, the dependence becomes different [17]. The ordering trend (see above) leads to a noticeable modification of the AB dependence. Indeed, as was noted above, the pair-breaking effect leads to depression of T_c, critical field,etc. The order parameter which determines the amplitude of the Josephson current is also depressed. However, because of the ordering trend, the spin-flip relaxation time $\tau_s = \Gamma_s^{-1}$ depends on temperature (see above). As a result, one can observe a temperature dependence of the amplitude different from that for usual junctions. Particularly interesting is the appearance of the upturn in the low temperature region. This upturn is the manifestation of the "recovery" effect .

The ordering trend can be provided not only by decrease in temperature, but an external magnetic field. Its presence frustrures the spin-flip scattering. An interesting phenomenon was observed in [18]. Namely, the microwave losses decrease in an external magnetic field. In addition, a decrease in the penetration depth was observed. This strengthening of superconductivity is caused by frustration of the spin-flip scattering (see above) and corresponding weakening of the pair-breaking effect.

"INTRINSIC" T_C ; UPPER LIMIT OF T_C IN THE CUPRATES

An analysis of the overdoped state, and the dependence $H_{c2}(T)$, in particular, allows us to introduce an important new parameter, the so-called "intrinsic" T_c. The dependence $H_{c2}(T)$ can be used in order to evaluate $\Gamma_{cr.}$ (see Eq.(1)).

One can obtain the value $\Gamma_{cr.}=137K$ for the 2201 Tl-based cuprate studied in [1]. Using the relation $T_{c;m}^0=2\gamma\Gamma_{cr.}/\pi$ [3], where $T_{c;m}^0$ is the value of critical temperature in the absence of the impurities, one can obtain $T_{c;m}^0=160K$ for the same layered compound. It is remarkable that this value greatly exceeds the experimental value $T_{c;m}=90K$, observed at the optimum ambient pressure doping. Therefore, the material contains magnetic impurities even at optimum doping, and the value of the critical temperature $T_{c;m}^{dop.}$ is depressed relative to the "intrinsic" value $T_{c;intr.}=T_{c;m}^0$. The value of tc can be raised if another metod of doping (e.g.pressure, see [19]), different from the chemical method, is used.

The layered superconductors are characterized by a large ratio of the intrinsic critical temperature $T_{c;m}^0$ which corresponds to the absence of the magnetic impurities (they act as a depressing mechanism) to the observed value of T_c.

Therefore, the 2201 Tl-based cuprate contains magnetic impurities even at optimum doping; the critical temperature is depressed relative to its "intrinsic" value, which is equal to $T_{c;intr.}$ =160K. One cam carry out a similar analysis for the 2201 Bi-based cuprate with use of the data [2]. A detailed calculation [13] leads to a similar value $T_{c;intr.}$ =160K.

Recently measurements of $H_{c2}(T)$ for the underdoped Y-Ba-Zn-Cu-O system were described in [12c]. The behavior of the critical field appears to be similar to that of the overdoped cuprates studied in [1,2]. This case corresponds to the "dirty" case (the mean free path l=12A). The value of the "intrinsic" T_c has been also evaluated [13]. It is remarkable that for the YBCO compound the value of $T_{c;intr.}$ also appears to be close to that for other cuprates.

We conclude that all cuprates has a similar value for $T_{c;intr.}$. This is due to the fact that all cuprates contain the same basis structural unit, namely, the Cu-O plane. In some sense, there is one high T_c superconductor, namely, the Cu-O plane, and this superconductor, depending on many factors, such as the doping level, structure of the sample, phonon spectrum, oxygen content, etc. can have a different value of critical temperature in the range from T_c=0K up to $T_{c;intr.}$=160K.

Of course, the value of this upper limit is characterized by some uncertainty. It could be affected by strong coupling effects. In addition, there is the possibility of reaching a doping level (in the absence of magnetic impurities), which exceeds the maximum level for the overdoped cuprates. But our main conclusion is that there is a maximum value of T_c for the cuprates, and this value is in the region near 160-170K.

The maximum value of T_c for the Hg-based cuprate is not far from our estimated limit. One can ask: what makes this material almost ideal? We speculate that this is related to the fact that the Hg-based cuprate , even at maximum T_c , is characterized by a small amount of oxygen in the Hg-O layer, and, according to our scenario, by a small concentration of magnetic pair-breakers. In addition, the distance between the apical oxygen and the Cu-O plane is larger in this material relative to other cuprates.

NOVEL ISOTOPE EFFECTS

In this section we focus on the novel isotope effect which are not affected by lattice dynamics [19-21]. Let us consider a superconductor which contains magnetic impurities. This case is related to recent experiments on isotope substitution with the high-T_c oxides (see below). Note, however, that the effect we are discussing, can be observed in conventional superconductors as well.

Magnetic Scattering; Zn Substitution.

As was noted above,the presence of magnetic impurities leads to decrease of the critical temperature, T_c, relative to the intrinsic value T_{c0} ,because of the pair-breaking effect. This depression is described by a well-known equation [3a]:

$$\ln (T_c^0 /T_c) = \Psi[0.5+ \gamma_s] - \Psi(0.5) \tag{2}$$

Here $\gamma_s= \Gamma_s/2\pi T_c$, $\Gamma_s =\tau_s^{-1}$ is the spin-flip scattering amplitude (see above); $\Gamma_s \propto n_M$, n_M is the concentration of magnetic impurities. Eq.(2) is valid in the weak coupling approximation. Note that there is a non-linear coupling between T_{c0} and T_c.

The isotope substitution $M \rightarrow M^*$ for the sample without magnetic impurities allows one to observe the shift in T_{c0} and measure the isotope coefficient α_0 which is described by the relation :

$$\alpha_0=- (M/\Delta M)(\Delta T_{c0}/T_{c0}) \tag{3}$$

Here $\Delta M = M^* - M$, $\Delta T_{c0} = T_{c0}^* - T_{c0}$. The presence of magnetic impurities leads to a change in T_C and in the isotope coefficient. One can see directly from Eqs. (2) and (3) that the shift in T_C and the new value of the isotope coefficient

$$\alpha = -(M/\Delta M)(\Delta T_c/T_c) \tag{4}$$

differ from $\Delta T_{c0}/T_{c0}$ and the value of α_0. Calculating the shift ΔT_c from Eqs.(2)-(4), one can arrive at the following equation:

$$\alpha = \alpha_0 [1 - \psi'(0.5 + \gamma_s) \gamma_s]^{-1} \tag{5}$$

Increase in n_M leads to increase in γ_s. Eq.(5) was obtained in [22].

The presence of magnetic impurities leads to an increase of the isotope coefficient ($\alpha > \alpha_0$), since $\psi' > 0$; this can be seen directly from the expression $\psi'(0.5 + x) = \Sigma(k + 0.5 + x)^{-2}$. One can study the dependence α on the concentration of magnetic impurities. For small γ (small values of n_M) $\Delta\alpha \propto n_M$. Therefore, near T_{c0} the critical temperature displays a linear decrease with increasing in n_M, whereas the isotope coefficient increases linearly as a function of n_M. In the region $\gamma \gg 1$ one can use an asymptotic expression for the digamma function, and we obtain $\Delta\alpha \propto n_M^2$, i.e. the dependence becomes strongly non-linear. This picture is in very good agreement with experimental data [4,5].

Therefore, the effect of isotope substitution can be greatly affected by the presence of magnetic impurities. This effect should be observed for conventional as well for high T_c superconductors. To the best of our knowledge, this effect has not been studied experimentally for conventional superconductors, and it would be interesting to carry out these measurements on simple mono-atomic superconductors.

With regard to the high-T_c oxides, we think that the experimental data [4,5] are directly related to the present theory. The value of the critical temperature for the YBCO compound was modified by Zn and Pr substitutions. For both types of the substitution one can observe a decrease in T_c and an increase in the value of the isotope coefficient.

Consider first the case of Zn-substitution . The substitution of Zn for Cu in the YBCO compound leads to a decrease in T_c. This decrease can be explained by the pair-breaking effect. Indeed, according to [23], this substitution leads to the appearance of additional magnetic moments on the Cu site in the Cu-O plane. It is important to note that this decrease in T_C is also accompanied by a peculiar temperature dependence of the critical field H_{c2} observed in [12c] (positive curvature, sharp upturn as $T \to 0K$, see above).

The isotope coefficient on T_C is described by Eqs.,(2),(5); $n_M = n_{Zn}$. There is a very good agreement with the data [4]. It is essential that the described analysis has been carried out without any adjustable parameter.

Proximity Effect

The pair-breaking can be also provided by the proximity effect. The behavior of T_C is described by equation similar to (2), see [9,24]. As a result, one can obtain the following equation for the isotope coefficient for the S-N proximity system:

$$\alpha = \alpha_O [1 + (\upsilon_N L_N)/(\upsilon_S L_S)] \tag{6}$$

One can see that, indeed, the value of the isotope coefficient is modified by the proximity effect. Moreover, $\alpha > \alpha_0$. Therefore, a decrease in T_c which is a well-known feature of the proximity effect, is accompanied by an increase in the isotope coefficient. It is interesting that one can modify the value of α by changing the thickness of the films. For example, the increase in the thickness of the normal film L_N leads to decrease in T_c, but the value of α increases.

Non-Adiabatic Isotope Effect

In our paper [19] we introduced the" non-adiabatic" isotope which occurs in the presence of the Jahn-Teller crossing of the electronic terms and which is manifested in the dependence of the carrier concentration on isotope substitution. In the perfectly realistic case when we are dealing with the degeneracy of the terms (for example, this leads to an appearance of several close minima for the apical and in-plane oxygen in YBCO), and presence of magnetic impurities, one should use Eq.(5) with

$$\alpha_0 = \alpha_{na} = \beta(n/T_c)(\partial T_c/\partial n) \, , \quad \beta = const.$$

The change of the isotope coefficient caused by the Pr-substitution [4] can be described by this equation. Indeed, the Pr-substitution occurs on the Y site. The value of Tc decreases with increasing Pr content, whereas the value of the isotope coefficient increases. As is known (see e.g. [25]), the effect of Pr on T_c is two-fold. First, the presence of Pr leads to a pair-breaking effect, similar to Zn. Secondly, the mixed valence state of Pr leads to depletion of holes from the CuO plane and, correspondingly, to an additional decrease in T_c. This second channel is related to the charge transfer between the CuO plane and Pr and involves dynamics of the non-adiabatic in-plane oxygen. The analysis of the data [4], based on Eq.(5) with $\alpha_0 = \alpha_{na}$, leads to good agreement between the theory and the data.

ORIGIN OF HIGH T$_C$; PHONON-PLASMON MECHANISM

This section is concerned with problem of the origin of high-T_c in the cuprates. The superconducting state in conventional superconductors is caused by the electron-phonon interaction (BCS theory). At the same time it is known that the pairing can be mediated by other excitations.

We think that the high T_c is due to strong coupling of the carriers with low energy excitations (generalized phonon or phonon-plasmon mechanism), namely with phonons and a peculiar phonon-like electronic acoustic branch ("electronic" sound) [26,27].

Phonons play a key role in high T_c superconductivity. Such statements may sound old-fashioned, because it means a similarity with conventional superconductors. However, we want to stress that we are dealing with an exotic phonon system in the cuprates. These materials contain soft optical modes. The lattice dynamics is very peculiar and differs drastically from that in conventional metals.

The concept of plasmons is not a new phenomenon in solid state physics. Plasmons describe the collective motion of the carriers relative to the lattice. Usual metals are characterized by the following dispersion relation for the plasmons: $\omega = \omega_0 + aq^2$; therefore, the plasmon spectrum has a finite value ω_0 at q=0. The situation in layered materials is quite different. Note that the term layered electron gas (LEG) describes the basic physical concept,namely an infinite set of two-dimensional layers of carriers,described as a 2D electron gas,separated by electronically inactive (insulating) layers.

In the LEG model the Coulomb interaction between charge carriers on different layers is included exactly, while the polarization of the system is treated by including the electron-hole pair responce calculated for a single layer with 2D plane-wave wave-functions for the carriers.

The plasmon spectrum in the layered conductors such as cuprates represent the plasmon band .The density of states is peaked near the upper ($q = 0$) and low ($q = \pi/d_c$) boundaries. Qualitatively, one visualizes the plasmon band as a set of two branches , the upper (U) branch is similar to that in usual metals [27]. A very important feature of the layered metals is the appearance of the lower (L) branch, which has an acoustic dispersion law and can be called "electronic" sound. One should stress also that the slope of this acoustic branch is of order of v_f (not the sound velocity as for usual phonons), however, the value of v_f is small [28], and therefore, we have an additional phonon-like branch with large phase space. This branch makes an additional contribution to the pairing.

It is very important to determine the strength of the coupling between the carriers forming the paired state. Indeed, the value of T_c is determined by two parameters: the strength of the coupling λ and the energy scale W, so that $T = T(\lambda,W)$. For example, in the BCS model W is a scale of the phonon energies.

The determination of electron-phonon coupling is not a trivial task.The use of conventional technique,tunneling spectroscopy, has been frustrated by a short coherence length. In the absence of high quality tunnelling spectroscopy data, which is the best way to determine λ, the electron phonon coupling parameter, we have presented an alternative method based on the analysis of a fundamental bulk property,heat capacity [29]. Electron phonon coupling (EPC), along with temperature dependence of the phonon distribution lead to the dependence of the Sommerfeld constant γ ($\gamma=C_{el}/T$) on temperature [30]. This manifests itself in a deviation of the electronic specific heat from a simple linear law. This dependence is described by the equation

$$\gamma(T)=\gamma(0)\{1+\rho[[\kappa(T)/\kappa(0)]-1]\}$$

(7)

where $\gamma(0) = \gamma^0(1+\lambda)$ and γ^0 is the band value of the Sommerfeld constant, $\lambda=2\int d\Omega g(\Omega)\,\Omega^{-1}$;$\rho=\lambda[1+\lambda]^{-1}$ and $\kappa(T) = 2\int d\Omega g(\Omega)\,\Omega^{-1}Z(T/\Omega)$ where Z is a universal function that has been derived in [30] and Ω is a phonon frequency. One can see that at low temperatures $(T\to 0)$ the Sommerfeld constant approaches the band value renormalized by the EPC; i.e $\gamma(0) = \gamma^0(1+\lambda)$; it means that the carriers are "dressed" by phonons and other excitations. Then we obtain

$$\lambda^* = [\gamma(0)/\gamma(T_c)]-1$$

(8)

where $\lambda^* = \lambda_{ph} / (1+ \lambda_{pl})$; therefore λ^* represents the minimum value of the electron- phonon coupling constant. Using Eq.(8) ,one can evaluate the value of λ^*. The value of the carrier-phonon coupling constant is even larger.

Thus, if we can use experimental data to extract the values of the low and high temperature Sommerfeld constant then we can get an estimate the minimum value of the electron-phonon coupling constant λ_{ph}. As a result,we obtained the value $\lambda^* = 2.5$, and it means a strong carrier-phonon coupling (let us remind,that for Pb, which is conventional superconductor with strong coupling, $\lambda = 1.4$).

The critical temperature can be calculated from the general equation valid for any value of the coupling constant [31] :

$$T_c= 0.25<\Omega>/(e^{2/\lambda_{eff}} -1)^{1/2}$$

(9)

$\lambda_{eff} = (\lambda-\mu^*)/ [1+ 2\mu^*+\lambda\mu^* t (\lambda)]^{-1}$; the function $t(\lambda)$ is introduced in [31]. The presence of the low frequency plasmons leads to an effective decrease in μ^*, so that $\mu^* = 0,-0.1$. Then we obtain high value of the critical temperature.

Our analysis leads to the conclusion that the unusual lattice dynamics as well as the layered structure in the cuprates leads to an exotic version of the BCS theory with a strong interaction between low energy excitations (phonons and phonon-like plasmon branch) and the carriers.

CONCLUSION

The main results can be summarized as follows:

1. One can observe a "two-gap" spectrum, that is, the density of states is characterized by presence of two peaks.

2. The cuprates contain magnetic impurities ; they act as pair-breakers, and their presence leads to several substantial effects, including depression in T_c in the overdoped compounds, gapless behavior, etc. The temperature dependence of the spin-flip scattering amplitude leads to an unusual temperature dependence of the upper critical field.

3. The cupraters are characterized by an "intrinsic" T_c, which corresponds to the absence of magnetic impurities. It appears that $T_{c;intr.}$ is in the range160-170K ; this value appears to be a universal parameter for all cuprates and represents an effective upper limit of critical temperature for this class of materials.

4. One can observe unconventional isotope effects not related to the pairing mechanism.

5. High T_c is due to strong electron-phonon coupling and weakening of the Coulomb repulsion by peculiar "acoustic" plasmon branch.

REFERENCES

1   A.P.Mackenzie et al., *Phys.Rev.Lett.* **71** , 1938 (1993); A.Carrington et al. ,*Phys.Rev.* B **49**, 13243 (1994).

2. M.Osofsky et al., *Phys.Rev.Lett.* **71**, 2315 (1993).

3. L.Gor'kov, *Sov. Phys. JETP* **10**, 593 (1960); E.Helfand, and N.R.Werthamer, *Phys.Rev.Lett.* **13**, 686 (1964); *Phys.Rev.* **147**, 288 (1966).

4. J.P.Franck et al., *Phys.Rev.* B **44**, 5318 (1991); J.P.Franck et al., in *High-Tc Superconductivity, Physical Properties, Microscopic Theory and Mechanisms*, p.411, J.Ashkenazi et al., Eds., Plenum, NY (1991).

5. D. Zech et al., *Physica B* **219&220**,136 (1996).

6. V.Kresin, and S.Wolf, *Phys.Rev.B* **46**, 6458 (1992); **51**, 1229 (1995).

7. V.Kresin, S.Wolf, and G.Deutcher, *Physica C* **191**, 9 (1992).

8. X.Geerk, J.Xi, and G.Linker, *Z.Phys. B* **73**, 329 (1988).

9. W.McMillan, *Phys.Rev.* **175**, 537 (1968).

1 0 a)A.Abrikosov and L.Gor'kov, *Sov.Phys.-JETP* **12**,1243 (1961);
    b) P.de Gennes, *Phys. Condens. Matter* **3**, 79 (1964); P.de Gennes,
    Superconductivity in Metals and Alloys, Benjamin, NY (1966);
    c) S.Skalski, O.Betbeder, and P.Weiss, *Phys.Rev.* **136**, 1500 (1964).

11. a) D.Saint-James, G.Sarma, and E.Thomas, *Type II Superconductors*,
    Pergamon, Oxford (1969); b) K.Maki, in *Superconductivity*, R.Parks
    ed, Marcel Dekker, New York, (1969), p.1035; c) A.Abrikosov,
    Fundamentals of the Theory of Metals, North-Holland, Amsterdam
    (1988).

12. a) Y.Dalichaouch et al., *Phys.Rev.Lett.* **64**, 599 (1990); b) M.Suzuki
    and M.Hikita, *Phys.Rev.B* **44**, 249 (1991); c) D. Walker et al.,
    Phys.Rev. B **51**, 9375 (1995).

13. Yu.Ovchinnikov and V.Kresin, *Phys.Rev.B* **54**,1251 (1996).

14. H.Homma et al., *Phys.Rev.B* **33**, 3562 (1986).

15. a) N.Phillips, R.Fisher, J.Gordon, in *Proress in Low-Temperature
    Physics* 13,p.267, D.Brewer Ed., North-Holland, The Netherlands
    (1992); b) C. Niedermayer et al., *Phys.Rev.Lett.* **71**,1764(1993);
    c) J.Tallon, J.Loram, *J. of Supercond.* **7**,15 (1994); J.Tallon et al.
    Phys.Rev.Lett. **74**, 1008 (1995).

16. V.Ambegaokar and A.Baratoff, *Phys.Rev.Lett.* **10**,486 (1963).

17. V.Kresin et al., *J.of Low Temp.Phys.* **106**,159 (1997).

18. M.Hein et al., *J.of Supercond.,* to be published.

19. V.Kresin, S.Wolf,Yu.N. Ovchinnikov, *Phys.Rev.B* **53**,11831 (1996).

20. A.Furrer et al. , *Physica C* **235-240**, 101 (1994) ; *J.of Supercond.,*
    (to be published).

21. a)V.Kresin and S.Wolf, *Phys.Rev. B* **49** , 3652 (1994);
    b)V.Kresin, A.Bill, S.Wolf, Yu.Ovchinnikov, *Phys.Rev.* **56**,107
    (1997).
    c)A.Bill et al., preprint.

22. J.Carbotte et al., *Phys.Rev.Lett.* **66**,1789 (1991);
    S.Singh et al., *J.of Supercond.* **9**, 269 (1996).

23. a) H.Allout et al., *Phys.Rev.Lett.* **67**, 3140 (1991); b) T.Miyatake et
    al., *Phys.Rev.B* **44**, 10139 (1991); c) A.Mahajan et al., *Phys. Rev.
    Lett.* **72**, 3100 (1996); d) S. Zagoulaev, P. Monod, J. Jegoudez, *Phys.
    Rev. B* **52**, 10474 (1995); *Phys. C* **259**, 271 (1996).

24. V.Kresin , *Phys.Rev.B* **25**, 157 (1982).

25. a) J.Neumeier et al., *Phys.Rev.Lett.* **63**, 1516 (1990); b) B.Maple et al., *J.of Supercond.* **7**, 97 (1994).

26. V.Kresin , *Phys.Rev.B* **35**, 8716 (1987).

27. V.Kresin and H.Morawitz, *Phys.Rev.B* **37**,7854 (1988).

28 V.Kresin, H.Morawitz, S.Wolf, *Mechanisms of Conventional and High-Tc Superconductivity*, Oxford, NY (1993).

29. M.Reeves et al., *Phys.Rev.B* **47**, 6065 (1993).

30. V.Kresin and G.Zaitsev, *Sov.Phys.-JETP* **47**, 983 (1978).

31. V.Kresin, *Phys.Lett. A* **122**,434 (1987).

MOTT METAL-INSULATOR TRANSITION IN OXIDES

M. CYROT

Laboratoire Louis Néel,
CNRS, BP 166,
38042 GRENOBLE Cedex 9

The renewed interest in Mott metal-insulator transition came from the high temperature superconducting cuprates discovered by K.A. Muller and J.C. Bednorz. All the superconducting cuprates are very close to a parent phase which is insulating and antiferromagnetic. It raises the question of the link between high temperature superconductivity and the Mott transition i.e. the cuprates would be basically doped Mott insulator.

In this lecture, after a brief historical review, we will present the main hamiltonian which is used to study this Mott transition i.e. the so-called Hubbard model. We then review the main attempts to understand on this model the metal-insulator transition. First we describe Hubbard approach, then the Gutzwiller one, and finally what we can learn from a mean field approach in a space with infinite dimensions. Numerical computations will give also some other interesting results. In a second part, we will review the experiments on transitional metal oxides and compare it with the cuprates.

In a third part, we show how the proximity of the Mott transition can explain some properties of the cuprates.

A. MOTT TRANSITION

I. Historical background[1]

In 1937, de Boer and Verwey pointed out that nickel oxide should be metallic according to the Bloch Wilson band theory. It was pointed out by Peierls that it must be due to correlation. At that time Nickel oxide was not known as an antiferromagnet. Slater pointed out in 1951 that an antiferromagnetic lattice can split a band and make insulating something which is predicted

The Gap Symmetry and Fluctuations in High - T_c Superconductors
Edited by Bok *et al.*, Plenum Press, New York, 1998

metallic in a simple band theory. However antiferromagnetic insulator as the oxides retains their non-metallic behaviour above the Neel temperature. Thus this is not the whole story. Mott in 1949 described a metal insulator transition in a half filled band, which is metallic in a band theory, by considering a crystalline array of hydrogen-like atoms with a lattice constant **a** that could be varied. For large values of **a**, one should obtain an insulator and for small values, a metal as sodium shows this is the case. At what value of the lattice constant would a metal-insulator transition occur ? Mott assumed that this would occur when the screened potential round each positive charge was just strong enough to trap an electron i.e.the potential extracts a bound state from the bottom of the band. From that, he derived his famous criterion

$$n^{1/3} a_H = 0,2$$

where n is the number of electrons per unit volume and a_H the Bohr radius.

Later Anderson pointed out that such an insulator is antiferromagnetic by considering the hopping integral t as a perturbation. To second order perturbation, two electrons on neighbouring atoms will gain an energy $\frac{t^2}{U}$ to be in an antiferromagnetic configuration, U being the repulsion energy between the two electrons when they sit on the same atom.

II. The Hubbard model[2]

In band theory, the Coulomb repulsion between electrons is taken into account only by the Hartree-Fock self consistent potential. The many body wave function for the electrons is given by a Slater determinant. Two electrons with the same spin cannot be at the same point due to the antisymmetry of the wave function. Thus their repulsion is rather small. On the contrary two electrons of different spins can be at the same point. Thus the Coulomb repulsion in a band theory is important and not well taken into account. In 1963, Hubbard[3] proposed to add to the usual Hartree-Fock Hamiltonian a repulsive term between electrons of different spin when they sit on the same atom. His model Hamiltonian is written in second quantification form as

$$H = \sum_{ij\sigma} t_{ij}\, c_{i\sigma}\, c_{j\sigma} + hc + \sum_{i} U\, n_{i\uparrow}\, n_{i\downarrow} \qquad (1)$$

$c_{i\sigma}^{+}$ creates an electrons of spin σ on atom i.

The first term describes the possibility for an electron to hope from site j to site i. If the last term is absent, this hamiltonian can be solved exactly in term of Bloch waves

$$H = \sum_{k} \varepsilon_k\, c_{k\sigma}^{+}\, c_{k\sigma} \qquad (2)$$

74

Where $c_{k\sigma}^{+}$ creates an electron in a Bloch state of wave vector k and the energy of this state is

$$\varepsilon_k = \sum_j t_{ij} \, e^{ik(R_j - R_i)} \quad (3)$$

The second term is the repulsion between two electrons of different spin when they sit on the same site. Hubbard model neglects repulsion between electrons when they sit on different sites. In a metallic ground state the potential due to an electron is screened by the others and this is a valid approximation. However for an insulating ground state this would not be a good starting point.

We have some parameters in this model : the hopping term t_{ij} which is often taken only between nearest neighbours and we set $t_{ij} = -t < 0$ and the coulomb repulsion U in fact only U/t i.e. the Coulomb repulsion measured in unit of t enters into the problem. The second parameter is the number if electrons per site n. If $n = 1$ it means one electron per site or per atom. If $n \neq 1$, it means that we have doped the system with holes if $n<1$, or with electrons if $n>1$.

This Hamiltonian describes a metal to insulator transition when U/t increases for $n = 1$. Indeed if $U = 0$ one obtain a band which is half filled.

For a square lattice

$$\varepsilon_k = -2t \, (\cos k_x a + \cos k_y a) \quad (4)$$

The electrons kinetic energy is

$$E_k = \sum_{k<k_F} 2 \, \varepsilon_k < 0 \quad (5)$$

this ground state has lower energy compared to an insulating state where we put a non mobile election on each site. In that case the energy is

$$E_k = 0$$

If now we consider the Coulomb repulsion, we add to the kinetic energy of the metallic state a positive term. For the insulating ground state, the second term does not give any contribution as we have only one electron of a given spin per site. Thus the metallic state becomes unstable if its total energy becomes positive. Thus there is a critical value of U, where the minimisation of the energy gives an insulating ground state. The kinetic energy of an electron in a metallic state is of the order of the band width W ($W = 8t$ for the square lattice). The repulsive energy is of order U thus a metal non-metal transition occurs for

This hamiltonian, if we were able to solve it, would describe a metal non-metal transition and would predict that with one electron per atom, one can obtain an insulating state contrary to the band theory.

As it is directly related to the lattice constant, varying the lattice constant changes the kinetic energy compared to the Coulomb repulsion and can induce a Mott transition. As at first sight, this approach looks different from the historical Mott one, one often called this transition a Mott-Hubbard transition.

III. Hubbard approximation[3]

Unfortunately although Hubbard Hamiltonian looks simple, nobody has been able to solve it except in the one dimensional case. As the physic of a one dimensional problem is always very different from the two or three dimensional one, we cannot learn too much from this case and we do not describe it. Hubbard proposed an approximate solution for this model which is based on an alloy analogy.

He first considered the case $n = 1$ i.e. one election per site and he supposed that the total number of up spin electrons is equal to the total number of down spin electrons. Let us prevent to move the down spin electrons, they are fixed on $\frac{N}{2}$ sites if we have N sites in the lattice. The up electrons which can move, feel a potential zero is they are on a site with no electron and a potential U if they are on a site where there is a down spin electron. The up spin electrons move on a lattice where the sites have energy 0 or U as in a disordered alloy. Some general theorem in alloys states that the permitted energy for the alloys lies in two bands of width W of the pure compound centred at 0 and U.

It means that if $U < W$ the alloys has one band which is half filled and if $U > W$, the alloy has two separate bands which accommodates one electron. The lower band is filled and the upper band is empty. A gap separates the two bands. Thus we have a metal for $U < W$ and an insulator for $U > W$. The lower band is called the lower Hubbard band and the upper, the upper Hubbard band.

This result stems from the approximation of down spin kept fixed. This is not true as they can moved when one up spin electron answer. A refined approximation has been given by Cyrot[4] to try to take into account this effect. One writes the repulsive term as

$$U n_i n_i = \frac{U}{4}(n_{i\uparrow} + n_{i\downarrow})^2 - \frac{U}{4}(n_{i\uparrow} - n_{i\downarrow})^2 \quad (6)$$

the number of electrons on each site is taken constant thus only the second term can vary if we make a mean field approximation on the second term we write it as

$$-\frac{U}{2}\,\mu_i\,(n_{i\uparrow} - n_{i\downarrow}) \qquad (7)$$

$$\text{with } \mu_i = <n_{i\uparrow} - n_{i\downarrow}> \qquad (8)$$

The change of the potential between site is not U as in Hubbard approximation but $U\mu$ where μ is the average difference between up and down spin on a site. This rewriting of the Coulomb repulsion is interesting as it permits to treat charge and spin fluctuations separately and to introduce antiferromagnetic ground states in a simple way even if the state is metallic and has a reduced moment[5].

IV. Gutzwiller approximation[6]

Gutzwiller proposed another way of approach. He noticed that the Coulomb repulsion U will prevent double occupation on a site. Thus he used a variational method to reduce in the Slater determinant for uncorrelated electron the double occupancy.
If $|\psi_0>$ is the many body wave function of the kinetic part of the Hamiltonian i.e. a Slater determinant, he proposed the following wave function

$$|\psi> = g^D\,|\psi_0> \qquad (9)$$

where $D = \sum_i n_{i\uparrow} n_{i\downarrow}$ is the number of doubly occupied sites and g is a variational parameter.

Unfortunately it is not possible to solve exactly this variational problem and he has to use further approximations which give a simple renormalisation of the Hamiltonian

$$<\psi|\, c_{i\sigma}^{+}\, c_{j\sigma}\,|\psi> = q <\psi_0|\, c_{i\sigma}^{+}\, c_{j\sigma}\,|\psi_0> \qquad (10)$$

with $q = \dfrac{2\delta}{1+\delta}$ in the large U limit $(U \to \infty)$

δ is the deviation from the half filled case $n = 1$

$$\delta = 1 - n$$

It means that in the large U limit the hopping parameter t is renormalised in $\dfrac{2\delta}{1+\delta}$ t.

It has a very simple interpretation. Indeed δ is the probability to have a hole (no electron on a site) and in the large U limit, an electron can hope on a nearest neighbour site only if there is a hole. The possibility to move is reduced by a factor δ thus the kinetic energy. In this large U limit the kinetic energy of the electrons become zero for $\delta = 0$ i.e. one obtains an insulator. In the general case the energy of the ground state E_g is given by

$$E_g = \sum_\sigma q_\sigma (d, n\uparrow, n\downarrow) \; \overline{\varepsilon}_\sigma + Ud \quad (11)$$

$$\text{with } \overline{\varepsilon}_\sigma = \sum_{|k|<k_F} \varepsilon(k) < 0 \quad (12)$$

and d is the number of doubly occupied sites. In the case of a non magnetic half filled band, i.e. $n = 1$ $q\uparrow = q\downarrow = q$ $\overline{\varepsilon\uparrow} = \overline{\varepsilon\downarrow} = \dfrac{\overline{\varepsilon}}{2}$, Brinkman and Rice[7] have obtained from the general equation

$$d = \frac{1}{4} \left(1 - \frac{U}{U_c} \right) \quad (13)$$

$$q = 1 - \left(\frac{U}{U_c} \right)^2 \quad (14)$$

$$E_g = - |\overline{\varepsilon}| \left(1 - \frac{U}{U_c} \right)^2 \quad (15)$$

Thus there exists a critical value of U where the number of doubly occupied sites becomes zero. It is given by

$$U_c = 8 \, |\overline{\varepsilon}| \quad (16)$$

It means that for $U < U_c$ the ground state is metallic, the number of doubly occupied sites is reduced and the kinetic energy too. For $U > U_c$ the ground state is insulating with no doubly occupied site and a zero energy. Gutzwiller approximation also permits to describe the Mott transition.

V. Hubbard and Gutzwiller approximation compared to Fermi liquid theory

It is instructive to compare the results of both approximations knowing some general results on a gas of interacting particles called Fermi liquid theory[8]. The Fermi liquid theory starts from a gas of non interacting particles which is completely solved and considers what happens if the interaction are introduced. In a gas of non interacting particles the electron states are labelled by the wave vector k and at zero temperature, all the states with $|k| < k_F$ are filled and states with $|k| > k_F$ are empty. There is a discontinuity of n (k), the occupation number of the k states, at k_F. In the Fermi liquid theory, this discontinuity still occurs at k_F but it is reduced in amplitude. Thus one can still define a Fermi surface which has the same volume as the non interacting Fermi surface (this is known as Luttinger theorem). Contrary to the gas of non interacting particles, the states k are well defined only close to the Fermi surface. They are quasiparticle states and their life-time is

$$\frac{1}{\tau} \propto (E - E_F)^2 \qquad (17)$$

At the Fermi energy E_F the life time of the states k_F is infinite. Far from E_F the state k are not well defined and the states are said incoherent and not described by their wave vector.

In Hubbard approximation, k does not label the states close to E_F due to the alloy analogy. Thus there is no discontinuity of $n(k)$ and this is a serious drawback of this approximation. This is not the case of Gutzwiller approximation[9]. We can calculate the discontinuity of $n(k)$

$$n_{k\sigma} = <c^+_{k\sigma} c_{k\sigma}> = \sum_{ij} <c^+_{i\sigma} c_{j\sigma}> e^{ik \, [R_i - R_j]} \qquad (18)$$

using equation 10, and taking into account that the term $i = j$ is not renormalised and give the average number of electrons n_σ per site, we have

$$n_{k\sigma} = n_\sigma + \left(n^0_{k\sigma} - n_\sigma \right) q_\sigma \qquad (19)$$

where $n^0_{k\sigma}$ is the value for non interacting particle. The Gutzwiller approximation shows that the discontinuity at the Fermi surface disappears at U_c. Above U_c $n_{k\sigma}$ is a constant for all k in the band

$$n_{k\sigma} = \frac{1}{2}$$

This approximation fulfilled Luttinger theorem. The existence of a Fermi surface is linked with the metallic properties. Its absence means that the system is an insulator. The discontinuity in $n_{k\sigma}$ is related to the effective mass of the quasiparticle by Midgal theorem

$$\frac{m_{eff}}{m} = q^{-1} \qquad (20)$$

Thus, at the transition in the Gutzwiller approximation the effective mass becomes infinite and this explains the insulating properties. This has to be contrasted with Hubbard approximation where the transition occurs because the number of effective carriers is going to zero.

VI. Hubbard Model in infinite dimension[10]

The models in infinite dimensions are generally equivalent to a mean field approximation and this is the case for the Hubbard model. In fact it makes a link between Hubbard and Gutzwiller approximation which both capture some physics of the model.

From a theoretical point of view, this limit permits to the self energy to the Green function to be local i.e. to have no k dependence. The solution of Hubbard Hamiltonian reduces to a single site self consistent model. The problem is mapped on an Anderson impurity model in an electron bath provided the bath dispersion and the hybridisation coupling satisfy a self consistent equation. This makes the problem amenable to various analytical approximations or make numerical computations possible.

Results[11] of numerical computations for n = 1 show that the density of states in the metallic state $U \leq U_c$ has the two Hubbard gap (fig. 1).This peak is narrow and the effective mass of the carriers is going to infinity as in the Gutzwiller approximation[12]

$$\frac{m^{eff}}{m} \propto [1 - \left(\frac{U}{U_c}\right)^2]^{-1} \qquad (21)$$

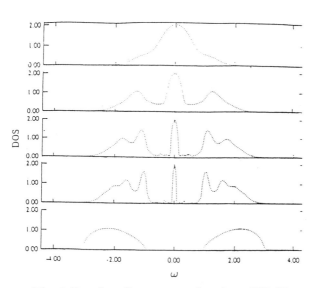

DOS

ω

Fig. 1 Density of states as a function of U, U increases from top to bottom

U_c is also of the order of 8 $\overline{\varepsilon}$ as in the Gutzwiller approximation. Thus it connects Hubbard and Gutzwiller approximation. The differences with the last approximation are not only the existence of the two Hubbard band but also a paramagnetic susceptibility which does not diverge as $\frac{m^{eff}}{m}$. There is also a discontinuous opening of the gap at U_c and also contrary to Gutzwiller approximation a reduced magnetic moment on the insulating side.

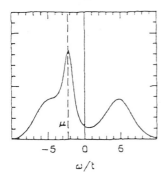

ω/t

Fig. 2 Density of states for n= 0.87

In the non half filled case, when doped with holes $\delta = 1 - n$, the quasi particle peak is at the top of the lower Hubbard band. Similar results are obtained with quantum Monte Carlo simulations[13] Fig. 2.

VII. Different types of Mott metal-insulator transition

Up to now, we consider only a Mott transition in a Hubbard Hamiltonian where the state on an atom is non degenerate. In that case the transition can be experimentally studied either by varying the doping or by pressure which increases the hopping parameter letting U unchanged. If the state on an atom is degenerate, new behaviour can arise. For instance[14] in a doubly degenerate e_g states with states α and β and with one electron per atom, one can construct orbital super lattice where the state of an electron is either α on a sublattice or β on the other. In that case ferromagnetism can occur and different behaviour can be predicted. We will not consider this case.

Finally another band due for instance to the oxygen can be close to the Fermi level. This is the case at the end of the first transitional series. In that case, an oxygen band can lie in the middle of the Hubbard gap and the gap of the material is between this band and the upper Hubbard band. One usually call it charge transfer gap[15] by opposition to the Hubbard gap. The superconducting cuprates are in that class of materials. We will not consider this case in details.

B. METAL INSULATOR TRANSITION IN PEROVSKITES

Soon after the discovery of the superconducting cuprates, Anderson proposed that they can be considered as doped Mott insulator. This renewed the study of the Mott metal insulator transition in other perovskites. As the cuprate band stems from a d^9 copper, compounds with a d^1 transitional metal were considered. T_i^{3+} is a d^1 ion and the two compounds $LaTiO_3$ and $YTiO_3$ are Mott insulator. By doping the first with Strontium[16] and the second with Calcium[17] a metal insulator transition is obtained. Their behaviour close to the transition were compared with the transition obtained in $La_{2-x}Sr_xCuO_4$. However at the start, there are some differences between these systems. First, the titanate are three dimensional system compared to the two dimensional cuprates. Second, the titanate are real Mott-Hubbard compounds where the gap is between the lower and upper Hubbard band while the cuprates have a charge transfer gap between the p band of oxygen and the upper Hubbard band. Third, the tetragonal distortion of the octahedron around the transitional metal left the e_g degeneracy for copper but not the whole e_g degeneracy for Titanium.

I. Specific heat and Hall effect

In order to know if the transition is due to the vanishing of the number of carriers or to the divergence of the effective mass of the carriers, one has to study the electronic specific heat and the Hall coefficient.

Indeed we have $\quad C_V = \gamma T \quad$ with $\quad \gamma \propto \dfrac{m^*}{m} \quad$ and $\quad R_H^{-1} \propto n$

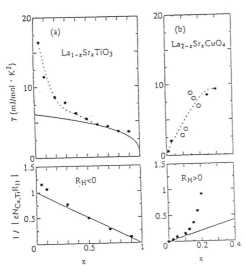

Gutzwiller approximation as in the infinite dimensional case gives a diverging effective mass and n constant. Hubbard approximation gives vanishing of the number of carriers and constant effective mass. Fig. 3 gives experimental results for $La_{1-x}Sr_xTiO_2$[16,18] and $La_{2-x}Sr_xCuO_4$. The data shows clearly that the behaviour close to the transition is very much different.

Fig.3 Electronic specific heat and inverse Hall coefficient as a function of hole concentration.

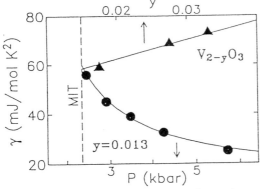

In figure 4, we show new results on the V_2O_3 materials which has served as a prototype for Mott Hubbard physics. These authors[19] found a different behaviour if the transition is obtained with pressure or with doping.

Fig.4 γ versus vacancy concentration and pressure

II. Photoemission

The main results of integrated photoemission experiment[20-22] and inverse photoemission are first the appearance of ingap states when doping. Contrary to predictions, the chemical potential lies in the Hubbard gap where states develop with doping. In other words, E_F does not move into states present in the insulator and does not jump across the gap if the doping is changed from holes to electrons but has roughly at the same position relative to the valence band maximum for both holes and electrons.

This finding does not correspond to any predictions as the chemical potential is thought to move in the lower Hubbard band with hole doping.

82

Thus the paradoxical behaviour of μ still needs a theoretical explanation. The second main feature of the high Tc materials and the doped Titanium perovskite is a strong doping dependence of spectral distribution and a shift of intensity from high to low energies. This is in contrast to the behaviour of a doped semiconductor and is a characteristic of doped strongly correlated system. There is a simple picture which can be used to understand this large spectral-weight transfer in Mott-Hubbard system[23]. Consider N atoms with one electron in an insulating state. The total electron removal (photoelectron) and electron addition (inverse photo electron) spectral weight is equal to the number of occupied and empty levels respectively. Thus each has intensity N. Let us now consider a one hole doped case. There are now N - 1 way of removing an electron. At the same time there are N - 1 and not N way of occupying the upper Hubbard band. This leaves two states at the site with the missing electron. Both of these are electron addition states. This shows that the electron addition spectral weight goes as $a\delta$ where $a = 2$ and δ is the doping. One of these states comes from the upper Hubbard band so there is a transfer of spectral weight from high to low energy. A correlated system behaves quite differently from a semiconductor with respect to doping[23].

III. Optical conductivity[24] and Raman

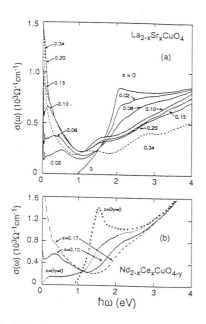

The main conclusion from the optical conductivity is that upon doping weight appears inside the charge transfer gap of the undoped compounds defining the so-called mid infrared band. Fig. 5 shows the optical conductivity of two cuprates[25]. A clear feature centred below 0.5 eV appears.

Fig.5 Optical conductivity

The same mid infrared band is observed in the Titanium perovskite[26]. The d infinite model qualitatively explains the existence of weight in the gap[27] but they failed to explain the mid infrared location except if a very small value of U is taken which seems contradictory. Numerical calculations[28] on 4 x 4 clusters also give qualitative explanation. However the mid infrared band seems to appears only in the two dimensional case.

Another feature to be contrasted with the free election behaviour is the decay of the Drude peak in $1/\omega$ rather than in $1/\omega^2$. Or the linear dependence[29] of $1/\tau$ as a function of ω or kT that can be extract from the data. This is also a puzzling experimental result.

The anomalous characteristic of the electronic Raman spectra is a Raman scattering continuums which is nearly independent of energy[30] up to eV. This is contradictory to all theoretical theories[31] which predicts Raman only at very low energy i.e. below qv_F or $1/\tau$ if v_F is the Fermi velocity, q an inverse length which can be taken as the wave number of the light and $1/\tau$ the scattering time of the excitation.

C. A POSSIBLE SCENARIO FOR SUPERCONDUCTING CUPRATES

I. Scenario for a Mott metal-insulator transition

If we assume that the Fermi liquid theory holds in the metallic state close to the Mott transition, the scenario is the one given by the Gutzwiller approximation. The discontinuity at the Fermi momentum, in the momentum distribution of occupied single particle states decreases continuously to zero as the interaction is increased to a critical value. The coherent states close to the Fermi Level should be roughly at energy given by the Hartree-Foch theory so close to $\frac{U}{2}$

i.e. in the middle of the Hubbard gap for a nearly half filled materials. Thus the picture given by the infinite dimension mean field approximation should be qualitatively valid. The weight of the peak can be given in a simple picture. The spectral weight of the lower Hubbard subband measured for instance by photoemission spectroscopy is given by the probability of singly occupied site. This probability is

$$n\uparrow (1 - n\downarrow) + n\downarrow (1 - n\uparrow) = (n\uparrow - n\downarrow)^2 \quad (22)$$

where we have used the fact that $n_\sigma^2 = n_\sigma$

This instantaneous magnetic moment is given by

$$(n_i\uparrow - n_i\downarrow)^2 = (n_i\uparrow + n_i\downarrow)^2 - 4d \qquad (23)$$

d being the probability of double occupation. If the total number of electrons is $1-\delta$

$$(n_i\uparrow - n_i\downarrow)^2 = (1 - \delta)^2 - 4d \qquad (24)$$

the spectral weight of the upper Hubbard subband measured by inverse photo emission spectroscopy is the same. Thus the spectral weight of the quasi particle band is

$$2 - 2(1 - \delta)^2 + 8d = 8d + 4\delta + 2\delta^2 \quad (25)$$

If we use Gutzwiller result for the number of doubly occupied sites, we have

$$d = A\,\delta \quad (U > U_c)$$

A being a constant, the weight of the quasiparticle peak is going to zero for zero doping if $U > U_c$.

We now discuss the range of energy where the concept of quasiparticle hold. In the Green function formalism, quasiparticles of the Landau theory appear as poles in the complex energy plane. The real part of the pole position gives the quasiparticle energy, and the imaginary part gives the quasiparticle damping or inverse lifetime. The discontinuity q is the residue at the pole evaluated at the Fermi momentum.

The one particle Green function may be written as

$$G(k,\varepsilon) = \frac{1}{\varepsilon - \varepsilon(k) - \Sigma(k,E)} \quad (26)$$

$\Sigma(k,E)$ arises from interactions. Near the Mott transition, it is argued that $\Sigma(k,E)$ depends only weakly on k. We neglect the k dependence in the following.

If the transition is given by the disappearance of the discontinuity in the probability $n(k)$, it means that the real part of the self energy is going to infinity. Close to E_F, the self energy can be written

$$\Sigma(\varepsilon) = \alpha\varepsilon + i\beta\varepsilon^2 \quad (27)$$

with $\varepsilon = (E - E_F)$, the energy ε is measured with respect to the Fermi energy. A quasiparticle is well defined if its energy is much smaller than its width

$$\varepsilon \ll \frac{\beta}{\alpha}\,\varepsilon^2 \quad (28)$$

which is true usually and in most cases $\dfrac{\beta}{\alpha} \approx \dfrac{-1}{E_F}$ and the quasiparticle concept is valid in the whole band. Close to the Mott transition α is going to infinity and it can be shown that β behaves as α^2. Thus the range of validity of the concept of quasiparticle is restricted to

$$\varepsilon < \frac{\alpha}{\beta} \approx \frac{1}{\alpha} \quad (29)$$

The range of validity shrinks to zero at the transition. The kinetic energy of the quasi particles are reduced by a factor given by q in the Gutzwiller approximation. Thus the width in energy of the quasiparticle states are of the order of qw with w the band width of non interacting particles. As d and q diverge in the same manner, the density of states at the Fermi energy should be roughly constant and drops to zero discontinuously at the Mott transition. In this scenario γ is roughly constant and the inverse of the Hall constant is going to zero.

Description of the incoherent states[32]

The incoherent states of the lower Hubbard subband correspond to singly occupied sites. We now show that, in some particular cases, they can be described as nearly localised states. Indeed let us create a hole on this site let us consider its motion. The problem is analog to the motion of a particle creating an attractive potential and moving in a Fermi liquid. This problem is a many body problem which has an answer[33] which depends on the phase shift created by the perturbing potential on the electrons at the Fermi level. It has to be large enough. However Friedel sum rule usually prevents this possibility. Thus it is impossible to localise a particle in a Fermi liquid except in one case where two holes are linked together. This is exactly the situation here because the hole created into the lower Hubbard subband is a hole for the up and down spin electrons. In a two dimensional system, phase shifts are large and it is possible to localise a hole. In that case the width of the Hubbard subband represents the lifetime of this hole. Thus there exist excitations which can be called virtual exciton in a strongly correlated system. They are quasilocalised excitations created with a hole on a site and the excited electron is in a virtual bound state around it. This picture permits an approximate description of the lower Hubbard band in a two dimensional materials.

A very important consequence of the quasilocalisation of this excited state is that the energy of the excitation is not the Hartree-Fock energy[34]. The Hartree-Fock energy, which relies on Koopman theorem would predict a large excitation energy of order of U. If the excitation is quasilocalised, Koopman theorem does not hold and there is a shift of the scattering states due to the attractive potential. This shift is not small as computation shows. This can be understood as the shift in energy is

$$\Delta E = \frac{2}{\pi} \int^{E_F} \eta(E) \, dE \quad (30)$$

where η is the phase shift at energy E. This shift is large because the phase shift at E_F is large of order of π. This excitation would be the explanation of the mid infrared band in the cuprates.

We emphasise that these results do not follow directly from the Hubbard model. Indeed use of the Friedel sum rule in order to calculate phase shift at the Fermi energy relies on long range Coulomb interaction. Thus our results differ substantially from that of a simple Hubbard model. The doped Mott insulator would behave as an "excitonic" metal where exist low energy excitations of the Fermi liquid type and high energy excitations of excitonic type.

86

II. Scenario the mechanism of superconductivity[35]

The existence of excitations of the type of virtual exciton creates the possibility of excitonic superconductivity in this system. Such possibility has been studied by Bardeen et al[36]. They have calculated a rough estimate of the electron-electron coupling λ. We have

$$\lambda \approx \mu\, f \frac{\omega_p^2}{\omega_{ex}^2 \varepsilon} \quad (31)$$

where μ is the usual average of electrons-electrons repulsive parameter. f is the fraction of the spectral weight of particle-hole excitations in the virtual exciton. ω_{ex} is the energy of the virtual exciton. f is roughly the spectral weight of the lower Hubbard subband, thus is close to one near half filling. As ω is also very much reduced compared to U, this would permit excitonic mechanism in the cuprates.

III. Properties superconductivity

In this scenario, differences with BCS theory lie in the very small range in energy where the Fermi liquid theory holds. The inequality

$$W \gg \omega_D$$

is now replaced by $qW \leq \omega_{ex}$.

The whole band of quasiparticles is concerned by the attractive interaction. The main qualitative result[37] is that Tc is proportional to the number of quasiparticles and the ratio $\frac{\Delta}{2k_B Tc}$ is greater than 4. This seems to be the case for the superconducting cuprates.

CONCLUSION

The idea that superconducting cuprates are doped Mott insulator first proposed by Anderson permits to explain many unusual properties of the metallic phase. The discrepancy sometimes obtained or the lack of explanation mainly rely on the absence of exact theory of the Mott transition obtained by doping. Even the qualitative physical picture is not completely set up. We propose a model of "excitonic" metal where the range of energy in which the Fermi liquid theory holds shrinks near the transition. Excitonic superconductivity could explain the unusual superconducting properties.

REFERENCES

1. For this review, see N.F. Mott, Metal-Insulator transitions edited by Taylor and Francis (1974).
2. The main papers in the subject are published in a reprint volume The Hubbard Model edited by A. Montorsi, World Scientific (1992). For a review see also M. Cyrot, Physica 91B 141 (1977).
3. J. Hubbard, Proc. Roy. Soc. A 276, 238 (1963), A 281, 401 (1964).
4. M. Cyrot, Phys. Rev. Lett. 25, 871 (1970).
5. M. Cyrot, J. de Physique 33,125 (1972).
6. M.C. Gutzwiller, Phys. Rev. Lett. 10, 159 (1963). Phys. Rev. A 134, 923 (1964).
7. W.F. Brinkman and T.M. Rice, Phys. Rev B2, 4302 (1970).
8. L.D. Landau, Sov. Phys. JETP 3, 920 (1957), 5, 101 (1957).
9. For a study of the relation between Gutzwiller approximation and Fermi liquid theory see D. Volhardt , Rev. Mod. Phys. 56, 99 (1984) and Ph. Nozières, Lectures notes at College de France (1986) unpublished.
10. W. Metzner and D. Vollardt, Phys. Rev. Lett. 62, 324 (1989). For a review see D. Vollhardt, Physica B, 169, 277 (1991) and E. Muller-Hartmann, Int. J. of Mod. Physic 3, 2169 (1989).
11. A. Georges and W. Krauth, Phys. Rev. Lett. 69, 1240 (1992).
12. X.Y. Zhang, M.J. Rozenberg and G. Kotliar, Phys. Rev. Lett. 70, 1666 (1993).
13. For a review see E. Dagotto, Rev. Mod. Phys. 66, 763 (1994).
    N. Bulut and D.J. Scalapino, J. Phys. Chem. Solids, 56, 1597 (1995).
14. M. Cyrot and C. Lyon-Caen, J. Phys. (Paris) 36, 253 (1974).
    J.L. Garcia-Munoz, J. Rodriguez-Carvagal and P. Lacorre, Europhys. Lett. 20, 241 (1992).
15. J. Zaanen, G.A. Sawatzky and J.W. Allen, Phys. Rev. Lett. 52, 418 (1985).
16. Y. Tokura et al, Phys. Rev. Lett. 70, 2116 (1993).
17. Y. Taguchi, Y. Tokura, T. Arima and F. Inaba, Phys. Rev. B 48, 511 (1993).
18. K. Kumagai et al, Phys. Rev. B 48, 7636 (1993).
19. S.A. Carter et al, Phys. Rev. B, 48, 16841 (1993).
20. A. Fujimori et al, Phys. Rev. Lett. 69, 1796 (1992).
21. K. Morikawa et al, Phys. Rev. B 54, 8446 (1996).
22. A. Fujimori et al, Phys. Rev. B 46, 9841 (1992).
23. H. Eskes, M.B.J. Meinders and G.A. Sawatzky, Phys. Rev. Lett. 67, 1035 (1991).
24. For a review see Tanner and Timusk (1992).
25. S. Uchida et al, Phys. Rev. B 43, 7942 (1991).
26. Y. Taguchi, Y. Tokura, T. Arima and F. Inaba. to be published.
27. M.J. Rozenberg et al, Phys. Rev. Lett. 75, 105 (1995).
28. E. Dagotto, Review of Mod. Phys. 66, 763 (1994).
29. L.D. Rotter et al, Phys. Rev. Lett. 67, 2741 (1991).

30 For a review see S. Sugay in Mechanism of HTS, H. Kemimura and A. Oshiyama, Editors Springer Verlag (1989).

31. A. A. Abrikosov and L.A. Falkovskii, Soviet Phys. JETP 13, 179 (1961). L.A. Falkovskii, Sov. Phys. JETP 68, 661 (1989). A. Zuwakowski and M. Cardona, PR B 42, 10732 (1990).

32. M. Cyrot, Sol. State Comm. 97, 639 (1996).

33. Y. Yamada et al, Prog. Theoret. Phys. 70, 73 (1983).

34. U. Muschelknautz and M. Cyrot, Phys. Rev. B 54, 4316 (1996).

35. M. Cyrot, Modern Physic Letters B 6, 383 (1992).

36. D. Allender, J. Bray, J. Bardeen, Phys. Rev. B 7, 1020 (1973).

37. D.J. Thouless, Phys. Rev. 117, 1256 (1960).

SCALING BEHAVIOR OF THE NORMAL STATE PROPERTIES AND THE SUPERFLUID DENSITY IN METALLIC YBa$_2$Cu$_3$O$_x$ CUPRATES

Victor V. Moshchalkov, Bart Wuyts, Annemie Steegmans,
Rik Provoost, Roger E. Silverans, Yvan Bruynseraede

Laboratorium voor Vaste-Stoffysica en Magnetisme,
Katholieke Universiteit Leuven,
Celestijnenlaan 200D,
B-3001 Leuven, Belgium

ABSTRACT

We have studied the effect of reduced oxygen content x on the temperature dependence of resistivity $\rho(T)$ and the Hall number $n_H(T)$ of metallic YBa$_2$Cu$_3$O$_x$ (YBCO) epitaxial films. These results have been analyzed in terms of the Hall angle $\cot\theta_H$, which demonstrates a quadratic-like temperature dependence for all x values, with the systematic deviations at high and low doping levels. We have shown that the $\rho(T)$, $n_H(T)$, $\cot\theta_H(T^2)$ data for all metallic YBCO compositions fall onto single curves by scaling the temperature with T_0, the temperature above which resistivity becomes linear with temperature. The comparison with the transport data for other cuprates reported in literature strongly indicates that the observed scaling behavior is universal for underdoped cuprates. Furthermore we show that the NMR Knight shift data for oxygen deficient YBCO can also be mapped on a single scaling curve, by using the same scaling parameter $T_0(x)$ derived from our transport measurements. These findings demonstrate that the spin correlations, which determine magnetic properties, are also governing the transport of the high-T_C cuprates. Therefore, it is then most likely that T_0 is related to the opening of a spin pseudo gap. Finally, by using the NMR decoration technique we have studied the temperature dependence of the superfluid density $n_S(T) \propto \lambda^{-2}(T)$ in YBCO powders in the superconducting state as a function of the oxygen content x. The evolution of the behavior of the $\lambda(T)$ with doping is discussed in the framework of the d-wave pairing model and is also compared with the theory based on the solution of the Eliashberg equations which takes into account anisotropy and the presence of an intermediate electron-phonon coupling.

The Gap Symmetry and Fluctuations in High - T$_c$ Superconductors
Edited by Bok *et al.*, Plenum Press, New York, 1998

91

1. SCALING BEHAVIOR OF THE NORMAL-STATE PROPERTIES
OF YBa$_2$Cu$_3$O$_x$

1.1. Introduction

Systematic studies of the normal-state properties as a function of doping from an insulator to a high-temperature superconductor (HTSC) may reveal some interesting insights concerning the evolution of the exotic metallic state of HTSC's from the antiferromagnetic insulating state. Quite important in this respect is the study of the electrical transport properties as a function of doping. The anomalous temperature dependences of the resistivity ρ and the Hall coefficient R_H, which seem to be a fingerprint of the HTSC,[1,2] remain difficult to explain. Moreover, the available experimental data to indicate that both $\rho(T)$ and R_H depend delicately on the doping level.[3-9] Since these properties are intimately related to the scattering mechanism of the charge carriers, a clear insight in their behavior is highly desirable.

In section 1.2 of this paper we present the results of a systematic study of the normal-state transport properties of YBa$_2$Cu$_3$O$_X$ (YBCO) epitaxial thin films as a function of the oxygen content x. The results for $\rho(T)$ and $R_H(T)$ will be described together with an analysis in terms of the Hall angle θ. A quadratic-like temperature dependence is obtained for all oxygen content values, but systematic deviations seem to question the universality of the T^2 dependence of $\cot\theta(T)$. We show that the changes induced by a diminishing hole doping level in the behavior of $\rho(T)$, $R_H(T)$, and $\cot\theta(T)$ can be understood as the result of a changing temperature (or energy) scale. For all the studied properties it will be demonstrated that the obtained data can be *scaled onto universal curves*, as a function of T/T_0.

We present evidence for a remarkable resemblance between the characteristic features found in transport and magnetic properties. This correlation strongly indicates the magnetic origin of the dominant charge carrier scattering mechanism in the cuprates and allows us to consider the scaling parameter as the width of the spin pseudo gap.

1.2. Scaling Behavior of Transport Properties

The c-axis-oriented epitaxial thin YBCO films with a thickness of 1000-1100 Å were grown on MgO (100) substrates by a single target 90° off-axis magnetron sputtering technique.[10] X-ray diffraction characterization shows that the as-prepared films are c-axis oriented and of single-phase character. The as-prepared films have a high critical temperature ($T_C \cong 89$ K) and a narrow normal-to-superconducting transition ($\Delta T_C < 2$ K). Patterning of the films into a (2×10) mm^2 structure, necessary for Hall effect measurements, is done by classical photolitographic and wet-etching techniques, without deterioration of the film properties.

The oxygen depletion is carried out *ex situ* using a special heat treatment procedure reported previously.[11] The temperature dependence of the resistivity and Hall coefficient are measured inside a temperature-stabilized He flow cryostat, using a four-point ac lock-in technique, with a current density $J \leq 10^2$ A/cm^2. All electrical contacts are made by direct wire bonding onto the film surface using AlSi wires. The Hall coefficient is obtained from the transverse voltage between two opposing contacts, measured using a classical field inversion technique, in a magnetic field $B = 0.72$ T. For all the films the Hall voltage varies linearly with applied field and current.

1.2.1. Resistivity. Using the method described above, we have systematically varied the oxygen content in several c-axis-oriented YBCO films and measured the temperature dependence of the resistivity. Figure 1 presents a compilation of $\rho(T)$ data from three different films, showing very similar and reproducible results.

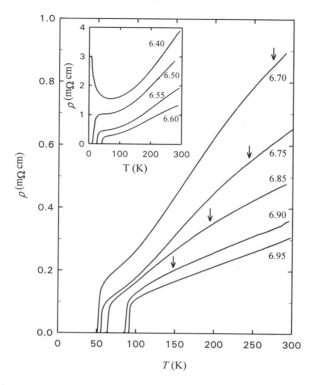

Figure 1. Resistivity ρ versus temperature T as a function of the oxygen content x in c-axis-oriented YBCO films. The arrows show the scaling temperature T_0.

According to Fig. 1, the $\rho(T)$ curve for $x = 6.95$ shows the well-known linear temperature dependence. However, reducing x below its optimum doping value produces a clear change in the temperature dependence of ρ.[3,4] The temperature interval, in which ρ is linear, systematically shifts to higher temperatures, and a downward bending develops at lower temperatures, eventually followed by an upward bending becoming more pronounced at the lowest x values. Finally, for the nonsuperconducting $x = 6.3$ sample, we observe only the upward bending, indicating the onset of localization effects.

A similar change in temperature dependence, including the development of a kind of "S-shape," has been reported also for oxygen-deficient YBCO single crystals[6,12] and thin films,[13-15] for Co-doped YBCO single crystal,[8] and also for underdoped $La_{2-y}Sr_yCuO_4$ (LSCO) ceramics[16] and single crystals.[17]

Hence the following consistent picture emerges: a pronounced nonlinear $\rho(T)$ behavior develops systematically in cuprates with reduced hole densities, while a linear behavior with constant slope but increased magnitude of ρ is obtained in cuprates in which T_C is suppressed by other means than hole doping (e.g., by disorder).

1.2.2. Hall Effect. For the same samples on which we measured the resistivity and in the same measuring run, we have studied the behavior of the Hall number $n_H(T) =$

$1/[R_H(T)e]$. The main results of $R_H(T)$ as a function of x are shown in Fig. 2. Note that the Hall effect behavior in the vicinity of T_C, i.e., in the "mixed state," is mainly governed by the movement of magnetic flux lines. This regime will not be treated here since it goes beyond the scope of the present work.

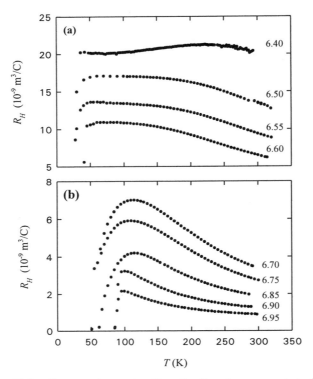

Figure 2. Hall coefficient R_H versus temperature T as a function of oxygen content x in c-axis oriented YBCO : (a) $x < 6.7$, (b) $x \geq 6.7$.

1.3. Discussion

1.3.1. Scaling Behavior. Although the reduction of the oxygen content in YBCO gives rise to different temperature dependences of the transport properties, the changes take place in a very systematic way.[5-7] As an example, we refer back to the $\rho(T)$ behavior, where we saw that the T-linear part systematically shifts towards higher temperatures when x is lowered and simultaneously a pronounced S-shape develops at lower temperatures. In order to quantify this evolution, we can define a characteristic temperature T_0 as the temperature above which ρ is linear in T. Then T_0 corresponds to the temperature above which the derivative $d\rho/dT$ is constant and is approximately equal to twice the temperature of the maximum in $d\rho/dT$. By deriving T_0 in this way for all measured samples, we estimate that an error of $\pm 20K$ has to be taken into account.

Using the characteristic temperature T_0, we can now scale the temperature axis and simultaneously scale the resistivity with the value of ρ at T_0.[6] The result of this operation for all the available thin film resistivity data (15 data sets in total) is shown in Fig. 3a. Very interestingly, we can see in this figure that to a good approximation *all the $\rho(T)$ curves collapse onto one universal curve.*[4,5] Given the considerable overlap in the reduced-temperature scale between the 15 resistivity curves, it is very unlikely that this scaling result is a fortuitous coincidence or an artificial effect of splicing separate curves together.

The universal resistivity curve consists of an S-shaped part for $T/T_0<1$, followed by a linear part for $T/T_0>1$, which approximately extrapolates to zero for $T=0$. Hence this scaling curve clearly demonstrates the limited validity of the often-cited T linearity of the resistivity. Note that the curves tend to separate slightly at $T/T_0>1.5$. This may be due to small differences in residual resistivity between the different samples, which was not included as an extra parameter in the scaling.

In order to check whether the scaling behavior of the resistivity is an intrinsic property of YBCO, we have also applied the same scaling procedure to reported data on oxygen deficient YBCO *single crystals* ($x = 6.45$, 6.58, 6.68, 6.78, 6.85, 6.9) by Ito *et al.*[6] and on *Co-doped* YBCO single crystals (Co doping level $y = 0.0\%$, 1.1%, 4.1%, 9.6%) by Carrington *et al.*[8] and included them in Fig. 3a. Hence a combination of the thin film data with the studies cited above compares samples of different nature (thin film versus single crystal) and hole doping mechanisms from different origins (O deficiency versus Co doping). It can be clearly observed in Fig. 3a that the scaled data points for the single crystals agree very well with the thin film data. The scaled curves from the oxygen deficient YBCO crystal tend to saturate at a slightly smaller low-temperature value, which may be due to a lower residual resistivity of these single crystals.

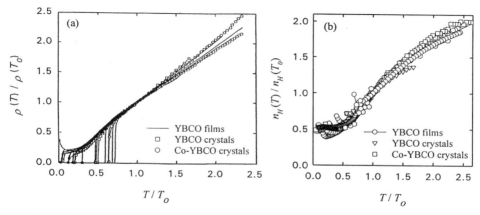

Figure 3. Scaled in-plane resistivity $\rho/\rho(T_0)$ (a), and Hall number $n_H/n_H(T_0)$ (b) versus normalised temperature T/T_0 for 15 sets of thin film data together with reported data for oxygen-deficient (Ref. 6) and Co-doped (Ref. 8) YBCO single crystals.

The fact that the scaling behavior of $\rho(T)$ is observed in different YBCO samples with a variety of hole densities strongly indicates that the main scattering mechanism in YBCO remains unchanged when the carrier concentration is reduced from its optimum value down to the metal-insulator transition.[4,6] The characteristic energy scale on which the scattering occurs and which is determined by the temperature T_0 decreases monotonically with increasing oxygen content or decreasing Co doping lever, as shown in fig. 4.

After having demonstrated the scaling behavior of the resistivity, it becomes particularly interesting to make the same kind of analysis for the Hall effect data. Using *the same characteristic temperature T_0* as for the resistivity, we have plotted in Fig. 3b the Hall number versus temperature scaled with $n_H(T_0)$ and T_0, respectively. Included are the 15 sets of thin film data together with the reported data on oxygen-deficient[6] and Co-doped[8] YBCO single crystal. It can be seen that, within the experimental scatter on the data, again one universal Hall number versus temperatures curve emerges. This behavior follows an S-like shape which is approximately linear in only a very limited temperature region $0.75 < T/T_0 < 1.25$. At higher temperatures the curve bends off, tending towards a saturation, while at lower temperature the scaling curve levels off more sharply at $T/T_0 \cong 0.5$ to a constant value. The inflection point of the transition from a positive to a negative curvature occurs exactly at the characteristic temperature T_0.

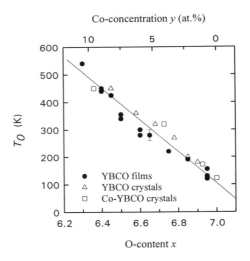

Figure 4. Characteristic temperature T_0 versus oxygen content x and Co doping level y, for the oxygen-deficient c-axis YBCO films and single crystals (Ref. 6) and for the Co-doped YBCO single crystals (Ref.8). A typical error bar is indicated; the line is a guide for the eye.

The observation of scaling behavior for the Hall number, using the same scaling parameter as the one derived from the resistivity data, is quite surprising. It strongly indicates that both properties are governed by the same characteristic energy.

In view of the scaling observations for the separate resistivity and Hall effect data, a scaling behavior is expected to appear also for the Hall angle $\cot\theta_H = \rho/R_H.B$. Again, using the same scaling parameter T_0 derived from the resistivity data, we plot in Fig. 5 the scaled cotangent of the Hall angle $\cot\theta_H/\cot\theta_H(T_0)$, versus the square of the reduced temperature, $(T/T_0)^2$. As expected, all the data are mapped onto a single curve.

Figure 5 clearly shows that the often-cited quadratic temperature dependence of $\cot\theta_H$ is only valid in a limited low-temperature region $[0.3 \le (T/T_0)^2 \le 1.5]$. At higher reduced temperature a good fit to the data is obtained using the relation $\cot\theta_H = \alpha'T^p + C'$ with $p = 1.69$. The latter value is in good agreement with $p = 1.63$ reported by Hofmann et al.[18] for nearly optimum doped YBCO thin films.

1.3.2. Comparison with Magnetic Measurements. In order to get more insight into the meaning of the characteristic temperature T_0, we compare in this section the *electrical* transport results with published data on normal-state *magnetic* properties. There exists nowadays growing evidence for the intimate relation between magnetic and transport properties,[6,19,20] which may be the clue to the understanding of HTSC.[21]

In the past few years, a very intensive effort has been developed to study the spin excitations in HTSC's using inelastic neutron scattering (INS)[1,2]. It was found that in underdoped YBCO samples a so-called *spin gap* or *pseudogap* opens up in the spin excitation spectrum at a temperature T_{SG}, well above the superconducting transition temperature T_C, below which the spins become highly correlated.

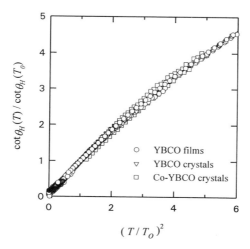

Figure 5. Cotangent of the Hall angle $\cot\theta_H$ divided by $\cot\theta_H(T_0)$ versus the square of the reduced temperature $(T/T_0)^2$, for 15 sets of thin film data together with reported data for oxygen-deficient (Ref. 6) and Co-doped (Ref. 8) YBCO single crystals. The same $T_0(x)$ data were used as those shown in Fig. 4.

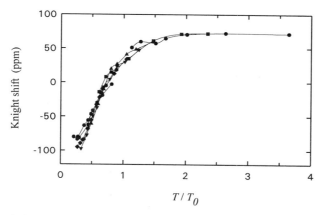

Figure 6. Knight shift data from Alloul (Ref. 22) for $(7 \leq x \leq 6.48)$ plotted versus reduced temperature T/T_0. The scaling temperature $T_0(x)$ is the same as the one used for the transport properties scaling (see fig. 4).

The opening of a spin gap has also been confirmed by NMR measurements of the Knight shift ΔK and of the spin-lattice relaxation time (T_1). It was shown before[6] that the Knight shifts data closely resemble the behavior of ρ/T versus T. Hence the latter suggests that the spin gap temperature T_{SG} determined by NMR Knight shift measurements may be related to the characteristic temperature T_0 determined from transport measurements.

In order to check this suggestion, we have assigned a characteristic temperature T_0 value to the NMR data $(7 \leq x \leq 6.48)$ from Alloul,[22] using the same $T_0(x)$ relation found from transport data and presented in Fig. 4 and plotted the Knight shift versus reduced temperature T/T_0. The result of this operation is shown in Fig. 6. Clearly, all the temperature dependent Knight shift data for the samples with different oxygen contents collapse onto a single curve.

This scaling observation shows that also for the magnetic properties the physics does not change with decreasing hole doping, but rather the temperature scale does. Moreover, the fact that this scaling behavior is obtained using the *same characteristic temperature T_0 derived from transport measurements* strongly indicates that transport and magnetic properties in YBCO are closely related. Similar scaling behavior has also been reported on *underdoped* $La_{2-y}Sr_yCuO_4$.[9,23] From this it becomes clear that there exists a close correlation between the electrical transport and the magnetic properties in several HTSC compounds. These results clearly indicate that both transport and magnetic properties are governed by the same characteristic temperature, which may be related to the opening of a spin gap. Because of the apparent universality of the characteristic temperature, we believe that the latter should be included in the generic phase diagram of the HTSC's, as shown schematically in Fig. 7. Very similar phase diagram has been recently obtained for YBCO films from the analysis of their high frequency properties.[24]

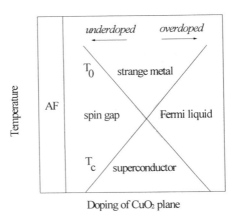

Figure 7. Schematic presentation of the mean field phase diagram for the CuO_2 plane as proposed by Nagaosa and Lee (Ref. 25).

The available experimental data on YBCO indicate that the spin gap temperature of the resistivity (T_0) coincides with the onset temperature of the spin gap in the Knight shift, while the spin gap opening in ^{63}Cu $(T_1T)^{-1}$ apparently occurs at about $T_0/2$.

Besides the scaling behavior of the transport and magnetic properties of metallic high-T_c cuprates, the scaling behavior of the localization length and conductivity has also been reported at the insulator-metal transition in $Bi_2Sr_2(Ca_ZY_{1-Z})Cu_2O_8$.[26]

In spite of the growing experimental evidence for existence of the scaling behavior of transport and magnetic properties, the fully consistent theoretical description of this phenomena is still lacking. For possible theoretical models, reflecting the personal point of view of one of the authors (V.V.M.) of the present paper ($T_0 \cong W$), we refer to Ref. 27, where the narrow feature with metallic conductivity with the width W and the treatment of high-T_C cuprates as 2D quantum antiferromagnets have been considered.

1.4. Scaling Behavior of the Normal State Properties : Conclusions

In the previous section we have presented and discussed experimental normal-state transport data on oxygen-deficient YBCO films. From the resistivity curves a characteristic temperature T_0 is derived which separates the linear part at high temperature from the S-shaped part at low temperature. By scaling T with T_0 and ρ with $\rho(T_0)$, all the thin film data collapse onto one universal curve. In addition, resistivity data taken from the literature for underdoped YBCO single crystals are mapped on the universal resistivity curve as well. Using the same characteristic temperature T_0, a scaling behavior is also observed for the Hall effect and Hall angle data.

These results indicate that a reduction of the hole doping level in YBCO does not affect the dominant scattering mechanism, but rather modifies the energy scale on which the scattering occurs. Furthermore, the universal scaling curves clearly show that the transport properties obey the often cited linear or quadratic temperature dependences only within limited temperature intervals. From a comparison of the transport behavior with published data on the magnetic normal-state properties of the cuprates, it becomes clear that both types of properties are governed by the same physics. As the most prominent feature, we note that we have mapped published NMR Knight shift data for oxygen-deficient YBCO onto a single curve by scaling the temperature axis with the *same characteristic temperature T_0 derived from our transport measurements*. These findings indicate that the spin correlations, which are believed to determine the anomalous behavior of the magnetic properties, are also governing the transport properties in the cuprates. It is then most likely that T_0 is related to the opening of a spin gap in the spin excitation spectrum, which is observed in NMR and neutron scattering experiments. It is in fact not only a gap in the spin excitations, but also in the charge channel. Therefore its a partial gap in the entire excitation spectrum, e.g. a pseudogap.[28]

2. SUPERFLUID DENSITY IN METALLIC YBa$_2$Cu$_3$O$_x$ CUPRATES

2.1. Introduction

In part 2 of this paper, we focus on the evolution of the superfluid density with doping in YBCO. The HTSC's are highly anisotropic type-II superconductors with high critical temperature, and with a very short superconducting coherence length ξ. The London penetration depth λ, another fundamental superconducting parameter, is directly related to the superfluid density n_s, given by the square of the modulus of the superconducting order parameter $|\Psi(\vec{r})|^2 = n_s(\vec{r})$. Therefore, the behavior of λ *as a function of temperature T* can reveal important information on the *superconducting pairing mechanism*, since the latter strongly influences the $\lambda(T)$ dependence. As a consequence, the temperature dependence of the London penetration depth $\lambda(T)$ has been the topic of many experimental and theoretical studies.

Nuclear Magnetic Resonance (NMR),[29,30] Muon Spin Relaxation (μSR),[31,32] Electron Spin Resonance (ESR)[33] and the direct AC method[34] are only a few of the different techniques applied to measure $\lambda(T)$ directly or derive it indirectly from reliable model calculations. Several groups measured the London penetration depth by μSR in oriented polycrystalline YBa$_2$Cu$_3$O$_7$ samples.[31,32] The $\lambda(T)$ behavior was well described by the two-fluid model, from which the authors[31,32] concluded that the conventional BCS s-wave pairing mechanism is valid. Direct measurements of $\lambda(T)$ on single crystals of YBa$_2$Cu$_3$O$_7$[35] led to a similar conclusion. Other groups observed linear and power law temperature dependences on YBa$_2$Cu$_3$O$_7$ thin films and Bi$_2$Sr$_2$CaCu$_2$O$_8$ single crystals.[36,37,38] These temperature dependencies are predicted for an unconventional pairing state, gapless or d-wave superconductivity.[39] A transition from conventional to gapless superconductivity with varying carrier concentration has also experimentally been reported.[40,41]

Besides the pairing mechanism there are also other parameters that can have an important influence on the behavior of $\lambda(T)$. For instance, the anisotropy due to the layered structure of the HTSC can cause a substantial modification of the $\lambda(T)$ curves.[42] The temperature dependence of the London penetration depth is also known to be very sensitive to the hole doping level in the superconducting layers.[37,41]

All these different factors have led to some controversy concerning the temperature variation of $\lambda(T)$ in the high-T_C oxides. The closely related essential question - which mechanism is responsible for the superconductivity in these materials - remains open. Therefore further systematic studies of the $\lambda(T)$ behavior by different methods are necessary.

Recently we have developed the NMR decoration technique to measure $\lambda(T)$ from the additional broadening of the NMR signal in a thin decorating layer formed around superconducting grains. In contrast to the conventional NMR method, which uses the intrinsic nuclei as the NMR probe[29], we monitor ^{1}H-NMR nuclei in a thin layer at the surface of the high-T_C material. The fields required to measure intrinsic nuclei, for instance Y or O, are much higher than the field needed for ^{1}H-nuclei. Our method is based on the idea proposed by Bontemps et al.[33] to study the field distribution near the surface of a superconductor using ESR spin label layers. Whereas in continuous wave ESR a varying magnetic field is used, in pulsed NMR we use a constant field which gives us the advantage of being able to perform measurements in a broader temperature range, whereas the cw-ESR method is limited to the reversible area in a narrow temperature interval close to T_C. Compared to the implantation of muons needed for the μSR experiments, the preparation of the decorated samples, based on wetting of silicon oil on the YBCO grains, is much easier.

Our NMR decoration method was calibrated with the classical superconductor Nb$_3$Sn[43]. Recently we reported on a NMR decoration study of the field distribution in different high-T_C cuprates[44]: YBa$_2$Cu$_3$O$_7$, Bi$_2$Sr$_2$CaCu$_2$O$_8$ and Bi$_2$Sr$_2$Ca$_2$Cu$_3$O$_{10}$. We found that not only the low temperature value $\lambda(0)$, but also the temperature behaviour $\lambda(T)$ is specific to each compound. Here we report on a systematic study of the temperature dependence of the penetration depth in YBCO powder as a function of hole doping and anisotropy. A specific behavior of $\lambda(T)$ for each oxygen content is observed.

2.2. Temperature Dependences of the Superfluid Density in YBa$_2$Cu$_3$O$_x$

The NMR decoration measurements have been carried out on oxygen deficient YBCO powder. We prepared the oxygen deficient powders with $x = 6.5 - 7$ using the method developed by Osquiguil et al.[11] The parent compound, obtained from Seattle Specialty Ceramics, Inc., is powdered material and has a grain radius of about 0.5 µm. Since single crystalline material has a very low oxygen diffusion rate, it is more straightforward to use powdered material for the experiments with the oxygen deficient YBCO.

The magnetic moment m as a function of temperature T was measured for all the oxygen deficient YBCO samples using a SQUID magnetometer (MPMS-Quantum Design). In Fig. 8 the $m(T)$ data are shown for YBa$_2$Cu$_3$O$_x$ with $x = 6.8, 6.7, 6.6$ and 6.5, for clarity we do not show the results for all nine samples. The measurements were performed in an applied magnetic field of 5×10^{-4} T. All samples show a transition typical for single phase material, which is important for our NMR decoration measurements. If a sample consists of different phases, then each phase demonstrates a different behavior of the NMR line broadening as a function of temperature. As a consequence, all the different phase contributions will be mixed and the interpretation of the additional line broadening becomes impossible.

We used a Bruker 2.2 T electromagnet and a Bruker CXP-90 pulse spectrometer for the NMR experiments. A solid echo pulse sequence was applied to generate the NMR signal. For a detailed description of the experimental setup we refer to Steegmans *et al.*[43,44] The YBCO powders were covered by a thin layer of silicone oil, which is used as the NMR probe. The silicone oil gives a ^{1}H-NMR spectrum with a single line. For temperatures $4\ K < T < 200\ K$ the frozen oil has a temperature independent line width Δw_i of about 32 kHz, mainly due to dipole-dipole interaction.

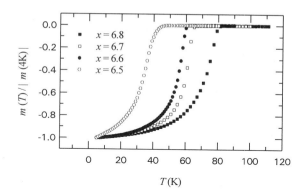

Figure 8. Magnetic moment m for $B_a = 5\times10^{-5}$ T as a funtion of temperature, normalized to the value at 4 K, for YBa$_2$Cu$_3$O$_x$ with x = 6.5, 6.6, 6.7 and 6.8 and T_C = 46 K, 61 K, 67 K and 81 K, respectively.

Below T_C the width of the NMR line of the mixture oil and superconducting powder increases due to field inhomogeneities at the surface of the YBCO grains. The additional line broadening is defined as:

$$\Delta w = \sqrt{\langle\Delta w\rangle^2_{meas} - \langle\Delta w\rangle^2_i} \tag{1}$$

This broadening is mainly related to the presence of the vortex lattice.[43] Because of this the second moment data can be analyzed using Brandt's formula[45] for a triangular vortex lattice:

$$\Delta B^2 = 0.00371\frac{\Phi_0^2}{\lambda^4(T)} \tag{2}$$

and the relationship $\gamma(\Delta B) = 2\pi(\Delta w)$, with $\gamma = 2.675\ 10^8$ rad Hz/T the gyromagnetic ratio for ^{1}H nuclei. Demagnetization effects can also cause a contribution in the additional line broadening. In this case one should have a line broadening which is dependent on the grain size of the superconducting powder.[46] Our experimental data (Figure 9) show that the line broadening as a function of temperature for YBa$_2$Cu$_3$O$_7$ powders remains the same for different grain sizes. This proves that demagnetisation effects can be ignored. Therefore the additional line broadening is stemming from field inhomogeneities due to the flux line lattice and this justifies the use of Eq. (2). The main assumption made in deriving Eq. (2) is the presence of a static flux line lattice with vortices parallel to the applied field B_a.

The $\lambda(T)$ behavior can be characterized by the simple power law :

$$\lambda(T) = \frac{\lambda(0)}{\sqrt{\left(1-(T/T_c)^n\right)}} \tag{3}$$

The temperature dependence of the penetration depth in the two fluid model is given by Eq. 3 with n = 4.

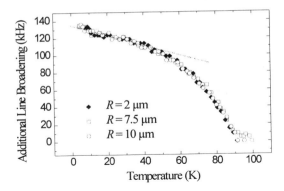

Figure 9. The NMR line broadening for $YBa_2Cu_3O_7$ as a function of temperature for different grain sizes in an applied field of 1.7 T.

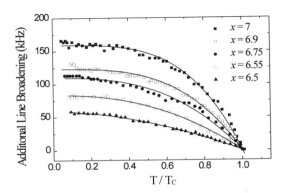

Figure 10. The additional line broadening as a function of reduced temperatures for $YBa_2Cu_3O_x$ with x = 7, 6.9, 6.75, 6.55, 6.5 and T_C = 92 K, 89 K, 65 K, 51 K, 44 K, respectively. The solid lines are the best fits to the power law given by Eq. (3), the fit parameters $\lambda(0)$ and n are listed in Table 1.

In Figure 10 we show some examples of the additional line broadening as a function of the reduced temperature T/T_C. The solid lines are the fits with Eq. (3). For a decrease in oxygen content the low temperature value of the additional line broadening becomes smaller, which means an increasing $\lambda(0)$. The parameters n and $\lambda(0)$ obtained from the fittings are listed in Table 1 for each oxygen content x. Since we used powdered samples we have to take into account that the $\lambda(0)$ value is an average over the different grain orientations and it can be defined as $\lambda(0) = \sqrt[3]{\lambda_{ab}^2(0)\lambda_c(0)}$. With decreasing carrier concentration we observe an increase of the London penetration depth $\lambda(0)$. Not only the penetration depth at 0 K depends on the doping level but also the deviation of $\lambda(T)$ from the two-fluid model (n = 4) for isotropic material varies in a systematic way with the oxygen stoichiometry x. Thus it is obvious that not only the low-temperature value $\lambda(0)$, but also the temperature behavior $\lambda(T)$ is specific for each compound.

x	$\lambda(0) \pm 10\text{Å}$	n
7	2590	3.7 ± 0.1
6.9	2950	3.27 ± 0.09
6.85	2990	2.9 ± 0.1
6.8	2980	2.36 ± 0.05
6.75	3125	2.8 ± 0.1
6.7	3480	2.89 ± 0.08
6.6	3400	3.0 ± 0.1
6.55	3580	2.33 ± 0.06
6.5	4270	1.83 ± 0.04

Table 1. The fit parameters $\lambda(0)$ and n derived from the fit to $\lambda(T) = \lambda(0) / \sqrt{1 - (T/T_c)^n}$.

Our $\Delta w(T/T_c)$ data as a function of carrier density show a very similar behavior as that reported by Zimmerman et al.[41] based on µSR studies. They also found a stronger deviation of $\lambda(T)$ from the conventional two-fluid dependence and a reduced exponent n for lower carrier concentrations.

Recently a remarkable empirical relation between T_C and $\lambda_{ab}^{-2}(0)$ was derived from µSR results.[47] This relation seems to be universal for most high-T_C cuprates and other extreme type II superconductors[48,49,50]. This so-called Uemura-plot (T_C versus $\lambda_{ab}^{-2}(0)$) exhibits the following feature : with increasing carrier doping, T_C initially increases linearly to saturate at optimum doping and finally T_C is suppressed for higher carrier concentration. For anisotropy ratios λ_c/λ_{ab} greater than five the relation $\lambda = 1.23 \lambda_{ab}$ is valid[46]. From this relation we calculated λ_{ab} and plotted it versus T_C (Fig. 11), which gives us the linear part, which agrees with the underdoped regime of the Uemura plot.[50]

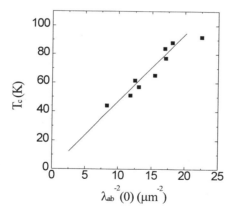

Figure 11. T_C versus $\lambda_{ab}^{-2}(0)$ for $YBa_2Cu_3O_x$ ($6.5 \leq x \leq 7$)

We can conclude that with decreasing oxygen content x in the $YBa_2Cu_3O_x$ cuprates there is a systematic trend towards weaker variation of the additional line broadening Δw as a function of reduced temperature T/T_C. The crossover from the underdoped ($x < 6.9$) regime to optimal doping is accompanied by the distinct evolution of the penetration depth $\lambda(T)$ as a function of temperature. We should also note that the values derived from our

$\Delta w(T)$ measurements for the oxygen deficient samples are in a very good agreement with the values reported in the literature.[41]

2.3. Discussion

We have presented a convincing experimental evidence for a substantial deviation of the $\lambda(T)$ dependences in YBa$_2$Cu$_3$O$_x$ from the two-fluid behavior. There are several possiblities to explain this unusual behavior of the high-T_C cuprates. We consider three different theories: the Bose-gas theory, anisotropy and interlayer coupling, and s- or d-wave pairing.

We will begin with the Bose-gas theory proposed by Nagaosa and Lee,[25] who argued that a strongly underdoped system corresponds to a dilute Bose-gas, which can be described by an ideal Bose-gas model. The temperature dependence of the penetration depth in case of an ideal Bose-gas is given by Eq. (3) with an exponent $n = 3/2$ (fig. 12a). Increasing the carrier density up to optimal doping leads to a crossover to the dense limit for which $\lambda(T)$ can be approximated by the two-fluid model with $n = 4$. Thus by changing the carrier density it is possible to cover the whole range of powers between 1.5 and 4. In Fig. 12b we compare the experimental additional line broadening of the optimum doped sample YBa$_2$Cu$_3$O$_7$ and the two-fluid like behavior ($n = 4$). The best fit is achieved for $n = 3.7$. This is not far but still distinctly different from the two-fluid value. At low temperatures (see fig. 9) the tendency towards a linear λ versus T dependence shows up, though the two-fluid model predicts the saturation of $\lambda(T)$ at T→0. In Fig. 12a a comparison is made between the data obtained on YBa$_2$Cu$_3$O$_{6.5}$ ($n = 1.83$), which is

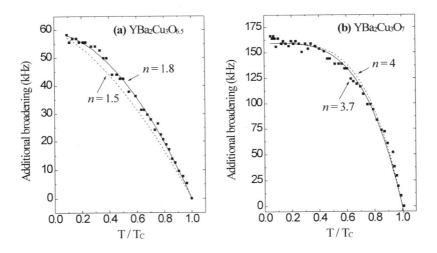

Figure 12. a) The additional line broadening for YBa$_2$Cu$_3$O$_{6.5}$ is presented together with the temperature dependence of $\lambda(T)$ for an ideal Bose-gas ($n = 1.5$) (dashed line). The solid line is the best fit to Eq. (3) with $n = 1.8$ and $\lambda(0) = 4270 \pm 10$ Å. b) The additional line broadening for YBa$_2$Cu$_3$O$_7$ is compared to the $\lambda(T)$ behavior predicted by the two-fluid model ($n = 4$) (dashed line). The solid line is the best fit to Eq. (3) with n = 3.7 and $\lambda(0) = 2590 \pm 10$ Å.

strongly underdoped, and the $\lambda(T)$ behavior predicted for an ideal Bose-gas ($n = 1.5$). Thus in this case our measurements on the YBCO samples are not far from the prediction of the Bose-gas theory. In general underdoped samples exhibit the $\lambda^{-2}(T) \propto \Delta w(T)$ behavior with n

close to 1.5, while the optimal doped sample shows certain similarities with the two-fluid like behavior.

The Bose-gas scenario, used above, is not the only one which can be used to interpret our experimental observations. According to the theoretical calculations by Nicol and Carbotte,[51] the behavior of $\lambda(T)$ can also be influenced by the strength of the electron-phonon (e-ph) coupling. This theory, based on the Eliashberg equations, predicts a reduced n value for a decreasing coupling strength. In this case the power $n = 4$, which corresponds to the two-fluid model, would agree with a strong e-ph coupling. In our $\Delta w(T/T_C) \propto \lambda^{-2}(T/T_C)$ measurements on the oxygen deficient YBCO, we see a trend towards a temperature dependence with smaller n for decreasing oxygen content. The reduction of the power n from 3.69 for $x = 7$ to 1.83 for $x = 6.5$ reflects this trend, which could mean a decreasing e-ph coupling strength with reduced oxygen content. The samples with the highest oxygen deficiency ($x = 6.55$ and 6.5) have a $n \cong 2$, which is in agreement with the weak-coupling BCS theory which predicts $n \cong 2.4$ in Eq. (3).[41] For the sample with the lowest oxygen deficiency we derived $n = 3.69$ which is clearly the variation towards the two-fluid like behavior ($n = 4$). According to Nicol and Carbotte,[51] this would mean a stronger e-ph coupling. The possibilities of a strong e-ph coupling in samples with high oxygen content is also supported by recent work of Reeves[52] (specific heat) and Neminsky[53] (AC susceptibility). We think that these arguments should still be kept in mind as an alternative to the d-wave scenario which is based on the rapidly growing experimental evidence.[54]

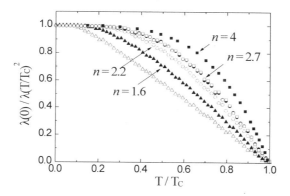

Figure 13. The normalized $\lambda^{-2}(T/T_C)$ curves (Ref. 56) are substantially changed as one goes from low anisotropy ($\square$) to high anisotropy ($\triangle$). For their calculations the authors[56] assumed an intermediate e-ph coupling strength.

An alternative explanation for the variation of the $\lambda^{-2}(T)$ curves with x could also be based on the increasing anisotropy. Since it is known that the anisotropy grows with decreasing oxygen content,[55] we can say again that a larger anisotropy gives us a stronger deviation of $\Delta w(T/T_C) \propto \lambda^{-2}(T/T_C)$ from the two-fluid model. However from the calculations of Nicol and Carbotte it is clear that the strength of the influence of the anisotropy on the $\lambda^{-2}(T)$ variation depends also on the e-ph coupling strength. Taking in the calculations the electron-phonon interaction typical for the strong e-ph coupling, reduces the dependence of $\lambda^{-2}(T)$ on the anisotropy. In figure 13 we show the normalized $\lambda^{-2}(T)$ reported by Jiang et al.[56] for an intermediate e-ph coupling strength. The solid squares correspond to the two-fluid model. All the other curves have a fitting parameter n in the range 2.7 - 1.6 when anisotropy is decreased. If now we compare the calculated curves to the ones we obtained a very similar variation of n is observed. This means that the calculations based on the Eliashberg equations could, in principle, explain the $\Delta w(T/T_C) \propto \lambda^{-2}(T/T_C)$ behavior found in our NMR decoration experiments.

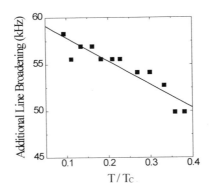

Figure 14. Additional line broadening ($\blacksquare$) as a function of reduced temperature in the low temperature range with $T_C = 44$ K. The solid line gives the fit to a linear temperature dependence.

The results obtained on oxygen deficient YBCO can be interpreted also in the framework of s- and d-wave superconductivity. The calculations by Adrian *et al.*[40] show that depleting the chains of oxygen leads to a series of functional forms for $\lambda(T)$ going from two-fluid like s-wave behavior to a linear dependence pointing towards d-wave pairing. This would mean that by reducing the oxygen content in the $YBa_2Cu_3O_7$ sample, an admixture of the s-wave contributions can be varied. Our measurements of $\Delta w(T/T_C)$ for $YBa_2Cu_3O_7$ show an almost two-fluid like behavior, but with a tendency toward the linear $\lambda(T)$ variation at low temperatures (see fig. 9). The dependence of $\Delta w(T/T_C)$ for $YBa_2Cu_3O_{6.5}$ deviates strongly from the two-fluid model and at low temperatures demonstrates more clearly the linear behavior (Fig. 14). It is worth noting that a similar change of $\lambda(T)$ was also predicted for the case when magnetic impurities are added to the N layer of the S/N model. The S-layers are the CuO_2-planes and the N-layers are the layers between the supercon-ducting planes.

By now, there is growing experimental evidence for a superconducting order parameter of mainly d-wave symmetry, which leads, among others to a linear $\Delta\lambda(T)/\lambda(0)$ dependence.[36] There are always several reasons why the pure d-wave density of states is smeared out, for example due to the presence of impurities and inhomogeneities. In this case an admixture of s-component should be present.

2.4. Summary

In NMR decoration studies a substantial modification of $\Delta w(T/T_C) \propto \lambda^{-2}(T/T_C)$ was observed for oxygen deficient YBCO, with a transition from an almost two-fluid like behavior for $YBa_2Cu_3O_7$ ($n = 3.7$) to a power law behavior (Eq. (3)) with $n = 1.83$ for $YBa_2Cu_3O_{6.5}$. For all x the linear contribution to $\lambda^{-2}(T/T_C)$ is seen at low temperatures, and it is noticably enhanced for lower oxygen concentrations. Three possible theories to interpret these results are presented. The first is based on the Bose-gas picture where one would expect a power n between 4 and 1.5 with decreasing oxygen content. The second interpretation relies upon the calculations using the Eliashberg equations. In this case the anisotropy and the electron-phonon coupling strength play an important role. The third possible interpretation requires an enhancement of the d-wave superconducting character on the way from the optimally doped to underdoped YBCO. The final choice in favour of only one specific model among those three considered here can not be done in the framework of the analysis of the $\lambda^{-2}(T)$ data and requires further detailed studies of $YBa_2Cu_3O_x$ using other experimental techniques, which are indicating more and more in the direction of the d-wave superconductivity.

ACKNOWLEDGMENTS

We would like to thank V.V. Metlushko for the magnetization measurements on the YBCO powders, M. Maenhoudt and G. Jacob for the preparation of the thin films, and E. Osquiguil and H. Adrian for useful discussions. The YBCO powders with different grain sizes were provided by Hoechst AG Germany. This research has been supported by the Fund for Scientific Research Flanders (FWO), the Concerted Research Action programmes (G.O.A.), the Interuniversity Poles of Attraction programmes (I.U.A.P.) -Belgian State, Prime Minister's Office- Federal Office for Scientific, Technical and Cultural Affairs. A.S. is a research assistant of the Belgium Interuniversity Institute for Nuclear Sciences (I.I.K.W).

REFERENCES

1. For a review, see N.P. Ong, in : *Physical Properties of High Temperature Superconductors II*, edited by D.M. Ginsberg, World Scientific, Singapore (1991), p. 459.
2. For a review, see Y. Iye, in : *Physical Properties of High Temperature Superconductors III*, edited by D.M. Ginsberg, World Scientific, Singapore (1993), p. 285.
3. B. Wuyts, E. Osquiguil, M. Maenhoudt, S. Libbrecht, Z.X. Gao, and Y. Bruynseraede, Influence of the oxygen content on the normal-state Hall angle in $YBa_2Cu_3O_x$ films, *Phys. Rev. B* 47 5512 (1993).
4. B. Wuyts, E. Osquiguil, M. Maenhoudt, S. Libbrecht, Z.X. Gao, and Y. Bruynseraede, Relation between the Hall angle slope and the carrier density in oxygen-depleted $YBa_2Cu_3O_x$ films, *Physica C* 222, 341 (1994).
5. B. Wuyts, V.V. Moshchalkov, and Y. Bruynseraede ,Scaling of the normal-state transport properties of underdoped $YBa_2Cu_3O_x$, *Phys. Rev. B* 51, 6115 (1995); B. Wuyts, V.V. Moshchalkov, and Y. Bruynseraede, Resistivity and Hall effect of metallic oxygen-deficient $YBa_2Cu_3O_x$ films in the normal state, *Phys. Rev. B* 53, 9418 (1996).
6. T. Ito, K. Takenaka, and S. Uchida, Systematic deviation from T-linear behavior in the in-plane resistivity of $YBa_2Cu_3O_{7-y}$: Evidence for dominant spin scattering, *Phys. Rev. Lett.* 70, 3995 (1993).
7. B.Batlogg, H.Y. Hwang, H. Takagi, R.J. Cava, H.L. Kao, and J. Kwo, Normal State Phase Diagram of $(La,Sr)_2CuO_4$ from Charge and Spin Dynamics, *Physica C* 235-240, 130 (1994).
8. A. Carrington, A.P. Mackenzie, C.T. Lin, and J.R. Cooper, Temperature dependence of the Hall angle in single crystal $YBa_2(Cu_{1-x}Co_x)_3O_{7-\delta}$, *Phys. Rev. Lett.* 69, 2855 (1992).
9. H.Y. Hwang, B. Batlogg, H. Takagi, H.L. Kao, J. Kwo, R.J. Cava, J.J. Krajewski, and W.F. Peck, Jr., Scaling of the temperature dependent Hall effect in $La_{2-x}Sr_xCuO_4$, *Phys. Rev. Lett.* 72, 2636 (1994).
10. B. Wuyts, Z.X. Gao, S. Libbrecht, M. Maenhoudt, E. Osquiguil, and Y. Bruynseraede, Growth of particle-free $YBa_2Cu_3O_7$ films by off-axis sputtering, *Physica C* 203, 235 (1992).
11. E. Osquiguil, M. Maenhoudt, B. Wuyts, and Y. Bruynseraede, Controlled preparation of oxygen deficient $YBa_2Cu_3O_x$ films, *Appl. Phys. Lett.* 60, 1627 (1992).
12. J. Harris, Y.F. Yan, and N.P. Ong, Experimental test of the T^2 law for the Hall angle from T_C to 500K in oxygen-reduced $YBa_2Cu_3O_{6+x}$ crystals, *Phys. Rev. B* 46, 14293 (1992).
13. P. Xiong, G. Xiao, and X.D. Wu, Hall angle in $YBa_2Cu_3O_{7-\delta}$ epitaxial films: Comparison between oxygen reduction and Pr doping, *Phys. Rev. B* 47, 5516 (1993).
14. E.C. Jones, D.K. Christen, J.R. Thompson, R. Feenstra, S. Zhu, D.H. Lowndes, J.M. Phillips, M.P. Siegal, and J.D. Budai, Correlations between the Hall coefficient and the superconducting transport properties of oxygen-deficient $YBa_2Cu_3O_{7-\delta}$ epitaxial thin films, *Phys. Rev. B* 47, 8986 (1993).
15. A. Carrington, D.J.C. Walker, A.P. Mackenzie, and J.R. Cooper, Hall effect and resistivity of oxygen-deficient $YBa_2Cu_3O_{7-\delta}$ thin films, *Phys. Rev. B* 48, 13051 (1993).

16. R. Decca, E. Osquiguil, F. de la Cruz, C. D'Ovidio, M.T. Malachevski, and D. Esparza, Non linear temperature dependence of the resistivity of La-Sr-Cu-O: Effects of Sr content, *Solid State Commun.* 69, 355 (1989).

17. H. Takagi, B. Batlogg, H.L. Kao, J. Kwo, R.J. Cava, J.J. Krajewski, and W.F. Peck, Jr., Systematic evolution of temperature dependent resistivity in $La_{2-x}Sr_xCuO_4$, *Phys. Rev. Lett.* 69, 2975 (1992).

18. L. Hofmann, K. Harl, and K. Samwer, Normal-state Hall effect and resistivity measurements of $YBa_2Cu_3O_x$ films with different oxygen content, *Z. Phys. B* 95, 173 (1994).

19. B. Bucher, P. Steiner, J. Karpinski, E. Kaldis, and P. Wachter, Influence of the spin gap on the normal state transport in $YBa_2Cu_4O_8$, *Phys. Rev. Lett.* 70, 2012 (1993)

20. T. Nakano, M. Oda, C. Manabe, Y. Mompno, Y. Miura, and M. Ido, Magnetic properties and electronic conduction of superconducting $La_{2-x}Sr_xCuO_4$, *Phys. Rev. B* 49, 16000 (1994).

21. T.M. Rice, in : *The Physics and Chemistry of Oxide Superconductors*, edited by Y. Iye and H. Yasuoka, Springer-Verlag, Berlin (1992).

22. H. Alloul, Comment on "Nature of the conduction-band states in $YBa_2Cu_3O_7$ as revealed by its yttrium Knight shift", *Phys. Rev. Lett.* 63, 689 (1989).

23. T. Nishikawa, J. Takeda, and M. Sato, Anomalous temperature dependence of the Hall coefficient in $La_{2-x}Sr_xCuO_4$ above room temperature, *J. Phys. Soc. Jpn.* 62, 2568 (1993).

24. I. François, C. Jaekel, G. Kyas, R. Heeres, D. Dierickx, H. Rokos, V.V. Moshchalkov, Y. Bruynseraede, H. Kurz, and G. Borghs, Influene of Pr-doping and oxygen deficiency on the scattering behavior of $YBa_2Cu_3O_7$ thin films, *Phys. Rev. B* 53, 12502 (1996).

25. N. Nagaosa and P.A. Lee, Ginzburg-Landau theory of the spin-charge-seperated system, *Phys. Rev. B* 45, 966 (1992).

26. C. Quitmann, D. Andrich, C. Jarchow, M. Fluster, B. Beschoten, G. Güntherodt, V.V. Moshchalkov, G. Mante, and R. Manzke, Scaling behavior at the insulator-metal transition in BiSrCaYCuO, *Phys. Rev. B* 46, 11813 (1992).

27. V.V. Moshchalkov, Transport phenomena and magnetic susceptibility of highly correlated charge carriers, *Physica B* 163, 59 (1990); V.V. Moshchalkov, High-T_C cuprates as two-dimensional quatum antiferromagnets, *Solid State Commun.* 86, 715 (1993)

28. B. Batlogg, in the proceedings of this NATO/ASI in Cargèse, September 1-13, 1997, Plenum Press.

29. H.B. Brom and H. Alloul, The flux pattern in the high-T_c superconductor $YBa_2Cu_3O_7$ studied by ^{89}Y-NMR, *Physica C* 185-189, 1789 (1991).

30. P. Pincus, A.C. Gossard, V. Jaccarino, and J.H. Wernick, NMR measurements of the flux distribution in type II superconductrors, *Phys. Lett.* 13, 21 (1964).

31. B. Pümpin, H. Keller, W. Kündig, W. Odermatt, I.M. Savić, J.W. Schneider, H. Simmler, and P. Zimmermann, Measurement of the London penetration depths in $YBa_2Cu_3O_x$ by means of spin rotation (μSR) experiments, *Physica C* 162-164, 151 (1989).

32. D.R. Harshman, L.F. Schneemeyer, J.V. Waszczak, G. Aeppli, R.J. Cava, B. Batlogg, L.W. Rupp, E.J. Ansaldo, and D.Ll. Williams, Magnetic penetration depth in single-crystal $YBa_2Cu_3O_{7-\delta}$, *Phys. Rev. B* 39, 851 (1989).

33. N. Bontemps, D. Davidov, P. Monod. and R. Even, Determination of the spatial length scale of the magnetic-field distribution in the $YBa_2Cu_3O_7$ ceramic by surface EPR, *Phys. Rev. B* 43, 11512 (1991).

34. N. Athanassopoulou, J.R. Cooper, and J. Chrosch, Variation of the magnetic penetration depth of $YBa_2Cu_3O_{7-\delta}$ with oxygen depletion, *Physica C* 235-240, 1835 (1994).

35. L. Krusin-Elbaum, R.L. Greene, F. Holtzberg, A.P. Malozemoff, and Y. Yeshurun,Direct measurement of the temperature-dependent magnetic penetration depth in Y-Ba-Cu-O crystals, *Phys. Rev. Lett.* 62, 217 (1989).

36. W.N. Hardy, D.A. Bonn, D.C. Morgan, R. Liang, and K. Zhang, Precision Measurements of the temperature dependence of λ in $YBa_2Cu_3O_{6.95}$: Strong evidence for nodes in the gap function, *Phys. Rev. Lett.* 70, 3999 (1993); W.N. Hardy, in the proceedings of this NATO/ASI in Cargèse, September 1-13, 1997, Plenum Press.

37. N. Klein, N. Tellmann, H. Schulz, K. Urban, S.A. Wolf, and V.Z. Kresin, Evidence of two-gap s-wave superconductivity in $YBa_2Cu_3O_{7-x}$ from microwave surface impedance measurements, *Phys. Rev. Lett.* 71, 3355 (1993).

38. M.R. Beasley, Recent penetration depth measurements of the high-T_C superconductors and their implications, *Physica C* 209, 43 (1993).

39. P.J. Hirschfeld and N. Goldenfeld, Effect of strong scattering on the low-temperature penetration depth of a d-wave superconductor, *Phys. Rev. B* 48, 4219 (1993).

40. S.D. Adrian, M.E. Reeves, S.A. Wolf, and V.Z. Kresin, Penetration depth in layered superconductors: Application to the cuprates and conventional multilayers, *Phys. Rev. B* 51, 6800 (1995).

108

41. P. Zimmermann, H. Keller, S.L. Lee, I.M. Savić, M. Warden, D. Zech, R. Cubitt, E.M. Forgan, E. Kaldis, J. Karpinski, and C. Krüger, Muon-spin-rotation studies of the temperature dependence of the magnetic penetration depth in the $YBa_2Cu_3O_x$ family compounds, *Phys. Rev. B* 52, 541 (1995).

42. A.A. Abrikosov and R.A. Klemm, The dependence of Δ and T_C on hopping and the temperature variation of Δ in a layered model of HTSC, *Physica C* 191, 224 (1992).

43. A. Steegmans, R. Provoost, V.V. Moshchalkov, R.E. Silverans, S. Libbrecht, A. Buekenhoudt and Y. Bruynseraede, Study of the vortex-state field distribution in superconducting Nb_3Sn by the NMR decoration technique, *Physica C* 218, 295 (1993).

44. A. Steegmans, R. Provoost, V.V. Moshchalkov, H. Frank, G. Güntherodt and R.E. Silverans, NMR decoration study of the mixed state in high-T_c cuprates, *Physica C* 259, 245 (1996).

45. E.H. Brandt, Flux distribution and penetration depth measured by muon spin rotation in high-T_C superconductors, *Phys. Rev. B* 37, 2349 (1988).

46. W. Barford and J.M.F. Gunn, The theory of the measurement of the London penetration depth in uniaxial type II superconductors by muon spin rotation, *Physica C* 156, 515 (1988).

47. C. Bernhard, Ch. Niedermayer, U. Binninger, A. Hofer, Ch. Wenger, J.L. Tallon, G.V.M. Williams, E.J. Ansaldo, J.I. Budnick, C.E. Stronach, D.R. Noakes, and M.A. Blankson-Mills, Magnetic penetration depth and condensate density of cuprate high-T_c superconductors determined by muon-spin-rotation experiments, *Phys. Rev. B* 52, 10488 (1995).

48. Y.J. Uemura *et al.*, Systematic variation of magnetic-field penetration depth in high-T_C superconductors studied by muon-spin relaxation, *Phys. Rev. B* 38, 909 (1988).

49. Y.J. Uemura *et al.*, Universal correlations between T_c and n_s/m^* in high-T_C cuprate superconductors, *Phys. Rev. Lett.* 62, 2317 (1989).

50. Y.J. Uemura *et al.*, Basic similarities among cuprate, bismuthate, organic, Chevrel-phae, and heavy-Fermion superconductors shown by penetration-depth measurements, *Phys. Rev. Lett.* 66, 2665 (1991).

51. E.J. Nicol and J.P. Carbotte, Penetration depth in phenomenological marginal-Fermi-liquid model for CuO, *Phys. Rev. B* 43, 1158 (1991).

52. M.E. Reeves, D.A. Ditmars, S.A. Wolf, T.A. Vanderah, and V.Z. Kresin, Evidence for strong electron-phonon coupling from the specific heat of $YBa_2Cu_3O_{7-\delta}$, *Phys. Rev. B* 47, 6065 (1993).

53. A.M. Neminsky and P.N. Nikolaev, Temperature dependence of anisotropic penetration depth in $YBa_2Cu_3O_7$ measured on aligned fine powder, *Physica C* 212, 389 (1993).

54. D. Pines, in the proceedings of this NATO/ASI in Cargèse, September 1-13, 1997, Plenum Press; B. Batlogg, idem.

55. B. Janossy, D. Prost, S. Pekker, and L. Fruchter, Magnetic study of oxygen-deficient $YBa_2Cu_3O_{7-\delta}$, *Physica C* 181, 51 (1991).

56. C. Jiang and J.P. Carbotte, Penetration depth in layered high-T_c superconductors, *Phys. Rev. B* 45, 10670 (1992).

THE SPIN FLUCTUATION MODEL FOR HIGH TEMPERATURE SUPERCONDUCTIVITY: PROGRESS AND PROSPECTS

David Pines

Physics Dept., University of Illinois at Urbana-Champaign
1110 West Green Street, Urbana, IL 61801-3080

I. INTRODUCTION AND OVERVIEW

Following a decade of work, there is now an experimental and theoretical consensus that the behavior of the elementary excitations in the Cu-O planes provides the key to understanding the normal state properties of these cuprate superconductors, and that essentially no normal state property (save one) resembles that found in the normal state of a conventional, low T_c, superconductor. As may be seen in Table 1, both the charge response (measured in transport and optical experiments), and the spin response (measured in static susceptibility, nuclear magnetic resonance (NMR) experiments and inelastic neutron scattering (INS) experiments),of the high T_c materials are dramatically different from their low T_c counterparts, as is the single particle spectral density measured in angle-resolved photoemission studies (ARPES).

Moreover, essentially no property of the superconducting state is that of a conventional superconductor, in which BCS pairing takes place in a singlet s-wave state, and the quasiparticle energy gap at low temperatures is finite and isotropic as one moves around the Fermi surface. Despite the fact that something quite new and different is required to understand normal state behavior, there is also a consensus that BCS theory, suitably modified, will provide a satisfactory description of the transition to the superconducting state, and the properties of that state.

There is a near consensus as well on the basic building blocks required to understand the high temperature superconductors. These can be summarized as follows.

*The action occurs primarily in the Cu-O planes, so that it suffices, in first approximation, to focus both experimental and theoretical attention on the

The Gap Symmetry and Fluctuations in High - T_c Superconductors
Edited by Bok *et al.*, Plenum Press, New York, 1998

Table 1
Some ways in which the normal state of high T_c materials is anomalous.

	Conventional	High T_c
Resistivity	$\rho \sim T^2$	$\rho \sim T$
Quasiparticle lifetime, $1/\tau(T,\omega)$	$aT^2 + b\omega^2$	$aT + b\omega$
Spin excitation spectrum	Flat	Peaked at $\mathbf{Q}_i \sim (\pi/a, \pi/a)$
Maximum strength of spin excitations	~1 state/eV	20-300 states/eV
Characteristic spin excitation energy	~E_f	$\omega_{sf} < T << E_f$
AF correlations	None	strong, with $\xi_{AF} \geq 2a$
Uniform susceptibility, $\chi_o(T)$	Flat	varies with temperature, possesses a maximum at $T_{cr} > T_c$ for magnetically underdoped systems

behavior of the planar excitations, and to focus as well on the two best-studied systems, the 1-2-3 system ($YBa_2Cu_3O_{7-x}$) and the 2-1-4 system ($La_{2-x}Sr_xCuO_4$).
*At zero doping ($YBa_2Cu_3O_6$; La_2CuO_4) and low temperatures, both systems are antiferromagnetic insulators, with an array of localized Cu^{2+} spins which alternate in sign throughout the lattice
*One injects holes into the Cu-O planes of the 1-2-3 system by adding oxygen; for the 2-1-4 system this is accomplished by adding strontium. The resulting holes on the planar oxygen sites bond with the nearby Cu^{2+} spins, making it possible for the other Cu^{2+} spins to move, and, in the process, destroying the long range AF correlations found in the insulator.
*If one adds sufficient holes, the system changes its ground state from an insulator to a superconductor.
*In the normal state of the superconducting materials,the itinerant, but nearly localized Cu^{2+} spins form an unconventional Fermi liquid, with the quasiparticle spins displaying strong AF correlations even for systems at doping levels whch exceed that at which T_c is maximum, the so-called overdoped materials.

There is, however, no consensus among theorists as to how to develop a more detailed theoretical description of the cuprates. The approaches which have been tried can be classified as top-down or bottom-up. In a top-down approach, one chooses a model early on (the Hubbard model and the recent SO5 model are typical examples), develops solutions for alternative choices of model parameters, and then sees whether the solutions lead to results consistent with experiment. In a bottom-up approach one begins with the experimental results, and attempts to devise a phenomenological description of a subset of the experimental results. One then explores alternative scenarios which appear consistent with this description, working out the microscopic consequences of each scenario, until one arrives at a scenario and associated microscopic calculations which are consistent with experiment. Then, and only then, does one search for a model Hamiltonian whose solution might provide the ultimate microscopic theory. It was this second approach which John Bardeen followed in his work on conventional superconductors, and guided by his example, it was the approach our research group in Urbana followed for high T_c. We arrived in this fashion at a spin fluctuation model and a *nearly antiferromagnetic Fermi liquid* (NAFL) description of the effective quasiparticle interaction responsible for the strange normal state properties and the superconducting transition at high T_c.

The spin fluctuation model for the high temperature superconductors was proposed in 1989;[1] it was then developed independently by theoretical groups in Tokyo and Urbana.[2] Both groups assumed that the magnetic interaction between the planar quasiparticles was responsible for the anomalous normal state properties and found a superconducting transition at high temperatures to a $d_{x^2-y^2}$ pairing state. Moriya et al.[2] used a self-consistent renormalization group approach to characterize the dynamic spin susceptibility. The resulting effective magnetic interaction between the planar quasiparticles was then used to calculate the superconducting transition temperature, T_c, and normal state resistivity. Monthoux et al.[2] did not attempt a first-principles calculation of the planar quasiparticle interaction; rather they turned to experiment, using quasiparticles whose starting spectra was determined by fits to ARPES experiments and an effective magnetic interaction between these which was assumed to be proportional to a mean field expression for the dynamic susceptibility which had been developed by Millis et al., [3] and which had been shown to provide an excellent fit to NMR experiments on the YBaCuO system.[3,4]

Both groups followed up their initial weak coupling calculations (which gave high values for T_c) with strong coupling (Eliashberg) calculations[5,6] which enabled them to take into account lifetime effects and other changes in the quasiparticle behavior brought about by the strong magnetic inter-action, and which showed that superconductivity at high T_c was a robust phenomena. Both groups found that the pairing state would be $d_{x^2-y^2}$ in agreement with the Hubbard model calculations of Bickers et al.[7] However, in the latter calculations it was found that the superconducting transition temperature was comparatively low (<40K) under what seemed to be optimal conditions, hole concentrations close to half-filling, with T_c decreasing as one increased the planar hole concentration (in contradiction to experiment). Indeed it could be argued that these Hubbard model results, together with the early pentration-depth experiments which supported an s-wave pairing state, were responsible for the fact that the magnetic mechanism and $d_{x^2-y^2}$ pairing had been abandoned by most of the high temperature superconductivity community by the end of 1989.[8]

In our Urbana work,[2,6] we were encouraged by the fact that it seemed possible to get significantly higher transition temperatures (than those achieved using the Hubbard model) by using our experimentally-based magnetic interaction. Indeed, when Monthoux and I found[6] that we could obtain a quantitative account of the temperature-dependence of the planar resistivity of optimally-doped YBaCuO using the same coupling constant (and parameters which characterized the quasiparticle and spin spectrum) which had yielded a T_c of 90 K, we concluded we had established a "proof of concept" for a spin fluctuation mechanism based on the Urbana quasiparticle interaction. Since our calculations unambiguously predicted a $d_{x^2-y^2}$ pairing state, experimental detection of that state was a necessary condition for our spin fluctuation mechanism to be viable. In effect we argued that an accurate determination of the pairing state was central to the determination of the mechanism for high T_c: were the pairing state anything else, the spin fluctuation mechanism could be ruled out, and one would have to look to other mechanisms for an explanation of high T_c.

At the time we challenged the experimental community to find our predicted pairing state, only NMR Knight shift and spin-lattice relaxation rate supported $d_{x^2-y^2}$ pairing;[9] however within the next year or so, the tide turned dramatically away from s-wave pairing, with ARPES,[10] penetration depth,[11] and NMR experiments[12] on the oxygen spin-lattice relaxation time and the anisotropy of the copper spin-lattice relaxation time[2] all supporting a $d_{x^2-y^2}$ state. The decisive experiments were the direct phase-sensitive tests of pairing symmetry carried out by Van Harlingen and his group in Urbana[13] which were confirmed by Kirtley, Tseui, and their collaborators in subsequent work at IBM,[14] and which will be discussed at this Institute by John Kirtley.

For theorists who had advocated other high T_c mechanisms (and pairing states), identification of the pairing state has served as a stimulus to adjust their models and model-dependent parameters in such a way as to obtain a d-wave pairing state; in some cases such efforts have been successful. Thus the community has begun to look to the anomalous normal state behavior as a testing ground for the models and mechanisms which have been proposed, and it is natural to inquire whether the Urbana spin fluctuation model is equally capable of explaining the remarkable changes in spin and charge behavior found when one moves from overdoped to underdoped systems, as well as characterizing and explaining quasiparticle pseudogap behavior in the underdoped systems. In these lectures I will discuss the extent to which it can do so.

I begin with an overview of the Urbana spin fluctuation model and then discuss in some detail NMR measurements of the low frequency magnetic response in the normal state, and the normal state phase diagram to which these give rise. I next consider the extent to which the results obtained in NMR experiments are consistent with inelastic neutron scattering (INS) experiments on similar samples, and show how what, at first sight, appear to be conflicting results can be reconciled. Following a brief qualitative discussion of the relationship between changes in quasiparticle behavior and changes in the low frequency magnetic response, I present a brief summary of the detailed microscopic calculations based on the Urbana spin-fluctuation model of normal state properties and the extent to which these provide a quantitative account of the experimental results. I conclude with a discussion of some of the remaining open questions in high temperature superconductivity.

II. OVERVIEW OF THE URBANA SPIN FLUCTUATION MODEL

In common with many other models, in calculations using the Urbana spin fluctuation model, the planar quasiparticles are assumed to be characterized by a starting spectrum which reflects their barely itinerant character, and which takes into account both nearest neighbor and next nearest neighbor hopping, according to

$$\varepsilon_p \sim -2t(\cos p_x a + \cos p_y a) - 4t' \cos p_x a \cos p_y a \qquad (2.1)$$

where t, the nearest neighbor hopping term, ~ 0.25 eV, while the next nearest neighbor hopping term is taken to be t' $\sim$ -0.45 t for the 1-2-3 system, and is assumed to be somewhat smaller (t' $\leq$ -0.25 t) for the 2-1-4 system. Where the Urbana model differs from its counterparts (Hubbard, SCRG et al.) is in the ansatz that effective magnetic interaction between these planar quasiparticles mirrors their dynamic spin susceptibility $\chi(\mathbf{q},\omega)$, which is taken from fits to NMR and INS experiments which show that it is strongly peaked at wave vectors in the vicinity of $\mathbf{Q} = (\pi,\pi)$ as a result of the close approach of even optimally doped systems to antiferromagnetism. Although, as discussed below, the spin spectrum may in general possess four incommensurate peaks at wave vectors $\mathbf{Q}_i$ near $\mathbf{Q}$ (as seen, for example, in the 2-1-4 system), in most of the microscopic calculations of normal state properties carried out to date, the dynamic spin susceptibility is modeled by the phenomenological expression introduced by Millis et al. (hereafter MMP),[3]

$$\chi_{NAFL}(\mathbf{q},\omega) = \frac{\chi_\mathbf{Q}}{1+(\mathbf{Q}-\mathbf{q})^2\xi^2-i\omega/\omega_{SF}} . \qquad (2.2)$$

From the detailed fits to NMR and INS experiments discussed in the following sections, it is found that as T approaches T_c, the static staggered spin susceptibility,

$$\chi_\mathbf{Q} = \alpha \, \xi^2, \qquad (2.3)$$

is large compared to the uniform spin susceptibility, χ_0, ξ, the antiferromagnetic correlation length measured in units of the lattice spacing, a, satisfies $\xi >> 1$, and ω_{SF}, the frequency of the relaxational mode, is small compared to the planar quasiparticle band width or Fermi energy; indeed, for optimally doped and underdoped systems one finds that over a considerable regime of temperatures,

$$\omega_{SF} << \pi \, T, \qquad (2.4)$$

and it is only as T approaches T_c that ω_{SF} becomes comparable to πT. Thus the effective interaction between quasiparticles is typically modelled as

$$V_{eff}(\mathbf{q},\omega) = g^2 \, \chi_{NAFL}(\mathbf{q},\omega) \qquad (2.5)$$

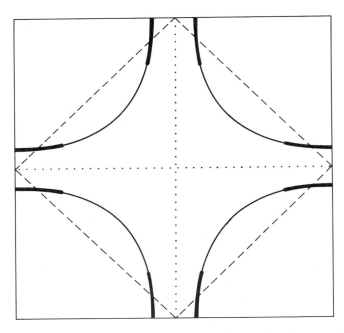

Fig. 1 Hot (thick lines) and cold (thin lines) quasiparticles on the Fermi surface. Hot quasiparticles on the Fermi surface, which lie a distance apart in momentum space of $Q\pm1/\xi$, feel the maximum consequences of the NAFL interaction, Eq. (2.5); the remaining (cold) quasiparticles are much more weakly coupled.

where in many situations the coupling constant g is the only free parameter. Thus the doping level is assumed known, the parameters which enter ε_p are taken from fits to ARPES experiments, while the parameters which enter χ_{NAFL} are determined by fits to NMR and, where appropriate, INS experiments.

Since, $\chi(\mathbf{q},\omega)$ is highly peaked in the vicinity of $\mathbf{Q}$, it follows that the full effects of this interaction will be felt by only those quasiparticles which are located a distance of order $\mathbf{Q}$ away from each other on the Fermi surface. The situation will be approximately that shown in Fig 1. Clearly quasi-particles located near the four singular points (those connected by $\mathbf{Q}$) will feel a markedly different interaction than those which lie a distance away of, say, $|(\mathbf{Q}\text{-}\mathbf{q})\xi| = 2$. This highly peaked quasiparticle interaction produces, in essence, a "two-class" society: *hot* quasiparticles (the "elite") interact very strongly because they feel the peaks of the interaction; in contrast, *cold* quasiparticles (the "underclass") feel only the valleys, i.e. an interaction which corresponds to a normal Fermi liquid interaction. The stronger the AF correlations, the more striking this difference becomes. It is this strong anisotropy of quasiparticle behavior as one moves around the Fermi surface, as illustrated by this two-class society of hot and cold quasiparticles which is the defining characteristic of the Urbana spin fluctuation model. We show below how it provides a natural explanation for the measured doping and temperature dependence of the anomalous normal state properties.

116

Because the interaction responsible for the anomalous normal state properties (and high T_c) is essentially purely electronic (electron-phonon coupling plays little or no role in determining normal state properties), feedback effects can play a significant role in the high temperature superconductors. Quasiparticles which move in response to the effective magnetic interaction in turn give rise to it. Feedback effects are most pronounced for the hot quasiparticles, since not only are these the quasi-particles most strongly affected by $V_{eff}(\mathbf{q},\omega)$, but it is these same hot quasiparticles which determine the response function, $\chi(\mathbf{Q},\omega)$, and thence the effective interaction, $V_{eff}(\mathbf{Q},\omega)$ to which they respond.

Consider, for example, how $\chi_{NAFL}(\mathbf{Q},\omega)$ arises in a microscopic theory. Quite generally we may write:

$$\chi_{NAFL}(\mathbf{Q},\omega) = \frac{\tilde{\chi}(\mathbf{Q},\omega)}{1 - J_{\mathbf{Q}}\tilde{\chi}(\mathbf{Q},\omega)} \tag{2.6}$$

where $\tilde{\chi}(\mathbf{Q},\omega)$ is the irreducible "hot" quasiparticle-quasihole susceptibility, and $J_{\mathbf{Q}}$ is the effective restoring force which connects these irreducible contributions. Through $\tilde{\chi}$ (and possibly $J_{\mathbf{Q}}$ as well) any changes in hot quasiparticle behavior brought about by $V_{eff}(\mathbf{q},\omega)$ are fed back into V_{eff}. This feedback will in general be non-linear; it may be either negative (and so tend to keep matters near equilibrium) or positive (in which case crossover behavior occurs).

Matters are otherwise for the cold quasiparticles, since their much weaker interaction, (which, to the extent it can be approximated as the tail of the interaction, Eq. (2.5), will be independent of the strength of the antiferromagnetic correlations as soon as $(\mathbf{Q}-\mathbf{q})^2\xi^2 \gg 1$), will not change appreciably with doping or temperature. Cold quasiparticles will therefore not contribute appreciably to changes in $\tilde{\chi}(\mathbf{Q},\omega)$. Of course, in fact one has a continuum of both anisotropic quasiparticle behavior and feedback as one moves around the Fermi surface; characterizing quasiparticles as either hot or cold, while often a useful simplification, misses an important class of quasiparticles (luke-warm?) which are neither hot or cold, and which in practice may contribute to various phenomena under study.

Since their effective interactions are markedly different, the lifetimes of hot and cold quasiparticles must differ in both magnitude and temperature dependence. This is easily verified both experimentally and theoretically. For example, Stojkovic and Pines[15] have calculated quasiparticle lifetimes around the Fermi surface using the effective interaction, Eq. (2.5) and second-order perturbation theory. For hot regions they find:

$$\left(\frac{1}{\tau_k}\right)_{hot} \cong \frac{a\, g^2 T\xi}{4v_f} \tag{2.7}$$

while for cold quasiparticles located a distance Δk away from the hot regions they find

$$\left(\frac{1}{\tau_k}\right)_{cold} \cong \frac{a g^2}{8v_f(\Delta k)}\frac{T^2}{T_0+T} \tag{2.8}$$

where $T_0 \sim \omega_{SF}(\Delta k)^2/\pi$. Stojkovic and Pines show that in the limit of large anisotropy, which is present for all but quite overdoped systems at very low

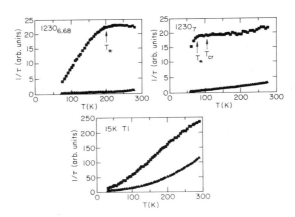

Fig. 2 Hot and cold quasiparticle lifetimes, deduced from measurements of the resistivity and Hall effect, for three cuprate superconductors (after Stojkovic and Pines); the squares denote the hot quasiparticles. The crossover at T_* found in the magnetic behavior of the underdoped systems is clearly visible in the 1-2-3 $O_{6.68}$ and 1-2-3 O_7 results.

temperatures, these lifetimes can be extracted directly from a self-consistent analysis of measurements of the longitudinal resistivity and the Hall conductivity, using the following simple expressions:

$$\sigma_{xx} = ne^2 \left[\frac{\tau_{hot}\tau_{cold}}{m_{hot}m_{cold}} \right]^{1/2} \tag{2.9}$$

$$\text{ctn } \theta_H = \frac{2}{\omega_c \tau_{cold}}. \tag{2.10}$$

The results of that analysis is given in Fig. 2 for three representative systems: optimally-doped and underdoped $YBa_2Cu_3O_{7-x}$, and overdoped Tl 2201. Note that the temperature dependence of the scattering rates for the hot and cold quasiparticles is close to that obtained using simple perturbation theory. Note, too, that the hot quasiparticle scattering rate for $YBa_2Cu_3O_{6.68}$ exhibits a crossover at T_*, a result which follows directly from the measured crossover behavior of the longitudinal resistivity, [16] and which reflects a crossover in the low frequency magnetic behavior found in NMR experiments.

From the simple expressions, Eq. (2.9) and (2.10) we see that the "classic" result, first emphasized by Anderson, [17] that at high temperatures the resistivity is linear in T while ctn θ_H varies as T^2, follows directly from the spin fluctuation model, since in this model one finds (see Eqs. (2.7) and (2.8)) that for temperatures above T_* where $\xi \sim T^{-1}$, τ_{hot} is independent of temperature, while τ_{cold} varies as T^{-2}. As discussed in detail in SP, and summarized briefly below, the spin fluctuation model is capable as well of explaining the departures from these simple results as one varies temperature and doping, and has quite recently been shown to provide a simple explanation for the measured magnetotransport properties of optimally doped and overdoped Tl samples in magnetic fields of up to 60 T.[18]

A further check on the predictions of the spin fluctuation model comes from Raman experiments. As discussed by Hackl in these proceedings, in the B_{2g} channel one measures cold quasiparticle lifetimes, while the B_{1g} channel picks up primarily the behavior of the hot quasiparticles. The Raman results show that the scattering intensities in these two channels are quite different in magnitude, and exhibit an overall temperature and doping dependence consistent with that deduced above from transport measurements.[19,20]

Perhaps the most striking consequence of the anisotropic quasiparticle interaction is found in the pseudogap behavior of the underdoped super-conductors measured in ARPES experiments. Quite recently, as discussed in detail below, Schmalian et al. (hereafter SPS) have carried out micro-scopic calculations based on the spin fluctuation model which are exact in the high temperature limit, Eq. (2.4).[21] These show that for underdoped systems the anisotropy in the effective quasiparticle interaction gives rise to a striking anisotropy in quasiparticle spectral density evolution as the temperature is lowered. Thus SPS find that below a critical temperature, T_{cr}, a pseudogap associated with an appreciable transfer of spectral weight from low energies to energies ~200 meV, opens up for hot quasiparticles, while the spectral weight of cold quasiparticles remains peaked at the Fermi surface. This is just what is seen in the ARPES experiments.[22] SPS calculate the changes in $\bar{\chi}(\mathbf{Q},\omega)$ and in $\bar{\chi}_0$ brought about by this hot quasiparticle pseudogap, and show that these explain the changes in the low frequency magnetic response and uniform susceptibility seen at T_{cr} in the NMR experiments I discuss in detail in the following section.

A final, and especially important consequence of a highly peaked quasi-particle interaction of the form, Eq. (2.5), is that it always brings about a transition to a superconducting state with d_{x2-y2} pairing. To see why this occurs, consider a configuration space description of this interaction between planar quasiparticles almost localized on a 2D lattice.[23] It is straightforward to show that it is strongly repulsive on-site, attractive for the four nearest neighbors, then alternates between repulsion and attraction as one proceeds to consider next nearest neighbors etc. It is this on-site repulsion, plus the effective attraction between the nearest neighbors, which is responsible for d-wave pairing; in this pairing state quasiparticles, by virtue of their $\ell = 2$ relative angular momentum, avoid sampling the on-site repulsion while taking advantage of the attraction between nearest neighbors to achieve superconductivity. That the pairing state must be d_{x2-y2} follows from the fact that along the diagonals (where $x^2 = y^2$) in configuration space, the effective quasiparticle interaction is always repulsive; hence it is energetically favorable to place the nodes of the gap parameter,

$$\Delta(k,T) = \Delta(T)[\cos k_x a - \cos k_y a] \tag{2.11}$$

along the k-space diagonals, $k_x^2 = k_y^2$.

III. LOW FREQUENCY MAGNETIC RESPONSE AND THE MAGNETIC PHASE DIAGRAM

Of the various anomalous aspects of normal state behavior of the super-conducting cuprates, their low frequency magnetic response measured in NMR and uniform susceptibility experiments is perhaps the most unusual, in that one finds nearly antiferromagnetic behavior and three distinct magnetic phases (above T_c!) in all but highly overdoped systems. Let me begin by summarizing what we have learned from NMR experiments.

Nuclear spins probe the local environment. The Knight shift provides a direct measure of the uniform magnetic susceptibility seen at a particular nuclear site, while measurements of the ^{63}Cu spin-lattice relaxation rate, $^{63}T_1^{-1}$, provide information on the very low frequency behavior of $\chi''(\mathbf{q},\omega)$, the imaginary part of the dynamic spin-spin susceptibility, again at a particular nuclear site. Measurements of the spin-echo decay rate $^{63}T_{2G}^{-1}$, tell us about $\chi'(\mathbf{q},0)$, the real part of that susceptibility. [24] Knight-shift experiments on ^{63}Cu and ^{17}O (and where relevant, ^{89}Y and ^{205}Tl) nuclei show that the coupled system of planar Cu^{2+} spins and holes forms a single component, with the magnetism residing on the Cu^{2+} spins. [24a] Although the Cu^{2+} spins become itinerant upon doping, with the coupled system changing its behavior below the superconducting transition, to a remark-able extent the system turns out to exhibit behavior which is close to that of localized Cu^{2+} spins, as demonstrated by the fact that the basic Hamiltonian proposed by Shastry[25] and Mila and Rice[25] to describe the coupling of a nuclear spin to the Cu^{2+} spin system in the insulating state of La_2CuO_4 or $YBa_2Cu_3O_6$ remains valid, subject to the minor modifications discussed below, at doping levels up to $La_{1.75}Sr_{0.25}CuO_4$ or $YBa_2Cu_3O_7$, with many of the hyperfine coupling constants which describe system behavior changing by less than 5%.

Specifically, the Shastry-Mila-Rice (hereafter SMR) Hamiltonian for nuclei coupled to Cu_2 spins, $S(r)$, in the 1-2-3 system takes the form

$$H = {}^{63}\mathbf{I}(\mathbf{r}_i)\bullet[A\Sigma\mathbf{S}(\mathbf{r}_i)+B\sum_j^{nn}\mathbf{S}(\mathbf{r}_j)] + {}^{17}\mathbf{I}(\mathbf{r}_i)\bullet\mathbf{C}\bullet\sum_j^{nn}\mathbf{S}(\mathbf{r}_j)$$

$$+ {}^{89}\mathbf{I}(\mathbf{r}_i)\bullet\mathbf{D}\bullet\sum_j^{nn}\mathbf{S}(\mathbf{r}_j) \tag{3.1}$$

where $\mathbf{A}$ is the tensor which describes the direct, on-site-coupling of the ^{63}Cu nuclei to the Cu_2 spins nearly localized there, and B is the strength of transferred hyperfine coupling of a ^{63}Cu nuclear spin to the four nearest neighbor Cu^{2+} spins. The ^{17}O and ^{89}Y nuclei see only their nearest neighbor Cu^{2+} spins, through transferred hyperfine interaction tensors of strength C and D respectively, i.e. no spin component resides directly on these sites in the crystal. As a result, these different nuclei probe different regions in momentum space of $\chi''(q,\omega)$; specifically, one finds for the spin lattice relaxation time, $(^{\alpha}T_1)_\gamma$ for a nucleus α responding to a field in the γ direction,

$$\frac{1}{(^aT_1)_\gamma T} = \frac{k_B}{2\mu_B^2} \sum_{\substack{q \\ \omega \to 0}} {}^\alpha F_\gamma(q) \frac{\chi''(\mathbf{q},\omega)}{\omega} \tag{3.2}$$

where the filter functions, ${}^\alpha F_\gamma(\mathbf{q})$, are given by[3]:

$$^{63}F_{\|}(q) = [A_\perp + 2B(\cos q_x + \cos q_y)]^2 \tag{3.3}$$

$$^{63}F_\perp(q) = \frac{{}^{63}F_{\|}(q) + {}^{63}F^{eff}(q)}{2} \tag{3.4}$$

$$^{17}F_{\|}(q) = 2C^2(1 + \cos q_x a) \tag{3.5}$$

and the subscripts, $\|$ and $\perp$, refer to the direction of the applied magnetic field with respect to the axis perpendicular to the Cu-O planes, $(^{17}T_1)\|$ is the isotropic average of the ^{17}O spin-lattice relaxation rates for a field applied perpendicular to the Cu-O plane and the form factor

$$^{63}F^{eff}(q) = [A_{\|} + 2B(\cos q_x + \cos q_y)]^2 \tag{3.6}$$

is the filter function for the ^{63}Cu spin-echo decay time, $^{63}T_{2G}$,[26]

$$\frac{1}{^{63}T_{2G}} = \left(\frac{0.69}{128}\right)^{1/2} (^{63}\gamma_n)^2 \left\{ \frac{1}{N} \sum_q F_{eff}^2(q)\,[\chi'(q)]^2 - \left[\frac{1}{N} \sum_q F_{eff}(q)\,\chi'(q)\right]^2 \right\}^{1/2} \tag{3.7}$$

The hyperfine coupling constants, $A_{\|}$, $A_\perp$, B, and C may then be determined from fits to the experimental measurements of the anisotropic ^{17}O and ^{63}Cu Knight shifts and spin-lattice relaxation times, the spin-echo decay time, and the measured frequency of the antiferromagnetic resonance for $YBa_2Cu_3O_6$, by combining the phenomenological

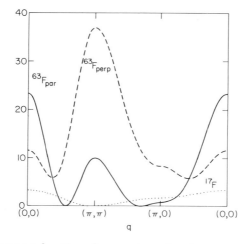

Fig. 3 Form factors as a function of momentum, for planar oxygen and copper sites, in units of the hyperfine coupling constant B.

one-component expression, $\chi_{NAFL}(\mathbf{q},\omega)$, Eq. (2.2), for the dynamical spin susceptibility in the vicinity of the commensurate wave vector, $\mathbf{Q} = (\pi/a,\pi/a)$, which reflects the close approach of the system to antiferromagnetism, with a Fermi-liquid-like expression, $\chi_{FL}(\mathbf{q},\omega)$ for the susceptibility at wave vectors far removed from $\mathbf{Q}$. i.e. for $(\mathbf{Q}-\mathbf{q})^2\xi^2 \gg 1$.[3,4,26] In the latter regime, the dynamical susceptibility may be parametrized by a pheno-menological expression appropriate to a Fermi liquid,

$$\chi_{FL}(\mathbf{q},\omega) = \frac{\chi_{\mathbf{q}}}{1-i\omega/\Gamma_q} \cong \frac{\chi_0(T)}{1-i\omega/\Gamma_0} \tag{3.8}$$

where Γ_0 is comparable to the bandwidth.

The resulting momentum dependence of the filter functions determined using hyperfine constants which fit NMR experiments on YBa$_3$Cu$_3$O$_{7-x}$ is shown in Fig. 3; there one sees that both $^{63}F_{\parallel}$ and $^{63}F_{\perp}^{eff}$ are peaked at $(\pi/a,\pi/a)$, and thus act to reinforce the peaking associated with $\chi_{NAFL}(\mathbf{q},\omega)$, while because $^{17}F_{\parallel}(\mathbf{q})$ vanishes at $\mathbf{Q}$, $^{17}T_1$ misses, to first approximation, the antiferromagnetic "action" present in $\chi_{NAFL}(\mathbf{q},\omega)$.

The striking difference between the ^{63}Cu relaxation rates and the ^{17}O relaxation rates measured for all the superconducting cuprates is explained by combining the filter functions with the phenomenological expression, $\chi_{NAFL}(\mathbf{q},\omega)$. Experiment shows that for the fully and over-doped systems, both $^{17}T_1^{-1}$ and $^{89}T_1^{-1}$ obey a modified Korringa relation, $^{17}T_1 T \chi_0(T)$ const, while $^{63}T_1^{-1}$ is more than an order of magnitude larger, and quite "non-Korringa" in its temperature dependence. Both the SMR Hamiltonian and the strong antiferromagnetic correlations present in $\chi_{NAFL}(\mathbf{q},\omega)$ for $\xi \geq a$, are absolutely essential if one seeks to obtain a quantitative fit to the degree of AF enhancement required by the $^{63}T_1$ data, while a temperature dependent $\xi(T)$ provides a natural explana-tion of its striking non-Korringa temperature dependence.

The indispensability of the SMR Hamiltonian and strong antiferro-magnetic correlations becomes even more evident when one compares the $^{17}T_1$ measurements with the results for $^{63}T_1$. These show that a nuclear spin just a few Å away from the Cu^{2+} spin system (or a ^{63}Cu nucleus) sees a completely different low frequency magnetic behavior; the strength of the magnetic response it picks up is an order of magnitude smaller, and it displays no trace of the striking non-Korringa temperature dependence found using a ^{63}Cu probe. Part of the answer to this dilemma comes from the SMR filter function, $^{17}F(\mathbf{q})$, shown in Fig. 3. However, close exami-nation shows that this filtering is not in itself sufficient; it must be accompanied by the strong AF correlations described by $\chi_{NAFL}(\mathbf{q},\omega)$. Specifically one finds for YBa$_2$Cu$_3$O$_7$ that unless $\xi \geq 1.5a$ for $T \leq 250K$, even with the filter function, the leakage of the Cu^{2+} spin fluctuation spectrum onto the ^{17}O site will be so large that the resulting $^{17}T_1$ will display a non-Korringa temperature dependence which is not seen experimentally. A second consequence of the SMR filtering action is that the resulting magnitude of the ^{17}O relaxation rate is too small; thus, as MMP first argued, the ^{17}O nuclei spin-lattice relaxation rate is determined in part

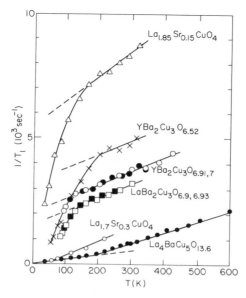

Fig. 4 Spin lattice relaxation rates for superconducting and non-superconducting cuprate oxides (Imai[27]).

by the long wave length, Fermi-liquid like, part of the spin fluctuation spectrum, $\chi_{FL}(\mathbf{q},\omega)$, Eq. (3.8). Hence in fitting to experiments involving ^{17}O or ^{89}Y nuclei, one should use the more general expression

$$\chi(\mathbf{q},\omega) = \chi_{NAFL}(\mathbf{q},\omega) + \chi_{FL}(\mathbf{q},\omega), \tag{3.9}$$

while for ^{63}Cu NMR experiments, the contribution from wave vectors near $\mathbf{Q}$ is so large that the contribution from χ_{FL}, Eq. (3.8), can safely be neglected.

The early experiments on $^{63}T_1$ for a series of cuprates established the doping-dependence of the antiferromagnetic correlations. [27] As may be seen in Fig. 4, it is what one would expect; at a given temperature (say 200K) as the concentration of planar holes increases, the strength of the antiferromagnetic peak present in $\chi_{NAFL}(\mathbf{q},\omega)$, measured by $^{63}T_1^{-1}$, decreases. These experiments also suggest an intimate relation between antiferromagnetic behavior and superconductivity, since $^{63}T_1$ for the over-doped non-superconducting cuprates, $La_{1.7}Sr_{0.3}CuO_4$ and $La_4BaCu_5O_{13}$ is Korringa-like, and shows no sign of AF enhancement.

The early experimental results for the spin component of the bulk susceptibility, $\chi_0(T)$ in $YBa_2Cu_3O_{6-x}$, (which can be directly obtained from the ^{89}Y Knight shift) by Alloul et al., [28] showed (see Fig. 5) that even for the optimally-doped system, in striking contrast to a Landau Fermi liquid, χ_0 was temperature dependent, and that the lower the doping, the greater its temperature dependence. These measurements led Friedel[29] to conclude that the physical origin of the unusual behavior of χ_0 was the near for-mation of a spin density wave, SDW, and, by analogy to charge density wave

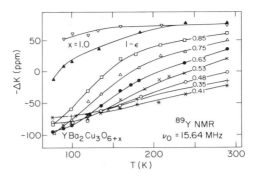

Fig. 5 ^{89}Y Knight shift measurements in the 1-2-3 system (from Alloul et al.[28]).

behavior, to introduce the term "pseudogap" to describe the changes in quasiparticle behavior (brought on by SDW precursor behavior) which are responsible for $\chi_0(T)$. Subsequent measurements on $La_{2-x}Sr_xCuO_4$ showed that the same phenomenon appeared there; indeed, for doping levels $x \lesssim 0.25$, $\chi_0(T)$ possesses a maximum at some temperature, T_{cr}, greater than T_c, while for $x \gtrsim 0.25$, the behavior of $\chi_0(T)$ is monotonic. Similar results are found in the Bi, Tl, and Hg based systems. These results led Barzykin and Pines[30] to suggest that since T_{cr} marks the onset of pseudogap behavior, the distinction between a *magnetically* underdoped and overdoped system is whether one finds $T_{cr} > T_c$. On this basis, optimally doped systems, which are conventionally defined on the basis of the doping dependence of T_c (as corresponding to the doping level, x_{opt}, for which T_c is maximum), turn out,

Table 2
Systems for which detailed NMR results presently exist.

System	$^{63}T_1$	$^{63}T_{2G}$	$^{17}T_1$	^{63}Ka	^{17}Ka	^{89}Ka	T_c
$YBa_2Cu_3"O_7"$	o	o	o	o	o	o	93K
YBa_2Cu4O_8	o	o	o	o	o	o	80K
$YBa_2Cu_3O_{6.63}$	o	o	o	o	o	o	60K
$La_{2-x}Sr_xCuO_4$	o		o	o	o	o	$\leq$38K
$Tl_2Ba_2CuO_{6+y}$	o		o				0,40K,72K
$Tl_2Ba_2Ca_2Cu_3O_{10}$	o	o	o				$\leq$127K
$HgBa_2Ca_2Cu_3O_{6+y}$	o	o	o				$\leq$160K
$Bi_2Sr_2CaCu_2O_8$	o		o				77K;90K
$(Bi,Pb)_2SrCa_2Cu_3O_y$	o		o	o			120K

in fact, to be magnetically underdoped. Put another way, _all_ optimally doped systems display some degree of pseudogap behavior, and it is only when one reaches doping levels $x > x_{opt}$ that one finds monotonic behavior for $\chi_0(T)$ and no trace of pseudogap behavior in the normal state.

An incomplete list of NMR experiments on the superconducting cuprates systems is given in Table 2. These show that as a result of the opening of the quasiparticle pseudogap, both $^{63}T_1$ and $^{17}T_1$ for the under-doped and optimally doped systems display a quite different temperature dependence from that measured for magnetically overdoped systems. Moreover, NMR experiments on quantities dominated by $\chi_{NAFL}(\mathbf{q},\omega)$, the ^{63}Cu spin-lattice relaxation rate, $^{63}T^{-1}_1$ and the spin-echo decay time, $T_{2G} \sim \xi^{-1}$, show that $\chi_{NAFL}(\mathbf{q},\omega)$ varies dramatically with doping and temperature through changes in $\chi_{\mathbf{Q}_i}$, ω_{SF}, ξ, and the dependence of ω_{SF} on ξ. The T_{2G} measurements, which have now been carried out for $YBa_2Cu_3O_7$, $YBa_2Cu_3O_{6.63}$, and $YBa_2Cu_4O_8$, make it possible to determine directly the temperature dependence of the AF correlation length, $\xi(T)$ for these materials, and, when combined with $^{63}T_1$ and $^{17}T_1$ measurements, set at least a lower limit on the magnitude of $\xi(T)$.

In the long correlation length limit, the theoretical expressions for both $^{63}T_1$ and $^{63}T_{2G}$ obtained using the MMP expression for χ_{NAFL} simplify considerably. Specifically, one finds,[30]

$$^{63}T_1 T \sim \left[\frac{\omega_{SF}\xi^2}{\chi_Q} \equiv \frac{\omega_{SF}}{\alpha} \right] \tag{3.11}$$

$$\frac{1}{^{63}T_{2G}} = \left(\frac{.69}{512\pi} \right)^{1/2} F_{eff}(Q) \frac{\chi_Q}{\xi} \sim \alpha\xi \tag{3.12}$$

where α, the scale factor which relates χ_Q to ξ^2 is assumed to be temperature independent, and we adopt the convention, $^{63}T_1 = (^{63}T_1)_{\parallel}$. Hence up to the scale factor α, $^{63}T_1 T$ provides a direct measurement of ω_{SF}, while $^{63}T_{2G}$ provides a direct measurement of ξ; together with α, these are the three quantities which determine the behavior of $\chi_{NAFL}(\mathbf{q},\omega)$. Moreover, the ratios $(^{63}T_1 T/^{63}T_{2G})$ and $^{63}T_1 T/(^{63}T_{2G})^2$ tell us about the scaling laws, if any, which relate ω_{SF} to ξ, since

$$(^{63}T_1 T/^{63}T_{2G}) \sim \omega_{SF}\,\xi \tag{3.13}$$

$$(^{63}T_1 T/^{63}T^2_{2G}) \sim \chi_Q\omega_{SF} \equiv \alpha\,\omega_{SF}\,\xi^2. \tag{3.14}$$

It is these expressions which Barzykin and Pines (30) used to determine crossover behavior in $\chi_{NAFL}(\mathbf{q},\omega)$ from $^{63}T_1$ and $^{63}T_{2G}$ measurements and to relate this in turn to crossover behavior in $\chi_0(T)$. Specifically, they concluded, as shown in Fig. 6, that above the _same_ temperature, T_{cr}, at which $\chi_0(T)$ displays a maximum χ_{NAFL} displays mean field or RPA behavior, with both ω_{SF}, and ξ^{-2}, varying linearly with temperature in such a way that

$$\omega_{SF}(T) \sim \xi^{-2}(T) \sim a + bT, \tag{3.15}$$

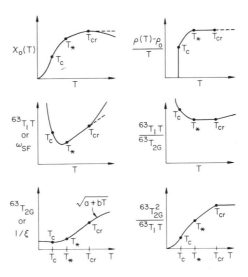

Fig. 6 A schematic depiction of the temperature crossovers measured for various observable quantities in the magnetically underdoped cuprates.

while below T_{cr} the relationship between ω_{SF} and ξ changed, in such a way that between T_{cr} and a second crossover temperature, T_*, $\chi_{NAFL}(\mathbf{q},\omega)$ would display z=1 dynamical scaling behavior, with

$$\omega_{SF}(T) = \widehat{c}/\xi \sim c + dT, \qquad (3.16)$$

Because the scaling behavior found between T_{cr} and T_* turns out not to be universal, it is best described as, "pseudoscaling."

BP concluded that the crossover at T_{cr} seen in both $\chi_0(T)$, and in NMR experiments which probe $\chi_{NAFL}(\mathbf{q},\omega)$, must be of quasiparticle origin, rather than representing a "spin only" phenomenon. The microscopic calculations of Schmalian et al.,[21] which show the common origin of both phenomena is the onset of weak pseudogap behavior in the hot quasiparticle spectrum support this reasoning. What determines T_{cr}, i.e. what is the physical origin of this crossover from mean field to pseudoscaling behavior in the dynamical spin susceptibility? Monthoux and Pines[31] suggested that T_{cr} is associated with a critical (and doping independent) value of ξ. From their analysis of the behavior of $\chi_0(T)$ and $^{63}T_1$ in optimally doped 1-2-3, Barzykin and Pines suggested that at T_{cr} for this system, one had

$$\xi(T_{cr}) \cong 2 \qquad (3.17)$$

and that Eq. (3.17) provides a general criterion for the doping dependence of T_{cr} as one goes to the underdoped systems. Thus it could be argued that the physical origin of the crossover at T_{cr} is the approach to an SDW instability of the "hot" portion of the Fermi surface, the conjecture of Monthoux and Pines which was verified in the subsequent calculations of Schmalian et al.[21] Put another way, strongly *magnetically* correlated quasiparticles will not respond to an external field as effectively as those which are weakly

126

correlated; hence the reduction in $\chi_0(T)$ below T_{cr}, while the SPS calculations show that this reduction reflects, in fact, the development of a high energy peak (~200 meV) in the hot quasiparticle spectral density. Experimental confirmation of the BP criterion, Eq. (3.17), for T_{cr} comes from INS experiments on $La_{1.86}Sr_{0.14}CuO_4$ discussed below, and from the Raman experiments discussed in these proceedings by Hackl. The latter experiments show that below T_{cr} one finds the high energy peak expected for a system with AF correlations sufficiently strong that $\xi \geq 2$, while above T_{cr} that peak is absent; T_{cr} determined from Raman experiments agrees with the magnetic determination.

In most magnetically underdoped systems, pseudoscaling behavior does not persist all the way down to T_c; rather $^{63}T_1$ experiments show that in the vicinity of the temperature, T_*, at which $^{63}T_1T$ ceases to be linear in T, ω_{SF} reaches a minimum, and then increases rapidly as the temperature is further decreased. This behavior suggested that some kind of further gap in the spin excitation spectrum opens up at T_*. Inspection of the behavior of $\chi_0(T)$ shows that between T_{cr} and T_* there is a comparatively gentle fall-off in $\chi_0(T)$, but that around T_*, $\chi_0(T)$ begins to decrease more rapidly, as though a significant gap had opened in the quasiparticle spectrum as well. Thus below T_*, one finds strong quasiparticle gap-like behavior for both χ_0 and χ_{NAFL}. Because this gap-like behavior is not accompanied by the long range order of an antiferromagnet or a super-conductor it is best described by the term introduced by Friedel, *pseudogap*.

It is convenient to distinguish between these two rather distinct regimes of pseudogap behavior by characterizing the behavior of the system between T_{cr} and T_* as *weak* pseudogap behavior, reflected in the slow fall-off in $\chi_0(T)$ as T decreases, and the pseudoscaling behavior of $\chi_{NAFL}(\mathbf{q},\omega)$, while between T_* and T_{cr}, the system exhibits *strong* pseudogap behavior seen in both χ_0 and χ_{NAFL}; ARPES experiments discussed by Campuzano in this volume, show that in the vicinity of T_*, hot quasiparticles [those near $(\pi,0]$ develop a "leading-edge" gap, so that it is reasonable to assume that it is this *further* transfer of hot quasiparticle spectral density (over an energy region, $2\Delta_{hot}$, which does not appear correlated with T_c), which is responsible for the measured changes in the low frequency dynamical response.

As may be seen in Fig. 7, rather similar phase diagrams emerge from an analysis of the low frequency magnetic behavior of the 1-2-3 and 2-1-4 systems. For both systems, $T_{cr}(x)$ increases rapidly as the doping decreases on the magnetically underdoped side; $T_*(x)$ on the other hand increases less rapidly with decreased doping and may, in fact lie below T_c for optimally and overdoped systems while possessing a rather broad maximum for underdoped systems. Indeed, as befits crossover behavior, how one defines T_* is somewhat arbitrary: if one defines it as the temperature at which $^{63}T_1T$ reaches a minimum, (rather than following the BP criterion that T_* is the temperature at which $^{63}T_1T$ begins to exhibit sublinear behavior), and then further requires that $^{63}T_1T$ increase substantially at still lower temperatures, then an analysis of existing experiments suggests that optimally doped 2-1-4 does not display strong pseudogap behavior over any appreciable temperature range above T_c.

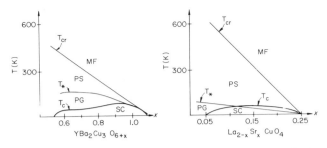

Fig. 7 Candidate phase diagrams for two families of cuprate superconductors. In both the 1-2-3 and 2-1-4 systems one finds in NMR experiments the crossovers at T_{cr} and T_* from mean field (MF) to pseudoscaling (PS) to pseudogap (PG) behavior discussed in the text, before the transition at Tc to the superconducting (SC) state. Note the similarities in the doping dependence of T_{cr}, T_*, and T_c.

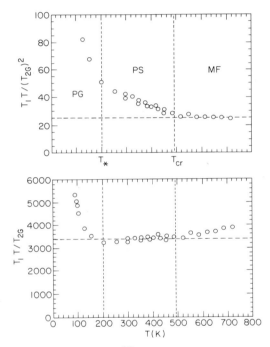

Fig. 8 The measurements by Curro et al.[33] which show the crossovers in low frequency magnetic response at T_{cr} and T_*.

This conclusion is supported by the results of the inelastic neutron scattering experiments of Aeppli et al. [32] on $La_{1.86}Sr_{0.14}CuO_4$ discussed in the following section. It is bolstered by the results of the recent Raman experiments of Naeini et al. [20] on $La_{1.78}Sr_{0.22}CuO_4$, which again find no evidence for strong pseudogap behavior.

The NMR experiments of Curro et al. [33] on $YBa_2Cu_4O_8$ at temperatures from T_c up to 700K provide striking confirmation of the BP pseudogap scenario. As may be seen in Fig. 8, these show that pseudoscaling behavior in $\chi_{NAFL}(\mathbf{q},\omega)$ and weak pseudogap behavior in $\chi_0(T)$ both begin at ~500K, which is therefore the appropriate value of T_{cr} for this system. Pseudoscaling and weak pseudogap behavior persist down to temperatures, T_*,

~180±20K, while $^{63}T_1T$ possesses a well-defined minimum at ~160K, below which the strong pseudogap behavior of the hot quasiparticles brings about a rapid increase in $^{63}T_1T$ and $\omega_{SF}(T)$.

IV. RECONCILING NMR AND INS EXPERIMENTS

As both sample quality and INS (inelastic neutron scattering) techniques have improved, there have emerged a number of apparent contradictions between the spin fluctuation spectrum measured in INS experiments on the 1-2-3 system (about which Philippe Bourges reports elsewhere in this volume) and that deduced from NMR experiments. As Zha et al. [34] (hereafter ZBP) have shown, such contradictions are not present for the 2-1-4 system; indeed the reconciliation of the NMR and INS experiments on $La_{1.86}Sr_{0.14}CuO_4$ provides significant clues to how the corresponding experiments on the 1-2-3 system may be reconciled.

For the 2-1-4 system, INS experiments reveal the existence of four incommensurate peaks at wavevectors $\mathbf{Q}_i$ close to $\mathbf{Q}$, whose position is $\mathbf{Q}_i = [\pi,\pi(1\pm\delta)], [\pi(1\pm\delta);\pi]$ where δ depends on the doping level. For the $Sr_{0.14}$ system studied by Aeppli at al., [32] one has $\delta = 0.245$. With this amount of incommensuration, one finds, using the SMR hyperfine Hamiltonian, Eq. (3.1), that there is so much leakage of the strong spin fluctuation peak seen in the INS experiments onto the nearby oxygen site that one gets completely the wrong magnitude and temperature dependence for the measured ^{17}O spin-lattice relaxation rate. ZBP showed that if one supplements the SMR hyperfine Hamiltonian with a term which couples the oxygen nucleus to the next nearest neighbor Cu^{2+} spins, then a comparatively modest amount of next nearest neighbor coupling provides sufficient filtering action that the contribution coming from χ_{NAFL} to $^{17}T_1$ is reduced to the point that the Fermi liquid contribution, Eq. (3.8) dominates, and agreement between the INS result and experiment is restored.

ZBP went on to compare the results of Aeppli et al.,[32] which was the first experiment in which it proved possible to obtain the magnitude of the INS peaks, with the results of the Cu NMR experiments, which are, of course, primarily sensitive to χ_{NAFL}. They showed that on taking the Lorentzian form, Eq. (2.2) for χ_{NAFL}, one could combine the INS results at 35K ($\chi_{\mathbf{Q}_i} = 350$ states/eV, and $\omega_{SF} = 8.75$ meV) with the NMR results of Ohsugi et al., [35] for $^{63}T_1$ to fix the scale factor, α, and the AF correlation length ξ. On taking hyperfine coupling constants chosen to get an anisotropy in $^{63}T_1$, (for fields parallel and perpendicular to the Cu-O plane) which agreed with experiment, they found in this way, $\alpha \simeq 24$ states/eV and, on making use of Eq. (2.3), $\xi \simeq 7.6$. The resulting value of ξ agrees remarkably well with the low frequency limiting value of ξ obtained directly by Aeppli et al.[32] from their measurement of the intrinsic line width, $\xi = 7.7$. Thus at 35K (with these hyperfine coupling constants) the 35K NMR and INS experiments are consistent with one another.

ZBP then went on to establish the consistency of the two sets of measurements at higher temperatures. Since all the free parameters in χ_{NAFL} have been determined, the issue was whether consistency could be found if one made the assumption, inherent in the expression, (2.3), that α

is independent of temperature, and used only the NMR results to predict the results of the INS experiment. The excellent agreement which was found for the magnitude of the peak, $\chi_{\mathbf{Q_i}}$, the frequency of the relaxational mode, ω_{SF}, and the correlation length, ξ, at temperatures up to 300K, demonstrates that with a temperature independent value of α, consistency can be achieved.

Two important additional results emerge from the INS experimental results. The first is the explicit demonstration of pseudoscaling. Consider the frequency dependence of the peak behavior of one of the four incommensurate peaks,

$$\chi''_{NAFL}(\mathbf{Q}_i,\omega) = \frac{\chi_{\mathbf{Q_i}}\omega/\omega_{SF}}{1+\omega^2/\omega_{SF}^2} \tag{4.1}$$

At any given temperature one measures $\chi_{\mathbf{Q_i}}$, ω_{SF}, and ξ. Moreover, with $\chi_{\mathbf{Q_i}} = \alpha \xi^2$ and $\omega_{SF} = \widehat{c}/\xi$, one finds from Eq. (4.1), that in the pseudoscaling regime the slope of χ'' obeys the scaling relationship,

$$\lim_{\omega\to 0} \left(\frac{\chi''(\mathbf{Q}_i,\omega)}{\omega}\right) = \frac{\chi_{\mathbf{Q_i}}}{\omega_{SF}} = \left(\frac{\alpha}{\widehat{c}}\right)\xi^3 \tag{4.2}$$

This is just what Aeppli et al. find between 35K and 300K. The second important result is the direct determination of the temperature at which $\xi = 2$. This turns out to be ~325K, which, within experimental error, is identical to the crossover temperature, T_{cr}, determined by BP from their analysis of Knight shift experiments. Thus the INS experiment provides direct experimental confirmation of the prediction by BP that $\xi(T_{cr}) = 2$.

The above results were obtained using experimental data kindly provided by the Aeppli group in advance of publication of their important work, a publication which turned out to be delayed for almost two years. In the quite recently published version of their work, Aeppli et al. [32] use a somewhat different expression for $\chi''(\mathbf{Q}_i,\omega)$ than the MMP form, Eq. (2.2), to analyze their data; Schmalian and Morr[36] have shown that the two forms are, however, consistent and have shown as well that with $\xi^{-1}(T) = a + bT$ the simple MMP expression provides a fit to their data on the frequency and temperature dependence of the half-width, κ, of the neutron peaks which is indistinguishable, within experimental error, from that employed by Aeppli et al. [$\kappa^2 = \kappa_0^2 + (T/\widehat{c})^2 + (\omega/\widehat{c}))^2$], with $\widehat{c} = 47$ meV); the latter expression corresponds to assuming that the temperature dependence of ξ is given by $\xi^{-2}(T) = a + bT^2$. Which expression does a better job of describing the temperature dependence of $\xi(T)$ is thus an open question.

Let us turn now to the INS experiments on the 1-2-3 system. Here, in the normal state, with the exception of the very recent experiments by Mook et al. (37), no incommensurate spin fluctuation peaks have been seen. Moreover, the width of the commensurate peaks appears to be, at most, weakly temperature dependent, and suggests correlation lengths of order unity. ZBP point out that this apparent discrepancy would be resolved if the broad lines observed in INS experiments in the normal state originate in four incompletely resolved incommensurate peaks. They explored this possibility using a simple model for the underlying incommensurability: four peaks located along the zone diagonals, at $\mathbf{Q}_i = \mathbf{Q}_i = [\pi(1\pm\delta),\pi(1\pm\delta)]$. On making use of the parameters which fit the NMR data, at 100K,

$$\omega_{SF} = 17 \text{ meV}; \quad \xi = 2.1; \quad \alpha = 15 \text{ states/eV} \tag{4.3}$$

they find, on taking $\delta = 0.1$

$$\chi''(\mathbf{Q},\omega) = \frac{34\omega/\bar{\omega}_{SF}}{1+\omega^2/\bar{\omega}_{SF}^2} \text{ states/eV} \tag{4.4}$$

where $\bar{\omega}_{SF}$ reflects the degree of incommensuration, being given by

$$\bar{\omega}_{SF} = \omega_{SF} \times [1+(\mathbf{Q}_i-\mathbf{Q})^2\xi^2] = 1.85 \ \omega_{SF} = 31 \text{ meV} \tag{4.5}$$

for the incommensuration assumed in this simple model. We see that incommensuration has the effect of shifting the position of the maximum in ω space, and that for a correlation length, ξ - 2, a quite modest degree of incommensuration brings about a substantive shift.

Moreover, the value of α and the comparatively short correlation length required to fit the NMR experiments both mitigate against being able to pick up a signal against the always present background in an INS experiment on the normal state. Under optimal conditions one is seeking to resolve peaks whose maximum strength is some 17 states/eV for an energy transfer of some 30 meV; the peak is down by an order of magnitude from that seen for the $Sr_{0.14}$ 2-1-4 sample. *Thus from fits to the NMR experiments we can conclude that until one gets almost an order of magnitude improvement in resolving signals from noise, one will not be able to pick up useful information about the spin fluctuation spectrum in the normal state of optimally doped 1-2-3 from an INS experiment.* Matters are improved if one goes to underdoped systems, but even for $O_{6.6}$ at 70K preliminary estimates using the ZBP parameters, and a correlation length of 4, suggest that the maximum intensity is about 30% of that found by Aeppli et al. at 35K in their 2-1-4 sample.

It should be noted that even though one may not be able to resolve underlying incoherent peaks, a study, such as Bourges has carried out, of the frequency dependence of an assumed unresolved set of peaks at $\mathbf{Q}$ can, through application of Eq. (3.4), provide useful information, albeit somewhat indirect, on the degree of underlying incommensuration, provided one has a reasonable estimate of ω_{SF} from fits to NMR experiments.

A corollary of the above discussion is the conclusion that while one can, by introducing incommensuration, use spin fluctuation parameters deduced from NMR experiments to explain, or even predict, the results of INS experiments, the reverse procedure is in general not feasible. Thus if one tries to use the comparatively short and temperature independent correlation lengths measured in INS experiments on say, $YBa_2Cu_3O_{6.6}$ to explain the ^{63}NMR results on $^{63}T_1$ and $^{63}T_{2g}$ for this system, one is immediately led to a contradiction with experiment.

In concluding this discussion, I would like to call attention to an unexpected bonus of the incommensuration hypothesis for the 1-2-3 or 1-2-4 system. ZBP show for optimally doped 1-2-3 that incommensuration plus next nearest neighbor coupling of the oxygen nucleus to its Cu+ neighbors, provides a natural explanation, within the one component model, of the dependence on field orientation of the temperature dependence of the anisotropic ^{17}O relaxation rates recently found by Martindale et al. [38] for both the optimally-doped and $O_{6.6}$ members of the 1-2-3 system, while the Brinkmann group in Zurich finds this works as well for the measured anisotropies in the 1-2-4 system. [39]

V. THE MAGNETIC SCENARIO FOR NORMAL STATE PROPERTIES

Mean field and weak pseudogap behavior

Since the dramatic changes in low frequency magnetic behavior we have considered in Sec. III must, in the spin fluctuation model, reflect changes in quasiparticle behavior, without doing any detailed calculations (or experiments) the qualitative nature of those changes is clear. Consider first the mean-field regime, where as one reduces the temperature, hot quasiparticle life-time effects, which act to reduce $\tilde{\chi}(\mathbf{Q},\omega)$, from its non-interacting value, $\tilde{\chi}_0(\mathbf{Q},\omega)$, play less of a role. Hence one expects that at all wavevectors, the static susceptibility, $\chi_{\mathbf{q}}$, will increase as the temperature decreases, and the Eliashberg calculations of Monthoux and Pines[6] show that this is indeed the case. The feedback effects associated with the hot quasiparticles are negative in this regime; for example, self-consistent one-loop calculations (c.f. Pao and Bickers, and Monthoux and Scalapino[40]) show that if one starts with an effective interaction, $J_{\mathbf{Q}}$, which is, say, so strong that its product with the non-interacting susceptibility, $\tilde{\chi}_0(\mathbf{Q},\omega)$, would be greater than unity, no instability in fact occurs, since en route to that instability the interaction so reduces $\tilde{\chi}_0(\mathbf{Q},\omega)$, that the product, $J_{\mathbf{Q}} \tilde{\chi}(\mathbf{Q},\omega)$, remains less than unity.

Matters are clearly otherwise in the pseudoscaling, or weak pseudo-gap regime, since below T_{cr}, the hot quasiparticle spectral density must change in such a way that its contribution to χ_0 overrides the lifetime corrections and causes χ_0 to decrease as the temperature decreases. In the spin fluctuation scenario this must reflect positive feedback which, for the hot quasiparticles, begins at T_{cr} or about $\xi = 2$. Positive feedback brings about a crossover to a regime in which for the hot quasiparticles there must be a transfer of spectral weight from low energies to higher energies which acts to override lifetime effects. Specifically, the strength z_{hot} of the coherent low energy part, which above T_{cr} had been increasing as the temperature decreased (because of the reduced effectiveness of lifetime effects), must now decrease as the temperature decreases, while the strength of the higher frequency incoherent part increases. As discussed in the following section, this is just what is both seen experimentally in ARPES experiments, and is found in microscopic calculations based on the spin fluctuation model.

Monthoux and Pines[31] pointed out that by combining the results of measurements $^{63}T_1$ and $^{63}T_{2G}$ one gets a direct measure of changes in the irreducible particle-hole susceptibility which occur at T_{cr}. To see how this works, let us parametrize the low frequency behavior of $\tilde{\chi}(\mathbf{Q},\omega)$ by the following simple expression:

$$\tilde{\chi}(\mathbf{Q},\omega) = \frac{\tilde{\chi}_{\mathbf{Q}}}{1-i\omega/\tilde{\Gamma}_{\mathbf{Q}}} \tag{5.1}$$

where $\tilde{\chi}_{\mathbf{Q}}$ is the static susceptibility and $\tilde{\Gamma}_{\mathbf{Q}}$ characterizes the low frequency energy spectrum. On rewriting Eq. (2.6) in the form

$$\frac{1}{\chi(\mathbf{Q},\omega)} = \frac{1}{\tilde{\chi}(\mathbf{Q},\omega)} - J_{\mathbf{Q}} \tag{5.2}$$

it follows at once that $\text{Im } \chi^{-1}(\mathbf{Q},\omega) = \text{Im } \tilde{\chi}^{-1}(\mathbf{Q},\omega)$, and

$$\chi_{\mathbf{Q}}\,\omega_{SF} = \tilde{\chi}_{\mathbf{Q}}\,\tilde{\Gamma}_{\mathbf{Q}} \tag{5.3}$$

from Eq. (3.14) it follows that the ratio, $^{63}T_1/^{63}T_{2G}^2$, provides direct information on the product $\tilde{\chi}_{\mathbf{Q}}\tilde{\Gamma}_{\mathbf{Q}}$. Below we shall see that the microscopic calculations of this quantity are in good agreement with the experimental results on $YBa_2Cu_4O_8$ in the weak pseudogap regime.

Nonlinear feedback and strong pseudogap behavior

Below the lower crossover temperature, T_*, further changes in quasiparticle behavior must be responsible for the observed changes in the fermion relaxational frequency, ω_{SF}, and the correlation length, ξ. Again without detailed calculations, we can see what the character of those changes must be. First, not far below T_*, ω_{SF} reaches a minimum, and then increases as the temperature decreases further. Such an increase requires a rather more considerable change in the hot quasiparticle behavior than is found above T_*; it is reminiscent of what happens when a gap in the quasiparticle spectrum opens up, for example as a result of superconductivity. Since, however there is no long-range order, this hot quasiparticle gap cannot be complete; i.e. it is a pseudogap, but one which is quite effective at bringing about a shift in the low frequency part of the quasiparticle spectral weight. In similar fashion, the considerably more rapid fall-off in the uniform susceptibility found below T_* plausibly reflects this strong pseudogap behavior, which, it may be argued also is responsible for a freezing of the build-up of the antiferromagnetic correlations (seen as the flattening of the measured values of T_{2g}). Both the specific heat experiments of Loram[41] and the recent direct measurements in ARPES[42] and STM[43] experiments of the presence of a leading edge gap for hot quasiparticles which opens up rapidly below T_* provide significant experimental support for this general physical picture, and we return to this question below.

VI. MICROSCOPIC CALCULATIONS

I now present a brief summary of some of the detailed microscopic calculations of normal state properties which have been carried out using the Urbana effective quasiparticle interaction, $V_{eff}(\mathbf{q},\omega)$, Eq. (2.5). For each I indicate the key results and their possible applicability to experiment In these calculations the starting quasiparticle spectrum was specified in terms of the tight binding parameters, t, and t', and the dependence of the results on the planar hole density and spin fluctuation parameters was explored. Wherever possible the parameters chosen for the spin fluctuation and quasiparticle spectra were taken from fits to NMR and ARPES experiments. The use of an experiment-based specific interaction and quasiparticle spectrum possesses the advantage that one is dealing with a well-posed problem in which key parameters are specified within a comparatively narrow range; it possesses the disadvantage that one is not carrying out a fully self-consistent calculation of all relevant quantities, but is, in a very real sense, starting in the middle, which means that unless one succeeds in getting results which are close to those seen experimentally, the approach may not be taken seriously. The approach has the further disadvantage that one has comparatively little "wiggle room"; since the basic parameters which enter the model interaction and the quasiparticle spectrum are constrained by experiment; for a given hole density, there is, with these ground rules, only one "free" parameter, the coupling constant, g^2. Once this is chosen to obtain agreement with experiment for a given calculated system property, one is essentially dealing with a situation in which all the other calculated properties must follow if the model interaction is to be be regarded as credible by the broader community.

Weak coupling calculations of the resistivity , T_c, and pairing state[2]

These found high T_c, $d_{x^2-y^2}$ pairing, a rapid opening of the energy gap below T_c, and a comparatively large gap ratio, $(\Delta/T_c)>3$, for comparatively modest values of the dimensionless coupling constant. Our results showed that an interaction which is strongly peaked at large wavevectors is much more effective in bringing about $d_{x^2-y^2}$ pairing and superconductivity at high temperatures than the comparatively featureless Hubbard interaction. We also demonstrated that one obtained a considerably higher T_c, and a considerably larger (Δ/T_c) ratio by keeping the full structure of the interaction, rather than employing, at an early stage, approximations based on averaging procedures borrowed from theoretical treatments of phonon-induced superconductivity. Because lifetime effects, which are known to reduce T_c dramatically, were not taken into account, the results should be regarded as qualitative and suggestive, rather than quantitative results to be compared directly to experiment.

Strong Coupling Calculations[6]

These were the first Eliashberg calculations which took into account the full structure of the spin-fluctuation induced interaction between the planar quasiparticles. As with the weak coupling calculations, for a given coupling constant, doing so led to a considerably higher value of T_c.

Lifetime effects were found to reduce T_c dramatically from the weak coupling value, as had been widely anticipated, but $d_{x^2-y^2}$ pairing persists, and a T_c of 90K was found for comparatively modest values (~1/2) of the dimensionless coupling constant. Quite importantly, once the coupling constant was fixed in this way, we found a normal state resistivity that varied linearly with temperature, and possessed a magnitude comparable to that seen experimentally; the calculated optical properties were also in agreement with experiment. For Monthoux and me, this was the "proof of concept" for the spin-fluctuation model, and accordingly we had considerable confidence that the pairing state our theory required would, in due course, become firmly established.

Explicit calculations of the real and imaginary parts of the quasi-particle energy at various points on the Fermi surface were carried out. These first calculations of the difference between "hot" and "cold" quasi-particles (the term was introduced subsequently by Hlubina and Rice[44]) established the very considerable anisotropy of quasiparticle behavior, and demonstrated that when self-consistent calculations were carried out, the singularities which appear in first-order calculations of the influence of van Hove singularities are washed out by lifetime effects. The extent to which calculations of the irreducible particle-hole susceptibility in the Eliashberg approximation enable one to obtain the measured spin fluctuation spectrum was explored.

In the second paper in this series, the influence of potential impurity scattering on T_c was studied. Zn was found to be a superunitary scatterer, while Ni was a sub-unitary scatterer; these results support the spin-fluctuation model, since Zn not only scatters at the unitary limit, but because it changes the local magnetic order, it changes the effective inter-action responsible for superconductivity and so produces an even greater change in T_c than is obtained in the unitary (strong scattering) limit. Explicit numerical calculations showed that the influence of Ni impurities on the imaginary part of the self-energy was a minor add-on to the much larger contribution from inelastic spin-fluctuation scattering. The sensitivity of the calculated T_c to changes in doping, hopping terms, and spin fluctuation parameters was studied in some detail. It was shown that by using a more accurate spin-fluctuation spectrum and taking vertex corrections into account, one could obtain considerably better quantiutative agreement betwen theory and experiment for both the resistivity and optical properties of optimally doped 1-2-3.

Longitudinal and Hall Conductivities[15]

The spin-fluctuation model interaction with spin fluctuation and quasiparticle parameters once more taken from fits to experiment was used to calculate the doping and temperature dependent resistivity and Hall conductivity for both overdoped and underdoped systems. Direct numerical (nonvariational) solutions of the Boltzmann equation for representative members of the 1-2-3 and 2-1-4 systems were carried out, with results in quantitative agreement with experiment. The same numerical approach was used to calculate the influence of chains on the a-b plane anisotropy, and agreement with experiment was again obtained. This work thus demonstrated that the spin-fluctuation model could, with no additional parameters, explain in detail not only the "classic" behavior of optimally doped systems, a linear in T longitudinal resistivity and a quadratic dependence on temperature of the cotangent of the Hall angle, but was capable of explaining the departures from this behavior as one changes the doping level or goes to low temperatures. Thus, as discussed in some detail in Ref. 15, experiments using single crystals[16,45] show that $\rho_{xx}(T)$ displays a downturn from linear in T behavior as T approaches T_* from above, and that such downturns are a common feature of all magnetically underdoped materials. As may be seen in Fig. 9, the character of the departure of ρ_{xx} from linear in T behavior depends on doping level; optimally doped systems display the least departure from T-linear behavior, while magnetically overdoped systems display an upturn at comparatively high temperature.

In the magnetically underdoped cuprates, the transverse conductivity, σ_{xy} is $\sim T^{-3}$ at high temperatures, with significant deviations occurring at temperatures below T_{*2} As a result, in the pseudoscaling regime, the Hall

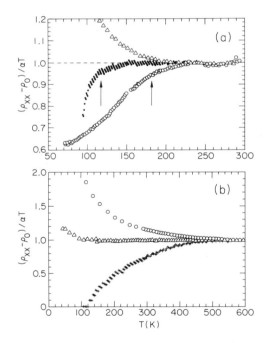

Fig. 9 The measured resistivity as a function of temperature, showing the deviation from linearity in T in underdoped, overdoped, and optimally doped cuprates. Panel (a) shows the results obtained in (top to bottom) 15-K Tl 2201, $YBa_2Cu_3O_7$, and $YBa_2Cu_3O_{6.68}$ compounds. The dashed line is a guide to the eye and the arrow marks the crossover from pseudoscaling to pseudogap behavior at T_* in the under-doped and optimally doped materials. Panel (b) shows $La_{2-x}Sr_xCuO_4$ at three doping levels (top to bottom) x=0.22, 0.15, and 0.10. The quantity plotted is $[\rho_{xx}(T)-\rho_0]/\alpha T$, where ρ_0 and α are obtained by fitting the high-T, linear part of the resistivity (from Stojkovic and Pines.[15]

resistivity, $\sigma_{xy} \cong (\sigma_{xy}/\sigma_{xx})$, is temperature dependent ($\sim$ T^{-1}), while $ctn\theta_H \equiv (\rho_{xx}/\rho_{xy})$ displays T^2 behavior. Departures from this behavior are found in the pseudogap regime and may occur as well in the mean field regime. For example, in YBa$_2$Cu$_3$O$_{6.63}$, one finds σ_{xy} is $\sim$ T^{-4} below 200K ($\sim$ T$_*$ for this material). Hwang et al.[46] finds that in the La$_{2-x}$Sr$_x$CuO$_4$ system, for $x \gtrsim 0.15$, the Hall resistivity becomes temperature dependent below a characteristic temperature $\sim$T$_{cr}$. Thus the crossovers at T$_{cr}$ and T$_*$ seen in the spin response possess direct counterparts in the charge response of the planar quasiparticles in the cuprate superconductors. Indeed, as discussed in Ref. 30, a phase diagram for the underdoped systems which is quite similar to the BP magnetic phase diagram was constructed by Hwang et al.,[46] Batlogg et al.,[46] and Nakano et al.[47] using crossovers in charge response as a principal ingredient.

Magnetotransport in Fields up to 60T

Quite recently, Stojkovic and Pines[18] have used the spin fluctuation model to calculate the shift, $\Delta\rho_{xx}$, in the longitudinal resistivity, ρ_{xx}, and the Hall angle, θ_H, brought about by strong magnetic fields and compared their results with the experiments of Harris et al.,[48] who found that $(\Delta\rho_{xx}/\rho_{xx}) \sim \theta_H^2$, and Tyler et al.[49] who find a departure from this modest field, B^2 behavior which takes the form

$$\frac{\Delta\rho_{xx}}{\rho_{xx}} = \frac{\alpha B^2}{1+\beta^2 B^2},$$

(6.1)

a departure which occurs at higher threshold fields for optimally doped or underdoped systems than for overdoped cuprates. Stojkovic and Pines find that in the limit of strong anisotropy ($\tau_{cold} \gg \tau_{hot}$) both α and β are determined by cold quasiparticle behavior, with

$$\frac{\Delta\rho_{xx}}{\rho_{xx}} \cong \frac{1}{8} \, \omega_c^2 \tau_{cold}^2$$

(6.2)

and

$$\beta^2 B^2 \cong \frac{5}{16} \, \omega_c \tau_{cold} \, .$$

(6.3)

On making use of their earlier calculations of τ_{cold} for optimally and over-doped Tl 2201 samples,[15] they are able to obtain a quantitative fit to the results of Harris et al. and Tyler et al., while explaining both the doping-dependent onset of strong magnetic field corrections to transport, and the specific values of $\omega_c\tau$ deduced by Tyler et al. for overdoped and optimally doped samples. Thus Stojkovic and Pines find that the spin fluctuation model, with no additional adjustable parameters, gives a consistent description of the existing magnetotransport data.

Optical Conductivity[50]

The highly anisotropic scattering rate in different parts of the Brillouin zone was calculated as a function of frequency and temperature, and shown to yield an average scattering rate of the Marginal Fermi liquid form for overdoped and optimally doped systems, and for underdoped systems at high temperatures. Numerical calculations of the optical conductivity were carried out for several compounds for which the spin fluctuation parameters were known, with results in quantitative agreement with experiment for both optimally doped and overdoped systems.

Raman Scattering

Deveraux and Kampf[19] have used the spin-fluctuation model to calculate the doping and temperature dependence of Raman scattering in the B_{1g} and B_{2g} channels, while a detailed comparison between the resulting theoretical calculations and experiments on $La_{1.78}Sr_{0.22}CuO_4$ ($T_c = 30K$) has been carried out by Naeini et al.[20] The theoretical calculations of the quite different response functions associated with hot and cold quasiparticles, measured in the B_{1g} and B_{2g} channels respectively, both support and complement the quasiparticle lifetime calculations reported by Stojkovic and Pines.[15] Naeini et al.[20] find a crossover in the hot quasiparticle scattering rate at ~160K which they identify with the meanfield to weak pseudogap crossover at T_{cr}, an identification which is consistent with the doping dependence of T_{cr} deduced by Barzykin and Pines[30] from NMR data. No evidence for strong pseudogap behavior is found, which leads Naeini et al. to conclude that $T_* < T_c$ for 2-1-4 $Sr_{0.11}$.

Weak Pseudogap Behavior in the Underdoped Superconductors[21]

An exact solution of the spin-fluctuation model, obtained in the limit of $\pi T \gg \omega_{sf}$, which is valid throughout the weak pseudogap (pseudo-scaling) regime, demonstrates that the broad high energy features found in ARPES measurements of the spectral density are determined by strong antiferromagnetic correlations and precursor effects of an SDW state. The onset temperature, T_{cr}, for pseudogap behavior is determined by the strength of the AF correlations, in agreement with the proposal of Monthoux and Pines. Moreover, as may be seen in Fig. 10 the generic changes in low frequency magnetic behavior below T_{cr} are obtained

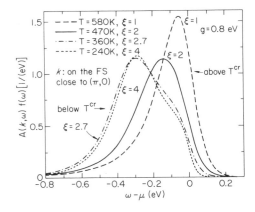

Fig. 10 Evolution of the hot quasiparticle spectral density as a function of correlation length calculated by Schmalian et al.[21]

138

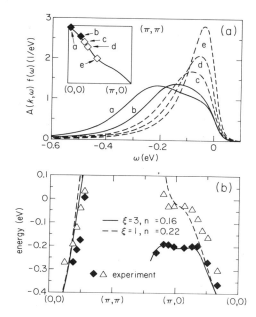

Fig. 11 (a) Spectral density multiplied with Fermi function on the Fermi surface for ξ=3. The distinct behavior of hot and cold quasiparticles is visible. The inset shows the corresponding Fermi surface. (b) Momentum dependence of local maxima of the spectral density as function of ξ and hole doping concentration n_h is compared with experiments of Ref. 22 for $Bi_2Sr_2Ca_{1-x}Dy_xCu_2O_{8+\delta}$ with x=0.01 (triangles) and x=0.175 (diamonds). Only maxima with relative spectral weight <10% are shown [from Schmalian et al.[21]].

beginning at ξ = 2, thus confirming the BP crossover criterion. As may be seen in Fig. 11, the distinct behavior of the evolution of the spectral density of hot and cold quasiparticles found in ARPES experiments is obtained, and quantitative agreement between theory and the ARPES experiments of Marshall et al.[22] is found.

Strong Pseudogap Behavior

Chubukov[51] has recently used the spin fluctuation model to develop a theory of the leading edge gap found below T_* in the underdoped cuprates. He focusses on the behavior of the hot quasiparticle spectral density in the temperature regime close to T_c in which the leading edge gap is maximum, and shows how this maximum gap transforms into a true superconducting gap below T_c. He finds the leading edge gap is brought about by pairing correlations in the $d_{x^2-y^2}$ channel associated with the large incoherent component of the hot quasiparticle spectral density, a d-wave precursor scenario which is quite different from conventional scenarios which focus on the coherent component of the quasiparticle spectral density. In subsequent work, Chubukov and Schmalian[52] find that this d-wave precursor behavior is destroyed as πT approaches ω_{SF}, by the thermal spin fluctuations which dominate system behavior in the weak pseudogap regime (T > T_*).

VII. CONCLUDING REMARKS

The success of the Urbana spin fluctuation model in explaining the anomalous results of transport, magnetotransport, optical, Raman, and ARPES measurements in the normal state of both underdoped and over-doped cuprates coupled with its unambiguous prediction of the $d_{x^2-y^2}$ pairing state, make it a strong candidate for a starting point in developing a microscopic theory of high temperature superconductivity. The calculations which have been described in these lectures provide answers to some of the key questions about high T_c. Thus the physical origin of the anomalous normal state behavior of the cuprate superconductors may reasonably be assumed to be a highly anisotropic effective magnetic inter-action between planar quasiparticles which are hybrids of holes and the "almost localized" Cu^{2+} spins. Even overdoped systems are never far from being antiferromagnetic, and the planar quasiparticles are best described as a nearly antiferromagnetic Fermi liquid, a system whose properties differ in almost every way from the Landau Fermi liquids found in conventional superconductors. The mechanism for high T_c is spin-fluctuation exchange, an electronic mechanism, which produces a quasiparticle interaction which mirrors the dynamic spin susceptibility measured in NMR experiments. The superconducting order parameter and pairing state is predominantly the $d_{x^2-y^2}$ pairing state.

We are, however, far from possessing a complete understanding of these fascinating materials. For example, we do not have as yet microscopic finite temperature calculations of the strong pseudogap behavior found below T_*, or of the doping dependence of either T_* or the superconducting transition temperature, T_c. Understanding why intense magnetic fields produce an insulating ground state for underdoped 2-1-4 samples[53] is a major challenge, as is developing a microscopic theory of c-axis conductivity which takes into account the role played by hot and cold quasiparticles. Still another fascinating problem is the doping dependence of the superfluid density, ρ_s, which appears to track the quasiparticle density (~1-x) for optimally and overdoped systems, and then tracks the hole density, x, for underdoped systems. Does this mean that hot quasiparticles, being already pseudogapped in the normal state, do not contribute appreciably to ρ_s, and that their average density is ~1-2x? Finally, once these and other open questions are settled, there remains the fundamental problem of developing, from first principles, a microscopic theory of the dynamic spin susceptibility and the effective quasiparticle interactions whose consequences we have considered here.

ACKNOWLEDGEMENTS

It gives me pleasure to acknowledge my collaborators, A. Balatsky, V. Barzykin, A. Chubukov, J.P. Lu, A. Millis, H. Monien, P. Monthoux, D. Morr, J. Schmalian, C. P. Slichter, A. Sokol, B. Stojkovic, M. Takigawa, and D. Thelen who have contributed significantly to the research described here, and to thank the National Science Foundation and STCS for the support provided through NSF Grant DMR91-20000 to the research described herein, and to thank the Center for Nonlinear Systems and the Center for Materials Science at Los Alamos National Laboratory for their support and warm hospitality during the past year. Some of the material presented here is taken in part from my earlier reviews on this and related topics.[54,55]

REFERENCES

1. D. Pines, *Physica B* 163:78 (1990).
2. T. Moriya, Y. Takahashi, and K. Ueda, *J. Phys. Soc. Jpn.* 59:2905 (1990); Physica C 185-189:114 (1991); P. Monthoux, A.V. Balatsky, and D. Pines, *Phys. Rev. Lett.* 67:3448 (1991); *Phys. Rev. B* 46:14803 (1992).
3. A. Millis, H. Monien, and D. Pines, *Phys. Rev. B* 42:167 (1990).
4. H. Monien, D. Pines, and M. Takigawa, *Phys. Rev. B* 43:258 (1991).
5. K. Ueda, T. Moriya, and Y. Takahashi, *J. Phys. Chem. Solids* 53:1515 (1992).
6. P. Monthoux and D. Pines, *Phys. Rev. Lett.* 69:961 (1992); P. Monthoux and D. Pines, *Phys. Rev. B* 47:6069 (1993); P. Monthoux and D. Pines, *Phys. Rev. B* 49:4261 (1994).
7. N.E. Bickers, D. Scalapino, and S. White, *Phys. Rev. Lett.* 62:961 (1989).
8. See papers and discussion in *High Temperature Superconductivity*, K.S. Bedell et al. eds, Addison-Wesley (1990).
9. W.W. Warren et al., *Phys. Rev. Lett.* 61:1860 (1987); R.E. Walstedt et al., *Phys. Rev. B* 36:5727 (1987); M. Takigawa et al., *Phys. Rev. B* 39:7371 (1989); S. Barrett et al., *Phys. Rev. B* 41:6283 (1990); H. Monien and D. Pines, *Phys. Rev. B* 41:6297 (1990).
10. Z.-X. Shen et al., *J. Chem. Phys. Solids* 54:1169 (1993).
11. W. Hardy et al., *Phys. Rev. Lett.* 70:399 (1993).
12. J. Martindale et al., Phys. Rev. B 47:9155 (1993); N. Bulut and D. Scalapino, *Phys. Rev. Lett.* 68:705 (1992); D. Thelen, D. Pines, and J.P. Lu, *Phys. Rev. B* 47:951 (1993).
13. D. Wollman et al., *Phys. Rev. Lett.* 71:2134 (1993).
14. C.C. Tsuei et al., *Phys. Rev. Lett.* 73:593 (1994).
15. B. Stojkovic and D. Pines, *Phys. Rev. B* 55:8576 (1997).
16. T. Ito et al., *Phys. Rev. Lett.* 70:3995 (1993).
17. P.W. Anderson, *Phys. Rev. Lett.* 67:2092 (1991).
18. B. Stojkovic and D. Pines, Cond-matt 97-11269.
19. T. Deveraux and A. Kampf, cond-mat 97-11039
20. J.G. Naeini et al., cond-mat 97 11272
21. J. Schmalian, D. Pines, and B. Stojkovic, cond-mat 97 08238
22. D. Marshall et al., *Phys. Rev. Lett.* 76:4841 (1996).
23. P. Monthoux and D. Pines, *J. Phys. Chem. Solids* 56:1651 (1995).
24. C. Pennington and C.P. Slichter, *Phys. Rev. Lett.* 66:381 (1991).
24a. M. Takigawa et al., *Phys. Rev. B* 43:247 (1991).
25. B. Shastry, *Phys. Rev. Lett.* 63:1288 (1989); F. Mila and T.M. Rice, *Physica C* 157:561 (1989).
26. D. Thelen and D. Pines, *Phys. Rev. B* 49:3528 (1994).

27. T. Imai, *Physica C* 162-164:169 (1989).

28. H. Alloul, *Phys. Rev. Lett.* 63:1700 (1989).

29. J. Friedel, *J. Phys. Condens. Matter* 1:7757 (1989).

30. V. Barzykin and D. Pines, *Phys. Rev. B* 52:13585 (1995).

31. P. Monthoux and D. Pines, *Phys. Rev. B* 50:16015 (1994).

32. G. Aeppli et al., *Science* 278:1432 (1997).

33. N.J. Curro et al., *Phys. Rev. B* 56:877 (1997).

34. Y. Zha, V. Barzykin, and D. Pines, *Phys. Rev.* 54:7561 (1996).

35. S. Ohsugi et al., *J. Phys. Soc. Jpn.* 63:700 (1994).

36. J. Schmalian and D. Morr, private communication.

37. H. Mook, Proc. of 1997 SPNS Conference, to appear in *J. Chem. Phys. Solids.*

38. J. Martindale et al., preprint.

39. D. Brinkmann, private communication.

40. C.H. Pao and N.E. Bickers, *Phys. Rev. Lett.* 72:1870 (1994); P. Monthoux and D. Scalapino, *Phys. Rev. Lett.* 72:1874 (1994).

41. J. Loram et al., *Phys. Rev. Lett.* 71:1740 (1993); *J. Supercond.* 7:243 (1994).

42. A.G. Loeser et al., *Science* 273:325 (1996); H. Ding et al., *Nature* 328:51 (1996); J. Campuzano, these proceedings.

43. O. Fischer, these proceedings.

44. R. Hlubina and T.M. Rice, *Phys. Rev. B* 51:9253 (1995); *ibid.* 52:13043 (1995).

45. B. Bucher et al., *Phys. Rev. Lett.* 70:2012 (1993)

46. H.Y. Huang et al., *Phys. Rev. Lett.* 72:2636 (1994); B. Batlogg et al., *Physica C* 235-240:130 (1994).

47. T. Nakano et al., *Phys. Rev. B* 49:16000 (1994).

48. J. Harris et al., *Phys. Rev. Lett.* 75:1391 (1995).

49. A. Tyler et al., cond-mat 97 10032

50. B. Stojkovic and D. Pines, *Phys. Rev. B* 55:8576 (1997).

51. A. Chubukov, preprint.

52. A. Chubukov and J. Schmalian, cond-mat 97-11041.

53. T. Ando et al., *Phys. Rev. Lett.* 75:4662 (1995); G.S. Boebinger et al., *Phys. Rev. Lett.* 77:5417 (1996).

54. D. Pines, in *High Temperature Superconductors and the C^{60} family*, S. Feng and H.C. Ren eds., Gordon & Breach, pp. 1-52 (1995).

55. D. Pines, in *Zeit. f Physik B* 103:129 (1997).

A THEORY OF THE NON-FERMI-LIQUID PROPERTIES AND SUPERCONDUCTIVITY IN COPPER-OXIDES

C.M. Varma

Bell Laboratories, Lucent Technologies, Murray Hill, NJ., USA

In my lectures, I described the theory developed in the last several years to understand the anomalous normal state and the superconducting properties of Copper-oxide metals. The theory has been presented in full detail elsewhere [1,2].

The starting point of the theory is based on two basic assumptions:
(i) Breakdown of Landau Fermi-liquid theory, clearly obvious in the transport properties of copper-oxide metals near the optimum composition, requires scale-invariant low energy fluctuations.
(ii) The solid-state chemistry of copper-oxide is special, and responsible for its unique properties. In particular, a one-band Hubbard model cannot describe the physics of the copper-oxides. The minimum necessary model must explicitly include both copper and oxygen orbitals with hopping between them, local repulsions, as well as ionic interactions.

Experimental results severly delimit the form of the allowed scale-invariant low-energy fluctuations. The fact that the transport relaxation rates have the same frequency and temperature dependence as the single particle relaxation rate implies that the fluctuations should be essentially momentum independent. The fact that these relaxation rates are proportional to $\max(\omega, T)$ imply that the dependence on frequency has the marginal-Fermi-liquid form [3]. The fact that NMR properties are anomalous implies that the fluctuations have a magnetic character. The fact that these anomalies are confined to copper and are absent on oxygen, independent of electron density, is a severe constraint on the spatial properties of the fluctuations within each unit cell.

Scale-invariant fluctuations in more than one-dimension appear only at quantum-critical points. So one has to search for a broken symmetry terminating as a function of electron-density at T=0 near the density for the highest T_c. A Systematic theory of the general model mentioned above has revealed a new form of broken symmetry in which a periodic pattern of current flows in each unit cell, as illustrated in the Figure 1. This pattern breaks time-reversal symmetry as well as four-fold rotation but the product of time-reversal and four-fold rotation is preserved. The transition temperature (in the pure limit) to such a phase is shown to rise from 0 as one changes the hole or electron density below those optimum for T_c. The quantum fluctuations near the quantum critical point satisfy the marginal-Fermi-liquid requirements as well as promote superconductivity of d-wave symmetry.

The Gap Symmetry and Fluctuations in High - T$_c$ Superconductors
Edited by Bok *et al.*, Plenum Press, New York, 1998

143

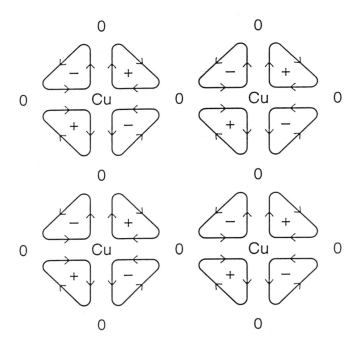

Figure 1: A schematic diagram of a periodic pattern of current that flows in each unit cell in the copper-oxide plane.

There are several predictions of the theory. The most significant is that in the underdoped phases, the current pattern of Figure 1 should be observable by polarised neutron or x-ray scattering through the periodic magnetic field it produces. The polarisation of angle-resolved photoemission should also show a four-fold variation at otherwise equivalent points in the Brillouin zone.

It should also be mentioned that $Ba_xK_{1-x}BiO_3$ has many of the same non-Fermi-liquid properties as the copper-oxides and at half filling is the negative U analog of the copper-oxides [4]. A look at their phenomenology reveals that if their electron density were similar to the copper-oxides T_c would be substantially higher. It would be very worthwhile to synthesise higher density valence-skipping isomorphs of such compounds with Sb or other elements substituting for Bi. If valence skipping persisits they should have a much higher T_c.

REFERENCES

[1] C. M. Varma, Phys. Rev. B55, 14554 (1997)
[2] C. M. Varma, Phys. Rev. Lett. **77**, 3431, (1996)
[3] C. M. Varma et al., Phys. Rev. Lett. **63**, 1996 (1989)
[4] C. M. Varma, Phys. Rev. Lett. **61**, 2713 (1988)

CURRENT RESEARCH ISSUES FOR THE ELECTRON-DOPED CUPRATES

P. Fournier, E. Maiser * and R.L. Greene

Center for Superconductivity Research and Department of Physics
University of Maryland, College Park, MD 20742

INTRODUCTION

It is widely accepted that the key building block of the cuprate superconductors is the copper-oxygen (CuO_2) plane. The general belief is that a mixture of itinerant electrons (holes) together with some important antiferromagnetic correlations (or fluctuations) are the important ingredients for the emergence of high temperature superconductivity. The surrounding environment, usually called the charge reservoirs, plays somehow a passive role as suppliers of charge carriers for the planes. The electron-doped cuprates are part of the "single-layer" subgroup, in which the CuO_2 planes are all equidistant (although, the unit cell can contain two CuO_2 planes : see below). However, most of their normal-state (NS) and superconducting (SC) properties are very different. For example, it was observed that the resistivity of $Nd_{1.85}Ce_{0.15}CuO_4$ (optimally doped) is very close to quadratic (T^2) [1], much different than the optimally doped $La_{1.85}Sr_{0.15}CuO_4$ (LSCO) where the resistivity is linear over an extended temperature range above the critical temperature (T_c) [2]. Moreover, the zero-temperature (residual) resistivity (ρ_o), often associated to impurity scattering, is in the range of 60 - 100 $\mu\Omega$-cm, a fairly high value compared to the values extrapolated from the linear resistivity in LSCO and other cuprates [3]. What made it even more difficult initially to assess any clear systematics of transport in this materials was the difficulty of preparing single phase polycrystalline materials (with a single T_c) [4]. Large contributions from grain boundaries make the interpretation difficult. In the case of single crystals and thin films, the materials appear to show reproducibility for the magnitude (and sign) of the resistivity and the Hall coefficient [5-7]. These material issues are going to be explored briefly below.

Another intriguing observation is made in the superconducting state of these materials as their penetration depth follows an exponential dependence at low temperatures, consistent with an isotropic (s-wave) order parameter [8]. That distinguishes the electron-doped from the other cuprates, most of them showing, in some part of their doping range (roughly from underdoped to optimally doped), various signatures of a very anisotropic gap function, in particular the linear or quadratic temperature dependence of the penetration depth [9,10].

These disparities alone make the electron-doped cuprates the most puzzling materials, a sort of "black sheep" challenging theories proposed to explain the mechanism

The Gap Symmetry and Fluctuations in High - T_c Superconductors
Edited by Bok *et al.*, Plenum Press, New York, 1998

of superconductivity in the cuprates. In this paper, we review in parallel the structural and doping behaviors of the normal-state properties of electron-doped $Pr_{2-x}Ce_xCuO_4$ (PCCO) thin films and the hole-doped $La_{2-x}Sr_xCuO_4$ (LSCO). The objective here is not to give an exhaustive listing of all the papers published on every theme mentioned in the paper. This kind of review can be found already in the literature [11,12]. A careful choice was made, the only purpose being to illustrate the main points of our arguments. The details of the PCCO thin film growth and the related systematics of the normal-state properties will be published elsewhere [13-15]. It must be mentioned however that a battery of measurements confirms that the materials are single-phase, that the doping dependence of the structure and the transport is well behaved (see below). In a second part, we explore several superconducting properties as we emphasize the symmetry of the order parameter. Finally, we try to understand the possible origin of a different order parameter in the electron-doped cuprates, and argue that sample quality is probably not a direct major factor.

STRUCTURE, DOPING AND SUPERCONDUCTIVITY

The discovery of electron-doped SC cuprates by Tokura *et al.* [16] was an exceptional achievement as these materials, unlike the other hole-doped cuprates, have to be doped with electrons in order to get them superconducting. The same group reported that the material crystallizes into the T'-structure (Fig. 1 (a)), fairly different from the LSCO crystal T-structure (Fig. 1(b)) in many aspects.

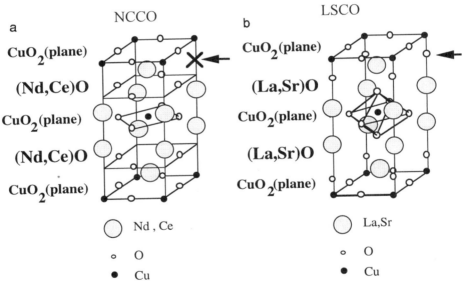

Figure 1 : Crystal structures of (a) $Re_{2-x}Ce_xCuO_4$ in the T'-structure [16]; (b) $La_{2-x}Sr_xCuO_4$ in the T-structure. The arrow and X in (a) indicate the approximate position of the empty apical site in the T'-structure.

First, there is no apical oxygen (see arrow in Fig. 1), as the Re and O ions are offset vertically (not found on the same layer). This implies that the Cu ions are part of a planar structure shared with only four O ions as nearest neighbors, contrary to the octahedral and tetrahedral coordination in LSCO and $YBa_2Cu_3O_7$ (YBCO), respectively. Moreover, the lattice parameters can be varied for ReCeCuO as a function of the radius of the Re ions [17]. This has a substantial effect on the volume of the unit cell, decreasing the amount of maximum Ce doping allowed (solubility limit) for Pr, Nd, Sm and Eu, respectively [17]. It is also believed that crystal structure static distortions limit the range of

Re ion radius that stabilizes the T'-structure: the smallest lanthanide ion that can be used is Gd [18] (GdCeCuO has to be synthesized under high pressure).

Electron-doping is achieved by substitution of the Re^{3+} ions by a substantial percentage of Ce ions : Ce is expected to carry a valence of 4+ (Ce^{4+}). Tokura *et al.* [16] showed that the material becomes superconducting in the doping range of x = 0.13 to 0.18 (see Figure 2) in $Re_{2-x}Ce_xCuO_4$ (with Re = Nd and Pr). This range is much smaller than the one observed for LSCO for example (see also Fig. 2), where SC is observed for x ~ 0.05 to 0.28 [19]. T_c in NCCO increases so sharply between x = 0.13 to 0.15, that it is very difficult to get samples with a well defined intermediate T_c, surely underlying the severe material issues one has to face in these system. After research by several groups, it was clear that this phase diagram (T_c vs Ce concentration) could not be changed drastically, even by playing around with the substituants (Ce and Th) or the rare earth, or both [11] : this phase diagram remains robust. However, there has been some mention of an extended doping range for superconductivity either by the partial substitution of Re by La [20], or by strong oxygen reduction [21]. In both these cases, one has to question the homogeneity of the materials.

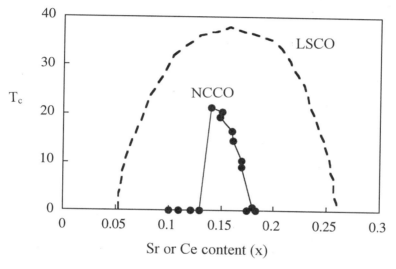

Figure 2 :   Schematic of the phase diagram determined by Tokura et al [16] for Re = Nd (circles). The dashed line shows the approximate T_c vs x in $La_{2-x}Sr_xCuO_4$.

In a very crude picture, the additional electrons added to the structure are transferred to the CuO_2 planes (at least in the metallic state). Then, the appearance of SC in the electron-doped cuprates with a maximum T_c at x ~ 0.15 suggests an electron-hole symmetry in most of the properties [11]. Of course, this fact has implications for the microscopic theory to describe SC in the cuprates, and for the normal-state transport properties. However, a closer look at the existing phase diagram reveals : 1) the AF doping range extends up to x = 0.12 - 0.14 [22], while for LSCO, it disappears at x ~ 0.05 ; 2) the carriers involved in the NS transport seem to show both polarities (electrons, but also holes) [5] ; 3) the materials become fully hole-type beyond x ~ 0.20 [5]. Moreover, there are some suggestions that the electron-doped cuprates need to have hole-type carriers in order to become superconducting [23].

The extent of the antiferromagnetic (AF) phase in the phase diagram [11] can be qualitatively explained if one assumes that electron doping dilutes the Cu^{2+} (spin 1/2) system, decreasing the Néel temperature fairly slowly. In comparison, hole doping could frustrate the AF order. This frustration would lead to a strong suppression of the AF order, and would occur over a much smaller doping range.

Another important feature observed very distinctively in the electron-doped system, and rarely mentioned, is the large magnetic moment developed in the crystal electric field (CEF) by the core $4f$ electrons of the rare earth ions. The large CEF originates from the low positional symmetry at the Re site. This splits strongly the energy levels of the $4f$ states, and applying Hund's rule results in an electronic ground state with a large magnetic moment [24]. Ordering of these moments at low temperatures adds to the already rich phase diagram [25,26], and as we will show later on, might be playing an important role in the low temperature behavior of the penetration depth in the electron-doped cuprates.

TRANSPORT ABOVE T_c

For our comparison with LSCO, the reader should refer to B. Batlogg's paper in these proceedings. Figure 3 shows the temperature dependence of the resistivity of PCCO thin films for four different dopings. We first observe that the doping range for SC is a bit wider than originally published [16]. As shown in Figure 4, an almost discontinuous rise from x = 0.12 to 0.15 is found, followed by a smoother decrease beyond optimal doping (x = 0.15) [14]. Thin films with doping as high as x = 0.20 are still superconducting [14], unlike what was shown by Tokura *et al.* [16] in polycrystals : x-ray diffraction and microprobe analysis indicate that the cerium content corresponds to the target composition [14], and we can discard inhomogeneity problems. The slightly expanded doping range for SC in thin films might be a result of better controlled stoichiometry. In Fig. 3 (a), the resistivity decreases gradually with increasing Ce doping. At optimal doping (x = 0.15), the resistivity just above T_c is of the order of 70 $\mu\Omega$-cm, a value fairly comparable to what is observed in LSCO (x = 0.15). However, the temperature dependencies above T_c are very different. While it is well known that the resistivity is linear over an anomalously large range of temperature in LSCO (and YBCO), it is very close to $\rho(T) = \rho_o + AT^2$ in PCCO, with both ρ_o and A being functions of the Ce doping. This is shown in Fig. 3 (b) with the resistivity plotted as a function of T^2, showing a very close agreement, except at higher temperatures. Immediately, one would conclude that this temperature dependence is con-

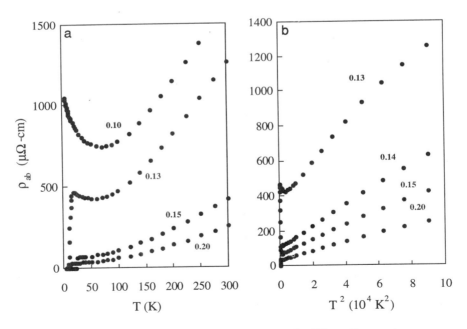

Figure 3 :   Resistivity as a function of temperature for different Ce contents : (a) vs T ; (b) vs T^2.

sistent with electron-electron scattering, added to a large impurity scattering contribution at low temperatures. The zero-temperature intercepts of the resistivity are quite different in LSCO and PCCO, one being close to zero (LSCO), while the other basically unchanged from the value of ρ at 30K (PCCO). This observation might actually be biased by the unknown NS transport properties for temperatures below T_c, which can actually be unraveled by quenching SC with a high magnetic field as we discuss below.

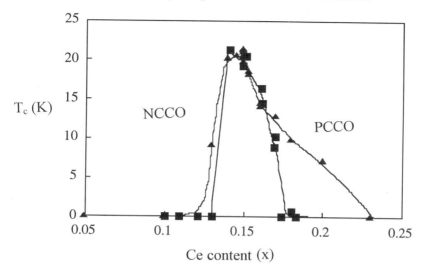

Figure 4 :   Phase diagram of NdCeCuO polycrystals (squares) from Tokura et al. , [16] and our PrCeCuO thin films (triangles).

In Figure 5, we show the Hall coefficient for four different doping levels. Using the fact that the Hall coefficient is inversely proportional to the density of free carriers in a simple single carrier description, the evolution with increasing Ce content toward smaller values is not puzzling, being consistent with the fact that the electron density increases with Ce content. However, it becomes more interesting to observe, for x = 0.15, a low temperature upturn toward positive values. This upturn is uncharacteristic of the hole-doped materials (which show $R_H \sim 1/T$ for optimal doping) and might be suggesting the presence of two types of carriers (holes and electrons) in the system [23]. For large enough x (x = 0.20), the materials has a completely positive Hall coefficient and a temperature dependence very comparable to the one measured on LSCO with x = 0.28 [27] : the electron-like character has basically disappeared, while the material is still superconducting.

The frequency dependence of the ab-plane optical conductivity of the hole-doped cuprates has been extensively studied recently [28,29]. In a narrow frequency range, the conductivity first develops a Drude-like peak at large frequency, accompanied by a fairly broad mid-infrared band. To describe the frequency dependence of $\sigma(\omega)$, Schlesinger et al [30] suggested the use of a frequency-dependent relaxation time and effective mass. Looking at $1/\tau(\omega)$ in particular, it was found to be linear in frequency for $\omega \geq 1000$ cm^{-1}, with a distinctive dip at lower frequencies (see schematics in Fig. 6). The extrapolation of $1/\tau$ at zero frequency for various temperatures was found to follow closely the resistivity. The same signature has been seen recently by Homes et al. [31] in the ab-plane conductivity of NCCO, x=0.15, single crystals. These authors find the same linear frequency dependence at large frequencies (beyond $\omega = 2000$ cm^{-1}), and a similar low frequency dip. Although $1/\tau(\omega)$ at low frequency looks fairly similar to the observations in the hole-doped systems, the extrapolation of $1/\tau$ at $\omega = 0$ leads to a temperature dependence consistent with the measured T^2 dependence of resistivity in NCCO. This underlines that the very details of transport (temperature dependence of resistivity for example) are determined uniquely by the low energy electronic spectrum for both the electron- and hole-doped cuprates, and there is little influence from high energy excitations. The common

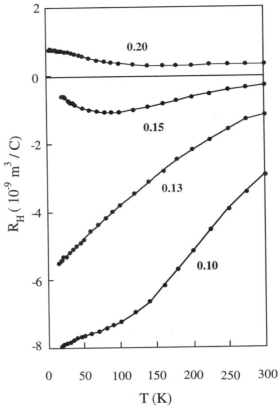

Figure 5 :   Hall coefficient of $Pr_{2-x}Ce_xCuO_4$ as a function of temperature for
different Ce contents.

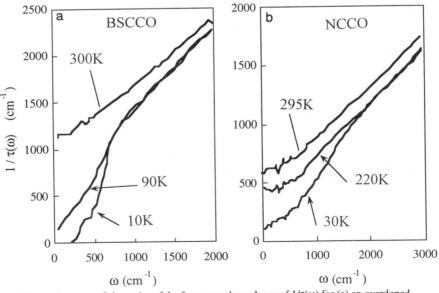

Figure 6 :   Schematics of the frequency dependence of $1/\tau(\omega)$ for (a) an overdoped
BSCCO [29] and (b) a NCCO, $x = 0.15$ [31].

linear frequency dependence of $1/\tau(\omega)$ at high ω found for both hole- and electron-doped cuprates is probably a fundamental characteristic of the CuO_2 planes, as examined recently by Anderson [32] with the hole-doped cuprates. The fact that the high ω behavior does not care if carrier doping is done with holes or electrons, but the low ω behavior does care is an interesting fact that needs further investigation.

Previous reports of angle-resolved photoemission spectroscopy (ARPES) in BSCCO [33] have shown the unusual nature of the electronic excitation spectrum for the hole-doped cuprates. During this summer school, J.C. Campuzzano, M. Onellion and G. Margaritondo (see these proceedings) have shown *solid* evidence for behavior uncharacteristic of Fermi liquids (FL) by ARPES. In particular, Campuzzano *et al.* showed that the lineshapes for BSCCO suggest the absence of quasiparticle excitations in some parts of the Brillouin zone close to intersections of the Fermi surface with the *magnetic* Brillouin zone (see the dashed line in Figure 7). These points are also <u>very close in energy</u> to the saddle-point regions around $(0,\pi)$ and $(\pi,0)$ (often mentioned as a Van Hove singularity), and are believed, in several theories, to play a major role for the determination of many properties [34], including transport. Two early papers of ARPES in NdCeCuO (x = 0.15 and 0.22) show very similar results (see the schematic on the same Fig. 7) [35,36] with a lesser energy resolution. The trend with the increasing Ce content is even consistent with the increase of the electron density as the hole pocket seems to shrink gradually around the (π,π) point [35]. However, we must recall here that NCCO, x = 0.15, has a negative Hall coefficient, not consistent with a hole pocket. It is even more difficult to reconcile the abrupt change in the Hall coefficient from negative to positive (Fig. 4) with minute variations of the Fermi surface with doping from x = 0.15 to 0.22 (as seen in Fig. 6). The data from Ref. [35] suggest that the Fermi surface for x = 0.22 <u>does not</u> cross the magnetic Brillouin zone, unlike x = 0.15 and BSCCO (see these proceedings). This might point out the unusual properties of the quasiparticles in the vicinity of the magnetic Brillouin zone and be consistent with the non-Fermi-liquid behavior of the cuprates in the vicinity of optimal doping. After a closer look at the data of Refs. [35,36], we also realize that the Fermi surface is <u>further away in energy</u> (around 300 meV) from the saddle-point regions than in BSCCO [33]. One could then speculate that the singularity is not influencing much the properties in NCCO in comparison to the hole-doped cuprates (at least BSCCO) which have the singularity within 30 - 50 meV from the Fermi energy. A more detailed study of the electron-doped cuprates using ARPES (with more energy resolution) is needed to shed some light on this issue.

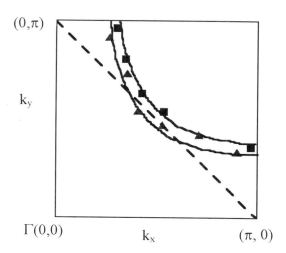

Figure 7 :   Schematic (extraction) of the Fermi surface (crossings) as determined by ARPES on NCCO, x = 0.15 (triangles) [35,36] and x = 0.22 (squares) [35]. The dashed line is the magnetic Brillouin zone. Solid lines are guides to the eye.

SYMMETRY OF THE GAP

Only microwave measurements of the penetration depth and Raman scattering have been convincingly done on the electron-doped system to determine the symmetry of the gap function (order parameter). The initial ab-plane penetration depth $\lambda_{ab}(T)$ of NCCO x = 0.15, measured using a microwave resonant cavity, was shown to be consistent with an exponential temperature dependence, giving a ratio $2\Delta_0/k_BT_c = 4.1$ [8]. One could argue that the data was over too limited range of temperature (the lowest temperature reached was 4.5K, giving $T/T_c \sim 0.2$). However, another set of data collected by Andreone et al. [37], still on NCCO, x = 0.15, and done to lower temperatures (2K), confirms the previous interpretation with basically the same $2\Delta_0/k_BT_c$. These results (extracted from the above mentioned papers) are compared in Figure 8. The two independent experiments are consistent with a s-wave gap, with no evidence of anisotropy or d-wave contribution. The penetration depth was also determined from AC-susceptibility at 100 kHz on thin films of NCCO by Schneider et al. [38] leading to the same conclusion. Raman scattering is another sensitive experiment which has revealed the anisotropic order parameter in hole-doped cuprates [39,40]. Displacement of low frequency spectral weight to higher frequencies as the sample is cooled down below T_c is attributed to the opening of the gap. Using two different polarization configurations (called B_{1g} and B_{2g}), Raman can infer a superconducting gap and evaluate its anisotropy. However in NCCO, Stadlober et al. find that the spectra are consistent with an s-wave order parameter [41].

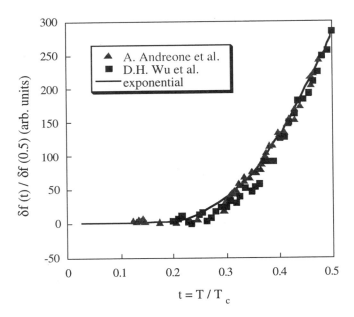

Figure 8 :   Microwave penetration depth from D.H. Wu et al. [8] (squares) and A. Andreone et al. (triangles) [37]. The solid line is a fit using the s-wave exponential model.

Tunneling is another interesting avenue to look at these materials. The few attempts [42] known by the present authors suggest that material issues are likely to affect the reproducibility of the spectra : a good description of this problem can be found in Ref. [43] (in particular Fig. 7 of that paper showing the junction characteristics with two different preparation routes). It was also shown recently [44] that NCCO does not show the zero-bias conductance peak (ZBCP) observed in YBCO, pointing out another difference between the hole- and electron-doped cuprates. The absence of the ZBCP might be just another indication that electron-doped cuprates are s-wave-like since the origin of the ZBCP in the hole-doped cuprates was attributed recently to the d-wave order parameter [45].

Phase-sensitive experiments giving directly the phase of the order parameter in hole-doped cuprates [46] have not yet been reported on the electron-doped family, probably because of material issues (once again!) yet to be understood. The difficulty of making good junctions with thin films of NCCO on bicrystal substrates, as emphasized by Kirtley in this present summer school and showed in the Ph.D. thesis of S.N. Mao [47], is quite puzzling, and hinders the study of the phase of the order parameter. It is surprising not to get good junctions on a bicrystal substrate with a material that has a coherence length of 80 Å (NCCO), while YBCO is successfully done with a coherence length of 10 - 20 Å .

Finally, it would be interesting to measure the nuclear magnetic resonance (NMR) of these materials. NMR results have been mentioned as evidence for d-wave symmetry, signature of a pseudo-gap, etc..., and look like a very powerful tool to study the spin dynamics at well defined lattice sites. However, the large magnetic response due to the presence of the Re ions in ReCeCuO is likely to affect strongly the measurability of any relaxation rate of nuclear moments.

To summarize the parallel between the electron-doped and the hole-doped cuprates, a list of the most important experiments applied to the determination of the symmetry of the order parameter is shown in Table 1. The direct conclusion from this listing is that there is as yet no clear evidence of a gap anisotropy or a d-wave symmetry in the electron-doped cuprates. Most of the experiments are consistent with an s-wave order parameter with $2\Delta_0/k_B T_c \sim 3.5 - 4.5$.

Table 1 :   Some experiments used to determine the anisotropy and symmetry of the order parameter, with results found.

Experiments	hole-doped	electron-doped
NMR, NQR, Knight shift	d-wave	No data because of large Re^{3+} moment in CEF
Penetration depth	$\sim T$ Pure d-wave $\sim T^2$ Dirty d-wave	Exponential with $2\Delta_0/k_B T_c \sim 4$
Phase sensitive (SQUID-type)	d-wave van Harlingen (Ref. [45])	No experiments due to difficulties with Josephson junction fabrication
ARPES	Anisotropic gap (pseudogap with same symmetry)	Not enough resolution !!! ($2\Delta_0 \sim$ 8 meV)
Raman Scattering	In-plane anisotropy	Not much anisotropy
Tunneling	Zero-bias conductance peak d + s from c-axis tunneling [48]	More data needed ... But no zero-bias conductance peak

J.R. Cooper suggested recently [49] by solving Maxwell's equations for a system having a large (normal-state) susceptibility (or permeability, μ_r), that the microwave response of superconductors in a resonant cavity (in particular the frequency shift, δf) can be affected when μ_r diverges at low temperatures, as was reported for NCCO just above T_c. This fast divergence is caused by the large Nd moments in the CEF and their tendency to order antiferromagnetically at low temperature. A Néel temperature of 0.5 to 1.5K (for x = 0.15) has been inferred for that ordering [26], although no clear signature of it in specific heat or magnetization was reported. One can calculate in fact that δf (T) ~ $(\mu_r(T))^{1/2} \delta\lambda_{SF}(T)$, where $\delta\lambda_{SF}(T)$ is the real value to evaluate (renamed here as the superfluid penetration depth), $\mu_r = 1 + \chi_{ac}$ is the magnetic permeability of the unscreened material (within $\lambda_{SF}(T)$), and $\delta f(T)$ is the experimentally measured frequency shift (usually taken as equal to $\delta\lambda_{SF}(T)$ within a constant). Using this expression and the measured μ_r from the DC susceptibility, Cooper shows that the extracted value of $\delta\lambda_{SF}(T)$ is consistent with the behavior found in thin films of the hole-doped cuprates ($\delta\lambda_{SF}(T)$ ~ T^2 : dirty d-wave). A very important detail, still under debate, is the fact that the susceptibility at microwave frequencies (10 GHz) might not be as significant as the zero frequency (DC) value. If one uses ReCeCuO materials with much lower susceptibility, like for Re = Pr ions instead of Nd, then one expects a much smaller effect. Our preliminary results obtained with PCCO crystals (x = 0.15) [50] do not show significant differences in the temperature dependence, and an exponential fit to the PCCO data reveals a $2\Delta_o/k_BT_c$ ~ 3.6 ratio. More work on $\lambda(T)$ in PCCO is in progress [50].

OTHER ISSUES

One very difficult issue with the electron-doped cuprates is the homogeneity of the materials. To give a good example, we can mention the determination of the Ce homoge-

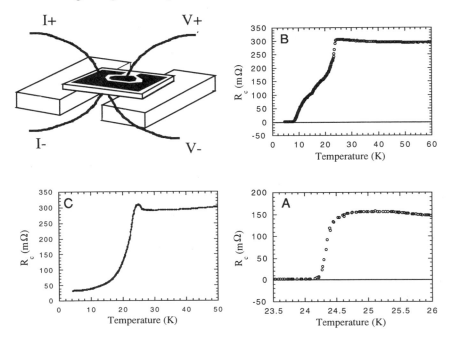

Figure 9 :   Measurement of the c-axis resistivity in NCCO single crystal. In the upper left corner, we show the usual contact configuration. In A, a very sharp transition is observed on a 10 μm thick crystal. In B, multiple transitions are observed, while in C, the material never becomes fully superconducting. The crystals in B and C have thicknesses of approximately 50 μm

neity as a function of position along the thickness (c-axis) of a thick crystal using a micron-sized x-ray beam by Skelton *et al.* [51]. The authors observe phase segregation over a thickness of the order of 100μm, indicating that the material grows layer by layer with smoothly varying Ce for Nd substitution. One realizes that, because T_c is sharply peaked around x = 0.15, the transport properties measured along the ab-plane of such a crystal might not represent exactly the properties of the material with the nominal concentration (say, x = 0.15). A prescription for further measurements is to select crystals with thicknesses of the order of 10 μm . In Figure 9, we show a clear result of inhomogeneity problems, where the c-axis resistivity ($\rho_c(T)$) is measured for crystals of different thickness. The best results are always obtained with the thinnest samples (example A of Fig. 9), whereas very often, the measured $\rho_c(T)$ of thick crystals show broad, multiple (example B of Fig. 9), and in some cases, incomplete (example C of Fig. 9) transitions. Most important, all of these crystals had a very sharp transition with no anomalies from the ab-plane resistivity ($\rho_{ab}(T)$). Brinkmann *et al.* [21] reported recently a wider doping range of SC than shown in the phase diagram of Fig. 2 following very intensive reductions (post-annealing in inert atmospheres at unusually high temperatures to remove oxygen) for thick samples (50 to 100 μm). These authors found superconductivity with $0.05 \leq x \leq 0.10$ in PrCeCuO crystals which reached values of $T_c \sim 25 - 28K$ [21]. It is likely that an inhomogeneous Ce distribution over the thickness could lead to such observations. Moreover, our latest work on PrCeCuO thin films with x = 0.05 and 0.10 post-annealed under similar conditions to Brinkmann *et al.* does not show this behavior [14]. Instead, we find that the phase diagram determined by Tokura *et al.* [16] (see Fig. 4) is very robust.

Two other questions to be answered are : 1) the impurity level in the materials and their impact on the normal-state properties (they show usually large residual resistivity) and, 2) the possible electron-hole doping symmetry of the cuprates. To answer these questions simultaneously, one wants to explore the NS transport at low temperatures below T_c, above H_{c2}, such that the samples are in the normal state. Boebinger *et al.* [52] reported resistivity measurements in the low-temperature normal state of LSCO using pulsed magnetic fields, and find a metal-insulator (MI) transition at low temperature at optimal doping (x = 0.15). In Figure 10, we show the ab-plane resistivity of PCCO thin films for several Ce contents around optimal doping. It is interesting to note that the resistivity for PCCO, x = 0.15 mimics the one observed in LSCO, x = 0.15, and even more important, that the magnitude is comparable to, /and even lower than LSCO. This suggests that the residual resistivity of the electron-doped cuprates is not as bad as one would believe, when comparing to LSCO (this means that the zero intercept of the linear resistivity in LSCO is not a good reference!), and that perhaps impurity scattering has nothing to do with the large residual resistivity : how else would one explain that for both LSCO and NCCO, the impurity scattering term is decreasing with increasing Sr or Ce content (respectively) ? The more you dope, the less impurity you get ? There are many theories that can describe the MI transition (minimum in resistivity). For example, Moshchalkov [53] suggests that the Fermi energy lies within k_BT of a mobility edge (at E_c), and that mobile carriers are obtained by thermal activation. A crossover from metallic to hopping-like resistivity is then expected at low temperatures (roughly when $k_BT = | E_c - E_F |$). This interesting avenue should be investigated more closely using other transport properties as the Hall effect and the magnetoresistance (in the normal state). One has to remark also that the electron-doped cuprates present the same metal-insulator transition of the low temperature NS resistivity at <u>optimal doping</u>. This suggest, once again, an electron-hole doping symmetry for the cuprates. In Figure 11, we show the phase diagram as determined by our measurements. The MI transition (crossover) has been defined as the temperature where the resistivity reaches a minimum.

Here, we did not mention another subtle ingredient in the complexity of the electron-doped cuprates : oxygen doping. In reality, for every single Ce content, the oxygen content in the crystal structure can be tuned to maximize the critical temperature. However, the amount removed to get a sample superconducting is very small, of the order of 0.01 per unit formula. Intuitively, this amount is expected to have little influence on the carrier concentration. Some neutron scattering experiments indicate that the excess oxygen removed during the reduction process is sitting on the apical site, usually unoccupied in the T'-structure [54]. Obviously, the presence of such interstitial oxygen is likely to distort locally the crystal structure and to disturb profoundly the electronic states in the nearby CuO_2 planes. Several recent papers have been dealing with this problem and suggest that oxygen is acting more as a scattering impurity than as a provider of additional carriers [6,7,55,56].

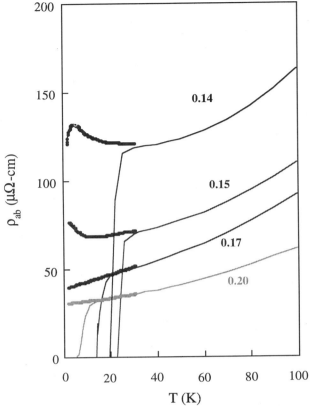

Figure 10 :   Resistivity as a function of temperature for the PrCeCuO thin films for various Ce contents. The applied field is 8.7 Tesla.

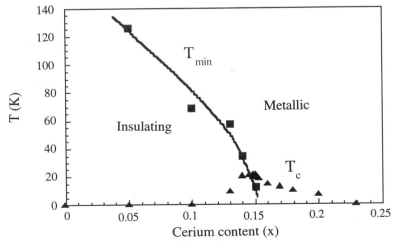

Figure 11 :   Normal-State phase diagram of the electron-doped cuprate $Pr_{2-x}Ce_xCuO_4$. The solid line is a guide to the eye.

156

SUMMARY

This work attempted to summarize very briefly the current status of some of the research on the electron-doped cuprates, in particular $Nd_{2-x}Ce_xCuO_4$ and $Pr_{2-x}Ce_xCuO_4$. Normal-state and superconducting properties were explored in parallel with the hole-doped single-layer cuprate $La_{2-x}Sr_xCuO_4$. The differences between the two systems were emphasized. One of the main conclusions related to the symmetry of the gap was the fact that there is no clear evidence of a d-wave component (even an anisotropic gap) in the electron-doped cuprates, most of the measurements being consistent with an s-wave BCS gap. However, it is important to mention that many of the experiments used to check the symmetry of the gap in the hole-doped systems have never been attempted successfully in the electron-doped materials. Most of them probably failed because of detrimental material issues. Phase-sensitive experiments should be attempted more seriously in the future to clarify the nature of the superconducting state in the electron-doped cuprates.

ACKNOWLEDGEMENTS

We would like to acknowledge the hard work of S. Darzens, F.M.Araujo-Moreira, D. Strachan, Z.Y.Li, J.L. Peng, S.N. Mao, D. Kokales, D.H.Wu, S.P. Pai, W. Jiang, X. Jiang, W. Liu, A. Smith, T. Clinton, X.X. Xi, X. Xu, S. Hagen (and all the students and post-docs of the thin film deposition group) over the many years of involvement of our group in the study of the electron-doped cuprates. Moreover, we would like to thank our faculty colleagues at Maryland for their constant inputs : C.J. Lobb, S.M. Anlage, F. Wellstood, J.W. Lynn, V. Venkatesan, H.D. Drew ; and also in Karlsruhe : G. Czjzek. This work was supported by the National Science Foundation, Division of Condensed Matter Physics, through grant DMR 9510475.

REFERENCES

*- Also at : Forschungszentrum Karlsruhe, Institut für nukleare Festkörperphysik, Postfach 36 40 , D- 76021 Karlsruhe, Germany.

1- C.C. Tsuei *et al* . , Physica C **161**, 415 (1989).

2- T. Kimura *et al.* , Phys. Rev. B **53**, 8733 (1996).

3- N.P. Ong, in *Physical Properties of High -Temperature Superconductors II*, ed. D.M. Ginsberg (World Scientific, Singapore, 1990) p. 459.

4- G. Spinolo *et al.* , Physica C **254**, 359 (1995); P. Ghigna *et al.* , Physica C **268**, 150 (1996).

5- S.J. Hagen *et al.* , Phys. Rev. B **43**, 13606 (1991).

6- W. Jiang *et al.* , Phys. Rev. Lett **73**, 1291 (1994); Phys. Rev. B **47**, 8151 (1993).

7- M. Brinkmann *et al.* , Physica C **235-240**, 1365 (1994).

8- D.H. Wu *et al.* , Phys. Rev. Lett. **70**, 85 (1993); S. M. Anlage *et al.* , Phys. Rev. B **50**, 523 (1994).

9- W.N. Hardy *et al.* , Phys. Rev. Lett. **70** , 3999 (1993).

10- Z.Ma *et al.* , Phys. Rev. Lett. **71**, 781 (1993).

11- M.B. Maple, Mat. Res. Bull. **15**, 60 (1990).

12- J. Fontcuberta *et al.*, in *Studies of High Temperature Superconductors* , ed. A.V. Narlikar, Vol. 16 (1995), p. 185.

13- J.L. Peng *et al.* , Phys. Rev. B **55**, R6145 (1997).

14- E. Maiser *et al.*, submitted to Physica C.

15- E. Maiser *et al.* , unpublished.

16- Y. Tokura *et al.* , Nature **337**, 345 (1989); H. Takagi *et al.* , Phys. Rev. Lett. **62**, 1197 (1989).

17- T.H. Meen *et al.* , Physica C **260**, 117 (1996).

18- Y.Y. Xue *et al.* , Physica C **165**, 357 (1990).

19- H. Takagi *et al.* , Phys. Rev. Lett. **69**, 2975 (1992).

20- M. Suaaidi *et al.* , Physica C **235-240**, 789 (1994).

21- M. Brinkmann *et al.* , Phys. Rev. Lett. **74**, 4927 (1995).

22- G.M. Luke *et al.* , Phys. Rev. B **42**, 7981 (1990).

23- J.E. Hirsch, Physica C **243**, 319 (1995).

24- A.T. Boothroyd *et al.* , Phys. Rev. B **45**, 10075 (1992).

25- S. Skanthakumar *et al.* , J. Appl. Phys. **73**, 6326 (1993), and references therein.

26- M.F. Hundley *et al.* , Physica C **158**, 102 (1989); S. Ghamaty *et al.* , Physica C **160**, 217 (1989).

27- B. Batlogg *et al.* , J. Low Temp. Phys. **95**, 23 (1994).

28- A. El Azvak *et al.* , J. Alloys Compd. **195**, 663 (1993).

29- A.V. Puchkov *et al.* , Phys. Rev. Lett. **77**, 3212 (1996).

30- Z. Schlesinger *et al.* , Phys. Rev. Lett. **65**, 801 (1990).

31- C.C. Homes *et al.* , Phys. Rev. B **56**, 5525 (1997).

32- P.W. Anderson, Phys. Rev. B **55**, 11785 (1997).

33- D.S. Dessau *et al.* , Phys. Rev. Lett. **71**, 2781 (1993); A.G. Loeser *et al.* , Science **273**, 325 (1996); see also the present proceedings for Campuzzano, Onellion and Margaritondo papers.

34- This is still a controversial issue, and the reader is referred to the theory papers of this summer school debating on that problem.

35- D.M. King *et al.* , Phys. Rev. Lett. **70**, 3159 (1993).

36- R.O. Anderson *et al.* , Phys. Rev. Lett. **70**, 3163 (1993).

37- A. Andreone *et al.* , Phys. Rev. B **49**, 6392 (1994).

38- C.W. Schneider *et al.* , Physica C **233**, 77 (1994).

39- T.P. Devereaux *et al.* , Phys. Rev. Lett. **72**, 396 (1994); C. Kendziora *et al.* , Phys. Rev. Lett. **77**, 727 (1996).

40- See also R.Hackl and A. Sacuto papers in these proceedings.

41- B. Stadlober *et al.* , Phys. Rev. Lett. **74**, 4911 (1995).

42- For example : Q. Huang *et al.* , Nature **347**, 369 (1990).

43- H. Yamamoto *et al.* , Phys. Rev. B **56**, 2852 (1997).

44- J.W. Ekin *et al.*, unpublished.

45- M. Covington *et al.* , Phys. Rev. Lett. **79**, 277 (1997).

46- D.J. Van Harlingen, Rev. Mod. Phys. **67**, 515 (1995).

47- S.N. Mao, *Fabrication, Characterization, and Properties of $Nd_{2-x}Ce_xCuO_{4-y}$ Superconducting Thin Films and Heterostructures*, Ph.D. thesis, U. of Maryland, College Park (1995).

48- J. Clarke *et al.* , Proceedings of *Spectroscopies in Novel Superconductors* , Cape Cod, Sept. 1997.

49- J.R. Cooper, Phys. Rev. B **54**, R3753 (1996).

50- D. Kokales *et al.* , unpublished.

51- E.F. Skelton *et al.* , Science **263**, 1416 (1994).

52- G.S. Boebinger *et al.* , Phys. Rev. Lett. **77**, 5417 (1996).

53- See the proceedings of this summer school.

54- P.G. Radaelli *et al.* , Phys. Rev. B **49**, 15322 (1994); A.J. Shultz *et al.* ., Phys. Rev. B **53**, 5157 (1996).

55- M. Brinkmann *et al.* , Physica C **269**, 76 (1996).

56- P. Fournier *et al.* , Phys. Rev. B , Dec. 1st 1997.

SUPERCONDUCTIVITY OF HEAVY-ELECTRON COMPOUNDS - COMPARISON WITH CUPRATES

H.R. Ott

Laboratorium für Festkörperphysik
ETH Hönggerberg
8093 Zürich
Switzerland

INTRODUCTION

Considering the results of low temperature measurements of the electrical resistivity, the magnetic susceptibility or the specific heat of heavy-electron materials, it is readily recognized that the characteristic features of these physical properties are distinctly different from those of ordinary metals[1]. Figure 1 is intended to demonstrate these differences schematically. First we note that all electronically dominated quantities, ρ, χ and C_p, are two to three orders of magnitude larger for heavy-electron metals than for ordinary metals. The figure emphasizes the distinct differences in the temperature variations of these properties and we briefly comment the features of heavy-electron materials shown in the right part of fig.1. The resistivity at room temperature is usually of the order of 100 $\mu\Omega$cm and more or less temperature independent. With decreasing temperature $\partial\rho/\partial T$ may be positive or negative, depending on the individual material. What is common, however, is a distinct loss of resistivity at low temperatures, sometimes by two orders of magnitude in a relatively narrow temperature regime, without a cooperative phase transition. At very low temperatures, $\rho(T)$ varies as T^2, with unusually large prefactors. Instead of a constant and rather small magnetic susceptibility, one observes a Curie-Weiss type behaviour of $\chi(T)$ due to rather well defined local moments of not completely filled atomic electron shells. At very low temperatures, the $\chi(T)$ behaviour is not universal but often reveals a tendency to temperature independence. Finally, we note that the electronic specific heat is by factors of 10^2 to 10^3 larger than for ordinary metals. The increasing C_p/T ratio with decreasing temperature is one of the essential features indicating the formation of a heavy-electron state. Again this happens without a distinct phase transition.

The phenomenon of heavy-electron formation is most often observed for intermetallic compounds where one of the constituents appearing in the chemical formula is a rare-earth or an actinide element with a partially filled 4f- or 5f-electron shell. In the rare-earth series it is either Cerium with one electron or Ytterbium with one hole in the 4f-electron shell that form the most obvious cases of heavy-electron compounds. In the

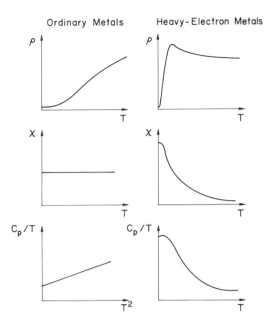

Fig. 1 - Comparison of the temperature dependences of various properties of ordinary metals and heavy-electron compounds.

actinide series such compounds certainly exist with Uranium as the cation, possibly also with Neptunium or Plutonium. There is no clear preference for the ligand atoms, as many heavy-electron compounds in combination with a large variety of metallic elements have been reported.

All of the above mentioned properties at very low temperatures are commonly ascribed to the development of anomalously large effective masses m^* of the conduction electrons in these metals as T approaches zero. Hence the name heavy-electron compounds. A large effective mass implies automatically a low Fermi velocity v_F and a small Fermi energy E_F. In this sense, the itinerant electrons that provoke the above mentioned features in these systems may also be regarded as slow.

Many of the puzzling experimental results concerning the normal state of copper oxides do not have an obvious answer when analyzed in terms of concepts that are commonly employed for describing non-insulating condensed matter. Although various experiments, mostly using angle resolved photoemission[2], seem to have identified a Fermi surface of the itinerant charge carriers in some superconducting cuprates, the theoretical debate is still on whether new schemes are needed to describe the metallic state of these oxides[3].

For both types of materials, the discovery of their superconductivity came as a great surprise. In heavy-electron materials[4,5], the superconductivity occurs under, conventionally viewed, very unfavourable conditions of inherently strong magnetic interactions. Hence it has been speculated that both the superconducting state and the mechanism inducing it might be different from those known in all conventional superconductors. As we shall demonstrate below, the superconducting state of some heavy-electron metals indeed reveals quite unusual features which may fairly well be explained by invoking an unconventional pairing of the heavy electrons. The discovery of superconductivity of copper oxides[6] is probably one of the most unexpected and, because of the magnitude of the critical temperatures T_c, one of the most exciting events in the history of condensed matter physics. The anomalous features of both the normal and the superconducting state of cuprates have led to early speculations[7] on unconventional superconductivity of these materials and many experiments have since corroborated this view[8].

PROBING UNCONVENTIONAL SUPERCONDUCTIVITY

As pointed out above, unusual superconductivity is tied to materials for which already the normal state properties are anomalous and therefore the exploration of possible unconventional superconductivity of these materials has to include investigations of the electronic properties above T_c. As usual there are a number of ways of how to probe the nature of a superconducting state. The formation of a gap in the electronic excitation spectrum of a superconductor implies that the temperature dependences of physical properties that are dominated by excitations of itinerant quasiparticles, follow fairly strict predictions. In the most simple case, which is believed to apply for conventional superconductors, the energy gap around E_F is non zero in all directions of the reciprocal space and the rate of quasiparticle excitations vanishes exponentially as the temperature drops well below T_c and approaches 0 K [9]. Thus in this range, the expected temperature variation of relevant thermal-, transport- or microscopic properties is exponential in $-1/T$, which may be verified by suitable experiments at $T \ll T_c$. Unconventional superconducting states are usually related with nodes of the gap function due to additional symmetry breaking. This implies that the density of electronic states is not simply zero within the range of the maximum gap energy but explicitly depends on energy. This in turn leads to power-law-type temperature dependences of physical parameters, again for $T \ll T_c$.

Another way is offered by searching for anomalous responses with respect to external perturbations, such as magnetic fields, high pressure or changes in the chemical composition of the materials. For conventional superconductors, the influence of these perturbations on important parameters characterizing the superconducting state such as T_c, the critical field H_c, the penetration depth λ, and others, has been studied in great detail and is believed to be fairly well understood. For example, in the phase diagram of a classical type II superconductor in the [H,T] plane, two important boundaries, the upper (H_{c2}) and the lower (H_{c1}) field are experimentally and theoretically well established. Any deviation from these conventional diagrams may be regarded as an indication for unconventional superconductivity and deserves to be thoroughly investigated.

A third and very elegant method is to probe the gap function directly. For common superconductors the sign of the gap function's amplitude does not change across the entire Fermi surface. This may not be true for more complicated pairing arrangements and intrinsic gap nodes in the form of points or lines may appear. In particular cases, these gap nodes also imply a sign change of the gap amplitude. As was first pointed out by Geshkenbein and Larkin[10], these sign changes may be identified by probing the response of specially tailored superconducting loops to external magnetic fields.

SUPERCONDUCTIVITY OF HEAVY-ELECTRON COMPOUNDS

The occurrence of superconductivity in some compounds of this type provides an unambiguous proof that the large low temperature specific heat is indeed due to thermal excitations of electronic quasiparticles with enormous effective masses m^*. This is particularly evident from fig.2, where the low temperature specific heat of UBe_{13} is plotted in the form C_p^{el}/T versus T [11]. The massive quasiparticles are manifest in the very large and, with decreasing temperature, increasing ratio C_p^{el}/T of the electronic specific heat. Upon entering the superconducting state, the entropy of these

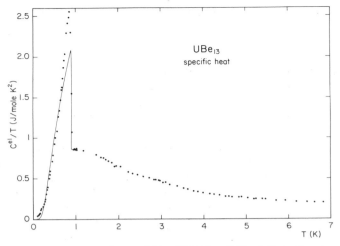

Fig. 2 - Electronic specific heat of UBe$_{13}$ between 0.06 and 7 K. The solid line indicates the universal BCS curve for the superconducting state and the discontinuity at T$_c$.

quasiparticles is lost as indicated by the anomaly of the specific heat at T$_c$ and below. The discontinuity ΔC at T$_c$ is large and compatible with the electronic specific heat parameter γ of the normal state and hence confirming that the heavy quasiparticles form the superconducting state.

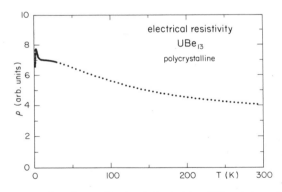

Fig. 3 - Temperature dependence of the electrical resistivity of UBe$_{13}$ between 1 and 300 K.

As may be seen in fig.3, also the electrical resistivity ρ(T) of UBe$_{13}$ below room temperature is quite different from those of common metals. No evidence for a coherent Fermi-liquid state above 1 K may be recognized. The drop of ρ(T) just below the distinct maximum around 2.5 K simply manifests the onset of some coherence. This decrease of the resistivity, still of the order of 100 $\mu\Omega$cm, is intercepted at approximately 0.9 K by the transition to the superconducting state. This is displayed in fig.4, where resistive transitions are shown in various external magnetic fields[12]. Apart from a large negative magnetoresistance, the normal state just above T$_c$ is characterized by a linear

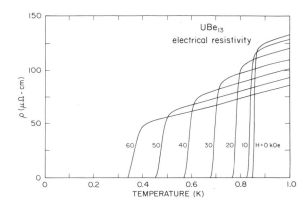

Fig. 4 - Influence of external magnetic fields on the low-temperature resistivity of UBe$_{13}$.

decrease of ρ with T, demonstrating that complete coherence has not been reached yet in this temperature range. The ρ(T) curves in fig.4 also demonstrate that the resistive transition remains fairly narrow also in external magnetic fields, contrary to what is observed in cuprate superconductors[13]. It may also be seen that the upper critical field H$_{c2}$(T) is relatively large, implying that UBe$_{13}$ is an extreme type II superconductor.

Experimental evidence for unconventional superconductivity in these materials has been accumulated over a number of years and only a few examples of relevant results of different measurements can be presented here. Various experiments aiming at providing evidence for the presence of nodes of the energy gap of the electronic excitation spectrum have been made. Most of the low temperature properties that were studied are dominated by excitations of heavy quasiparticles, lattice effects may usually be neglected. This is a clear advantage over the situation with cuprate superconductors. Although, in the latter case it is easier to fulfill the condition T<<T$_c$, necessary for this type of studies, the electronic response of these cuprates is often drowned in background effects of uncontrolled origin. Below we present a few examples where the difference between properties of the superconducting state of heavy-electron compounds and those of common superconductors or predictions of the basic BCS theory is particularly obvious.

Measurements of the specific heat C$_p$(T) should in principle pose the least problems with respect to their analysis and interpretation. In fig.5 we show the electronic part C$_p^{el}$ of the specific heat of UBe$_{13}$, well below its critical temperature T$_c$ = 0.85 K [14]. Clearly the temperature variation is not of exponential type. The C$_p$/T versus T^2 plot illustrates that the temperature dependence is close to a T^3 power law. The solid line is a fit assuming an odd parity type of pairing and point-like nodes of the gap function, a so called axial state. The calculations include the non negligible influence of resonant impurity scattering on such a state [14]. Regardless of details, the difference to simple BCS-type behaviour of the form
C$_p^{el}$ = a exp(- bT$_c$/T) is quite striking. Equally anomalous is the temperature variation of the electronic specific heat below T$_c$ of UPt$_3$ displaying two consecutive transitions separated by approximately 60 mK [15]. A T^2 decrease of C$_p^{el}$ well below T$_c$ is intercepted by an anomaly in the form of a strongly increasing C$_p$/T ratio with decreasing temperature[16,17]. The origin of this anomaly has not yet been identified.

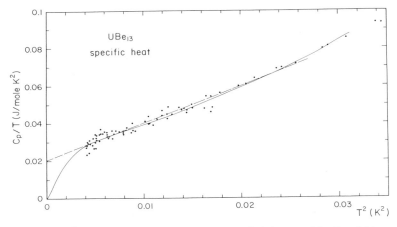

Fig. 5 - C_p/T vs. T^2 for UBe_{13} at $T \ll T_c$. The solid line is a calculation as explained in ref. 14.

Power law rather than exponential temperature dependences have been identified absorption-type experiments, such as ultrasound absorption and relaxation rates in nuclear magnetic resonance (NMR). Fig.6 shows data that were obtained on UPt_3 where it has been found that the T dependence of the ultrasound absorption coefficient changes upon simply altering the polarization of the ultrasound waves propagating in a chosen direction of the crystal lattice. This phenomenon indicates an additional broken symmetry, inconsistent with the expected features of a conventional superconducting state[18].

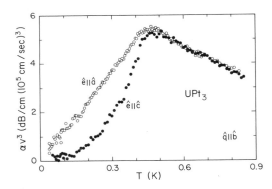

Fig. 6 - Attenuation of ultrasound in superconducting UPt_3. Note the dependence on wave polarization in the same direction of propagation (see ref. 18).

Equally anomalous is the temperature dependence that was found for the NMR spin-lattice relaxation rates T_1^{-1} in the superconducting state of heavy-electron materials. As an example we show the results of measurements made with UBe_{13} [19] in fig. 7. Apart from the non exponential decay of the relaxation rate T_1^{-1} well below T_c we note a remarkable absence of the so called Hebel-Slichter peak in $T_1^{-1}(T)$ [20] just below T_c,

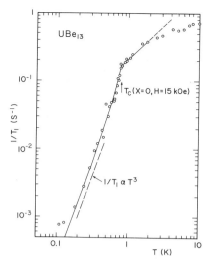

Fig. 7 -   Temperature dependence of the ^{11}B NMR relaxation
in superconducting UBe$_{13}$.

expected for conventional superconductors because of coherence-factor effects. This lack of a substantially enhanced relaxation rate just below T_c is manifest both for UBe$_{13}$ and UPt$_3$ [21], as well as for all so far investigated cuprate superconductors [22].

One phenomenon that is intimately related with the onset of superconductivity is the expulsion of magnetic flux from the bulk of the superconducting material. As is well known, the magnetic flux is not discontinuously reduced to zero at the surface of the superconductor but disappears exponentially within the so-called penetration depth. Measurements of this penetration depth, given by

$$\lambda_L = (m^* c / 4\pi \, n_s e^2)^{1/2}$$

probe directly the density n_s of the superconducting condensate, whose temperature dependence is influenced by the symmetry of the energy gap function, i.e., the presence of gap nodes. In fig.8 we present the direct comparison of the temperature variation $\Delta\lambda_L(T)$ of the heavy-electron superconductor UBe$_{13}$ with that of a conventional superconductor (Sn), both measured with the same technique and in the same reduced temperature range[23]. This obvious difference in behaviour provides clear evidence for the different symmetries of the gap functions of these two superconductors. The T^2 power law dependence of $\lambda_L(T)$ of UBe$_{13}$ is compatible with gap zeroes on points of the Fermi surface. Numerous attempts have been made to measure $\lambda_L(T)$ of cuprate superconductors. For temperatures $T \ll T_c$, the penetration depth of YBa$_2$Cu$_3$O$_7$, most likely the best studied copper oxide material, has been found to vary linearly with temperature, compatible with lines of nodes of the gap function[24].

At least circumstantial evidence for unconventional superconductivity of UPt$_3$ is obtained by mapping its superconducting phase diagram in the [H,T] plane, shown in fig.9. The phase boundaries, which have no counterparts in the [H,T] phase diagrams of conventional superconductors, have mainly be established via measurements of the

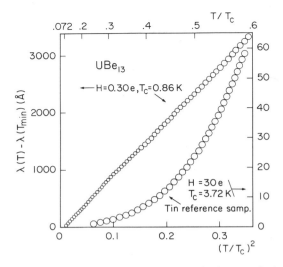

Fig. 8 - Incremental magnetic-field penetration depth of superconducting
UBe$_{13}$ and Sn as a function of reduced temperature T/T$_c$.

specific heat[25] and ultrasound absorption[26]. The exact location of the phase boundary separating the B- and the C phase depends on the orientation of the external magnetic field with respect to the c axis of the hexagonal crystal structure of UPt$_3$. Distinctly different behaviour in some physical properties has also been established for the A- and the B phase, respectively. Measurements of the local magnetization using μSR techniques[27] have indicated that the B phase might be of odd parity, i.e., a superconducting phase in which the time reversal symmetry is broken.

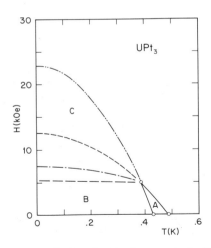

Fig. 9 - Schematic [H,T] phase diagram of superconducting UPt$_3$ (see text).

A second example of an unusual phase diagram for superconductivity has been established for U$_{1-x}$Th$_x$Be$_{13}$ by plotting the critical temperature T$_c$ versus x. In fig. 10 we display the plot that resulted from measurements of the specific heat[28]. The phase

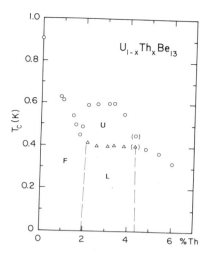

Fig. 10 - [x,T] phase diagram of superconducting $U_{1-x}Th_xBe_{13}$.

boundary between the U- and the L phase is derived from the observation of a second distinct specific heat anomaly at temperatures below the superconducting phase transition at Tc, depending on x. The vertical boundaries are dictated by thermodynamic considerations. The magnetic features of phases U and L, respectively, are quite different, as has been observed in µSR experiments. In a theoretical attempt[29] to reproduce a phase diagram of this kind, a discussion based on the Ginzburg-Landau theory of phase transitions has revealed that the assumption of odd-parity pairing leads to phase diagrams of the kind that is shown in fig.10 and it was concluded that the L phase ought to be of non-unitary character, again with broken time reversal symmetry. Subsequent measurements employing the µSR-technique gave clear support for the actual realization of this situation in ther L phase[30].

Unusual phase diagrams, in the sense demonstrated above, for the superconducting state of cuprates are much less easy to establish because in these copper oxides the superconducting state is overwhelmingly influenced by particular features of the vortex state. Investigations addressing these questions have grown into an industry of its own[31].

No reliable results of experiments probing the symmetry of the order parameter directly with phase coherence experiments using specially tailored superconducting loops, which have been so successful in the case of cuprate superconductors[32-34] and are described in the contribution of J. Kirtley et al. to this volume[35], are yet available for heavy-electron compounds. This is mainly due to experimental problems in the sense that first, it is difficult to obtain reliable tunnel contacts with heavy-electron materials and second, all experiments have to be done well below 1 K.

SUMMARY AND CONCLUSIONS

This brief review of experimental attempts to probe and identify the character of the superconducting state of heavy-electron materials should show that there is growing evidence for unconventional features of superconductivity in some of these materials.

Although heavy-electron metals and metallic copper oxides are different in many ways, they share the common aspect that both their normal and their superconducting states are still not well understood, i.e., many experimental observations are waiting for their convincing and correct interpretation.

Acknowledgement

I wish to thank many colleagues and friends for fruitful collaborations over the years, in particular E. Felder, Z. Fisk, J.L. Smith, T.M. Rice, K. Ueda, and M. Sigrist. Much of the work presented here has been supported financially by the Schweizerische Nationalfonds zur Förderung der wissenschaftliche Forschung.

References

1. see, e.g., H.R. Ott, in *Progress in Low Temperature Physics*, vol. XI, ed. D.F. Brewer (North Holland, Amsterdam 1987) p. 215
2. Z.X. Shen, W.E. Spicer, D.M. King, D.S. Dessau, and B.O. Wells, *Science* 267:343 (1995); and references therein
3. see, e.g., *Proc. of the 10th Anniversary of HTSC Workshop*, eds. B. Batlogg, C.W. Chu, K.W. Chu, D.U. Gubser and K.A. Müller (World Scientific, Singapore 1996)
4. F. Steglich, J. Aarts, C.D. Bredl, W. Lieke, D. Meschede, W. Franz, and H. Schäfer, *Phys. Rev. Lett.* 43:1892 (1979)
5. H.R. Ott, H. Rudigier, Z. Fisk, and J.L. Smith, *Phys. Rev. Lett.* 50:1595 (1983)
6. J.G. Bednorz and K.A. Müller, *Z. Phys. B* 64:189 (1986)
7. P.W. Anderson, *Science* 235:1196 (1987)
8. D.J. Scalapino, *Physics Reports* 250:331 (1995)
9. J. Bardeen and J.R. Schrieffer, *Progress in Low Temperature Physics*, vol. III, ed. C.J. Gorter (North Holland, Amsterdam 1961) p. 170
10. V.B. Geshkenbein and A.I. Larkin, *Pis'ma Zh. Eksp. Teor. Fiz.* 43:306 (1986) [*JETP Lett.* 43:395 (1986)]
11. H.R. Ott, H. Rudigier, Z. Fisk, and J.L. Smith, *Physica* 127B:359 (1984)
12. M.B. Maple, J.W. Chen, S.E. Lambert, Z. Fisk, J.L. Smith, H.R. Ott, J.S. Brooks, and M.J. Naughton, *Phys. Rev. Lett.* 54:477 (1985)
13. T.T.M. Palstra, B. Batlogg, L.F. Schneemeyer, and J.V. Waszczak, *Phys. Rev. Lett.* 61:288 (1988)
14. H.R. Ott, E. Felder, C. Bruder, and T.M. Rice, *Europhys. Lett.* 3:1123 (1987)
15. R.A. Fisher, S. Kim, B.F. Woodfield, N.E. Phillips, L. Taillefer, K. Hasselbach, J. Flouquet, A.L. Georgi, and J.L. Smith, *Phys. Rev. Lett.* 62:332 (1989)
16. E.A. Schuberth, H.B. Strickler, and K. Andres, *Phys. Rev. Lett.* 68:117 (1992)
17. J.P. Brison, N. Keller, P. Lejay, J.L. Tholence, A. Huxley, N. Bernhoeft, A.I. Buzdin, J. Fåk, J. Flouquet, L. Schmidt, A. Stepanov, R.A. Fisher, N.E. Phillips, and C. Vettier, *J. Low Temp. Phys.* 95:145 (1994)
18. B.S. Shivaram, Y.H. Jeong, T.F. Rosenbaum, and D.J. Hinks, *Phys. Rev. Lett.* 56:1078 (1986)
19. D.E. MacLaughlin, C. Tien, W.G. Clark, M.D. Lan, Z. Fisk, J.L. Smith, and H.R. Ott, *Phys. Rev. Lett.* 53:1833 (1984)
20. L.C. Hebel and C.P. Slichter, *Phys. Rev. Lett.* 3:1504 (1959)
21. Y. Kohori, T. Kohara, H. Shibai, Y. Oda, Y. Kitaoka, and K. Asayama, *J. Phys. Soc. Jap.* 56:867 (1987)
22. see, e.g., H. Yasuoka, T. Imai, and T. Shimizu, in *Strong Correlation and Superconductivity*, eds. H. Fukuyama, S. Maekawa, and A.P. Malozemoff (Springer, Berlin 1989) p. 254

23. F. Gross, B.S. Chandrasekhar, D. Einzel, K. Andres, P.J. Hirschfeld, H.R. Ott, J. Beuers, Z. Fisk, and J.L. Smith, *Z. Phys. B* 64:175 (1986)
24. see, e.g., W.N. Hardy, D.A. Bonn, D.C. Morgan, R. Liang, and K. Yhang, *Phys. Rev. Lett.* 70:3999 (1993)
25. K. Hasselbach, L. Taillefer, and J. Flouquet, *Phys. Rev. Lett.* 63:93 (1989)
26. A. Schenstrom, M.F. Xu, Y. Heng, D. Bein, M. Levy, B.K. Sarma, S. Adenwalla, Z. Zhao, T. Tokuyasu, D.W. Hess, J.B. Ketterson, J.A. Sauls, and D.G. Hinks, *Phys. Rev. Lett.* 62:332 (1989); G. Bruls, D. Weber, B. Wolf, P. Thalmeier, B. Lüthi, A. de Visser, and A. Menovsky, *Phys. Rev. Lett.* 65:2294 (1990)
27. G.M. Luke, A. Keren, L.P. Le, W.D. Wu, Y.J. Uemura, D.A. Bonn, L. Taillefer, and J.D. Garret, *Phys. Rev. Lett.* 71:1466 (1993)
28. H.R. Ott, H. Rudigier, E. Felder, Z. Fisk, and J.L. Smith, *Phys. Rev. B* 33:126 (1986)
29. M. Sigrist and T.M. Rice, *Phys. Rev. B* 39:2200 (1989)
30. R.H. Heffner, J.L. Smith, J.O. Willis, P. Birrer, C. Baines, F.N. Gygax, B. Hitti, E. Lippelt, H.R. Ott, A. Schenck, E.A. Knetsch, J.A. Mydosh, and D.E. MacLaughlin, *Phys. Rev. Lett.* 65:2816 (1990)
31. G. Blatter, M.V. Feigel'man, V.B. Geshkenbein, A.I. Larkin, V.M. Vinokur, *Reviews of Modern Physics* 66:1125 (1994)
32. D.A. Wollman, D.J. van Harlingen, W.C. Lee , D.M. Ginsberg, and A.J. Leggett, *Phys. Rev. Lett.* 71:2134 (1993)
33. D.A. Brawner and H.R. Ott, *Phys. Rev. B* 50:6530 (1994)
34. C.C. Tsuei, J.R. Kirtley, C.C. Chi, L.S. Yu-Jahnes, A. Gupta, T. Shaw, J.Z. Sun, and M.B. Ketchen, *Phys. Rev. Lett.* 73:593 (1994)
35. J.R. Kirtley et al., this volume

THE INHOMOGENEITY OF HIGH -T$_c$ SUPERCONDUCTORS

Z.X. Zhao and X.L. Dong

National Laboratory for Superconductivity
Institute of Physics & Center for Condensed Matter Physics
Chinese Academy of Sciences, Beijing 100080, China

I. INTRODUCTION

At present, the topic of inhomogeneity is getting warm. The reason lies in that inhomogeneity is associated not only with the quantities of materials (samples) but also with the mechanism of high temperature superconductivity, especially after the discovery of electronic phase separation and stripe phase.

Unlike conventional superconductor, high temperature superconductor has a rather short coherence length which is comparable with the size of unit cell and a strong two-dimensional character which is much remarkable in Bi-based superconductors[1,2], so the strong fluctuation exists in most cuprate superconductors. The more significant feature is that the critical temperature of high-temperature superconductor depends on carrier density. Either sharp transition or ideal Meissner effect hence is difficult to obtain. In early years, majority effort was paid on the quantity of sample. Now, single crystal has been successfully grown, and rich physical properties have been obtained. However, these single crystals should be called quasi-single crystals since the carriers in high-temperature superconductor come from defects. Inhomogeneity is unavoidable provided that the defects exist there. Since absolute homogeneous sample has no way to exist, the key point thus lies in that to what extent the inhomogeneity affects the experimental data. Only after well understand various inhomogeneities could people analyze and utilize the data intelligently, thereby we should know that what kind of "homogeneous" sample is obtainable. For instance, to wisely discuss the gap symmetry in high temperature superconductors, i.e., the topic of this workshop, ideal sample is required because anisotropic superconductivity is solidly sensitive to disorder as well known.

Different theory models suggested different phase diagrams. First, is there a so called spin glass part between antiferromagnetic and superconductive regions in underdoped regime? The recently proposed SO(5) theory[3] predicts a direct first order phase transition from antiferromagnetic(AF) phase to superconducting(SC) phase and a bi-critical point

The Gap Symmetry and Fluctuations in High - T$_c$ Superconductors
Edited by Bok *et al.*, Plenum Press, New York, 1998

171

where both T_N and T_c merge. Under the regime for such a low carrier density, of course it is rather difficult to synthesize a "homogeneous" sample. Hence the hinge lies in that what a homogeneous sample is capable of providing reasonable data. Another phase diagram suggested that in the non-overdoped region, even at optimal doing level, the sample will be insulate when T=0K. In normal state region, the line of pseudo gap also needs to determine. Physical measurements on "homogeneous" sample are necessary in order to both give a description in details and a precise outline of the phase diagram. Even in overdoped region, some remarkable physical features such as supercooling and electronic phase separation were observed as well. Similar question thereby arises: how homogeneous the sample should be therefore it can be guaranteed that the results from local measurement such as STM are consistent with that from statistic one for example specific heat measurement. Moreover, it is well known that high temperature superconductivity depends on carrier density. Therefore, regarding the superconductive part in phase diagram, the question, i.e., whether the dependence of T_c on carrier density is of "bell-jar" type or a step one, remains open. At least in La-214 system, several physicists have observed the step-type dependence. Both 30K "plateau" and 15K "plateau" in "homogeneous" samples were reported[4]. Very recently, Jogensen[5] reported a plateau at optimal doping region in Hg-1201 sample, and this plateau is supposed to be the reflection of the effects caused by two types of oxygen defects. In fact, precisely description of phase diagram will provide an important clue for the study on high temperature superconductivity mechanism as well.

The observation of stripe phase reported by Tranquada et al.[6] undoubtedly provided a support to some theory models. Meanwhile, new theory models thereby were put forward, for instance, Zaanan's dynamical stripe correlation model[7] and Emery's spin-paring model[8], etc. Since stripe phase revealed the intrinsic inhomogeneity of high temperature superconductors, above models suspected that intrinsic inhomogeneity is essential for high temperature superconductivity. In addition, stripe phase is also associated with electronic phase separation. So far, to solve above questions, investigations on some typical materials are expected. With regarding to chemical inhomogeneity, stripe phase and electronic phase separation, etc., inhomogeneity in high temperature superconductors certainly is a topic that requires continuously thorough investigation.

Therefore, in this paper, we will mainly discuss the typical inhomogeneous phenomena in both Bi-based and La-214 superconductors.

II. CHEMICAL INHOMOGENEITIES

It is well known that high temperature superconductors are often synthesized under somewhat complicated processes containing three or more different components, therefore, to synthesize a single-phased sample without deviation of chemical composition is rather difficult. Furthermore, the parents of high temperature superconductors are insulated antiferromagnets[9], they cannot exhibit superconductivity without enough carriers. Since the carries are major caused by either intercalating anions or substituting cations, i.e., introducing defects, it is difficult to avoid the non-uniformity of composition and anti-site among elements. The structure distortion or quenched disorder, therefore, might also be involved. The variation of oxygen content and its inhomogeneous distribution probably play a role in producing chemical inhomogeneities as well. In fact, it has been concluded that considerable disorder in the arrangement of oxygen atoms in buffer layers connecting the conducting CuO_2 planes was a feature of all the layered copper-oxide superconductors[10]. Therefore, it is understandable that chemical inhomogeneities can affect various properties of high temperature superconductors to some extent. Here, we would like to give an example of chemical inhomogeneous phenomena in Bi-based superconductors.

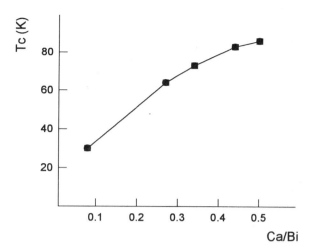

Figure 1. T_c vs. the Ca/Bi ration in $Bi_{2+x}Sr_2Ca_{1-x}Cu_2O_y$ single crystal.

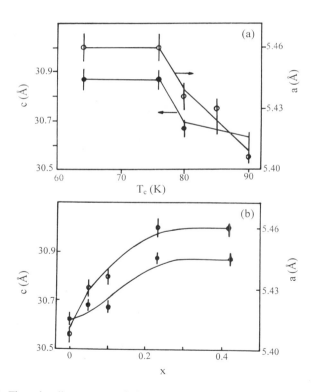

Figure 2. The unit cell parameters of $Bi_{2+x}Sr_2Ca_{1-x}Cu_2O_y$ as a function of (a) T_c and (b) x

The experiment performed on $Bi_{2+x}Sr_2Ca_{1-x}Cu_2O_y$ single crystal gave a strong evidence of composition deviation[11]. A large Bi-2212 single crystal was smashed. The structure of each piece was investigated by electron diffraction and single crystal x-ray diffraction, the corresponding composition was determined with element analysis(EDX), and the T_c was measured simultaneously. It was found that the x is not the same in different pieces, and T_c decreased rapidly with the increase of x. Meanwhile, lattice parameters were increased with the increase of x, i.e., with the decrease of T_c. These results are illustrated in Fig.1 and Fig.2, respectively. Obviously, nonstoichiometry in Bi-2212 superconductors is inevitable despite the samples are all from a single crystal, and the superconductivity directly associates with chemical composition and structural features. This is why it is supposed that the smaller or thinner the Bi-2212 single crystal is, the better it will be.

In fact, chemical inhomogeneity is undoubtedly a common phenomenon in all known high temperature superconductors. It has been experimentally ensured that the chemical inhomogeneity in Tl- and Hg-based superconductors is beyond all question, moreover, recent investigations on YBCO single crystal and thin films[12] showed that though the samples were claimed to be "optimal" considering their very good transport properties, gross structural inhomogeneities were revealed.

A question hence is posed: could the various properties drawn from the measurements really reflect the intrinsic properties of samples taking into account the existence of chemical inhomogeneity? In fact, various inhomogeneous phenomena such as incommensurate modulations, phase separation and stripe phase, all of which are associated with the unperfect CuO_2 plane and non-uniform distribution of carrier density, are also difficult to eliminate even though the chemical inhomogeneity is overcome.

III. INCOMMENSURATE MODULATIONS

One-dimensional incommensurate modulation is the common feature of Bi-based superconductors[13].

There are three major types in Bi-based $Bi_2Sr_2Ca_{n-1}Cu_nO_y$ superconductors, they are called 2201, 2212 and 2223 phases corresponding to n = 1, 2 and 3, and the T_cs are 20K, 85K and 110K, respectively.

A group in the Institute of Physics (Chinese Academy of Sciences) has done a systematic study on Bi-based superconductors. The modulated structures were determined by multi-dimensional electron crystallographic methods and/or x-ray diffraction direct method. The main results show that there are both positional and occupational modulations in Bi-based superconductors. Hence, at least, the inhomogeneity can be found in CuO_2 plane.

3.1 Bi-2223 Crystal

Bi-2223 has the highest T_c (110K) among Bi-based superconductors. The structure analysis on a Pb-doped Bi-2223 sample was carried out in 4-dimensional space[14,15]. The symmetry of the sample belongs to the superspace group P[B bmb] 1-11 with the 3-dimensional unit cell a = 5.49Å, b = 5.41Å, c = 37.1Å; $\alpha = \beta = \gamma = 90°$ and the modulation wave vector q = 0.117b*. The phases of the main reflections $0kl0$ were calculated from the known average structure, while phases of the satellites $0klm$ were derived by the multidimensional direct method. Modulation waves of all metal atoms were measured directly through calculating a Fourier map in multidimensional space.

The incommensurate modulated structure in the 3-dimensional physical space can be obtained by cutting the 4-dimensional Fourier map with a 3-dimensional hyperplane. Fig.3

is the sketch of the incommensurate modulation of the Pb-doped Bi-2223 phase in the 3-dimensional physical space projected along the a-axis. Ten unit cells of the average structure are plotted along the b-direction. In conclusion, in the Pb-doped Bi-2223 high-T_c (104K) phase, positional modulations exist for all metal atoms except Cu(1), and for some oxygen atoms. On the other hand, occupational modulations exist simultaneously for all atoms.

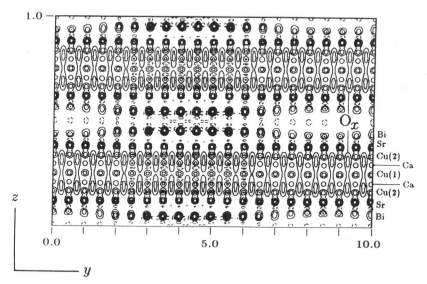

Figure 3.   Contour potential map of Pb-doped Bi-2223 phase projected along the **a** axis.

Both occupational and positional modulations are evident for Bi atoms. The strong occupational modulation of Bi implies a large number of Bi vacancies disordered on the planes normal to the b-axis. The same feature is also found for Ca and Sr atoms. However, no positional modulations are found for Cu(1). Another prominent feature that can be seen in Fig.3 is that oxygen atoms of the Cu-O layers move towards the Ca layer, forming a disordered oxygen bridge across the layers of Ca-Cu(1)-Ca-Cu(2). In addition, occupational and positional modulations along the b-axis are also found for the disordered oxygen atoms. Furthermore, some parts of the Cu(2)-Ca-Cu(1)-Ca-Cu(2) layers are bridged by disordered oxygen atoms arranged on lines perpendicular to the b axis, which obviously is an evidence of the non-uniform distribution of carriers.

Trokiner et al.[16] investigated the $(Bi,Pb)_2Sr_2Ca_2Cu_3O_y$ powder with $T_c \sim 110K$ using ^{17}O NMR spectra analysis. In the same compound, two types of CuO_2 planes with different carrier density were revealed. This is clearly consistent with the results derived from the Pb-doped Bi-2223 crystal. The conclusion was that the CuO_2 planes between Sr-O and Ca-O planes had a high density of holes, and no pronounced manifestations of antiferromagnetic correlations, while the central CuO_2 plane which embedded in the two vicinal Ca-O planes was lack of holes, and presented antiferromagnetic correlations.

3.2 Bi-2212 Crystal

Bi-2212 crystal belongs to the superspace group N [B bmb] 1-11 with a = 5.42Å, b = 5.43Å, c = 24.6Å, $\alpha = \beta = \gamma = 90°$. F.H. Li et al.[17] has obtained the [100] electron

175

diffraction pattern and structure image of light Pb-doped Bi2212 with modulation wave vector $q \sim 0.22b^* + c^*$, and the incommensurate modulation of the structure was clearly revealed by combining the electron micrograph and the electron diffraction techniques. The [100] electron diffraction data of Bi-2212 with modulation wave vector $q \sim 0.117b^*$ were collected at 200K and 40K as well, the two projected potential maps are shown in Fig.4. Both positional and occupational modulations were observed below and above T_c. However, the modulations in superconducting state became weaker. The position of Ca with a high occupancy corresponds to those of Bi also with the high occupancy in the case above T_c but corresponded to positions of Bi with the low occupancy in the case below T_c. In addition, the high occupancy of Ca always led to a displacement of two adjacent Cu-O and Sr-O layers towards the Bi-O layers in both cases above and below T_c. Besides, a prominent, and perhaps more important, feature was that the 'extra' oxygen atoms, which distributed on the Bi-O, Sr-O and Cu-O layers at 200K, moved towards the Cu-O layer at the temperature (40K) below T_c[15].

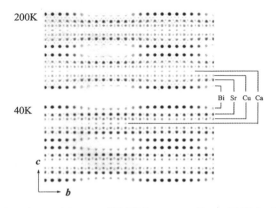

Figure 4. [100] projected potential maps of Bi-2212 corresponding to 200K (top) and 40K (bottom).

In addition, modulation in Sm-doped Bi-2212 single crystal, which was observed by single-crystal x-ray diffraction direct method[18], seemed to be weaker than that without Sm-doped sample. However, the corresponding electron density map of Cu-O layer (Fig.5) clearly showed that in Sm-doped case, the CuO_2 plane deviated from the ideal square shape. This means that perfect CuO_2 plane is hardly obtained in Bi-2212 single crystal even though the modulation is weakened.

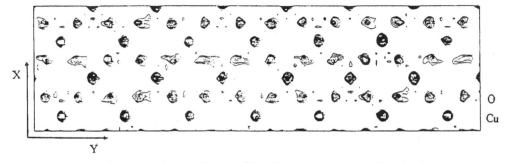

Figure 5. The electron density map of Cu-O layer in Sm-doped Bi-2212 structure.

3.3 Bi-2201 Crystal

$Bi_{2+x}Sr_{2-x}CuO_{6+y}$ is a one CuO$_2$ plane Bi-based superconductors. Its one-dimensional incommensurate modulation varies with the different composition. The c* component of the modulation wave vector decreases with the increasing of x. Fig.6 is the potential map which provides a precise structure image of Bi-2201[19], its incommensurate modulation wave vector $q \sim 0.2b* + 0.58c*$. Both positional and occupational modulations can be seen, and the trace of extra oxygen in both Bi-O and Sr-O layer is also observed.

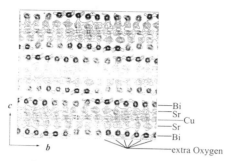

Figure 6. Contour potential map of Bi-2201 single crystal projected along the **a** axis.

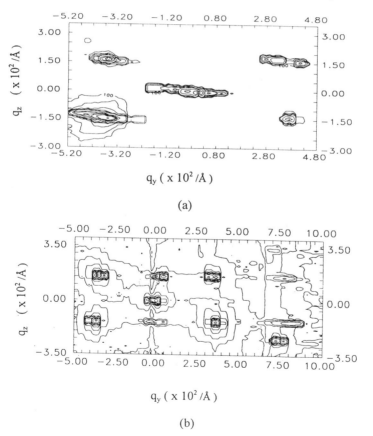

(a)

(b)

Figure 7. Reciprocal lattice mapping of 0 0 16 reflection, (a) sample BS4 and (b) BS5.

Recently, a new modulation structure associated with the superconductivity in $Bi_{2.13}Sr_{1.87}CuO_{6+\delta}$ single crystal was observed by using x-ray scattering reciprocal space mapping (plotted on a two-circle, triple-crystal diffractometer using a GX21 high-brilliance rotation-anode source with Cu K_α radiation operating at 3 kW) combined with grazing incidence diffraction (performed on a SLX-1 diffractometer using a 18kW Rigaku rotating-anode generator and Cu $K_{\alpha 1}$ radiation) under total external reflection conditions[20]. The reciprocal space mapping of 0 0 16 reflection for two samples is plotted in Fig.7, sample BS4 (Fig.7(a)) is non-superconducting while sample BS5 (Fig.7(b)) exhibits superconductivity with $T_c \sim 6K$. In Fig.7 (a) only a main reflection and four oblique first-order modulation satellites are observed. While in Fig.7 (b) the modulation has independent satellite peaks along c-axis and four new satellites align with the original first-order satellites along b-axis. The new modulation structure of the superconducting Bi2201 is two dimensional which needs two based wave vectors, $q_{//} \sim 0.19b^*$ and $q_{\perp} \sim 0.38c^*$ to describe. Therefore, there exists structural complexity and inhomogeneity in Bi-2201 crystal. Hence, it is difficult to obtain high quality single crystal.

Theoretical calculation[21] has inferred the electronic inhomogeneity in layered high temperature superconductors. By analogy with graphite intercalation compounds, it was predicted that the charge distribution of holes among the various CuO_2 planes in the unit cell was strongly inhomogeneous (for the number of CuO_2 planes per unit cell, n > 3) with a depletion in the central layers. Furthermore, the decrease of T_c at large n was supposed to result from the depletion of the hole density in the central layers. This inhomogeneity of the charge distribution among the various CuO_2 planes is suspected to influence various properties of high temperature superconductors.

From the structural and electronic details discussed above, it can be suspected that inhomogeneities associated with the non-perfect CuO_2 plane and non-uniform carrier distribution at least exists in Bi-based superconductors.

IV. PHASE SEPARATION, STRIPE PHASE AND RELATED MODULATIONS

Phase separation, a phenomenon that two or more different electronic phases coexist in the same compound, is a common feature of La-214 superconductors though it was also observed in other superconducting copper oxides. In general, phase separation in La-214 system manifests itself as the coexistence of hole-rich (superconducting) and hole-poor (antiferromagnetic) phases at lower doping level or the coexistence of different superconducting phases at higher doping level. Since it might be related to the origin of high temperature superconductivity, phase separation has become a hot topic on studying the mechanism of high temperature superconductivity[22]. Recently observed stripe phase in La-214 system has drawn remarkable interests because it was assumed to directly relate to superconductivity and phase separation.

4.1 Phase separation in La-214 System

La_2CuO_4-based superconductor has the simplest structure among high temperature superconductors. Moreover, it has extremely broad doping region and can exhibit various properties upon doping. Therefore, remarkable interests have been attracted in La_2CuO_4 system.

Upon oxygen doping either by carefully annealing processes or by high oxygen pressure synthesis or by electrochemical or chemical oxidization, $La_2CuO_{4+\delta}$ can become superconducting. At low doping level, the holes introduced by excess oxygen, unlike that doped by cations, are mobile. Therefore, with the promotion of antiferromagnetic exchange interaction, holes tend to segregate thus phase separation takes part in. In consequence, hole-rich and hole-poor domains coexist and optimal superconductivity ($T_c \sim 30K$) reaches though the average carrier density is much lower than that of the optimal doping. Extensive work has been done on this topic[22].

Z.G. Li et al.[23] studied the excess oxygen content and hole concentration in $La_2CuO_{4+\delta}$ prepared by electrochemical oxidation for $0 < \delta < 0.12$. They found that there existed two distinct sites with doping efficiency, which were described by two linear regimes: (1) $p = 2\delta$ up to $\delta \sim 0.03$; and (2) $p = 1.3\delta + 0.019$ for $0.03 < \delta < 0.11$. The change in doping efficiency from 2 holes per excess oxygen atom to 1.3 holes per excess oxygen atom occurred at a critical hole concentration $p_c \sim 0.06$, which was characterized by a sudden increase of chemical potential of doped holes. Gas effusion spectra of $La_2CuO_{4+\delta}$ for $\delta = 0.015$, 0.055 and 0.075, which revealed two oxygen peaks, demonstrated that there existed two bonding configurations of excess oxygen. Obviously, the results from gas effusion spectra are consistent with the doping efficiency data. Hence, oxygen-doped $La_2CuO_{4+\delta}$ can be thought as a typical electronic inhomogeneity.

Different oxygen bonding configurations are not the unique feature of $La_2CuO_{4+\delta}$, quite the contrary, oxygen defects are considered to be a remarkable factor that affects various properties of high temperature superconductors. For instance, Jorgensen et al.[5] investigated Tl-2201 and Hg-1201 superconductors by using neutron powder diffraction. They found that oxygen defects affected T_c. Furthermore, a plateau at maximum T_c was observed in Hg-1201. This plateau was suggested to result from the different effects caused by the two types of oxygen defects at different doping level.

Phase separation was also reported in other high temperature superconductors[22]. Photogeneration experiments[24] in $YBa_2Cu_3O_{6.3}$, however, give a direct evidence of pure electronic phase separation. Resistivity measurements showed that the photoinduced charge carriers in YBCO spontaneously became inhomogeneously distributed after photogeneration, i.e., that phase separation occurred. The phase separation manifested itself as disconnected metallic (superconducting) regions separated by insulating regions.

Generally speaking, phase separation is definitely a strong proof of intrinsic electronic inhomogeneity.

4.2 Stripe Phase and Related Modulations in La-214 System

Recently, stripe phase in CuO_2 plane and related modulations related to phase separation were observed in high-temperature superconductors[6,25-28]. They are suggested to be the consequence of charged stripes embedded in antiferromagnetic correlated spin framework, and be a reflection of the inhomogeneous distribution of carriers associated with high temperature superconductivity.

Tranquada et al.[6] reported an evidence for stripe correlations of spins and holes along Cu-O-Cu direction in the CuO_2 plane of $La_{1.6-x}Nd_{0.4}Sr_xCuO_4$ ($x = 0.12$) crystal obtained by neutron diffraction below 70K. The anomalous suppression of superconductivity at a dopant concentration near 1/8 that was first observed in $La_{2-x}Ba_xCuO_4$, i.e., the "1/8 problem", was assumed to result from a distortion of the lattice from the usual low-temperature orthorhombic (LTO) to the low-temperature tetragonal (LTT) structure. Since the horizontal charge stripes were likely pinned in the LTT phase, but not in the LTO phase, hence superconductivity might be suppressed due to the distortion from LTO phase to LTT phase. While in LTO phase, dynamic stripe correlations may exist.

The Cu-rich $La_2CuO_{4.0033}$ with La : Cu = 2:1.06 was investigated by magnetization measurements and electron microscopy technique[25]. It was found that excess Cu could promote phase separation in $La_2CuO_{4.0033}$. Considering that the excess oxygen of 0.0033 in

La$_2$CuO$_{4+\delta}$ without excess Cu is too small to phase-separate[22], it was suspected that the excess Cu^{2+} ion, which posses a spin of 1/2, could enhance the antiferromagnetic interaction, thus promote phase separation through "driving" doped holes to segregate. Meanwhile, the temperature to phase separation was raised as well. Besides the phase separation, i.e., the coexistence of superconducting phase with T$_c$ ~ 35K and antiferromagnetic phase with Neel temperature T$_N$ ~ 243K, a stripe-like structure along with Cu-O-Cu direction with period around 12Å was observed in ab-plane (Fig.8) at room temperature. After Ar-annealing at 800°C for 8 hours, most of these stripes disappeared. The above phenomena imply that both phase separation and stripes are introduced by excess oxygen, while the excess Cu enables them to be stable even at room temperature and with a small excess oxygen content. However, whether the stripe we observed is similar to that Tranquada discovered remains ambiguous.

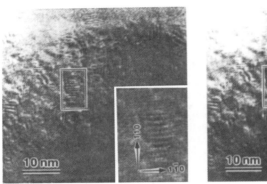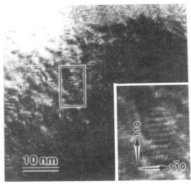

Figure 8. Room temperature lattice image of ab-plane for Cu-rich La$_2$CuO$_{4.0033}$.

In addition, a modulation of the type, q ~ 0.246b* +(1/3) c* in the orthorhombic notation at 45° from the Cu-O-Cu direction was also observed in the Cu-rich La$_2$CuO$_{4.0033}$ at room temperature[30]. The corresponding electron diffraction pattern and high resolution electron microscopy (HREM) image are demonstrated in Fig.9. Furthermore, we found that the modulation was enhanced and its wave vector slightly changed when the temperature decreased. Fig.10 is the electron diffraction pattern of Cu-rich La$_2$CuO$_{4+\delta}$ at 93K, the modulation seemed stronger than that at room temperature, and its wave vector is q ~ 0.192b* + (1/2)c*. Besides, these modulations, like the stripes in ab-plane, also disappeared together with phase separation after the Ar-annealing. These modulations are clearly associated with excess oxygen, and are suspected to be the reflection of the ordering of excess oxygen in the hole-rich regions created by phase separation. Due to the enhancement of antiferromagnetic exchange interaction by excess Cu, phase separation temperature was raised. This implies that the doping holes can segregate at higher temperature comparing with the case without excess Cu. Therefore, the incommensurate modulation observed at room temperature is understandable. Furthermore, the thermal fluctuation will be weakened when the temperature is lowered, hence the incommensurate modulation is strengthened from room temperature to 93K, as shown in Fig. 9 and Fig. 10.

Lanzara et al.[26] studied the La$_2$CuO$_{4.1}$ single crystal with T$_c$ ~ 38K by in-plane-polarized Cu K-edge extended x-ray-absorption fine structure (EXAFS), and found minority domains characterized by a long Cu-O (planar) bound length at T < 150K. The neutron scattering data and x-ray diffraction on the crystal showed a superstructure of the type, giving a one-dimensional modulation with period λ ~ 24.3Å. It was suggested that the CuO$_2$

plane was decorated by stripes of distorted lattice containing relatively trapped charge carriers with width ~9.7Å and stripes of undistorted lattice containing itinerant charge carriers with width ~14.5Å.

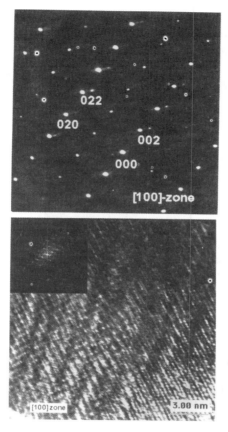

Figure 9. Room temperature [100] zone-axis diffraction pattern (top) and the corresponding HREM image (bottom) of Cu-rich $La_2CuO_{4.0033}$.

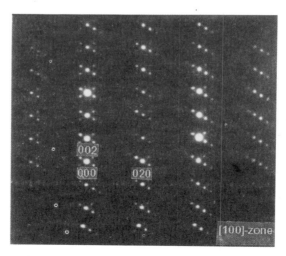

Figure 10. [100] zone-axis diffraction pattern of Cu-rich $La_2CuO_{4.0033}$ at 93K.

181

Bianconi et al.[27] measured the Cu-O distances by a local and fast probed, polarized Cu K-edge extended x-ray absorption fine structure (EXAFS) in $La_{1.85}Sr_{0.15}CuO_4$ crystal ($T_c \sim$ 35K). Two different conformations of the CuO_6 octahedra below 100K assigned to two types of stripes with different lattice were shown. The stripe width was measured to be L ~ 16Å and W ~ 8Å, respectively. The authors suggested that this experiment supported the "two component" models for the interpretation of the anomalous electronic and transport properties of cuprate superconductors.

Theories of doped Mott-Hubbard insulators[29] also suggested that the stripe phase must exist though the proposed mechanisms for producing stripe phases were not quite consistent. Therefore, both experiments and theories support that at least in the underdoping region, stripe phase in CuO_2 plane is a common feature for high temperature superconductors, which implies that charge inhomogeneity is intrinsic to the high temperature superconductors.

VI. CONCLUSIONS

Inhomogeneous phenomena in high temperature superconductors have been discussed in terms of chemical inhomogeneity, incommensurate modulation and electronic inhomogeneity. It is found that the inhomogeneity associated with the incommensurate modulation in Bi-based superconductors, the phase separation and stripe phase in La-214 superconductors is of general characteristic. In other words, non-perfectness of CuO_2 planes and/or the non-uniformity of carrier distribution are the common phenomena in high temperature superconductors. While the chemical inhomogeneity in high temperature superconductors is also hardly avoided.

In fact, it is well understood that no absolute homogeneity exists in any material. However, we believe that a criterion should be put forward in order to judge the quality of the high temperature superconductors. Following this criterion, we can decide whether the data we obtained can be attributed to the intrinsic properties. Therefore, to find out the criterion of the inhomogeneity in high temperature superconductors is absolutely of importance.

Acknowledgment We would like to thank Prof. H.F. Fan, Prof. F.H. Li, Prof. B.R. Zhao, Prof. X.F. Duan, Prof. L.M. Peng, Dr. Z.F. Dong, Dr. Y. Li, Dr. Z.H. Wan and Dr. S.F. Cui for the helpful discussions and allowing us to use their new results. We also thank Prof. E. Kaldis, Dr. K. Conder and Dr. Trokiner for sending their new results to us.

REFERENCES

1. D.M. Ginsberg, *Physical Properties of High Temperature Superconductors I*, D. M. Ginsberg ed., World Scientific, Singapore, 1989, 1

2. S. Martin, A.T. Fiory, R.M. Fleming, L.F. Schneemeyer and J.V. Waszczak, *Phys. Rev. Lett.* 60, 2194 (1988)

3. S.C. Zhang, *Science*, 275 (1997)

4. H.H. Feng, Z.G. Li, P.H. Hor, S. Bhavaraju, J.F. DiCarlo and A.J. Jacobson, *Phys. Rev. B* 51, 16499 (1995)

5. J.D. Jorgensen, O. Chmaissem, J.L. Wagner, W.R. Jensen, B. Dabrowski, D.G. Hinks and J.F. Mitchell, invited report in *the 5th International M²-HTSC conference*, to be published in *Physica C*

6. J.M. Tranquada, B.J. Sternlieb, J.D. Axe, Y. Nakamura and S. Uchida, *Nature* 375, 561(1996), and references therein

7. J. Zaanen and W. van. Saarloos, invited report in *the 5th International M²-HTSC conference*, to be published in *Physica C*

8. V.J. Emery, S.A. Kivelson and O. Zachar, invited report in *the 5th International M²-HTSC conference*, to be published in *Physica C*

9. N.M. Plakida, *High-Temperature Superconductivity*, Springer-Verlag Berlin Heidelberg 1995

10. A.W. Hewat, E.A. Hewat, P. Bordet, J.J. Capponi, C. Chaillout, J. Chenavas, J.L. Hodeau, M. Marezio, P. Strobel, M. Francois, K. Yvon, P. Fischer, J.L. Tholence, *Physica B* 156-157, 874(1989)

11. Z.X. Zhao, S.L. Jia, Y.M. Ni and J.Q. Li, *Physica C* 185-189, 651(1991)

12. S.B. Qadri, E.F. Skelton, M.S. Osofsky, V.M. Browning, J.Z. Hu, P.R. Broussard, M.E. Reeves and W. Prusseit, invited report in *the 5th International M²-HSTC conference*, to be published in *Physica C*

13. See, for examples, D.Y. Yang, J.Q. Li, F.H. Li, Y.Q. Zhou, L.Q. Chen, Y.Z. Huang, Z.Y. Ran and Z.X. Zhao, *Supercond. Sci, Technol.* 1, 100(1988); S.Ikeda, K. Aota, T. Hatano and K. Ogawa, *Jpn. J. appl. Phys.* 27, L2040(1988); Y.Matsui and S. Horiuchi, *Jpn. J. Appl. Phys.* 27, L2306(1988)

14. Y.D. Mo, T.Z. Cheng, H.F. Fan, J.Q. Li, B.D. Sha, C.D. Zheng, F.H. Li and Z.X. Zhao, *Supercond. Sci. Technol.*5, 669(1992)

15. FAN HAI-FU, WAN ZHENG-HUA, LI JIAN-QI, FU ZHENG-QING, MO YOU-DE, LI YANG, SHA BING-DONG, CHENG TING-ZHU, LI FANG-HUA AND ZHAO ZHONG-XIAN, *Multi-Dimensional Electron Crystallography of Bi-Based Superconductors*, to be published

16. A. Trokiner, L. Le Noc, J. Schneck, A.M. Pougnet, R. Mellet, J. Primot, H. Savary, Y.M. Gao and S. Aubry, *Phys. Rev. B* 44, 2426(1991)

17. F.H. Li, H.F. Fan and Z.X. Zhao, invited report in *the 5th International M²-HSTC Conference*, to be published in *Physica C*

18. Y. Li, *Doctoral Degree Thesis* (1997) and references therein

19. Z.H. Wan, *Doctoral Degree Thesis* (1997)

20. C.F. Cui, C.R. Li, Z.H. Mai, A.J. Zhu, Z.X. Zhao, J.W. Xiong, Y.T. Wang, X.M. Jiang and P.D. Hattron, preprint (1997)

21. M. Di Stasio, K.A. Müller and L. Pietronero, *Phys. Rev. Lett.* 64, 2827(1990)

22. See, for examples, *Proceedings of the Workshop on Phase Separation in Cuprate Superconductors*, K.A. Müller and G. Benedek eds, World Scientific, Singapore, 1992; *Proceedings of the 2nd International Workshop on Phase Separation in Cuprate Superconductors*, E. Sigmund and K.A. Müller eds, Springer Verlag, Berlin Heidelberg, 1993; *International Workshop on Phase Separation, Electronic Inhomogeneities, and Related Mechanisms in High T_c superconductors*, D. DiCastro and E. Sigmund eds, J. Supercond. vol.9, no.4, 1996

23. Z.G. Li, H.H. Feng, Z.Y. Yang, A. Hamed, S.T. Ting and P.H. Hor, *Phys. Rev. Lett.* 77, 5413(1996)

24. G. Yu, C.H. Lee and A.J. Heeger, *Proceedings of the 2nd International Workshop on Phase Separation in Cuprate Superconductors*, E. Sigmund and K.A. Müller eds., Springer-Verlag, Berlin Heidelberg, 17(1993)

25. B.R. Zhao, X.L. Dong, Z.X. Zhao, W.Liu, Z.F. Dong, X.F. Duan, B. Xu, L. Zhou, G.C. Che, S.Q. Guo, B. Yin, H. Chen, F. Wu, L.H. Zhao and Z.Y. Xu, *Proceedings in the 5th M2-HSTC Conference*, to be published in *Physica C*

26. A. Lanzara, N.L. Saini, A. Bianconi, J.L. Hazemann, Y. Šoldo, F.C. Chou and D.C. Johnston, *Phys. Rev. B* 55, 9120(1997)

27. A. Bianconi, N.L. Saini, A. Lanzara, M. Missori and T. Rossetti, *Phys. Rev. Lett.* 76, 3412(1996)

28. A. Bianconi, N.L. Saini, T. Rossetti, A. Lanzara, A. Perali, M. Missori, H. Oyanagi, H. Yamaguchi, Y. Nishihara and D.H. Ha, *Phys. Rev.B* 54, 4310, 12018 (1996)
29. See, for examples, V.J. Emery and S.A. Kivelson, *Physica C* 263, 44(1996); H.J. Schulz, *J. Phys. France* 50, 2833(1989); A.H. Castro Neto and Daniel Hone, *Phys. Rev. Lett.* 76, 2165(1996); J. Zaanen, M.L. Horbach and W.van Saarloos, *Phys. Rev. B* 53, 8671(1996)
30. X.L. Dong, Z.F. Dong, B.R. Zhao, Z.X. Zhao, X.F. Duan, L.-M. Peng et al., in preprint.

APPLICATIONS OF HTS TO SPACE AND ELECTRONICS

Stuart A. Wolf

Naval Research Laboratory,
Washington DC, 20375
and
DARPA/Defense Sciences Office
Arlington VA, 22203

INTRODUCTION

In this review article, I will briefly describe some of the electronic applications of HTS technology that have been supported by the DoD since 1987. Most of these applications have important military potential as well as commercial implications. These efforts that I will describe have had substantial support from DARPA. The areas in which these applications fall are space electronics, communications, radar, countermeasures, medical applications and cryo-electronics. In addition, I will describe several efforts to provide reliable, affordable cryocoolers which are an enabling technology for any superconductivity application.

THE HIGH TEMPERATURE SUPERCONDUCTIVITY SPACE EXPERIMENT (HTSSE)

The High Temperature Superconductivity Space Experiment (HTSSE) was initiated in late 1988 to focus the attention of the High Temperature Superconductivity (HTS) community on the potential space applications of the then recently discovered materials. The ultimate goal of the program was to demonstrate the viability and system advantages that might result from the use of HTS technology in space systems. In order to demonstrate these goals, the program was formulated to have three satellite launches at the device, subsystem and system level.

The HTSSE-I experiment extended from 1989 to 1992 when a fully integrated and space qualified satellite payload was delivered for integration on a

The Gap Symmetry and Fluctuations in High - T_c Superconductors
Edited by Bok *et al.*, Plenum Press, New York, 1998

185

Department of Defense satellite. A total of twenty-three different HTS devices were promised for the HTSSE-I program with only 18 delivered to NRL for electrical testing and space qualification. Fifteen of these devices, mostly microwave components, were integrated into a payload consisting of a long-lifetime cryogenic refrigerator and a space qualified microwave network analyzer which could be commanded with signals from the ground to monitor the devices while in orbit. The satellite launch, onto which HTSSE-I was manifested, did not achieve orbit and the HTSSE-I payload was lost. Despite the failure to achieve orbit, HTSSE-I experimental payload demonstrated the robustness of HTS technology and that HTS technology could yield components with electrical and physical characteristics that could provide significant improvements compared to normal-conducting technologies.

The HTSSE-II experiment, which started in 1992 is focused on the design, fabrication, testing and space qualification of advanced HTS devices and subsystems which could have significant impact on future space communications and surveillance systems. A total of thirteen deliverables were proposed for the HTSSE-II experiment. They ranged from complicated passive components such as demultiplexes to high performance analog-to-digital convertors fabricated using HTS Josephson junction devices. In the following sections, a brief summary will be presented on each type of component followed by a brief comparison of HTS technology and conventional technology performing the same functions. An artist's sketch of the HTSSE II outboard deck is shown in Figure 1 illustrating the complexity of this project. To get an idea of the scale, the A-Frame is approximately 18 inches long.

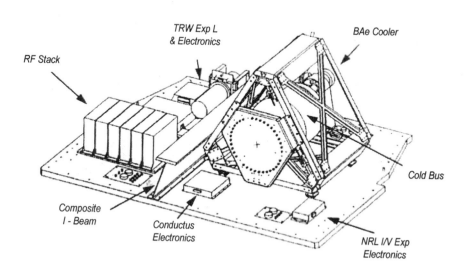

Fig 1. HTSSE II Outboard Deck

PASSIVE MICROWAVE MULTIPLEXERS AND DEMULTIPLEXERS

Three multi-channel analog demultiplexers were delivered and space qualified. For HTSSE-II, the providers had to design multi-channel filterbanks with specified center frequencies, channel bandwidths and band-pass characteristics. One system, delivered by COM DEV (Cambridge, Ontario, Canada), was a four channel dielectrically loaded multiplexer with circulator coupling. A second channelizer, built by Westinghouse Science and Technology Center (now Northrup Grumman--soon to be Lockheed Martin) was a four channel, thin film edge coupled resonator filterbank using thin-film hybrid couplers. The final demultiplexers, built by the Naval Research Laboratory, consisted of four channels of thin-film, end-coupled log-periodic resonators, a common thin-film input manifold and GaAs MMIC mixer chips to form the equivalent of a front end of a channelized receiver.

For the same volume, superconducting resonators will have 10 to 30 times higher Q-values. For resonators with the same Q-values, superconducting resonators will be 10 to 100 times smaller in volume than normal conducting resonators (without cryocooler).

DELAY LINES

The very low insertion loss of superconducting technology can be used to provide very wide signal bandwidth, very low insertion loss delay lines. In the HTSSE-II program, Westinghouse Science and Technology Center (now Northrup-Grumman, soon to be Lockheed-Martin) proved two delay lines whose total delay was 44 nanoseconds. (HTS delay lines with delays exceeding 100 nanoseconds have been reported in the open literature.)

Superconducting delay lines are physically compact with very LOW insertion Loss and high linearity and large dynamic range, much larger than can be achieved with delay lines of the same volume but using conventional metalization techniques.

LOW NOISE RECEIVER AND DOWNCONVERTERS

Another deliverable to HTSSE-II was a low noise receiver and downconvertor built by NASA Jet Propulsion Laboratory (Pasadena) and NASA Lewis Research Center (Cleveland). The receiver consisted of an HTS pre-selector filter, a two-stage GaAs amplifiers, and a GaAs mixer with a HTS resonator-stabilized local oscillator. The input signal was near 7.4 GHz and the IF near 1 GHz. This type of receiver built with conventional technologies operating at room temperature would have a noise temperature near 154 K while the HTS version delivered to HTSSE-II exhibited a noise temperature of 44 K with the potential of achieving noise temperatures as low as 32 K.

HTS hybrid receivers and downconvertors can provide very LOW noise performance which can increase data rates of communications systems by a factor of two or more or increase the range of surveillance systems by at least a factor of four.

CUEING TYPE RECEIVERS

In many surveillance and communications systems used in space, there is need for receivers which can identify the frequency of incoming signals. One approach for identifying the frequency components of incoming signals is a cueing receiver which has a frequency-dispersive element for resolving the incoming signal into its frequency components. The HTSSE cueing receiver had an effective signal bandwidth of 3 GHz, limited by the characteristics of the GaAs circuits employed in the back-end of the receiver.

Another form of frequency-dispersive receiver is known as a Digital Instantaneous Frequency Measurement (DIFM) System which uses a series of HTS delay lines, whose relative delays are precise multiples of a basic unit to provide information about the incoming signal. The number of delay lines and the precision of their delays determine the resolution of such a system.

Hts receivers using low-loss delay lines can be used to produce frequency sensitive equipment with much larger signal bandwidths and larger dynamic range than can be achieved with comparable conventional technologies.

HTS DIGITAL CIRCUITRY

Superconductivity can provide very high speed, very low power dissipation signal processing circuits such as analog-to-digital (A/D) convertors, digital multiplexers etc. which are of interest in a variety of satellite systems. In 1992, when contracts were being awarded for HTSSE-II deliverables, a contract for an A/D convertor was awarded to Conductus, Inc. (Sunnyvale, CA) and another one for a high-speed digital multiplexer to TRW (Redondo Beach, CA). Unfortunately, the technology has not matured as rapidly as had been hoped and, at this time, there still is no manufacturable HTS device technology, that is, a technology in which the characteristics of the devices and circuits are predictable, controllable, and can be fabricated with very tight margins and acceptable yields.

The analog-to-digital convertor was much too complex a circuit to build at that time and the effort was terminated. The digital multiplexer that was to be provided by TRW, was drastically reduced in complexity and a very simplified digital multiplexer circuit, mounted on a small Oxford-heritage cryogenic refrigerator also developed by TRW, was delivered to HTSSE-II.

HTS digital technology has the potential to have a major impact on space systems. However, a controllable, reproducible, and manufacturable device technology must be developed before such digital circuits become available for possible insertion into space

PACKAGING, INTEGRATION AND SPACE QUALIFICATION

Once the above equipment were received at NRL, they were tested electrically and, they were space-qualified. The main cold bus assembly was mounted on the HTSSE-II experimental deck and then interfaced with a British Aerospace Oxford cryogenic refrigerator (the so-called BAe "80-K" cooler) via a flexible thermal clutch. In addition, the ambient temperature electronics for supplying the incoming signal to the various HTS devices and to process the output signals from these devices were mounted on the deck. The TRW digital multiplexer integrated with the TRW cryogenic refrigerator was also mounted on the deck and interfaced with the signal routing circuits and the common power supply. This assembly was tested and space qualified.

The HTSSE-II payload is will be integrated on the Advanced Research and Global Observation Satellite (ARGOS). When in orbit and in sight of an NRL ground station, a signal will be sent up to the satellite, it will be received by an antenna and then supplied sequentially to the HTS devices. The output signals will be amplified, digitized and then sent back to the ground station, a so-called "bent-pipe" configuration commonly employed in satellite systems. Once HTS technology has a record of on orbit operation, spacecraft designers will be more willing to consider insertion of this technology into future operational satellite systems.

CRYOGENIC RADAR

One of the most pressing problems faced by the Navy is the threat of cruise missiles launched from a hostile shore. In this littoral environment, the clutter that is associated with various land masses can mask the missile and render it nearly impossible to detect. The very loss surface resistance of superconducting materials and the complementary low loss of high quality sapphire have enable the development of a highly stable reference oscillator (STALO) that can enhance lower the phase noise floor of a ship defense radar by a significant amount enough so that it is projected that a cruise missile can be picked out of the doppler clutter. The reason for this is illustrated in Fig. 2 which demonstratges that the radar return from large land masses is has frequency noise addedto it by the phase noise of the STALO. This in turn rasies the overeall noise floor burying the targets in the clutter. The low phase noise of the superconducting resonator allow the targets to be detected. A STALO built by Northrup-Grumman (soon to be Lockheed-Martin) with these characteristics is currently under test with a specially upgraded prototype ship defense radar. Preliminary results indicate about an order of magnitude improvement in the ability of the radar to detect small signals in a highly land cluttered background. This is a major breakthrough for HTS technology. In addition to the STALO a switched filterbank will eventually be incorporated at the front end, behind the dish to filter any unwanted signals from overloading the radar. This can be very effective against a jammer attempting to defeat the radar set.

The reduction in the noise floor of the radar with a superconducting front end puts an additional burden on the A/D converter that is essential in any radar to digitize the waveforms for processing. In fact the measured improvement would have been much better had there been an better A/D converter in the test radar. Superconducting digital technology offers a unique solution for the very high resolution A/D converters that are being targeted for this application (20 bits at 100 MHz). This will allow the full dynamic range of the radar to be processed.

Recently it was realized that a very low phase noise waveform generator can be built using superconducting technology. This waveform generator will allow the production of many single tones as well as the production of complicated waveforms that will significantly enhance the capabilities of advanced radar sets and push superconductivity into other modalities of radars. This may be one of the most important military applications of superconducting electronic technology.

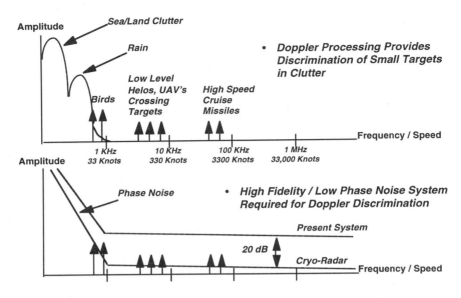

Fig. 2 - Why low phase noise is important for clutter rejection near land masses.

CRYO-COMMUNICATIONS

Interference in communications is a significant problem for the military as well as commercial communications. The ability to build very low loss, extremely high performance filters will push superconductivity into many markets. One of the first commercial applications of superconducting filters and cryogenic low noise semiconducting amplifiers will be for cellular and PCS systems. Several superconductivity vendors have demonstrated system performance that significantly exceeds the performance of conventional filters. The remaining issues are reliability and cost. The ability to reduce the interference from competing systems has very significant military implications as well. There are a plethora of military communication systems spanning many decades of frequency space. Often these systems are co-located and they can and <u>do</u> interfere with each other since small non-linearities present in one system cause spillover into a neighboring band causing interference. Superconductivity offers a the ability to provide excellent filtering on small, mobile platforms as well as on aircraft and ships. This interference problem exists within all of the military services and often is very severe. Recently, a prototype communication receiver utilizing HTS filters in the front end has been deployed by the services. The preliminary reports from these field tests are extremely encouraging and these superconducting front ends may be part of a major upgrade of the communications receivers. Also, several cellular providers have begun to deploy superconducting filters and cryogenic low noise amplifiers in some of their more suburban locations to enhance the performance of these base stations without having to expand the number of base stations. This may expand into significant business for the HTS vendors in the near future.

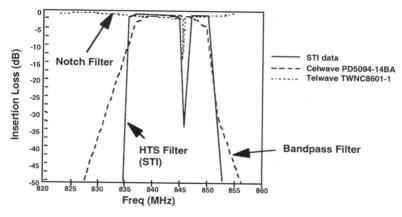

- **Receive channel:**
  - High Qs (25,000 to 40,000) provide sharp skirts, notch
  - Low insertion loss
- **Transmit channel:**
  - Low intermodulation products
- **Cryocooler can accommodate many filters**

**Fig. 3. Comparison of Conventional Cavity Filter
with Superconducting Filters**

COUNTERMEASURES

Many of our advanced aircraft suffer from an interoperability problem between the radar and the radar warning receivers. The newest generation of radars hop in frequency so that they avoid being jammed. However the radar warning receivers that are currently in operation do not have the ability to notch out the rapidly hopping radars and are thus overloaded by the radar on the same platform or from a neighbor in the formation. Again, superconductivity provide a very convenient solution to this problem. Very compact and switchable HTS filters can be built that can effectively notch out the frequency hopping radar and enhance the survivability of the aircraft. Simulations show that many more types of missions can be flown if this capability becomes operational. Currently there are programs targeting upgrades for the B1-B and the F-15.

MEDICAL APPLICATIONS

Superconducting pickup coils can reduce the noise inherent in nuclear magnetic resonance systems. When the signal to noise ratio is limited by the instrument rather than the sample than superconductivity can play a role. MRI applied to limbs or breasts are an example of when the signal is limited by the pickup coil rather than the object being measured. For whole body MRI, it is usually the body that is the dominant noise source. Thus, in the former case a superconducting pickup coil can enhance the quality of image or reduce the time necessary to produce an image. For NMR spectroscopy and or pathology superconductivity can also significantly enhance the performance in the same way. Either smaller samples can be measured or they can be scanned more rapidly.

CRYO-ELECTRONICS

Early in the DARPA (ARPA) HTS program, a study pointed to an opportunity to utilize superconducting interconnects on a multi-chip-module (MCM) to provide an unprecedented level of interconnection between very high performance semiconductor chips. A program was established to develop all of the required infrastructure to build such a superconducting MCM. This program was a true precompetitive consortium involving many of the HTS vendors, universities and other labs. It reached a significant level of maturity, however the missing link was the semiconductor chips to operate at 77 K and the cryo-cooler. The effort punted and now is focused on the low temperature optimization of the semiconductor chips. In this regard a potential breakthrough was uncovered which, if realized would allow the development of a new type of FET which would be highly optimized a cryogenic temperature and would offer an order of magnitude higher speed at two orders of magnitude lower power. Thus even if the power to cool the chip were included there would still be a significant overall gain.. If this all comes to fruition then the opportunities presented by superconducting interconnects will again come to the forefront.

CRYO-COOLER DEVELOPMENT

The cryocooler development that has been sponsored by DARPA is extremely important for the future of superconducting electronics. Without reliable, affordable cryocoolers it would be impossible for superconducting or cryo-electronic systems to be commercialized and it would make the military applications more difficult in the current military procurement environment where

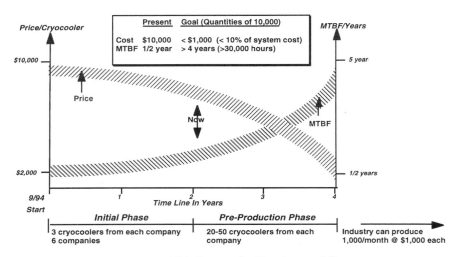

Fig. 4. DARPA/NRL Cryocooler Development Program

leveraging the commercial marketplace is considered highly desirable if not essential. In this regard DARPA has sponsored a program run through NRL that has contracted with six of the major cooler manufacturers to develop a series of cryocoolers that have greater than a three year mean time to failure, will cost about $1000.00 and are targeted at several applications including the ones that have been mentioned above. This program started about one and a half years ago and several of the vendors are in the process of delivering the first set of improved coolers.. The cost goal seems to have been too ambitious for many of the vendors but significant cost reductions over the current state-of-the-art coolers will certainly be achieved. A new effort has just begun to further the cryocooler developments as well as to expand into the area of thermoelectric cooler development. This thermoelectrics program will initially focus on materials improvements with the goal of finding new compounds or structures that will enable thermoelectrics to provide cooling for superconducting electronics.

ACKNOWLEDGMENTS

This article would not be possible without the efforts of APD Cryogenics, Com Dev, Conductus, Cryomech, CTI, Dupont, Hughes Aircraft, Lincoln Laboratory, Loral, MMR, NASA-Lewis, NRL, JPL, SAI, SCT, STI, TRW, Westinghouse, Wright Laboratory, and many others. Some of these institutions have not been specifically named in the places where they have made major contributions when the description was meant to be generic.

PHOTOEMISSION AS A PROBE OF SUPERCONDUCTIVITY

G. Margaritondo

Institut de Physique Appliquée, Ecole Polytechnique Fédérale,
CH-1015 Lausanne, Switzerland
and
Sincrotrone Trieste SCpA, Trieste, Italy.

FOREWORD

Photoemission, since the advent of high-temperature superconductors, has become a leading experimental technique in superconductivity. Most superconductivity researchers, however, are still unaware of its background (beyond the most elementary version of the Einstein model) -- and therefore unable to fully understand its results and their fundamental implications. In order to remove this obstacle and pave the way to more specialized presentations, we present here a simple discussion of photoemission techniques for superconductivity research. The discussion also deals with the remaining conceptual problems of this approach.

INTRODUCTORY REMARKS

Photoemission techniques emerged, in the early days of high-temperature superconductivity (HTSC), as very effective probes of this novel phenomenon[1,2] Its impact has steadily increased, as demonstrated for example by the contributions presented at this school. But one difficulty remains: such techniques are not widely understood, and some of the aspects relevant to HTSC are not fully appreciated even by specialists.

It is thus desirable to present a background discussion, to understand and exploit the most recent photoemission achievements in HTSC: such is the objective of this short review. Note that our presentation is not an overview of recent results. Such results are indeed presented by other authors at the school, and we invite the readers to consult their articles.[3,4]

In many cases, the knowledge of photoemission is limited to a primitive version of the Einstein-Fermi model.[5,6] This is an excellent starting point, but not sufficient to understand the use of photoemission in HTSC physics. The problem is the collective nature of the superconducting state.

In fact, the simplest description of the photoelectric effect[5,6] is based on three individual particles: the photon, the electron inside the system, and the free electron or photoelectron. The Einstein equation:

$$E(\mathbf{k}) = E_i(\mathbf{k_i}) + h\nu - \Phi , \tag{1}$$

relates the energy E of the free photoelectron to the sum of the photon energy $h\nu$ and of the energy E_i of the electron inside the system; $\mathbf{k}$ and $\mathbf{k_i}$ are the k-vector outside and inside the solid and Φ is the work function of the material. Note (Fig. 1a) that Eq. 1 adopts the Fermi level E_F as the reference energy for the E_i-scale, and the vacuum level $E_V = E_F + \Phi$ as the reference energy for the E-scale.

Suppose now that the photoelectron-emitting material is a metal. It is quite clear from Eq. 1 and Fig. 1a that photoelectrons can have energies up to $E = E_F + h\nu - \Phi$. In a plot of the energy distribution of the photoelectrons (integrated over all emission directions), there will be an edge at the kinetic energy $E_F + h\nu - \Phi$. Note, by the way, that in plotting the photoelectron energy distribution, one normally does not use the kinetic-energy E-scale, but the "binding energy" E_i-scale; thus, the edge is at $E_i = 0 = E_F$ (see Fig. 1a).

Suppose now that the metal becomes a superconductor, and that one can describe what happens in terms of the BCS (Bardeen-Cooper-Schrieffer) theory. As it is known, the BCS theory explains conventional superconductivity by assuming a phonon-mediated interaction that results in the formation of Cooper pairs. However, many of the BCS predictions are interaction-independent, and can then be used even without knowing the nature of the interaction that leads to superconductivity.

One the most relevant of such predictions is shown in Fig. 1b: the individual-particle energy spectrum is drastically modified in a region near the Fermi energy E_F, whose magnitude Δ is related to the basic energy kT_c (in the BCS theory, $\Delta \approx 3.5\ kT_c$; T_c is the critical temperature). Whereas in the metallic state the spectrum as a function of E_0 continuously extends up to $0 = E_F$, for the superconducting state a 2Δ-wide forbidden gap is created, with its center at E_F (the lower portion of the gap is visible in Fig. 1b)

This explains the interest in photoemission experiments: after measuring the energy spectrum of the emitted electrons, i.e., their flux as a function of the energy, one derives using Eq. 1 the individual-particle (occupied-state) energy spectrum. For the metallic state, the spectrum is extended up to E_F. For the superconducting state, no signal is detected for the E_i-region closer than Δ to E_F (see Fig. 1b)

Thus, a comparison of photoemission spectra taken in the normal and superconducting state reveals the opening of the gap. Furthermore, it can be used to quantitatively analyze subtle properties of the gap, and test the corresponding predictions of different models of HTSC.[1,2]

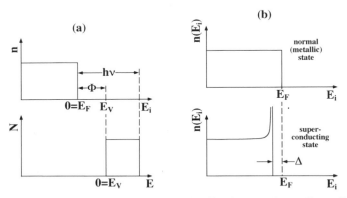

Figure 1. (a) Simple energy scheme for the photoelectric effect in a metal according to Eq. 1. The upper part shows the individual-particle energy distribution $n(E_i)$ of electrons in an electron gas near the Fermi level, E_F. E_i = electron energy inside the material; E_V = vacuum level; $h\nu$ = photon energy; Φ = work function. In the lower part, the photoelectron energy distribution $N(E)$ as a function of E, the photoelectron kinetic energy. In the usual practice, N is plotted as a function of E_i rather than of E. (b) Electron energy distribution for a normal state and for a superconducting state; note the opening of the superconducting gap, and the consequent empty region Δ below the Fermi level.

The feasibility of this approach primarily depends[1,2] on two factors:

1.   Sufficient energy resolution: most photoemission experiments are performed with limited energy resolution not exceeding 100 meV, and cannot detect effects on a small scale like Δ. For a conventional (low-temperature) superconductor, Δ does not exceed a fraction of a millivolt, thus photoemission studies of the gap are exceedingly difficult. For HTSC materials, Δ is typically of a few ten millivolt. Photoemission studies become feasible with a good resolution, typically in the 10-20 meV range. This requires the combination of a good resolution of both the photon beamline and the electron analyzer.

2.   Good surface quality: photoemission is a surface phenomenon, quite sensitive to even small levels of contamination.[5,6] In fact, photons can penetrate quite deeply in the solid, and lose their energy over distances including many atomic planes. But excited electrons can travel only for a short distance (of the order of the interatomic spacing) before losing at least part of the energy absorbed from a photon. Thus, only excitations occurring near the surface can produce the emission of photoelectrons. Thus, typical HTSC gap studies by photoemission must be performed on high-quality single crystals, cleaved and kept under ultrahigh vacuum to slow down contamination.

If these two requirements are met, reliable photoemission spectra can be taken.[1,2] But their interpretation is still problematic, because of the conceptual problems discussed in the next section.

THEORETICAL BACKGROUND

Angle-Resolved Photoemission

The practical experience with photoemission analysis of the superconducting gap[7] indicates that its most attractive feature is k-resolution. In fact, a photoemission experiment[5] can analyze the emitted photoelectrons one by one, and identify the direction of emission for each of them with an angle-resolved analyzer. This is perhaps the most important advantage of photoemission with respect to more traditional superconductivity-research techniques such as tunneling.

Angle-resolved photoemission, however, is not a straightforward technique. When a photoelectron is captured by an angle-resolved energy analyzer, its (free-electron) quantum state $|\mathbf{k}\rangle$ is completely identified by measuring the magnitude and direction of the k-vector. But the problem is to convert this information into k-resolved information for the electron state *inside* the solid.

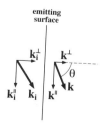

Figure 2. Link between the k-vectors inside and outside the photoelectron-emitting surface, Note the conservation of the parallel component.

This problem is not only difficult to solve, but even difficult to define, since the translation symmetry conditions for defining a three-dimensional k-vector are not necessarily present for the electron-emitting system. If such conditions do exist at least in the directions parallel to the electron-emitting surface, then one can - under suitable assumptions and approximations - relate the measured photoelectron k-vector **k** to the "internal" k-vector $\mathbf{k_i}$.

For example (see Fig. 2), assume that in the directions parallel to the photoelectron-emitting surface, the Bloch theorem is valid. In the same parallel direction the k-vector is conserved during the photoemission process -- whereas this is not true in the perpendicular direction. The parallel k-conservation implies that umklapp processes involving G-vectors of the bulk crystal are taken into account by reduction to the first Brillouin zone . For a complete analysis, the possibility of umklapp processes involving surface G-vectors (e.g., due to surface reconstruction) must also be considered.[5]

Neglecting this possibility, **k** and $\mathbf{k_i}$ are linked by:

$$k_i{}^{\parallel} = k^{\parallel} = k \sin q = [(2mE)^{1/2}/\hbar] \sin\theta \; ; \tag{2a}$$

$$k_i{}^{\perp} \neq k^{\perp} \; . \tag{2b}$$

In three dimensions, one must find the relation between the perpendicular components $k_i{}^{\perp}$ and $k^{\perp}$ with adequate empirical approaches.[5] In the case of essentially two-dimensional materials like the HTSC's, one usually confines[5] the analysis to the parallel components $k_i{}^{\parallel}$ and $k^{\parallel}$, for which a straightforward link is given by Eq. 3a.

This approach has been widely used, for example, to investigate band structures -- which are the $E_i(\mathbf{k_i})$ dispersion curves.[5,6] Suppose that the same, simplified approach can be extended to HTSC materials. In the normal state, the $E_i(\mathbf{k_i})$ dispersion curves derived with Eq. 3 must include at least one band that crosses E_F (see Fig. 3), thereby guaranteeing the metallic character of the system.

At temperatures below T_c, however, the energy region closer than Δ to E_F becomes forbidden and the angle-resolved spectra reflect the presence of the superconducting gap. Are the corresponding photoemission measurements of Δ equal in all directions? The standard BCS theory would give a positive answer, since its features - in particular the gap Δ-parameter - are direction-independent. This is the so-called "s-wave" superconductivity.

Most theorists, however, do not believe that s-wave superconductivity applies to HTSC materials.[7,8] Many HTSC theories are based on anisotropic gaps -- for example, "d-wave" superconductivity. The corresponding Δ-parameter depends on the direction, and this dependence can be observed by measuring the gap with angle-resolved photoemission spectra.

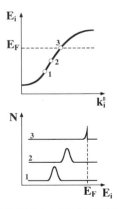

Figure 3. Top: in a metal. there must be at least one band crossing the Fermi level. Bottom: angle resolved photoemission spectra reflecting the points 1, 2 and 3 in the band, with the Fermi-level crossing.

This approach has been widely used to analyze the gap of Bi-containing HTSC materials.[7-13] The results are clearly inconsistent with the hypothesis of an entirely isotropic s-wave gap.

A similar approach can be used gaps related to phenomena other than superconductivity.[14,15] Figure 4, for example, schematically illustrates the changes in the angle-resolved photoemission spectra, induced by a metal-insulator transition related, for example, to the creation of charge-density waves. Such phenomena, of course, may prove to be relevant to HTSC.[14]

Collective Phenomena

In a sense, the analysis adopted in the previous sections is self-defeating. In fact, it is based on the notion that the electrons inside the system behave like individual particles. But superconductivity is precisely due to interactions between electrons, so the approach could appear groundless. Similarly, strong interactions could invalidate the individual-particle analysis for the normal state.

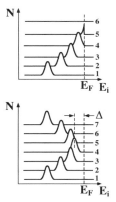

Figure 4. Difference between the angle-resolved photoemission spectra of a specimen before (top) and after (bottom) a metal-insulator transition. After the transition, there is a gap Δ below the Fermi level.

As we will see, this negative conclusion is not necessarily valid. But the analysis cannot be kept at the naïve level of Eq. 1. A complete description of the photoelectric effect for a system of N interacting electrons can be found, for example, in the literature cited by Ref. 15. We will qualitatively outline here the most important conclusions.

Essentially, angle-resolved photoemission measures the probability $P(E_i,\mathbf{k}_i)$ of observing at the detector a photoelectron with energy E and k-vector $\mathbf{k}$, related to an electron with energy E_i and k-vector $\mathbf{k}_i$ inside the solid. Such a probability is proportional to the square of the matrix element M for the optical excitation and to the so-called spectral function $A(E_i,\mathbf{k}_i)$:

$$\text{Angle-resolved photoemission spectrum} \approx |M|^2 A(E_i,\mathbf{k}_i) . \tag{3}$$

In turn, the spectral function $A(E_i,\mathbf{k}_i)$ is proportional to the imaginary part of the Green function $G(E_i,\mathbf{k}_i)$ for the process that creates the photoelectron.

The simplest case is photoemission from a gas of non-interacting electrons -- the so-called Fermi gas. In that case, $A(E_i,k_i)$ is a delta function:

$$A(E_i,\mathbf{k_i}) = \delta[E_i(k_i) + h\nu - \Phi - E] \,, \tag{4}$$

corresponding to the Einstein Eq. 1.

An angle-resolved photoemission spectrum consists then in the delta-function corresponding to the initial state of the extracted electron in the solid, as schematically shown in Fig. 5a. An angle-integrated photoemission spectrum corresponds to the integral of the delta-function of Eq. 4 over all k-vectors, which corresponds by definition to the density of states.

Thus, except for the modulating role of the matrix element in Eq. 3, an angle-integrated photoemission spectrum of a Fermi gas shows the density of occupied states. At $T = 0K$, the spectrum has a discontinuity at $E_i = E_F$, the Fermi edge, and is zero at higher energies. The same discontinuity can be seen in the so-called momentum distribution function $n(k_i)$, as shown in Fig. 6a.

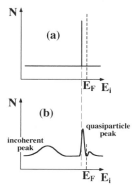

Figure 5. (a) Angle-resolved photoemission spectrum of a Fermi gas, consisting of a delta-function. (b) Deformation of the same spectrum due to electron-electron interactions, as described by Landau's Fermi liquid model. The delta-function becomes a quasiparticle peak with finite width and reduced intensity, and an incoherent peak appears at lower energies.

One consequence of this individual-particle picture is that the elementary excitation left in the system after emitting the photoelectron is a "hole" with energy determined by that of the initial electron state. Due to the lack of interactions, the hole has an infinite lifetime, consistent with the infinitely narrow spectral function -- indeed, the delta-function of Eq. 4 and Fig. 5a.

. What happens when the electron-electron interactions are taken into account? First of all. it is no longer true that the angle-integrated spectrum shows the density of states. Second, the spectral function is deformed and has a finite linewidth, consistent with a finite lifetime. Third, the discontinuity of $n(k_i)$ decreases and tends to disappear for very strong interactions.

The simplest way to analyze these effects is Landau's Fermi-liquid model.[15] The theory is valid for relatively weak interactions, and its major feature is that it retains the notion of individual particles, replacing the free electrons of the Fermi gas with quasiparticles.

Assume that the electron-electron interaction corresponds to a (complex) electron self energy $\Sigma(E_i,\mathbf{k_i}) = \Sigma_R(E_i,\mathbf{k_i}) + i\Sigma_I(E_i,\mathbf{k_i})$. Then, according to the Fermi-liquid approach, the spectral function becomes:

$$A(E_i,\mathbf{k_i}) = (1/\pi)\ \frac{-\Sigma_I}{[E_i(k_i) + h\nu - \Phi - E - \Sigma_R]^2 + \Sigma_I^2} \,. \tag{5}$$

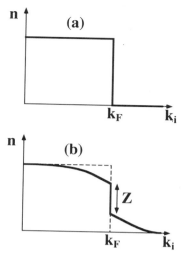

Figure 6. (a) Momentum distribution function of a Fermi gas; (b) the same function taking into account electron-electron interactions; the reduced step height is equal to the renormalization parameter Z.

Note that for $\Sigma(E_i, k_i) = 0$ Eq. 5 becomes the delta function of Eq. 4, and one finds again the Fermi-gas limit. But when $\Sigma(E_i, k_i) \neq 0$, there are significant modifications as schematically shown in Fig. 5b. In particular:

- There is still a well-defined and sharp peak reminiscent of the Fermi-gas delta function This means that one can still think of the excitations left by the emission of a photoelectron as individual particles or quasiparticles. But the peak is closer to the Fermi energy than for the Fermi gas.

- This quasiparticle peak is sharp but it is no longer an infinitely narrow delta-function: Fig. 5b shows that it has a finite width. This corresponds to a finite lifetime for the quasiparticle. The Fermi liquid predicts a quadratic E_i^2 dependence for the lifetime broadening.

- The integrated intensity of the quasiparticle peak is weaker than that of the Fermi-gas delta-function. Part of the intensity is transferred to the so-called "incoherent peak" (on the left-hand side of Fig. 5b). This implies that the removal of the electron can leave the system in an excited state not identifiable with a quasiparticle.

The integrated intensity of the quasiparticle peak is proportional to the so-called renormalization factor Z (which also determines the renormalization of the electron mass due to interactions with the other electrons).[15] For a Fermi gas, $Z = 1$ and all the spectral intensity remains in the delta-function. For a Fermi liquid, $Z < 1$ and intensity is transferred in part to the incoherent peak.

Furthermore, as seen in Fig. 6b, for a Fermi liquid the step of the $n(k_i)$ function decreases as the electron-electron interactions increase. The step height is given by the renormalization factor Z, and $Z = 1$ for a Fermi gas, but $Z < 1$ for a Fermi liquid. Yet, a Fermi liquid still exhibits a clear discontinuity at the Fermi level, reminiscent of the Fermi surface for a Fermi gas.

The limit Case: the Luttinger Liquid

The Fermi-liquid approach, however powerful, breaks down when the electron-electron interactions are too strong.[15] This can be realized by extrapolating some of the features derived within its framework. For example, the step of the $n(k_i)$ function tends to vanish, making it impossible to think in terms of a Fermi surface. And the quasiparticle peak in the spectrum also vanishes, making it impossible to think in terms of quasiparticles.

Forced to leave the safe harbor of the Fermi-liquid model, the analysis of photoemission travels in largely uncharted waters. The most solid reference point is the so-called Luttinger-liquid model.[16] This is an exactly solvable model in one dimension, whose major feature is that the excitations are no longer quasiparticles with spin and charge like a hole. There are in fact two kinds of excitations: those carrying spin or "spinons" and those carrying charge or "holons".

The charge-spin separation has been proposed as a background for the onset of HTSC - notably by Phil Anderson[17] - but the tests are inconclusive.[2,9,18,19] For one-dimensional systems, as one can imagine from Fig. 6, one of the fingerprints of the Luttinger liquid and of the consequent spin-charge separation would be a vanishing photoemission signal at the Fermi level.[15] And a vanishing signal near E_F seems indeed a general feature of quasi-one-dimensional metallic crystals.[14] However, not all the experimental features can be explained by the Luttinger liquid model, so its applicability to these results is still in question.[14]

What happens in higher-dimensionality systems? Photoemission from a three-dimensional system whose elementary excitations are spinons and holons was analyzed by Huber, reaching[18] similar conclusions as the one-dimensional Luttinger model: the photoemission signal near E_F should vanish.

Extensive searches for this phenomenon in the normal state of HTSC materials failed to produce results.[9,19] And the possibility still remains of yet other exotic non-Fermi-liquid states, perhaps with intermediate features between the two cases. The least one can say is that this issue is only partially explored, and therefore quite exciting.

Other Theoretical Issues

The possibility of collective effects does not exhaust the list of potential problems that suggest prudence while interpreting HTSC photoemission data. Photoemission has been very extensively and very successfully used[5,6] to analyze other classes of materials -- and the corresponding assumptions are more or less taken for granted. But this may prove to be dangerous in the case of new materials like the HTSC crystals.

For example, the matrix element M for the excitation is calculated assuming a perturbation hamiltonian of the $\mathbf{A} \cdot \mathbf{p}$ type, as it is done for optical excitations in the visible.[5] But the Dirac hamiltonian also gives a $\mathbf{p} \cdot \mathbf{A}$ term, which is clearly negligible for visible optical excitations because of the weak spatial variation of the vector potential. This might not be true near a surface, and the term $\mathbf{p} \cdot \mathbf{A}$ can give specific photoemission phenomena.[5]

What could happen near the surface of a superconductor, in particular considering the exceedingly small value of the correlation length of HTSC crystals? To the best of my knowledge this issue has not been analyzed, and might produce some surprises.

Also about the matrix element: its energy dependence is normally neglected, again as an extrapolation of what is done in optical spectroscopy.[5] But in optical spectroscopy this is justified by the small energy range involved. Deviations from this assumption are well known in photoemission, and constituted one of the early justifications for the use of energy-tunable synchrotron photon sources.[5] There was a time when it was fashionable to arbitrarily label "matrix element effect" all unexplained phenomena in synchrotron photoemission.

Now the tendency is opposite, and matrix element effects are largely ignored. One might suspect that this is not a problem for superconducting gap spectroscopy, since its energy domain is quite small. But the analyzed effects are also very small, so prudence is not out of place.

Note that the matrix element properties can be exploited in conjunction with the polarization of the photon beam to gain useful information about the state of the analyzed system.[5] This approach is presented in detail[3] elsewhere in this book, so I will only quickly review its implementation. Basically, M is zero for certain combinations of the symmetry of the final state, of the initial state and of the $\mathbf{A} \cdot \mathbf{p}$ operator. Thus, by manipulating the $\mathbf{A} \cdot \mathbf{p}$ symmetry and the final-state symmetry, one can obtain straightforward information on the initial-state symmetry.[5,20]

Consider for example the conditions of the so-called[20] Hermanson's rule: the angle-resolved analyzer is in plane of mirror symmetry for the crystal, thus the final state |f⟩ has even mirror-reflection symmetry; if the photon is linearly polarized parallel to the plane, the $\mathbf{A} \cdot \mathbf{p}$ operator is even, $\mathbf{A} \cdot \mathbf{p}$ |f⟩ is even, and to have a non-zero M the initial-state must also be even. If the photon polarization is perpendicular to the mirror-symmetry plane, $\mathbf{A} \cdot \mathbf{p}$ is odd, $\mathbf{A} \cdot \mathbf{p}$ |f⟩ is odd, and a non-zero M requires an odd initial state.

The implementation of this approach requires of course linearly-polarized synchrotron radiation.[5] Otherwise, it is quite simple and straightforward.

Although photoemission studies of HTSC are still affected by several difficulties, some of the old issues were resolved and removed. One of such issues was the quasiparticle lifetime.[21] We have seen that the Fermi liquid model predicts a quadratic energy dependence for the lifetime broadening. But the results on Bi-containing HTSC seemed initially quite puzzling, with a clearly non-quadratic behavior.[21]

As it turned out, many other classes of materials - including simple metals like aluminum - exhibited similar or even more dramatic "anomalous" lifetime effects. And there was really nothing mysterious, since other factors can contribute to the lifetime broadening besides the one giving the quadratic dependence. So, this issue is no longer an issue.[21]

But there exists a more fundamental difficulty: no complete theory has been developed for the photoelectric effect in a superconductor. This issue was raised when, in the early days of this field, the magnitude if the gap parameter Δ was commonly derived by attributing a BCS lineshape to the experimental spectra.[1] The risky character of this procedure was enhanced by the singularity of the BCS lineshape at $E_i = -\Delta$.

The problem was tackled by Huber[2] with a weak-interaction approach, leading to reassuring results. Huber proved that the photoelectron spectral lineshape was indeed given by the BCS function, so the procedure to extract the gap parameter seemed well-founded. Quite interestingly, the result did not appear rooted in some fundamental physical point, but just a mathematical coincidence.

The problem is that Huber's result was derived for weak interactions, and no one - to the best of my knowledge - has developed a complete treatment for strong interactions. Once again, this is an unresolved issue that should stimulate some interesting theoretical work.

EARLY HISTORICAL EVOLUTION

The first actual experiments on HTSC materials were preceded several years before by precursor studies of instability gaps in layered chalcogenides. Specifically, extensive experiments[22] were performed on $TiSe_2$, for which a debate occurred about the nature of the ground state (metal or insulator with small gap). During these experiments, the spectral modifications due to the onset of a charge-density-wave instability were clearly observed and revealed the creation of the gap.[22] Subsequently, the gap was carefully mapped in k-space.[22]

These precursor experiments, of course, were facilitated by the large size of the gap, for which the standard photoemission energy resolution of that time was sufficient. But extending the approach to conventional superonductors was out of the question. The surprising discovery of HTSC materials with their large gaps created then much excitement: for the first time, electron spectroscopy could be used to study superconductivity.[1]

The first steps of this direction were both bold at the limit of irresponsibility.[1] The early HTSC were very questionable candidates for photoemission experiments: polycrystalline aggregates with large chemical fluctuations, unknown chemical stability, unknown surface stability. In fact, it took some time before reaching the first reliable results.

The initial puzzle was the lack of a Fermi edge in the normal-state spectra of YBCO.[1,2] As we have seen, a weakening of the photoemission signal near E_F could be the exciting fingerprint of some exotic state. But in many cases it is simply due to spurious effects like contamination (by and large, contaminants are insulators with a gap). And the absence of a Fermi edge for the normal state was somewhat disappointing since it made it impossible to observe the spectral evidence for the opening of a superconductivity gap.

The first significant step out of this impasse was the use of high quality single crystals.[23] Yet, the puzzling result for YBCO survived this step. But the decisive event was the discovery of BCSCO and the consequent observation[24] of a photoemission Fermi edge a few days later -- see Fig. 7. This suggested that the absence of the Fermi edge for YBCO was a spurious effect - as it was - and opened the way to the photoemission observation of the gap. Shortly afterwards, good-quality BCSCO single crystals[25] became available, facilitating the development of gap spectroscopy.

The first glimpse of the opening of the superconductivity gap was an experiment by Y. Chang et al.[26,27]; the resolution, however, was still in the 100-meV range, thus the symptoms of the gap opening were not easy to observe. These first steps remained controversial for a few months, until better resolution was achieved and the results unmistakably revealed the gap opening.[9,28-30] Figure 8 shows two of the early high-resolution experiments.

For some time, the experiments were primarily feasibility tests. But some of the spectral features almost immediately attracted the attention of theorists; this was the case, for example, of the minimum ("dip") that follows the main peak below the gap in the superconducting-state.[30]

The full power of photoemission, however, became clear when issues related to the direction dependence were raised for HTSC. There were two classes of such issues. First, those concerning the normal state. For example, many theorists proposed that HTSC is

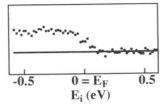

$$-0.5 \qquad 0 = E_F \qquad 0.5$$
$$E_i \text{ (eV)}$$

Figure 7. First evidence of a Fermi edge in the photoemission spectra of HTSC materials. Data for BCSCO from Ref. 24.

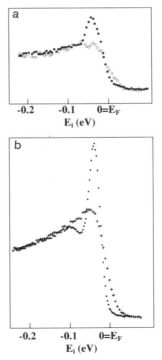

Figure 8. Some of the early photoemission data which reveal the opening of the superconductivity gap in BCSCO by comparing normal-state and superconducting-state data. (a) from Ref. 9; (b) from Ref. 30.

related to a peculiarity of the normal-state band structure, the Van Hove singularity that enhances the density of states at the Fermi level.[31] This type of issue stimulated a very extensive analysis of the normal-state electronic structure with angle-resolved photoemission, as a function of doping, disorder and other parameter -- an analysis which is still very active.

The second class of issues concerned the superconducting state, and specifically the symmetric s-wave or non-symmetric character of the gap. This was a question in which photoemission could fully show its power. But the approach was not immune from problems, and for some time it was full of controversies (most of which were later resolved).[7,8]

One of the problems was the identification of the specific symmetry for the gap parameter. The experiments clearly demonstrated that a symmetric s-wave gap could be ruled out.[7,8] But hat was the specific symmetry? The most likely candidate was d_{x2-y2}. However, proving this symmetry required measuring a zero value of the gap (a node) in certain directions. The experimental accuracy, almost by definition, does not allow to measure a zero gap with infinite accuracy, because of finite resolution, lineshape modeling and other factors. The most one could conclude was that the measured gap was, in the relevant directions, consistent with zero within the experimental accuracy (see Fig. 9).[7,8]

To complicate the issue, the gap symmetry - as it was discovered later - depended on doping.[8] Divergent results triggered a controversy between different groups, which fortunately was later removed. In essence, it is now clear that reliable results can be obtained about the gap symmetry if (1) sufficient resolution (energy and angle) is used; (2) the limitations of the data analysis are taken into account, without making extravagant assumptions about accuracy; (3) the crystal quality is carefully controlled (see Fig. 10), and in particular the effects of doping are systematically assessed.

Photoemission emerged from the early and controversial period of gap symmetry analysis as a generally accepted, powerful and reliable probe of the superconducting state. Its wild youth was over, and a productive maturity initiated, whose results are evident from other presentations in this school.

FUTURE DIRECTIONS

The easiest prediction is that photoemission studies of the superconductivity gap will continue and expand in the future. In particular, they will be progressively extended to other

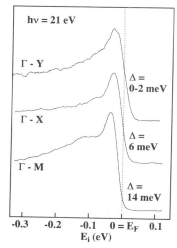

Figure 9. Superconducting-state photoemission spectra for BCSCO, revealing the gap in different directions of the Brillouin zone. A pure d_{x2-y2} symmetry would require a zero gap in the top curve. However, the accuracy of this conclusion is limited by the noise, by the experimental broadening and by the data analysis assumptions. Results from Ref. 7.

Figure 10. Early large-size, single-crystal BCSCO specimen grown by H. Berger at the Institut de Physique Appliquée of the Ecole Polytechnique Fédérale de Lausanne.

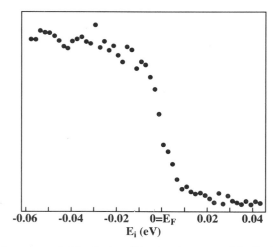

Figure 11. Metallic edge measured at the Center of Spectromicroscopy of the Ecole Polytechnique Fédérale de Lausanne. The unpublished data by M. Grioni et al. demonstrate the excellent energy resolution; after correcting for the temperature broadening, the instrumental Gaussian broadening is 9 meV.

HTSC families, whereas the majority of the past experiments were on BCSCO. New materials will be investigated both as high-quality single crystals and thin films.

Part of this expansion is related to instrumentation improvements. Basically, the experiments require very high energy and angle resolution, which in turn reduces the signal level and makes the experiments quite time consuming, Furthermore, most experiments are performed with synchrotron light, and suitable synchrotron beamtime is affected by a worldwide shortage.

But the situation is improving from several different points of view. First of all, new and advanced synchrotron facilities are being implemented.[5,6] This increases the available beamtime. Furthermore, the new facilities are characterized[5,6] by very high brightness, thus they increase the signal level and shorten the data taking time, enhancing the effectiveness of every hour of beamtime.

On the other hand, significant instrumentation improvements make it possible to use laboratory sources for part of these experiments. Figure 11 shows, for example, a Fermi-edge spectrum demonstrating very high energy resolution for the Scienta ESCA 300 system of the Center of Spectromicroscopy in Lausanne.

The increase in source brightness[5,6] brought by the recently commissioned third generation of synchrotron facilities - e.g., ELETTRA in Trieste, Max-Lab in Sweden, the ALS in Berkeley, SRRC in Taiwan and PAL in Korea - is really tremendous. And a further increase is on the way, with the recent approval of the first fourth-generation source SLS in Switzerland.[32] Such sources not only make high-resolution photoemission experiments easier and faster, but also open the way to entirely new techniques.

Consider for example lateral resolution. This is an important issue for HTSC materials, that are often subject to fluctuations in the chemical composition and in other properties. Standard photoemission tests average the properties over an area of the order of 1 mm^2, thus they are blind to fluctuations on a smaller scale. But it is now possible to overcome this barrier with the so-called photoelectron spectromicroscopy, which exploits the brightness of the new synchrotron sources to achieve lateral resolutions down to 0.1 micron or even less.[33]

Unfortunately, the insufficient signal level does not allow a combination of high lateral resolution and high energy and angle resolution. This situation is likely to change with the advent of better sources and better instrumentation. In fact, high-energy-resolution spectromicroscopy is one of the objectives of the CERVIN facility, proposed for the SLS.[34]

Similarly, high brightness synchrotron sources decrease the data taking time to the point that time-resolved photoemission analysis becomes feasible.[33] This approach is being used to analyze the dynamic evolution of surface chemical properties. Once again, this possibility is quite interesting for HTSC research -- but the signal level is not sufficient to combine it with high energy resolution. And fourth-generation sources could remove this limitation.[34]

In essence, photoelectron spectroscopy has emerges as a leading probe of HTSC properties. This is not surprising, since the photoelectric effect is one of the most fundamental phenomena in nature, and carries plenty of information on the involved electronic states. In its history, the photoelectric effect includes outstanding milestones such as the foundation of quantum mechanics. In the past decades. it was exploited as a basic electronic probe for materials. The combination of larger gaps and better resolution extended its domain to superconductivity in the late 1980's. The subsequent expansion is still underway, and its results are among the most important in the general field of superconductivity.

ACKNOWLEDGMENTS

The results used as examples in this overview were produced in collaboration with many scientists, including all authors of Refs. 1, 2, 3, 7, 8, 11, 12, 19, 21, 22-27 and 30-34. Support for the author is provided by the Fonds National Suisse de la recherche Scientifique, the Ecole Polytechnique Fédérale de Lausanne, the Sincrotrone Trieste SCpA and the European Commission. I am particularly grateful to Marshall Onellion for many stimulating discussions during our collaboration, and to Alex Müller for stimulating ideas.

REFERENCES

1. See, for example: G. Margaritondo, Robert Joynt and M. Onellion, Eds. : *High T_c Superconducting Thin Films, Devices and Characterization*, (American Institute of Physics, New York 1989).
2. G. Margaritondo, D. L. Huber and C. G. Olson, Science 246:770 (1989), and the references therein.
3. M. Onelliom these proceedings.
4. J. C. Campuzano, these proceedings.
5. See, for example: G. Margaritondo, *Introduction to Synchrotron Radiation*, Oxford University Press, New York (1988)
6. G. Margaritondo, Riv. Nuovo Cimento 18:1 (1995), and the references therein.
7. Jian Ma, C. Quitmann, R. J. Kelley, H. Berger, G. Margaritondo and M. Onellion, Science 267:862 (1995), and the references therein.
8. R. J. Kelley, C. Quitmann, M. Onellion, H. Berger, P. Alméras and G. Margaritondo, Science 271:1255 (1996), and the references therein.
9. C. G. Olson et al. Science 245:731 (1989).
10. Z.-X. Shen et al., Phys. Rev. Lett. 70:1553 (1993).
11. Y. Hwu et al., Phys. Rev. Lett. 67:2573 (1991).
12. R. J. Kelley, Jian Ma, G. Margaritondo, M. Onellion, Phys. Rev. Lett. 71:4051 (1993).
13. D. S. Dessau, et al., Phys. Rev. Lett. 66:2160 (1991).
14. G. Margaritondo, Phys. Low. Dim. Struct. 1:11 (1994).
15. D. Malterre, M. Grioni and Y. Baer, Adv. Phys. 45:299 (1996), and the references therein.
16. J. M. Luttinger, Phys. Rev. 119:1153 (1960).
17. P. W. Anderson, Science 235:1196 (1987).
18. D. L. Huber, Solid State Commun. 68:459 (1988).
19. Y. Chang et al., Phys. Rev. B39:7313 (1989).
20. J. Hermanson, Solid State Commun. 22:9 (1977).
21. Y. Hwu, L. Lozzi, S. La Rosa and M. Onellion, P. Alméras, F. Gozzo, F. Lévy, H. Berger and G. Margaritondo, Phys. Rev. B45:5438 (1992); Phys. Rev. B48:624 (1993), and the rereferences therein.
22. M. M. Traum, G. Margaritondo, N. V. Smith, J. E. Rowe and F. J. DiSalvo, Phys. Rev. B17:1836 (1978); G. Margaritondo, C. M. Bertoni, J. H. Weaver, F. Lévy, N. G. Stoffel and A. D. Katnan, Phys. Rev. B23:3765 (1981) N. G. Stoffel, F. Lévy, C. M. Bertoni and G. Margaritondo, Sol. State Commun. 41:53 (1982).
23. N. G. Stoffel, Y. Chang, M. K. Kelly, L. Dottl, M. Onellion, P. A. Morris, W. A. Bonner and G. Margaritondo, Phys. Rev. (Rapid Commun.) B37:7952 (1988).
24. M. Onellion, Ming Tang, Y. Chang, G. Margaritondo, J. M. Tarascon, P. A. Morris, W. A. Bonner and N. G. Stoffel, Phys. Rev. (Rapid Commun.) B38:881 (1988).
25. R. Zanoni, Y. Chang, Ming Tang, Y. Hwu, M. Onellion, G. Margaritondo, P. A. Morris, W. A. Bonner, J. M. Tarascon and N. G. Stoffel, Phys. Rev. B38:11832 (1988).
26. Y. Chang, Ming Tang, R. Zanoni, M. Onellion, Robert Joynt, D. L. Huber, G. Margaritondo, P. A. Morris, W. A. Bonner, J. M. Tarascon and N. G. Stoffel, Phys. Rev. (Rapid Commun.) B39:4740 (1989).
27. Y. Chang, Ming Tang, R. Zanoni, M. Onellion, Robert Joynt, D. L. Huber, G. Margaritondo, P. A. Morris, W. A. Bonner, J. M. Tarascon and N. G. Stoffel, Phys. Rev. Letters 63:101 (1989).
28. J. M. Imer et al., Phys. Rev. Letters 62:336 (1989); ibid. 63:102 (1989).
29. R. Manzke, T. Buslap, R. Claesen and J. Fink, Europhys, Lett 9:477 (1989).
30. Y. Hwu, L. Lozzi, M. Marsi, S. La Rosa, M. Winokur, P. Davis, M. Onellion, H. Berger, F. Gozzo, F. Lévy and G. Margaritondo, Phys. Rev. Letters 67:2573 (1991).
31. Jian Ma, C. Quitmann, R. J. Kelley, P. Alméras, H. Berger, G. Margaritondo and M. Onellion, Phys. Rev. B51:3832 (1995), and the references therein.
32. H. J. Weyer and G. Margaritondo, Synchrotron Radiation News 8:19 (1995).
33. G. Margaritondo, A. Savoia, S. Bernstoff, M. Bertolo, G. Comelli, F. De Bona, W. Jark, M. Kiskinova, G. Paolucci, K. Prince, A. Santaniello, G. Tromba, R. Walker and R. Rosei, Acta Phys. Polonica 91:631 (1997), and the references therein.
34. G. Margaritondo, E. Kapon, K. Kern and W. Schneider, unpublished.

ELECTRONIC STRUCTURE AND DOPING IN CUPRATE SUPERCONDUCTORS

Marshall Onellion

Physics Department
University of Wisconsin,
Madison, WI 53706

I. INTRODUCTION TO BCS SUPERCONDUCTIVITY AND TO PHOTOEMISSION

A. BCS SUPERCONDUCTIVITY

To understand the significance of electronic structure in the cuprates, we compare the low temperature (low-T_c) and high temperature (high-T_c) superconductors. Most low-T_c superconductors have a superconducting coherence length, $\xi \geq 100$ nm. Further, the pairing interaction, due to a retarded electron-phonon interaction, is spatially very short ranged. Consequently, the pairing interaction connects essentially all parts of the Fermi surface with almost the same strength. In addition, low-T_c superconductors are typically good metals with a screening length much less than the coherence length at all frequencies. In three dimensions, BCS theory [1] predicts the superconducting gap, Δ, and the superconducting transition temperature, T_c, both depend on the same, single energy scale. This energy scale depends on the angle-integrated density of electronic states, $N(E_F)$, at the Fermi energy, E_F. In weak-coupling BCS theory, the ratio of ($2\Delta_{max}/kT_c$) is 3.52;[1] for strong coupling it increases to ~4.3. For a weak-coupling d-wave order parameter the ratio ($2\Delta_{max}/kT_c$) is 4.3.[2] However, in two dimensions with a circular Fermi surface, neither Δ nor T_c depend on $N(E_F)$, nor on the Fermi wavevector, k_F. BCS theory also predicts an upper bound on T_c of ~30K.[1]

High-T_c materials exhibit several contrasts. Materials with T_c values above 160K have been discovered. The coherence length $\xi \sim 2$nm is barely large enough to include the two carriers that form a Cooper pair. Further, based on a bandwidth of ~400 meV, the DeBroglie wavelength of the Fermi surface electrons is ~2 nm as well, comparable to rather than much less than the coherence length. High-T_c materials act as good metals at high frequencies, but are poor metals at low frequencies,[3] so the screening length at low frequencies is not negligable compared to the coherence length. Also, in-plane resistivity measurements are anomalous. Low-T_c materials exhibit a normal state resistivity (ρ) dominated by phonons, and a characteristic change in $\rho(T)$ from below to above the Debye temperature. However, high-T_c materials exhibit $\rho(T) \sim T$ over two decades of temperature.[4] A further anomaly is the size of the transport mean free path. The concept of a quasiparticle wavepacket becomes tenuous when the transport mean free path (λ_L) becomes comparable to the DeBroglie wavelength (λ_D) of the wavepacket. However, the high-T_c materials exhibit no substantial change in $\rho(T)$ between $\lambda_L \gg \lambda_D$ and $\lambda_L \sim \lambda_D$. Several theories argue that the exchange boson is a paramagnon, or other boson due to an electron-electron interaction, rather than a phonon.[5] Such electron-electron pairing interactions link different parts of the Fermi surface is qualitatively different ways.

Because in BCS theory superconductivity arises from the electron-phonon interaction, the superconducting order parameter has the symmetry of the crystallographic lattice. This leads to the well-known observation that the only symmetry difference between the normal and superconducting states is that gauge invariance is broken in the superconducting state. Such an order parameter is known as "conventional."

B. ANGLE-RESOLVED PHOTOEMISSION

Until the discovery of the cuprate superconductors, angle-resolved photoemission measurements did not have a major impact on the field of superconductivity. There were two very good reasons for this. First, the

The Gap Symmetry and Fluctuations in High - T$_c$ Superconductors
Edited by Bok *et al.*, Plenum Press, New York, 1998

superconducting gaps were so small that photoemission techniques could not measure the gap. Second, BCS theory predicted that the pairing interaction averaged over the entire Fermi surface. Consequently, although angle-integrated photoemission might be of some small interest- based on measuring the total density of states- angle-resolved photoemission would not be of interest.

The situation changed dramatically with the cuprate superconductors. Several factors lead to angle-resolved photoemission becoming more important. The traditional techniques for obtaining information about the gap and excitation spectrum- tunneling and optical absorption- encountered unexpected difficulties and have not yet proven as decisive as they were for the low temperature superconductors. Also, the superconducting gap could be measured- indeed, directly observed- using angle-resolved photoemission.[6] Further, the electronic band structure, and superconducting gap, were found to be highly anisotropic, which meant that wavevector-resolved information was essential to our understanding. Angle-resolved photoemission is virtually the only technique that yields unambiguous wavevector-resolved information. For all these reasons, angle-resolved photoemission has become much more important with the advent of cuprate superconductors.

Figure One illustrates the fundamentals of angle-resolved photoemission measurements.[7] A source of photons, generally in the extended ultraviolet range of the electromagnetic spectrum, with photon energies between 5-100 eV, illuminates the sample. The polarization of the electric field of the light can be used to obtain some information on the symmetry of the initial state wavefunction, as discussed below in some detail.

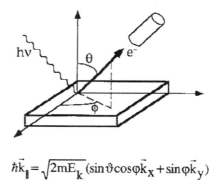

$$\hbar \vec{k}_{\parallel} = \sqrt{2mE_k}\,(\sin\vartheta\cos\varphi \vec{k}_x + \sin\varphi \vec{k}_y)$$

Figure One: Typical angle-resolved photoemission setup, including photon energy (hv), photoelectron, detector, and sample.

A single electron absorbs a single photon. For photon energies of 5-100 eV, the change of linear momentum that accompanies photoabsorption ($\leq 10^{-2}$ Å^{-1}) is negligable compared to the electron linear momentum in the sample (1 Å^{-1}). Photoabsorption is thus viewed as a transfer of energy, with negligable change of linear momentum. If the photoelectron possesses sufficient kinetic energy, and linear momentum parallel to the sample surface normal, it emerges from the sample.[8] For a well-ordered sample, the electrostatic potential parallel to the surface is periodic, and specifically has translational invariance. Consequently, the component of photoelectron wavevector (linear momentum) parallel to the sample surface are conserved when the photoelectron emerges from the sample. By contrast, the component of photoelectron wavevector parallel to the surface normal is not a conserved quantity during the photoemission process.

Fortunately for using angle-resolved photoemission, the cuprate electronic band structure is quasi-two-dimensional. This means that angle-resolved photoemission, which measures the kinetic energy and wavevector components of the photoelectron, also measures the wavevector components and kinetic energy of the electron before the photoemission process. Specifically, angle-resolved photoemission provides the following information about the electrons in the material:

$$E_i = E_f - hf + \Phi \qquad (1)$$

where E_i is the kinetic energy of the initial electron (before photoemission), hf is the photon energy, and Φ is the work function of the material;

$$k_{||, \text{out}} = k_{||, \text{in}}$$

(2)

where $k_{||, \text{in}}$ is the component of electron wavevector parallel to the sample surface before photoabsorption, and $k_{||, \text{out}}$ is the component of electron wavevector parallel to the sample surface after absorption, when the photoelectron emerges from the sample;

$$<\Psi_f \,|\, \mathbf{p} \cdot \mathbf{A} \,|\, \Psi_I >$$

(3)

must be symmetric with respect to any operator that commutes with the Hamiltonian, where (Ψ_I) is the electron initial state wavefunction, $\mathbf{p} \cdot \mathbf{A}$ is the photoabsorption operator and includes the electron linear momentum $\mathbf{p}$ and the photon vector potential $\mathbf{A}$, and Ψ_f is the electron final state wavefunction. By using synchrotron-radiation, which is highly linearly polarized (degree of linear polarization >95%), and noting that Ψ_f must be totally symmetric with respect to any operator that commutes with the Hamiltonian, we are able to measure the reflection symmetry of the initial electronic state, Ψ_I. Figure Two illustrates the experimental setup. Note that by measuring in the horizontal plane, we measure initial states that have only even reflection symmetry, while by measuring in the vertical plane we measure initial states that have only odd reflection symmetry. If Ψ_I has pure even (or pure odd), reflection symmetry with respect to a particular plane of reflection symmetry in the lattice, then we will observe Ψ_I with one measurement and it will disappear with the opposite orientation. If Ψ_I has components of both even and odd reflection symmetry (which can happen), then the initial state is observed in both orientations.

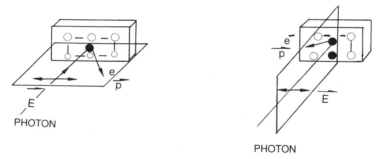

Figure Two: Experimental setup for even and odd reflection symmetry angle-resolved photoemission measurements.

The discussion so far is based on a single electron picture, and implicitly the idea is that photoemission measures the Hartree-Fock single electron energy levels.[9] This is not true: Koopmans theorem [10] does not hold, and the system responds to both the photoelectron and the photohole left behind by adjusting energy levels. Further, since photoemission leaves the system behind in an ionic state, which is not an eigenstate of the Hamiltonian, there is more than one configuration (eigenstate) that can be observed. The full details are complicated. However, what saves the day for the study of the cuprate superconductors is the close relation between photoemission and the one-particle Green's function.[11] For an interacting system, the Green's function has the general formalism:

$$G_i(E) = \frac{1}{E - E_i^0 - \sum_i(E)}$$

(4)

leading to the spectral function:

$$A_i(E) = \frac{1}{\pi} \mathrm{Im}\{G_i(E)\} = \frac{1}{\pi}\left(\frac{\mathrm{Im}\{\sum_i(E)\}}{\left[E - E_i^0 - \mathrm{Re}\{\sum_i(E)\}\right]^2 + \left[\mathrm{Im}\{\sum_i(E)\}\right]^2} \right)$$

(5)

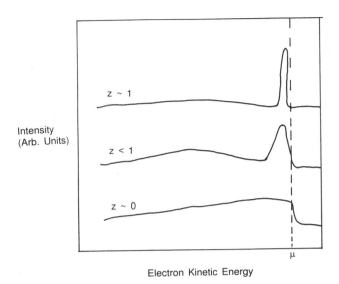

Figure Three: Angle-resolved photoemission spectra for renormalization factor Z from ~1 to ~0.

It is important to note that photoemission measures the spectral function. Consequently, in view of (5), it is common to say that photoemission measures the one-particle Green's function- specifically the imaginary part. There is another important consequence of (5). The spectral function has a maximum for a vanishing:

$$E - E_i^0 - \sum_i(E)$$

(6)

For a non-interacting system, the peaks in the spectral function are the quasiparticle eigenstates of the system, at energy values corresponding to the Hartree-Fock energy eigenvalues. In general, for the idea of a quasiparticle is well defined, the spectral function must have a maximum. To find the spectral intensity of such a maximum, make a Taylor expansion of (6):

$$E - E_i^0 - \mathrm{Re}\{\sum_i(E)\} \cong \left(\frac{1}{Z_i(E_{max})} \right)(E - E_{max})$$

(7)

212

$$Z_i(E) = \left(1 - \frac{\partial}{\partial E}\left[\text{Re}\{\textstyle\sum_i(E)\}\right]\right)^{-1}$$

(8)

$$A_i(E) \cong \frac{Z_i(E_{\max})}{\pi} \frac{\Gamma_{\max}/2}{(E - E_{\max})^2 + (\Gamma_{\max}/2)^2}$$

(9)

where $\Gamma(E) = 2\,Z_i(E)\,\text{Im}\{\textstyle\sum_i(E)\}$. The renormalization factor, $Z_i(E)$, is quite important. For non-interacting quasiparticles, $Z=1$ exactly. For a Fermi liquid, $Z\sim1$, and the idea of a quasiparticle is well-defined. However, as $Z\to0$, the idea of a quasiparticle becomes undefined. In particular, as $Z\to0$, the spectral function has no well defined quasiparticle of energy (E). Instead, the spectral intensity (i.e., carrier concentration) is spread over a range of energies as incoherent spectral area. Figure Three illustrates the difference the occurs as $Z\to0$.

The important point is that disorder leads to a transfer of spectral weight from the coherent to the incoherent part of the spectral function, and this is observed directly by using photoemission.

II. TWO ENERGY SCALES IN SUPERCONDUCTING PAIRING

Soon after the discovery of cuprate superconductors, reports using several experimental techniques estimated the size of the gap (Δ) as incongruously large for the values of T_c, with the ratio of ($2\Delta/kT_c$) as 7-10, depending on experimental method.[12] More recent reports, including tunneling,[13] Raman,[14] and photoemission[6] have produced estimates of the ratio ($2\Delta/kT_c$) between 7-9. This result, as is widely recognized, stands in marked contrast to BCS theory. In BCS theory, the ratio is 3.52 in the limit of weak coupling and increases to ~4.3 for strong-coupling. Recently, in an 18 month collaboration that includes C.G. Olson, G. Margaritondo, C. Kendziora and the author,[15] overdoped samples have been studied in four different laboratories and with unparalleled photoemission energy resolution. We find that the ratio ($2\Delta/kT_c$) decreases from 7-8 for nearly optimally doped samples to 3.2 ±0.9 for overdoped samples with T_c =56-62K. This marked change, much greater than explicable based on the strength of a single coupling constant, indicates that there are two energy scales involved in the superconductivity of cuprates. It is, after all, the presence of a single energy scale (the strength of the electron-phonon coupling) that leads both to the BCS value for ($2\Delta/kT_c$) and limits the variation of this ratio when the strength of the coupling changes. In view of the ratio for overdoped samples, 3.2±0.9, the data indicate that one of the energy scales is the BCS energy scale.

It is worth noting that the overdoped samples exhibiting the ($2\Delta/kT_c$) ratio of 3-4 also exhibit a significant T^2 component to the ab-plane resistivity, unlike the optimally doped samples that exhibit T-linear behavior. This indicates that with doping the system is changing toward a "normal" Fermi liquid.[4]

In addition to this new result, our study of more heavily overdoped samples confirms the earlier report in Ref. 16: the gap in the $\langle\pi,\pm\pi\rangle$ directions is not zero. Instead, for the overdoped samples studied, it is 3±1 meV in the Γ - Y direction and 5±1 meV in the Γ - X direction. This result, as with the earlier results, is inconsistent with a pure d-wave order parameter symmetry.

III. NORMAL STATE

A. SYMMETRY OF QUASIPARTICLE STATES IN THE NORMAL STATE

I discuss four aspects of the electronic band structure in the normal state. The first is an evolution of the quasiparticle states from the parent antiferromagnetic insulator to the overdoped cuprate. Since the symmetry of the initial electronic states changes with doping, without a structural phase transition, there must be a non-lattice part of the Hamiltonian changing with doping. The electronic band structure results indicate that short-range antiferromagnetic order persists well into the metallic phase, and in particular into the overdoped part of the phase diagram.

There are three general, noteworthy, observations that we use as the basis for our data analysis and interpetation.

1) As Bloch first reported,[17] itinerant carriers exhibit the symmetry of the potential energy term they experience. If the potential energy is solely the electrostatic potential of the lattice, then the carriers exhibit the periodicity of the lattice. Thus, if one studies electronic states on the CuO_2 planes as a function of doping- with no structural change occurring- then the symmetry of the quasiparticle states should be constant with doping;

2) The dispersion of electronic states must be symmetric about high symmetry points. For example, along a $\Gamma-X-\Gamma$ line in the Brillouin zone, the electronic states must be symmetric about the X point, because it does not matter from which Γ point one starts. This means that a dispersion relation exhibit symmetric behavior about high symmetry points, but only high symmetry points;

3) If there is a superlattice periodicity to the system, then there can be band folding which is visible in photoemission spectra.[7]

We use these general observations in simple, logical ways.

1) We measure the reflection symmetry of the itinerant, quasiparticle, electronic states as a function of doping.[18] Figure Four illustrates the reflection symmetry in the $<0,\pi>$ direction in the Brillouin zone for oxychloride samples. Table One provides a summary of the reflection symmetry in the three directions for all doping levels. We find that there is a change in reflection symmetry in both the $<\pi,\pi>$ and $<\pi-\pi>$ directions, while no change with doping is observed in the $<0,\pi>$ direction. There is no indication of a structural phase transition with doping. Consequently, we know (a) the potential energy term experienced by the itinerant carriers is changing with doping, but (b) the electrostatic potential of the lattice is not changing with doping. We conclude that the spin-dependent potential energy term, specifically the short-range antiferromagnetic ordering, is changing with doping, and that this change is reflected (pun intentional) in the symmetry of the itinerant carriers.

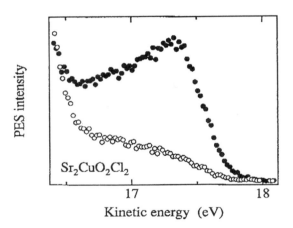

Figure Four: Angle-resolved photoemission spectra of the same Brillouin zone location for oxychloride $Sr_2CuO_2Cl_2$ sample, including even (closed circle) and odd (open circle) reflection symmetry.

2) We measured the dispersion relation of both oxychloride and cuprate samples.[18] In all instances, the electronic states nearest the chemical potential are on the CuO_2 planes, and these planes have the same structure for all sample types. For the oxychloride samples, the dispersing quasiparticle states exhibit a symmetric dispersion about the $(\pi/2,\pm\pi/2)$ points. This is not surprising; the oxychloride samples are antiferromagnetic insulators and the magnetic Brillouin zone makes the $(\pi/2,\pm\pi/2)$ points symmetry points due to the antiferromagnetic order. What of the cuprate superconductors? One common method for changing the carrier concentration is to add or remove oxygen. Recently, we have, in two different laboratories, succeeded in removing sufficient oxygen to render the cuprate samples insulating.[19] Further, we have done so reversibly- we can replace the oxygen and restore the superconductivity. We have further irradiated $Bi_2Sr_2CaCu_2O_{8+x}$ single crystal samples with thermal neutrons.[20] The as-irradiated samples are insulating, but gentle annealing is adequate to partially or totally restore superconductivity. The main effect of the thermal neutrons is to disorder- but not remove- oxygen. We find that the insulating $Bi_2Sr_2CaCu_2O_{8+x}$ samples obtained by oxygen removal and by oxygen disorder exhibit the same spectral function. This leads to a natural question: are the $(\pi/2,\pm\pi/2)$ points symmetry points for different doping

214

levels of cuprate samples? Figure Five illustrates the results, which are quite similar in both the <π,π> and <π−π> directions. The (π/2,±π/2) points are symmetry points for the oxychloride, for the insulating cuprate (obtained by neutron irradiation), and for the T_c =30K underdoped cuprate samples. For the T_c =60K and optimally doped cuprates, the (π/2,±π/2) points are not symmetry points as such, but the dispersion between (0,0) and (π/2,±π/2) is quite similar for all the samples. The data in Fig. 5 indicate that for insulating cuprates and heavily underdoped cuprates obtained by oxygen removal, the (π/2,±π/2) points are indeed symmetry points. The only interaction in the problem leading to this symmetry is short-range (or, for the insulator, perhaps long-range) antiferromagnetic order.

Table One: Reflection symmetry of quasiparticle state in high-symmetry directions of Brillouin zone versus doping level, including underdoped (UD), optimally doped (O) and overdoped (OD) superconducting samples.

sample/direction	$\Gamma\bar{M}$	ΓY	ΓX
insulator	even	even	even
UD Bi-2212	even	even	odd
O Bi-2212	even	mostly even	odd
OD Bi-2212	even	mostly odd	even + odd

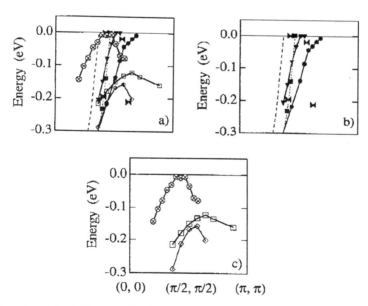

Figure Five: Dispersion relation in the <π,π> direction, including (a) all sample types, (b) superconducting samples only, and (c) insulating samples only. Sample types include optimally doped (dotted line), 60K superconductor (closed squares), 30K superconductor (closed facing triangles), oxychloride (circle with x), low dose irradiated insulator (open square), irradiated and annealed 43K superconductor (closed circles), high dose irradiated insulator (open diamond) and irradiated and annealed 83K superconductor (closed single triangle).

3) The report by P. Aebi and colleagues at Friborg, Switzerland of "shadow" features in the Fermi surface was immediately recognized as important.[21] Figure Six illustrates the data from Ref. 21 that exhibit the shadow features. The results were important for several reasons, including the possibility that the shadow features were due to short-range antiferromagnetic fluctuations. Such features, although due to long-range antiferromagnetism, had been predicted earlier by Kampf and Schrieffer.[22] Subsequent work suggested that the shadow features might be due to lattice (Umklapp) processes, rather than a superlattice periodicity due to magnetic order.[23] Recent work by my colleagues and me have established that the shadow features are due to short-range antiferromagnetic fluctuations.[24] Figure Seven illustrates the experimental data that exhibit the shadow feature, using traditional angle-resolved photoemission. The question is: how to distinguish between the two types of superlattice periodicities, lattice and spin? A. Chubukov obtained an answer,[25] by realizing that Umklapp processes are insensitive to electron binding energy, while spin processes are much more sensitive to electron binding energy. The calculation of Ref. 25 indicates that by taking the ratio of the spectral intensity of the main quasiparticle state to that of the shadow feature at the same binding, it is possible to distinguish the origin of the shadow feature. Figure Eight illustrates the results. If the shadow feature is due to Umklapp processes, then the ratio is virtually independent of electron binding energy. However, if the shadow feature is due to spin-based superlattice periodicity, then the ratio changes quadratically with electron binding energy. As Fig. 8 demonstrates, the data are consistent with the shadow features due to spin-based superlattice periodicity, and rule out Umklapp processes as the origin of the shadow features.

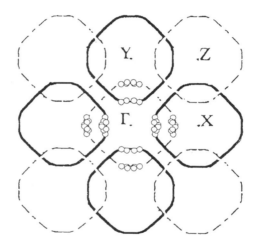

Figure Six: Main (solid line) and shadow (dotted line) Fermi surface as reported in Ref. 21. Open circles indicate our measurement of the main and shadow features at the Fermi energy.

We conclude from the change in quasiparticle symmetry with doping, and the ratio between main and shadow quasiparticle features, that there is a part of the Hamiltonian that involves short-range antiferromagnetic order, and that this persists well into the overdoped part of the phase diagram.

B. FERMI SURFACE TOPOLOGY

One of the long-standing puzzles in cuprate superconductors concerns the evolution of the Fermi surface with doping. In a doped semiconductor, in which carriers occupy the bottom of the conduction band (electron majority) or the top of the valence band (hole majority), the size of the Fermi surface scales with the number of carriers. In fact, this relation between carrier density and size of Fermi surface is enshrined in Luttinger's theorem, which provides the mathematical argument that a Fermi liquid should "always" behave in this fashion. A very different picture emerges for the cuprate superconductors. The topology of the Fermi surface changes qualitatively with doping. Highly underdoped samples exhibit a very small Fermi surface, and in particular there are pockets around the ($\pi/2$,$\pm\pi/2$) points. As the doping level increases, the Fermi surface becomes large, and the large parts roughly encircle the (π,$\pm\pi$) points. This is important since since several models predict qualitatively different behavior of the Fermi surface with doping. In particular, our data largely agree with the model proposed by A. Chubukov,[26] that there should be a topological phase transition in the underdoped part of the phase diagram.

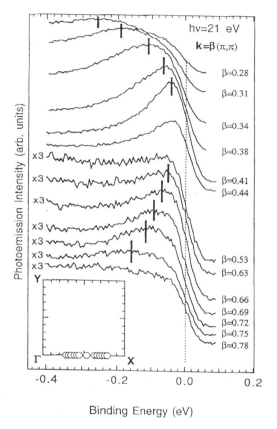

Figure Seven: Angle-resolved photoemission data in the $\langle\pi,\pi\rangle$ direction showing both main and shadow features. Inset: locations in the Brillouin zone at which spectra were taken.

The astute reader will have noted a peculiarity in Fig. 5: the underdoped, $T_c = 30K$, samples exhibit two crossings. This is part of a larger story, that the Fermi surface changes qualitatively with doping. Figure Nine illustrates the data from a recent study of the Fermi surface for $T_c = 30K$ underdoped samples. From these data, there is a crossing in the Γ-Y direction, that is, the direction in which there is no Bi-O superlattice modulation to complicate the analysis with Umklapp band folding. Figure Ten illustrates the Fermi surface measured for optimally doped samples, for $T_c = 60K$ underdoped samples, and for $T_c = 30K$ underdoped samples. The change in carrier concentration is from about 0.20 carriers per CuO_2 unit (optimally doped) to 0.15 ($T_c = 60K$) to 0.11 ($T_c = 30K$). That is, a change in carrier concentration of ~x2. However, the Fermi surface changes by significantly more than x2. Specifically, the Fermi surface decreases by ~x3 for the $T_c = 60K$ samples, and decreases by ~x30 for the $T_c = 30K$ samples. These data indicate that the size of the Fermi surface does not scale with the carrier concentration. In addition, there is

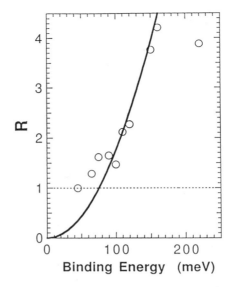

Figure Eight: Ratio of intensity of main feature to that of the shadow feature versus binding energy. The ratio, R, has been normalized to 1.0 at a binding energy of 50 meV.

a change in the shape of the Fermi surface, from "large" portions centered around the $(\pi,\pm\pi)$ points to a "small" portion centered around the $(\pi/2,\pm\pi/2)$ points. Recalling the earlier discussion about symmetry points, these results indicate a qualitative change from the $(\pi/2,\pm\pi/2)$ points as symmetry points for the antiferromagnetic insulator parent compound to the $(\pi/2,\pm\pi/2)$ points not acting as symmetry points for sufficient carrier concentration.[27]

In a recent paper, H. Ding et.al. [28] have argued that the "large" Fermi surface is retained for the entire underdoped part of the phase diagram. Their data analysis is, however, flawed in a way both fundamental and informative. Figure 11 illustrates the essential features of our data and that of Ref. 28. In Ref. 28 the quasiparticle state does not cross the Fermi level. Instead, the minimum binding energy is between 100-200 meV, depending on the temperature and wavevector. This is illustrated in exaggerated form in Figure 12. In analyzing photoemission data in which the quasiparticle peak is not at the Fermi energy, most authors conclude that the wavevector is not on the Fermi surface. The authors of Ref. 28, however, propose a different criterion for what "on the Fermi surface" means. They argue, as illustrated in Fig. 11, that if the leading edge of the spectral function is resolution limited, then they are "on" the Fermi surface. I am not aware of any theoretical underpinning to this argument. For example, if I apply this argument to the spectra of a disordered gold film deposited to obtain a Fermi energy reference, I would conclude that every point in the Brillouin zone is on the Fermi surface. However, the authors of Ref. 28 have raised two very important points: (1) the spectral function changes qualitatively between optimal doping and heavily underdoped, and (2) there is spectral intensity, and a "sharp" leading edge, at the chemical potential for heavily underdoped samples. These two points are discussed- in fact form the topic for- section III(C).

However, if one studies the actual quasiparticle states that form a Fermi surface, then our data indicate an evolution of the Fermi surface with doping. As Figure 10 illustrates, there is a qualitative change from a small Fermi surface with pockets around the $(\pi/2,\pm\pi/2)$ points for heavily underdoped samples to a large Fermi surface with pockets around the $(\pi,\pm\pi)$ points for optimally and overdoped samples. Such an evolution of the Fermi surface has been proposed theoretically by Chubukov, who argues that there is a topological phase transition with doping.[26] The practical importance of this idea is that for heavily underdoped samples the Fermi surface does not support d-wave superconductivity very well, because d-wave superconductivity involving spin fluctuations is enhanced when parts of the Fermi surface differing by a wavevector $\mathbf{k} = \langle\pi,\pi\rangle$ can be connected. A small Fermi surface, such as we have measured, does not support a d-wave order parameter very well.

C. SPECTRAL FUNCTION

The underdoped part of the phase diagram is of great current interest. Using a variety of experimental

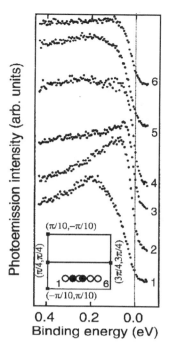

Figure Nine: Angle-resolved photoemission spectra parallel to the <π,π> direction for underdoped, T_c =30K, superconductor samples.

methods, several researchers have reported "pseudogaps" in the excitation spectrum. There are two fundamental reasons for the interest. One is the possibility that there are preformed d-wave pairs above T_c, that is, a non-zero superconducting gap without phase coherence. The other is that the gap in the excitation spectrum leads to a non-Fermi-liquid in the normal state. Implicit in the analysis, particularly of the

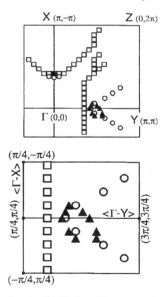

Figure Ten: Fermi surface measured for almost optimally doped (open square), underdoped T_c =60K (open circle) and underdoped T_c =30K (closed triangle) superconductor samples. Included is the full Fermi surface (top) and expanded view in the <π,π> direction (bottom).

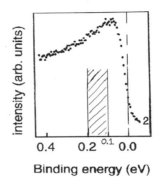

Binding energy (eV)

Figure Eleven: Angle-resolved photoemission spectrum from Fig. 9 compared to (dashed area) smallest binding energy of quasiparticle state reported in Ref. 28.

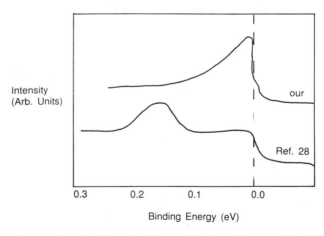

Figure Twelve: Schematic comparison of angle-resolved photoemission data, both ours and Ref. 28.

photoemission data, is that the pseudogap- change in the excitation spectrum- is a rigid shift of the same one particle Green's function and spectral function.[23,29] As I discuss in detail below, this is wrong. Instead, disorder, including the underdoped part of the phase diagram with oxygen defects, leads to qualitative changes in the spectral function. The experimental results indicate that some of the properties attributed to a pseudogap are, to put it kindly, overstated.

Figure 13 illustrates schematically three important aspects of the pseudogap in photoemission, including (a) the ideal pseudogap measurement, with perfect energy resolution, (b) a pretty good pseudogap measurement, with imperfect energy resolution, and (c) close to the actual situation, at least for heavily underdoped samples. In Fig. 13(a), there is only a quasiparticle, coherent, part to the spectral function, and this δ-function is shifted rigidly by some value (Δ). In Fig. 13 (b), there is both a coherent and an incoherent part to the spectral function, but this spectral function is also shifted rigidly by some value (Δ).

In Fig. 13 (c), there are changes to the spectral function. When there are changes to the spectral function, the idea of a pseudogap- in analogy to a semiconductor bandgap- no longer has any meaning.

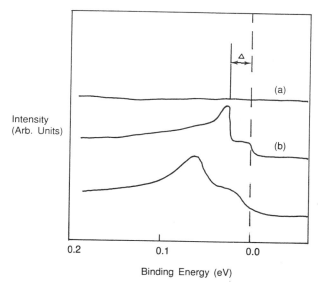

Figure Thirteen: Pseudogap (Δ) developed for (a) perfect energy resolution, (b) imperfect energy resolution, and (c) close to the experimental situation.

We have studied both underdoped and disordered (by neutron-irradiation) samples. In both instances, the spectra correspond to Fig. 13(c): there are significant changes to the spectral function. We observe three changes, including:

(1) the coherent part of the spectral function becomes broader, that is, the full width half maximum increases, with disorder. This along can cause an apparent shift in the centroid of the coherent part of the spectral function;

(2) the intensity of the coherent part of the spectral function, relative to the incoherent part, decreases with disorder, which means that the idea of a rigid shift (Fig. 13(b)) is invalid;

(3) the leading edge of the spectral function changes. In perfectly ordered samples, the leading edge of the spectral function is determined by the Fermi-Dirac distribution function. In such a circumstance, the width of the leading edge is limited by the energy resolution of the experimental equipment. However, for sufficient disorder, the width of the leading edge is not resolution-limited. In particular, there is a transfer

of spectral weight to higher binding energy. As a result, the midpoint of the leading edge also shifts to higher binding energy. Since recent reports have used the midpoint as the reference point, the result, "automatically," is a pseudogap. However, such a pseudogap reflects the change in the spectral function, not as such a gap in the excitation spectrum.

Now let me turn from idealization to actual data.[30] We measured the leading edge of the spectral function for four sample types. In all four instances, care was taken to keep the oxygen stoichiometry the same. The samples have been irradiated with thermal neutrons and, for two types, subsequently annealed at low temperatures. The only difference among the photoemission spectra taken is the amount of oxygen disorder. The samples include two that are superconducting, with T_c values of 83K and 43K, and two that are insulating, with a difference of a factor of two in the neutron dose levels. The ab-plane resistivity for the four sample types includes metallic-like resistivity (above T_c)for the superconducting samples, an almost constant resistivity for the low dose insulator, and a two-dimensional hopping conductivity for the high-dose insulator.

The photoemission spectra of all four sample types were taken at wavevectors $k > k_F$. The T_c =83K and T_c=43K samples exhibits a leading edge that is limited by the total energy resolution of the system. However, both insulating samples, which exhibit spectral intensity up to the chemical potential, exhibit leading edges that are not resolution-limited. This result indicates that the increasing disorder results in a changes in the spectral function, including a change in the sharpness of the leading edge. Previous work by my colleagues and myself on Co-doped and Ni-doped $Bi_2Sr_2CaCu_2O_{8+x}$ single crystal samples [31,32] also

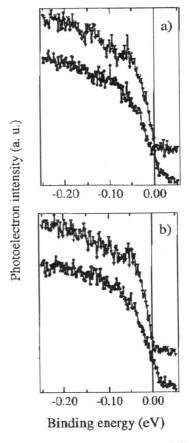

Figure Fourteen: Angle-resolved photoemission spectra in (a) the $<0,\pi>$ and (b) the $<\pi,\pi>$ directions for Pr-doped BSCCO-2212 samples. Data in the normal state (closed circles) and in the superconducting state (closed triangles) both exhibit a pseudogap.

demonstrates that for cation doping, with T_c values ~60K, the leading edge is not resolution-limited. If we were to use the criterion advocated in Refs. 23 and 29, we would conclude that there is a pseudogap of 15 meV for the cation-doped samples, and higher values for the lower dose insulating and higher dose insulating samples. For the cation-doped samples, the pseudogap of ~15 meV is also isotropic in the three high-symmetry directions. We have further studied Pr-doped $Bi_2Sr_2CaCu_2O_{8+x}$ single crystal samples, which have a T_c of 60K .[30] These samples have been studied in both the normal and superconducting states. In the normal state, there is a pseudogap of 15 meV, and again this pseudogap is isotropic in the three high-symmetry directions. In the superconducting state, as illustrated in Figure 14, there is a further change in the spectral function, a very weak superconducting feature, and no further gap opening. Thus, for Co-, Ni- and Pr-doped samples, one observes almost the same value of both T_c and of the **isotropic** pseudogap. However, Pr-doped samples are viewed as changing T_c via changes in carrier concentration (doping) while Co- and Ni-doped samples are viewed as changing T_c due to disorder. I argue from these data that in fact attributing a pseudogap as related to superconductivity is wrong. Instead, the data indicate that the pseudogap observed in photoemission arises primarily from disorder, and that it functions much as a mobility edge, visible at lower temperatures and not visible at higher temperatures.[33]

D. VAN HOVE SINGULARITY

Early in the development of electronic band structure calculations, the idea of singularities in the electronic density of states was found.[34] Because the electronic density of states diverges- at least mathematically- these places in the Brillouin zone were particularly important for physical properties that depend on the electronic density of states, e.g., photoemission signal. However, at the time, BCS theory prevailed in superconductivity, and BCS theory argued that the superconducting pairing interaction averaged over the electronic density of states around the entire Fermi surface. In reduced dimensions, however, these van Hove singularities become more important. As Abrikosov [35] and Bok [36] have particularly emphasized, the

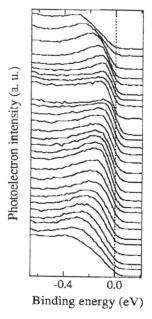

Figure Fifteen: Angle-resolved photoemission spectra in the <0,π> direction. The spectra exhibit an extended van Hove singularity near but below the chemical potential.

enhanced density of states can, under rather undemanding restrictions, lead to a larger value of T_c. The increase in T_c is greater provided that (a) the van Hove singularity is not merely a single point in the Brillouin zone, but rather a line of singular points, the so-called extended singularity, and (b) the van Hove singularity has a particular topology, that of a horse saddle, in which there is no dispersion in one axis but dispersion up through the chemical potential in the perpendicular axis. There remain two important experimental questions:
1) Do the cuprates exhibit such an extended, saddle-point, van Hove singularity?
2) Does this van Hove singularity change, with doping and disorder, in a way that make sense in explaining the values of T_c ?

By now, the first question is answered in the affirmative. Several of the cuprates exhibit a saddle point van Hove singularity.[37-40] This was first reported by J.C. Campuzano and colleagues,[37] and has been studied and reported by several research groups. The van Hove singularity is extended, and is a saddle-point type of singularity.

The second question remains in dispute. On the one hand, different materials with markedly different values of T_c do indeed exhibit changes in the van Hove singularity as expected if the van Hove singularity was the controlling factor for T_c.[37-40] Specifically, the van Hove singularity is closest to the chemical potential for the samples with the highest value of T_c, and there is a well-ordered relaion between proximity

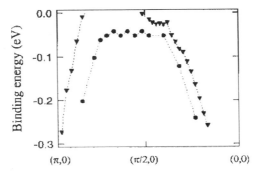

Figure Sixteen: Dispersion relation of the T_c =43K (closed triangle) and the T_c =83K (closed circle) superconducting samples in the <0,π> direction. The peak positions are shown, rather than the best fit including the finite energy resolution.

of the van Hove singularity to the chemical potential and the value of T_c. However, the doping dependence within a single system, while not yet settled, does not thus far provide any particular support for the importance of a van Hove singularity in determining T_c. As of this writing, however, the doping dependence studies are scattered and not conclusive.[41]

If the doping dependence studies are not yet conclusive, studying the effects of disorder has barely begun. The first such study is illustrated in Figure 15.[30] Fig. 15 illustrates angle-resolved photoemission data for a neutron-irradiated $Bi_2Sr_2CaCu_2O_{8+x}$ single crystal sample, subsequently annealed to restore superconductivity with a T_c =83K. We have also taken angle-resolved photoemission data for a neutron-irradiated $Bi_2Sr_2CaCu_2O_{8+x}$ single crystal sample, subsequently annealed at a lower temperture to restore superconductivity with a T_c =43K. Figure 16 illustrates the dispersion relation in the <0,π> direction for both samples. There are two major differences between the samples. The T_c =43K sample exhibits a dispersion through the chemical potential at a wavevector of ~0.85 Å^{-1} and subsequent dispersion back below the chemical potential at a wavevector of ~1.33 Å^{-1}. By contrast, the T_c =83K sample exhibits a van Hove singularity that remains below the chemical potential. The lower T_c sample thus exhibits (a) a less extended van Hove singularity below the chemical potential, and (b) greater dispersion, which further weakens the effect of the van Hove singularity. These differences are, qualitatively, as expected if the van Hove singularity controls T_c. However, in the absence of a serious calculation no further conclusion can be drawn.[30]

The single most serious criticism of using the van Hove singularity to explain T_c is that systems with very low values of T_c (~1K) also exhibit a van Hove singularity. Recently, T. Cummins and colleagues,[42] and T. Takahashi and colleagues,[43] have reported on the presence of a van Hove singularity in Sr_2RuO_4. Sr_2RuO_4 becomes superconducting at ~1K. However, the angle-resolved photoemission data show an extended, saddle-point type van Hove singularity. The van Hove singularity for the ruthenate differs from that of the cuprate in two respects. First, the van Hove singularity in the ruthenate is less extended in the bond-axis direction, and hence more two-dimensional in character. Second, the atomic orbitals from which the van Hove singularity is comprised are the $Ru4d(\varepsilon)$-$O2p(\pi)$ antibonding states, while in the cuprate it is the $Cu3d(x^2-y^2)$-$O2p(\sigma)$ antibonding orbitals.[43] There are three possible explanations for this. The most widely believed is that the van Hove singularity simply does not affect T_c in a significant way. However, this rush to judgment ignores two other possibilities. The symmetry of the electronic states that comprise the van Hove singularity is different for Sr_2RuO_4 as compared to the cuprates. It may be that this different combination of atomic orbitals alters the effect of the van Hove singularity. Further, the difference in the extent of the van Hove singularity may be enough to explain the difference in T_c. So at this point, the van Hove singularity explanation for high values of T_c is viewed as inconsistent with some experimental data with several crucial questions remaining unanswered.

IV. SUPERCONDUCTING STATE

A. SUPERCONDUCTING GAP ANISOTROPY AND SYMMETRY

In 1989, several investigators reported observing a superconducting gap using angle-resolved photoemission;[6,44,45] of these, the most definitive was by C.G. Olson and colleagues[6] Another remarkable result was the report by B.O. Wells and colleagues that the superconducting gap is anisotropic.[46,47] The latter result was particularly important because the error bars of the gap size in the $<\pi,\pi>$ direction overlap with zero, meaning that the data are consistent with a $d(x^2 - y^2)$ order parameter. These results have lead to several other important studies. One, by R.J. Kelley and colleagues,[16] reports that the superconducting gap anisotropy changes with doping. For optimally doped samples, the ratio of the gap in the $<0,\pi>$ direction (the Cu-O-Cu bond axis direction in real space) to that in the $<\pi,\pi>$ direction as large as 20:1. For overdoped samples, the ratio $\Delta(<0,\pi>):\Delta(<\pi,\pi>)$ is reduced to ~2:1 Equally important, the gap size in the $<\pi,\pi>$ direction is definitely not zero for overdoped samples, which means that the order parameter is not pure d-wave.[16] This result is in marked contrast to BCS theory, in which the carrier concentration does not affect the gap anisotropy at all. Another important consequence, however, is that a pure d-wave order parameter does not explain the gap size data for overdoped samples.

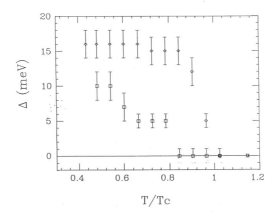

Figure Seventeen: Superconducting gap (Δ) versus normalized temperature in the $<0,\pi>$ direction (open diamond) and the $<\pi,\pi>$ direction (open square).

It has been widely argued in the superconductivity community that photoemission can only measure the size of the superconducting gap (order parameter), and photoemission only provides information about the symmetry of the order parameter by where the nodes (zeros) in the gap are. Since no experimental technique can prove that the gap is actually zero, this provides endless fun and arguments for scientists. In fact, this argument is wrong. Photoemission, while it cannot provide **all** symmetry information, can provide limited information on the symmetry of the order parameter. The argument goes as follows: In sect. I(B), I discussed the idea of using angle-resolved photoemission to obtain information on the symmetry of the quasiparticle state. In sect. II(A), using this idea, my colleagues and I measured the evolution of reflection symmetry of the normal state quasiparticle states with doping. Below T_c, as first shown in Ref. 6, there is a narrow feature characteristic of the superconducting state. If this feature has the same reflection symmetry as the lattice, then nothing can be said about the order parameter symmetry from photoemission data. However, if this feature, present only due to superconductivity, has a reflection symmetry **different** from the normal state quasiparticle states, then the order parameter is by definition unconventional. R.J. Kelley et.al. [48] reported that for nearly optimally doped $Bi_2Sr_2CaCu_2O_{8+x}$ samples, the reflection symmetry of the superconducting feature was different from the normal state quasiparticles. This difference proves direct evidence that the order parameter is not s-wave, that is, purely conventional. Because the measurement is one of symmetry, it does not depend on the size of the superconducting gap and is not limited by experimental energy resolution.

B. TEMPERATURE DEPENDENCE OF GAP

Ref. 6 also noted that the size of the superconducting gap changes with temperature. This report was followed by Jian Ma et.al.,[49] who reported on the size of the superconducting gap for somewhat overdoped samples. Figure 17 illustrate the experimental results. The data lead to three important conclusions:

(1) The gap in the $<0,\pi>$ direction does not follow the temperature dependence expected of a BCS gap. Instead, the gap opens much faster as the temperature decreases below T_c;

(2) In a BCS superconductor, such as lead, that has a small gap anisotropy at low temperatures, the anisotropy disappears as the temperature approaches T_c from below. The behavior of the cuprate superconductors is qualitatively different: the gap anisotropy increases markedly with increasing temperature;

(3) Close to T_c, the gap is consistent with a pure d-wave order parameter, while well below T_c the gap is inconsistent with a pure d-wave order parameter. This changes suggests that the order parameter might have two components.

C. GAP AND DISORDER

As noted in section II, there is a qualitative change to the spectral function with increasing disorder. A natural question is whether there is any change to the photoemission spectra observed for such materials, materials in which disorder- localization- is not negligable. The recent, beautiful, work by Bobinger et.al. [50] makes this issue even more important. In Ref. 55, when the superconducting phase transition is suppressed, the ground state of optimally doped and underdoped $La_{2-x}Sr_xCuO_4$ is found to be localized, while that of overdoped $La_{2-x}Sr_xCuO_4$ is metallic. The analysis of photoemission data in the superconducting state has thus far assumed- as per BCS theory- that pairs of quasiparticles form Cooper pairs and condense at T_c. In BCS theory, disorder is entirely bad, and electrons localized to atomic sites cannot participate in superconductivity. Our results indicate that at temperatures well below T_c one observes a superconducting condensate, and gap, even for samples without quasiparticle states observed above T_c .[51] Recently, A. Chubukov proposed the first theoretical model that provided an explanation for such observations.[52] The essential argument that Chubukov makes is that there is an energy scale that increases as the disorder decreases. From the chemical potential to this energy scale, the system behaves as a Fermi liquid. Below T_c, disorder- scattering- decreases markedly. The energy scale thus increases. If the energy scale is larger than the gap, then a "regular" superconducting gap is observed, even if in the normal state, due to increased scattering, there is no coherent quasiparticle state observed. The data of Ref. 50 indicate this behavior.

As noted previously, we have also studied neutron-irradiated samples in which T_c ranges from 83K to non-superconducting, depending on the level of disorder. We have measured the normal state photoemission spectra at several cuts in the Brillouin zone, including the three high-symmetry directions.[20] We find that the value of T_c depends on the coherent, quasiparticle spectral intensity in or near the $<0,\pi>$ direction, that is, the van Hove singularity direction. T_c does not depend on the quasiparticle spectral intensity in the $<\pi,\pm\pi>$ directions. The value of T_c changes monotonically with the quasiparticle spectral intensity in the $<0,\pi>$ direction. In the $<\pi,\pm\pi>$ directions, by contrast, the quasiparticle spectral intensity for an insulating sample is **greater** than for a sample with T_c =43K. We conclude that it is the itinerant carriers in the van Hove singularity that predominantly determine T_c .[20]

D. OTHER MEASUREMENTS

In a field with over 30,000 refereed journal articles, one can find support for practically any view. That being said, there are some recent measurements that are particularly noteworthy considering the data and interpretations that I have presented. The idea that the superconducting order parameter might have two components is not new; it was predicted theoretically by Kotliar [53] and by Joynt.[54] On the experimental side, the recent work by C. Rossel and colleagues,[55] using torque magnetometry- a bulk measurement technique- also leads to a two-component order parameter.

The non-BCS temperature dependence of the gap in the $<0,\pi>$ direction was first reported by C.G. Olson et.al.[6] The idea that the ratio of the two order parameter components changes with temperature is also found in the torque magnetometry data of C. Rossel and colleagues.[55]

The argument that the gap becomes closer to isotropic with overdoped samples has been concluded by experiments in four different laboratories and by both photoemission and Raman measurements;[14] the latter is bulk sensitive. The idea that the ratio of the maximum gap to T_c decreases for overdoped samples has also been reported by both photoemission and Raman measurements.[14,15]

V. CONCLUSIONS

The experimental work presented in this tutorial lead to several noteworthy conclusions. In the normal state, photoemission provides strong evidence that short-range antiferromagnetic fluctuations remain well into the overdoped part of the phase diagram. There is an evolution of the Fermi surface with doping, and the evolution observed is from a very small Fermi surface at low doping to a large Fermi surface for optimally doped and overdoped samples. The photoemission data give clear evidence that underdoped samples are also disordered samples, and the pseudogap almost certainly arises from this change in spectral function, having nothing to do with superconductivity, pre-formed d-wave pairs, or other such exotica. The van Hove singularity exists for several cuprates- all of those studied- but explaining the relation between the van Hove singularity and T_c has substantial uncertainty at present.

The superconducting gap studies using photoemission are, for near optimally doped samples, consistent with a d-wave order parameter. By the same token, photoemission studies of overdoped samples are **inconsistent** with a pure d-wave order parameter. The superconducting gap anisotropy changes markedly with doping, and the temperature dependence of the gap is quite different from a BCS superconductor. The ratio of the maximum superconducting gap to T_c changes markedly with doping, indicating that there are two energy scales in the cuprate superconductivity problem, rather than the single energy scale in a BCS superconductor. There is evidence that a superconducting gap can be observed well below T_c without observing an itinerant, quasiparticle feature above T_c . There is some evidence that carriers need not be Bloch states in order to participate in superconducting pairing, but this complicated situation is not fully understood.

Acknowledgements

Parts of the work presented here were performed in collaboration with P. Almeras, H. Berger, A. Chubukov, M. Grioni, R. Joynt, A. Karkin, R.J. Kelley, Jian Ma, G. Margaritondo, S. Misra, C.G. Olson, C. Quitmann, I. Vobornik, M. Zacchigna and F. Zwick. Financial support was provided by EPFL, Fonds National de Recherche Scientifique, Deutsche Forschungsge-meinschaft, and the U.S. NSF (DMR-96-32527). The author has benefitted from conversations with and the work of many colleagues, particularly J. Bok, H. Ding, J.C. Campuzano, and T. Takahashi. Prof. Takahashi and the author discussed Ref. 43 in some detail. I. Vobornik and S. Misra assisted in preparing the figures.

REFERENCES

1. J. Bardeen, L.N. Cooper and J.R. Schrieffer, Phys. Rev. 108 , 1175 (1957).
2. K.A. Musaelian et.al., Phys. Rev. B 53, 3598 (1996).
3. J. Hamlicek et.al., Physica C 206 , 345 (1993).
4. See, e.g., B. Batlogg et.al., Physica C 235-240 , 130 (1994) and the contribution by B. Batlogg to this NATO Summer School.
5. D. Pines in "High T_c and the C-60 family," H.C. Ren, ed., Gordon and Breach, NY, 1995, p. 1.
6. C.G. Olson et. al., Science 245, 732 (1989).
7. E.W. Plummer and W. Eberhardt, Adv. Chem. Phys. 49, 533 (1982).

8. A. Einstein, Ann. Phys. (Leipzig) $\underline{17}$, 132 (1905).

9. "Quantum Mechanics," L.I. Schiff, 3rd Ed., McGraw-Hill, New York, 1968.

10. T. Koopmans, Physica $\underline{1}$, 104 (1934).

11. "Breakdown of the One Electron Pictures in Photoelectron Spectra," Gorin Wendin, Springer-Verlag, Berlin, 1981.

12. See,e.g., J. Bok and J. Bouvier, SPIE $\underline{296}$, 131 (1996) and references therein.

13. K. Kitazawa, Science $\underline{271}$, 313 (1996); H.J. Tao, A. Chang, F. Lu and E.L. Wolf, Phys. Rev. B $\underline{45}$, 10622 (1992).

14. C. Kendziora, R.J. Kelley and M. Onellion, Phys. Rev. Lett. $\underline{77}$, 727 (1996).

15. M. Onellion et.al., submitted.

16. R.J. Kelley et.al., Science $\underline{271}$, 1255 (1996).

17. E. Bloch, Z. Physik $\underline{57}$, 545 (1929).

18. S. LaRosa et.al., Phys. Rev. B $\underline{58}$, R595 (1997).

19. C. Kendziora, R.J. Kelley, E. Skelton and M. Onellion, Physica C $\underline{257}$, 74 (1996).

20. C. Quitmann et.al., submitted.

21. P. Aebi et.al., Phys. Rev. Lett. $\underline{72}$, 2757 (1994).

22. A.P. Kampf and J.R. Schrieffer, Phys. Rev. B $\underline{42}$, 7967 (1990).

23. H. Ding et.al., Nature $\underline{382}$, 51 (1996).

24. S. LaRosa et.al., Solid State Commun., in press (1997).

25. A. Chubukov, Phys. Rev. B $\underline{52}$, R3840 (1995).

26. A. Chubukov et.al., Phil. Mag. $\underline{74}$, 563 (1996).

27. S. LaRosa et.al., submitted.

28. H. Ding et.al., Phys. Rev. Lett. $\underline{78}$, 2628 (1997).

29. A.G. Loeser et.al., Science $\underline{273}$, 325 (1996).

30. I. Vobornik et.al., submitted.

31. P. Alméras et.al., Solid State Commun. $\underline{91}$, 535 (1994).

32. C. Quitmann et.al., Phys. Rev. B $\underline{53}$, 6819 (1996).

33. B. Beschoten et.al., Phys. Rev. Lett. $\underline{77}$, 1837 (1997).

34. L. Van Hove, Phys. Rev. $\underline{89}$, 1189 (1953).

35. A.A. Abrikosov, Physica C $\underline{222}$, 191 (1994).

36. J. Bok and L. Force, Physica C $\underline{185\text{-}189}$, 1449 (1991).

37. A.A. Abrikosov, J.C. Campuzano and K. Gofron, Physica C $\underline{214}$, 73 (1993).

38. K. Gofron et.al., Phys. Rev. Lett. $\underline{73}$, 3302 (1994).

39. D.S. Dessau et.al., Phys. Rev. Lett. $\underline{71}$, 2781 (1993).

40. Jian Ma et.al. Phys. Rev. B $\underline{51}$, 3832 (1995).

41. E. Dagotto et.al., Phys. Rev. Lett. $\underline{73}$, 728 (1994).

42. D.H. Lu et.al., Phys. Rev. Lett. $\underline{76}$, 4845 (1996).

43. T. Yokoya et.al., Phys. Rev. Lett. $\underline{76}$, 3009 (1996).

44. Y. Chang et.al., Phys. Rev. B $\underline{39}$, 4740 (1989); Y. Chang et.al., Phys. Rev. Lett. $\underline{63}$, 101 (1989).

45. J.-M. Imer et.al., Phys. Rev. Lett. $\underline{62}$, 336 (1989).

46. B.O. Wells et.al., Phys. Rev. B $\underline{46}$, 11830 (1992).

47. Z.-X. Shen et.al., Phys. Rev. Lett. $\underline{70}$, 1553 (1993).

48. R.J. Kelley, Jian Ma, G. Margaritondo and M. Onellion, Phys. Rev. Lett. $\underline{71}$, 4051 (1993).

49. Jian Ma et.al., Science $\underline{267}$, 862 (1995).

50. G.S. Boebinger et.al., Phys. Rev. Lett. $\underline{77}$, 5417 (1996).

51. Jian Ma et.al., Solid State Commun. $\underline{94}$, 27 (1995).

52. A. Chubukov, unpublished.

53. G. Kotliar, Phys. Rev. B $\underline{37}$, 3664 (1988); G. Kotliar and J. Liu, op. cit. $\underline{38}$, 5142 (1988).

54. J. Betouras and R. Joynt, Europhys. Lett. $\underline{31}$, 119 (1995).

55. C. Rossel et.al., unpublished and private communication.

THE ELECTRONIC STRUCTURE OF THE HIGHT_C SUPERCONDUCTORS OBTAINED BY ANGLE-RESOLVED PHOTOEMISSION

Juan-Carlos Campuzano,[1,2], Mohit Randeria[3], Michael Norman[2], and Hong Ding[1,2]

[1]Physics Department, University of Illinois at Chicago,
Chicago, IL 60607, USA
[2]Material Science Division, Argonne National Laboratory,
Argonne, IL 60439, USA
[3]Theoretical Physics Group, Tata Institute of Fundamental Research,
Mumbai, India

INTRODUCTION

Angle-resolved photoemission spectroscopy (ARPES)[1,2] has played a major role in understanding the electronic structure of the high temperature superconductors (HTSCs). This has occurred because of three factors, a) the two-dimensional nature of the HTSCs, b) the large energy scales involved, and c) the progress in energy resolution of current spectrometers. Some of the important milestones are: the first observation of the superconducting gap in angle-integrated PES by Imer, et al.[2], the first observation of photoemission in the HTSCs by Arko, et al.[3], the subsequent angle-resolved study of the gap by Olson, et al.[4], followed by the observation of a normal state with a Luttinger Fermi surface by Campuzano, et al.[5] and Olson et al.[6], flat bands in the dispersion by Gofron, et al.[7] and Dessau, et al.[8], and the anisotropy of the SC gap by Shen, et al.[9]; see also a review by Shen and Dessau[10]. Here we will concentrate on the tremendous progress of the past three years in which, we believe, there has been a qualitative change in thinking about ARPES data and its analysis. As a consequence of this change, many new physics results have shed very important new light on the high Tc superconductors.

Most of the results described here are from the $Bi_2Sr_2CaCu_2O_{8-\delta}$ (Bi2212) compound, shown in Fig. 1. Although the structures of the HTSCs are well known, here we point out a few important points which are relevant to the photoemission experiments. Bi2212 is unique among the HTSCs in that it has two BiO layers, which are believed to be van der Waals coupled, leading to the longest bond length in all the cuprates. This results in extremely smooth surfaces, which are crucial for ARPES, since this is a surface-sensitive technique due to the short escape depth (10 Å) of the outgoing electron. Furthermore, the fact that there is no charge transfer means that when a sample is cleaved to expose a fresh

The Gap Symmetry and Fluctuations in High - T$_c$ Superconductors
Edited by Bok *et al.*, Plenum Press, New York, 1998

229

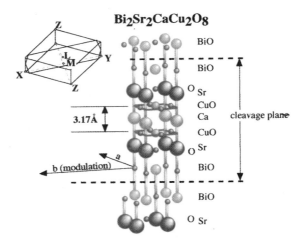

Bi₂Sr₂CaCu₂O₈

BiO
BiO
O Sr
CuO
Ca   cleavage plane
CuO
O Sr
BiO
3.17Å
b (modulation)   BiO
O Sr

Fig. 1. Structure of Bi₂Sr₂CaCu₂O₈. Of note are the two *BiO* layers with a superlattice of period (0.21π, 0.21π) along the *b*-axis.

surface for an experiment, no surface states are created, since the potential below the surface is minimally disturbed. Another important consideration is the presence of a structural superlattice in the *BiO* planes with a periodicity of (0.21π, 0.21π) along the *b*-axis of the structure. This superlattice diffracts the outgoing photoelectrons which originated in the CuO_2 planes below, creating additional structure in the ARPES spectra. Fortunately, as we show below, the *BiO* superlattice apparently does not disturb the potential in the CuO_2 layers.

Nevertheless, there are important issues which have to be addressed. We have little prior experience in analyzing ARPES data on the energy scale of few 10's of meV. Much of the rest of these lectures will focus on recent efforts towards properly analyzing such data, and developing an understanding of low frequency information contained in it. We will focus here on the conceptual issues leaving many of the technical details to the papers referred to in the text.

WHAT DOES ARPES MEASURE?

In principle, the theoretical interpretation of ARPES spectra is complicated by the fact that photoemission measures a *nonlinear* response function. Shortly after Hertz discovered photoemission in 1887, the experiments of Lenard in 1990 showed that the photo-electron current at the detector is proportional to the number of incident photons, i.e., to the *square* of the vector potential. Shaich and Ashcroft[11] first emphasized that the relevant correlation function is a three current correlation.

It is instructive to briefly review their argument. Using the standard Kubo response function calculations, lets look at an expansion of the current at the detector, the response, in powers of the applied vector potential (incident photons). Let $\mathbf{R}$ be the location of the detector in vacuum, and $\mathbf{r}$ denote points inside the sample. The zeroth order piece $\langle 0|j_\alpha(\mathbf{R},t)|0\rangle$ vanishes as usual; there are no currents flowing anywhere in the absence of the applied field. Here $|0\rangle$ is the ground state of the unperturbed system. The linear response also vanishes. $\langle 0|j_\alpha(\mathbf{R},t)j_\beta(\mathbf{r},t')|0\rangle = 0$ and $\langle 0|j_\alpha(\mathbf{r},t')j_\beta(\mathbf{R},t)|0\rangle = 0$, since there are no particles at the detector, in absence of the e.m. field, and $j_\beta(\mathbf{R},t)|0\rangle = 0$. Thus the leading term which survives is

$$\left\langle j_\gamma(\mathbf{R},t) \right\rangle \propto \int \, d\mathbf{r}' \, dt' \, d\mathbf{r}'' \, dt'' \, A_\alpha(\mathbf{r}',t') A_\beta(\mathbf{r}'',t'')$$
$$\langle 0 | j_\alpha(\mathbf{r}',t') j_\gamma(\mathbf{R},t) j_\beta(\mathbf{r}'',t'') | 0 \rangle \tag{1}$$

where only current operators *inside* the sample act on the unperturbed ground state on either side and the current at the detector is sandwiched in between.

The three current correlation function can be represented by the triangle diagrams[12] of Fig. 2, where the line between the two external photon vertices is the Greens function of the "initial state" or "photo-hole" and the two lines connecting the photon vertex to the current at the detector represent the "photo-electron" which is emitted from the solid.

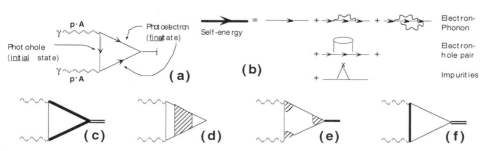

Fig. 2. a) The photoemission Feynman diagram; b) Possible self-energy corrections to each fermion line; c) Self-energy corrections to the photoelectron lines; d) photoelectron inelastic scattering corrections; e) vertex corrections; f) photohole self-energy

There is a large literature[13] on the ab-initio evaluation of the bare triangle diagram (a), incorporating the realistic electronic structure and surface termination, together with multiple-scattering effects in the photo-electron final state. Such studies are useful for understanding the photoemission intensities, final state and surface effects, but not the line-shape and the many-body aspects of the problem

Important propagators are shown in (b), and all possible renormalizations -- vertex corrections and self energy effects -- of the bare triangle diagram are shown in (c) through (f) of Fig. 2. Although it is impossible to evaluate these diagrams in any controlled calculation, they are useful in understanding, qualitatively, what the various processes are and in estimating their importance. Diagram (f) represents the many-body renormalization of the occupied initial state, which is what we are interested in; (c) and (d) represent final state line-width broadening and inelastic scattering, and (e) is a vertex correction that describes the interaction of the escaping photo-electron with the photo-hole in the solid. [An additional issue in a quantitative theory of photoemission is related to the modification of the external vector potential inside the medium, i.e., renormalizations of the photon line. This problem is somewhat well understood[13]].

What we wish to know is under which experimental conditions --if any-- can the many-body interactions shown in diagrams (c) to (e) be minimized, so that one is left with mostly diagram (e). This is known in the literature as the "sudden approximation", where it is assumed that the photoelectron leaves the sample so quickly that its interaction with the photohole left behind is negligible. Let us first discuss the validity of the sudden approximation, for 15 - 30 eV (ultraviolet) incident photons, by making some simple time scale estimates. The time t spent by the escaping photo-electron in the vicinity of photo-hole is the time available for interactions ("vertex corrections") which would invalidate the sudden approximation. A photoelectron with a kinetic energy of (say) 20 eV has a velocity

$v = 3 \times 10^8$ cm/s. The relevant length scale, which is the smaller of the screening radius (of the photo-hole) and the escape depth, is ~10 Å. Thus $t = 3 \times 10^{-16}$ s. This is to be compared with the time scale for electron-electron interactions (which are the dominant source of interactions at the high frequencies of interest): $t_{ee} = 2\pi / \omega_p = 4 \times 10^{-15}$ s, using a plasma frequency $\omega_p = 1$ eV for the cuprates (this would be even lower if c-axis plasmons are involved). If $t < t_{ee}$, then we can ignore vertex corrections. Our very crude estimate is $t / t_{ee} = 0.1$; all that we can say is that the situation with regard to the validity of the impulse approximation is not hopeless, but clearly a better estimate or a different approach is needed. We have taken the approach of verifying the validity of the impulse approximation experimentally, as discussed below.

Very similar estimates can be made for renormalizations of the outgoing photo-electron due to its interaction with the medium (diagrams c and d); again e-e interactions dominate at the energies of interest. The relevant length scale here is the escape depth, which leads to a process of self-selection: those electrons that actually make it to the detector with an appreciable KE have suffered no collisions in the medium. Such estimates indicate that the "inelastic background" must be small -- although its precise dependence on $\mathbf{k}$ and ω is uncertain.

To summarize: 1) the corrections to the sudden approximation are probably small and we shall test its validity further below. 2) Final state line-width effects are negligible. A clear experimental proof for Bi2212 is the fact that deep in the SC state a resolution limited spectral peak is seen, as discussed below in Section 6. 3) While the additive extrinsic background due to inelastic scattering is small, its precise form remains an important unresolved problem.

SPECTRAL FUNCTIONS AND SUM RULES

We will now discuss the experimental test of the validity of the impulse approximation. The strategy is as follows[14]: We assume the validity of the impulse approximation, and by using some sum rules, we derive some general consequences. We then test the consequences experimentally, which in turn indicates that the sudden approximation is a reasonable one. If the sudden approximation is correct, and ignoring the extrinsic background, the intensity of the ARPES spectrum, or the energy distribution curve (EDC), is then given by

$$I(\mathbf{k}, \omega) = I_0(\mathbf{k}) f(\omega) A(\mathbf{k}, \omega) \qquad (2)$$

where $\mathbf{k}$, the in-plane momentum, gives the location in the 2D Brillouin zone, and ω is the energy of the initial state measured relative to the chemical potential. (Experimentally we measure ω relative to the Fermi level of Pt, a good metal in electrical contact with the sample). $I_0(\mathbf{k})$ includes all the kinematical factors and the dipole matrix element (squared). It depends, in addition to $\mathbf{k}$, on the incident photon energy and polarization. The only general constraints on $I_0(\mathbf{k})$ come from dipole selection rules obeyed by the matrix elements. We will discuss these later on.

The spectral line-shape (ω dependence) of the EDC and its T dependence, at the low frequencies and temperatures of interest to us, are entirely controlled by $f(\omega)A(\mathbf{k}, \omega)$. Here $A(\mathbf{k}, \omega)$ is the initial state or "photo-hole" spectral function $A(\mathbf{k}, \omega) = -(1 / \pi) \operatorname{Im} G(\mathbf{k}, \omega + i0^+)$, and the Fermi function ensures that we are only looking at the *occupied* part of this spectral function. This can formally be seen as follows: the spectral function consists of two pieces $A(\mathbf{k}, \omega) = A_+(\mathbf{k}, \omega) + A_-(\mathbf{k}, \omega)$, which are spectral weights to add and to remove an electron from the system. In ARPES, where one extracts an electron one is measuring $A_-(\vec{k}, \omega) = Z^{-1} \sum_{m,n} e^{-\varepsilon_m} |\langle n|c_k|m\rangle|^2 \delta(\omega + \varepsilon_n - \varepsilon_m)$, which can be rewritten, using standard manipulations, as $A_-(\mathbf{k}, \omega) = f(\omega)A(\mathbf{k}, \omega)$.

There are some general sum rules that apply to the spectral function. The well-known sum rule $\int_{-\infty}^{+\infty} d\omega A(\mathbf{k},\omega) = 1$, which simply expresses the conservation of probability is not very useful for ARPES, since it is a sum-rule for angle-integrated photoemission (PES) A_- and *inverse* PES (A_+). The density of states (DOS) sum rule $\sum_{\mathbf{k}} A(\mathbf{k},\omega) = N(\omega)$ is also not directly useful since there is the $\mathbf{k}$-dependent matrix element factor $I_0(\mathbf{k})$. The important sum rule is

$$\int_{-\infty}^{+\infty} d\omega f(\omega) A(\mathbf{k},\omega) = n(\mathbf{k})$$

(3)

which directly relates the energy-integrated ARPES intensity to the momentum distribution $n(\mathbf{k})$. Somewhat surprisingly, its usefulness has never been exploited in the ARPES literature. We will use (3) to get experimental information on $n(\mathbf{k})$.

We first focus on the Fermi surface $\mathbf{k} = \mathbf{k}_F$. One of the major issues that will occupy us in the rest of these lectures, is how to define $\mathbf{k}_F$ in a strongly interacting system, which may not even have well-defined quasiparticles -- at finite temperatures -- and how to determine it experimentally. At this point, we simply define the Fermi surface to be the $\mathbf{k}$-space locus of gapless excitations in the normal state, so that $A(\mathbf{k},\omega)$ has a peak at $\omega = 0$. To make further progress with eqn. (3), we need to make a weak particle-hole symmetry assumption: $A(\mathbf{k}_F, -\omega) = A(\mathbf{k}_F, \omega)$ for "small" ω, where small means those frequencies for which there is significant T-dependence in the spectral function. It then follows that[14] $\partial n(\mathbf{k}_F)/dT = 0$, i.e., *the integrated area under the EDC at $\mathbf{k}_F$ is independent of temperature*. To see this, rewrite eqn. (3) as $n(\mathbf{k}_F) = 1/2 \int_0^\infty d\omega \tanh(\omega/2T) [A(\mathbf{k}_F,\omega) - A(\mathbf{k}_F, -\omega)]/2$, and take its T-derivative. It should be emphasized that we cannot say anything about the *value* of $n(\mathbf{k}_F)$, only that it is T-independent. (A much stronger assumption, $A(\mathbf{k}_F, -\omega) = A(\mathbf{k}_F, \omega)$ for *all* ω, is sufficient to give $n(\mathbf{k}_F) = 1/2$ independent of T). We emphasize that the $\mathbf{k}_F$-sum-rule is only approximate, since there is no exact symmetry that enforces it.

We note that we did not make any use of any properties of the spectral function other than the weak p-h symmetry assumption, and to the extent that this is also valid in the SC state, our conclusion $\partial n(\mathbf{k}_F)/dT = 0$ holds equally well below T_C. There is the subtle issue of the meaning of "$\mathbf{k}_F$" in the SC state. In analogy with the FS as the "locus of gapless excitations above T_C," we can define the "minimum gap locus" below T_C. We will describe this in great detail in Section 10 below; it suffices to note here that $\mathbf{k}_F$ is independent of temperature, within experimental errors, in both the normal and SC state of the systems studies thus far.

EXPERIMENTAL DETAILS

We now describe the experiments that first test the above ideas and then use them to get new information. Most of the data to be discussed in these lectures is on very high quality single crystals of $Bi_2Sr_2CaCu_2O_{8+x}$ (Bi2212), grown by the traveling solvent floating zone method with an infrared mirror furnace, with low defect densities and sharp x-ray diffraction rocking curves with structural coherence lengths ~1250Å. The near optimally-doped samples (which we shall focus on, except in the last part of the lectures) have T_C = 87K with a transition width of 1K as determined by a SQUID magnetometer. The samples are cleaved in-situ at 13 K in a vacuum of < 5x10^{-11} Torr, and have optically flat surfaces as measured by specular laser reflections. Another measure of the sample quality, within ARPES, is the observation of "umklapp" bands in the electronic structure (described below) due to the presence of a structural superlattice distortion.

The experiments were performed at the Synchrotron Radiation Center, Wisconsin, using a high resolution 4-m normal incidence monochromator with a resolving power of 10^4 at 10^{11} photons/s. The samples are carefully oriented in the sample holder to an accuracy 1° by Laue diffraction, and the orientation is further confirmed by the observed symmetry of sharp PES features around high symmetry points. Various experiments have been done using 17 -- 22 eV photons, with an energy resolution (FWHM) in the range of 15 -- 25 meV and a typical momentum window of angular range ±1°.

For the Brillouin zone of Bi2212 (see Fig. 1), we use a square lattice notation with $\Gamma - \overline{M}$ along the Cu-O bond direction. $\Gamma = (0, 0)$, $\overline{M} = (\pi, 0)$, $X=(\pi, -\pi)$ and $Y=(\pi, \pi)$ in units of $1/a^*$, where $a^* = 3.83$Å is the separation between near neighbor Cu ions. (The orthorhombic a axis is along X and b axis along Y).

EXPERIMENTS ON SUM RULE AND $n(\mathbf{k})$

Fig. 3(a) shows ARPES spectra for slightly overdoped Bi2212 (T_C =87 K) at the FS crossing along $(\pi, 0)$ to $(\pi, 0)$ at two temperatures: T=13 K, which is well below T_C, and T=95 K, which is in the normal state. The two data sets were normalized in the positive

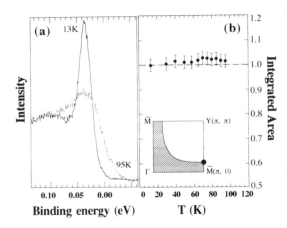

Fig. 3. (a) ARPES spectra for Bi2212 T_C =87 K at $\mathbf{k} = \mathbf{k}_F$ (FS crossing along $(\pi, 0)$ to $(\pi, 0)$) at T=13 K and T=95 K; (b) Integrated intensity vs. temperature.

energy region[16], which after normalization was chosen to be the common zero baseline. For details, see ref. 14. Remarkably, even though the spectra themselves are very strongly T-dependent, their integrated intensity in Fig. 3(b) is constant within experimental error bars (arising from the normalization), as predicted by the $\partial n(\mathbf{k}_F) / dT = 0$ sum rule. The sum rule has also been checked at other FS crossings, but is much less informative along the $(0, 0)$ to (π, π) crossing where the observed line-shape is not too strongly T-dependent.

An important application of (3) would be to use it to experimentally determine the momentum distribution, particularly since no other methods have successfully addressed this problem for the cuprates (e.g., positron annihilation in $YBa_2Cu_3O_{6.9}$ apparently only yields information about the chain bands). There are several caveats to keep in mind here, before discussing the data. First, we do not have an absolute scale for the integrated intensity, and, the unknown scale factor is $\mathbf{k}$-dependent, since from (2) and (3):

$\int d\omega I(\mathbf{k}, \omega) = I_0(\mathbf{k})n(\mathbf{k})$. (In principle, electronic structure theory can provide useful input on the $\mathbf{k}$-dependence of matrix elements[13]). Second, we do not know what the "zero" for the integrated intensity is, in view of the unknown "extrinsic background". Finally there is the question of the integration limits in (3): while the Fermi function cutoff makes the upper limit irrelevant, the lower limit may be more problematical. Thus, quantitative studies on $n(\mathbf{k})$ are not possible at the moment, but important qualitative information can be extracted as shown below.

In view of the above discussion, we illustrate the idea of measuring $n(\mathbf{k})$ on Bi2212. We had earlier measured $n(\mathbf{k})$ on $YBa_2Cu_4O_8$ (Y124)[7], where the low background and the loss of emission once $\mathbf{k}_F$ is crossed gives a clean result. The situation in Bi2212 (see Fig. 4) is not as clear cut, both as regards the background, since there is considerable emission after crossing $\mathbf{k}_F$, even though its much smaller than in the occupied states, and as regards the lower limit of integration. Even with these limitations, the integrated intensity shown in Fig. 4 is very informative[16]. (Note that the integrated intensity for $\mathbf{k}$ way past $\mathbf{k}_F$, i.e. deep on the unoccupied side, is set to zero, by hand). To minimize the effects of the matrix elements and the slowly varying additive background, it is useful to look at peaks in $|\nabla_{\mathbf{k}} n(\mathbf{k})|$. As seen from Fig 4, these correlate very well with the FS crossing inferred from the dispersion data.

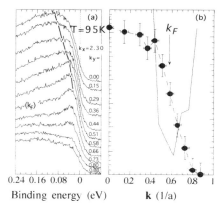

Fig. 4. (a): Normal state (T=95K) Bi2212 spectra for a set of $\mathbf{k}$ values corresponding panel. (b): Integrated intensity (black dots) from data in (a); its derivative is shown by a solid curve (arbitrary scale).

ARPES SPECTRA: QUALITATIVE FEATURES

At this stage, having gained some confidence in interpreting ARPES spectra in terms of the (occupied part of the) one-particle spectral function, let us discuss some of the important qualitative features of the data. The first thing to emphasize is that the peak of the experimental spectrum, the EDC, is *not* necessarily that of the spectral function. This is obvious from eqn. (2), $I \sim f(\omega)A(\mathbf{k}, \omega)$, and directly seen from the data in Fig. 5. In the normal state, the EDC peak is produced by the Fermi function $f(\omega)$ cutting off the spectral function $A(\mathbf{k}, \omega)$, which would presumably peak at $\omega = 0$ for $\mathbf{k} = \mathbf{k}_F$. We note, in passing, that recently we have succeeded in developing methods for "dividing out the Fermi function" in the data, which is nontrivial because of the convolution with the energy resolution. This gives very useful direct information about the low frequency $A(\mathbf{k}, \omega)$, as we will discuss elsewhere[15]

An important consequence is that the normal state spectral function is extremely broad, the observed full width of the EDC being less than the actual half-width of $A(\mathbf{k}, \omega)$!

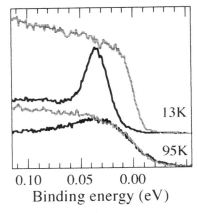

Fig. 5. SC (T=13K) and normal state (T=95K) Bi2212 spectra (solid curves) and reference Pt spectra (dashed curves) at the same temperatures.

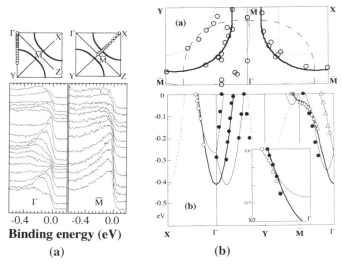

Fig. 6. (a) Normal state (T=95K) spectra for Bi2212 along two symmetry lines at values of the momenta shown as open circles in the upper insets. The photon polarization, **A**, is horizontal in each panel. (b) Fermi surface (top panel) and dispersion (bottom panel) obtained from normal state measurements. The thick lines are obtained by a tight binding fit to the dispersion data of the main band with the thin lines (0.21π, 0.21π) umklapps and the dashed lines (π, π) umklapps of the main band. Open circles in the top panel are the data. In the bottom panel, filled circles are for odd initial states (relative to the corresponding mirror plane), open circles for even initial states, and triangles for data taken in a mixed polarization geometry. The inset in the bottom panel is a blowup of ΓX.

Does this anomalous normal state spectrum imply a breakdown of Fermi liquid theory, as suggested by numerous transport experiments? In principle, one should be able to answer this question by analyzing the ARPES data using

$$A(\mathbf{k}, \omega) = \frac{\Sigma''(\mathbf{k}, \omega)/\pi}{(\omega - \varepsilon_{\mathbf{k}} - \Sigma'(\mathbf{k}, \omega))^2 + \Sigma''(\mathbf{k}, \omega)^2} \qquad (4)$$

where Σ' and Σ'' are the real and imaginary parts of the self energy. In practice, the number of parameters involved in Σ, coupled with uncertainties about the extrinsic background, lead to serious questions about the uniqueness of such fits. Instead of trying to extract the ω-dependence of Σ it is more useful, at the present time, to focus on the **k**-dependence of the line-shape as one approaches the FS. For a Fermi liquid, with well defined quasiparticles, the spectrum should sharpen up as k_F is approached. However, as the normal state data in Fig. 7 clearly show, and simple fits[18] corroborate, this does not happen. *There are no well defined quasiparticles above T_C!* It is very important to emphasize that the large linewidths observed in ARPES are not extrinsic, or artifacts of any analysis. As we will see next, when quasiparticles do exist (for $T<<T_C$) they are clearly seen in the experiment.

The remarkable T-dependent changes in the line-shape in Figs. 3 and 5 may be understood as follows. For $T < T_C$ the SC gap opens up and spectral weight at k_F shifts from $\omega = 0$ (in the normal state) to either side of it, of which only the occupied side ($\omega < 0$) is probed by ARPES. At the lowest temperature the EDC peak *is* the peak of the spectral function (unlike the normal state) since, as is obvious from Fig. 5, the Fermi function has now become sharper and spectral weight has moved down to below the gap energy. A detailed analysis of the SC gap data will described in Section 10 below.

Another striking feature of the data is the sharpening of the peak with decreasing T in the SC state. This is *not* a "BCS pile up" in the density of states, a description frequently used in the early literature, since we are not measuring a DOS. With a rapid decrease in linewidth below T_C, the only way the conserved area sum rule can be satisfied is by having a large rise in intensity. The dramatic decrease in the linewidth (Σ'') below T_C is a consequence of the SC gap leading to a suppression of electron-electron scattering which was responsible for the large linewidth above T_C. *Thus coherent quasiparticle (q.p.) excitations do exist for $T<<T_C$* [19]. The rapid T-dependence of the line width is in qualitative agreement with the results of various transport measurements[20]. A quantitative extraction of the scattering rate from ARPES data is an important open problem.

It is worth emphasizing that every aspect of this data, from the broad normal state spectrum to the highly non-trivial SC state line shape, points to the importance of e-e interactions. The strong T-dependence of the linewidth is very unusual, and would not occur in conventional metallic SC's where the e-e interaction contribution to the scattering rate is weak. We will argue below that e-e interactions are also responsible for the non-trivial dip and hump structure (see Fig. 3) present beyond the sharp q.p. peak in the SC state; see Section 9. Finally, the fact that we see spectral shifts in the same data all the way down to 100 meV ~1000 K, for a temperature change of 100 K, also suggests that e-e interactions are at work.

NORMAL STATE OF OPTIMALLY DOPED BI2212

We now briefly summarize the main results of a very detailed study[21] of the electronic excitations in the normal state ($T = 95K$) of near-optimal Bi2212 ($T_C = 95K$). We begin with a discussion of the dispersion of the electronic excitations and the Fermi Surface.

Two representative data sets are plotted in Fig. 6a: the left panel shows dispersing peaks along the diagonal $(0, 0)$ to (π, π), while the right panel shows data along the zone boundary $(\pi, 0)$ to $(\pi, -\pi)$. Spectral peak positions as a function of **k** are plotted in the bottom panel of Fig. 6(b), and the corresponding Fermi surface (FS) crossings in the top panel of Fig. 6b..

In addition to the symbols in Fig. 6b, there are also several curves, which we now describe; these curves make clear the significance of all of the observed features. The thick curve is a 6-parameter tight-binding fit[22] to the Y-quadrant data; this represents the main

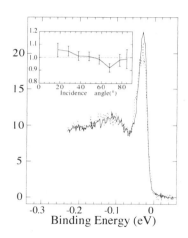

Fig. 7. Low temperature (T=13K) EDC's of Bi2212 at $\overline{M}$ for various incident photon angles. The solid (dashed) line is 18° (85°) from the normal. The inset shows the height of the sharp peak for data normalized to the broad bump, at different incident angles.

CuO_2 band. The two thin curves are obtained by shifting the main band fit by $\pm\mathbf{Q}$ respectively, where $\mathbf{Q} = (0.21\pi, 0.21\pi)$ is the superlattice (SL) vector known from structural studies[23]. We also have a few data points lying on a dashed curve, which is a (π, π) foldback of the main band fit; this "shadow band" will be discussed below. The Fermi surfaces corresponding to the main band fit (thick line), the SL umklapps (thin lines) and the shadow band (dashed) are plotted as curves in the top panel of Fig. 6b. We note that the main FS is a large hole-like barrel centered about the (π, π) point whose enclosed area corresponds to (approximately) 0.17 holes per planar Cu.

One of the key questions is why only one CuO main band is found in Bi2212 which is a bilayer material. We will discuss this in depth in the Section 9. The next important point relates to the "shadow bands" first observed in ARPES experiments[24] done in a very different mode (roughly, those experiments measure $\int_{\delta\omega} d\omega A(\mathbf{k},\omega) f(\omega)$ over a small range $\delta\omega$ near $\omega = 0$. The shadow bands were not seen earlier in the EDC mode experiments probably because of their sensitive photon energy dependence and the absence of a strong feature near E_F. These "shadow bands" were predicted early on to arise from short ranged antiferromagnetic correlations[25]. An alternative explanation, which needs to be tested further, is that they are of structural origin: Bi2212 has a face-centered orthorhombic cell with two inequivalent Cu sites per plane, which by itself could generate a (π, π) foldback.

We now turn to the effect of the superlattice (SL) on the ARPES spectra. This is very important, since a lack of understanding of these effects led to incorrect conclusions regarding such basic issues as one versus two Femi surfaces (see Section on Bilayer Splitting), and the anisotropy of the SC gap (see Section on Superconducting Gap and its Anisotropy). All of the experimental evidence is in favor of interpreting the SL umklapp bands as arising from a final state effect in which the exiting photo-electron scatters off the structural SL superlattice distortion (which lives primarily) on the Bi-O layer[26]

We use the polarization selection rules to disentangle the main and SL bands in the X-quadrant where the main and umklapp FSs are very close together; see the insert in Fig. 6b. The point is that ΓX (together with the z-axis) and, similarly ΓY, are mirror planes, and an initial state arising from an orbital which has $d_{x^2-y^2}$ symmetry about a planar Cu-site is odd under reflection in these mirror planes. With the detector placed in the mirror plane the final state is even, and one expects a dipole-allowed transition when the photon polarization $\mathbf{A}$ is perpendicular to (odd about) the mirror plane, but no emission when the polarization is parallel to (even about) the mirror plane. While this selection rule is obeyed along ΓY it is violated along ΓX. In fact this apparent violation of selection rules in the X quadrant, was a puzzling feature of all previous studies[10] of Bi2212. It was first pointed out in ref. 27, and then experimentally verified in ref. 21, that this "forbidden" $\Gamma X \|$ emission originates from the SL umklapps. We will come back to the $\Gamma X \|$ emission in the superconducting state below.

8. EXTENDED SADDLE POINT SINGULARITY

Some aspects of the normal state dispersion plotted in Fig. 6b deserve special mention: while the dispersion along the diagonal $(0, 0)$ to (π, π) is very rapid, that near the $(\pi, 0)$ point is very flat. In particular, along $(0, 0)$ to $(\pi, 0)$ there is an intense spectral peak which disperses towards E_F but stays just below it at a binding energy of (approximately) -30 meV. This is often called the "flat band" or "extended saddle point", and appears to exist in all of the cuprates, though at different binding energies in different materials; see 7,8,10.

In our opinion this flat band is not a consequence of the bare electronic structure but rather a many-body effect. The argument for this is that a tight-binding description of such a dispersion requires fine-tuning which would be unnatural even in one material, let alone many. Another important issue is whether this flat band leads to a singular density of states.

It is very important to recognize that, while Fig. 6b *looks like* a conventional band structure, the dispersing states whose "centroids" or "peak positions" are plotted are extremely broad, with width comparable to binding energy, and these simply cannot be thought of as quasiparticles. This general point is true at all **k**'s, but specifically for the flat band region it has the effect of spreading out the spectral weight over such a broad range that any singularity in the DOS would be washed out.

BILAYER SPLITTING?

On very general grounds, one expects that the two CuO_2 layers in a unit cell of Bi2212 should hybridize to produce two electronic states which are even and odd under

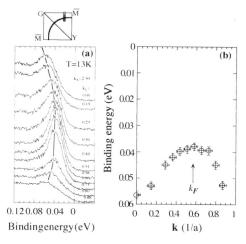

Fig. 8. Superconducting state EDCs for Bi2212 for the set of **k**-values (1/a units) which are shown at the top. (For corresponding normal state data, see Fig. 4). (b) SC state peak positions (white dots) versus **k** for data of part (a). The $\mathbf{k}_F$ marked is the same as that determined from the normal state analysis of Fig. 4.

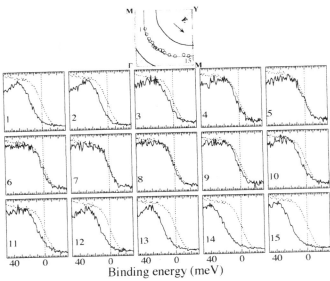

Fig. 9 Bi2212 spectra (solid lines) for a 87K T_C sample at 13K and Pt spectra (dashed lines) versus binding energy (meV) along the Fermi surface in the Y quadrant. The photon polarization and BZ locations of the data points are shown in inset to Fig 10.

reflection in a mirror plane mid-way between the layers. Where are these two states? Why did we find only one main "band" and only one FS in Fig. 6b?

We have carefully checked the absence of a FS crossing for the main band along $\Gamma\overline{M}$ by studying the integrated intensity and its derivative $|\nabla_k n(\mathbf{k})|$ and found no sharp feature in $n(\mathbf{k})$. Further the FS crossing that we do see near $(\pi, 0)$ along $\Gamma\overline{M}$ in Fig. 6a is clearly associated with a SL umklapp band, as seen both from the dispersion data in Fig. 6b and its polarization analysis. This FS crossing is only seen in the $\Gamma\overline{M}\perp$ (odd) geometry both in our data and in earlier work[8] (where it was erroneously identified as part of a second FS closed round around Γ). Emission from the main $d_{x^2-y^2}$ band, which is even about $\Gamma\overline{M}$, is dipole forbidden, and one only observes a weak SL signal crossing E_F. This clearly demonstrates that the bilayer splitting of the CuO_2 states does not lead to two experimentally resolvable Fermi surfaces. It should be emphasized that this, by itself, is not in contradiction with expectations from electronic structure calculations[28]. Whether or not the two Fermi surfaces are resolvable depends sensitively on the exact doping levels and on the presence of Bi-O pockets, which are neither treated accurately in the theory nor observed in the ARPES data. However, there *is* a clear prediction from band theory: at $\overline{M} = (\pi, 0)$, where both states are occupied the bilayer splitting is the largest, of order 0.25 eV.
The normal state spectrum at $\overline{M}$ is so broad that it may be hard to resolve two states. However, for $T \ll T_C$, when a sharp quasiparticle peak is seen, the bilayer splitting should be readily observable. For this one needs to interpret the non-trivial line shape at $\overline{M}$ shown in Fig. 7: with a dip[10,29] in between the q.p. peak and a broad bump at 100 meV.

Probably the simplest interpretation would be (I) where the bump is the second band, which is resolved below T_C once the first band becomes sharp. The other alternative (II) is determined in this way, gives information about the underlying FS (which is, of course, gapped below T_C).
that non-trivial line shape is due to many-body effects in a single spectral function $A(\mathbf{k}, \omega)$. To choose between these two hypotheses, we exploit the polarization dependence of the matrix elements. In case (I) there are two independent matrix elements which, in general, should vary differently with $\mathbf{A}$, and thus the intensities of the two features should vary independently. While for case (II), the intensities of the two features should scale together.

240

In ref. 21 we found, by varying the z-component of **A**, evidence supporting hypothesis (II): the q.p. peak, dip and bump are all part of a single spectral function for Bi2212. The same conclusion can be quite independently reached from the dispersion data in the SC and normal states shown in ref. 30. Additional experimental evidence against a two band interpretation of the dip structure comes from tunneling[31].

There are two important questions arising from this conclusion. First, what causes this non-trivial line shape? The answer is the non-trivial ω-dependence of the self energy: at low ω, Σ'' is suppressed by the opening of the gap which leads to the q.p. peak, but at $\omega >> \Delta$, Σ must recover its normal state behavior. This effect is qualitatively able to account for the dip-bump structure[32]. A more quantitative description of the SC line-shape is lacking at the present time; for some recent progress in this direction, see 30.

The second question to ask is: what conspires to keep the two states degenerate? Anderson[33] had predicted that many-body effects within a single layer would destroy the coherent bilayer splitting *in the normal state*. But why it should not be visible in the SC state is not so clear, at least to us. Finally, it should be mentioned that, in contrast to the Bi2212 case, there is some evidence for bilayer-split bands in YBCO[5,34,7].

SUPERCONDUCTING GAP AND ITS ANISOTROPY

In this Section, we will first establish how the SC gap manifests itself in ARPES spectra, and then directly map out its variation with **k** along the FS. Since ARPES is the only available technique for obtaining such information, it has a played an important role[9,35] in establishing the *d*-wave order parameter in the high Tc superconductors[36].

In Fig. 8 we show SC state spectra for Bi2212 for a sequence of **k**'s. In the normal state these **k**'s go from the occupied (top) to unoccupied (bottom) states, through k_F, as shown in Fig. 4. However, in the SC state the spectral peaks do not disperse through the chemical potential, rather they first approach $\omega = 0$ and then recede away from it, as can be clearly seen from Fig. 8b. In comparing the normal and SC state data in Figs. 4 and 8 (which have different energy scales!), it is important to bear in mind the discussion based on Fig. 5 that in the normal state the EDC peak is caused by the Fermi function cut-off while for a gapped spectrum, the EDC peak is that of the spectral function.

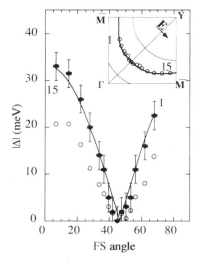

Fig. 10. *Y* quadrant gap in meV versus angle on the Fermi surface (filled circles) from fits to the data of Fig. 9. Open symbols show leading edge shift with respect to Pt reference. The solid curve is a d-wave fits to the filled symbols.

There are several important conclusions to be drawn from Fig. 8. First, the bending back of the spectral peak, for **k** beyond $\mathbf{k}_F$, is direct evidence for particle-hole mixing in the SC state; for details see ref. 17. The energy of closest approach to $\omega = 0$ is related to the SC gap that has opened up at the FS, and a quantitative estimate of this gap will be described below. The location of closest approach to $\omega = 0$ ("minimum gap") coincides, within experimental uncertainties, with the $\mathbf{k}_F$ obtained from the normal state $n(\mathbf{k})$ analysis of Fig. 4. It is important for later purposes to note that the "minimum gap locus",

In Fig. 9, we show the $T = 13K$ EDCs for the 87K T_C sample for various points on the main band FS in the Y-quadrant. Each spectrum shown corresponds to the minimum observable gap along a set of **k** points normal to the FS, obtained from a dense sampling of **k**-space[37]. We used 22 eV photons in a $\Gamma Y \bot$ polarization, with a 17 meV (FWHM) energy resolution, and a **k**-window of radius $0.045\pi / a^*$.

The simplest gap estimate is obtained from the mid-point shift of the leading edge of Bi2212 relative to Pt in electrical contact with the sample. This has no obvious quantitative validity, since the Bi2212 EDC is a spectral function while the polycrystalline Pt spectrum (dashed curve in Fig. 9) is a weighted density of states whose leading edge is an energy-resolution limited Fermi function. We see that the shifts (open circles in Fig. 10) indicate a highly anisotropic gap which vanishes in the nodal directions, and these results are qualitatively similar to one obtained from the fits described below.

Next we turn to modeling[35,38] the SC state data in terms of spectral functions. It is important to ask how can we model the non-trivial line shape (with the dip-bump structure at high ω) in the absence of a detailed theory, and, second, how do we deal with the extrinsic background? We argue as follows: in the large gap region near $(\pi, 0)$, we see a linewidth collapse for frequencies smaller than $\sim 3\Delta$ upon cooling well below T_C. Thus for estimating the SC gap at the low temperature, it is sufficient to look at small frequencies, and to focus on the coherent (resolution limited) piece of the spectral function. (Note this argument fails at higher temperatures, e.g., just below T_C). We model this coherent piece by the BCS spectral function $A(\mathbf{k},\omega) = u_\mathbf{k}^2 \Gamma / \pi[(\omega - E_\mathbf{k})^2 + \Gamma^2] + v_\mathbf{k}^2 \Gamma / \pi[(\omega + E_\mathbf{k})^2 + \Gamma^2]$ where the coherence factors are $v_\mathbf{k}^2 = 1 - u_\mathbf{k}^2 = \frac{1}{2}(1 - \varepsilon_\mathbf{k}/E_\mathbf{k})$ and Γ is a phenomenological linewidth. The normal state energy $\varepsilon_\mathbf{k}$ is measured from E_F and the Bogoliubov quasiparticle energy is $E_\mathbf{k} = \sqrt{\varepsilon_\mathbf{k}^2 + |\Delta_\mathbf{k}|^2}$, where $\Delta_\mathbf{k}$ is the gap function. Note that only the second term in $A(\mathbf{k},\omega)$, with the $v_\mathbf{k}^2$-coefficient, makes a significant contribution to the ARPES spectra.

The effects of experimental resolution are taken into account via

$$\tilde{I}(\mathbf{k},\omega) = I_0 \int_{\delta k} d\mathbf{k}' \int_{-\infty}^{\infty} d\omega' \, R(\omega - \omega') f(\omega') A(\mathbf{k}',\omega') \tag{5}$$

where $R(\omega)$, the energy resolution, is a normalized Gaussian and $\Delta_\mathbf{k}$ is the **k**-window of the analyzer. In so far as the fitting procedure is concerned, all of the incoherent part of the spectral function is lumped together with the experimental background into one function which is added to the $\tilde{I}$ above. Since the gap is determined by fitting the resolution-limited leading edge of the EDC, its value is insensitive to this drastic simplification. To check this, we have made an independent set of fits to the small gap data where we do not use any background fitting function, and only try to match the leading edges, not the full spectrum. The two gap estimates are consistent within a meV. Once the insensitivity of the gap to the assumed background is established, there are only two free parameters in the fit at each **k**: the overall intensity I_0 and the gap $|\Delta|$; the dispersion $\varepsilon_\mathbf{k}$ is known from the normal state study, the small linewidth Γ is dominated by the resolution.

The other important question is the justification for using a coherent spectral function to model the rather broad EDC along and near the diagonal direction. We have found that such a description is self-consistent[35,38] (though perhaps not unique), with the entire width

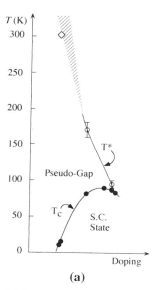

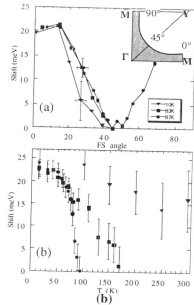

Fig. 11. (a) Schematic phase diagram of Bi2212 as a function of hole doping. The filled symbols are the measured T_C's for the superconducting phase transition from magnetic susceptibility. The open symbols are the T* at which the (maximum) gap seen in ARPES closes; for the T_C 10K sample the symbol at 301K is a lower bound on T*. **(b)** Momentum and temperature dependence of the gap estimated from leading edge shift (see text). a) **k**-dependence of the gap along the "minimum gap locus'" (see text) in the 87K T_C, 83K T_C and 10K T_C samples, measured at 14K. b) T-dependence of the (maximum) gap in a near-optimal 87K sample (circles), underdoped 83K (squares) and 10K (triangles) samples. Note smooth evolution of gap from SC to normal state for 83K sample.

of the EDC accounted for by the large dispersion (of about 60 meV within our **k**-window) along the zone diagonal.

The gaps extracted from fits to the spectra of Fig. 10 are shown as filled symbols in Fig. 12. For a detailed discussion of the error bars (both on the gap value and on the FS angle), and also of sample-to-sample variations in the gap estimates, we refer the reader to ref. 35. The angular variation of the gap obtained from the fits is in excellent agreement with $\cos k_x - \cos k_y$ form. The ARPES experiment cannot of course measure the phase of the order parameter, but this result is strongly suggestive of $d_{x^2-y^2}$ pairing. Such an order parameter arises naturally in theories with strong correlations and/or antiferromagnetic spin fluctuations[39].

For completeness, we add few lines clarifying the earlier observation of two nodes in the X-quadrant[38], and the related non-zero gap along ΓX in the ΓX‖ geometry[38,40]. It was realized soon afterwards that these observations were related to gaps on the superlattice bands[27], and not on the main band. To prove this experimentally, the X-quadrant gap has been studied in the ΓX⊥ geometry[35] and found to be consistent with Y-quadrant $d_{x^2-y^2}$ result described above.

PSEUDOGAP IN THE UNDERDOPED MATERIALS

We finally turn to one of the most fascinating recent developments -- pseudogaps -- in high T_C superconductors in which ARPES has again played a major role[41,42,43].

Up to this point we have discussed optimally doped Bi2212. We now contrast this with the remarkable behavior of the underdoped materials, where T_C is suppressed by lowering the carrier (hole) concentration. See Fig. 11a for a schematic phase diagram[41]. Underdoping was achieved by adjusting the oxygen partial pressure during annealing the float-zone grown crystals. These crystals also have structural coherence lengths of at least 1,250Å as seen from x-ray diffraction, and optically flat surfaces upon cleaving, similar to the near-optimally doped T_C samples discussed above. (Actually, those samples are now believed to be slightly overdopded, optimal doping corresponding to T_C = 92K). We denote the underdoped samples by their onset T_C: the 83K sample has a transition width of 2K and the highly underdoped 15K and 10K have transition widths > 5K. (Other groups have also studied samples where underdoping was achieved by cation substitution[43].

The first point to note about the high temperature ARPES spectra of underdoped Bi2212 is that they become progressively broader with underdoping. While the excitations of the optimally doped material were anomalously broad (non-Fermi liquid behavior), there was nevertheless an identifiable spectral peak in the normal state. In contrast, the underdoped spectra above T_C are so broad that there is no identifiable peak at all. One might question: how do we know that these featureless EDCs are spectral functions? There are two reasons: first, even above T_C there is observable dispersion, and second, way below T_C a coherent (almost resolution-limited) quasi-particle peak emerges (in the 83K T_C samples in which this regime is accessible).

In fact the SC state spectra in the underdoped regime look very similar to those at optimal doping, with the one difference that the spectral weight in the coherent q.p. peak diminishes rapidly with underdoping. This can be seen from Fig. 12a, which shows the EDCs along the $\overline{M}\,Y$ crossing. The important point is that there are always two features in the spectra, a sharp quasiparticle peak at low binding energy, and a broad hump at higher binding energy. In underdoped samples, such as the 60 K fresh shown in Fig. 12b, the quasiparticle peak is small, but nonetheless a clear break in the spectra signals its presence. As the sample ages in the spectrometer, the doping increases, with a consequent increase of the quasiparticle peak. This reversibility of the doping effects give some confidence in the interpretation of the results.

It is not possible at the present time to quantify this important relationship between doping and q.p. peak size. However, the q.p. exhibits almost the same "minimum gap locus" in the SC state (see Section on Superconducting Gap) as the overdoped samples, which suggests a large underlying Fermi surface, satisfying the Luttinger count of (1+x) holes per planar Cu[42], as shown in Fig. 13. This "minimum gap locus" coincides with the high temperature FS, which is the locus of gapless excitations[42].

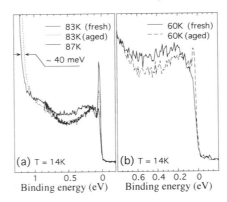

Fig. 12a. EDCs of the 83K sample at $\overline{M}$ for a newly cleaved surface (blue line) and after 41 hours (red line), with the EDC of the 87K sample (black line) plotted for comparison. The 40 meV shift is between the leading edges of the main valence bands of the two 83K spectra. **Fig. 12b.** EDC of the 60K sample near $\overline{M}$ for an unaged surface (blue line) and an aged surface (red line). Note the 200 meV hump moves to lower binding energy, and the low binding energy leading edge evolves into a quasiparticle peak.

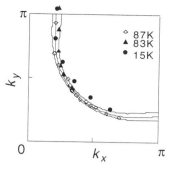

Fig. 13. Fermi surfaces of the 87K, 83K, and 15K samples. All surfaces have a large volume. The solid lines are tight binding estimates of the Fermi surface at 18\%, 13\%, and 6\% doping assuming rigid band behavior.

The major difference with the optimally doped sample is evident upon heating through T_C. While the gapless excitations along the diagonal remain gapless, the large gap along the $(\pi, 0)$ to (π, π) crossing does not close above T_C, as seen from Fig. 13a. One has to go to a (crossover scale) T^* which is much higher than T_C before this gap vanishes and a closed contour of gapless excitations (the FS) is recovered. Note that in Fig. 13 we use the leading edge shift to estimate the gap, since except for $T<T_C$ we do not know enough about the line-shape to make any quantitative fits (as explained in section on gap analysis).

It is important to emphasize that our understanding of the 83K T_C sample is the best amongst all the underdoped materials. In this sample all three regimes -- the SC state below T_C, the pseudogap regime (83K= $T_C < T < T^*$=170K) and the gapless "normal" regime above T^* -- have been studied in detail. In contrast, the 10K and 15K samples have such low T_C's and such high T^*s that only the pseudogap regime is experimentally accessible. Nevertheless, the results on the heavily underdoped samples appear to be a natural continuation of the weakly underdoped materials and the results (same magnitude of gap, higher value of T^*,) on the low T_C samples are in qualitative agreement with those obtained from other probes (see ref. 44). Perhaps the most controversial of the results on the heavily underdoped samples is the inference about a large underlying FS from the "minimum gap locus" in the pseudogap regime as opposed to small hole pockets. While this is certainly a tricky issue, and there may also be materials problems in the very low T_C sample, we did not find any evidence for either the closure of a hole pocket (concave arc about the Γ point) or for shadow bands which are (π, π)-foldbacks of the observed state.

To summarize the ARPES results in the underdoped regime: a highly anisotropic SC gap is found in the underdoped samples which is essentially independent of the doping level both in its magnitude and in its **k**-dependence. Thus in this respect the underdoped samples are very similar to optimally doped Bi2212. The key differences in the SC state are first, the value of T_C, and second, the spectral weight in the coherent q.p. peak at $T<T_C$, both of which drop rapidly with underdoping. Above T_C the ARPES spectra in the underdoped state are qualitatively different from optimal doping. ARPES continues to show a gap which evolves smoothly through T_C and has the same anisotropy as the SC gap below T_C. This gap-like suppression of spectral weight, called the pseudogap, persists all the way to a much higher scale T^* at which a locus of gapless excitations (Fermi surface) is recovered.

CONCLUSIONS

In conclusion, we hope that we have been able to convey to the readers the exciting new physics that has come out of ARPES studies of the high T_C superconductors. What is really astonishing is the range of issues on which ARPES has given new insights: from non-Fermi liquid behavior with a Fermi surface, to the symmetry of the superconducting order parameter, to the development of a Fermi surface in a doped Mott-insulator and the pseudo-gap phenomena in the underdoped cuprates.

ACKNOWLEDGMENTS

We gratefully acknowledge the contribution of all our collaborators. JCC acknowledges support of the National Science Foundation DMR 9624048, JCC and MRN acknowledge support by the U. S. Dept. of Energy, Basic Energy Sciences, under contract W-31-109-ENG-38, and the National Science Foundation DMR 91-20000 through the Science and Technology Center for Superconductivity.

REFERENCES

1. S. Hüfner, *Photoelectron Spectroscopy*, Springer Verlag, Berlin, (1995); *Photoemission in Solids* I, edited by M. Cardona and L. Ley, Springer Verlag, Berlin, (1978).
2. J-M. Imer, etal, Phys. Rev. Lett. 62, 336 (1989).
3. A.A. Arko, et al. Phys. Rev.
4. C.G. Olson, *et al.*, Science 245, 731 (1989).
5. J. C. Campuzano, *et al.*, Phys. Rev. Lett. 64, 2308 (1990).
6. C.G. Olson, *et al.*, Phys. Rev. B42, 381 (1990).
7. K. Gofron, *et al.*, J. Phys. Chem. Sol. 54, 1193 (1993) and Phys. Rev. Lett. 73, 3302 (1994).
8. D. S. Dessau *et al.*, Phys. Rev. Lett. 71, 2781 (1993).
9. Z. X. Shen,*et al.*, Phys. Rev. Lett. 70, 1553 (1993).
10. For a review of the first five years of work on the cuprates, see Sec. 4 and 5 of Z. X. Shen and D. S. Dessau, Phys. Repts. 253, 1 (1995).
11. W. L. Shaich and N. W. Ashcroft, Phys. Rev. B3, 2452 (1971).
12. C. Caroli, D. Lederer-Rozenblatt, B. Roulet, and D. Saint-James, Phys. Rev. B8, 4552 (1973).
13. R. Pendry, Surf. Sci. 57, 679 (1976). For application of these methods to the cuprates, see: A. Bansil, M. Lindroos, K. Gofron, J.C. Campuzano, R. Liu, H. Ding, and B. W. Veal.A. Bansil and M. Lindroos, J. Phys. Chem. Solids 53 (1992) 1541.
14. M. Randeria, H. Ding, J. C. Campuzano, A. Bellman, G. Jennings, T. Yokoya, T. Takahashi, H. Katayama-Yoshida, T. Mochiku, and K. Kadowaki, Phys. Rev. Lett. 74, 4591 (1995).
15. Work in progress with M.R. Norman and H. Ding.
16. The essentially ω-independent emission at positive frequencies, which is not cut-off by the Fermi function, is due to higher harmonics of the incident photon beam ("second order light").
17. J. C. Campuzano, H. Ding, M. R. Norman, M. Randeria, A. Bellman, T. Yokoya, T. Takahashi, H. Katayama-Yoshida, T. Mochiku, K. Kadowaki, and G. Jennings, Phys. Rev. B53 R14737, (1996).
18. We have found that simple ways of analyzing the data are adequate to establish this, unlike the low frequency behavior of the self energy. Ding \etal (unpublished).
19. Actually, this has been established only for **k**'s around $(\pi, 0)$. We cannot answer the question of whether quasiparticles exist for $T < T_C$ along the (0, 0) to (π, π) FS crossing because of the rapid dispersion with our **k**-window (see Fig. 8 and also Section 10).
20. M. C. Nuss *et al.*, Phys. Rev. Lett. 66, 3305 (1991); W.N. Hardy *et al.*, Phys. Rev. Lett. 70, 399 (1993); K. Krishana, J.M. Harris, and N.P. Ong, Phys. Rev. Lett. 75, 3529 (1995).
21. H. Ding, A. Bellman, J. C. Campuzano, M. Randeria, M. R. Norman, T. Yokoya, T. Takahashi, H. Katayama-Yoshida, T. Mochiku, K. Kadowaki, G. Jennings, and G. P. Brivio, Phys. Rev. Lett. 76, 1533 (1996).
22. M. R. Norman, M. Randeria, H. Ding, and J. C. Campuzano, Phys. Rev. B52, 615 (1995).
23. R. L. Withers *et al.*, J. Phys. C21, 6067 (1988).
24. P. Aebi *et al.*, Phys. Rev. Lett. 72, 2757 (1994).
25. A. Kampf and J. R. Schrieffer, Phys. Rev. B42, 7967 (1990).
26. This would allow us to understand (1) why polarization selection rules are obeyed for the $\Gamma\overline{M}$ mirror plane, and (2) qualitatively, why the intensities for odd and even polarizations along ΓX are

comparable. This point was not fully appreciated at the time \onlinecite{DING_96} was published

27. M. R. Norman, M. Randeria, H. Ding, J. C. Campuzano, and A. F. Bellman, Phys. Rev. B52, 15107 (1995).

28. S. Massida, J. Yu, and A.J. Freeman, Physica C152, 251 (1988); O. K. Andersen, *et al.*, Phys. Rev. B49, 4145 (1994).

29. D.S. Dessau {\it et al.}, Phys. Rev. Lett. 66, 2160 (1991).

30. M. R. Norman, H. Ding, J. C. Campuzano, T. Takeuchi, M. Randeria, T. Yokoya, T. Takahashi, T. Mochiku, and K. Kadowaki, cond-mat/9702144

31. There is striking evidence in point-contact tunneling data on a variety of cuprates (including some one-layer per unti cell materials) that the dip-hump structure scales with the gap; see J. Zasadzinski, *et al.* (unpublished) in Fig. 1 of D. Coffey, J. Phys. Chem. Solids 54, 1369 (1993).

32. C. M. Varma and P. B. Littlewood, Phys. Rev. B46, 405 (1992); L. Coffey and D. Coffey, Phys. Rev. B48, 4184 (1993).

33. P. W. Anderson, Science 256, 1526 (1992).

34. R. Liu, *et al.*, Phys. Rev. B45, 5614 (1992) and B46, 11056(1992).

35. H. Ding, M. R. Norman, J. C. Campuzano, M. Randeria, A. Bellman, T. Yokoya, T. Takahashi, T. Mochiku, and K. Kadowaki, Phys. Rev. B54 R9678 (1996).

36. B. Goss-Levi, Physics Today, 49, 1, p. 19 (1996).

37. The large dispersion along ΓY, of about 60 meV within our momentum window $\delta \mathbf{k}$, makes it hard to locate $\mathbf{k}_F$ accurately and to map out the nodal region. To this end the we use a step size of $\delta k/2$ normal to the FS and $\delta \mathbf{k}$ along it.

38. H. Ding, J. C. Campuzano, A. Bellman, T. Yokoya, M. R. Norman, M. Randeria, T. Takahashi, H. Katayama-Yoshida, T. Mochiku, K. Kadowaki, and G. Jennings, Phys. Rev. Lett. 74, 2784 (1995) and 75, 1425(E) (1995).

39. See: D. Pines and P. Monthoux, J. Phys. Chem. Solids 56, 1651 (1995) and D. J. Scalapino, Phys. Rep. 250, 329 (1995).

40. R. J. Kelley et al., Phys. Rev. B50, 590 (1994).

41. H. Ding, T. Yokoya, J.C. Campuzano, T. Takahashi, M. Randeria, M.R. Norman, T. Mochiku, K. Kadowaki, and J. Giapintzakis, Nature 382, 51 (1996).

42. H. Ding, M.R. Norman, T. Yokoya, T. Takuechi, M. Randeria, J.C. Campuzano, T. Takahashi, T. Mochiku, and K. Kadowaki, Phys. Rev. Lett. 78, 2628 (1997).

43. D.S. Marshall *et al.*, Phys. Rev. Lett. 76, 4841 (1996). A. Loesser *et al.*, Science, 273, 325 (1996); J. Harris *et al.*, Phys. Rev. B 54, R15665 (1996).

44. M. Randeria, Varenna Lectures on "Precursor Pairing and Pseudogaps'"(1997).

LIGHT SCATTERING FROM CHARGE AND SPIN EXCITATIONS IN CUPRATE SYSTEMS

Rudi Hackl

Walther-Meissner-Institut
Bayerische Akademie der Wissenschaften
D-85748 Garching, Germany

Results from Raman scattering experiments in differently doped cuprate systems as well as the theoretical background for the analysis of the data are examined with the main focus placed on electronic excitations and their interactions. The response is studied as a function of polarization and temperature in both the normal and the superconducting state. Anisotropies in the CuO_2 plane increase with decreasing doping level suggesting an increasingly strong $\mathbf{k}$ dependence of electronic properties. In the normal state, below approximately 200 K, intensity anomalies are found in the spectra of underdoped materials being indicative of a pseudogap in direction of the principal axes. At all doping levels studied the superconducting gap has predominantly $d_{x^2-y^2}$ symmetry with an amplitude of $2\Delta_0/kT_c = 8$ independent of the transition temperature T_c.

INTRODUCTION

The study of excitations in solids requires in general both high momentum and high energy resolution making neutron scattering a favorable experimental technique as any combination of these quantities can be accessed. As usual however, real life is a little more complicated: One sticks to relatively big samples, in most of the cases single crystals, with masses of several tens of milligrams containing elements with a sufficiently large coherent scattering cross section. In addition, the study of carrier exitations is not quite feasible since the coupling is indirect via the spin degrees of freedom. In spite of the problems many of the puzzles of the former high-T_c materials, i.e. superconductors with transition temperatures of the order of 20 K such as refractory compounds like NbN[1] and materials with A15 structure like Nb_3Sn[2-4], have been unraveled with neutrons. This is related to the fact that phonons and their interaction with electrons play a crucial role for the superconductivity in these compounds. According to what we know to date, the importance of vibrational excitations seems to be smaller or negligible in the new copper-oxygen based high-temperature superconductors discovered in 1986.[5] Here the careful study of carrier and magnetic properties shifted into the main focus of interest[6-12]. Therefore, in comparison to the "classical" era of superconductivity other spectroscopic techniques attracted increasing attention.

Angle-resolved photoemission spectroscopy (ARPES), for instance, helped to get insight into the Fermi-surface topology[13,14] and the variation of the energy gap Δ with momentum $\mathbf{k}$[15-18]. In the latter case limits are imposed by the relatively poor energy resolution of 5 - 10 meV, and materials with T_c less than some 50 K such as electron-doped cuprates cannot be studied in the superconducting state. On the other hand, the momentum resolution is particularly good in two-dimensional materials, and this is the reason why ARPES became so popular in the layered cuprate systems. The situation is reversed in tunneling spectroscopy where almost no momentum but very good energy resolution can be achieved[19-22]. Here, the single-particle density of states is directly measured. However, in a superconductor with a strong momentum dependence of the energy gap it is unfavorable to get only an average from an integration over the whole Fermi surface. In addition, both ARPES and tunneling are extremely sensitive to the surface, since in the first case the electrons cannot escape ballistically from a depth larger than approximately 10 Å and since in the latter case the superconducting coherence length ξ does not exceed 20 Å. For this reason consistent results could only be obtained in bismuth-based compounds such as $Bi_2Sr_2CaCu_2O_8$ (Bi2212). Here, atomically flat and stable surfaces consisting of BiO_2 layers can easily be produced by cleaving. In most of the other compounds, the surface is unstable at ambient air against chemical reactions with water or carbon dioxide or even in vacuum as oxygen diffuses out at temperatutes well below 300 K.

The probing depth is much larger in optical experiments. Here, due to the relatively low carrier concentration of only a few 10^{21} cm^{-3} the photons penetrate approximately 1000 Å in contrast to some 100 Å in conventional metals. For the study of the carrier dynamics over a broad frequency range infrared (IR) spectroscopy is well established and has returned several important results in classical[23] and cuprate superconductors[24-28]. In the new high-T_c materials the interpretation of the results in the superconducting state is still under debate[24,26,29] since according to the theory by Matthis and Bardeen[30] an energy gap should be observable only in "dirty" materials with $\Delta \leq \hbar/\tau$, τ being the average relaxation time for excited carriers. It has been argued, and we shall see the justification of this assumption below, that the cuprates are in the clean limit, $\Delta \gg \hbar/\tau$, at least well below T_c and at a doping level where T_c is maximal (optimal doping). In the normal state however, there is general agreement that predominantly carrier properties are probed in the far infrared range between 2 and 100 meV (corresponding to 16 and 800 cm^{-1}, respectively) although details of the interpretation are developing continuously[31-33]. In any case all quantities measured by unpolarized IR spectroscopy are averages over the entire Fermi surface. By polarization a weight is introduced. Then, as one effect the response to certain types of carriers can be switched on and off. For example the CuO chains running along the b axis in $YBa_2Cu_3O_7$ (Y123) crystals can be seen only with the photons polarized parallel to the b direction while with perpendicularly polarized light carrier excitations in the planes can be selected out[34]. Other types of implications have not yet been explored.

Summarizing the preceding brief overview, it seems that there exists another exclusion principle: Energy and momentum cannot be accessed simultaneously with satisfactory resolution in a single experiment. It will be one of the purposes of the following discussion to demonstrate that Raman scattering leads partly out of this dilemma. In recent years it has been shown that properties of the normal and the superconducting state can indeed be studied by inelastic scattering of light as a function of $\mathbf{k}$ since the specific interaction between photons and electrons characterized by the Raman vertices $\gamma_\mathbf{k}$

allows to weigh out different parts of the Fermi surface by different polarisation combinations of the incoming and outgoing photons[35]. As a prominent result, the analysis of the superconducting spectra taken from several different types of compounds with one, two, and three copper-oxygen layers showed consistently that the energy gap is generally strongly **k** dependent and has predominantly $d_{x^2-y^2}$ symmetry[36-47]. Unconventional pairing was already conjectured much earlier[48] from the linear increase of the Raman intensity at low energy and temperature[49], but a quantitaive description of the strongly channel dependent continua[50-57] by establishing a relation between light polarizations and electronic momentum[36] was delayed by several years. Of course, the momentum resolution is by far not as good as in ARPES but the resolution in energy is considerably better, and half an meV can be achieved easily. At low energies of approximately 1 meV the accuracy of the Raman experiment is still very high in contrast to IR. In addition, the desired response function $Im\chi$ is measured directly while in most of the cases the conductivity $Re\sigma$ must be calculated from the IR reflectivity by Kramers-Kronig transformation[24,26,58]. In the superconducting state, Raman is particularly useful in the clean limit, just opposite to IR, and is still working in not too strongly disordered materials[59,60]. Due to the specific two-particle correlation function involved[61-64] light scattering is sensitive to coherence effects of the condensate comparable to e.g. the famous Hebel-Slichter peak in the temperature dependence of the the spin-lattice relaxation rate $1/T_1T$ right below T_c[65].

Light does not only scatter inelastically from carriers but from any other excitation in a solid with finite coupling to the electrons. Scattering from phononic and magnetic excitations has indeed been studied extensively in the cuprates.[66-68] Among the remarkable effects are the strong coupling of certain phonon modes to charge excitations in Y123[49,51,69] and the related theoretical work[70-72], the anharmonicity of the phonon originating from the oxygen in apex position[73-75] and the various investigations aiming to monitor the material properties, e.g. the stoichiometry and the quality of the samples[76]. The phononic properties depend strongly on the respective material class and do not seem to be tied to properties common to all cuprates. In other words, the lattice dynamics do not show systematics which can be related to the unique doping phase diagram that varies only little with the occasionally very complicated crystallographic structure of some high-T_c families. In contrast, the spin and charge dynamics are much closer related to general properties of the cuprates. Since all undoped CuO_2 systems are antiferromagnetically (AF) ordered Mott-Hubbard insulators and since the superconducting metal evolves from the insulator upon doping with either holes or electrons, the question for the magnetism of the CuO_2 systems continuously played a dominant role in the theoretical discussion.[8-10,77-84] The experiments showed that AF correlations are present not only in the insulator but also in the superconductor, most probably at fairly high carrier concentrations beyond optimal doping.[67,68,85-95] In this context, Raman results can provide useful information[96,97] and scattering from magnons has been studied from the beginning and revealed important information such as the magnitude of the exchange coupling J[89] or an approximate value for the magnetic correlation length ξ_{mag} as a function of temperature[94,98].

In what follows I will first describe some experimental aspects and then give an overview over the theoretical framework of Raman scattering in metals and super-conductors touching briefly scattering from spin excitations. I will confine myself to the minimum neccessary for an understanding of the experiments to be discussed in the third part. For a closer study of the continuously growing theoretical work the interested reader is refered to the literature. A relatively elementary review and a fairly recent compilation of references can be found in.[35] Finally I will try to contrast conclusions and open questions and discuss the Raman results in relation to other spectroscopic methods.

SAMPLES AND EXPERIMENTAL

Most of the cuprate systems have a fairly complicated crystal structure (Fig. 1) with a minimum of four different elements involved.[99] The main subunits, however, are simple and consist only of copper and oxygen atoms arranged in planes parallel to the crystallographic a and b axes. Units of one or more adjacent CuO_2 planes are separated by other constituents and stacked along the c axis like a pack of cards. The separating layers are mainly a charge reservoir for the CuO_2 planes. By slightly changing their composition while leaving the CuO_2 planes untouched the properties of most of the compounds can be

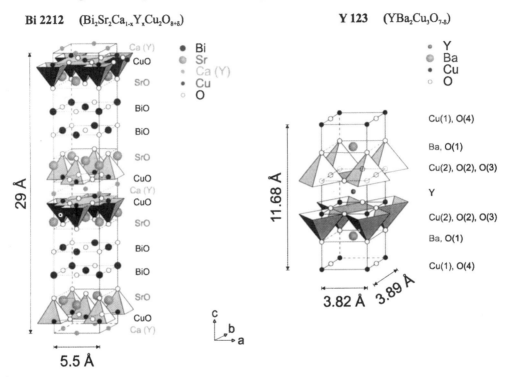

Figure 1: Crystal structures of $Bi_2Sr_2CaCu_2O_8$ and $YBa_2Cu_3O_7$.

varied considerably. An idealized doping phase diagram is shown in Fig. 2. $La_{2-x}Sr_xCuO_4$ (LSCO) can indeed be doped from the AF insulating phase through the superconducting to the normal metallic one. In many cases such as in Y123 or Bi2212 the structure is only stable in a limited doping range. Due to the layered structure the transport is very anisotropic and the ratio of the out-of-plane to the in-plane resistivities ρ_c/ρ_{a-b} varies between 10 for Y123 and 10^5 for Bi2212.[6] Since the Cu $3d^9$ spins in the planes show AF order not only in undoped but also in metallic samples anisotropies can also be expected in the a-b plane. It is this type of anisotropy which will be studied with light scattering in the following.

In a Raman experiment (Fig. 3) the sample is attached with good thermal contact to the cold finger of a cryostat which, in the case of the cuprate superconductors, should cover a temperature range from 4 K to at least 300 K. At higher temperatures the stability of the sample should be checked regularly as in most of the cases the mobility of the oxygen is very high. The quality of the surface is crucial since the Raman effect in a metal is weak in

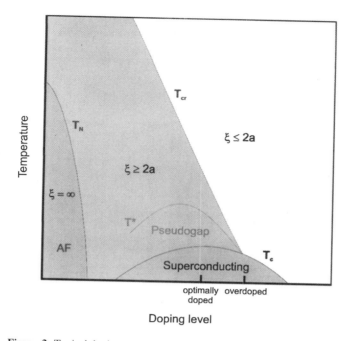

Figure 2: Typical doping-temperature phase diagram of CuO_2 compounds.

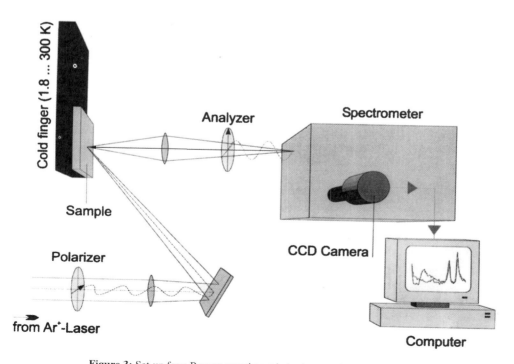

Figure 3: Set-up for a Raman experiment in back-scattering configuration.

253

contrast to scattering from most of the insulating contaminants. Hence, high or ultra-high vacuum is desirable. The exciting monochromatic, linearly polarized photons come from various types of lasers which normally provide several lines throughout the visible and near infrared or ultraviolet spectral range. In opaque materials the sample surface is oriented perpendicular to the optical axis of the system. To prevent directly reflected light from entering the spectrometer a large angle of incidence is used. Due to the relatively high refraction index of typically $\sqrt{\varepsilon} = 2.0 + 0.5i$ [58,100] the exciting photons propagate under a small angle to the surface normal. For optimal efficiency the beam is focused to some 100 µm diameter (at least in one direction). Even at low laser powers of the order of 1 mW corresponding to 10 W/cm^2 the spot temperature may be well above that of the holder and should be determined experimentally (see below). Typical values for ΔT are between 1 and 10 K/mW. The backscattered light is collected with an objective lens with large aperture and focused on the entrance slit of a highly dispersive and selective grating spectrometer. In spite of the good selectivity of double and triple monochromators elastically scattered laser light cannot completely be suppressed and reduces the experimental accuracy at low frequencies. Different polarization states of the scattered photons can be selected. The polarizations are given with respect to the crystal axes. Most widely used is the Porto notation $\mathbf{k}^I(\mathbf{e}^I;\mathbf{e}^S)\mathbf{k}^S$ which uniquely determines polarization ($\mathbf{e}$) and propagation directions ($\mathbf{k}$) of incoming (I) and scattered (S) photons. Here we are only interested in the properties of the CuO$_2$ planes being parallel to the sample surface, hence the scattering geometry is unambiguously given by the $\mathbf{e}^{I,S}$ with x = [100], y = [010], x' = [110]/$\sqrt{2}$, and y' = [$\bar{1}$10]/$\sqrt{2}$. We will always use the Cu-O bonds as a coordinate system and indicate the polarizations symbolically in the figures. The momenta of the photons are also important as they determine the momentum $\mathbf{q}$ of the created (+) or annihilated (-) elementary excitation $\hbar\omega$ in the sample. Since the change in the photon energy is small, $\omega^I = \omega^S \pm \omega \approx \omega^S$, $\mathbf{q} \approx \pm 2|\sqrt{\varepsilon}|\mathbf{k}^I$ for backscattering conditions. With $\lambda^I = 500$ nm, $|\mathbf{q}| = q$ is of the order 0.01 π/a with a the lattice spacing. Therefore, in a clean material only phonons in the center of the Brillouin zone can be observed. In case of scattering from electrons the conservation of momentum defines a frequency cut-off, $\omega_c = \mathbf{v}_F\mathbf{q} \approx 2\omega^I|\sqrt{\varepsilon}|(v_F/c)$, with $\mathbf{v}_F$ the Fermi velocity v_F its magnitude, and c the speed of light. For the scattering configuration described above $\hbar\mathbf{v}_F\mathbf{q}$ is of order 1 meV in the cuprates and therefore much smaller than all important energy scales. In conventional metals the cut-off may be much larger.

THEORETICAL FRAMEWORK

In an ideal Raman experiment a plane wave of generally linearly polarized, monochromatic light with a photon flux density $\dot{n}^I = I(\omega^I)/\hbar\omega^I$ with $I(\omega^I)$ being the power per unit area, $I(\omega^I) = P/A = \varepsilon_0 cE^2$, is hitting a sample volume V. Then the number of photons $\Delta\dot{N}^S$ scattered per unit time and energy interval $\Delta\hbar\omega^S$ into a solid angle $\Delta\Omega$ is given by

$$\Delta\dot{N}^S = \dot{n}^I V \frac{\partial^2\sigma}{\partial\omega^S\partial\Omega} \Delta\omega^S \cdot \Delta\Omega. \qquad (1)$$

This defines the spectral differential photon cross section $\dfrac{\partial^2\sigma}{\partial\omega^S\partial\Omega}$ per unit volume inside of the sample. The number of photons I_R counted in the detector is reduced considerably by surface losses, internal absorption, and the system response. Apart from frequency dependent factors to be determined carefully I_R is proportional to $\Delta \dot{N}^S$, somewhat inconsistently called Raman intensity, and usually plotted as a function of energy transfer, $I_R(\omega = \omega^I - \omega^S)$. It is straightforward but tedious to calculate the details if one is interested in absolute cross sections.[45,61] In a typical experiment the energy resolution is of the order of 1 meV (8 cm^{-1}) and the solid angle outside of the sample is 0.2 sr. The internal cross section is calculated via Fermi's golden rule[101-103] or the principle of detailed balance[61] giving

$$\frac{\partial^2 \sigma}{\partial \omega^S \partial \Omega} = \frac{\omega^S}{\omega^I} r_0^2 S(\mathbf{q}, \omega, T) \tag{2}$$

with $r_0 = e^2/4\pi\varepsilon_0 mc^2$ the classical radius of the electron and S the generalized structure function at momentum and energy transfer $\hbar\mathbf{q}$ and $\hbar\omega$, respectively, and temperature T. The principle of detailed balance[104,105] relates the energy loss (Stokes, $\omega^S < \omega^I$) and the energy gain (anti-Stokes $\omega^S < \omega^I$) spectra,

$$\left(\frac{\partial^2 \sigma}{\partial \omega^S \partial \Omega}\right)_{AS} = \left(\frac{\omega^I + \omega}{\omega^I - \omega}\right)^2 \left(\frac{\partial^2 \sigma}{\partial \omega^S \partial \Omega}\right)_{ST} e^{\left(-\frac{\hbar\omega}{kT}\right)} \tag{3}$$

and is equivalent to the fluctuation-dissipation theorem relating the structure function with the response function $\chi(\mathbf{q},\omega,T)$[106],

$$S(\mathbf{q}, \omega, T) = -\frac{\hbar}{\pi}\{1 + n(\omega, T)\} \operatorname{Im} \chi_{\gamma,\gamma}(\mathbf{q}, \omega, T), \tag{4}$$

with the Bose-Einstein thermal function $n(\omega,T) = [\exp(\hbar\omega/kT) - 1]^{-1}$. Relation (3) is very useful as the temperature of the illuminated spot can be obtained by just comparing energy gain and loss spectra (Fig. 4). In addition, if the two spectra agree reasonably well one has some evidence that the intensity comes from a spontaneous process with no allowed intermediate states involved, meaning, for instance, sufficiently far away from resonances. Then, the analysis in terms of linear response theory has also some experimental justification. It is not clear however, which type of excitations give rise to the observed scattering and in many cases careful experimental work is neccessary to single out the different contributions.

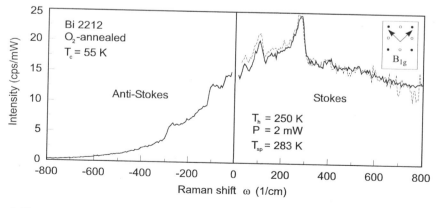

Figure 4: Energy gain (anti-Stokes) and energy loss (Stokes) spectra of Bi2212 at B_{1g} symmetry (full lines). The dashed line is the AS spectrum converted with Eq. 3.[43]

Eqs. (1)-(4) are completely general. For the explicit calculation of the Raman response function χ one has to specify which type of excitation gives rise to the inelastic scattering and how the interaction with the light works. Here we will focus on scattering from carriers in an anisotropic superconducting metal. The interaction Hamiltonian is obtained if in $\mathbf{H} = \Sigma \mathbf{p}_i^2/2m$ the canonical momentum is replaced by the kinematic one, $\mathbf{p}_i \rightarrow \mathbf{p}_i - e\mathbf{A}$ with $\mathbf{A} = \mathbf{A}^I + \mathbf{A}^S$ the vector potentials of the photons. Then, a non-zero amplitude for light scattering is found in first order perturbation theory for the term proportional to $\mathbf{A}^2$ corresponding to direct scattering on charge-density fluctuations and in second order for the $\mathbf{p} \cdot \mathbf{A}$ term via intermediate electronic states. It turns out that in a metal both contributions are important. This means that one scatters from an effective density,

$$\tilde{\rho}_\gamma = \sum_{k,\sigma} \gamma_k c^+_{k+q,\sigma} c_{k,\sigma} \qquad (5)$$

with electron creation and annihilation operators c^+ and c, respectively, and the Raman vertex γ_k taken in the limit $\mathbf{q} = 0$,

$$\gamma(\mathbf{k}, \mathbf{q} = 0; \omega^I, \omega^S) = \mathbf{e}^I \cdot \mathbf{e}^S + \frac{1}{m} \sum_v \left[\frac{\langle n\mathbf{k}|\mathbf{e}^S \hat{\mathbf{p}}|v\mathbf{k}\rangle \langle v\mathbf{k}|\mathbf{e}^I \hat{\mathbf{p}}|n\mathbf{k}\rangle}{\varepsilon_n(\mathbf{k}) - \varepsilon_v(\mathbf{k}) + \hbar\omega^I} + \frac{\langle n\mathbf{k}|\mathbf{e}^I \hat{\mathbf{p}}|v\mathbf{k}\rangle \langle v\mathbf{k}|\mathbf{e}^S \hat{\mathbf{p}}|n\mathbf{k}\rangle}{\varepsilon_n(\mathbf{k}) - \varepsilon_v(\mathbf{k}) - \hbar\omega^S} \right], \qquad (6)$$

where m is the free electron mass, $\hat{\mathbf{p}} = -i\hbar\nabla$ and $\varepsilon_n(\mathbf{k})$ and $\varepsilon_v(\mathbf{k})$ are Bloch conduction (n) and intermediate (v) state energies, respectively.[61-64,101-103,107] In order to complete the derivation of the cross section, the response function χ is to be calculated,

$$\chi^0_{a,b}(\mathbf{q}, \omega) = \frac{1}{V} \langle [\tilde{\rho}_a(\mathbf{q}), \tilde{\rho}_b(-\mathbf{q})] \rangle(\omega), \qquad (7)$$

where a,b = γ,1, < > denotes the thermal average and [ ] is the commutator.[37,106,107] Instead of using diagrammatical techniques $\chi^0_{a,b}$ can alternatively be obtained from Boltzmann transport theory.[35,37] In this approach, one considers deviations from the equilibrium distribution function as a result of an external perturbation being characterized by the Raman vertex $\gamma_k \delta \varepsilon_\gamma$ with $\delta \varepsilon_\gamma = r_0 |\mathbf{A}^I||\mathbf{A}^S|$. The different susceptibilities for specific perturbations, i.e. experimental probes, can then be obtained by just selecting the respective vertex functions and transport and Raman can be calculated from the same starting point.

Eq. (7) neglects Coulomb screening effects which are important in charged systems. The long-range Coulomb interaction leads to a separation of the total Raman response into transverse and longitudinal contributions and the full result reads[35,62-64]

$$\chi_{\gamma,\gamma} = \chi^0_{\gamma,\gamma} - \frac{(\chi^0_{\gamma,1})^2}{\chi^0_{1,1}} \left(1 - \frac{1}{\varepsilon}\right). \qquad (8)$$

The first two terms on the right hand side correspond to the transverse response and are of order q^0 and hence finite at $q = 0$, the last longitudinal one is suppressed by the dielectric function $\varepsilon \approx 1 - (\omega_{pl}/\omega)^2$. With the characteristic frequency $\hbar v_F q = 1...10$ meV and a plasma energy of 1 eV the difference in intensity is at least a factor of 10^4, and the last term can safely be neglected here. To make the response gauge invariant the vertex has to be dressed (vertex corrections). This can change the spectra seriously under special conditions.[63,64] In many cases such as d-wave superconductors the influence of vertex corrections is only mild[37] so that we will not discuss them further.

For the calculation of the response as given in Eq. (8) the vertex (Eq. 6) is more of academic use since it requires the knowledge of the full band structure in an energy window around the Fermi surface given by the photons involved. In addition, the band structure of the cuprates is still a matter of debate. Far away from resonances, $|\varepsilon_n(\mathbf{k}) - \varepsilon_v(\mathbf{k})| \gg \omega^{I,S}$, (6) can be simplified enormously, and the vertex is given by the curvature of the conduction band or, more precisely, by the tensor of the inverse effective mass[103],

$$\gamma_{\mathbf{k}} = \frac{m}{\hbar^2} \sum_{\alpha,\beta} e_\alpha^S \frac{\partial^2 \varepsilon_n(\mathbf{k})}{\partial k_\alpha \partial k_\beta} e_\beta^I . \qquad (9)$$

If there exists an analytic approximation for the conduction band, $\gamma_{\mathbf{k}}$ can be calculated immediately. For cuprate systems a 2-D tight-binding conduction band on a quadratic lattice with spacing a is often used, $\varepsilon(\mathbf{k}) = -2t[\cos(k_x a) + \cos(k_y a)] + 4t' \cos(k_x a) \cos(k_y a)$ with t and t' the nearest and next nearest neighbor hopping integrals.[108-110] It is much more instructive, however, to exploit the symmetry properties of the vertex, since polarized Raman scattering projects out its different symmetry components μ. To this end one decomposes $\gamma_{\mathbf{k}}$ into a set of orthonormal functions.[111] Then, weighted averages over the Fermi surface can be calculated for each polarization configuration. In this way a relation between polarization, i.e. configuration space and certain directions in $\mathbf{k}$-space can be established,

$$\gamma_{\mathbf{k}}^\mu = \sum_{L=0,2,4,\ldots} \gamma_L^\mu(\omega^I,\omega^S) \Phi_L^\mu(\mathbf{k}) \qquad (10)$$

with μ = A_{1g}, B_{1g}, B_{2g}, and A_{2g} for a tetragonal lattice, $\Phi_0^\mu(\mathbf{k}) \equiv 1$, and $\sum_{\mathbf{k}} \Phi_L^\mu(\mathbf{k}) \Phi_{L'}^{\mu'}(\mathbf{k}) = \delta_{L,L';\mu,\mu'}$. It is noted that A_{2g} is not Raman active. For a cylinder-like Fermi surface parallel to the c-axis expansion (10) is particularly simple, and the first nontrivial basis functions are given by $\cos(4\varphi)$, $\cos(2\varphi)$, $\sin(2\varphi)$, and $\sin(4\varphi)$ for the symmetries specified above and φ the azimuthal angle. If the response should be evaluated in the whole Brillouin zone, appropriate basis functions might be [const. + $\cos(k_x a)$ + $\cos(k_y a)$], [$\cos(k_x a)$ - $\cos(k_y a)$], and [$\sin(k_x a)\sin(k_y a)$] for the three Raman active symmetries A_{1g}, B_{1g}, and B_{2g}. Independent of the choice of the basis functions, the sensitivity for the B_{1g} component projected out with x'y' polarizations is predominantly along the principal axes, while B_{2g} (xy) emphasizes the diagonals. A_{1g} has always admixtures of the B components, is generally seen with parallel polarizations (xx, x'x', etc.), and is not specifically sensitive for a certain direction (Fig. 5). If the expansion is truncated after L = 2 and plugged into Eq. 8 the response reads,

$$\lim_{q \to 0} \chi^\mu = -\left(\gamma_2^\mu\right)^2 \left\{ \left\langle \left(\Phi_2^\mu(\mathbf{k})\right)^2 \Theta_{\mathbf{k}} \right\rangle - \frac{\left\langle \Phi_2^\mu(\mathbf{k}) \Theta_{\mathbf{k}} \right\rangle^2}{\left\langle \Theta_{\mathbf{k}} \right\rangle} \right\}. \qquad (11)$$

Here $\Theta_{\mathbf{k}}$ is the response kernel for either the normal or the superconducting state and does not depend on the selected symmetry, <...> repesents the $\mathbf{k}$-sum over the Brillouin zone. Apparently all $\mathbf{k}$-independent contributions to the Raman vertex drop out. This means, among other things, that no electronic Raman scattering can be observed for purely

Polarization	FS-Harmonics		BZ-Harmonics

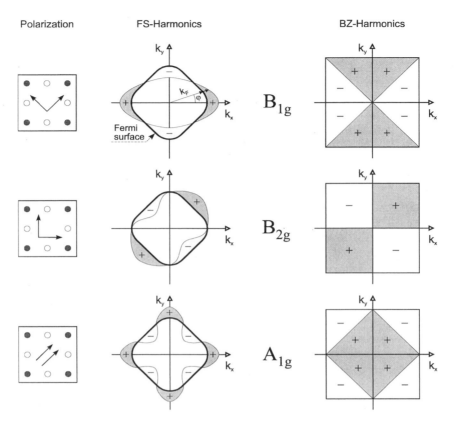

Figure 5: Fermi-surface and Brillouin-zone harmonics projected out for different polarizations. The bold line is the Fermi surface. A_{1g} components are seen generally with parallel polarizations, however, with contributions from B_{1g} and B_{2g} depending on the angle. With x'x' shown in the figure, $A_{1g} + B_{2g}$ is projected out.

parabolic dispersion. In the normal state, Θ_k is finite only in the collision-limited regime and is then given by a Drude-like response v_k characterized by a lifetime τ_k,[106,112]

$$v_k = \phi_k \frac{1}{1 - i\omega\tau_k} \qquad (12)$$

with $-\phi_k$ the derivative of the equilibrium distribution function with respect to the quasiparticle energy $\xi_k = \epsilon(\mathbf{k}) - \mu$, μ the chemical potential. ϕ_k measures the density of states for a given $\mathbf{k}$, and the $\mathbf{k}$-integral over the Fermi surface gives N_F, the total density of states at the Fermi level. In the superconducting state and $\tau_k \to \infty$, Θ_k can be expressed in terms of the Tsuneto function λ_k,[113]

$$\lambda_k = \frac{\tanh\left(\dfrac{E_k}{2kT}\right)}{2E_k} \frac{4|\Delta_k|^2}{4E_k^2 - (\hbar\omega + i\delta)^2} \qquad (13)$$

with $E_k^2 = \xi_k^2 + |\Delta_k|^2$ and T the temperature. After energy integration, the more popular form

$$\text{Im}\lambda_k = N_F \frac{\pi}{2\hbar\omega} \tanh\left(\frac{\hbar\omega}{4kT}\right) \text{Re} \frac{4|\Delta_k|^2}{\sqrt{(\hbar\omega)^2 - 4|\Delta_k|^2}} \qquad (14)$$

is recovered which shows the threshold and the singularity at $2\Delta_k$ the latter one giving rise to the pair-breaking peak. Using (11) - (13) the full $\mathbf{k}$ sums have to be performed explicitly taking into account the band structure. Normally the $\mathbf{k}$ sums are replaced by an energy and surface integral, $\sum_k \rightarrow N_F \int d\xi d\Omega_k$, which returns the well known results, e.g. (14) for the superconducting state[63,64]. Then however, details of the band structure such as multi-band effects or the van Hove singularity close to $(\pi,0)$,[114,115] or a $\mathbf{k}$-dependent temperature variation of φ_k, as we will discuss in context with the pseudogap scenario, cannot be handled any more. For further details see Ref.[37].

Usually the magnitude and $\mathbf{k}$ dependence of τ_k and Δ_k are factorized, e.g. $\Delta_k = \Delta_0 g(\mathbf{k})$ with $|g(\mathbf{k})| \leq 1$. For all practical purposes $g(\mathbf{k})$ is expanded according to Eq. (10) in the same way as γ_k. This means that the symmetry related functions have to be selected carefully whenever a certain variation for τ_k or Δ_k is being studied. If expanded to different order, problems arise with the orthogonality of the respective basis functions selected for the response kernel and the vertex. Both τ_k and $|\Delta_k|^2$ have the full symmetry of the underlying lattice, transforming as A_{1g}. Therefore, in most of the cases, the second term in (11) is finite only for A_{1g}-type vertex functions and may then change the spectral shape and the overall intensity completely.[37] Accordingly, for B_{1g} and B_{2g} spectra the first term gives normally the full response. As in most of the other response functions only the magnitude of the gap shows up in (13) and (14), and the Raman experiment is not sensitive to the phase of the order parameter. It should also be noted that (12) and (13) are only compatible if $\hbar/\tau_k << \Delta_0$.[35,115] With increasing $\hbar/\tau_k$ the singularity at $2\Delta_k$ is more and more suppressed and the effect of the superconducting transition in the spectra becomes weaker.[59,60,116] It turns out that the cuprates are in the collision-limited regime in the normal state and in the opposite one below T_c as the relaxation of the electrons is obviously dynamic and strongly temperature dependent. This justifies to combine the results for certain limites.[35]

I will consider now a few limits of the functions presented above. In the low frequency limit of the normal state one finds

$$\lim_{\omega \to 0}(\text{Im} \, v_k) = \varphi_k \omega \tau_k \qquad (15)$$

showing that the slope $\partial \text{Im} v_k / \partial \omega$ in the d.c. limit is proportional to the relaxation time. The full response for $\mu = B_{1g}, B_{2g}$ and $\varphi_k = \text{const.}$ is then

$$\lim_{\omega \to 0}(S^\mu(\omega)) = \frac{\hbar}{\pi}\left\{\frac{kT}{\hbar\omega}\right\}(\gamma_2^\mu)^2 N_F \omega \langle \Phi_2^\mu(\mathbf{k})\tau_k \rangle = R^\mu \cdot kT \cdot \tau_0^\mu. \qquad (16)$$

Eqs. (15) and (16) demonstrate that a symmetry-specific τ_0^μ at $\omega = 0$ can be obtained from the spectra if the "intensity" factor R^μ is known. One either can determine the absolute intensity or try to find an approximation for R^μ from the other parts of the spectra. To this end a τ has to be specified. If τ is constant, as in the case of impurity scattering, v assumes a Lorentzian shape[112] peaking at $(1/\tau; R^\mu/2)$, hence $R^\mu = 2S^\mu(\omega=1/\tau)$. Whenever the electronic lifetime is limited by inelastic scattering temperature and frequency dependences come into play. In the nested Fermi liquid model[117] which describes the normal state response of the high-T_c cuprates fairly well an interpolation formula for the scattering rate can be found,

$$\Gamma(\omega, T) \equiv \frac{\hbar}{\tau(\omega, T)} = \sqrt{\left(\Gamma_0(T)\right)^2 + (\alpha\hbar\omega)^2},$$

with (17)

$$\Gamma_0(T) = \frac{\hbar}{\tau_{imp}} + \beta kT,$$

and $0.5 < \alpha < 1.5$ and $1 < \beta < 2$. Here, in addition to the original version a small impurity scattering rate is added. At large energies $S''(\omega \rightarrow \infty) = R''\alpha/(1+\alpha^2) \approx R''/2$. This means that an estimate for τ in the d.c. limit can be obtained independent of a specific assumption for details of the response if the right spectral range is selected for the determination of R''.

In the superconducting state the response at small energies can also be analysed further. This is particularly interesting since in this limit band structure effects and vertex corrections drop out completely: For symmetry reasons and independent of the number L of basis functions the B_{1g} vertex always changes sign on the diagonals and can be considered linear in the vicinity. The B_{2g} vertex, on the other hand, is constant here. Therefore, whenever the gap has nodes on the main symmetry lines, frequency power laws result which depend only on the selected polarization. For a gap with $d_{x^2-y^2}$ symmetry one obtains for zero temperature[36]

$$\text{Im}\,\chi^{B_{1g}} \propto \left(\frac{\hbar\omega}{2\Delta_0}\right)^3$$

$$\text{Im}\,\chi^{B_{2g}} \propto \left(\frac{\hbar\omega}{2\Delta_0}\right)$$ (18)

$$\text{Im}\,\chi^{A_{1g}} \propto \left(\frac{\hbar\omega}{2\Delta_0}\right)$$

with Δ_0 the gap maximum. In the limit $\omega = 0$ superconducting correlation effects vanish for systems with a gap being finite on the entire Fermi surface. They can also be neglected for gap functions with point and line nodes which obviously have zero area. Hence the number of thermally excited quasiparticles which determine the intensity in the limit considered depend only on the magnitude of the gap. In an isotropic system the result comes out in the same way as for a semiconductor and holds in the clean limit[118-120] and with non pair-breaking impurities.[59,60] The ratio of the superconducting (sc) to the normal (nc) response at the same temperature reads

$$\frac{I_{sc}(\omega=0, T)}{I_{nc}(\omega=0, T)} \equiv \frac{\text{Im}\,\chi_{sc}(\omega=0, T)}{\text{Im}\,\chi_{nc}(\omega=0, T)} = f\{|\Delta(T)|\}.$$ (19)

Here f is the Fermi function. The normal metallic response for the superconducting state can easily be calculated since it is temperature independent in this approximation. In an anisotropic system with nodes, impurities are generally pair breaking and a similarly simple expression as (19) can only be obtained for the clean limit,[36,37]

260

$$\left[\frac{\operatorname{Im}\chi_{sc}(\omega=0,T)}{\operatorname{Im}\chi_{nc}(\omega=0,T)} \right]^{\mu} = \frac{2\left\langle f\{|\Delta_{\mathbf{k}}(T)|\}|\Phi_2^{\mu}(\mathbf{k})|^2 \right\rangle}{\left\langle |\Phi_2^{\mu}(\mathbf{k})|^2 \right\rangle} \tag{20}$$

where, for simplicity, first order expansion functions are used. Once again the result depends on the symmetry selected, and power laws can be predicted for low temperature. For a given pairing state they are identical to those found for the frequency with $\hbar\omega \to kT$ (Fig. 6). With impurities (isotropic potential) the ratio is no longer zero for $T = 0$ because the response is always normal metallic at $\omega = 0$ even at the lowest temperatures due to the quasiparticles created by pair breaking.[115] In the cuprates the situation is even more complicated since the finite lifetime of the carriers does not predominantly come from impurity scattering but from inelastic and strongly temperature-dependent processes.

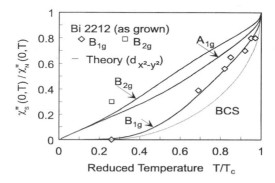

Figure 6: Theoretical prediction for the normalized d.c. Raman response in a clean d-wave superconductor with a cylinder-like Fermi surface.[37] The experimental points are from refs.[36,56]. The low temperature results can be visualized in Fig. 10 below.

Therefore, the intensity below T_c cannot be calculated prior to an understanding of the scattering process which is very likely related to the mechanism responsible for the superconductivity. This, on the other hand, allows to test different model predictions. However, the preceding considerations show at least qualitatively that the temperature variation of the d.c. intensity may also depend on the selected channel and provide supplemental information.

Finally, I will briefly address scattering from spin excitations. The spin does not directly couple to the magnetic field of the light but to excited electrons via spin-orbit coupling[96,97] in a similar way as the phonons. For the determination of the dispersion and the energy scales involved one considers the Heisenberg Hamiltonian

$$\hat{H} = J\sum_{i,j}\hat{\mathbf{S}}_i \cdot \hat{\mathbf{S}}_j \tag{21}$$

with J the exchange interaction and $\hat{\mathbf{S}}_i$ a spin operator. For an antiferromagnet J is positve for a ferromagnet negative. The sum runs over all nearest neighbors. One obtains a dispersion for the magnons which is similar to that of an acoustic phonon with an energy of JzS at the zone boundary with z the number of nearest neighbors and S the spin quantum number. In the cuprates, we have z = 4 and S = 1/2. As soon as the system is anisotropic there is a gap in the spectrum at q = 0. The interaction J between spins is mediated by exchange of electrons. For finite coupling to light, the connections $\mathbf{r}_{i,j}$ between spins have

261

to have a component parallel to the polarization vectors. In lowest order the Hamiltonian is then given by[96]

$$\hat{H} = \alpha \sum_{i,j} \left(\mathbf{e}^I \cdot \mathbf{r}_{i,j} \right) \left(\mathbf{e}^S \cdot \mathbf{r}_{i,j} \right) \left(\hat{\mathbf{S}}_i \cdot \hat{\mathbf{S}}_j \right). \qquad (22)$$

Using this Hamiltonian the selection rules can be found. The strongest scattering is expected and observed for B_{1g} symmetry[90] although some intensity, weaker by approximately an order of magnitude, is also found at other polarizations.[121] Here two adjacent spins are flipped and the energy transferred to the system is approximately 3 J by just counting the bonds. A complete calculation gives 2.8 J for the position of the two-magnon peak[97], hence J can be determined from the Raman spectrum. For a prediction of the full line shape the two magnon correlation function has to be calculated. So far only approximations could be obtained.[122-124]

QUALITATIVE COMPARISON OF CONVENTIONAL AND CuO_2 SYSTEMS

Raman spectra of a conventional metal (Fig. 7) and a cuprate superconductor (Fig. 8) consist of a broad almost structureless continuum and superimposed bands with various widths. In most of the cases the bands come from phonons. In $Nd_{1.84}Ce_{0.16}CuO_4$ (NCCO, see Fig. 30) some of the lines originate from crystal-field excitations[127,128]. If one is lucky the different types of excitations can be distinguished from the selection rules such as in NCCO. Sometimes a lattice-dynamical calculation is required to approximately determine the frequencies. The situation is additionally complicated by disorder since in this case the selection rules are relaxed and more phonon lines than those allowed by symmetry can be present, for instance the lines at 590 cm^{-1} in NCCO[128,129]. The origin of the continuum is not completely clarified. However, most of the intensity at low energies is due the

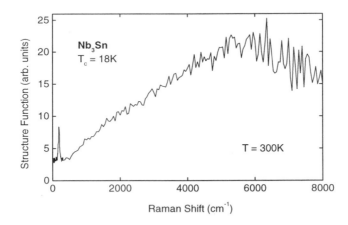

Figure 7: Raman spectrum for Nb_3Sn at E_g symmetry. Part of the increasing background is coming from luminescence since the maximum is always at the same absolute energy of approximately 2 eV independent of the exciting frequency.[125] The narrow line at 200 cm^{-1} is the E_g phonon which is strongly coupled to the carriers.

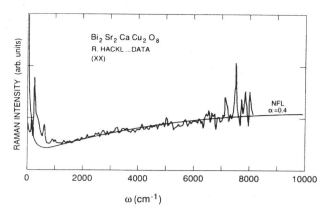

Figure 8: Raman spectrum for Bi2212 at $A_{1g} + B_{2g}$ symmetry.[126] The fit (smooth line) is according to the nested-Fermi-liqid model[117] which exhibits the same phenomenology as the marginal Fermi liquid[194].

scattering from carriers as is evident from the spectra in the superconducting state. In addition, the Fano-type line shape of certain phonons can originate only from their strong interaction with Raman active electronic excitations.[48,49,51,69-72,130] This does not explain the origin of the continuum extending far beyond the cut-off energy. In A15 compounds and in cuprates with $\mathbf{k}^1$ parallel to the c axis $\hbar v_F \mathbf{q}$ is of the order of 10 cm^{-1} due to a Fermi velocity $\mathbf{v}_F$ of the order of 10^6 cm/s. Therefore the electronic lifetime must be limited and $\hbar v_F \mathbf{q} \ll \hbar/\tau$ in order to relax the momentum conservation and produce intensity at reasonably high energy transfers. So far elastic scattering from impurities, electron-phonon[131,132], electron-spin[133,134] and electron-electron[117] interactions have been studied theoretically and found to be candidates for explaining intensities at sufficiently high frequencies. Impurity scattering results in a Lorentzian-type frequency dependence falling off as $1/\omega$ at $\omega \to \infty$. All inelastic processes lead to the required $1/\tau \propto \omega$ implying more or less constant Raman intensity at $\omega \gg kT$. In high-T_c A15 compounds, because of the strong electron-phonon interaction with coupling constants λ between 1 and 2,[135] it is likely that the phonons carry away the excess momentum. There is however a gap in the experimental information which hinders a thorough understanding. In cuprates electron-electron scattering on a nested Fermi surface very successfully describes the Raman spectra at high energy transfers as shown in Fig. 8[117], but fails to explain the symmetry dependence of the relaxation rate at low frequency[136]. To date the study of electron-spin interactions is right at the beginning. Apparently, the high energy cross section is not uniquely linked to a certain type of interaction.[137,138]

In the superconducting state the spectra at low energy are rearranged (Fig. 9). The intensity decreases below a characteristic energy, piles up in the intermediate range and asymptotically approaches the normal state response from above at twice or three times the crossover point. In conventional superconductors the threshold is fairly well defined at all symmetries which is certainly the strongest evidence for a finite only slightly anisotropic gap. Nevertheless, there is considerable symmetry dependence (compare Fig. 9 a and b) the origin of which, however, is unlikely to be a gap anisotropy of some 30%. It rather comes from strong final state interactions[139] which have been discussed for the first time for Raman scattering from rotons in superfluid ^{4}He.[140,141] Though predicted correctly the importance for superconductors was not immediately acknowledged.[63] Later on it became

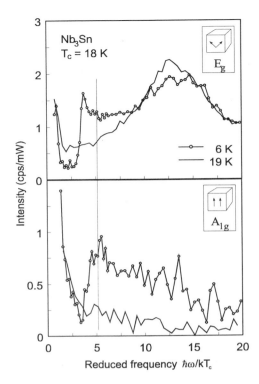

Figure 9: Normal- and superconducting Raman spectra for Nb$_3$Sn.[35]

clear that theory and experiment fit much better with final state interactions taken into account.[64] Indeed, after a deconvolution the structures in Nb$_3$Sn are very narrow[142,143] and can not at all be modelled by a square root singularity as given in Eq. (14). I note that the effects of final state interactions are not included in the above equations and the reader is referred to the literature[37,64] for details. The physical origin of final state interactions is a small residual electron-electron interaction with non-zero angular momentum l orthogonal to the Copper channel with lowest angular momentum ($l = 0$ for s-wave and $l = 2$ for d-wave pairing) responsible for the superconductivity. In A15 compounds the phonon with E$_g$ symmetry may be coupled strongly enough to open a channel orthogonal to $l = 0$ to explain the pole split off considerably from the gap edge.[144-146] Consistently, no narrow peak is observed for the two other symmetries, A$_{1g}$ (Figs. 9 b) and T$_{2g}$ (not shown, see Refs.[142,147]). Finally the finite intensity below the threshold will be addressed. At a temperature of T = 0.3 T$_c$ the gap of a conventional superconductor should be clean (see Fig. 6). Instead approximately 30% of the normal state intensity are seen in the experiment. The residual scattering can be traced back either to intrinsic luminescence[148] or a chemically changed composition of the sample surface which can affect layers as thick as some 20% of the penetration depth δ.[149] Without the appropriate UHV tools this question cannot be answered experimentally. Most likely however, the major contribution comes from a degradation of the surface since the residual intensity is very small in samples with a stable surface such as Bi2212 (see below). To summarize the results in conventional superconductors we note that a well-defined threshold is observed at all symmetries. The relatively low peak frequency in E$_g$ symmetry is specific to A15 compounds and unlikely to be a signature of a strong gap anisotropy since the line width is too small. It rather is a bound state below the gap. Therefore the gap energy is approximately given by the peak maxima in the A_{1g} and T_{2g} spectra, and 2Δ can be estimated to be 5.5 in units of kT_c. The residual scattering below the threshold is most likely not an intrinsic property of the electronic system, but comes from surface degradation. However, it cannot be excluded that the decay of long-lived intermediate states of the intrinsic sample material has some spectral weight in this energy range.

The spectra of Bi2212 look strikingly different (Fig. 10): There is no threshold, the maxima are at fairly different positions and not particularly pronounced, and the slope at low frequency depends on the symmetry. The intensity at the B_{1g} symmetry is found to be

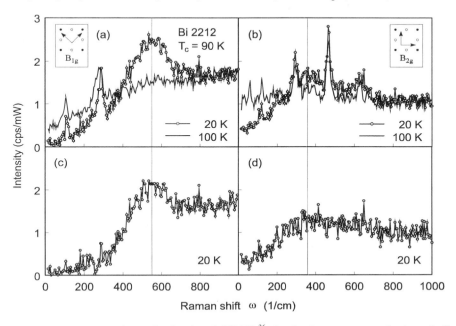

Figure 10: Raman spectra for optimally doped Bi2212.[36] In the lower two panels (c and d) the superconducting spectra are shown with the phonons subtracted.

very small, i.e. below 10% of the normal state value, at low temperature. This is unique for all superconductors studied so far. The reason is that Bi2212 can easily be cleaved between the weakly coupled BiO_2 layers (Fig. 1) which, in turn, are chemically inert at ambient air and preserve the intrinsic properties perfectly up to the surface. Therefore Bi-based compounds are most favorable for ARPES and tunneling probing only a few Ångstrøms. From this consideration the 30% left over at B_{2g} must be considered intrinsic. If one looks at samples close to the T_c maximum with the resistivity extrapolating to a value very close to zero at $T = 0$ the response in the low-T, small-ω limit varies as ω^3 at B_{1g} and as ω at the two other symmetries.[36,38,46,56] All this information taken together does not allow an isotropic or slightly anisotropic gap to account for the observed spectra. However, very good agreement can be achieved if a gap transforming as B_{1g} is used, $\Delta_k(T) = \Delta_0(T)\sum_L g_L^{B_{1g}}\Phi_L^{B_{1g}}(\mathbf{k})$ with $\Delta_k(T) = \Delta_0(T)\cos(2\varphi)$ being the simplest representaion of a $d_{x^2-y^2}$ gap on a cylinder-like Fermi surface (Fig. 11).[36] In particular, the different peak positions and shapes, the symmetry dependence of the slope (see Eq. 18) and that of the intensity in the d.c. limit (cf. Fig. 6) can be explained naturally. This concludes the considerations about the qualitative differences between the old an the new superconductors. For a deeper understanding of the cuprates the unique dependence of their properties on doping and temperature must be studied in detail. I will now compile the available Raman data.

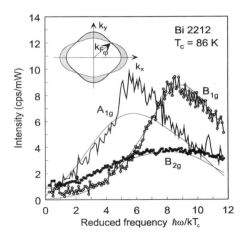

Figure 11: Electronic spectra for optimally doped Bi2212 at 20 K.[56] The light lines represent the theoretical prediction for unconventional $d_{x^2-y^2}$ pairing as sketched in the inset.[36]

RECENT EXPERIMENTAL RESULTS FOR HIGH-T_c MATERIALS

In this section the doping and symmetry dependence of the Raman spectra of normal and superconducting cuprate systems will be described starting with general trends and the d.c. limit in the spectra at $T > T_c$. Then the intermediate frequency range up to some 100 meV will be addressed in order to better understand the peculiarities related to the pseudogap state in underdoped samples at temperatures between 100 and 200 K. Finally I will present more details about the spectra below T_c, in particular, I will try to find some link to the normal state results and will conclude with a comparison of electron and hole-doped systems.

Normal State

Anisotropies in the normal state of optimally doped cuprate superconductors have been observed by Raman scattering[126,150,151] briefly after the study of those below T_c.[50,51] Since a few years the information from Raman spectroscopy about differently doped samples is growing continuously. This is closely related to the availability of well characterized high-quality single crystals. The first systematic study in the normal state was presented for LSCO[152] followed by results on Y123.[153] The main conclusion was that the ratio of the overall intensity of the B_{1g} to that in the B_{2g} channel decreaces by factors on the way from overdoped to underdoped still superconducting samples. The same trend is found in Bi2212. Three differently doped samples having T_c's of 57, 90, and 55 K in the underdoped, the optimally doped and the overdoped range of the phase diagram, respectively, will be compared. Underdoping was achieved by replacing 38% of Ca^{2+} by Y^{3+} thus filling in holes in the planes. For overdoping an optimally-doped as-grown crystal was annealed in 1600 bar oxygen at 450°C. The transition widths were between 2 and 5 K. In Fig. 12 the intensity ratios at 800 cm^{-1} ($\hbar\omega \gg kT$) are plotted as a function of doping δ.[154] To the obtainable accuracy of 5% rms of the absolute intensity they are independent of temperature. In the spectra shown in Fig. 10 the ratio can be visualized for the optimally doped sample. The absolut numbers in Y123 are larger by a factor of two[155] on account of the CuO chains being invisible for B_{2g} polarizations. Hence we have another property of

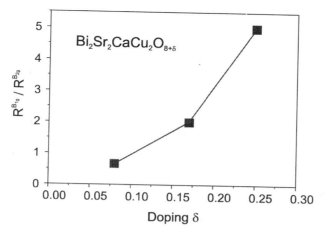

Figure 12: Ratio of the Raman intensities at B_{1g} and B_{2g} symmetry at 800 cm^{-1} as a function of doping.

the cuprates which depends only on the doping level but not on the specific sample material, and it is clear for symmetry reasons that predominantly the CuO_2 planes are probed.

For the study of the d.c. limit we will first look at the data for strongly overdoped Bi2212. B_{1g} spectra (phonons subtracted) at various temperatures are displayed in Fig. 13. In addition to the intensity $I_R(\omega)$, the response function $Im\chi(\omega)$ is shown. The change of the slope with temperature $\partial Im\chi(T,\omega)/\partial\omega$ is indicated by short dashes resembling directly the

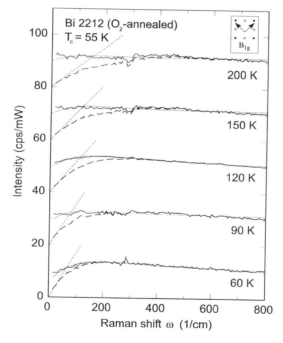

Figure 13: Electronic Raman spectra (full lines, phonons subtracted) and response functions $Im\chi$ (long dashes) for overdoped Bi2212.[43] The light lines are fits according to Eq. (4), (12), and (17).[117] The short dashes represent the slope at $\omega = 0$ which is proportional to the carrier lifetime. The spectra above 60 K are shifted by 20 units each.

temperature dependence of $\tau_0^{B_{1g}}$ without, however, pinning down the magnitude. As outlined above figures can be obtained in various ways. As the prediction for a Fermi liquid with nesting properties[117] (Eqs. (4), (12) and (17)) is found to fit the data quite well (mass renormalization is neglected here) the respective results can be used as well as those from the intensity intercept. Similar procedures apply at all doping levels. As one may anticipate the temperature dependence can be revealed with better accuracy than the magnitudes being subject to the model selected. The relaxation rates, $\Gamma_0^{B_{1g}}(T) = \hbar / \tau_0^{B_{1g}}(T)$, in the limit $\omega = 0$ for the three doping levels described above are compiled in Fig. 14. Obviously, there is a strong variation which, as far as comparable, is corroborated by the results in Y123.[95,155] Although Raman and transport relaxation rates cannot be compared directly due to the different vertices involved one can nevertheless expect similar temperature dependences. Actually, resistivity ρ and relaxation rate Γ should roughly be proportional according to the Drude relation,

$$\rho = \frac{1}{\varepsilon_0 \omega_{pl}^2} \frac{1}{\tau_{tr}}, \tag{23}$$

where ε_0 is the permittivity constant, ω_{pl} the plasma frequency and τ_{tr} the transport lifetime, or, rearranged for the calculation of numbers,

$$\Gamma \equiv \frac{\hbar}{\tau} = 1.08 \cdot \rho_\mu \left(\hbar \omega_{pl} \right)^2,$$

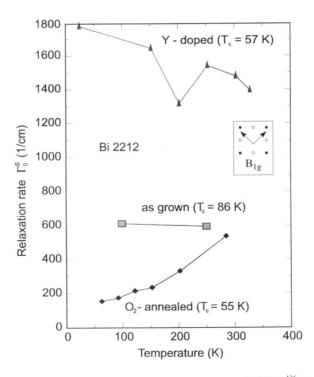

Figure 14: Relaxation rates $\Gamma = \hbar/\tau$ at B_{1g} symmetry for differently doped Bi2212.[136] The magnitude of Γ in the underdoped sample depends critically on the method of determination (see text). The accuracy is not better than ±50%.

with the relaxation rate in cm^{-1}, ρ_μ in $\mu\Omega$cm, and the plasma frequency in eV. While the overdoped sample fits satisfactorily, the discrepancies to the resistivity data for comparable samples at the T_c maximum and in the underdoped range of the phase diagram[156] become progressively severe. This concerns not only the magnitude but in particular the dependence on T. At low doping $\Gamma_0^{B_{1g}}(T)$ is very large implying a mean free path of the order of the lattice constant. This means that we are already at the brink of validity of the analysis. In clear contrast, the B_{2g} data are compatible with transport at all doping levels including the respective zero-temperature extrapolation values (Fig. 15). This means that the normal-state anisotropy is small in the overdoped range and continuously increases towards underdoping. More remarkably, the complete anisotropy comes from a split off of $\Gamma_0^{B_{1g}}(T)$ whereas $\Gamma_0^{B_{2g}}(T)$ remains closely related to ordinary transport. For completeness it is noted that Bi2201, the single layer compound in the Bi family, is very similar to overdoped Bi2212.[35] Phenomenologically, the variation of the relaxation with symmetry and doping can be understood by invoking an increasingly strong anisotropic scattering potential with little temperature dependence in the range studied varying, for instance, as $\left(\Phi_2^{B_{1g}}(\mathbf{k})\right)^2$.[45,115] As a physical origin the antiferromagnetic background of the Cu $3d^9$ spins can be imagined which interacts quite differently with spins of carriers moving along the

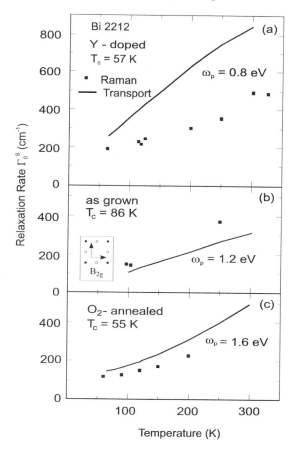

Figure 15: Relaxation rates $\Gamma = \hbar/\tau$ at B_{2g} symmetry for differently doped Bi2212.[136] The transport data and plasma frequencies for the calculation of the respective relaxation rates according to Eq. (23) are taken from the literature.[156,157]

principle axes or the diagonals. Towards low doping the magnetic correlation length and consequently the coupling increases.

This type of physics is very thoroughly (and less naively) persued in all models incorporating electron-spin interactions.[158-160] In the model of a nearly antiferromagnetic Fermi liquid (NAFL) the phenomenology has been pushed furthest including Raman response.[133,159,161] For the Fermi surface predicted by LDA calculations[108-110] and observed with ARPES the spin susceptibility is strongly peaked close to $\mathbf{Q} = (\pi,\pi)$, the antiferromagnetic (AF) wave vector. The peaks in the susceptibility have indeed been found by inelastic neutron scattering[86,87] At the points close to the principle axis where the Fermi surface crosses the AF Brillouin zone (hot spots, see Fig. 16) the electrons are scattered very effectively while being unaffected around the diagonals (of the large Brillouin zone). The coupling strength scales with the correlation length ξ_{mag} of the AF spin fluctuations providing a natural explanation for the doping dependence.[82,159] At low doping the lifetime and the spectral function of the electrons with momenta along the principle directions can be expected to be strongly reduced while in the overdoped regime the anisotropy should vanish. The suppression of the spectral function at the hot spots has been calculated explicitly[161] and has been observed in ARPES a while ago.[13,14] This can at least qualitatively explain the weak Raman intensity at B_{1g} symmetry in underdoped cuprates.

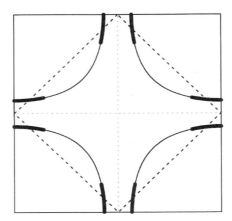

Figure 16: Brillouin zone (big square), AF Brillouin zone (dashed) and tight-binding Fermi surface[108] typical for cuprates over a wide doping range[162]. The AF wave vector $\mathbf{Q} = (\pi,\pi)$ connects the crossing points of the AF Brillouin zone with the Fermi surface close to $(\pi,0)$ (hot spots visualized by bold lines).[82]

For an explanation of the intensity variation alone one might look at the tight-binding band structure given above. Formally, the ratio of the intensities is proportional to $(t/t')^2$ and can be tuned easily.[35,152] The required parameter range, however, is unrealistic, and no clarification of the doping dependence of the scattering rates $\Gamma_0^\mu(T)$ can be obtained. Alternatively, Fermi surface evolution as a function of doping can be considered. Since electronic correlations are expected to gain influence with decreasing carrier concentration the large Fermi surface centered around (π,π) in optimally and overdoped samples[13,110,162] may become unstable against a transition to a spin-density-wave-like state at low doping.[8,10,84,110] Then small pockets around $(\pi/2,\pi/2)$ are expected to develop[163] and the Raman intensity will be suppressed strongly in A_{1g} and B_{1g} symmetry due to the **k** dependence of the vertex (see Fig. 5). In this case however, one has to clarify how the small Fermi surface evolves continuously from the large one since the intensity ratio does not seem to change discontinuously. The remaining intensity at A_{1g} and B_{1g} can be traced back to scattering from spin excitations with completely different relaxational behavior.[94,121] This scenario is certainly compatible with longitudinal transport. An explanation of the almost doping independent Hall angle[164,165] has not been attempted in this model.

Pseudogap

Not only the increasing anisotropy in underdoped cuprates is a matter of debate, but also anomalies in the temperature dependence of their Fermi surface-integrated normal state properties such as kinks in the resistivity[166], in the infrared conductivity,[31-33] and in the electronic specific heat[167] as well as a strong suppression of the ^{63}Cu spin-lattice relaxation rate $1/T_1$ at $T^* > T_c$.[168] The latter two quantities are directly related to the density of states at the Fermi level N_F which apparently is suppressed below a characteristic temperature T^*. Since the materials are still metallic and even superconducting considerable density of states must, however, be left over. The phenomena are therefore referred to as a pseudogap. The questions is as to whether or not there exists a general suppression of the quasiparticle density, for example as a consequence of preformed (isotropically coupled) electron pairs[169,170], or whether an other type of anisotropy manifesting itself in the density of states is showing up. Recently, ARPES results came up strongly supporting a $\mathbf{k}$-dependent normal-state gap with the maximum along the principle directions.[171,172] This is actually the same variation with $\mathbf{k}$ as that for the superconducting gap derived from the data at $T < T_c$. In addition, the same magnitude for the two quantities is found in this study. Consequently, preformed pairs with $d_{x^2-y^2}$ symmetry were invoked moving incoherently between T^* and T_c. At T_c the phase becomes coherent determining the transition to the superconducting state.[83] Surprisingly, the characteristic energy for the pseudogap is here some 30 meV and at least a factor of two smaller than that in the IR study but still fairly high in units of kT_c. On the other hand, the superconducting gap is not clearly or not at all seen by IR spectroscopy. It seems therefore highly desirable to study the phenomena with a method which is sensitive to the density of states, to coherence effects in the superconducting state, and to the respective $\mathbf{k}$ dependences.

B_{2g} Raman data taken on Bi2212 single crystals with various doping levels are plotted as a function of temperature in Fig. 17.[173] To clarify the evolution of the spectra with temperature, we focus first on the optimally doped sample (Fig. 17 b). As we have already seen in the preceding paragraph, the slope increases upon cooling corresponding to an increasing lifetime of the quasiparticles. Above approximately 800 cm^{-1} all spectra merge. Essentially the same behavior of the continuum is found for the overdoped sample (Fig. 17 a). At all doping levels some of the phonons gain intensity at low temperature. The effect is relatively strong in the overdoped sample but not negligible either in the underdoped one (Fig. 17 c) and may mask variations of the continuum in the respective energy range. Down to some 200 K the continuum in the underdoped material exhibits a similar variation with temperature as that at other doping levels. Then, as a new feature, spectral weight is lost below approximately 800 cm^{-1} while the slope at low shifts continues to increase as is shown explicitly in Fig. 15. Between 250 and 600 cm^{-1}, the spectrum at 80 K is approximately 10% below those at high temperatures. Is the effect a peculiarity of Bi2212 or a general property of the CuO_2 planes in cuprates? To answer this question, an Y123 sample with comparable doping was studied. The single crystal was grown in $BaZrO_3$ and had a purity of better than 99.995%.[174] For adjusting the doping level the crystal was annealed for 100 h at 784 °C in 1 bar oxygen, and subsequently quenched. According to the calibration of Lindemer et al. this thermal treatment results in an oxygen

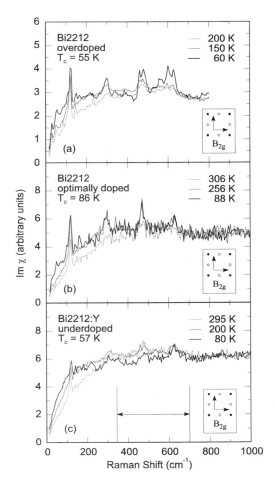

Figure 17: B_{2g} Raman reponse functions $Im\chi$ of differently doped Bi2212 single crystals in the normal state. In (c) the integration range is indicated for the determination of the spectral weight to bedisplayed in Fig. 21.[173]

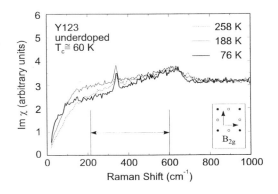

Figure 18: B_{2g} Raman reponse functions $Im\chi$ of an underdoped Y123 single crystals in the normal state. Due to illumination with intense light T_c increases in the laser spot by several degrees.[178] The arrows indicate the integration range for the determination of the spectral weight to be displayed in Fig. 21.[173]

content of 6.5.[175] The magnetically measured T_c was 53.5 ± 3 K. In this sample the very same variation with temperature is observed as for Bi2212 (Fig. 18). Due to the better thermal conductivity of this sample a higher laser power could be applied and, consequently, the statistics is improved as compared to Bi2212. Since the effects to be studied are very small one has to worry not only about the statistics but also about the stability. Therefore the continuum was measured at several temperatures up to 1500 cm⁻¹ to check possible variations. A systematic change was found neither in the slope nor in the integrated intensity in the whole temperature range. The intensity integral between 800 and 1000 cm⁻¹ calculated for Y123 is plotted as a function of temperature in Fig. 19. The variation is obviously purely statistical. Although the standard deviation of 5% is originating from instabilities in the experimental conditions between different runs and not from counting statistics with a typical error of 1%, it is safe to normalize the intensity of all spectra to the mean value. The same careful procedure has been applied to the B_{1g} symmetry. Here, for frequencies $\hbar\omega > 800$ cm⁻¹ the slope starts to change for both compounds above 250 K. Therefore the spectra cannot be normalized in the most inter-

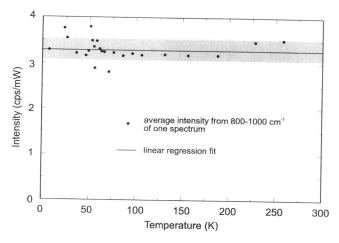

Figure 19: B_{2g} Raman reponse for Y123 at high energies as a function of temperature.[155]

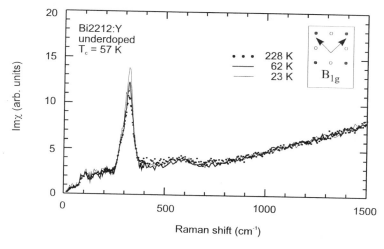

Figure 20: B_{1g} Raman reponse functions $Im\chi$ of underdoped Bi2212.[154] At 23 K (light line) the division by the Bose factor has only a neglegible effect, i.e. the spectrum shows essentially raw data.

273

esting temperature range. Since possible effects are much smaller than at B_{2g} symmetry we have to postpone the decision whether or not a signature of the pseudogap manifests itself also in the B_{1g} spectra. In a first approximation the B_{1g} spectra are therefore considered temperature independent excepting for the out-of-phase phonon at 320 cm^{-1} which gains almost a factor two in intensity from some 300 down to 20 K (Fig. 20). Since there are small changes in the energy range between 400 and 700 cm^{-1} we have to improve the reproducibility for a final decision.

In Fig. 21 the integrated spectral weights at B_{2g} symmetry for both Bi2212 and Y123 are plotted. The integration limits are set as indicated in Figs. 17 c and 18 and cut off the low-energy parts with trivial temperature dependence dominating the spectra of the optimally and the overdoped sample due to the increase of the carrier lifetime. In the underdoped materials the depletion of spectral weight starts at approximately 200 K and tends to saturate around 100 K. Obviously the maximal energy of some 800 cm^{-1} (100 meV) at which the depletion of weight occurs is temperature independent. If they exist at all the effects are much smaller at higher doping and are shifted to lower temperatures and, possibly, frequencies. So far, the body of data is too small to map out phase separation lines. However, the similarity in different but comparable samples is a strong case for an intrinsic feature. Further support is coming from NMR studies of the spin-lattice relaxation time in underdoped Y123[168] and IR and Raman experiments on phonons in Y123 and Y124 which shift slightly at approximately T*.[176] Here, in contrast to Bi2212, the phonons are relatively strongly coupled to the electrons because of the field gradient across the planes which has its origin in the asymmetric position of the differently charged Y^{3+} and Ba^{2+} atoms.[72] The observed variations of both $1/T_1T$ and the phonons are mediated by changes in the density of states. Actually, it would be interesting to study the

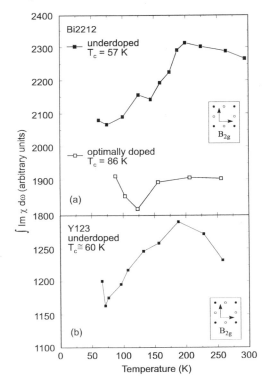

Figure 21: Integrated spectral weight of the B_{2g} Raman reponse of differently doped Bi2212 (a) and underdoped Y123 (b) as a function of temperature.[173]

274

symmetry properties of the pseudogap from the viewpoint of phonon renormalization in the same way as those of the superconducting gap.[72]

The discussion of the symmetry properties of the pseudogap from the electronic Raman response is aggravated by the missing information from the B_{1g} spectra. What can we extract from the B_{2g} data? As a matter of fact, the B_{2g} spectra look metallic at all temperatures and can be described by Eq. (12) over a fairly wide energy range. This implies that the B_{2g} response between T_c and T^* cannot be distinguished from that of an ordinary metal with finite lifetime of the carriers. However, the density of states in a certain range of energy and momenta is reduced. Therefore, in (π,π) direction we can exclude both line nodes such as in the superconducting state since the linear variation of the intensity is missing (c.f. Fig. 11) and a finite gap since there is no threshold in the spectrum. For an explanation of the missing spectral weight below $800 \, cm^{-1}$ and the simultaneously occurring metallic behavior at $\omega \to 0$ we rather have to invoke a reduced or vanishing density of states at the Fermi level away from the diagonals in the vicinity of the principal axes and, on the diagonals, extended areas on the Fermi surface with a finite density of states.

Are there alternative explanations and what might be the physical background? Hypothetically, one can imagine to decrease the density of levels in any $\mathbf{k}$ direction without opening a gap by changing the slope of $\varepsilon(\mathbf{k})$ at ε_F in a way right opposite to the effect of strong electron-phonon coupling,[103] for instance by switching off the interaction. Then, N_F would change by $1/(1 + \lambda)$ with λ the coupling strength. The energy range affected would actually fit as the biggest phonon frequency is roughly $100 \, meV$. It is very unlikely, however, that the temperature dependence and relative variation observed for the electronic spectral weight is exactly the same for the two compounds with quite different phonon spectra and coupling constants and that the coupling decreases upon cooling. In addition, the phonons should then harden below T^* as opposed to what is observed.[176] More realistically, the temperature dependence could be related to $\varphi_\mathbf{k}$ which determines the Raman intensity (see Eq. 12). At $T \neq 0$ $\varphi_\mathbf{k}$ has a Lorentzian type of shape with a width kT (and becomes a δ function at $T = 0$ in a Fermi liquid). Whenever the band structure has a region with very high density of states at a distance $\Delta\varepsilon$ to ε_F such as the van-Hove singularities along $(\pi,0)$[13,108-110] the intensity should start to decrease at $T \leq \Delta\varepsilon/k$.[12] $\Delta\varepsilon \leq 30 \, meV \approx 300 \, K$ is actually the appropriate range, however, it is not yet clear how the energy scale of $100 \, meV$ observed in Raman and IR experiments should come about. The characteristic energy J for spin excitations would fit better[89] and can be obtained from the B_{1g} Raman spectra at large energy transfer (Fig. 22). Here, at some $2400 \, cm^{-1}$, we see weak scattering from two-magnon excitations which, due to short range fluctuations, are still present in a superconducting sample far away from long-range AF order. In an antiferromagnet with four nearest neighbors the peak in the spectrum is at approximately $2.8 \, J$,[97] hence $J = 100 - 120 \, meV$. Since the AF wavevector $\mathbf{Q}$ connects parts of the Fermi surface close to $(\pi,0)$ we are now away from the diagonals and arrive again at the scenario described by the NAFL model with strong effects around $(\pi,0)$.

To reconcile ARPES and optical data a true gap close to $(\pi,0)$ has to be invoked. If one assumes that a band splits symmetrically with respect to the Fermi surface (for a not yet specified reason) the discrepancy in the energy scales basically vanishes: ARPES measures Δ^* from ε_F while the optical experiments, Raman and IR, need at least $2\Delta^*$ for an excitation. $2\Delta^*$ actually defines the onset of scattering and the range affected can be larger such as in the superconducting state where changes in the spectra are seen well above $2\Delta_0$ up to approximately $12 \, kT_c$ independent of polarization (see Fig. 10). With these

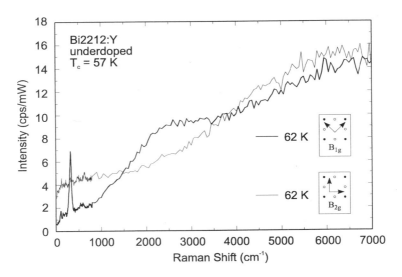

Figure 22: Raman spectra of underdoped Bi2212 at large energy transfers. The symmetries are indicated. The strong increase towards high energies stems from luminescence of so far unknown origin.[154] Note the broad band from two-magnon scattering centered at 2400 cm^{-1} in the B$_{1g}$ spectrum superimposed on the background.

arguments and the **k** dependence of the Raman vertices we can estimate $60 < 2\Delta^* < 100$ meV. As outlined above the gap Δ^* has to be confined to a very narrow region in **k** space. The minimal extension Δ**k** is approximately given by the band dispersion

$$\Delta\mathbf{k} = \frac{2\Delta^*}{\hbar \mathbf{v}_F}. \tag{24}$$

For an average in-plane Fermi velocity $\mathbf{v}_F \approx 10^7$ cm/s one obtains $\Delta\mathbf{k} \approx 0.1(\pi/a)$. This relatively small number is not incompatible with the ARPES results from underdoped Bi2212[171,172] which actually found deviations from $d_{x^2-y^2}$ symmetry in the right direction, i.e. a dirty d-wave gap. $\Delta\mathbf{k} \approx 0.1(\pi/a)$ fits also well to the results in B$_{2g}$ symmetry and could explain why the effect in B$_{1g}$ is not very pronounced either if this symmetry really samples carriers any more. It is noted that this scenario is quite similar to what has been discussed in context with CDW and SDW systems[177] like 2H-NbSe$_2$ and some organic compounds, respectively, and even A15 systems[144,145].

Superconducting state

There is another energy scale which becomes relevant below T$_c$. I will first address the results in the underdoped samples just to complete the discussion about the pseudogap started in the preceding section and to find a relation between Δ^* and Δ_0. Although the debate about the symmetry of the order parameter has converged for optimally doped cuprates, and the gross majority agrees that the $d_{x^2-y^2}$ contribution dominates, there is still a discussion on the other doping levels. I will therefore present also results for overdoped Bi2212, in particular for the low frequency-low temperature limit. A brief discussion about the influence of impurities and the electron-doped system will terminate the section.

In underdoped samples superconductivity-induced effects are only observed in B_{2g} symmetry (Fig. 23) and neither in B_{1g} (see Fig. 20) nor A_{1g} configurations. The B_{2g} spectra of the two underdoped compounds Y123 and Bi2212 are very similar to those at the T_c maximum (Fig. 10) and for the overdoped sample (see Fig. 25 below). Details of the shape can be influenced by band-structure effects[114] or by slightly different concentrations of impurities[116] The strong peak intensity in Y123 is in part certainly related to the extreme purity of this sample.[174] More important and significant is the slope at small energies. In both samples the intensity varies linearly over almost a decade clearly indicating a dominant admixture of d_{x2-y2} pairing. The zero frequency extrapolation value for the response is very small at the lowest temperatures (experimentally determined by comparing Stokes and anti-Stokes spectra). As a function of time the intensity off-set increased, obviously due to irreversibly adsorbed surface layers. If the temperature approaches T_c the pair-breaking peaks decrease and vanish at $T_c \pm \Delta T_c$. We take this as evidence that the structures are superconductivity induced although the expected shift of the peaks to lower energies is much smaller than predicted by BCS theory[179] and indeed observed in conventional systems.[50,180,181] The missing temperature dependence is not yet clarified but may be related either to the layered structure according to results in thin-film NbN[182], i.e. the dimensionality[56], or to the strong variation of the quasiparticle damping below T_c[59,60,183]. The peak frequencies at $T \rightarrow 0$ are in both cases 230 ± 10 cm^{-1} or 6 in units of kT_c. According to what we know from optimally doped samples (Fig. 10 and ref.[36]) the characteristic energy $2\Delta_0$ can be somewhat higher than the B_{2g} peak frequency. $2\Delta_0/kT_c = 8$ might be a reasonable estimate.

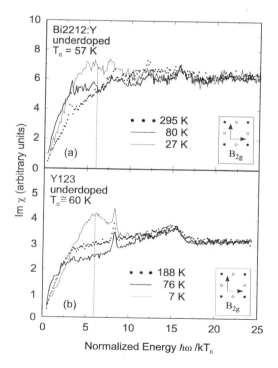

Figure 23: B_{2g} Raman response Imχ of underdoped Bi2212 (a) and Y123 (b) in the normal and in the superconducting states.[173] kT_c is approximately 40 cm^{-1} for both samples. The T_c indicated for Y123 (b) is higher than that measured magnetically according to what is expected for underdoped samples of this system under intense light.[178] Close to this temperature, the pair-breaking feature disappers.

In both samples kT_c is approximately 40 cm^{-1} or 5 meV. Then $2\Delta_0$ is 40 meV and is roughly a factor of 1.5 or even 2 smaller than $2\Delta^*$. The difference in the energy scales is directly seen in Fig. 23: The superconducting spectra and those right above T_c merge at $12\,kT_c$ while the two normal spectra taken below and above T* become identical between 18 and $20\,kT_c$. We recall that the energy of $12\,kT_c$ seems to indicate the maximum energy for superconducting correlations independent of polarization (Fig. 10) and conclude that the difference between Δ_0 and Δ^* is to be taken serious and really indicates two energy scales for the respective gaps above and below T_c. How can ARPES end up with a larger Δ_0 of 25 meV[171,172] which coincides with Δ^*? Since photoemission basically measures the distance from the Fermi edge always the larger gap is seen. At the momentum $\mathbf{k}$ where Δ^* has its maximum the effect of superconductivity at $T_c < T^*$ is limited to a change of the quasiparticle lifetime as can be seen as an increasingly strong pole below T_c. Since $\Delta_0 < \Delta^*$ one cannot expect a further shift of the "leading edge". Since the superconducting correlations affect the Raman spectra only up to $12\,kT_c < 2\Delta^*$ this may partly explain the missing pair-breaking peak in the B_{1g} spectrum. Thus, the spectra in the superconducting state additionally favor the opening of a normal state gap in the close vicinity of $(\pi,0)$ instead of an overall loss of density of states around the Fermi surface or a B_{2g}-type pseudogap. The most likely variation of the two types of gaps as a function of Fermi-surface length is shown in Fig. 24.

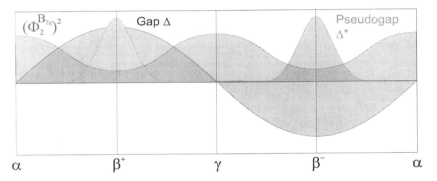

Figure 24: Variation of the B_{2g} Raman vertex, the pseudogap, and the superconducting gap on the Fermi surface. For the tight binding contour shown in Fig. 16 α is on the zone diagonal, β^- on the boundary close to $(\pi,0)$, and γ again on the diagonal. For a pocket around $(\pi/2,\pi/2)$[163] α is on the diagonal inside of the AF Brillouin zone β^+ is on the AF zone boundary, γ on the diagonal outside. Here α and γ are not on symmetrically equivalent points.

At optimal doping the interpretation of the data is straightforward, and quantitative agreement between the experiments and the theoretical prediction for $d_{x^2-y^2}$ pairing can be obtained (Fig. 11).[36] In underdoped samples the B_{1g} channel is not directly related to apparent carrier properties neither above nor below T_c, and a strong anisotropy becomes manifest. Overdoped material, on the other hand, looks pretty isotropic in the normal state. At the first glance, the superconducting spectra show the same trend (Fig. 25): All peak frequencies of the clearly pronounced pair-breaking maxima fall into the same energy range. From that it was concluded that the gap is almost isotropic in this compound.[184] As a striking difference to an isotropic superconductor (Fig. 9) however, the well defined threshold in the low energy part is missing. In this energy range the spectra fit much better to those obtained for cuprates at other doping levels. As observed there, the intensities increase continuously over the whole range from zero to the maximum. Since the slopes

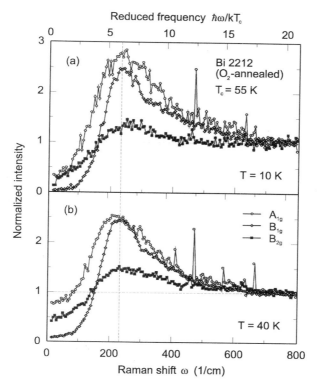

Reduced frequency $\hbar\omega/kT_c$

Bi 2212
(O_2-annealed)
$T_c = 55$ K

T = 10 K

(b)

—○— A_{1g}
—◇— B_{1g}
—■— B_{2g}

T = 40 K

Normalized intensity

Raman shift ω (1/cm)

Figure 25: Electronic Raman spectra (phonons subtracted) of overdoped Bi2212 at all three symmetries at 10 K (a) and at 40 K (b).[43] All spectra are normalized to the respective intensities at 800 cm^{-1}.

are obviously symmetry dependent an inhomogeneity of the sample can be excluded as an explanation. For more quantitative information, the intensity variation with energy is shown in a log-log plot (Fig. 26). Up to 2 kT_c an almost linear slope is found at all symmetries.[43,136] Then, the slope crosses over to a cubic variation at the B_{1g} symmetry while the close-to-linear increase continues at the two others. Of course, in a clean d-wave superconductor the B_{1g} intensity should vary cubically starting from zero energy. For a slightly disordered one, however, a crossover from linear to cubic variation at $\omega^* \approx \sqrt{\Delta_0 \cdot \Gamma}$ with Γ the impurity scattering rate is predicted.[116] The rate $\Gamma = 70$ cm^{-1} obtained from this relation with a crossover frequency of 100 cm^{-1} and a gap $\Delta_0 = 150$ cm^{-1} is actually very close to the zero temperature extrapolation value of the normal state relaxation rate and $\Gamma/\Delta_0 \approx 0.5$ is still sufficiently small to allow pronounced maxima.[116] Since the low frequency behavior is determined only by the symmetry properties of the vertices and the **k** dependence of the gap and not by details of the band structure, it is natural and safe to assign the same $d_{x^2-y^2}$ pairing state to overdoped Bi2212 as for the vast majority of other compounds and the doping levels studied so far. For additional support one may look at the temperature dependence of the scattering intensity in the limit $\omega \to 0$. In Fig. 25 a serious dependence on symmetry is obvious at the two temperatures which is plotted in detail in Fig. 27. Although a quantitative analysis is not yet possible, the pronounced channel dependence is nevertheless a clear indication of unconventional pairing. The peak frequencies, on the other hand, are much more subject to peculiarities of the band structure

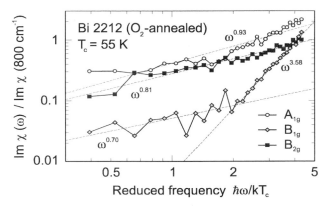

Figure 26: Low-energy part of the 10 K spectra shown in Fig. 25 a.[43]

and details of the selected basis functions for both the vertex and the gap.[107,114] Nevertheless, the 30% downwards shift of the B_{1g} peak remains a challenge as it is not yet quantitatively reproduced by theory. In addition, it is apparently an experimental fact that the B_{1g} pair-breaking maximum in a way resembles the trend observed for $\Gamma_0^{B_{1g}}(T)$, and more or less monotonically increases upon decreasing doping level instead of scaling with T_c (Fig. 28). In contrast, the B_{2g} peak frequency sticks to T_c and seems to be very robust against small changes of the sample quality. We therefore conclude that $2\Delta_0$ scales roughly with T_c. It has already been observed earlier that the peak frequency in B_{1g} symmetry may depend on the sample history and vary by some 20% in samples with the same T_c.[44,45] It has even been reported to increase below optimal doping.[185] In the latter samples however, the B_{2g} peak was lost earlier than the B_{1g} maximum which was seen even above the actual T_c. The missing B_{2g} pair-breaking feature in the work by Rosenberg et al.[185] can certainly not be reconciled with the data presented here. It is clear, on the other hand, that the B_{1g} maximum must vanish on the way from optimal doping to underdoped samples with $T_c \approx 0.5\ T_c^{max}$, and it is unlikely that the intensity is discontinuously switched off. Since it is more complicated to get homogeneous and chemically stable underdoped Bi2212 by just taking out oxygen,[186] it cannot be excluded that some of the samples were not single phase. This means that the studies in the underdoped range of the phase diagram are not yet finished.

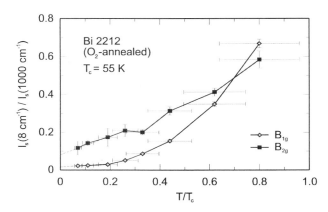

Figure 27: d.c. intensity of the electronic Raman spectra of overdoped Bi2212 for B_{1g} and B_{2g} symmetry. All intensities are normalized to the respective values at 1000 cm^{-1}, a temperature-independent constant.

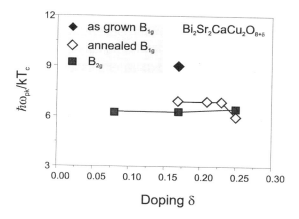

Figure 28: Frequencies of the pair-breaking peaks for B_{1g} and B_{2g} symmetry for various doping levels. At optimal doping an as-grown and a post-annealed sample of the same batch have been studied. Obviously, the B_{1g} frequency depends strongly on the sample history.

In overdoped Bi2212 we have a clear indication for a small concentration of impurities which can easily be determined in the superconducting state close to $T = 0$ by just finding the crossover from linearly to cubically varying intensity. This is much easier than the procedure in the normal state and allows to precisely measure the impurity scattering rate at very low temperature in the superconducting state. Recently, a new sample of Bi2212 close to optimal doping was studied which showed an offset of the zero temperature extrapolation value of both the resistivity and the inverse carrier lifetime in the normal state.[95] Consistently, a fairly long linear intensity variation for the B_{1g} spectrum was found with a crossover to a $I_R \propto \omega^3$ at 300 cm^{-1} at $T \rightarrow 0$. Still $\Gamma/\Delta_0 \leq 1$ and, consequently, the peaks were well resolved. Can the theoretically predicted suppression of the pair-breaking features indeed be observed? In single-layer Bi2201 there might be a chance since $\Gamma_0^\mu(T = 0)$ is larger than the expected magnitude of Δ_0. We actually looked at two different samples with $T_c = 18$ K and $T_c = 12$ K, respectively. The results are displayed in Fig. 29 and demonstrate that at sufficiently high disorder the gap-like features are really gone while at large Γ/Δ_0 the peak intensity is strongly reduced and the frequency is pushed beyond $2\Delta_0$ if one assumes a $2\Delta_0/kT_c$ ratio of 8 or 9 for this material similar as for Bi2212. More importantly, in the material with the higher T_c a linear increase of the intensity is found at all channels with no indication of a threshold in the spectrum in complete agreement with d-wave pairing. At this high impurity level a finite gap would show up for anisotropic s-wave pairing varying, for instance, as $|d_{x^2-y^2}|$. Since Raman is measuring only the magnitude of the gap $|d_{x^2-y^2}|$ is indistinguishable from $d_{x^2-y^2}$ in the clean limit.[116] On the other hand, the high impurity concentration could at least partly explain the low T_c in this material. The suppression of the pair breaking features by impurities should also be considered in context with underdoped Bi2212. The extremly short carrier lifetime at the B_{1g} symmetry (Fig. 14) may indeed be a reason for the missing pair-breaking structure. The increase of τ observed by ARPES[171,172] below T_c is probably too small to drive $\Gamma_0^{B_{1g}}(T = 0)/\Delta_0$ below 1 facilitating the formation of gap-like structures in the spectra.

So far we did not pay any attention to electron-doped systems such as NCCO. One of the problems is the supply with high-quality samples as the preparation efforts are usually directly proportional to T_c. Nevertheless this type of compounds is of enormous importance for the understanding. The single crystal used here had a T_c of 19.3 K.[187] Raman spectra for

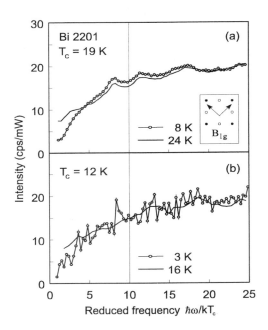

Figure 29: B_{1g} Raman spectra of two Bi2201 crystals with different transition temperatures.[35]

B_{1g} and B_{2g} symmetry are shown in Fig. 30 plotted in units of kT_c. In the superconducting state the differences to the well established properties of the hole-doped cuprates and the similarities with Nb_3Sn are obvious at the first glance: The peak frequencies are at the same position in units of kT_c and significantly below those in overdoped Bi2212 (Fig. 25), there is a threshold in the spectra, and the residual scattering is channel independent. The high residual intensity below the threshold is accompanied by a strong elastic line. Most likely it comes from a second phase besides the superconducting one. In fact, phase separation seems to be an intrinsic property of NCCO. In spite of these experimental shortcomings the clear messages is that NCCO is a slightly anisotropic intermediately coupled s-wave superconductor.[128] Owing to the similarity with Nb_3Sn we try to reproduce the data using a gap given by

$$\Delta(\mathbf{k}) = \frac{\Delta_{max} + \Delta_{min}}{2} - \frac{\Delta_{max} - \Delta_{min}}{2} \Phi_2^{A_{1g}}(\mathbf{k}) \tag{27}$$

with $2\Delta_{max} = 66$ cm^{-1} and $2\Delta_{min} = 56$ cm^{-1} or 4.9 and 4.1 kT_c, respectively. These figures are in quantitative agreement with tunneling and surface resistance measurements.[188,189] The result is unexpected[190] as NCCO was believed for a long time to be the electron-doped counterpart of the hole-doped cuprates. Even more surprising, the almost isostructural hole-doped LSCO fits well into the d-wave scenario.[38] There is no consensus yet about the mechanism of superconductivity in the cuprates, but it is another puzzle that NCCO is so different and looks like an ordinary intermediately coupled s-wave superconductor. On the other hand, the crystal structure of NCCO is similar to that of all other cuprates and has approximately the same electronic structure near the Fermi surface[191,192], in particular a similar carrier concentration and superconducting coherence length.[193] If in this system the superconducting state is indeed characterized by BCS-like behavior as can be concluded from the data, we have produced experimental evidence that a consistent use of a weak

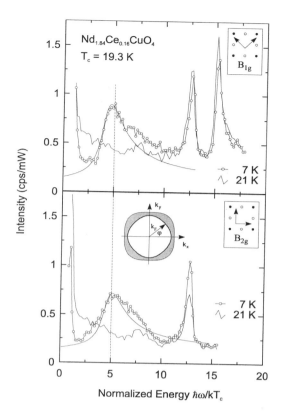

Figure 30: Raman spectra of NCCO at B_{1g} and B_{2g} symmetry.[128] Here kT_c is 13.5 cm^{-1}. The line at 200 cm^{-1} (15 kT_c) is a phonon, the one at 170 (12.5 kT_c) and the very weak one at 110 cm^{-1} (8 kT_c) are crystal-field excitations.[127,128] The variation of the gap with **k** used for the fit (light line in the spectra) is shown in the inset.

coupling BCS model for both the electron- and the hole-doped materials with high transition temperature is a reasonable approximation.

CONCLUDING REMARKS

Theoretical aspects and recent experimental results of Raman scattering experiments in cuprate metals with various degrees of disorder have been summarized. I have discussed both normal and superconducting state properties aiming predominantly at a better understanding of the electronic anisotropies in the CuO$_2$ planes.

The strength of electronic Raman scattering is closely linked to the symmetry selection rules. By a certain combination of the polarization vectors of the incoming and outgoing photons different symmetry components of the Raman vertex can be picked up. Since the respective Raman vertices are prefactors of the response kernel which has a symmetry related **k** dependence as well, different averages over the Brillouin zone or over the Fermi surface are weighed out for different polarization combinations. In this way one obtains a mapping of **k** space on configuration space or, less formal, a momentum resolution in electronic Raman scattering. The resolution is certainly not comparable to that of photoemission experiments as the leading order basis functions for the expansion of the Raman vertex are slowly varying with **k**. On a quadratic lattice, for instance, they have basically (modulo change of sign) fourfold symmetry. Therefore, Raman assumes some kind of an intermediate position between really **k** selective experiments like ARPES or neutron scattering (not suitable for carrier spectroscopy) and methods like IR spectroscopy, tunneling, calorimetry, magnetometry or NMR which measure averages over the entire Fermi surface. The poor **k** sensitivity in Raman is partly overcome by the fairly good energy resolution of better than 1 meV with the limit imposed by intensity. In addition, one can obtain reliable data down to very small energies of the order of 1 meV, i.e. close to the d.c. limit. As the Raman response is a two particle correlation function which is sensitive to coherence effects in a superconductor, not only perfect conductivity and the signature of a gap, but also the formation of coherently moving pairs is seen which manifest themselves as excess intensity close to the maximum gap energy. The superconducting state was actually in the main focus of theoretical work on Raman scattering in recent years, with the emphasis placed on anisotropic systems.[35,115] For unconventional pairing a rich variety of symmetry-related information has been predicted such as frequency and temperature power laws for the intensity which allowed a quantitative interpretation of the Raman data of high-T_c materials in terms of an order parameter with $d_{x^2-y^2}$ symmetry. Increasing effort is now focused on normal state properties[115] which, since a long time, are believed to be a link to the puzzle of superconductivity in the cuprates[194]. In this context the impact of different electronic interactions as suggested by the nested-Fermi-liquid[117], the Hubbard[160,161], or the NAFL model[133,161] on the Raman spectra is studied, in particular the symmetry related information.

One of the central experimental results, though unspectacular at the first glance, is the similarity of the symmetry selection rules for scattering from electronic excitations above and below T_c in all cuprates studied so far. The frequency, temperature and polarization dependences of the electronic continua for compounds with one, two or three adjacent CuO_2 planes from various families can almost be mapped onto each other, if the coordinate system of the Cu-O bonds is selected instead of that of the crystallographic axes. The selection rules for the other excitations such as phonons or crystal-field excitations are determined by the usual axes and are strongly material dependent. We have therefore clear evidence that the electronic scattering in all high-T_c materials originates from the planes (excepting for a small contribution from the chains in Y123). Obviously, the response from the planes is purely additive, and the single band model selected for the interpretation is a fairly good approximation.[36,107,114,195] In detail, for all compounds with comparable doping the same ratio of scattering intensities at the three symmetries is found. This ratio varies in a unique way with doping in that the B_{2g} spectrum dominates at low doping and the B_{1g} response is strongest in overdoped material. At all doping levels the B_{2g} response resembles carrier properties as obtained from other experiments. In the super-conducting state pronounced pair-breaking peaks are observed at approximately 6 kT_c independent of doping. The B_{1g} spectra vary qualitatively with doping. While in the overdoped range the same magnitude and temperature dependence for the relaxations rate can be derived from the B_{1g} and B_{2g} spectra in the normal state, the B_{1g} relaxation rate increases with decreasing doping level by almost a factor of 5 losing any relation to conventional carrier properties. At the same time, the intensity fades away and it is not yet clear whether or not carriers are seen any more at the lowest doping. In particular, no indication of super-conductivity, i.e. no pair-breaking peak is oberved in the B_{1g} in contrast to the B_{2g} spectra.

The interpretation of the superconducting spectra is qualitatively clear at all doping levels. On account of the low frequency part of the spectra, specifically the power laws for the intensity, isotropic or even anisotopic s-wave symmetry for the gap can be excluded. Rather a dominant $d_{x^2-y^2}$ component has to be invoked with a maximal magnitude of some 8 kT_c. Details of the spectral shape are still open such as the coincidence of all pair-breaking peaks in overdoped samples. Recently mixtures of s and d-wave pairing have been studied[196] which could at least qualitatively explain the down shift of B_{1g} peak. Small s components are actually to be expected due to the orthorhombic distortion of Y123 and Bi2212.[21,22] As an exception electron-doped NCCO seems to be an only slightly anisotropic s-wave superconductor. In all cases the temperature dependence of the pair breaking features deviates from the BCS prediction. So far there is no explanation on the market.

The normal state is much less clear. The small anisotropy in overdoped material can easily be understood in terms of the usual Fermi surface anisotropies and manifests itself also in magneto-transport.[197] Although Hall measurements[164,165] suggest stronger scattering on the faces of the Brillouin zone than along the diagonals, an order of magnitude difference in the respective scattering rates as found by Raman in underdoped samples is hardly explained in the framework of ordinary Boltzmann transport theory since the mean free path is of the order of the lattice constant. Strong temperature independent scattering has to be invoked as, for instance, proposed in the NAFL model. So far there is no theoretical background for a quantitative comparison. The same holds for the intensity anomalies or pseudogap observed between T^* and T_c. Qualitatively, the Raman spectra suggest the opening of gap in the vicinity of the $(\pi,0)$ point in **k** space close to the "hot spots" proposed by the NAFL model. If the bands would split symmetrically here, all spectroscopic results could be reconciled. Clearly, the one-electron picture is not a very good basis of argumentation in strongly correlated systems and becomes particularly questionable for regions in **k** space with strong inelastic scattering and, consequently, extremely short lifetime of the carriers. Since Raman is right at the beginning, phase separation lines cannot be mapped out. It is noted (see Fig. 2) that there is another characteristic temperature T^{cr} above T^* where the magnetic correlation length becomes smaller than twice the lattice constant so that even fluctuating antiferromagnetism vanishes. At this temperature light cannot any more scatter from spin excitations and therefore T^{cr} can basically also be measured by Raman by looking at the intensity of the two-magnon peak.

To summarize, there is a variety of symmetry dependent information in the electronic Raman spectra above and below T_c. Some features such as the **k** dependence of the superconducting gap are qualitatively clear. The temperature dependence however is already an unsolved problem. As soon as one starts to compare Raman with other methods the same holds for the magnitude of the gap which is at least 20% smaller than that found by electron tunneling.[20] There is a serious divergence of all experimental methods for underdoped materials. While the Raman indicates a constant $2\Delta_0/kT_c$, tunneling and ARPES are better consistent with constant Δ_0. In the normal state a pseudogap is observed. In pronounced disagreement with ARPES its magnitude is at least 1.5 times as large as that of the superconducting gap. The details of its **k**-dependence and its origin are open. As concerning Raman the B_{1g} response both in the normal and in the superconducting state is still hiding several puzzles particularly at low doping. Many of the open questions are indeed closely related to the underdoped range of the phase diagram which certainly deserves most of the experimental effort in the near future.

ACKNOWLEDGEMENTS

The experimental work has been done in collaboration with Christian Hoffmann, Gaby Krug, Peter Müller, Ralf Nemetschek, Matthias Opel, Richard Philipp, and Barbara Stadlober who all contributed a lot to the success of the research presented. In particular, I am indebted to Matthias Opel who prepared the figures for the paper, as usual in excellent quality. I profited a lot from the continuous collaboration with Tom Devereaux, Dietrich Einzel, István Tüttö, Attila Virosztek, and Fred Zawadowski. The samples we used came from various places, and I want to express my gratitude to H. Berger and L. Forró (Lausanne), J.L. Cobb and J.T. Markert (Austin), A. Erb and E. Walker (Genève), P. Müller (Erlangen), and J.J. Neumeier (FAU Florida). I would like to thank Klaus Andres and my collegues at the WMI for many useful discussions as well as expert technical support. Financial support by the Bayerische Forschungstiftung via the Forschungsverbund Hochtemperatursupraleiter (FORSUPRA) is gratefully acknowledged. Part of the collaboration with theory has been subsidized by the BMBF via the program "Bilaterale wissenschaftlich-technische Zusammenarbeit" under grant WTZ-UNG-052-96.

REFERENCES

1. For TaC and HfC see: H.G. Smith and W. Gläser, Phys. Rev. Lett. **25**, 1611 (1970).
2. G. Shirane and J.D. Axe, Phys. Rev. Lett. **27**, 1803 (1971).
3. B.P. Schweiß, B. Renker, E. Schneider, and W. Reichardt, in *Superconductivity in d- and f-Band Metals II*, D.H. Douglass ed., p. 189 (Plenum Press, New York, 1976).
4. L. Pintschovius, H. Takei, and N. Toyota, Phys. Rev. Lett. **54**, 1260 (1985).
5. J.G. Bednorz and K.A. Müller, Z. Phys. B **64**, 189 (1986).
6. B. Batlogg, in *High-Temperature Superconductivity: The Los Alamos Symposium-1989*, Eds. K. Bedell, D. Coffey, D. Meltzer, D. Pines, and J.R. Schrieffer, p. 37 (Addison-Wesley Publishing Company, 1990).
7. K. Levin, Ju H. Kim, J.P. Lu, and Qimiao Si Physica C **175**, 449 (1991).
8. E. Dagotto, Rev. Mod. Phys. **66**, 763 (1994).
9. H. Fukuyama and H Kohno, Czech. J. Phys. **46** Suppl. S6, 3146 (1996).
10. P.A. Lee J. Low Temp. Phys. **105**, 581 (1996).
11. K. Maki and H. Won, Ann. Phys **5**, 320 (1996).
12. J. Ruvalds, Supercond. Sci. Technol. **9**, 905 (1996).
13. For a review see: Z.-X. Shen and D.S. Dessau, Physics Reports **253**, 1 (1995) and Z.-X. Shen, W. E. Spicer, D. M. King, D. S. Dessau and B. O. Wells, Science **267,** 343 (1995).
14. P. Aebi, J. Osterwalder, P. Schwaller, L. Schlapbach, M. Shimoda, T. Mochiku, and K. Kadowaki, Phys. Rev. Lett. **72**, 2757 (1994).
15. Z.-X. Shen, D.S. Dessau, B.O. Wells, D.M. King, W.E. Spicer, A.J. Arko, D. Marshall, L.W. Lombardo, A. Kapitulnik, P. Dickinson, S. Doniach, J. DiCarlo, A.G. Loeser, and C.-H. Park, Phys. Rev. Lett. **70**, 1553 (1993).
16. R.J. Kelley, Jian Ma, G. Margaritondo, and M. Onellion, Phys. Rev. Lett. **71**, 4050 (1993).
17. H. Ding, J.C. Campuzano, A.F. Bellman, T. Yokoya, M.R. Norman, M. Randeira, T. Takahashi, H. Katayama-Yoshida, T. Mochiku, K. Kadowaki, and G. Jennings, Phys. Rev. Lett. **74**, 2784 (1995).
18. M.C. Schabel, C.-H. Park, A. Matsuura, Z.-X. Shen, D.A. Bonn, Ruixing Liang, W.N. Hardy, Phys. Rev. B **55**, 2796 (1997).
19. D. Mandrus, L. Forró, D. Koller, L. Mihaly, Nature **351**, 460 (1991).
20. Ch. Renner and Ø. Fischer, Phys. Rev. B **51**, 9208 (1995).
21. R.C. Dynes, Solid State Commun. **92**, 53 (1994).
22. R. Kleiner, A.S. Katz, A.G. Sun, R. Summer, D.A. Gajewski, S.H. Han, S.I. Woods, E. Dantsker, B. Chen, K. Char, M.B. Maple, R.C. Dynes, and John Clarke, Phys. Rev. Lett. **76**, 2161 (1996).
23. R.E. Glover and M. Tinkham, Phys. Rev. **104**, 844 (1956); D.M. Ginsberg and M. Tinkham, Phys. Rev. **118**, 990 (1960); L.H. Palmer and M. Tinkham, Phys. Rev. **165**, 588 (1968).
24. Review: T. Timusk and D. Tanner, in *Physical Properties of High Temperature Superconductors I*, D.M. Ginsberg ed., p. 339 (World Scientific, Singapore, 1989)
25. J. Orenstein, G.A. Thomas, A.J. Millis, S.L. Cooper, D.H. Rapkine, T. Timusk, L.F. Schneemeyer, and J.V. Wazczak, Phys. Rev. B **42**, 6342 (1990)
26. Review: D. Tanner and T. Timusk, in *Physical Properties of High Temperature Superconductors III*, D.M. Ginsberg ed., p. 363 (World Scientific, Singapore, 1992)
27. Z. Schlesinger, R.T. Collins, L.D. Rotter, F. Holtzberg, C. Field, U. Welp, G.W. Crabtree, J.Z. Liu, Y. Fang, K.G. Vandervoort, and S. Fleshler, Physica C **235-240**, 49 (1994).
28. S. Uchida, K. Tamasaku, K. Takenaka, and Y. Fukuzumi, J. Low Temp. Phys. **105**, 723 (1996).
29. K. Kamarás, S.L. Herr, C.D. Porter, N. Tache, D.B. Tanner, S. Etemad, T. Venkatesan, E. Chase, A. Inam, X.D. Wu, M.S. Hedge, and B. Dutta, Phys. Rev. Lett. **64**, 84 (1990).
30. D.C. Mattis and J. Bardeen, Phys. Rev. **111**, 412 (1958).
31. C.C. Homes, T. Timusk, Ruixing Liang, D.A. Bonn, and W.N. Hardy, Phys. Rev. Lett. **71**, 1645 (1993).
32. D.N. Basov, T. Timusk, B. Dabrovski, and J.D. Jorgensen, Phys. Rev. B **50**, 3511 (1994)

33. A.V. Puchkov, P. Fournier, D.N. Basov, T. Timusk, A. Kapitulnik, and N.N. Kolesnikov, Phys. Rev. Lett. **77**, 3212 (1996)

34. Z. Schlesinger, R.T. Collins, F. Holtzberg, C. Field, S.H. Blanton, U. Welp, G.W. Crabtree, Y. Fang, and J.Z. Phys. Rev Lett. **65**, 801 (1990).

35. For a review see: D. Einzel and R. Hackl, J. Raman Spectroscopy **27**, 307 (1996).

36. T. P. Devereaux, D. Einzel, B. Stadlober, R. Hackl, D. H. Leach and J. J. Neumeier, Phys. Rev. Lett. **72**, 396 (1994) and **72**, 3291 (1994)

37. T. P. Devereaux and D. Einzel, Phys. Rev. B **51**, 16336 (1995) and Phys. Rev. B **54**, 15547 (1996).

38. X.K. Chen, J.C. Irwin, H.. Trodahl, T. Kimura and K. Kishiiu, Phys. Rev. Lett. **73**, 3290 (1994).

39. A. Hoffmann, P. Lemmens, G. Güntherodt, V. Thomas, and K. Winzer, Physica C **235-240**, 1897 (1994).

40. B. Stadlober, G. Krug, R. Nemetschek, M. Opel, R. Hackl, D. Einzel, C. Schuster, T. P. Devereaux, L. Forró, J. L. Cobb, J. T. Markert and J. J. Neumeier, Journ. of Phys. Chem. Solids **56**, 1841 (1995).

41. A. Hoffmann et al., J. Low. Temp. Phys. **99**, 201 (1995).

42. A. Yamanaka, N. Asayama, T. Furutani, K. Inoue, S. Takegawa, SPIE (Bellingham) Vol. **2696**, 276 (1996).

43. R. Hackl, G. Krug, R. Nemetschek, M. Opel, and B. Stadlober, SPIE (Bellingham) Vol. **2696**, 194 (1996)

44. B. Stadlober, R. Nemetschek. O.V. Misochko, R. Hackl. P. Müller. J.J. Neumeier, and K. Winzer, Physica B **194-196**, 1539 (1994)

45. B. Stadlober, Doctoral thesis (Technische Universität München, 1996);

46. L.V. Gasparov, P. Lemmens, M. Brinkmann, N.N. Kolesnikov, and G. Güntherodt, Phys. Rev. B **55**, 1223 (1997).

47. A. Sacuto, R. Combescot, N. Bontemps, P. Monod, V. Viallet, and D. Colson, Europhys. Lett. **39**, 207 (1997). The data for $HgBa_2CaCu_2O_6$ are very similar to those found in the other compounds while the interpretation is different. In this paper, an extended s-wave gap is favored over d-wave. It is noted, however, that satisfactory agreement with the data can only be obtained if screening is not taken into account for the calculation of the A_{1g} response.

48. H. Monien and A. Zawadowski, Phys. Rev. Lett. **63**, 911 (1989).

49. S.L. Cooper, M.V. Klein, B.G. Pazol, J.P. Rice, and D. Ginsberg, Phys. Rev. B **37**, 5820 (1988)

50. R. Hackl, W. Gläser, P. Müller, D. Einzel and K. Andres, Phys. Rev. B **38**, 7133 (1988).

51. S.L. Cooper, F. Slakey, M.V. Klein, J.P. Rice, E.D. Bukowski, and D. Ginsberg, Phys. Rev. B **38**, 11934 (1988)

52. A. Yamanaka, F. Minami, and K. Inoue, Physica C **162-164**, 1099 (1989).

52. M. C. Krantz, H. J. Rosen, J. Y. T. Wie and D. E. Morris, Phys. Rev. B **40**, 2635 (1989).

54. A. Yamanaka, H. Takato, F. Minami, K. Inoue, and S. Takekawa, et al. Phys. Rev. B **46**, 516 (1992)

55. A. A. Maximov, A. V. Puchkov, I. I. Tartakovskii, V. B. Timofeev, D. Reznik and M. V. Klein, Solid State Commun. **81**, 407 (1992).

56. T. Staufer, R. Nemetschek, R. Hackl, P. Müller and H. Veith, Phys. Rev. Lett. **68**, 1069 (1992).

57. R. Nemetschek, O.V. Misochko, B. Stadlober, and R.Hackl, Phys. Rev. B **47**, 3450 (1993).

58. Ivan Bozovic, Phys. Rev. B **42**, 1969 (1990)

59. T. P. Devereaux, Phys. Rev. B **45**, 12965 (1992).

60. T. P. Devereaux, Phys. Rev. B **47**, 5230 (1993).

61. A. A. Abrikosov and L. A. Fal'kovskii, Zh. Eksp. Teor. Fiz. 40, 262 (1961) [Sov. Phys. JETP **13**, 179 (1961)].

62. A. A. Abrikosov and V.M. Genkin, Zh. Eksp. Teor. Fiz. 65, 842 (1973) [Sov. Phys. JETP **38**, 417 (1974)].

63. M. V. Klein and S. B. Dierker, Phys. Rev. B **29**, 4976 (1984).

64. H. Monien und A. Zawadowski, Phys. Rev. B **41**, 8798 (1990).

65. L.C. Hebel and C.P. Slichter, Phys. Rev. **107**, 901 (1957); ibid **113**, 1504 (1959); L.C. Hebel, ibid **116**, 79 (1959).

66. For reviews see: C. Thomsen and M. Cardona in in Physical Properties of High Temperature Superconductors I, D.M. Ginsberg ed., p. 509 (World Scientific, Singapore, 1989) and C. Thomsen in *Light Scattering in Solids VI*, Topics in Applied Physics **68**, p. 285 (Springer, Berlin Heidelberg, 1991)

67. K.B. Lyons and P.A. Fleury, J. Appl. Phys. **64**, 6075 (1988)

68. M. Pressl, M. Mayer, P. Knoll, S. Lo, U. Hohenester, and E. Holzinger-Schweiger, J. Raman Spectroscopy **27**, 343 (1996).

69. B. Friedl, C. Thomsen, and M. Cardona, Phys. Rev. Lett. **65**, 915 (1990)

70. R. Zeyher and G. Zwicknagl, Z. Phys. B **78**, 175 (1990)

71. T. P. Devereaux Phys. Rev. B **50**, 10287 (1994).

72. T. P. Devereaux, A.Virosztek, and A. Zawadowski, Phys. Rev. B **51**, 505 (1995).

73. R. Zamboni, G. Ruani, A.J. Pal, and C. Taliani, Solid State Commun. **70**, 813 (1989)

74. K.A. Müller, Z. Phys. B **80**, 193 (1990)
75. P. Knoll, C. Ambrosch-Draxl, R. Abt, M. Mayer, E. Holzinger-Schweiger, Physica C **235-240**, 2117 (1994).
76. For a review see: R. Feile, Physica C **159**, 1 (1989).
77. P.W. Anderson, Science **235**, 1196 (1987).
78. F.C. Zhang and T.M. Rice, Phys. Rev. B **37**, 3759 (1988)
79. D. Pines, Physica B **163**, 78 (1990)
80. A.J. Millis, H. Monien, and D. Pines, Phys. Rev. B **42**, 167 (1990).
81. J. Wagner, W. Hanke, and D Scalapino, Phys. Rev. B **43**, 10517 (1991) and Physica C **185-189**, 1617 (1991).
82. D. Pines, Z. Phys. B **103**, 129 (1997).
83. V. Emery and S.A. Kivelson, Nature **374**, 434 (1995).
84. J.R. Schrieffer and A.P. Kampf, J. Phys. Chem. Solids **56**, 1673 (1995).
85. J.M. Tranquada, P.M. Gehring, G. Shirane, S. Shamoto, and M. Sato, Phys. Rev. B **46**, 5561 (1992).
86. J. Rossat-Mignot, L.P. Regnault, C. Vettier, P. Bourges, P. Burlet, J. Bossy, J.Y. Henry, and G. Lapertot, Physica B **180&181**, 383 (1992).
87. T.E. Mason, G. Aeppli, and H.A. Mook, Phys. Rev Lett. **68**, 1414 (1992).
88. P. Bourges, L.P. Regnault, Y. Sidis, and C. Vettier, Phys. Rev. B **53**, 1 (1996).
89. K.B. Lyons, P.A. Fleury, J.P. Remeika, A.S. Cooper, and J.T. Negran, Phys. Rev. B **37**, 2353 (1987).
90. K.B. Lyons, P.A. Fleury, L.F. Scheemeyer, and J.V. Wazczak, Phys. Rev. Lett. **60**, 732 (1988).
91. S. Sugai, S. Shamoto, and M. Sato, Phys. Rev. B **38**, 6436 (1988).
92. I. Tomeno, M. Yoshida, K. Ikeda, K. Tai, K. Takamuku, N. Koshizuka, S. Tanaka, K. Oka, and H. Unoki, Phys. Rev. B **43**, 3009 (1991).
93. M. Rübhausen, N. Dieckmann, A. Bock, U. Merkt, W. Widder, H.F. Braun, J. Low Temp. Phys. **105**, 761 (1996)
94. M. Rübhausen, N. Dieckmann, K.-O. Subke, A. Bock, U. Merkt, Physica C **280**, 77 (1997)
95. R. Philipp, unpublished results.
96. P.A. Fleury and R. Loudon, Phys. Rev. **166**, 514 (1968).
97. J.B. Parkinson, J. Phys. C **2**, 2012 (1969).
98. M. Rübhausen, private communication
99. H. Shaked, P.M. Kaeane, J.C. Rodriguez, F.F. Owen, R.L. Hitterman, and J.D. Jorgensen, *Crystal Structures od High-T$_c$ Superconducting Copper Oxides* (Elsevier, Amsterdam, 1994)
100. H. Koch, H.P. Geserich, and Th. Wolf, Solid State Commun. **71**, 495 (1989)
101. W. Hayes and R. Loudon, *Scattering of Light by Crystals* (John Wiley & Sons, New York, 1978)
102. M.V. Klein, in in *Light Scattering in Solids III*, M. Cardona and G. Güntherodt eds., p. 136 (Springer, Berlin, 1982).
103. N.W. Ashcroft and N.D. Mermin, *Solid State Physics* (Holt-Saunders International Editions, 1981). The derivation of the cross section is exactly the same as for neutrons. Note, however, that second order perturbation theory is needed for Raman scattering. [62-64]
104. L.D. Landau and E.M. Lifshitz, *Statistical Physics* (Pergamon, Oxford, 1969).
105. L.D. Landau and E.M. Lifshitz, *Electrodynamics Of Continuous Media* (Pergamon, Oxford, 1960).
106. D. Forster, *Hydrodynamical Fluctuations, Broken Symmetry and Correlation Functions* (W.A. Benjamin Inc., Massachusetts, 1975).
107. T. P. Devereaux, A.Virosztek, and A. Zawadowski, Phys. Rev. B **54**, 12523 (1996).
108. J. Yu and A.J. Freeman, J. Phys. Chem. Solids **52**, 1351 (1991).
109. W.E. Pickett, Rev. Mod. Phys. **61**, 433 (1989).
110. O.K. Andersen, A.I. Liechtenstein, O. Jepsen, and F. Paulsen, J. Phys. Chem. Solids **56**, 1573 (1995).
111. P. B. Allen, Phys. Rev. B **13**, 1416 (1976).
112. A. Zawadowski and M. Cardona, Phys. Rev. B **42**, 10732 (1990).
113. Tsuneto, Phys. Rev. **118**, 1029 (1960).
114. T.P. Devereaux, SPIE (Bellingham) Vol. **2696**, 230 (1996).
115. T.P. Devereaux and A. Kampf, International Journal of Modern Physics B (1997)
116. T. P. Devereaux, Phys. Rev. Lett. **74**, 4313 (1995).
117. A. Virosztek and J. Ruvalds, Phys. Rev. Lett. **67**, 1657 (1991) and Phys. Rev. B **45**, 347 (1992).
118. D.R. Tilley, Z. Phys. **254**, 71 (1972).
119. D. Einzel, B. Stadlober, R. Nemetschek, T.Staufer, and R. Hackl, Physica B **194-196**, 1487 (1994).
120. C. Schuster, Diploma Thesis (Technische Universität München, 1995) unpublished.
121. P.E. Sulewski, P.A. Fleury, K.B. Lyons, and S.-W. Cheong, Phys. Rev. Lett. **67**, 3864 (1991).
122. R.R.P. Singh, P.A. Fleury, K.B. Lyons, and P.E. Sulewski, Phys. Rev. Lett. **62**, 2736 (1989).
123. P. Knoll, C. Thomsen, M. Cardona, and P. Murugaraj, Phys. Rev. B **42**, 4842, (1990).
124. M. Rübhausen, N. Dieckmann, A. Bock, and U. Merkt, Phys. Rev. B **54**, 14967 (1996).
125. R. Hackl, unpublished results.

126. T. Staufer, R. Hackl, and P. Müller, Solid State Commun. **75**, 975 (1990) and **79**, 409 (1991)
127. S. Jandl et al. Solid State Commun. **87**, 609 (1993).
128. B. Stadlober, G. Krug, R. Nemetschek, R. Hackl, J. L. Cobb and J. T. Markert, Phys. Rev. Lett. **74**, 4911 (1995).
129. E.T. Heyen et al. Phys. Rev. B **43**, 2857 (1991).
130. H. Wipf, M.V. Klein, B.S. Chandrasekhar, T.H. Geballe, and J.H. Wernick, Phys. Rev. Lett. **41**, 1752 (1978).
131. V.N. Kostur and G.M. Eliashberg, Pis'ma Zh. Eksp. Teor. Fiz. **53**, 373 (1991) [JETP Lett. **53**, 391 (1991) and V.N. Kostur, Z. Phys. B **89**, 149 (1992).
132. S.N. Rashkeev and G. Wendin, Phys. Rev. B **47**, 11603 (1993)
133. C. Jiang and J. Carbotte, Solid State Commun. **95**, 643 (1995) and Phys. Rev. B **53**, 11868 (1996).
134. T.P. Devereaux, priv. commum.
135. A. Junod, T. Jarlborg, and J. Muller, Phys. Rev. B **27**, 1568 (1983).
136. Hackl, M. Opel, P.F. Müller, G. Krug, B. Stadlober, R. Nemetschek, H. Berger, and L. Forró, J. Low Temp. Phys. **105**, 733 (1996).
137. R.T. Demers, S. Kong, M.V. Klein, R. Du, and C.P. Flynn, Phys. Rev. B **38**, 11523 (1988).
138. I. Bozovic, J.H. Kim, J.S. Harris Jr.,C.B. Eom, J.M. Phillips, and J.T. Cheung, Phys. Rev. Lett. **73**, 1436 (1994).
139. A. Bardasis and J.R. Schrieffer, Phys. Rev. **121**, 1900 (1961).
140. T.J. Greytag and J. Yan, Phys. Rev. Lett. **22**, 987 (1969).
141. A. Zawadowski, J. Ruvalds, and J. Solana, Phys. Rev. A **5**, 399 (1972).
142. R. Hackl, Doctoral Thesis (Technische Universität München, 1987).
143. R. Hackl, R. Kaiser, and W. Gläser, Physica C **162-164**, 431 (1989)
144. P.B. Littlewood and C.M. Varma, Phys. Rev. Lett. **47**,811 (1981).
145. D.A. Browne and K. Levin, Phys. Rev. B **28**, 4029 (1983).
146. I. Tüttö and A. Zawadowski, Phys. Rev. B **45**, 4842 (1992).
147. R. Hackl and R. Kaiser, J. Phys. C, **21**, L453 (1988).
148. A. Mooradian, Phys. Rev. Lett. **22**, 185 (1969).
149. H. Ihara, Y. Kimura, M. Yamazaki, and S. Gonda, Phys. Rev. B **27**, 551 (1983).
150. S. Sugai, T. Ido, H Takagi, S. Uchida, M. Sato, and S Shamoto, Solid State Commun. **76**, 365 (1990).
151. F. Slakey, M.V. Klein, J.P. Rice, and D.M. Ginsberg, Phys. Rev B **43**, 3764 (1991).
152. T. Katsufuji, Y. Tokura, T. Ido, and S. Uchida, Phys. Rev. B **48**, 16131 (1993).
153. X.K. Chen, J.C. Irwin, R. Liang, and W.N. Hardy, Physica C **227**, 113 (1994) and X.K. Chen, J.G. Naeini, K.C. Hewitt, J.C. Irwin, R. Liang, and W.N. Hardy, Phys. Rev. B **56**, R513 (1997)
154. P.F. Müller, Diploma Thesis (Technische Universität München, 1996) unpublished.
155. C. Hoffmann, Diploma Thesis (Technische Universität München, 1997) unpublished.
156. C. Kendziora, M.C. Martin, J. Hartge, L. Mihaly, L. Forró, Phys. Rev. **48**, 3531 (1993)
157. C. Kendziora, L. Forró, D. Mandrus, J. Hartge, P. Stephens, L. Mihaly, R. Reeder, D. Moecher, M. Rivers, and S. Sutton, Phys. Rev. **45**, 13025 (1992).
158. P.W. Anderson, Phys. Rev. Lett. **67**, 2091 (1991).
159. B.P. Stojkovic and D. Pines, Phys. Rev. B **55**, 8576 (1997).
160. J. Altmann, W. Brenig, and A. Kampf, preprint.
161. T.P.Devereaux and A. Kampf, preprint.
162. H. Ding, M.R. Norman, T. Yokoya, T. Takeuchi, M. Randeira, J.C. Campuzano, T. Takahashi, T. Mochiku, and K. Kadowaki, Phys. Rev. Lett. **78**, 2628 (1997).
163. D.S. Marshall, D.S. Dessau, A.G. Loeser, C.H, Park, A.Y. Matsuura, J.N. Eckstein, I. Bozovic, P. Fournier, A. Kapitulnik, W.E. Spicer, and Z.-X. Shen, et al. Phys. Rev. Lett. **76**, 4841 (1996).
164. T.R. Chien, Z.Z. Wang. and N.P. Ong, Phys. Rev. Lett. **67**, 2088 (1991).
165. A. Carrington, A.P. Mackenzie, C.T. Lin, and J.R. Cooper, Phys. Rev. Lett. **69**, 2855 (1992).
166. T. Ito, K. Takenaka, and S. Uchida, Phys. Rev. Lett. **70**, 3995 (1993).
167. J.W. Loram, K.A. Mirza, J.R. Cooper,and W.Y. Liang, Phys. Rev. Lett. **72**, 1740 (1993).
168. T. Imai et al., Physica C **162-164**, 169 (1989).
169. N.F. Mott, Philos. Mag. Lett. **63**, 319 (1991) and A.S. Alexandrov and N.F. Mott in *Polarons and Bipolarons* (World Scientific, Singapore, 1996).
170. J. Ranninger and J.M. Robin, Phys. Rev. B **53**, R11961 (1996).
171. A.G. Loeser et al., Science **273**, 325 (1996).
172. H. Ding et al. Nature **382**, 51 (1996)
173. R. Nemetschek, M. Opel, C. Hoffmann, P.F. Müller, R. Hackl, H. Berger, L. Forro, A. Erb, and E. Walker, Phys. Rev. Lett. **78**, 4837 (1997).
174. A. Erb, E. Walker, and R. Flükiger, Physica C **245**, 245 (1995).
175. Lindemer et al. J. Am. Ceram. Soc. **71**, 1775 (1992).
176. A.P. Litvinchuk, C. Thomsen, and M. Cardona, Solid State Commun. **83**, 343 (1992).

177. G. Grüner, Rev. Mod. Phys. **66**, 1 (1994).

178. E. Osquiguil et al., Phys. Rev. B **49** 3675 (1994).

179. J. Bardeen, L.N. Cooper, and J.R. Schrieffer, Phys. Rev. **108**, 1175 (1957).

180. R. Hackl, R. Kaiser and S. Schicktanz, J. Phys. C **16**, 1729 (1983).

181. R. Hackl, P. Müller, D. Einzel, and W. Gläser, Physica C **162-164**, 1241 (1989).

182. H.P. Fredericksen et al. Solid State Commun. **48**, 883 (1983).

183. D. A. Bonn, P. Dosanjh, R. Liang, and W. N. Hardy, Phys. Rev. Lett. **68**, 2390 (1992).

184. C. Kendziora, R.J. Kelley, and M. Onellion, Phys. Rev. Lett. **77**, 727 (1996).

185. C. Kendziora and A. Rosenberg, Phys. Rev. B **52**, 9867 (1995).

186. I. Bozovic, private communication.

187. Beom-hoan O and J.T. Markert, Phys. Rev. B **47**, 8373 (1993).

188. D. H. Wu et al., Phys. Rev. Lett. **70**, 85 (1993).

187. Q. Huang et al., Nature **347**, 369 (1990).

190. A.W. Sleight, Physica C **162-164**, 3 (1989).

191. D.M. King, Z.-X. Shen, D.S. Dessau, B.O. Wells, W.E. Spicer, A.J. Arko, D. Marshall, J. DiCarlo, A.G. Loeser, C.-H. Park, E.R. Ratner, J.L. Peng, Z.Y. Li, and R.L. Greene, Phys. Rev. Lett. **70**, 3159 (1993).

192. R.O. Anderson, J.W. Allen, C.G. Olson, C. Janowitz, L.Z. Liu, J.-H. Park, M.B. Maple. Y. Dalichaouch, M.C. de Andrade, R.F. Jardim, E.A. Early, S.-J. Oh, and W.P. Ellis, Phys. Rev. Lett. **70**, 3163 (1993).

193. M.C. de Andrade, Y. Dalichaouch, M.B. Maple, Phys. Rev. B **48**, 16737 (1993).

194. C.M. Varma, P.B. Littlewood, S. Schmitt-Rink, E. Abrahams, and A.E. Ruckenstein, Phys. Rev. Lett. **63**, 1996 (1989).

195. The interpretation in terms of d-wave pairing and in a single-band picture has been criticized, particularly because the A_{1g} response, in lowest order approximation, does not fit as well as those at the other symmetries: M.C. Krantz and M. Cardona, Phys. Rev. Lett. **72**, 3290 (1994); M.C. Krantz and M. Cardona, J. Low Temp. Phys. **99**, 205 (1995); M.C. Krantz, I.I. Mazin, D.H. Leach, W.Y. Lee and M. Cardona, Phys. Rev. B **51**, 5949 (1995); M. Cardona, T. Strohm, and J. Kircher, SPIE (Bellingham) Vol. **2692**, 182 (1996). As a matter of fact, the A_{1g} response, in contrast to the other symmetries, depends critically on the band structure and, consequently, on the order of the expansion (Eq. 10, see T.P. Devereaux and D. Einzel, Phys. Rev. **54**, 15547). Peak frequency and shape can be tuned in a wide range by only slightly changing the parameters. Therefore, it is at least questionable to use the A_{1g} response as a crucial test for the theoretical model proposed in Ref.[36] The universality of the symmetry selection rules found experimentally, specifically the independence of the response on the number of CuO_2 planes and, hence, the number of conduction bands further invalidates the basis of the critizism. It is nevertheless interesting to study multiband effects and different gap symmetries on different bands, as done in some of the references quoted here. The influence of a multi-sheeted Fermi surface has actually been calculated and is found to be mild for the B_{1g} and B_{2g} spectra.[107]

196. M.T. Beal-Monod, J.-B. Bieri, and K. Maki, Europhys. Lett. **40**, 201 (1997); T.P. Devereaux, priv. commun.

197. N.E Hussey, J.R. Cooper, J.M. Wheatley, I.R. Fisher, A. Carrington, A.P. Mackenzie, C.T. Lin, and O. Milat, Phys. Rev. Lett. **76**, 122 (1996).

ANALYSIS OF THE SUPERCONDUCTING GAP BY ELECTRONIC RAMAN SCATTERING IN HgBa$_2$Ca$_2$Cu$_3$O$_{8+\delta}$ SINGLE CRYSTALS

A. Sacuto[+] and R. Combescot[++]

[+]Physique de la Matière Condensée,
[++]Physique Statistique, Ecole Normale Supérieure, 24 rue Lhomond, 75231 Paris cedex 05, France

1. INTRODUCTION

Since the last few years, identifying the symmetry of the pairing state has been the expected major step towards an understanding of high T_c superconductivity. The controversy between s-wave or d-wave pairing is not yet resolved: some experiments appear to advocate strongly in favor of a d$_{x2-y2}$ symmetry [1], whereas others seem to show a significant s-wave contribution [2]. With respect to this problem, inelastic light scattering has been shown very early to be a powerful tool because besides probing the bulk (in contrast with photoemission and tunneling), the selection rules on the polarizations of the incident and scattered light make the spectra very sensitive to the wavevector of the electronic excitations [3-6]. A theoretical approach of electronic Raman scattering neglects other excitations such as phonons. The experimental major difficulty lies however in the fact that in La$_{2-x}$Sr$_x$CuO$_4$, YBa$_2$Cu$_3$O$_{7-\delta}$ (Y-123), Bi$_2$Sr$_2$CaCu$_2$O$_{8+\delta}$ (Bi-2212) and Tl$_2$Ba$_2$CuO$_{6+\delta}$ the phonons mask the electronic excitations, in particular at low frequency, which hampers an accurate determination of the shape of the electronic excitation spectrum in the superconducting state [7-13].

In this work, we report for the first time *pure - no phonon structures superimposed -* electronic Raman spectra in single crystals from the highest T_c cuprate family, namely HgBa$_2$Ca$_2$Cu$_3$O$_{8+\delta}$ (Hg-1223). The crystallographic structure is purely tetragonal (^{1}D$_{4h}$) [14], which allows an unambiguous comparison with theoretical calculations based on

The Gap Symmetry and Fluctuations in High - T$_c$ Superconductors
Edited by Bok *et al.*, Plenum Press, New York, 1998

291

tetragonal symmetry without the complications due to the orthorhombic distorsion in Y-123 [15]. Hg-1223 belong to the new family of mercury-based superconductors where the phonons do not mask the low energy electronic spectrum for an electric field within the planes [16]. No subtraction is thus necessary to obtain the electronic spectra for the various $A_{1g}+B_{1g}$, B_{1g} and B_{2g} symmetries. Hg-1223 is therefore particularly well suited for the study of the superconducting gap. Our most striking result is that the low frequency behavior of the B_{2g} but also the B_{1g} spectrum display an intrinsic linear term. The linearity of the B_{1g} spectrum have also been reported in Hg-1212 and Tl-2201 compounds [17-19]. Such an observation strongly suggests that the nodes exist outside the [110] and [1$\bar{1}$0] directions, a conclusion at odds with the generally accepted d_{x2-y2} gap symmetry. We believe that our new and unambiguous experimental data should be taken into account in any theoretical approach of high T_c superconductors.

2. ELECTRONIC RAMAN SCATTERING IN THE FRAMEWORK OF BCS THEORY

Electronic Raman scattering is a process of inelastic light scattering where an incident photon is absorbed by the crystal and a scattered one is emitted with the creation (Stokes) or the annihilation (anti-Stokes) of an electronic excitation. We first consider the scattering on one single electron before generalizing to an arbitrary number of electrons. The electron is in a Bloch eigenstate $\left|n_i, k_i\right\rangle$ with energy ε_{n_i, k_i}, and the electromagnetic fields are described by the vector potentials A_L and A_S of the laser and scattered light with

$$A_L = A_L^0 e_L a_{k_L} e^{ik_L \cdot r} + h.c. \quad \text{and} \quad A_S = A_S^0 e_S a_{k_S} e^{ik_S \cdot r} + h.c.. \qquad (1)$$

$a_{k_{L,S}}$ is the annihilation operators of the laser and scattered photons. The electron-photon interaction leads to the perturbative Hamiltonian with linear and quadratic terms in A :

$$H_A = (e/m)A \cdot P \quad \text{and} \quad H_{AA} = (e^2/2m)A^2. \qquad (2)$$

Under typical experimental conditions, the intensity is less than 100 Wcm^{-2}, the Fermi momentum $k_F \approx 10^9$ m^{-1} and therefore $eA/\hbar k_F \approx 10^{-5}$, justifying a standard perturbation calculation of the transition amplitude S_{fi} to a final state $\left|n_f, k_f\right\rangle$. Since we are interested in a two photon process, we must include the second order contribution of H_A but we can limit the calculation to first order in H_{AA}. The difference of the photon frequencies is noted $\omega = \omega_L - \omega_S$, and the difference of the photon momenta $q = k_L - k_S$.

Only the quadratic Hamiltonian H_{AA} contributes to the first order transition amplitude without intermediate virtual state. For an infinite time of interaction, we find

$$S_{fi}^{(1)} = \frac{e^2}{m\hbar} e_S \cdot e_L \left\langle A_S A_L \right\rangle \delta_{n_f, n_i} \delta_{k_f, k_i + q} \, \delta(\varepsilon_{n_f, k_f} - \varepsilon_{n_i, k_i} - \omega) \qquad (3)$$

where $\langle \ \rangle$ is the matrix element of the photons states. The conservation of momentum and energy is evident in eq.(3). The diagramatic representation of $S_{fi}^{(1)}$ is shown in Fig.1-a). Pure intraband is zero for crossed polarizations.

a)   b)

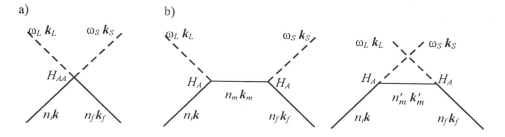

Figure 1. Feynman's diagrams for a) : the $S_{fi}^{(1)}$ scattering process, b) : the $S_{fi}^{(2)}$ scattering process ; photon state (dashed line) and electron state (full line)

The second order amplitude involves a summation over all possible intermediate virtual states. H_A generates two types of scattering processes : the annihilation of the incident photon followed by the creation of the scattered one or vice versa. For commodity, we will denote the contribution of the second process by $(L \leftrightarrow S)$ because it is obtained by the substitution $(\omega_L \leftrightarrow -\omega_S, k_L \leftrightarrow -k_S)$ in the first term. Momentum conservation contracts the summation over intermediate electron momenta, and one finds

$$S_{fi}^{(2)} = \frac{e^2}{m^2\hbar^2}\langle A_S A_L\rangle \delta_{k_f,k+q}\, \delta(\varepsilon_{n_f,k_f} - \varepsilon_{n_i,k_i} - \omega)\ \sum_{n_m}\Gamma_{fmi}(q,k) \quad (4)$$

with the matrix elements

$$\Gamma_{fmi}(q,k) = \frac{\langle n_f k_i + q | e^{-ik_S\cdot r} e_S \cdot P | n_m k_i + k_L\rangle\langle n_m k_i + k_L | e^{ik_L\cdot r} e_L \cdot P | n_i k_i\rangle}{\varepsilon_{n_i,k_i} - \varepsilon_{n_m,k_i+k_L} + \omega_L + i0} + (L \leftrightarrow S)$$

The Feynman's diagram of $S_{fi}^{(2)}$ is displayed in Fig.1-b).

We can take $\varepsilon_{n,k_i+k'} \approx \varepsilon_{n,k_i}$, since the momenta $k' = k_L, k_S, q \approx 10^7\,\mathrm{m^{-1}}$ are negligible compared to $k_F \approx 10^9\,\mathrm{m^{-1}}$. Using the identity $\exp(ik'\cdot r)|n,k_i\rangle = |n, k_i + k'\rangle$, we can write the matrix elements as

$$\Gamma_{fmi}(k) = \frac{\langle n_f k_i | e_S \cdot P | n_m k_i\rangle\langle n_m k_i | e_L \cdot P | n_i k_i\rangle}{\varepsilon_{n_i,k_i} - \varepsilon_{n_m,k_i} + \omega_L + i0} + (L \leftrightarrow S) \quad (5)$$

The transition induced is not necessarily intraband as in the first order, and there is now the possibility of resonant excitation for vanishing denominators. The ratio of intraband/interband matrix elements being of the order of ω / ω_L, the sum in eq.(4) can be restricted to $n_m \neq n_i$: for a typical excitation energy (2.5 eV), the intermediate electron state is preferentially situated in a band far the conduction band. The above results are equivalent to consider an effective perturbation Hamiltonian for the electron system :

$$H_R = \frac{e^2}{m} \langle A_S A_L \rangle e^{-i\omega t} \hat{\rho}_q \tag{6}$$

where we introduce the dimensionless perturbation density operator

$$\hat{\rho}_q = \sum_{n_f, n_i, k} \gamma_{n_f, n_i, k} \, c^+_{n_f, k+q} \, c_{n_i, k} \tag{7}$$

with the transition matrix element, known as the *Raman vertex,*

$$\gamma_{n_f, n_i, k} = e_S \cdot e_L \, \delta_{n_f, n_i} + \frac{1}{m\hbar} \sum_{n_m} \Gamma_{fmi}(k). \tag{8}$$

This result allows us to calculate the transition amplitude from a given initial to a final electronic state. By application of Fermi's golden rule to the effective perturbation Hamiltonian we calculate the *differential scattering cross section,* defined as the ratio of the transition rate and the flux of incident photons for a given frequency shift ω and momentum transfer q,

$$\frac{d^2\sigma}{d\Omega d\omega} = r_0^2 \frac{\omega_S}{\omega_L} S(\beta, \omega, q, \omega_L) \tag{9}$$

where $r_0 = e^2 / 4\pi \varepsilon_0 mc^2$ and the *generalized dynamical structure function*

$$S(\beta, \omega, q, \omega_L) = \sum_{f,i} \frac{1}{Z} e^{-\beta E_i} \left| \langle f | \hat{\rho}_q(\omega_{L,S}) | i \rangle \right|^2 \delta(\varepsilon_f - \varepsilon_i - \omega). \tag{10}$$

We recognize $e^{-\beta E_i} / Z$ as the probability of the initial state in thermodynamical equilibrium.

Within this framework two major simplifications are most often made when dealing with effective calculations of the Raman vertex. First a single band is assumed at the Fermi level. This excludes interband transitions for sufficiently weak energy transfer, so that $n_f = n_i$. More crucially the energy of all the states belonging to other bands are assumed to be far enough from the Fermi level, compared to the photon energies ω_L and ω_S, that the latter ones can be neglected in the denominator in the expression eq.(5) of Γ_{fmi}. It can then be shown that the Raman vertex is identical to the contraction of the k-dependent effective mass tensor with the polarization vectors of the incident and the scattered light. While these approximations are resonable for simple metals (one band near the Fermi level and the other bands are faraway), they are clearly not justified in high Tc superconductors. In particular it is known that the absorption is important at the incident frequency, linked to interband transitions. This invalidates the approximation of neglecting the photon energy in the expression of Γ_{fmi}, and leads instead to the existence of resonant terms. This is actually coherent with what we observe experimentally, since the electronic Raman scattering varies in an important way as a function of the incident photon energy (as we will see in section V). Moreover, in order to deal consistently with these quasi resonant terms, we have to take into account the finite lifetime of the highly excited intermediate states. We expect, on the basis of infrared data, that the relaxation rate is of order a few tenths of eV. All this makes it clear that the effective mass approximation is not valid in our experimental situation. Therefore we have to keep in mind that the Raman vertex that we consider is actually an effective quantity which depends not only on the wavevector k, but also on the incident and scattered photon energies. For simplicity we will not indicate explicitly this dependence in the following, but we will consider it in the analysis of our data.

The fluctuation-dissipation theorem (Kubo relation) allows us to relate the imaginary part $\chi'' = \text{Im} \chi$ of the susceptibility to the structure function in the differential cross section and to obtain

$$\frac{d^2\sigma}{d\Omega d\omega} = \frac{r_0^2}{\pi(1-\exp(-\beta\omega))} \frac{\omega_S}{\omega_L} \chi''(\beta,\omega,q).$$
(11)

In BCS theory with a coherence length much smaller than the optical penetration depth, the imaginary part of the response function for $T \to 0$ is

$$\chi''(\omega,q) = \pi \int \frac{d^3k}{(2\pi)^3} |\gamma_k|^2 \frac{EE' - \varepsilon_k \varepsilon_k' + \Delta_k^2}{2EE'} \delta(\omega - E - E')$$
(12)

where $\varepsilon_k' = \varepsilon_{k+q}$ (ε_k is zero at the Fermi energy) and the quasi-particle energy $E_k = (\varepsilon_k^2 + \Delta_k^2)^{1/2}$. Neglecting q compared to k, we have $\varepsilon_k' = \varepsilon_k$ and $E' = E$, and the Raman susceptibility can be rearranged to give for $q \to 0$ [3] :

$$\chi''(\omega) = \frac{2\pi N_0}{\omega} \text{Re} \left\langle \frac{|\gamma_k|^2 \Delta_k^2}{(\omega^2 - 4\Delta_k^2)^{1/2}} \right\rangle_k$$
(13)

N_0 is the density of states for both spin orientations at the Fermi level, and the brackets indicate an average over the Fermi surface. Δ_k stands for the superconducting, k-dependent gap.

3. CRYSTAL CHARACTERIZATION

The crystals were grown by a single step synthesis as previously described for Hg-1223 [20]. They are perfect little squares with typical 0.5 x 0.5 mm^2 cross section and thickness of 0.3 mm. The crystals were characterized by X-ray diffraction and energy dispersive X-ray analysis [20]. The [100] crystallographic direction lies at 45° of the edge of the square and the [001] direction is normal to the surface. The data concerning the lattice parameters have been collected with an Ennaf-Nonius CAD-4 diffractometer. The unit cell dimensions at 293 K are $a = 3.844 \pm 0.002$ Å, $c = 15.72 \pm 0.03$ Å. A weak proportion of intergrowth phases corresponding to different stacking of the CuO_2 planes along the c-axis has been detected. DC magnetic susceptibility measurements have been performed with a SQUID magnetometer on the different single crystals studied by Raman investigations. The Hg-1223 single crystals are underdoped with $T_c = 126 \pm 1K$.

4. EXPERIMENTAL PROCEDURE

Raman measurements were performed with a double monochromator using a single channel detection and the Ar^+ and Kr^+ laser lines. The spectral resolution was set at 3 cm^{-1}. The crystals were mounted in vacuum (10^{-5} mbar) on the cold finger of a liquid helium cryostat. The temperature was controlled by a Si diode located inside the cold finger. The incident laser spot is less than 100µm in diameter and intensity onto the crystal surface was below 100W/cm^2 in order to avoid heating of the crystal during the low-temperature runs. The temperature variation between the cooled face and the illuminated face of the crystals was estimated to 1 K which corresponds to a temperature of 13 K inside the scattering volume. The angle of incidence was 60° and the scattered light was collected along the normal to the crystal surface. Inside the crystal the incident wave vector is quasi-normal to the crystal surface. The polarizations of the incident and scattered lights are denoted in the usual way, x: [100] (a axis), y: [010], z: [001] (c axis), x': [110], y': [1$\bar{1}$0].

In order to compare our experimental data with the theoretical calculations [5, 6, 21], the pure B_{2g} (xy) and B_{1g} (x'y') symmetries are needed. This requires that the incident electric field lies within the xy plane. The incident angle being not zero, the crystal must be rotated by 45° after measuring the B_{1g} channel in order to get the B_{2g} one while the polarizor and the analysor remain unchanged (Fig.2). The $A_{1g}+B_{2g}$ symmetry is obtained in the (xx) polarization. The impact point of the laser beam onto the crystal surface was then precisely located before turning the crystal in order to probe the same area for these two symmetries. Finally, a very weakly diffusive spot was carefully chosen on each crystal surface to minimize the amount of spurious elastic scattering.

To conserve exactly the same laser spot for the B_{1g} and B_{2g} symmetries, we have the possibility to rotate the polarization instead of the crystal. After measuring the pure B_{1g} spectrum, both the polarizor and the analyzor are rotated of 45° (Fig.2). Unfortunately, in this configuration the incident electric field does not lie within the xy plane which induces symmetry mixing for B_{2g} with 5% admixture of $A_{1g}+B_{1g}$ (xx) and 16% of E_g(xz). Conversely, the $A_{1g}+B_{1g}$ spectrum contains 5% of B_{2g} (xy) and 16% of E_g(xz). Let us denote by B_{2g}' and $(A_{1g}+B_{1g})'$ these mixed symmetries.

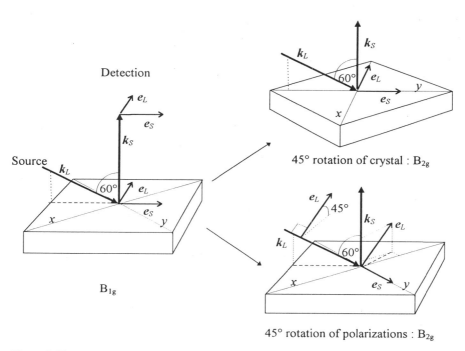

Figure 2. The two experimental configurations chosen to obtain the B_{1g} and B_{2g} symmetries.

To get a complete set of reliable data we have performed Raman measurements in both configurations : rotation of the crystal without changing the polarizations to get pure B_{1g} and B_{2g} symmetries and rotation of polarizations without moving the crystal to conserve the same spot. The Raman measurements have been performed on three different crystals coming from the same batch. On each crystal various spots have been probed. All the measurements obtained are consistent with each other.

5. EXPERIMENTAL RESULTS

Preliminary micro-Raman measurements [22] on the Hg-1223 single crystals at room temperature in (x'x'), (xx), x'y'), (xy), (zx') and (zz) polarizations using the 514.52 nm laser line are displayed in Fig.3. The incident and scattered light were emitted and collected by the same microprobe objective which do not allow an efficient rejection on contrary to the spectra obtained by conventionnal Raman technique (Fig.3). The most striking feature of these spectra at room temperature is the absence of peaks in x'x', xx, x'y' and xy polarizations. In the zx' polarization the peaks at 165 and 337 cm^{-1} have already been studied and assigned to the motion inside the CuO_2 planes of the coppers and the oxygens, respectively. In the zz polarization, the peak at 92 cm^{-1} has been attributed to the vertical motion along the c-axis of the barium. The high frequency modes at 533, 561, 581 and 592 cm^{-1} correspond to vibrations of the apical oxygen atoms and to the modes induced by the oxygen excess. They strongly depend on δ [16,23]. The very low intensity of the Raman peaks for exciting electric fields within the CuO_2 plane offer a unique opportunity to study the pure electronic excitations in the normal and superconducting state.

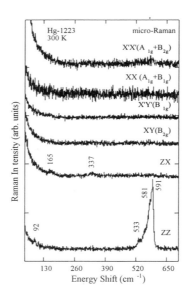

Figure 3. Micro-Raman spectra of the Hg-1223 single crystal at T=300 K in (x'x'), (xx), (x'y'), (xy), (zx') and (zz) polarizations with 514.52 nm. Rayleigh rejection is not efficient in the Micro-Raman technique used in comparison with the conventional Raman one.

We focus first on the low temperature limit ($T = 13$ K). The raw spectra in the three symmetries obtained from both configurations described in section 4 (rotation of the crystal or rotation of the polarization) are displayed in Figs. 4-a) and 4-b). For a given symmetry, the spectra obtained from both configurations are nearly the same. This confirms the reliability of our data and of the experimental procedure.

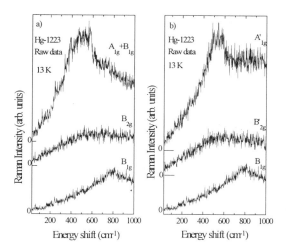

Figure 4. Raw Raman spectra of the Hg-1223 single crystals at T= 13 K obtained from both configuration described in Fig.2. a) pure B_{1g} and B_{2g} symmetries (rotation of the crystal). b) pure B_{1g} and the polluted B_{2g} symmetries (rotation of the polarization).

Two well marked maxima are observed around 530 cm^{-1} and 800 cm^{-1} for the $A_{1g}+B_{1g}$ and B_{1g} symmetries respectively. No clear maximum appears in the B_{2g} symmetry, for which actually the scattered intensity levels out smoothly around 530 cm^{-1}. However, the ratio of the peak energies is approximatively 1.5 while it was found nearly equal to 2 for Bi-2212 and Y-123 after subtraction of the phonon structures [9,13]. In the $A_{1g}+B_{1g}$ spectrum we can recognize the B_{1g} contribution by the presence of an electronic maximum around 800 cm^{-1}. This maximum is slightly lower in energy than in the pure B_{1g} spectrum because the additional A_{1g} electronic contribution decreases in this region. On the contrary, the B_{2g} and E_g contributions in the $(A_{1g}+B_{1g})'$ mask the B_{1g} electronic spectrum and the B_{1g} maximum is not observed. Note that the residual elastic scattering below 50 cm^{-1} is quite weak, due to the high quality of the crystal surface which allows the choice of laser spot with weak diffusion. Remarkably and in sharp contrast with Y-123 and Bi-2212, no Raman active phonon mode disturbs the B_{1g} and B_{2g} electronic spectra. Weak peaks appear at 388 and 580 cm^{-1} in the $A_{1g}+B_{1g}$ spectrum but their energy location and their intensity are such that no correction is actually needed to discuss the electronic spectra [23]. Therefore, we are in a position to turn immediately to the analysis of the data without dealing with the delicate handling of "phonon subtraction".

In contrast with previous data in Bi-2212 [9] the intensity ratios $I(50$ cm$^{-1}) / I(1000$cm$^{-1})$ from these raw data are nearly the same for the $A_{1g}+B_{1g}$, B_{1g} and B_{2g}

symmetries: 37%, 36% and 38% respectively. The intrinsic scattered intensities in $A_{1g}+B_{1g}$, B_{1g} and B_{2g} symmetries (after subtracting the dark current of the photomultiplier) drop close to a zero count rate whereas, in Bi-2212, a trend towards finite intensity at zero frequency was claimed to be observed in $A_{1g}+B_{2g}$ symmetry [9]. We note that in Bi-2212, the presence of numerous phonon structures at low frequency especially in x'x' polarizations (xx in the notation of ref.[9]), makes it difficult to estimate precisely the residual scattered intensity close to zero frequency. In this respect, we believe that our data are more reliable.

Fig.5 displays the imaginary part of the electronic response function at $T = 13\,\mathrm{K}$ associated to the $A_{1g}+ B_{1g}$, B_{1g} and B_{2g} symmetries with the 514.52 nm laser line. They were obtained from the raw spectra after correcting of the dark photomultiplier current, the response of the diffraction grating and the Rayleigh scattering. The Raman spectra have been divided by the thermal Bose-Einstein factor $1+ n(\beta, \omega) = (1 - \exp(-\beta \hbar \omega))^{-1}$ to obtain the imaginary part of the response function. The dashed lines in Fig. 5 are the imaginary part of the response functions at $T = 150\,\mathrm{K}$, displayed for clarity after smoothing the spectra. Note that the three response functions vanish at zero frequency.

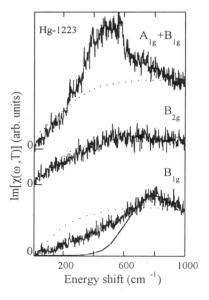

Figure 5. Imaginary parts of the response function $\mathrm{Im}[\chi(\omega, T=13K)]$ in the A_{1g}, B_{1g} and B_{2g}. The dashed lines represent $\mathrm{Im}[\chi(\omega, T=150K)]$. The solid line is a fit to the B_{1g} spectra of a Gaussian distribution of gaps (see text) $2\Delta_0=670\ \mathrm{cm}^{-1}$ and $\sigma=180\mathrm{cm}^{-1}$ for B_{1g} (Δ_0 and σ are the mean and the standard deviation of the Gaussian respectively).

The main new observations which emerge from Fig. 5 are: i) the B_{1g} response function exhibits a clear decrease of the electronic scattering rate at low energy with respect to the normal state, unlike the A_{1g} and the B_{2g} response functions where the difference is small. ii) the normal state intensity is recovered for all symmetries beyond 1000 cm^{-1}. iii) the scattered intensity in the B_{1g} channel exhibits a quasi-linear increase.

Indeed, as already pointed out, our experimental data are of special interest in the low frequency regime where we may analyze unambiguously the electronic scattering behavior as a function of the frequency ω. Let us focus first on the pure B_{1g} and B_{2g} symmetries, which are the ones which are expected to have the most different behavior (ω^3 and ω respectively) in the d_{x2-y2} model (as it will be explained in the next section). In order to characterize this frequency behavior, two fits have been performed for each symmetry: a first one to a power law ω^α and a second one to the simple polynomial function $b\omega + c\omega^3$. The α exponents calculated below 300 cm^{-1} for the B_{2g} and B_{1g} symmetries are 0.8 ± 0.2, 1.5 ± 0.5 respectively. The power law fit shows clearly that the low energy range of the B_{1g} spectrum does not display a ω^3 dependence, in contrast to the claim of [5,6,23] dealing with the Bi-2212 results. In particular a linear ω dependence in the low frequency regime is quite compatible with our B_{2g} and B_{1g} response function. The second fit provides quantitative evidence for the predominance of the linear part over the cubic part especially for the the B_{1g} symmetry. The ratio $\tau = b / c\omega^2$, which gives the relative weight of the linear and the cubic variation and calculated from the fit of various spectral ranges 50-300, 400, 500, 600 cm^{-1} is respectively $\tau = 2 \pm 1$, 4 ± 1, 3.3 ± 0.7, 3.4 ± 0.5. τ is much larger than an earlier estimate [15]. We suspect that the ω^3 fit to previous experimental measurements is less reliable because of the phonon background [9,13].

We have also carried out Raman measurements at T=13 K with different laser lines in order to test the linearity of the B_{1g} spectrum as function of the excitation sources. The imaginary parts of the electronic reponse functions in the B_{1g} symmetry for the 647.1, 568.2, 514.52, 488.0 and 476.5 nm lines are displayed in Fig. 6-a). The dark current have been substracted from the spectra and the spectra have been corrected from the response of the diffraction gratting and the thermal Bose factor. All the spectra have been normalized to have the same intensity at the 800 cm^{-1} maximum. The estimation of τ at 300 cm^{-1} gives $\tau \geq 8$, 52, 4, 3, 8 for 647.1, 568.2, 514.52, 488.0 and 476.5 nm respectively.

The τ ratio is much larger than one as a consequence the B_{1g} spectrum exhibits a linear ω dependence much larger than the cubic one. τ varies with the excitation line (this is illustrated by the τ variation between 514.52 and 568.2 nm), this changes the line-shape of the B_{1g} spectrum at intermediate energy (>400cm-1). The low frequency electronic excitations of the B_{1g} spectra are displayed in Fig. 6-b). Changes in the intermediate and low energy parts of the Raman spectrum with different excitation lines have also been reported in Tl-2201 compounds [19].

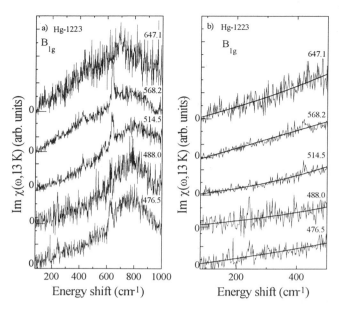

Figure 6. Imaginary parts of the response function $Im[\chi(\omega, T=13K)]$ in B_{1g} symmetries obtained from various excitation laser lines. a) between 80 and 1000 cm^{-1}, b) between 80 and 500 cm^{-1}.

6. ANALYSIS OF THE LOW RAMAN ENERGY EXCITATIONS

We now turn to the comparison of our data with existing theories. We start from eq.(13). Obviously, our results cannot be accounted for with an s isotropic gap combined with a cylindrical Fermi surface. Indeed for an s-wave gap $\Delta(k) = cst$, the Raman response function under any symmetry has to vanish below 2Δ which is not the case here. The s isotropic gap in k -space is displayed in Fig.7. Several Δ_k and γ_k distributions have been proposed to reproduce the electronic excitations. For instance, if we take a Gaussian distribution of Δ_k as used for Nb_3Sn and V_3Si [4] or YBCO [8] to simulate a gap anisotropy, we obtain for the B_{1g} channel the fits shown in Fig.5. We have added a $\omega^{1/2}$ function to the bracketed expression at energies greater than Δ_k, in order to mimic phenomenologically the asymptotic recovery of the normal state behavior. Indeed no Raman in the normal state $(\Delta_k \rightarrow 0)$ is expected if we refer to eq. 13. Clearly this part should be handled more quantitatively. However, this type of calculation does not yield any information on the gap symmetry and the low energy part of the response function in B_{1g} symmetry is not satisfactorily described.

We have attempted to fit our $A_{1g}+B_{1g}$, B_{1g} and B_{2g} response functions (obtained from the 514.52 nm which exhibit in B_{1g} symmetry the weaker τ value) by the calculations of Devereaux et al. for a d_{x2-y2} gap. The Devereaux's calculations were found to describe well the Bi-2212 results [21]. The d_{x2-y2} gap in k-space is shown in Fig.7.

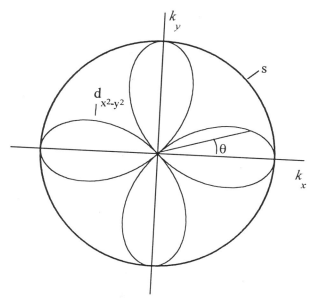

Figure 7. Representation of the isotropic and d_{x2-y2} gaps in the k-space.

We find a fair agreement with our data both for the relative energy location of the maxima in A_{1g}, B_{2g}, and B_{1g} symmetries, and for the linear dependence of the low energy part of the A_{1g} and B_{2g} symmetries (see Fig.8). However, the low frequency behavior of the theoretical B_{1g} spectrum is incompatible with our experimental result. This was already clear from our above discussion of the fitted exponents and of the ratio of the linear and cubic contribution with various excitation lines. We have also done the calculation for the simplest s-wave anisotropic gap $\Delta(\theta) = \Delta_0 + \Delta_1 \cos^2(2\theta)$, not finding a satisfactory agreement because no indication for a threshold corresponding to the minimum gap is seen in our data. Let us therefore consider the implications of our experimental findings and analyze first the B_{1g} and B_{2g} symmetries, which are easier to interpret due to simple symmetry considerations. For one thing, B_{1g} and B_{2g} symmetries are not affected by screening in contrast with the A_{1g} symmetry. We recall briefly their symmetry properties : in the B_{1g} case, the Raman vertex γ_k is zero by symmetry for $k_x = k_y$, so the electronic scattering is insensitive to the gap structure around 45° ; in contrast, the $k_x = 0$ and $k_y = 0$ regions do contribute, giving weight to the gap Δ_0 in these directions. Conversely, in the B_{2g} case, the Raman vertex γ_k is zero by symmetry for $k_x = 0$ or $k_y = 0$ and non zero elsewhere, hence provides weight in the $k_x = k_y$ direction. The Raman vertex

303

component in B_{1g} and B_{2g} symmetries act as a mask and probe selected parts of the Fermi surface shown in Fig.9. In the d_{x2-y2} pairing state, these considerations imply that the B_{1g} symmetry is insensitive to the nodes at 45° (hence the ω^3 dependence) and displays a maximum at $2\Delta_0$, whereas the B_{2g} symmetry exhibits a linear frequency dependence (because it probes the nodes) and a smeared gap (see Figs.7 and 9). An inescapable consequence of the linear observed frequency dependence is the presence of a finite density of states with the same dependence, most naturally due to nodes in the gap. Since these nodes are probed in both B_{1g} and B_{2g} symmetries, they cannot be located (only) in the 45° direction.

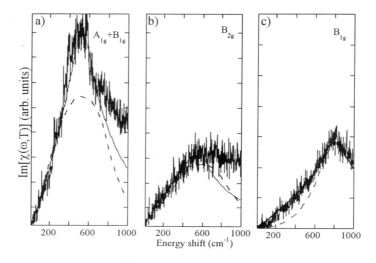

Figure 8. From left to right : $A_{1g}+B_{1g}$, B_{1g} and B_{2g} data and calculations for the d_{x2-y2} model (dashed line) and for a model with 8 nodes lying at $\theta_0=25°$ (see text).

A source of complication of the analysis comes from the dependence we have seen above (in section 2) of the Raman vertex on ω_L and ω_S. The dependence on ω_L is clear in our experiments since we see the strength of the Raman scattering vary with the laser frequency. The dependence on ω_S translates into a dependence on the Raman shift $\omega=\omega_L-\omega_S$. So we have to take into account that the Raman vertex depends also on ω. However, because this dependence comes essentially from interband transitions, we expect the typical energy scale for this variation to be of the order of 1 eV. Since the Raman shift is small on this scale, we can expand the Raman vertex (we omit the k dependence of γ_k for simplicity) to vary with the laser frequency as

$$\gamma(\omega_L,\omega) = \gamma(\omega_L,0) + \omega\,\gamma'(\omega_L,0) \quad \text{with} \quad \gamma'(\omega_L,0) = \left.\frac{\partial\gamma(\omega_L,\omega)}{\partial\omega}\right|_{\omega=0}$$

Therefore in the limit $\omega \to 0$, we can treat the Raman vertex as $\gamma(\omega_L, 0)$ that is independent of ω, which validates our analysis of the low frequency behaviour. On the other hand, for say $\omega \approx 1000$ cm^{-1}, it is not clear that the frequency dependence of γ can be neglected, which may explain the progressive change in the spectral shape of our results (Fig.6) as a function of ω_L.

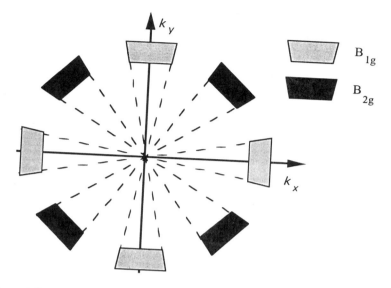

Figure 9. Fermi surface Area probed by the B_{1g} and B_{2g} components of the Raman vertex.

It could be argued that impurities are responsible for the observed low frequency density of states in the B_{1g} channel. Devereaux [25] has shown that, for a d_{x2-y2} gap, impurities induce for B_{1g} a linear rise of the electronic scattering at low frequency, crossing over to the ω^3 dependence at higher frequency. However the linear behavior in B_{1g} extends in our experiment up to 500 cm^{-1}, yielding a cross-over energy $\omega^* \sim (\Gamma\Delta)^{1/2}$ of order 500 cm^{-1}. In the most favorable case of unitary scatterers [25] , this would imply a scattering rate Γ of order of the gap Δ itself. For generic anisotropic scatterers such a high scattering rate should strongly reduce the critical temperature, in contradiction with $T_c = 126$ K of our single crystals. Only the consideration of very specific scatterers could allow to escape this problem. Therefore this explanation looks rather unlikely. We believe that our data are representative of a pure material and that the linear contribution found in the B_{1g} spectrum is intrinsic.

We are thus left with the conclusion that the nodes are shifted from the 45° direction. Hg-1223 being tetragonal, we cannot ascribe this to an orthorhombic distorsion as could be the case in Y-123. We therefore explore this shift more quantitatively. As a first approach, we take for the Raman vertices the simplest form compatible with symmetry $\gamma_{B1g} \sim \cos 2\theta$,

$\gamma_{B2g} \sim \sin 2\theta$ [21] and a cylindrical Fermi surface. We start by using a "toy model" where we compute the B_{1g} and B_{2g} spectra for an order parameter $\Delta(\theta) = \Delta_0 \cos(2\theta - 2\alpha)$ obtained by artificially rotating the $d_{x^2-y^2}$ one by an angle α. This is intended to check the sensitivity of the Raman spectra to the nodes location (note that a 22.5° rotation would naturally make B_{1g} and B_{2g} identical). We find that a rotation by $\alpha \approx 10°$ is barely noticeable in the spectra : the linear rise at low ω produced in B_{1g} by this rotation is comparable to the experimental accuracy. This suggests that the shift of the nodes away from 45° is quite sizeable.

Because of the tetragonal symmetry, one node located at θ_0 ($\neq 45°$ or $0°$) implies 8 nodes at $\pm\theta_0 + n\pi/2$ (n = 0,1,2,3) if we assume that tetragonal symmetry is not broken by the superconducting order. A very simple model corresponding to this situation is $\Delta(\theta) = \Delta_0 [\cos(4\theta)+s]$ with $0 < s < 1$, an order parameter which has the A_{1g} symmetry [26]. The gap has a maximum $\Delta_0(1+s)$ for $\theta=0$, seen in B_{1g}, and a secondary maximum $\Delta_0(1-s)$ for $\theta = \pi/4$, seen in B_{2g}, while the node lies at $\theta_0 = (1/4) \arccos(-s)$. This model could also be loosely called a g-wave model or a (super) extended s-wave model (in contrast to the d-wave model, which has the B_{1g} symmetry). With one more parameter we might decouple the position of the nodes and the size of the gap maxima. We choose the parameter s of our model to account for the relative peak position in B_{1g} and B_{2g} (inasmuch as we consider that B_{2g} has a very broad peak around 500 cm^{-1}). It can be seen in Fig.8 that B_{1g} and B_{2g} spectra calculated in this framework for s=0.2 and $\theta_0 = 25°$ are in good agreement with experiment for the low frequency behavior. In order to remove the singularities at the gap maxima, a smearing function with width proportional to frequency [25] has been incorporated in the calculation. We do not deal here with the finite electronic scattering in the normal state, which should be also accounted for, but which is beyond the scope of this work. This scattering is responsible for the discrepancy at high frequency, most conspicuous in B_{2g}, between our calculation and the experimental data.

Our model explains the different positions of the B_{1g} and B_{2g} peaks (also found in Bi-2212) by the gap anisotropy. This leads us to try and connect the position of the peak found in A_{1g} symmetry to the other peaks. In our calculations screening is fully taken into account. As noted above, because the A_{1g} symmetry is screened, its interpretation is more complicated and cannot be related directly to $\Delta(\theta)$. Since a constant vertex is completely screened, the simplest model giving a nonzero result for A_{1g} is $\gamma_{A1g} \sim \cos 4\theta$ for the A_{1g} Raman vertex. However, in our simple one parameter model, the position of the nodes is imposed by the relative peak position in B_{1g} and B_{2g}. In our case this leads to the simple $\gamma_{A1g} \sim \cos 4\theta$ being essentially zero at the nodes, and a very small low ω linear contribution in contrast with experiment. We have therefore included higher harmonics and made the calculation shown in Fig.8 with $\gamma_{A1g} \sim \cos 4\theta - 3 \cos 8\theta$. The agreement is quite satisfactory both with respect to the low ω behaviour and the peak position. However it is clear that this could even be improved in a more complicated two parameter model for $\Delta(\theta)$, decoupling the node position and the gap maxima, which could also include a θ dependence for the density of states.

306

7. CONCLUSION

In conclusion, we have presented for the first time pure electronic Raman spectra of Hg-1223 single crystals. Our most significant result is the observation of two gap maxima at 530 and 800 cm^{-1} in the A_{1g} and B_{1g} symmetry respectively as well as an intrinsic linear ω dependence - not only in the B_{2g} spectrum - but also in the B_{1g} spectrum. These experimental results clearly show evidence of a strong anisotropic superconducting gap with the existence of nodes. We have obtained indications on the location of the nodes, which were not reported sofar in other compounds, presumably because the experimental determination of the low frequency regime is usually hampered by phonons. Our observations advocate in favor of nodes existing outside the [110] and [1$\bar{1}$0] directions, which is inconsistent with the simple $d_{x^2-y^2}$ model.

8. ACKNOWLEDGEMENTS

We thank N.Bontemps, P.Monod, C.Müller, V.Viallet, D.Colson and M. Cyrot for very fruitful discussions.

REFERENCES

[1] D.A. Wollman et al, Phys. Rev. Lett. 74, 797 (1995), C.C. Tsuei et al. Nature 387, 481 (1997).

[2] J.R. Kirtley et al, Phys. Rev. Lett. 76, 1336 (1996).

[3] M. V. Klein and S. B. Dierker, Phys.Rev. B 29, 4976 (1984).

[4] S. B. Dierker, M. V. Klein, G. W. Webb and Z. Fisk, Phys. Rev. Lett. 50, 853 (1983).

[5] T. P. Deveraux et al, Phys. Rev. Lett. 72, 396, (1994).

[6] T.P. Devereaux and D. Einzel, Phys. Rev. B 51, 16336 (1995)

[7] S. L. Cooper, M. V. Klein, B. G. Pazol, J. P. Rice and D. M. Ginsberg, Phys. Rev. B 37, 5920 (1988); S. L. Cooper et al, Phys. Rev. B 38, 11934, (1988).

[8] R. Hackl, W. Gläser, P. Müller, D. Einzel and K. Andres, Phys. Rev. B 38, 7133 (1988).

[9] T. Staufer, R. Nemetschek, R. Hackl, P. Müller and H. Veith, Phys.Rev.Lett. 68, 1069 (1992).

[10] D. H. Leach, C. Thomsen, M. Cardona, L. Mihaly and C.Kendziora Solid. Stat. Comm. 88, 457 (1993).

[11] R.Nemetschek, O.V.Misochko, B.Stadlober and R.Hackl, Phys. Rev. B 47, 3450 (1993); L.V.Gasparov, P.Lemmens, M.Brinkmann, N.N.Kolesnikov and G.Güntherodt, unpublished.

[12] X.K.Chen, J.C.Irwin, H.J.Trodhal, T.Kimura and K.Kishio, Phys. Rev. Lett. 73, 3290 (1994).

[13] M.Krantz and M.Cardona, J.Low Temp. Phys.99, 205 (1995).

[14] M.Cantoni, A.Schilling, H.-U.Nisen and H.R. Ott, Physica C 215, 11 (1993).

[15] T.Strohm and M.Cardona, cond-mat/9609143.

[16] A. Sacuto et al, Physica C 259, 209 (1996).

[17] A. Sacuto, R. Combescot, N. Bontemps, P. Monod, V. Viallet and D. Colson, Europhys. Lett. 39, 207 (1997) ; proceeding of M2S-HTSC-V, Feb. 28-Mar.4, (1997) Beijing China, (1997).

[18] R.Nemetschek, O.V.Misochko, B.Stadlober and R.Hackl, Phys.Rev.B 47, 3450 (1997).

[19] Moonsoo Kang, G. Blumberg, M.V. Klein and N.N.Kolesnikov, Phys. Rev. Lett. 77, 4434 (1996).

[20] D. Colson, A. Bertinotti, J. Hammann, J.-F. Marucco, A. Pinatel, Physica C 233, 231 (1994); A. Bertinotti et al, Physica C 250, 213 (1995) ; « Studies of High Temperature Superconductors », A. Bertinotti, D. Colson, J-F. Marucco, V.Viallet, J. Le Bras, L.Fruchter, C. Marssenat, A.Carington, J.Hammann. Ed. by Narlikar, Nova Science Publisher (NY) (1997)

[21] T. P. Devereaux, Journal.of Superconductivity 8, 421 (1995); T. P. Devereaux et al., Phys. Rev. Lett. 72, 3291 (1994).

[22] A. Sacuto, C. Julien, V. A. Shchukin, C. Perrin and M. Mokhtari, Phys. Rev. B 52, 7619 (1995).

[23] Xingjiang Zhou, M.Cardona, C.W.Chu, Q. M. Lin, S.M.Loureiro and M.Marezio Phys. Rev. B 54, 6137 (1996).

[24] The 580 cm^{-1} peak shows a phonon asymmetric lineshape with an antiresonance on the high energy side (see Fig.4-a). This is characterisitic of the interference between an electronic continuum and a single phonon state. This Fano lineshape has already been detected in Y-123 and Bi-2212 for the B_{1g} normal mode in the vicinity of the gap structure maximum [7,10].

[25] T.P. Devereaux, Phys. Rev. Lett. 74, 4313 (1995).

[26] An order parameter with B_{1g} symmetry (changing sign under $\pi/2$ rotation) would imply 12 nodes.

EVIDENCE FOR GAP ASYMMETRY AND SPIN FLUCTUATIONS FROM NUCLEAR MAGNETIC RESONANCE (NMR)

S. Krämer and M. Mehring

2. Physikalisches Institut
Universität Stuttgart
D-70550 Stuttgart
Germany

INTRODUCTION

Nuclear Magnetic Resonance (NMR)[1-3] has been applied to conducting[4] and superconducting[5] solids from the beginning of its discovery about fifty years ago. It is therefore of no surprise that soon after the discovery of high-T_c superconductors by Bednorz and Müller[6] NMR played a key role in proving the significance of hole doping in the copper-oxygen plane for the normal state as well as for the superconducting properties of these materials. There are a number of abundant nuclei in the cuprate superconductors with different nuclear spins for NMR like ^{63,65}Cu (spin $I = 3/2$; quadrupole moment $Q = -0.222\ 10^{-24}\ \mathrm{cm}^2$, $Q = -0.195\ 10^{-24}\ \mathrm{cm}^2$), ^{89}Y ($I = 1/2$), ^{139}La ($I = 7/2$; $Q = 0.20\ 10^{-24}\ \mathrm{cm}^2$), ^{137}Ba ($I = 3/2$; $Q = 0.34\ 10^{-24}\ \mathrm{cm}^2$), ^{203,205}Tl ($I = 1/2$), ^{199}Hg ($I = 1/2$) and others which can be isotopically enriched like ^{17}O ($I = 5/2$; $Q = -0.026\ 10^{-24}\ \mathrm{cm}^2$). Because of the large quadrupole moments of ^{63,65}Cu and their non-cubic surrounding in the cuprate materials, their zero-field quadrupolar splitting corresponds to several megahertz (MHz) which lends itself to the rather simple experimental technique of pure Nuclear Quadrupole Resonance (NQR) in zero field. The same holds for ^{139}La, but not for ^{17}O because of its modest nuclear quadrupole interaction in the cuprates. All other nuclei are best investigated by NMR in high magnetic fields.

This contribution is not meant as a review of NMR/NQR in cuprate superconductors, but should be considered as a short presentation of some of the essential aspects NMR/NQR can contribute to the deeper understanding of correlations and gap symmetry in these materials. We therefore refer the interested reader to some of the earlier reviews[7-13], the special issue of Applied Magnetic Resonance on high-T_c superconductors[14] as well as more recent review articles[15-17] which summarize the essential observations. For introductory purposes we repeat some of the basics of reference 10.

We assume that the reader is familiar with the phase diagram of the cuprates which displays the reduction of the Néel temperature T_N of the antiferromagnetic parent compounds with increased hole doping until conductivity and superconductivity sets in at a hole concentration of about 0.02 per CuO_2 unit. The superconducting transition temperature increases with further hole doping (underdoping regime), reaches a maximum at about 0.15 per CuO_2 unit(optimum doping) and decreases with further doping (overdoping regime) until a non-superconducting metallic state is reached. The physical properties of the material

The Gap Symmetry and Fluctuations in High - T_c Superconductors
Edited by Bok *et al.*, Plenum Press, New York, 1998

change drastically in the different regimes and display a wealth of new physical phenomena like anti-ferromagnetic behaviour, spin-glass behaviour, highly correlated Fermi liquid (underdoping regime) and more weakly correlated Fermi liquid (overdoping regime) behaviour. The magnetic resonance properties change correspondingly in the different regimes.

The nuclei sense the different electronic states via their hyperfine interaction with the conduction electrons. This results in lineshifts (Knight shift) and relaxation rates (T_1^{-1}, T_2^{-1}).

PRINCIPLES OF NMR/NQR IN METALS AND SUPERCONDUCTORS

In this section we briefly summarize the consequences of the hyperfine coupling of the different nuclei to conduction electrons in metals and superconductors as far as the magnetic resonance properties Knight shift K and relaxation rates (T_1^{-1}, T_2^{-1}) are concerned. Although NQR techniques are applied in order to observe the magnetic resonance signal, we will not discuss here static or dynamic aspects of nuclear quadrupole interactions with conduction electrons, but will restrict ourselves to purely magnetic interactions. We will also not dwell on local magnetic fields caused by the vortex state of superconductors in magnetic fields.

Hyperfine interaction

Nuclear spin transitions between Zeeman levels in large magnetic fields B_0 lead to a single NMR line at the Larmor frequency $\omega_0 = \gamma_n B_0$, where the nuclear gyromagnetic ratio γ_n differs for different nuclei. If the orbital contribution of the surrounding electrons is taken into account, the resonance line is shifted according to the chemical shift Hamiltonian

$$\mathcal{H}_{cs} = \hbar\gamma_n \, \widehat{\boldsymbol{I}} \cdot \mathbf{S} \cdot \boldsymbol{B} \tag{1}$$

represented by the shift tensor $\mathbf{S}$. This Hamiltonian summarizes all orbital contributions including the van Vleck paramagnetic shift. Because of the second rank tensor $\mathbf{S}$ this orbital lineshift depends on the orientation of the principal axis with respect to the external field B_0. The chemical or orbital shift is non-zero in all conducting and non-conducting solids. In conducting solids an additional shift, the so-called Knight shift, named after its discoverer[18], appears which is the result of the spin-dependent paramagnetic hyperfine coupling Hamiltonian

$$\mathcal{H}_{hfc} = \hbar\gamma_n \sum_j \widehat{\boldsymbol{I}} \cdot \mathbf{A}_j \cdot \widehat{\boldsymbol{S}}_j \tag{2}$$

of the nuclei to the electron spins $\widehat{\boldsymbol{S}}_j$. The hyperfine tensor $\mathbf{A}$, defined here as a magnetic field (in Tesla units) at the nuclear site, contains both the isotropic Fermi contact term as well as the dipolar and spin-orbit interaction between electrons and nuclei. In cubic solids the hyperfine interaction is isotropic due to symmetry, but it is usually anisotropic in non-cubic solids like in the cuprate superconductors. This implies that the corresponding lineshift depends on the orientation of the external magnetic field with respect to the principal axis of this tensor. Both, the chemical shift and the hyperfine tensor often have the same principal axes as dictated by the local symmetry. We have reserved here the label *Knight shift* solely for the spin-dependent part which reflects the metallic behaviour, whereas in the solid state literature often all contributions to the NMR shift are called Knight shift.

Knight- and chemical shift

NMR and NQR spectra are today usually measured by fast Fourier transform (FFT) of either the timedomain response after a single pulse (free induction decay: FID) or Hahn spin echo (two pulse sequence). The lineshift is quoted with respect to a reference compound of known shift. The shift difference is normalized by the Larmor frequency, leading to a relative

310

shift δ which is usually quoted in ppm (10^{-6}) or for larger shifts in % (10^{-2}). The total shift δ contains contributions of the chemical shift δ_c and the Knight shift $K_s(T)$ according to

$$\delta(T) = \delta_c + K_s(T) + \frac{\boldsymbol{B_0}}{B_0^2} \cdot (1 - \boldsymbol{N}) \cdot \boldsymbol{M}(T) \tag{3}$$

where the susceptibility shift, caused by the bulk magnetization $\boldsymbol{M}(T)$ of the sample in combination with the demagnetizing factor $\boldsymbol{N}$, has also been included here. The Knight shift can be expressed as a product of the hyperfine interaction and the paramagnetic spin-susceptibility χ_s of the electrons as

$$K_{s,\alpha\alpha} = \frac{1}{g\mu_B} \sum_j A_{j,\alpha\alpha} \chi_s \quad \text{with} \quad \chi_s = \frac{\chi^{(molar)}}{\nu N_A \mu_0} \tag{4}$$

where summation over all hyperfine contributions, on-site, nearest neighbour and perhaps next-nearest neighbour sites, is performed. ν is the number of CuO_2-units in the unit-cell, N_A is Avogadros number and all other parameters have their usual meaning. The spin-susceptibility

$$\chi_s = g^2 \mu_B{}^2 \chi'(q = 0) \tag{5}$$

is defined through the real part of the static spin-susceptibility $\chi'(q = 0)$ which can be expressed within random phase approximation by

$$\chi'(q = 0) = \frac{1}{4} \frac{D(E_F)}{1 - \frac{1}{2}JD(E_F)} \tag{6}$$

where $D(E_F)$ is the density of states per CuO_2-unit and J is the exchange coupling constant of the electrons. Note that $\chi'(q = 0)$ includes a factor $S(S + 1)/3 = 1/4$ (for S=1/2) in our definition.

The orbital shifts depend predominantly on the filled valence bands and are only weakly or not at all temperature dependent. In a classical metal the Knight shift is as well temperature independent due to the temperature independent Pauli susceptibility. In correlated metals, however, the spin-susceptibility often changes with temperature and correspondingly the Knight shift. This allows to separate the spin dependent paramagnetic contribution to the total shift from the orbital shifts (Jaccarino-Clogston)[19, 20].

In particular in the superconducting state, where the superconducting gap opens below T_c, the paramagnetic susceptibility and therefore the Knight shift drops monotonically to zero, unless there is residual spin scattering at low temperatures. In the following analysis we follow the common assumption that the spin-paramagnetic Knight shift K_s equals zero at $T = 0$. However, if there are impurities in the material as for example in Zn-doped $YBa_2Cu_3O_7$, K_s does not vanish at $T = 0$ but approaches a constant value.

The functional form of the Knight shift decrease below T_c (Yosida function)[21] depends on the symmetry of the gap function. Grossly speaking K_s decreases rapidly below T_c and increases from $T = 0$ with an exponential law for an isotropic gap function(s-wave pairing). For an anisotropic gap function as for example anisotropic s-wave, antiferromagnetic s-wave(AFS)[22, 23] or d-wave, however, K_s increases from $T = 0$ proportional to T. In order to provide an intuitive understanding of these different behaviours we present in fig.1 the two different density of states near the Fermi energy E_F and the corresponding Knight shifts for s-wave and d-wave pairing. Note that thermal excitation across the gap with Boltzmann factor $\exp(-\Delta/k_BT)$ is required for s-wave pairing, whereas for d-wave pairing a distribution of energies is available for excitation leading to a linear increase of the Knight shift with T.

The following expression is a Yosida function result for the Knight shift in the superconducting state.

$$K_s = K_n 2 \int_{\Delta(T)}^{\infty} \frac{N_s(\epsilon)}{N_n(0)} \left(-\frac{\partial f(\epsilon)}{\partial \epsilon} \right) d\epsilon \quad \text{with} \quad \epsilon = E - E_f \tag{7}$$

311

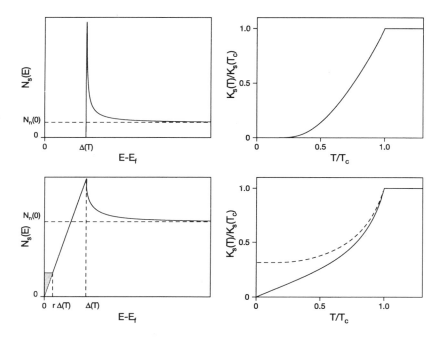

Figure 1. Density of states (left) and the corresponding normalized Knight shift (right) for different gap symmetries. Top: due to the isotropic gap function there are no states for $|\epsilon| < \Delta$. An exponential increase of K_s with T at low temperatures is observed. Bottom: anisotropic gap function with nodes. $N_s(\epsilon)$ is non-zero for $|\epsilon| < \Delta$ and K_s increases linearly with T at low temperatures. The dashed line of K_s corresponds with the shaded region in the density of states and displays the influence of impurities. See text for details.

$f(\epsilon)$ is the Fermi distribution function. $N_s(\epsilon)$ is the superconducting density of states which depends on the symmetry of the superconducting gap function. The latter is given for s-wave pairing by

$$\Delta(T) = \Delta_0 \tanh\left(\alpha\sqrt{\frac{T_c}{T} - 1}\right) \tag{8}$$

and for d-wave pairing by

$$\Delta_{\boldsymbol{k}}(T) = \frac{\Delta(T)}{2}\left(\cos k_x - \cos k_y\right) \tag{9}$$

where α is an adjustable parameter of order unity. The temperature dependence of the Knight shift differs significantly in particular at low temperatures as is shown in fig.1. We will come back to this point in the section on the superconducting state. However, one remark concerning the interpretation of magnetic shift data should be given here. For an extraction of the Knight shift from experimental data the other contributions have to be subtracted. The results for anisotropic pairing change dramatically at low temperatures at the presence of impurities (see fig. 1, bottom, dashed line): K_s does not longer increase linearly at low temperatures but shows a behaviour which resembles the s-wave case apart from its value at $T = 0$. Therefore a distinction between the two different pairing symmetries depends on a separate knowledge of all the contributions to the magnetic shift.

Spin-lattice relaxation

The nuclear spin levels approach the Boltzmann equilibrium with the spin-lattice relaxation rate T_1^{-1}. The recovery of the z-magnetization of the nuclear spins is measured a time t

after a perturbing rf-pulse by application of a $\pi/2$-pulse or a spin-echo sequence in order to transform the z-magnetization into measurable x,y-magnetization. Two different sequences are in use, namely the (i) $\pi - t - \pi/2$ (inversion recovery) and the (ii) $\pi/2 - t - \pi/2$ (saturation recovery) sequence, where in the latter sequence the initial $\pi/2$-pulse often consists of a series of pulses repeated rapidly with respect to T_1 in order to reach saturation. In simple cases the recovery curves are exponential, that is of the functional form $(1-a_0 \exp(-t/T_1))$ with $a_0 = 2$ in case (i) and $a_0 = 1$ in case (ii). In metallic and magnetic systems the general expression for the spin-lattice relaxation rate $T_{1,z}^{-1}$ goes back to Moriya[24]

$$\frac{1}{T_{1,z}} = \frac{1}{2}\hbar k_B T \gamma_n^2 (A_{xx}^2 + A_{yy}^2) I_{K,z} \quad \text{with} \quad I_{K,z} = \sum_q F_z(q) \frac{\Im m \chi^\pm(q,\omega_0)}{\hbar\omega_0}$$

$$\text{and} \quad F_z(q) = \frac{A_{xx}(q)A_{xx}(-q) + A_{yy}(q)A_{yy}(-q)}{A_{xx}^2 + A_{yy}^2}$$

$$\text{where} \quad A_{\beta\beta}(q) = \sum_j A_{j,\beta\beta} \exp(-iq r_j) \tag{10}$$

are the components of the hyperfine tensor(in Tesla units) perpendicular to the orientation of the z-axis of quantization, i.e. the magnetic field axis in NMR or the prominent electric field gradient axis in NQR.

Note that $\chi^\pm(q,\omega_0)$ is related to the electron spin correlation function $\langle S_-(0,0)S_+(r_j,t)\rangle$ by the fluctuation dissipation theorem, i.e. a factor of $2S(S+1)/3 = 1/2$ (for S=1/2) must be considered when calculating $\chi^\pm(q,\omega_0)$. We also note that ω_0 is essentially the Larmor frequency ω_e of the electron spins under the assumption that $\omega_n \ll \omega_e$[10]. Both frequencies are, however, rather small compared with the electronic excitation bandwidth so that the usual assumption of $\lim_{\omega_0 \to 0}$ is justified in most situations for the cuprate superconductors. The expression for the relaxation time T_1 quoted here is not what is usually measured in cuprate superconductors for nuclei with spin $I \neq 1/2$, because either selective excitation of the central $(-1/2 \to 1/2)$ transition is performed (NMR case) or the zero field transition $(\pm 1/2 \to \pm 3/2)$ (NQR case) is excited. In the NMR case the relaxation recovery curve is non-exponential with $S(t) = 1 - a_0((9/10)\exp(-6t/T_1) + (1/10)\exp(-t/T_1))$ and in the NQR case $S(t) = 1 - a_0 \exp(-3t/T_1)$[25].

In the case of isotropic hyperfine interaction the expressions reduce to the standard form with $A_{xx} = A_{yy} = A_{iso}$. In cases where the isotropic hyperfine interaction A_{iso} does not vanish it might be convenient to make the form factor $F_z(q)$ unit-free still by normalizing it by A_{iso}^2 as done here. In the rare cases where $A_{iso} = 0$ one can normalize $F_z(q)$ by another hyperfine component. We follow here the notation used earlier with a normalized formfactor which turned out to be convenient when comparing the experimentally determined quantity I_K for different superconductors[10].

For isotropic hyperfine interaction A_{iso} is directly connected with the isotropic Knight shift K_{iso} and a type of *sum-rule* can be formulated[26, 27] the Knight shift K_{iso} and the spin-lattice relaxation time T_1 measured at temperature T in the form

$$K_{iso}^2 T_1 T C_0 S_K = 1 \quad \text{with} \quad S_K = \frac{I_K}{4\pi\chi'(0)^2} \quad \text{and} \quad C_0 = 4\pi \frac{k_B}{\hbar}\left(\frac{\gamma_n}{\gamma_e}\right)^2 \tag{11}$$

where S_K is a unit-free scaling parameter which can be determined experimentally from the measured values of the isotropic Knight shift and the spin-lattice relaxation time and the known parameter C_0. We note that $S_K = 1$ for the 3D free electron gas. In this case the spin-lattice relaxation rate T_1^{-1} according to eq.(10) is proportional to T and K_{iso}^2 which again is proportional to the square of the density of states at the Fermi level. This type of behaviour is called Korringa relation after Korringa who first derived it[28]. In general $S_K \neq 1$ and K_{iso} might not be temperature independent. This leads to a breakdown of this simple form of the Korringa relation. In particular in low-dimensional systems[26] and correlated systems[10, 29] the parameter S_K can reach rather large values indicative of large fluctuations. Although anisotropic hyperfine interaction has been taken into consideration by multiplying the left hand side of eq.(11) by a factor $(1+\epsilon)$, where $\epsilon = A_D/A_{iso}$ represents the traceless anisotropic part of the hyperfine interaction[29] we present here a modified S_K sum-rule more relevant for oriented samples with the magnetic field oriented along the z-axis:

$$\frac{1}{2}(K_{xx}^2 + K_{yy}^2)T_1 T C_0 S_{K,z} = 1 \quad \text{with} \quad S_{K,z} = \sum_q F_z(q)\frac{\Im m\chi^\pm(q,\omega_0)}{\hbar\omega_0\, 4\pi\chi'(0)^2} \tag{12}$$

where the $F_z(q)$ are given according to eq.(10) and C_0 is given by eq.(11). In the case of only local (on-site) hyperfine interactions the form factor becomes q-independent and $F_z = 1$. If the $S_{K,z}$ is unfolded from the $F_z(q)$ factor $S_{K,z}$ finally contains only the ratio of the imaginary part of the dynamic susceptibility to the square of the static susceptibility. It provides therefore a numerical measure of the ratio of fluctuations (q-dependent dynamic susceptibility) to the static susceptibility i.e. of warping in q-space.

For completeness we present here also the expression relevant for the orbital contribution to the spin-lattice relaxation, which is considered not to be extremely relevant for the cuprate superconductors (see, however, C. M. Varma, *Phys. Rev. Lett.*, 77: 3431 (1996)).

$$\frac{1}{T_1} = \hbar k_B T\gamma_n^2 A_{orb}^2 I_{orb} \quad \text{with} \quad I_{orb} = \sum_q \frac{\Im m\chi_{orb}^\pm(q,\omega_0)}{\hbar\omega_0} \quad \text{and} \quad A_{orb} = 4\,g\,\mu_B\langle r^{-3}\rangle \tag{13}$$

Finally we want to discuss the standard behaviour of T_1^{-1} in the superconducting phase for a BCS type superconductor. In case of s-wave superconductivity the expression for T_1^{-1} is rather simple and includes the isotropic superconducting gap energy 2Δ

$$T_{1s}^{-1} = \frac{\pi\gamma_n^2}{\hbar}\int_\Delta^\infty \frac{1}{2}(A_{xx}^2 + A_{yy}^2)[N_s(\epsilon)^2 + M_s(\epsilon)^2]f(\epsilon)[1 - f(\epsilon)]\mathrm{d}\epsilon \tag{14}$$

where A_{xx} and A_{yy} are the perpendicular components of the the hyperfine interaction tensor and $N_s(\epsilon)$ and $M_s(\epsilon)$ are given by

$$N_s(\epsilon) = \frac{N(E_F)E}{\sqrt{\epsilon^2 - \Delta^2}} \quad \text{and} \quad M_s(\epsilon) = \frac{N(E_F)\Delta}{\sqrt{\epsilon^2 - \Delta^2}} \tag{15}$$

where $f(\epsilon)$ is the Fermi distribution function and $N(E_F)$ the density of states at the Fermi energy. The typical temperature dependence of T_{1s}^{-1} according to eq.(14) is displayed in

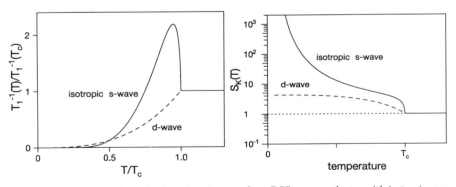

Figure 2. Left: Normalized spin-lattice relaxation rate for a BCS superconductor with isotropic gap function. The increase of the relaxation rate with decreasing temperature just below T_c is a result of the coherence factor in the BCS theory. The dashed line better describes the situation observed in high-T_c superconductors. At low temperature the rate follows a T^3 power law. Right: S_K in the superconducting state for a BCS like superconductor and a d-wave superconductor.

fig.2. Note the so-called Hebel-Slichter or coherence peak right below T_c which is a result of the coherence factor $M_s(\epsilon)$ in the BCS theory. Its observation was a hall-mark of the BCS theory[30]. Hebel and Slichter[31] as well as Redfield[32] and others observed such a behaviour in aluminum and other BCS superconductors. We note right here that such a Hebel-Slichter peak was never observed in the cuprate superconductors. On the right hand side of fig.2 we show the temperature dependence of the parameter S_K for a BCS like superconductor and a d-wave superconductor. In the first case S_K diverges for $T \to 0$, whereas in the latter case it remains finite.

Spin-spin relaxation

There is another spin relaxation rate, namely T_2^{-1} which reflects the phase fluctuations of the nuclear spins. The nuclear spin ensemble is first brought into a coherent superposition of states by a $\pi/2$-pulse, followed after a time τ by a π-pulse resulting in a Hahn-spin-echo at time $t = 2\tau$. Experimentally one observes that the spin echo decays in liquid samples exponentially as $\exp(-t/T_2)$ due to spin-lattice relaxation and phase fluctuations. In solids the decay is in general not exponential in particular if homonuclear spin-spin interactions like dipole-dipole or indirect-dipole interactions are involved. The spin-echo decay function $f(t = 2\tau)$ can often be expressed in the following form

$$f(t = 2\tau) = g_D(t)\, e^{-\frac{t}{T_{2R}}} \tag{16}$$

where the first factor represents the slowly fluctuating part which is usually non-exponential, whereas the second one, the Redfield spin-lattice relaxation, takes account of the fast fluctuations which also governs the spin-lattice relaxation time T_1. For a *flat* spectral density(rapid fluctuation limit) $1/T_{2R}$ is related to $1/T_1$ by the following relations:

$$\frac{1}{T_{2R}} = (\beta + R)\frac{1}{T_{1c}} \quad \text{with} \quad R = \frac{T_{1c}}{T_{1ab}} \tag{17}$$

and where $\beta = 2$ in the NQR case and $\beta = 3$ in the NMR case for $B_0 \parallel$ c-axis[25]. If $B_0 \perp$ c-axis the relation $1/T_{2R} = 1/2\, T_{1c} + 7/2\, T_{1ab}$ holds.

The first factor could include different slow motion contributions like atomic diffusion, site exchange, heteronuclear spin fluctuations and homonuclear spin-spin interaction, where some of these mechanisms lead to complicated non-exponential decays. The last two mechanism are indeed relevant for the cuprate superconductors.

If all other fast and slow motion contributions to the spin echo decay have been eliminated, one is left with $g_D(t)$ caused by homonuclear spin-spin interactions. In general it is not possible to calculate $g_D(t)$ rigorously. We note in passing that for the spin-spin Hamiltonian applied by Slichter and co-workers[33] in fact a rigorous expression for $g_D(t)$ can, after some algebra, indeed be obtained which has the same form as the free induction decay for heteronuclear spin-spin interactions as given by Mehring (p. 36, eq.(2.101a))[2]. Details will be published elsewhere. Van Vleck[34] has shown how to perform a moment expansion and calculated the first few moments rigorously when the interaction parameters are known. Usually one restricts oneself to the leading moment which is the second moment $M_2 = \langle \omega^2 \rangle$. Moreover one might assume a proper statistical distribution of spin orientations and interactions leading to a Gaussian decay function $g_D(t) = \exp[-t^2/(2M_2)]$ which can be expressed as $\exp[-t^2/(2T_{2G}^2)]$ where the Gaussian spin-spin relaxation rate $T_{2G}^{-1} = \sqrt{M_2}$ can be obtained via a second moment calculation. This can be performed rigorously, if the nuclear positions are known, for the dipole-dipole interaction, which is independent of temperature. The calculation of the indirect spin-spin coupling, mediated via the hyperfine interaction of the electrons, requires a calculation of the real part of the spin susceptibility $\chi'(q)$ as was pointed out by Slichter and co-workers[33,35] and Thelen and Pines[36]. The following relation is obtained for T_{2G}^{-1} for ^{63}Cu(2) in YBa$_2$Cu$_3$O$_7$

$$\left(\frac{1}{^{63}T_{2G}}\right)^2 = 0.69 \frac{A_{iso}^4}{8\hbar^2} \left[\frac{1}{N}\sum_{\boldsymbol{q}} {}^{63}F_{zz}(\boldsymbol{q})^2\chi'(\boldsymbol{q})^2 - \left(\frac{1}{N}\sum_{\boldsymbol{q}} {}^{63}F_{zz}(\boldsymbol{q})\chi'(\boldsymbol{q})\right)^2\right] \quad (18)$$

where $^{63}F_{zz} = A_{zz}(\boldsymbol{q})A_{zz}(-\boldsymbol{q})/A_{iso}^2$ is defined with respect to the quantization axis (z-axis) and where the isotopic abundance 0.69 of ^{63}Cu has been included in the prefactor. We have normalized here $^{63}F_{zz}$ with respect to A_{iso} because in the special case where the z-axis corresponds to the c-axis A_{zz} is nearly zero.

NMR OF THE CuO$_2$-LAYERS IN HIGH-T$_c$ SUPERCONDUCTORS

In this section we want to discuss the normal state properties of high-T_c superconductors as seen by NMR/NQR. They are already quite unusual and show intriguing features like temperature dependent Knight shifts, non-Korringa behaviour of the spin-lattice relaxation, spin-gap behaviour etc.. In order to obtain a consistent description we need to know the form of the anisotropic hyperfine interaction of the different nuclei in or near the copper-oxygen plane. Once an appropriate set of hyperfine interaction parameters is known the static susceptibility and the the dynamic susceptibility can be determined from Knight shift and relaxation measurements as outlined in the previous section. Standard expressions for the susceptibility can be applied and compared with model calculations of the correlated spin fluid representing the normal state properties of the high T_c superconductors. This will be called the *standard model* in the following.

Hyperfine Interaction à la Shastry-Mila-Rice (SMR)

Mila and Rice[37, 38] and independently Shastry[39] proposed a model for the hyperfine interaction ^{63,65}Cu and ^{17}O in the copper-oxygen plane and ^{89}Y between the CuO$_2$ planes. A single spin fluid is assumed where the electronic wavefunction of the Cu-O hybrid orbitals are assumed to be centered on the Cu sites[40]. Applying a simple tight-binding quantum chemical reasoning, Mila and Rice were able to provide a set of hyperfine interactions both on-site and nearest neighbour site as is summarized in the so-called Shastry-Mila-Rice Hamiltonian (SMR)

$$\mathcal{H}_{smr} = {}^{63}\gamma\hbar \left(\sum_i {}^{63}\widehat{\boldsymbol{I}}_i \cdot {}^{63}\mathbf{A} \cdot \widehat{\boldsymbol{S}}_i + {}^{63}\widehat{\boldsymbol{I}}_i \cdot \sum_{\delta=nn} {}^{63}B\,\widehat{\boldsymbol{S}}_{i+\delta}\right)$$
$$+ {}^{17}\gamma\hbar \sum_{j,\delta'=nn} {}^{17}\widehat{\boldsymbol{I}}_j \cdot {}^{17}\mathbf{A} \cdot \widehat{\boldsymbol{S}}_{j+\delta'} + {}^{89}\gamma\hbar \sum_{k,\delta''=nn} {}^{89}\widehat{\boldsymbol{I}}_k \cdot {}^{89}\mathbf{A} \cdot \widehat{\boldsymbol{S}}_{k+\delta''} \quad (19)$$

This Hamiltonian could readily be extended to e.g. the Cu(1) site in the CuO chains or to Tl nuclei in the TlO layers of the Tl cuprates[10]. It basically includes isotropic and anisotropic on-site hyperfine interactions represented by the tensor $\mathbf{A}$ and nearest neighbour Cu-site isotropic interaction ^{63}B. A pictorial representation of the coupling scenario is presented in fig. 3.

From a careful analysis of the temperature dependent Knight shift and spin-lattice relaxation measurements a number of authors have estimated the hyperfine parameters for different orientations of the magnetic field. Some representative values are summarized in Table 1[10]. However, the accuracy of the absolute values is rather poor and is about 20 %.

Once the hyperfine interactions and their spatial extensions are known, the Fourier transform from real space into q-space according to eq.(10) leads to the formfactors $F(\boldsymbol{q})$ used in the expression (eq.(10)) for the spin-lattice relaxation. If only local (on-site) hyperfine interactions would be present(δ-function in real space) the formfactor would be independent of q. The spin-lattice relaxation rate T_1^{-1} will in this case represent the integral of $\Im m\chi^{\pm}(\boldsymbol{q},\omega)$

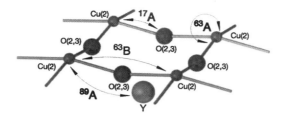

Figure 3. Standard model hyperfine couplings in the CuO_2-planes of $YBa_2Cu_3O_7$ according to the Shastry-Mila-Rice Hamiltonian (SMR).

over the whole q-space. Only if inter-site contributions to the hyperfine interactions occur, like in the SMR Hamiltonian, the formfactors assume some structure in q-space and provide a certain weighting of the different contributions of $\Im m \chi^{\pm}(\boldsymbol{q}, \omega)$ in q-space. This holds in particular for the three nuclei $^{63,65}Cu$, ^{17}O and ^{89}Y in $YBa_2Cu_3O_7$ where the formfactors are given by

$$
\begin{aligned}
^{63}Cu(2), B_0 \perp c: \quad ^{63}F_{ab}(\boldsymbol{q}) &= \frac{1}{A_c^2 + A_a^2}([A_{\parallel} + 2B(\cos q_x a + \cos q_y b)]^2 \\
&\quad + [A_{\perp} + 2B(\cos q_x a + \cos q_y b)]^2) \\
^{63}Cu(2), B_0 \parallel c: \quad ^{63}F_c(\boldsymbol{q}) &= \frac{1}{A_a^2}[A_{\parallel} + 2B(\cos q_x a + \cos q_y b)]^2 \\
^{17}O(2,3), B_0 \parallel c: \quad ^{17}F_c(\boldsymbol{q}) &= (1 + \frac{1}{2}(\cos q_x a + \cos q_y b)) \\
^{89}Y: \quad ^{89}F(\boldsymbol{q}) &= (1 + \cos q_x a)(1 + \cos q_x b)(1 + \cos q_z d)
\end{aligned}
\tag{20}
$$

d is the distance of the CuO_2-planes.

A pictorial representation of different formfactors for these nuclei for different orientations of the external field is displayed in fig. 4.

Note that the important $\boldsymbol{Q} = (\pi, \pi)$ fluctuations are weighted differently depending on the orientation of the magnetic field. We note that the formfactors for ^{17}O and ^{89}Y vanish for $\boldsymbol{Q} = (\pi, \pi)$ and have their maximum at $\boldsymbol{q} = (0,0)$. From these remarks it should be obvious that the Cu-relaxation rate $^{63}T_1^{-1}$ is sensitive to the (π, π) fluctuations, whereas the ^{17}O and ^{89}Y relaxation rates are not. They will instead be dominated by the low q fluctuations.

Table 1. Hyperfine couplings in $YBa_2Cu_3O_7$.

$YBa_2Cu_3O_7$ nuclear site	$A_{aa}[T]^a$	$A_{bb}[T]^a$	$A_{cc}[T]^b$	A_{iso} [T]	$A_{\parallel}$ [T] c	$A_{\perp}$ [T] c	B [T]
$^{63}Cu(2)$	36.8	36.8	-1.0	24.0	-33	4.8	8
$^{17}O(2,3)^d$	17.4	28.16	20.4	22.4			
^{89}Y	-3.12	-3.12	-3.12	-3.12			

[a] $A_{\perp} + 4B$ for the Cu nucleus.
[b] $A_{\parallel} + 4B$ for Cu nucleus.
[c] The subscripts $\parallel$ and $\perp$ denote the components of $\mathbf{A}$ relative to the c axis.
[d] For oxygen the Cu-O-Cu axis is parallel to the b-axis and labeled by Z, the other principal axis are labeled by X (parallel to c-axis) and Y (parallel to a-axis). For the other oxygen the system is rotated by $\pi/2$ about the X-axis.

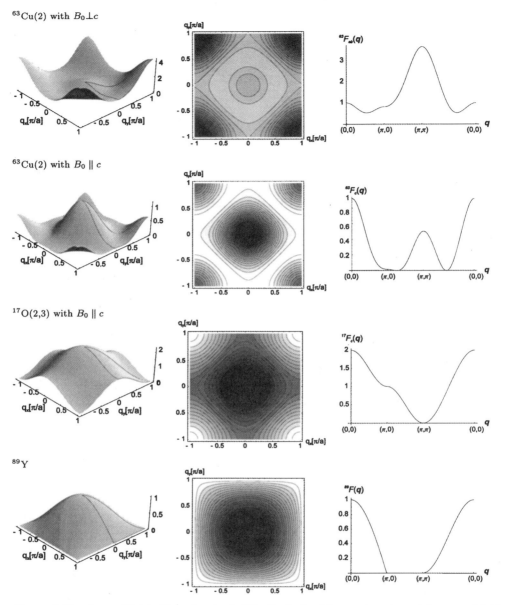

Figure 4. Formfactors $F(\mathbf{q})$ for T_1^{-1} relaxation in YBa$_2$Cu$_3$O$_7$ for different orientations of the external field B_0. From top to bottom: ^{63}Cu(2) with $B_0 \perp c$: $^{63}F_{ab}(\mathbf{q})$, ^{63}Cu(2) with $B_0 \parallel c$: $^{63}F_c(\mathbf{q})$, ^{17}O(2,3) with $B_0 \parallel c$: $^{17}F_c(\mathbf{q})$ and ^{89}Y. For each nucleus a 3D-plot, a contour plot and a trace along the path $(0,0)$-$(\pi,0)$-(π,π)-$(0,0)$ is displayed. Note that the different regions of the $\mathbf{q}$-space are weighted differently. ^{63}Cu(2) is sensitive to fluctuations at $\mathbf{Q} = (\pi, \pi)$ whereas the formfactors of ^{17}O(2,3) and ^{89}Y have a maximum at $\mathbf{q} = (0, 0)$.

Knight shifts

Since the Knight shifts are proportional to the static $\chi'(q = 0)$ spin susceptibility and the hyperfine interaction is a property of the electronic state it is assumed that the hyperfine interaction does not depend on temperature and any temperature dependence of K_s can be attributed to the spin susceptibility. Two representative examples are shown in fig. 5. We note a fairly temperature independent Knight shift above T_c in the highly doped samples, whereas

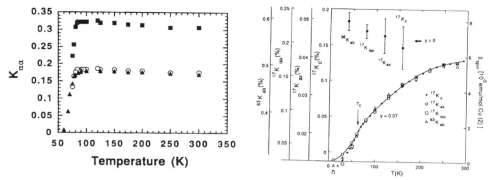

Figure 5. Left: Knight shift components of ^{17}O in $YBa_{1.92}Sr_{0.08}Cu_3O_7$ (overdoped) for different orientations of the external field: (■) K_{ZZ}, (○) K_{YY}, (▲) K_{XX}. Figure taken from reference 8. Reprinted from Appl. Mag. Res. 3, C. Berthier et al., pg. 453, 1992 with kind permission of Springer-Verlag, Wien, Austria. **Right:** Knight shift components of different nuclei at CuO_2-plane sites in $YBa_2Cu_3O_{6.63}$. The same temperature dependence of all components supports the single spin fluid model. For comparison the temperature independent data of $YBa_2Cu_3O_7$ are also marked ($y \simeq 0$). Figure taken from reference 41.

a strong temperature dependence of K_s even above T_c is observed in the underdoped samples. The characteristic temperature dependence of the Knight shift in underdoped samples was proposed to follow the temperature dependence[10, 42]

$$\chi(T) = \chi_0 \left[1 - \tanh^2 \left(\frac{E_g}{2T} \right) \right] \qquad (21)$$

for the $YBa_2Cu_4O_8$ compound and several Tl-cuprates. A slightly different expression, where the $\tanh(E_g/2T)$ appears without the square, was proposed by Takigawa[43] for underdoped $YBa_2Cu_3O_{6.6}$. Both approaches are connected with the proposal of Tranquada[44] who analyzed the spin gap behaviour observed in neutron scattering in a similar way. The appearance of a spin gap of total width $2E_g$ is one of the intriguing features of the normal state behaviour of high-T_c superconductors which is related to the unusual properties of this highly correlated Fermi liquid.

An interesting feature in underdoped cuprate superconductors is shown in fig. 5. Besides the strong temperature dependence even in the normal state, which can be represented by the spin-gap expression cited above, one observes the same behaviour for all three nuclei ^{63}Cu, ^{17}O and ^{89}Y. This has led to the conjecture of a single spin fluid. A note of caution might be appropriate here, however. The observed common temperature dependence of all three nuclei ^{63}Cu, ^{17}O and ^{89}Y intimately related to the CuO_2-subunit proves nothing but the fact that their Knight shift is dominated by the same spin-susceptibilities. This still leaves room for a three-band model with different partial susceptibilities which might be distinguishable only in the comparison of Knight shift and relaxation in particular in the high doping regime.

Another remarkable feature observed in the temperature dependence of the Knight shift K_s is an increase of K_s with decreasing temperature for overdoped compounds. This is related to the proximity of a van Hove singularity in the density of states. This was observed in our laboratory in Tl-cuprates some time ago[45]. The van Hove scenario in the cuprates was also discussed by J. Bok et al.[46, 47, 48] (see also his contribution to this volume).

Spin-lattice relaxation T_1

The discussion of the spin-lattice relaxation rate T_1^{-1} follows along the lines laid out in one of the previous sections and was discussed in detail before[10]. The essential quantity which governs the $(T_1T)^{-1}$ behaviour of the spin-lattice relaxation is I_K according to eq.(10). It reflects all the dynamical aspects of the electronic system. The different formfactors weight different parts of q-space and are responsible for the different temperature dependence of

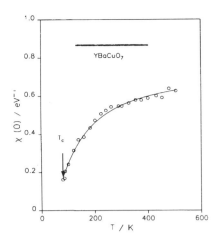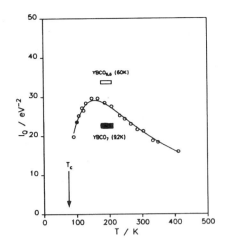

Figure 6. Left: Temperature dependence of $\chi(0)$ derived from Knight shift data of $YBa_2Cu_4O_8$. The temperature independent susceptibility of $YBa_2Cu_3O_7$ is also marked. Right: Antiferromagnetic spectral density I_Q in $YBa_2Cu_4O_8$. The I_Q values of $YBa_2Cu_3O_7$ (highly doped) and $YBa_2Cu_3O_{6.6}$ (underdoped) are also marked. The experimental sets of data are taken from Brinkmann et al.[49]. For each parameter the solid lines result from fits of the spin-gap expressions eq.(21) and eq.(22). From reference 10. Reprinted from Appl. Mag. Res. 3, M. Mehring, pgs. 395 and 409, 1992 with kind permission of Springer-Verlag, Wien, Austria.

^{63}Cu-relaxation on one hand and ^{17}O- and ^{89}Y-relaxation on the other hand. Since the ^{63}Cu-formfactors are sensitive for $Q=(\pi,\pi)$ fluctuations, whereas the ^{17}O- and ^{89}Y-formfactors are not, it might be useful to discuss the different parts of q-space separately, namely the q-space centered around (0,0), mainly seen by ^{17}O- and ^{89}Y-relaxation and partially also by ^{63}Cu-relaxation, and the q-space centered around $Q=(\pi,\pi)$ which affects only the ^{63}Cu-relaxation.

Whereas the $(^{89}T_1T)^{-1}$ is fairly temperature independent and corresponds to a Korringa type relaxation with S_K less than one[10], the $(^{63}T_1T)^{-1}$ behaviour and correspondingly I_K shows an increase with decreasing temperature, reaches a maximum and falls rapidly below a characteristic temperature T^* as shown in fig. (6) for the nominally underdoped $YBa_2Cu_4O_8$. This feature first noted by Walstedt and co-workers[50] was attributed to the spin gap phenomenon and was first simulated by one of us by using the expression[10]

$$I_K = \text{const} \left(\frac{T_0}{T}\right)^\alpha \left[1 - \tanh^2\left(\frac{E_g}{2T}\right)\right] \qquad (22)$$

where T_0 is some characteristic temperature where I_K becomes constant. In the case of $YBa_2Cu_4O_8$ as shown in fig. 6 the values $E_g = 280$ K and $\alpha = 1.5$ where obtained. Other functional forms to represent this so-called spin-gap behaviour have been proposed later[12, 51, 52, 53]. One should be aware, however, that these are just empirical functions in order to fit the experimental data. They do not explain the physics of the pseudo spin gap.

Simple Model Susceptibilities

From what we have discussed so far it should be evident that a proper physical model for the static as well as the dynamic susceptibility is required in order to relate the observed temperature dependences of the Knight shifts and the spin-lattice relaxation rates T_1^{-1}. At the phenomenological level it is tempting to use the following general and fairly simple expression for the dynamic susceptibility

$$\chi(\boldsymbol{q},\omega) = \frac{\chi'(\boldsymbol{q})}{1 - \dfrac{i\hbar\omega}{\Gamma_q}} \qquad (23)$$

320

where $\chi(\boldsymbol{q})$ represents the static q-dependent susceptibility which could be expressed, for example in terms of the usual RPA expression (eq.(6)) and Γ_q sets the energy scale for the spin fluctuations. The Knight shift follows immediately from $\chi'(0)$ and the relevant factor for the spin-lattice relaxation becomes

$$I_q = \frac{\Im m \chi^\pm(\boldsymbol{q}, \omega)}{\hbar \omega} = \frac{2\chi(\boldsymbol{q})}{\Gamma_q} \tag{24}$$

in the limit $\omega \ll \Gamma_q$. Within this phenomenological model the ^{89}Y- and ^{17}O-relaxation will be dominated by the typical quasiparticle contribution around $\boldsymbol{q} = (0, 0)$ and the ^{63}Cu-relaxation will be dominated by the $\boldsymbol{Q} = (\pi, \pi)$ fluctuations, where the different parts of q-space may be represented by the corresponding integrals around $\boldsymbol{q} = (0, 0)$ and $\boldsymbol{Q} = (\pi, \pi)$

$$I_0 = \frac{2\chi'(0)}{\Gamma_0} \quad \text{and} \quad I_Q = \frac{2\chi'(\boldsymbol{Q})}{\Gamma_Q} \tag{25}$$

Since the spin-gap features appear around $\boldsymbol{Q} = (\pi, \pi)$ only the ^{63}Cu-relaxation will show the corresponding characteristic behaviour of $(T_1 T)^{-1}$. Details of such an analysis have been presented some time ago[10].

One of the highly celebrated models which allows to simulate most of the normal state NMR features of the cuprate superconductors with only a few parameters is the Millis, Monien, Pines(MMP) model[54, 55]. Pines and co-workers have applied their expression of $\chi_{\mathrm{MMP}}(\boldsymbol{q}, \omega)$ to numerous experimental results and determined its characteristic parameters. The reader is referred to the papers by Pines and co-worker[56, 57, 58] and to his contribution in these proceedings for more details. For completeness we summarize here their essential expression. Following MMP the dynamic part of the spin-lattice relaxation can be summarized as

$$I_q = \frac{\Im m \chi^\pm(\boldsymbol{q}, \omega)}{\hbar \omega} = \pi \frac{2\chi(0)}{\Gamma} + \pi \frac{2\chi(Q)}{\omega_{\mathrm{SF}}} \frac{1}{(1 + \xi^2(\boldsymbol{q} - \boldsymbol{Q})^2)^2 + \omega^2/\omega_{\mathrm{SF}}^2} \tag{26}$$

where the important parameters are the antiferromagnetic spin-spin correlation length ξ and the spin-fluctuation energy ω_{SF}.

Extended 2D Hubbard model

Some of the normal state NMR properties have been simulated within the extended 2D Hubbard model by Scalapino and co-workers[62, 63], Lavagna and Stemmann[64] and in our group by Mack et.al.[59, 60].

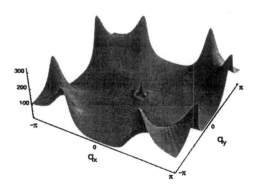

Figure 7. Extended 2D Hubbard calculation of the imaginary part of the dynamic susceptibility $\lim_{\omega \to 0} \Im m \chi^\pm(\boldsymbol{q}, \omega)/\omega$ for a typical parameters set. Note the peaks near the anti-ferromagnetic point at $\boldsymbol{Q} = (\pi \pm \delta, \pi \pm \delta)$ which corresponds to an incommensurate spin density wave (SDW). From 60.

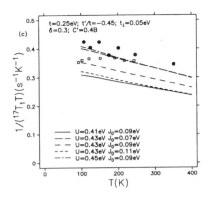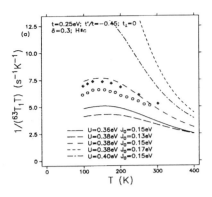

Figure 8. Spin-lattice relaxation data of ^{17}O (left) in $YBa_2Cu_3O_7$ ($\square$)[41], ($\bullet$)[61] and ^{63}Cu (right) in $YBa_2Cu_3O_{6.63}$ ($\blacklozenge$)[41] and $YBa_2Cu_3O_{6.52}$ ($\circ$)[61] . The data are shown together with theoretical calculations within the Extended 2D Hubbard model. From 60.

Our calculation starts from an extended 2D Hubbard model for a single and a double layer with transfer integrals t for nearest neighbour, t' for next-nearest neighbour and $t_\perp$ for transfer between neighbouring planes. The following dispersion relation is obtained after diagonalization of the tight binding part of the extended Hubbard Hamiltonian

$$\epsilon_\pm(\boldsymbol{k}) = -2t[\cos k_x a + \cos k_y a] - 4t' \cos k_x a \cos k_y a \pm t_\perp \qquad (27)$$

The coulomb correlation between the electrons is taken into account by an on-site Hubbard U and a nearest neighbour interaction J_0 as

$$U_{\boldsymbol{k}} = U - J_0[\cos k_x a + \cos k_y a] \qquad (28)$$

Temperature dependent Greens functions were calculated within RPA and by applying the Matsubara formalism[59, 60]. This procedure resulted in temperature dependent susceptibilities $\chi(\boldsymbol{q}, \omega)$ which could directly be inserted into the expressions for the Knight shift (eq.(4)), the spin-lattice relaxation T_1^{-1}(eq.(10)) and the spin-spin relaxation rate T_2^{-1}(eq.(18)). Typical parameter values ($t = 0.25$ eV, $t'/t = $ -0.45) were chosen which fit the experimentally observed Fermi surface best. Standard hyperfine tensors like those presented in Table 1 were used together with the already quoted formfactors.

Fig. 7 displays the imaginary part of the dynamic susceptibility $\lim_{\omega\to 0} \Im m\chi^\pm(\boldsymbol{q}, \omega)/\omega$ for a typical parameters set. Note the peaks near the anti-ferromagnetic point at $\boldsymbol{Q} = (\pi\pm\delta, \pi\pm\delta)$. Within this parameters set this dynamic susceptibility corresponds to an incommensurate spin density wave (SDW). Features of an incommensurate SDW have been observed in neutron scattering mainly in $LaCuO_4$[65] and more recently in $YBa_2Cu_3O_{6.6}$[66, 67]. Other values of the Hubbard U and J_0 give peaks at the anti-ferromagnetic point $\boldsymbol{Q} = (\pi, \pi)$.

Space does not permit to go into details here but we want to demonstrate briefly that the subtle differences in the temperature dependence of $^{63}T_1T$ and $^{17}T_1T$ can be accounted for by such a calculation. Fig. 8 displays the calculated temperature dependence of these quantities together with some experimental data. Note that the so-called spin-gap feature in $^{63}T_1T$, namely the appearance of a maximum can be reproduced as well as the flat temperature dependence of $^{17}T_1T$. In the latter case we have applied the extended formfactor for ^{17}O proposed by Pines and co-workers[68] which takes next-nearest neighbour hyperfine interaction into account and is given by

$$^{17}F_z(\boldsymbol{q}) = \frac{1}{2(A_{xx}^2 + A_{yy}^2)} \sum_{\beta=x,y} (1 + 2\frac{A'_{\beta\beta}}{A_{\beta\beta}})^{-2} \left[(1 + \cos q_y a)(A_{\beta\beta} + 2A'_{\beta\beta}\cos q_x a)^2 \right.$$
$$\left. + (1 + \cos q_x a)(A_{\beta\beta} + 2A'_{\beta\beta}\cos q_y a)^2 \right] \qquad (29)$$

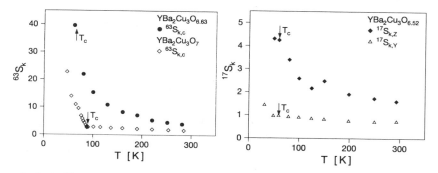

Figure 9. S_K for ^{63}Cu in YBa$_2$Cu$_3$O$_{6.63}$ and YBa$_2$Cu$_3$O$_7$ and ^{17}O in YBa$_2$Cu$_3$O$_{6.52}$. The experimental data are taken from reference 41, 69 and 70 for ^{63}Cu and from reference 61 for ^{17}O. See text for details.

This gave better agreement with the experimental data than without it. For more details see Mack et. al.[59, 60].

Limits of the standard model

In the previous subsections the standard model for an interpretation of NMR parameters in the normal state of high-T_c-superconductors was introduced and applied to experimental data. However, there are some experimental properties of these compounds which could not be explained within this model.

If one assumes a single spin fluid model the ^{17}O and ^{89}Y in YBa$_2$Cu$_3$O$_{6+x}$ nuclei are less sensitive to antiferromagnetic fluctuations compared to ^{63}Cu and rather probe the quasiparticle contributions to the magnetic susceptibility due to their formfactors. Within the standard model the S_K factors of these two nuclei should show a similar temperature dependence. Experimental results do not support this. Instead the ratio $(^{17}T_1^{-1}/^{89}T_1^{-1})$ increases with decreasing temperature although the ratio of the corresponding Knight shifts remains constant[71].

Moreover the absolute value of $(^{17}T_1^{-1})/(^{89}T_1^{-1})$ for YBa$_2$Cu$_3$O$_{6.63}$ and YBa$_2$Cu$_3$O$_7$ calculated according to the standard model results in a much larger value compared with the experimental data. This implies that either the spin-lattice relaxation at the yttrium site is enhanced or at the oxygen site is reduced[71].

The normal state spin-lattice-relaxation rate of ^{17}O itself shows for YBa$_2$Cu$_3$O$_{6.52}$ a temperature dependent anisotropy[72] which results in a different temperature dependence for the corresponding S_K-values as shown in fig. 9 (right). This behaviour can not be explained by using the Shastry-Mila-Rice Hamiltonian and the corresponding formfactors.

A recent comparison of ^{17}O Knight shift and spin-lattice relaxation rate measurements of several groups in YBa$_2$Cu$_3$O$_{6+x}$ by Martindale and Hammel[73] clearly shows that these parameters show no uniform behaviour although the samples vary only slightly in oxygen content. They consider a yet unknown physical parameter responsible for the ^{17}O NMR parameter in the YBa$_2$Cu$_3$O$_{6+x}$-compound.

There are several proposals which extend the standard model. As already mentioned before Zha et al.[68] introduced a modified ^{17}O formfactor, including next-nearest neighbour interaction, which screens the contributions of fluctuations in the vicinity of $Q = (\pi, \pi)$ in a more efficient way. Yoshinari[74] introduced an additional oxygen band in order to explain the anisotropy of the Knight shift at the O(2,3) site in YBa$_2$Cu$_3$O$_{6+x}$.

In order to extract the relevant physical parameters of the CuO$_2$-planes we apply the modified S_K sum-rule which has been developed in the previous section (eq.(12)). Figure 9 displays the results for ^{63}Cu and ^{17}O in YBa$_2$Cu$_3$O$_{6+x}$ for $x \simeq 1$ and underdoped compounds. In the latter S_K is strongly enhanced and shows a pronounced temperature dependence. This behaviour is due to strong antiferromagnetic fluctuations. The S_K-value of ^{63}Cu for

YBa$_2$Cu$_3$O$_7$, however, shows a weaker temperature dependence in the normal conducting state, but increases dramatically in the superconducting state[42]. Details of such an analysis will be presented elsewhere.

NMR IN THE SUPERCONDUCTING STATE

We have already discussed some general features of NMR in the superconducting state in the principles section. Here we want to take a closer look at the different behaviour of the Knight shift and the spin-lattice relaxation as well as the spin-spin relaxation in the superconducting state in particular with regard to s- and d-wave pairing. The essential aspect how the gap symmetry can influence the NMR parameters is through the corresponding change in the static and dynamic susceptibility. Basically the existence of a monotonic increase of density of states in the gap in the case of d-wave pairing in contrast to s-wave pairing where an isotropic gap with 2Δ occurs, leads to a power law behaviour for d-wave and an exponential temperature dependence for s-wave pairing.

Knight shift

The differences in the Yosida functions for s- and d-wave pairing which represent the temperature dependence of the Knight shift below T_c are not dramatic near T_c, unless extreme parameter sets are used. Only at low temperatures can both cases be distinguished. A characteristic feature of the d-wave Yosida function is its almost linear increase with T at low T in contrast to the s-wave function which increases exponentially.

In fig. 10 the temperature dependence of the Knight shift is shown as calculated within a 2D extended Hubbard model discussed already in the normal state section[59, 60]. Some experimental results are included for comparison. In order to include superconductivity in the extended 2D Hubbard model the following gap Hamiltonian with $\boldsymbol{k}$ dependent gap parameter $\Delta(\boldsymbol{k})$ was used

$$H = \sum_{\boldsymbol{k}\sigma} \Delta(\boldsymbol{k})a^\dagger_{\boldsymbol{k}\uparrow}a^\dagger_{-\boldsymbol{k}\downarrow} \tag{30}$$

Whereas $\Delta(\boldsymbol{k})$ is independent of $\boldsymbol{k}$ for s-wave pairing the following $\boldsymbol{k}$-dependence was used for d-wave pairing $\Delta(\boldsymbol{k}) = (\Delta_0(T)/2)(\cos k_x a - \cos k_y a)$.

We note that the more linear increase of the Knight shift with temperature calculated for d-wave pairing fits the experimental data better than the curves for s-wave pairing (fig. (10)). A note of caution might be appropriate, however. Because Knight shift measurements must

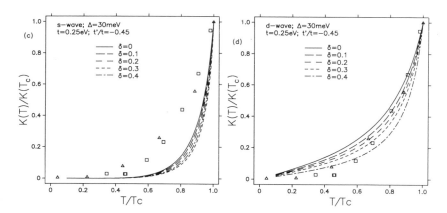

Figure 10. Extended 2D Hubbard calculations of the Cu(2)-Knight shift for $B_0 \perp c$ for s-wave (left) and d-wave pairing (right). The experimental data are taken from reference 69 (△) and reference 75 (□).

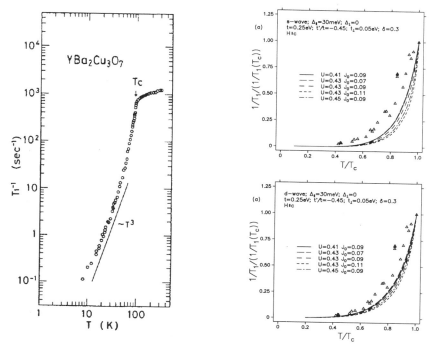

Figure 11. Left: Spin-lattice relaxation rate T_1^{-1} of $YBa_2Cu_3O_7$ in the superconducting state. For low temperatures a T^3 power law is observed. From reference 11. Reprinted from Appl. Mag. Res. 3, Y. Kitaoka et al., pg. 580, 1992 with kind permission of Springer-Verlag, Wien, Austria. Right: Temperature dependence of the normalized spin-lattice relaxation rate of ^{63}Cu for $B_0 \parallel c$. The lines are results of extended 2D Hubbard calculations for s-wave pairing (top) and d-wave pairing (bottom). The experimental data are taken from several references[61, 75, 76, 77].

be performed in rather large magnetic fields, the linebroadenings due to the vortex lattice and the diamagnetic lineshifts might prevent a precise determination of the Knight shift at low temperatures. Moreover the resolution is in general not great.

Spin-lattice relaxation T_1

The gap opening in the superconducting state finally leads to a drastic decrease of the relaxation rate T_1^{-1} below T_c with more or less exponential temperature dependence for s-wave pairing as was pointed out already in the principles section. For d-wave pairing the decrease is less dramatic and can in certain temperature ranges be described by different power laws as shown in fig. 11 (left). Due to the states in the gap the relaxation rate rises from $T = 0$ for d-wave pairing not as steep as for s-wave pairing. In order to demonstrate this we resort again to the extended 2D Hubbard calculations by Mack et.al.[59, 60].

Fig. 11 shows the calculated temperature dependence of $^{63}T_1^{-1}$ for s- and d-wave pairing together with some experimental data[59, 60]. The calculation reproduces the rapid decrease of the relaxation rate. However, agreement with the experimental data is rather poor. Nevertheless it is obvious that the decrease of $^{63}T_1^{-1}$ is much more dramatic for s-wave than for d-wave pairing. The fact that the experimental data appear beyond the d-wave calculation might be taken as support for d-wave pairing.

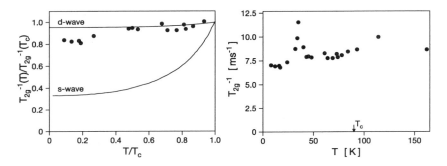

Figure 12. Normalized Gaussian part T_{2G} of the spin-spin relaxation rate of ^{63}Cu(2) in YBa$_2$Cu$_3$O$_7$ in the superconducting state (left). The fast fluctuating parts $(\exp(-t/T_{2R}))$ and the contribution of the direct dipole-dipole interaction are subtracted. The solid lines are recent calculations of Thelen and Pines[36] (s-wave pairing) and Pines and Wrobel[78] (d-wave pairing). Right: Overall temperature dependence of T_{2G}. Note the peak in the relaxation rate at 35 K which is omitted in the figure on the left.(See the next section for a discussion)

Spin-spin relaxation T$_2$

Slichter and co-workers[25] pointed out and demonstrated that measurement of the temperature dependence of the spin-spin relaxation rate T_2^{-1} can give some clues towards s- or d-wave pairing. Similar investigations were performed by Itoh[79] and Brinkmann and co-workers[80]. Scalapino and co-workers[81] as well as Pines and co-workers[36, 82, 78] performed calculations of the temperature dependence of T_2^{-1} in the superconducting state and demonstrated that there is an appreciable decrease of T_2^{-1} below T_c for s-wave pairing, whereas there is almost no decrease for d-wave pairing.

We present in fig. 12 our recent experimental data measured in YBa$_2$Cu$_3$O$_{6.9}$ in zero-field together with theoretical calculations based on s- and d-wave pairing[36, 78]. It is evident from fig. 12 that the experimental data strongly support d-wave pairing. This behaviour was also observed in YBa$_2$Cu$_4$O$_8$ by several groups[25, 79, 80]. In addition we note that NQR spin echo double resonance (SEDOR) experiments between ^{65}Cu and ^{63}Cu, performed in our laboratory some time ago[83], show very similar behaviour with almost no decrease of the heteronuclear spin-spin interaction below T_c. Both observations strongly support d-wave pairing or at least an anisotropic gap parameter. Calculations within the 2D extended Hubbard model[59, 60] support this conclusion also.

BEYOND THE STANDARD MODEL

In this final section we will address briefly some uncommon properties of the NMR/NQR in high-T_c superconductors. We have shown in the previous sections that Knight shift, T_1 and T_2 all show unusual temperature dependences which are incompatible with classical metallic behaviour or BCS type superconductivity. There is ample evidence from NMR that the normal state is dominated by strong electron-electron correlations and the superconducting gap is anisotropic. The most natural description of the NMR properties in the superconducting state seems to be by d-wave pairing. We want to contrast this conjecture in the following section by presenting the model of an antiferromagnetic s-wave superconductor.

The antiferromagnetic s-wave superconductor (AFS-model)

In the antiferromagnetic s-wave model(AFS) of Kulić et.al.[22, 23] the basic scenario is the following. Suppose spin 1/2 quasiparticles which undergo s-wave pairing in the superconducting state travel through a solid with localized antiferromagnetically correlated electron spins.

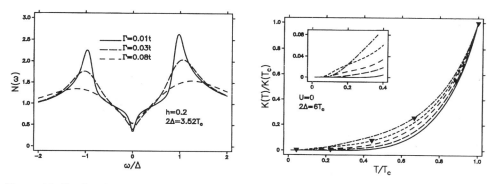

Figure 13. Density of states N_s (left) and Knight shift K_s (right) for an antiferromagnetic s-wave superconductor[23]. The different curves of N_s show the influence of the scattering parameter Γ. N_s shows gapless features caused by ferromagnetic interaction in certain directions. The different curves of K_s show the influence of the exchange field h. Experimental data are taken from reference 69. Reprinted from Physica C 252, M. L. Kulić et al., pgs. 34 and 37, 1995 with kind permission of Elsevier Science - NL, Sara Burgerhartstraat 25, 1055 KV Amsterdam, The Netherlands.

No antiferromagnetically ordered state is required, but rather long lived correlations on the timescale of the kinetic motion of the electron. The quasiparticles will feel an *alternating* field of strength h if they travel along the antiferromagnetic direction and feel a *ferromagnetic* field h when traveling in direction where the localized spins are ferromagnetically correlated. This leads to a gap in the quasiparticle spectrum which modifies the normal state properties and at the same time leads to d-wave like states in the gap in the superconducting state as is shown in fig. 13[23].

This is readily understood if one considers kinetic motion of the Cooper pairs along the antiferromagnetic direction which leads to an average field $\langle h \rangle = 0$, whereas in the ferromagnetic direction pair breaking takes place due to the ferromagnetic interaction with $\langle h \rangle \neq 0$. In the ferromagnetic direction the gap is therefore closed(see fig. 13). These *gapless* features cause a similar temperature dependence of the NMR parameters as d-wave pairing. As an example we present in fig. 13 the temperature dependent Knight shift for different strength of the exchange field h. Some of the results can be summarized as: (1) the Hebel-Slichter peak disappears, because of (a) the partially gapless character of the AFS model, (b) the large ratio $2\Delta/T \simeq 6$, (c) a finite scattering rate $\Gamma(T) = 0.8 T_c (T/T_c)^2$, (2) the Knight shift shows a linear temperature dependence at low T, (3) at low temperatures $1/T_1 \propto T^3$ and (4) T_{2G} is only weakly temperature dependent, depending on the magnitude of h. All these features appear under s-wave pairing within the AFS model and are usually accounted for by d-wave pairing. Further details may be found in reference 23.

Low temperature ordering transition

There is an intriguing feature observed by us and others[84, 85, 86] in the T_2-relaxation of ^{63}Cu(2) in YBa$_2$Cu$_3$O$_7$ well below T_c. Fig. 12 represents the data observed by us already quite some time ago and in more detail recently. The peak of T_2^{-1} around 35 K is reminiscent of a magnetic phase transition. We have found that it does not derive from the same mechanism as T_{2G}, but corresponds to local magnetic field or electric field gradient fluctuations resulting in an exponential decay. It is very pronounced and occurs in a very narrow temperature range. Moreover it cannot be a defect feature because **all** ^{63}Cu spins in the CuO$_2$-plane are affected in the same way. We speculate that it corresponds to the 3D low temperature ordering transition either of a spin density wave (SDW) or a charge density wave (CDW) state. When lowering the temperature an additional broadening of the spectral line is observed which increases with decreasing temperature. This would be expected if this feature is connected with the order parameter of the SDW/CDW state. However, the local field causing the broadening is rather low. If it is related to a SDW with local spins at the Cu-site this would imply that the amplitude of the SDW is extremely small and the distribution of local fields would point at an incommensurate SDW. One could speculate that the SDW spins reside on O(2,3), possibly in the π-bands including of course some hybridisation with Cu-orbitals. With nearly antiferromagnetic correlation we would expect rather small local field at the Cu sites. An alternative explanation would be due to a charge density wave (CDW) resulting in fluctuating electric field gradients causing quadrupolar relaxation and line broadening due to static electric field gradients with decreasing temperature related to the increase of the order parameter. Since the emergence of this feature does not disturb the superconducting condensate in first order the charge carriers responsible for the CDW should be distinct from the Cooper pairs and may be connected with the CuO chains or the apex oxygen ordering. Experiments are underway to clarify this point. Further details will be published elsewhere.

SUMMARY

We have discussed the characteristic NMR/NQR parameters Knight shift K_s, spin lattice relaxation rate T_1^{-1} and spin-spin relaxation rate T_2^{-1} in cuprate superconductors.

- The temperature dependence of K_s in the normal state shows a maximum at T_c for the optimally doped and overdoped compounds. The increase of K_s with decreasing temperature in these compounds is related to the van Hove singularity in their density of states. For underdoped compounds K_s shows a more or less uniform behaviour for all nuclei in the unit cell with a maximum at $T > T_c$ which could be explained within the single spin fluid model by the opening of a pseudo spin gap in the spin excitation spectrum near $q = 0$. In the superconducting state the decrease of the Knight shift can be explained within a d-wave pairing model but as well by an anisotropic s-wave pairing.

- The analysis of the spin-lattice relaxation rates T_1^{-1} of the nuclei of the unit cell shows an enhancement of $(T_1T)^{-1}$ with decreasing temperature at the ^{63}Cu sites due to strong antiferromagnetic correlations for underdoped and optimally doped compounds whereas the ^{17}O and ^{89}Y $(T_1T)^{-1}$ show a weaker temperature dependence but still distinct from a behaviour expected for a normal Fermi liquid.

- Measurement of the Gaussian part T_{2G}^{-1} of the spin-spin relaxation rate of ^{63}Cu gives access to the integrated real part of the spin susceptibility $\chi'(q, \omega)$. In the superconducting state the temperature dependence of T_{2G}^{-1} in $YBa_2Cu_3O_7$ and $YBa_2Cu_4O_8$ strongly supports a d-wave pairing model.

- In addition we have discussed alternative models, like the AFS model where s-wave pairing results in a gapless density of states and the calculated temperature dependence does indeed show similar behaviour as for d-wave pairing.

- Finally we have conjectured about a possible low temperature ordering transition in $YBa_2Cu_3O_{6+x}$ which could be due to a formation of a spin density wave (SDW) or a charge density wave (CDW) state. Evidence for this is provided by a T_2^{-1} peak at 35 K which is considered to be connected with the 3D ordering temperature of the SDW/CDW.

ACKNOWLEDGEMENTS

We would like to thank M. Kulić, F. Mack and E. Goreatchkovski for discussions on the theory and T. Stephany and R. Borenius for performing early SEDOR and T_2 measurements. A financial support by the Fonds der Chemischen Industrie is gratefully acknowledged.

328

REFERENCES

1. A. Abragam. *Principles of Nuclear Magnetism.* Oxford Univ. Press, Oxford, 1989.
2. M. Mehring. *Principles of High Resolution NMR in Solids.* Springer, Berlin, Heidelberg, 2nd edition, 1983.
3. C. P. Slichter. *Principles of Magnetic Resonance.* Springer, Berlin, Heidelberg, 3rd edition, 1990.
4. J. Winter. *Nuclear Magnetic Resonance in Metals.* Clarendom Press, Oxford, 1971.
5. D. E. MacLaughlin. Magnetic Resonance in the Superconducting State. In: *Solid State Physics vol. 31,* Seitz and Turnbull [eds.], p. 1. Academic Press , New York, 1976.
6. J. G. Bednorz and K. A. Müller. *Z. Phys.* B 64:189 (1986).
7. T. Imai. *J. Phys. Soc. Jpn.* 59:2508 (1990).
8. C. Berthier, Y. Berthier, P. Butaud, W. G. Clark, J. A. Gillet, M. Horvatić, P. Ségransan and J. Y. Henry. *Appl. Magn. Reson.* 3:449 (1992).
9. C. P. Slichter, S. E. Barret, J. A. Martindale, D. J. Durand, C. H. Pennington, C. A. Klug, K. E. O'Hara, S. M. DeSoto, T. Imai, J. P. Rice, T. A. Friedmann and D. M. Ginsberg. *Appl. Magn. Reson.* 3:423 (1992).
10. M. Mehring. *Appl. Magn. Reson.* 3:383 (1992).
11. Y. Kitaoka, K. Ishida, S. Ohsugi, K. Fujiwara, G. q. Zheng and K. Asayama. *Appl. Magn. Reson.* 3:549 (1992).
12. D. Brinkmann and M. Mali. *NMR Basic Principles and Progress,* p. 171. Springer, Berlin, Heidelberg, 1994.
13. A. Rigamonti, S. Aldrovandi, F. Borsa, P. Carretta, F. Cintolesi, M. Corti and S. Rubini. *Nuovo Cim.* 16:1743 (1994).
14. M. Mehring (guest editor). *Appl. Magn. Reson.* 3:383–747 (1992).
15. K. Asayama, Y. Kitaoka, G. Zheng and K. Ishida. *Prog. NMR Spectroscopy* 28:221 (1996).
16. C. Berthier, M. H. Julien, M. Horvatić and Y. Berthier. *J. Phys. France* 6:2205 (1996).
17. C. P. Slichter, R. L. Corey, N. J. Curro, S. M. DeSoto, K. O'Hara and T. Imai. *Phil. Mag.* B 74:545 (1996).
18. W. D. Knight. *Phys. Rev.* 76:1259 (1949).
19. A. M. Clogston, A. C. Grossard, V. Jaccarino and Y. Yafet. *Rev. Mod. Phys.* 36:170 (1964).
20. A. M. Clogston, V. Jaccarino and Y. Yafet. *Phys. Rev.* A134:650 (1964).
21. K. Yosida. *Phys. Rev.* 110:769 (1958).
22. M. L. Kulić, A. I. Liechtenstein, E. Goreatchkovski and M. Mehring. *Physica* C 244:185 (1995).
23. M. L. Kulić, E. Goreatchkovski, A. I. Liechtenstein and M. Mehring. *Physica* C 252:27 (1995).
24. T. Moriya. *J. Phys. Soc. Jpn.* 18:516 (1963).
25. R.L. Corey, N.J. Curro, K. O'Hara, T. Imai, C.P. Slichter, K. Yoshimura, M. Katoh and K. Kosuge. *Phys. Rev.* B 53:5907 (1996).
26. M. Mehring. In: *Low Dimensional Conductors and Superconductors,* D. Jerome and L.G. Caron [eds.], p. 185. NATO ASI Series 155,Plenum Press, NewYork and London, 1986.
27. M. Mehring, F. Rachdi and G. Zimmer. *Phil. Mag.* B 70:747 (1994).
28. J. Korringa. *Physica* 16:601 (1950).
29. M. Mehring, T. Kälber, U. Ludwig, K. F. Thier and G. Zimmer. In: *Physics and Chemistry of Fullerenes and Derivatives,* H. Kuzmany, J. Fink, M. Mehring and S. Roth [eds.], p. 360. World Scientific, Singapore, 1995.
30. J. Bardeen, L. N. Cooper and J. R. Schrieffer. *Phys. Rev.* 108:1175 (1957).
31. L. C. Hebel and C. P. Slichter. *Phys. Rev.* 113:1504 (1959).
32. Y. Masuda and A. G. Redfield. *Phys. Rev.* 125:159 (1962).
33. C. H. Pennington and C. P. Slichter. *Phys. Rev. Lett.* 66:381 (1991).
34. J. H. van Vleck. *Phys. Rev.* 74:1168 (1948).
35. T. Imai, C. P. Slichter, A. P. Paulikas and B. Veal. *Appl. Magn. Reson.* 3:729 (1992).
36. D. Thelen and D. Pines. *Phys. Rev.* B 49:3528 (1994).
37. F. Mila and T. M. Rice. *Phys. Rev.* B 40:11382 (1989).
38. F. Mila and T. M. Rice. *Physica* C 157:561 (1989).
39. B. Sriram Shastry. *Phys. Rev. Lett.* 63:1288 (1989).
40. F. C. Zhang and T. M. Rice. *Phys. Rev.* B 37:3759 (1988).
41. M. Takigawa, A. P. Reyes, P. C. Hammel, J. D. Thompson, R. H. Heffner, Z. Fisk and K. C. Ott. *Phys. Rev.* B 43:247 (1991).
42. N. Winzek and M. Mehring. *Appl. Magn. Reson.* 3:535 (1992).
43. M. Takigawa. *Appl. Magn. Reson.* 3:495 (1992).
44. J. M. Tranquada, P. M. Gehring, G. Shirane, S. Shamoto and M. Sato. *Phys. Rev.* B 46:5561 (1992).
45. N. Winzek. *NMR Untersuchungen zur Dotierungsabhängigkeit von Tl-Kuprat-Supraleitern.* Dissertation, Universität Stuttgart, 2. Phys. Inst., 1994.
46. J. Labbé and J. Bok. *Europhys. Lett.* 3:1225 (1987).
47. J. Bok and L. Force. *Physica* C 185-189:1449 (1991).
48. J. Bouvier and J. Bok. *Physica* C 249:117 (1995).
49. D. Brinkmann. *Appl. Magn. Reson.* 3:483 (1992).

50. W. W. Warren, R. .E. Walstedt, G. F. Brennert, R. J. Cava, R. Tycko, R. F. Bell and G. Dabbagh. *Phys. Rev. Lett.* 62:1193 (1989).
51. V. J. Emery and S. A. Kivelson. In: *Proceeding of the Workshop on Phase Separation in Cuprate Superconductors, Erice, Italy 1993.* K. A. Müller and G. Benedek [eds.], p. 1. World Scientific, Singapore, 1993.
52. M. Mehring, M. Baehr, P. Gergen, J. Groß and C. Kessler. *Physica Scripta* T 49:124 (1993).
53. M. Bankay, M. Mali, J. Roos and D. Brinkmann. *Phys. Rev.* B 50:6416 (1994).
54. A.J. Millis, H. Monien and D. Pines. *Phys. Rev.* B 42:167 (1990).
55. A.J. Millis and H. Monien. *Phys. Rev.* B 45:3059 (1992).
56. H. Monien, D. Pines and M. Takigawa. *Phys. Rev.* B 43:258 (1991).
57. V. Barzykin, D. Pines and D. Thelen. *Phys. Rev.* B 50:16052 (1994).
58. V. Barzykin and D. Pines. *Phys. Rev.* B 52:13585 (1994).
59. F. Mack. Der Einfluß von Elektron-Elektron Korrelationen auf NMR Parameter in Leitern und Supraleitern. Diploma Thesis, Universität Stuttgart, 2. Phys. Inst., 1996.
60. F. Mack, M.L. Kulic and M. Mehring. *Physica* C:1997 (submitted).
61. M. Horvatić, T. Auler, C. Berthier, Y. Berthier, P. Butaud, W. G. Clark, J. A. Gillet, P. Ségransan and J. Y. Henry. *Phys. Rev.* B 47:3461 (1993).
62. N. E. Bickers, D. J. Scalapino and S. R. White. *Phys. Rev. Lett.* 62:961 (1989).
63. N. Bulut and D. J. Scalapino. *Phys. Rev. Lett.* 64:2723 (1990).
64. M. Lavagna and G. Stemmann. *Phys. Rev.* B 49:4235 (1994).
65. R. J. Birgeneau, Y. Endoh, K. Kakurai, Y. Hidaka, T. Murakami, M. A. Kastner, T. R. Thurston, G. Shirane and K. Yamada. *Phys. Rev.* B 39:2868 (1989).
66. B. J. Sternlieb, J. M. Tranquada, G. Shirane, M. Sato and S. Shamoto. *Phys. Rev.* B 50:12915 (1994).
67. Pencheng Dai, H. A. Mook and F. Doğan. *preprint* (1997).
68. Y. Zha, V. Barzykin and D. Pines. *Phys. Rev.* B 54:7561 (1996).
69. S. E. Barrett, D. J. Durand, C. H. Pennington, C. P. Slichter, T. A. Friedmann, J. P. Rice and D. M. Ginsberg. *Phys. Rev.* B 41:6283 (1990).
70. P. C. Hammel, M. Takigawa, R. H. Heffner, Z. Fisk and K. C. Ott. *Phys. Rev. Lett.* 63:1992 (1989).
71. M. Takigawa, W. L. Hults and J. L. Smith. *Phys. Rev. Lett.* 71:2650 (1993).
72. M. Horvatić, C. Berthier, Y. Berthier, P. Ségransan, P. Butaud, W. G. Clark, J. A. Gillet and J. Y. Henry. *Phys. Rev.* B 48:13848 (1993).
73. J. A. Martindale and P. C. Hammel. *Phil. Mag. B Phys. Cond. Matt.* 74:573 (1996).
74. Y. Yoshinari. *Physica* C 276:147 (1997).
75. M. Takigawa, P. C. Hammel, R. H. Heffner, Z. Fisk, K. C. Ott and J. D. Thompson. *Physica* C 162-164:853 (1989).
76. S. E. Barrett, J. A. Martindale, D. J. Durand, C. H. Pennington, C. P. Slichter, T. A. Friedmann, J. P. Rice and D. M. Ginsberg. *Phys. Rev. Lett.* 66:108 (1991).
77. M. Horvatić, P. Butaud, P. Ségransan, Y. Berthier, C. Berthier, J. Y. Henry and M.Couach. *Physica* C 166:151 (1990).
78. D. Pines and P. Wrobel. *Phys. Rev.* B 53:5915 (1996).
79. Y. Itoh, H. Yasuoka, Y. Fujiwara, Y. Ueda, T. Machi, I. Tomeno, K. Tai, N. Koshizuka and S. Tanaka. *J. Phys. Soc. Jpn.* 61:1287 (1992).
80. R. Stern, M. Mali, J. Roos and D. Brinkmann. *Phys. Rev.* B 51:15478 (1995).
81. N. Bulut and D. J. Scalapino. *Phys. Rev. Lett.* 67:2898 (1991).
82. D. Thelen, D. Pines and Jian Ping Lu. *Phys. Rev.* B 47:9151 (1993).
83. R. Borenius. NQR Doppelresonanz an Hochtemperatursupraleitern. Diploma Thesis, Universität Stuttgart, 2. Phys. Inst., 1995.
84. M. Tei, H. Takei, K. Mizoguchi and K. Kume. *Z. Naturforsch.* 45a:429 (1990).
85. Y. Itoh, H. Yasuoka and Y. Ueda. *J. Phys. Soc. Jpn.* 59:3463 (1990).
86. C. H. Recchia, J. A. Martindale, C.H. Pennington, W. L. Hults and J. L. Smith. *Phys. Rev. Lett.* 78:3543 (1997).

330

NMR IN THE NORMAL STATE OF CUPRATES

Arlette Trokiner

Laboratoire PMMH, Groupe de Physique Thermique, UMR CNRS, ESPCI, 10 rue Vauquelin, 75231 Paris

INTRODUCTION

A common feature of all high-temperature superconducting compounds (HTSC) is the strong two dimensional (2D) character related to the presence of one or more CuO_2 planes in their unit cell, the CuO_2 planes playing a central role in the physics of these compounds. As HTSC are complex non stoechiometric compounds containing at least four chemical species, one may vary their composition. The superconducting phase is obtained from the parent compound, which is an antiferromagnetic (AF) insulator, by doping, that is, introducing charge carriers (holes) in the CuO_2 planes. Doping is obtained by increasing the oxygen content or by performing heterovalent substitution where the host ions are replaced by ions with a smaller valence. Starting from the parent compound, when the number of holes increases, the Néel temperature decreases sharply. When the number of holes is further increased the long range 3D antiferromagnetic order disappears and an insulator-metal transition occurs, the ground state at low temperature of the metal being the superconducting state. Due to the vicinity of the AF insulator-metal transition, it was early suggested that some magnetism still persists in the metallic range. Thus, one important issue was to understand the role of magnetism in the mechanism of superconductivity. In order to devise a microscopic theory describing the superconducting pairs formation, it appeared of prime importance to first understand the normal state since many of the temperature dependencies of the properties exhibited by HTSC in the normal state (magnetic susceptibility, resistivity, Hall coefficient ...) are unusual. That is why most of the NMR studies have concerned the metallic state demonstrating the importance of the short range AF correlations. Although there has been a huge amount of theoretical and experimental studies since the discovery of these compounds, no consensus has been yet reached on the microscopic description of the metallic state.

Nuclear Magnetic Resonance (NMR) provides valuable informations about the local environment of the probed nuclear spin through its coupling with its surrounding. Furthermore, NMR enables to study different chemical species in the material but also provides atomic site specific information. In HTSC, and more particularly in the conducting CuO_2 planes, the nuclear spins are mainly coupled to the electronic spins so that probing the oxygen or copper nuclei in the conducting layers or the nuclei (yttrium or calcium or thallium) separating the layers yields information about the static spin susceptibility $\chi'(0,0)$ as well as about the dynamical spin susceptibility, $\chi''(q,\omega)$. The former is deduced from the frequency shift of the NMR line and the latter from spin-lattice relaxation measurements.

NMR was a key experimental probe to show that HTSC are correlated systems and that AF electron spin correlations exist in the CuO_2 planes. The goal of this paper is to describe some of the unusual properties (compared to conventional metals) of these compounds, these anomalies being related to the presence of the AF correlations. The paper will be restricted to the normal state and to the NMR investigations concerning the CuO_2 planes.

The paper is organized as follows. In the first part some background needed to understand NMR studies will be described. In the following part, the appropriate spin Hamiltonian is introduced within the single spin fluid model. From this Hamiltonian, the expressions of the spin part of the NMR lineshift and of the spin-relaxation rate are then derived. In the third part, some of the most striking features concerning the underdoped regime will be reported. The last part will be devoted to the overdoped regime.

NMR BACKGROUND

A nuclear spin interacts with its electronic environment through electrostatic and magnetic hyperfine couplings. The Hamiltonian can be written as :

$$H = H_{Zeeman} + H_Q + H_{hf} \tag{1}$$

where $H_{Zeeman} = -^n\gamma \hbar H_0 I_z^n$ is the interaction between the applied magnetic field $\mathbf{H_0}$ and the nuclear spins $\mathbf{I}$, z is the quantization axis along $\mathbf{H_0}$ direction, n denotes the nuclear species and γ is the gyromagnetic ratio. When no perturbation is present, the NMR line is situated at the Larmor angular frequency $\omega_0 = {}^n\gamma H_0$. The two other terms H_Q and H_{hf} are respectively the quadrupolar and magnetic hyperfine interaction.

Although the quadrupolar interaction influences strongly the NMR spectrum of a nucleus with a spin I>1/2 (as is the case for ^{63}Cu, ^{17}O and ^{43}Ca in cuprates) when situated in a non-cubic symmetry site, we will not consider it but rather focus on the magnetic hyperfine interaction.

There are two contributions to the magnetic hyperfine Hamiltonian, a spin contribution which originates from the interaction between the nuclear spin and the electronic spins, and an orbital contribution due to the electron orbital motion. In other words, a local magnetic field is induced at the nuclear site by the electrons through their spin and orbital moments. The time-averaged part of the local magnetic field is responsible for the shift of the NMR line whereas the fluctuating part governs the nuclear spin-lattice relaxation.

The NMR line shift

The static part of the local magnetic hyperfine field gives rise to a shift of the NMR line from ω_0 to ω. The shift $\Delta\omega$ is proportional to the applied magnetic field, it obeys:

$$\omega_\alpha = \omega_0 + \Delta\omega_\alpha = {}^n\gamma H_0 (1 + {}^nK_{\alpha\alpha}) \tag{2}$$

where $\mathbf{H_0}$ lies in the α direction (α = x, y, z) and n denotes the nuclear species. The magnetic shift tensor ^nK can be decomposed into a spin and an orbital part. Its component parallel to the external field is :

$$^nK_{\alpha\alpha}(T) = {}^nK_{\alpha\alpha}^{spin}(T) + {}^nK_{\alpha\alpha}^{orb} \tag{3}$$

The first term is called Knight shift. In normal metals, it is related through the Pauli susceptibility to the density of states at the Fermi level and has no temperature dependence in the normal state. This is not the case in cuprates where $^nK_{\alpha\alpha}^{spin}$ exhibits a thermal dependence in the normal state. In the superconducting state, $^nK_{\alpha\alpha}^{spin}(T)$ is expected to vanish for T=0K, due to the singlet spin pairing. For copper and oxygen atoms, the orbital shift is due to both, the Van Vleck paramagnetism and the chemical shift (orbital contribution of the filled orbitals) whereas for yttrium or calcium atoms, it is only due to the chemical shift. $^nK_{\alpha\alpha}^{orb}$ is temperature independent. Hence, at T= 0K since $^nK^{spin}(T)=0$, the residual shift is mainly due to the orbital shift.

The spin part $^nK_{\alpha\alpha}^{spin}(T)$ can be expressed as a function of the component of the static spin susceptibility per site:

$$^nK_{\alpha\alpha}^{spin}(T) = \; ^nA_{\alpha\alpha}\, \chi_{\alpha\alpha}(T)/\, g\, \mu_B \tag{4}$$

where $^nA_{\alpha\alpha}$ is the magnetic hyperfine field, g is the Landé factor and μ_B is the Bohr magneton. Equation (4) shows that, in addition to the measurement of the electronic static susceptibility on different atomic sites, shift measurements also give information about the electronic structure through the determination of the magnetic hyperfine field.

SINGLE SPIN FLUID MODEL

In the case of YBCO or (LaSr)CuO systems, the superconducting phase is obtained from the antiferromagnetic phase by doping i.e. by introducing holes, these holes having a O(2p) character. The Cu configuration is approximately $3d^9$ (with a hole in the $3d_{x^2-y^2}$ orbital), so that each copper has an electron spin. There was an intense debate concerning the interaction between the local Cu(3d) holes and the doped O(2p) holes. Within the single band t-J model, Zhang and Rice[1] have proposed that the hole introduced by doping on each square of O atoms strongly hybridizes to the central Cu^{2+} ion to form a local singlet. This singlet then moves through the lattice of Cu^{2+} ions in a similar way to a hole. Many NMR studies have been devoted to answering the following question; is there only one single spin-fluid or does a second spin degree of freedom related an O(2p) band exist? In the latter case one should measure with NMR a spin susceptibility due to Cu electron spins and an independent contribution due to the O(2p) doped holes.

Knight shift studies

Before showing how NMR could answer to this question, it is first worth summarising, some of the early NMR results concerning the Knight shift tensor. The analysis of the anisotropy, sign and magnitude of the spin part of the shift tensor enables one to deduce information about the hyperfine coupling between the nuclear spins and the electronic spins.

Based on NMR results[2,3] in $YBa_2Cu_3O_7$ it was shown that the copper nuclear spin couples to the on-site Cu electron spin as well as to the four nearest neighbour Cu electron spins, so that in addition to a direct hyperfine interaction there is a transferred hyperfine interaction. In the latter case, the $3d_{x^2-y^2}$ orbitals of the four neighbouring sites are hybridized to the on-site Cu(4s) orbitals through O(2p) orbitals[4]. For oxygen, the axial part of ^{17}K is defined as $^{17}K_{ax} = [2\, ^{17}K_{//} - (^{17}K_{\perp}+^{17}K_c)]/6$ where $^{17}K_{//}$, $^{17}K_{\perp}$ and $^{17}K_c$ are respectively the components along the Cu-O axis, perpendicular to the bond in the CuO_2 plane and along the c axis. As was shown by various groups[5,6] the spin part of $^{17}K_{ax}$ couples to the O(2p) states through the spin density associated with the 2p states. On the other hand, the spin part of the isotropic shift, $^{17}K_{iso}$, is coupled to the 2s states, the spin density on O(2s) states being due to a non zero admixture Cu(3d)-O(2s). As for yttrium, the analysis of the isotropic shift[7] has shown that the ^{89}Y nuclear spins couple with the CuO_2 planes through the O(2p) orbitals. The same analysis was done later, for calcium atoms separating the CuO_2 planes in BiSrCaCuO compounds[8].

From these analyses it was thus inferred that ^{63}K and ^{17}K$_{iso}$ probe mainly the Cu spin susceptibility whereas ^{89}K$_{iso}$ and ^{17}K$_{ax}$ probe the spin density in the oxygen O(2p) orbitals (an extensive analysis can be found in ref. 9)

These results are an important step in order to distinguish between a two component model for which the doped holes have their own spin degree of freedom and a one component model for which the spin system of doped holes and the Cu electronic spins do not behave independently from each other.

The first NMR experiments showing that cuprates may be described by a single electronic spin system were performed in underdoped compounds for which, in the normal state, the spin susceptibility defined in eq (4), shows a strong decrease when the temperature is lowered. In $YBa_2Cu_3O_{6+x}$, ^{89}K$_{iso}$(T) was compared to the thermal variation of the macroscopic susceptibility χ_m[10]. The macroscopic susceptibility χ_m is the sum of various contributions : the spin susceptibility (Pauli-like contribution), a paramagnetic Van Vleck orbital term and a core diamagnetic term. Among these contributions, only the spin contribution is T-dependent. A good scaling was found between ^{89}K$_{iso}$(T) and χ_m(T). As χ_m(T) is dominated by the Cu spin contribution whereas ^{89}K$_{iso}$(T) probes the O(2p) spin polarization, these results are in agreement with a single spin fluid model. On the other hand, the T-dependencies of ^{63}K$_{ab}$(T), ^{17}K$_{iso}$(T) and ^{17}K$_{ax}$(T) were compared[11] in $YBa_2Cu_3O_{6.63}$. It was shown that ^{63}K$_{ab}$(T) and ^{17}K$_{iso}$(T) on the one hand and ^{17}K$_{ax}$(T) on the other hand, have the same thermal dependence. All these results demonstrate that the spin density on the Cu(3d) and O(2p) states behave as parts of a single spin system which can be associated with a strong hybridization of the Cu($3d_{x^2-y^2}$)-O(2pσ) orbitals. The same conclusion was drawn from Cu and O NMR shift measurements in $La_{1.85}Sr_{0.15}CuO_4$[12] and $YBa_2Cu_4O_8$[13].

It should be pointed out that some ^{17}O and ^{89}Y NMR studies[7,11] were devoted to try to obtain evidence of an independent spin susceptibility associated to the O(2p) doped holes. Up to now, the conclusion was that within the experimental accuracy, such a spin suceptibility was not detected. Nevertheless this issue is discussed in a recent publication where the author suggests to reconsider a second band associated with the O(2p$_\pi$) orbitals[14].

The nuclear Hamiltonian, spin-lattice relaxation and form factors

Considering that the electronic spins can be described within a single spin fluid model, since the extra holes go into O(2pσ) orbitals which are strongly hybridized with Cu $3d_{x^2-y^2}$, Mila and Rice[4] have proposed for Cu in the CuO_2 plane the following spin part for the magnetic hyperfine Hamiltonian where only the Cu electron spins S are considered:

$$^{63}H = \hbar\,^{63}\gamma\,^{63}I\,.\,[\mathbf{A.S} + \mathbf{B}.\sum_{i=1}^{4}\mathbf{S}_i]$$ (5.a)

The Hamiltonian was extended[15] to ^{17}O :

$$^{17}H = \hbar\,^{17}\gamma\,^{17}I\,.\,\mathbf{C}.\sum_{j=1}^{2}\mathbf{S}_j$$ (5.b)

and to ^{89}Y (or ^{43}Ca, or ^{205}Tl) nucleus separating the CuO_2 planes:

$$^{89}H = \hbar\,^{89}\gamma\,^{89}I\,.\,\mathbf{D}.\sum_{k=1}^{8}\mathbf{S}_k$$ (5.c)

where $\mathbf{A}$ is the ^{63}Cu on-site hyperfine coupling tensor, $\mathbf{B}$ is the ^{63}Cu isotropic transferred hyperfine coupling constant from the four nearest neighbour Cu spins, $\mathbf{C}$ is the ^{17}O transferred hyperfine coupling tensor from its two nearest neighbour Cu spins and $\mathbf{D}$ is the ^{89}Y (or ^{43}Ca, or ^{205}Tl) transferred hyperfine coupling tensor from its eight nearest neighbour Cu spins. From eqs. (4) and (5) the spin part of the shift, for instance in the case of ^{63}Cu, can be now expressed as:

$$^{63}K_c^{spin}(T) = (A_c + 4B)\,\chi/\,g\,\mu_B \qquad\qquad ^{63}K_{ab}^{spin}(T) = (A_{ab} + 4B)\,\chi/\,g\,\mu_B \qquad\qquad (6)$$

where the c component is obtained for $\mathbf{H_0}$ perpendicular to the CuO_2 plane and the ab component for $\mathbf{H_0}$ lying in the plane. Within the model, the spin susceptibility is assumed to be isotropic.

Now that the magnetic hyperfine coupling tensors are precisely defined by the Mila and Rice Hamiltonian, let us consider the spin-lattice relaxation. As mentioned above, the fluctuating part of the magnetic hyperfine field is responsible for the spin-lattice relaxation rate T_1^{-1}, where T_1^{-1} is the rate at which the disturbed nuclear magnetization relaxes back to its thermal equilibrium with its surrounding (the "lattice"). In cuprates, the nuclear relaxation rate is related to spin fluctuations through the spin part of the magnetic hyperfine couplings. After Moriya[16], T_1^{-1} divided by the temperature is given by:

$$^n\left(\frac{1}{T_1 T}\right)_\alpha = \frac{\gamma_n^2 k_B}{2\mu_B^2} \sum_{q,\,\alpha'\neq\alpha} |^n A_{\alpha\alpha'}(\mathbf{q})|^2\; \frac{\chi''_{\alpha\alpha'}(\mathbf{q},\omega)}{\omega} \qquad\qquad (7)$$

where α is the direction of $\mathbf{H_0}$ and α' denotes the directions perpendicular to α, ω is the resonance angular frequency of the nuclear species n. T_1^{-1} provides information about the imaginary part of the q-dependent dynamical electron spin susceptibility $\chi(\mathbf{q},\omega)$. The physical meaning of $\chi''(\mathbf{q},\omega)$ is the following. $\chi''(\mathbf{q},\omega)$ is proportional to the dynamic structure factor $S(\mathbf{q},\omega)$ through the fluctuation dissipation theorem where $S(\mathbf{q},\omega)$ is obtained by the double Fourier transform of the time-dependent correlation function of the electron spins $\mathbf{S}$:

$$S(\mathbf{q},\omega) = \int \exp i\,(\omega t - \mathbf{q.r}\,) <S(\mathbf{r},t)\,S(0,0)> d\mathbf{r}\, dt \approx kT\,\frac{\chi''_{\alpha\alpha'}(\mathbf{q},\omega)}{\omega} \qquad\qquad (8)$$

It is worth to note that $S(\mathbf{q},\omega)$ is directly related to the cross section that one measures in inelastic neutron scattering experiments.

The form factors $|^n A_{\alpha\alpha}(\mathbf{q})|^2$ entering in eq. 7 are the square of the Fourier transformed q-dependent hyperfine coupling field where $^n A_{\alpha\alpha}(\mathbf{q})$ are defined as:

$$^n A_{\alpha\alpha}(\mathbf{q}) = \sum_j\, ^n A_{j,\alpha\alpha}\, \exp(i\mathbf{q.r_j}) \qquad\qquad (9.a)$$

where α denotes the direction of the fluctuating field. In eq (9.a), $\mathbf{r_j}$ is zero for the direct (on-site) hyperfine coupling field and $\mathbf{r_j} \neq 0$ for the transferred hyperfine coupling field. For Cu, O and Y sites respectively, $^n A_{\alpha\alpha}(\mathbf{q})$ are expressed as follows:

$$^{63}A_{\alpha\alpha}(\mathbf{q}) = A_{\alpha\alpha} + 2B\left[\cos(q_x a) + \cos(q_y a)\right] \qquad\qquad (9.b)$$

$$^{17}A_{\alpha\alpha}(\mathbf{q}) = 2C_{\alpha\alpha}\cos\left(\frac{q_x a}{2}\right) \qquad\qquad (9.c)$$

$$^{89}A_{\alpha\alpha}(\mathbf{q}) = 8D_{\alpha\alpha}\cos\left(\frac{q_x a}{2}\right)\cos\left(\frac{q_y a}{2}\right)\cos\left(\frac{q_z d}{2}\right) \qquad\qquad (9.d)$$

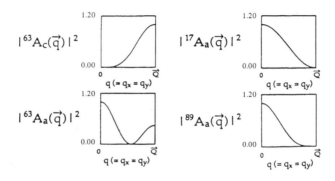

FIG. 1 The q-dependence of the form factors for ^{63}Cu, ^{17}O and ^{89}Y. From Slichter *et al.* (ref. 17)

where a is the lattice parameter in the CuO_2 planes and d is the distance between the two CuO_2 planes of the cell. As can be seen in Fig. 1(from ref. 17), the difference in the q-dependence of the form factors $\left|^nA_{\alpha\alpha}(q)\right|^2$ for Cu site on the one hand and for O and Y sites on the other hand, implies that T_1^{-1} measured at these sites will probe the same $\chi''(q,\omega)$ but at different parts of the q-space. Indeed at $q=Q_{AF}=(\pi/a,\pi/a)$ the form factor vanishes for O and Y sites but not for Cu site.

NMR STUDIES IN UNDERDOPED CUPRATES

We shall now describe some of the striking features which characterize the underdoped regime (for review papers see ref.18 and 19 and for a more recent review see ref. 20) i.e. the part of the phase diagram where T_c increases up to $T_{c,max}$ when the doping increases. $T_{c,max}$ corresponds to the optimal doping. The overdoped regime which corresponds to larger doping and for which T_c decreases when the doping increases will be described in the last part.

In the normal state of cuprates, the spin-lattice relaxation rates and shifts display unusual thermal behaviours compared to conventional metals.

A typical behaviour for the spin-lattice relaxation rates measured in underdoped compounds at Cu and O sites is shown for instance in $YBa_2Cu_3O_{6.63}$[11] with T_c= 62K on fig. 2. Both quantities $^{63}(T_1T)^{-1}$ and $^{17}(T_1T)^{-1}$ have a strong temperature dependence above T_c, contrary to the usually temperature independent behaviour found in metals where the Korringa relation $(T_1T)^{-1}$ =const. holds. When T is lowered, $^{17}(T_1T)^{-1}$ exhibits a monotonous decrease whereas $^{63}(T_1T)^{-1}$ first increases, with a nearly Curie-Weiss law, until a temperature $T^* \approx 140K$ and then decreases rapidly. No peculiar behaviour is evidenced at T_c= 62K. The same T-dependences was found for O and Cu relaxation respectively in $YBa_2Cu_4O_8$ where $T^* \approx 130K$ and T_c= 82K[21]. These results were interpreted as the existence of spatial antiferromagnetic correlations of the spin fluctuations at the wave vector $q=Q_{AF}$ (in short AF fluctuations) implying that $\chi''(q,\omega)$ has a peak at Q_{AF}. This is in agreement with neutron scattering experiments which evidence the presence of such short-range AF fluctuations. Thus, the very different T-dependences for O and Cu sites relaxation rate is due to the fact that O does not see the AF fluctuations whereas Cu does. In other words, $^{63}T_1^{-1}$ probes mainly $\chi''(Q_{AF},\omega)$ whereas for $^{17}T_1^{-1}$ and $^{89}T_1^{-1}$ it is mainly $\chi''(q=0,\omega)$.

As mentioned above, in the underdoped compounds, the Knight shift and thus the spin susceptibility $\chi'(q=0,\omega=0)$ is temperature dependent. It decreases when the temperature is

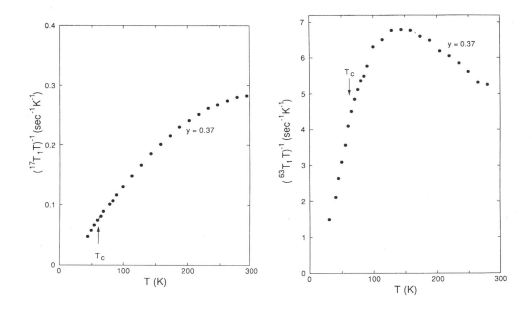

FIG. 2. Comparison of the T- dependence of $1/(^{17}T_1T)$ (left panel) and of $1/(^{63}T_1T)$ (right panel) at the O and Cu sites in CuO_2 layer in underdoped $YBa_2Cu_3O_{7-y}$. Reprinted with permission, from Takigawa *et al.* (ref. 11).

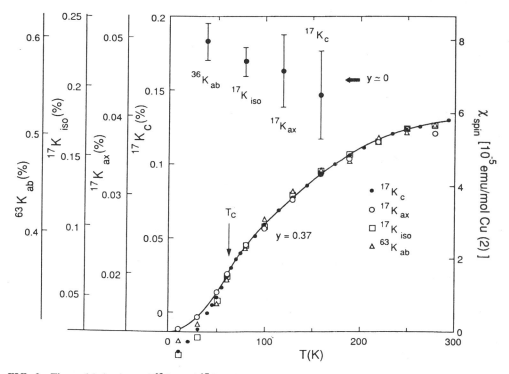

FIG. 3. Thermal behaviour of ^{63}Cu and ^{17}O Knight shift components in $YBa_2Cu_3O_{7-y}$. Reprinted with permission from Takigawa *et al.* (ref. 11).

337

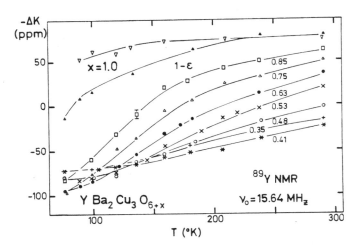

FIG. 4. Shift of the [89]Y line plotted versus T. Reprinted with permission from Alloul *et al.* (ref.10).

reduced. Typical behaviour is shown in fig. 3. for [17]O and [63]Cu line shifts in $YBa_2Cu_3O_{6,63}$[11]. From these measurements it can be deduced that the static spin susceptibility at $T=T_c$ is already reduced by a factor of about 3 compared to its value at room temperature. Furthermore, NMR studies of the T-dependence of the spin susceptibility as a function of the doping show that as the doping increases, the decrease of the spin susceptibility with T diminishes, and $\chi'(0,0)$ becomes constant at about the optimal doping as seen on fig. 4 .

For a quantitative analysis of NMR data, a theoretical model is needed in order to describe $\chi(q,\omega) = \chi'(q,\omega) + i \chi''(q,\omega)$. A model describing the spin fluctuations in nearly antiferromagnetic 2D itinerant electron systems, based on the self-consistent renormalization theory of spin fluctuations used in 3D systems, was proposed by T. Moryia *et al.*[22]. This model explains the Cu spin-lattice relaxation rate well. At the same time, Millis, Monien and Pines[23] proposed a phenomenological model (MMP) for $\chi(q,\omega)$ describing the system as a nearly antiferromagnetic Fermi liquid in which the dominant contribution comes from the antiferromagnetically correlated spins. The MMP model was widely used in order to extract detailed characters of the spin fluctuations such as the characteristic energy of the AF spin fluctuations or the static correlation length, ξ, from NMR data. The first quantitave analysis in an underdoped compound derived from this model is found in ref. 24. In the MMP model, a good fit is obtained for the experimental values of the relaxation rates at Cu, O and Y sites provided that the square of the static correlation length, ξ^2 , varies with temperature as $1/T$. This is in contradiction with inelastic neutron scattering studies[25] in which ξ is found to be T-independent, the origin of this discrepancy with the MMP model is still unclear.

Spin gap

The anomalous dependence of the Knight shift and the spin-lattice relaxation in underdoped compounds characterizes the existence of a gap in the spin excitation spectrum. Below T* the decrease of $(^{63}T_1T)^{-1}$ corresponds to the opening of the spin gap at Q_{AF} and may be interpreted as a transfer of low energy antiferromagnetic excitations of the spectral weight of $\chi''(q,\omega)$ to a high energy region. The opening of a gap in the spin excitations at

Q_{AF} was also discovered by inelastic neutron scattering[26]. The anomalous behaviour of $(^{63}T_1T)^{-1}$ was first shown[27,28] in $YBa_2Cu_3O_{6+x}$. Such a spin gap behaviour was also found in the underdoped two layered compounds $Bi_2Sr_2CaCu_2O_{8+x}$[29] (for which $T^* \approx 140K$ when $T_c = 90K$), $Tl_2Ba_2CaCu_2O_{8-\delta}$[30], $Y_2Ba_4Cu_7O_{15}$[31] and $TlSr_2(Lu,Ca)Cu_2O_y$[32] as well as in the three layered compound $HgBa_2Ca_2Cu_3O_{8+\delta}$[33] (with $T_c \approx 115K$ for which $T^* \approx 230K$). From all these measurements it has been observed that T^* decreases as the doping increases, getting very close to T_c at optimal doping.

Let us now consider the static spin susceptibility. The thermal behaviour generally observed in underdoped systems is as follows: $\chi'(0,0)$ starts to decrease slowly at a temperature usually called T_o ($T_c \leq T^* \leq T_o$) and then decreases more rapidly as the temperature is further reduced. The decrease at T_o is interpreted as the opening of a pseudogap at $q=0$. Such a pseudogap was observed in all the two and three layered compounds investigated. In the three layered compound $(Bi,Pb)_2Sr_2Ca_2Cu_3O_{10+x}$[34], the T-dependence of the ^{17}O Knight shift displays a pseudogap behaviour for the inner CuO_2 plane whereas for the outer ones ^{17}K is constant. This was interpreted as due to a smaller hole concentration in the inner plane. The same situation was observed in $(Tl,Pb)Sr_2Ca_2Cu_3O_{9-\delta}$[35] and same conclusion was drawn. In $HgBa_2Ca_2Cu_3O_{8+\delta}$[33] both types of layer display a pseudogap behaviour since there is about no difference in doping between the two types of CuO_2 layers. It is worth noting that the difference in hole concentration between the two types of copper planes disappears when the overall doping level of the compound decreases, as was shown in $Tl_2Ba_2Ca_2Cu_3O_{10-\delta}$ compounds[36].

In most cuprates T_o is too high to be measured without loosing some oxygen content. Nevertheless T_o was clearly evidenced ($T_o \approx 500K$) in $YBa_2Cu_4O_8$ by $^{63}K(T)$ measurements up to $700K$[37] as shown on fig. 5.

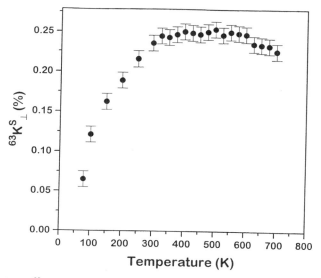

FIG. 5. ^{63}Cu Knight shift as a function of temperature in $YBa_2Cu_4O_8$. Reprinted with permission from Curro *et al.* (ref.37)

Analogously to the thermal behaviour of NMR shifts in underdoped cuprates, the T-dependence of the in-plane resistivity ρ_{ab} demonstrates a characteristic temperature T_ρ ($T_\rho > T_c$) above which ρ_{ab} has a linear T-dependence and below which $\rho_{ab}(T)$ deviates from linearity, its decrease with decreasing temperature becoming steeper[38]. This characteristic temperature T_ρ, derived from transport measurements is usually found as being close to T_0 determined by NMR. Nevertheless, in some other analyses T_ρ is considered to be close to T^* (in the case of $YBa_2Cu_4O_8$)[39]. As transport and magnetic properties are governed by the same characteristic temperature it is considered as indicative that the dominant charge carriers scattering mechanism can be of magnetic origin[40] and a pseudogap can be associated to the steeper decrease of $\rho_{ab}(T)$. The microscopic magnetic scattering mechanism of the charge carriers is still under debate.

It is worth noting that also by angle-resolved photoemission spectroscopy (ARPES) a pseudogap in the normal state was observed recently in underdoped $Bi_2Sr_2CaCu_2O_{8+x}$ compounds[41]. How the pseudogaps measured by magnetic or transport measurements or by ARPES are connected is still an open question.

Many theoretical works have been devoted to the spin gap nevertheless, its physical origin is not yet clearly understood. It has been described in terms of nesting properties of the band structure within the 2D Hubbard model[42] or within the t-J model[43]. On the other hand, within the frame of the scaling theory and the MMP approach, it was assumed that the electronic spin system exhibits two magnetic crossovers[44]. At high temperature the system crosses over from a non universal mean field regime (when $T>T_o$) to a quantum critical scaling regime (for $T^*<T<T_o$). The second crossover which occurs at $T=T^*$ corresponds to the onset of a quantum critical disorder regime.

One important issue concerning the spin gap was whether it is intrinsic to the underdoped systems or whether it is specific to bilayered and trilayered underdoped compounds. In the latter case, the magnetic coupling between CuO_2 layers in the bi- or trilayered materials would play a central role for the spin gap formation. Indeed, the spin gap was not clearly observed in underdoped $La_{2-x}Sr_xCuO_4$. Some $(^{63}T_1T)^{-1}$ NMR studies show an indication of the presence of the spin gap[45,46] and some show its absence[47]. On the other hand, the spin susceptibility measured by Cu NMR extrapolates to a positive value at $T=0K$[48] whereas it extrapolates to a negative value for other systems. This fact was interpreted as the absence of the pseudogap[49]. All these results were the only ones available in underdoped single layered cuprates until recently. Very recently, several groups have clearly demonstrated a spin gap behaviour in the single layer cuprates $HgBa_2CuO_{4+\delta}$. First, the T-behaviour of $(^{63}T_1T)^{-1}$, displays a broad maximum well above T_c[50,51], the values of T^* as a function of, δ, the oxygen concentration being $T^* \approx$ 260K, 200K and 160K for underdoped samples, with T_c=50K, 72K and 80K, respectively[50] and $T^* \approx$ 120K when T_c=96K[51]. It confirms also that T^* decreases, getting closer to T_c as the doping increases. Secondly, the existence of the $q=0$ pseudogap has been evidenced[52] by measuring the T-dependence of ^{17}O Knight shift, demonstrating that the spin susceptibility extrapolates to a negative value at $T=0K$, in contrast to $La_{2-x}Sr_xCuO_4$. Thus, it is now established that the spin gap behaviour is intrinsic to the underdoped regime irrespective of the number of CuO_2 planes in the unit cell demonstrating that the role of magnetic coupling between the CuO_2 planes in the unit cell is not dominant. This is consistent with Knight shift results in three layered compounds where both types of CuO_2 layers have not the same T-dependence for $\chi'(0,0)$ showing thus a weak magnetic interaction between the CuO_2 planes.

THE OVERDOPED REGIME

Compared to the underdoped regime, the overdoped one has been much less studied. Furthermore, there is even less work concerning the region at very high doping, near the superconducting-to-metal transition which can be observed only in few systems. Overdoped compounds show very different behaviours for shifts and relaxation rates compared to the underdoped ones.

340

For some lightly overdoped compounds, $Tl_2Ba_2CuO_{6+x}$ ($T_c=85K$)[53], $Bi_2Sr_2CaCu_2O_{8+x}$($T_c=86K$)[54] or $Tl_2Ba_2CaCu_2O_{8+x}$ ($T_c=103K$)[55], the static spin susceptibility has about the same behaviour as in the underdoped regime whereas for other lightly overdoped systems like $YBa_2Cu_3O_7$[56] or $Bi_2Sr_2CaCu_2O_{8+x}$($T_c=77K$)[29], $\chi'(0,0)$ increases linearly when T is reduced. For more heavily doped compounds the static susceptibility is either almost constant down to T_c as in $Tl_2Ba_2CuO_{6+x}$ ($T_c=42K$ and $T_c\approx0K$)[57] or it increases, in most of the cases linearly, when T is lowered as in $La_{1.8}Sr_{0.2}CuO_4$[58], $Bi_2Sr_2CuO_{6+x}$($T_c\approx10K$)[59] and $(Tl_{0.8}Pb_{0.2})Sr_2CaCu_2O_7$ ($T_c\approx0K$)[60]. Such behaviour is illustrated on fig. 6 and 7 which display the thermal dependence of ^{17}O and ^{43}Ca Knight shift respectively in the same $(Tl_{1-x}Pb_x)Sr_2(Ca_{1-y}Y_y)Cu_2O_7$ compounds. As can be seen in fig.6, the slope is about the same in the overdoped compound ($T_c\approx65K$) and in the strongly overdoped one ($T_c\approx0K$), the main difference being the abrupt decrease of $\chi'(0,0)$ at T_c (for the former sample) due to the singlet spin pairing in the superconducting state. The T-dependence of $\chi'(0,0)$ is illustrated in fig. 7 for a wide range of doping, from the underdoped to the overdoped regime, in the same system. These results show that as soon as the doping is large enough, one cannot define a temperature T_o and thus there is no pseudogap at $q=0$ for $T>T_c$.

In other respects, the single spin fluid model is still valid in overdoped materials as it was shown in lightly overdoped $Tl_2Ba_2CuO_{6+x}$[53] and $Tl_2Ba_2CaCu_2O_{8+x}$[55] as well as in strongly overdoped $(Tl_{1-x}Pb_x)Sr_2CaCu_2O_7$[61] compounds .

The spin gap at $q=Q_{AF}=(\pi/a,\pi/a)$ is also suppressed in overdoped compounds as can be seen in fig.8 from the monotonous increase of $(^{63}T_1T)^{-1}$ when the temperature is lowered down to T_c in $HgBa_2CaCu_2O_{6+x}$[63]. At T_c, $(^{63}T_1T)^{-1}$ exhibits a sharp decrease. A quite similar behaviour is measured for $(^{17}T_1T)^{-1}$ in overdoped $YBa_2Cu_3O_7$[56]. Nevertheless, for some lightly overdoped compounds $Tl_2Ba_2CuO_{6+x}$[53] and $LaBa_2Cu_2O_y$[64], the spin gap behaviour still exists indicating thus that the value of the doping where the spin gap behaviour disappears may not necessarily coincide with the optimal doping which corresponds to the maximum value of T_c.

The spin gap being suppressed in the overdoped regime, an open question was whether the AF fluctuations vanish or not. The strength of AF fluctuations can be estimated from the ratio of the measured spin-lattice relaxation rates at Cu and O sites. It is more convenient to introduce the experimental ratio : $^{63,17}R = (^{63}T_1^{-1}/ {^{17}T_1^{-1}}) (\gamma_{63}^2/\gamma_{17}^2)^{-1}$. In the limit where the AF fluctuations at $Q_{AF}=(\pi/a,\pi/a)$ dominate, by taking into account eqs. 7 and 9, one can approximate the ratio as :

$$\frac{\left(^{63}T_1\right)^{-1}}{\left(^{17}T_1\right)^{-1}}\left(\frac{\gamma_{63}^2}{\gamma_{17}^2}\right)^{-1} \approx \frac{\chi''(Q_{AF})}{\chi''(q=0)} \tag{10}$$

which measures the enhancement of $\chi''(Q_{AF})$ compared to $\chi''(q=0)$. In the opposite limit where the spins are uncorrelated, the form factors $|^nA_{\alpha\alpha}(q)|^2$ become q-independent as well as χ''. In this case, for instance for Cu, the form factors become $A_{ab,c}^2 + 4B^2$ and one may write:

$$^{63,17}A = \frac{\left(^{63}T_1\right)^{-1}}{\left(^{17}T_1\right)^{-1}}\left(\frac{\gamma_{63}^2}{\gamma_{17}^2}\right)^{-1} = \frac{A_{ab}^2 + 4B^2}{C_{//}^2 + C_\perp^2} \tag{11}$$

when H_0 is parallel to the c axis. Thus, the value of $(^{63,17}R / {^{63,17}A})$, where the hyperfine fields A_{ab}, B and $C_{//}$ and $C_\perp$ are deduced from the Knight shift analysis, enables an estimation to be made of the magnitude of the AF fluctuations. The ratio is equal to unity in the limit of uncorrelated spins. Such an analysis was done in the overdoped $Tl_2Ba_2CuO_{6+x}$ compound[53] yielding the thermal dependence of $(^{63,17}R / {^{63,17}A})$ shown in fig. 9. Although the AF fluctuations are weakened compared to the optimal $YBa_2Cu_3O_{6.96}$ they still exist in the overdoped regime.

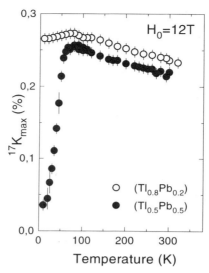

FIG. 6. ^{17}O shift vs temperature in overdoped $(Tl_{0.5}Pb_{0.5})Sr_2CaCu_2O_7$ with $T_c \approx 65K$ (open circle) and in strongly overdoped $(Tl_{0.8}Pb_{0.2})Sr_2CaCu_2O_7$ with $T_c \approx 0K$ (closed circle). Reprinted from Physica C, Bellot *et al.* (ref.60).

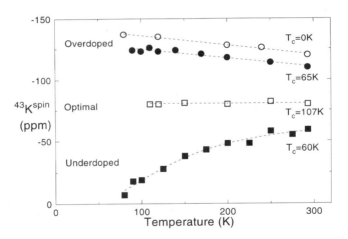

FIG. 7. Temperature dependence of the spin susceptibility measured by ^{43}Ca Knight shift from underdoped to strongly overdoped samples in the $(Tl_{1-x}Pb_x)Sr_2(Ca_{1-y}Y_y)Cu_2O_7$ system. From Bellot *et al.* (ref.62).

Other studies have demonstrated the persistence of the AF fluctuations in strongly overdoped compounds with no spin gap. Indeed, the presence of induced localized magnetic moments has been detected in the overdoped compounds $(Tl_{1-x}Pb_x)Sr_2CaCu_2O_7$ and $Bi_2Sr_2CaCu_2O_{8+x}$[65]. This fact has to be compared with the studies on substitutional impurities on the copper sites which mainly concern the underdoped regime[66-71]. One of the aspects of all these studies in correlated systems is to provide informations about the AF fluctuations. It was shown that non magnetic Zn^{2+} impurities induce a magnetic moment on its four nearest neighbours in underdoped compounds[68] and, as pointed out, the fact that the suppression of a spin (at Zn site) induces moments on the neighbouring Cu sites is a strong

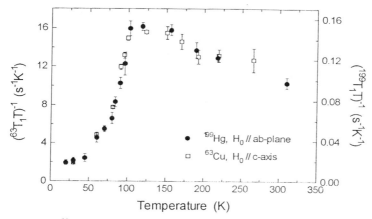

FIG. 8. $(^{63}T_1T)^{-1}$ vs. T in overdoped $HgBa_2CaCu_2O_{6+x}$. Reprinted from Physica C, Julien *et al.* (ref. 63)

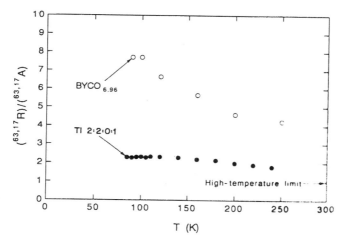

FIG. 9. T-dependence of $(^{63,17}R / ^{63,17}A)$ in overdoped $Tl_2Ba_2CuO_{\delta+x}$ and in $YBa_2Cu_3O_{6,96}$. Reprinted with permission, from Kambe *et al.* (ref. 53).

evidence that the CuO_2 planes are correlated electronic systems. This seems to be still valid for overdoped compounds. In overdoped $(Tl_{1-x}Pb_x)Sr_2CaCu_2O_7$ ($T_c=65K$ and $T_c \approx 0K$) and $Bi_2Sr_2CaCu_2O_{8+x}$ ($T_c=75K$) in which no magnetic impurities were included among the starting compounds during the preparation, the presence of localized magnetic moments was demonstrated from the thermal variation of the ^{17}O linewidth. As shown in fig.10, the linewidth corresponding to oxygen in CuO_2 layers can be well fitted with a C/T law. The Curie-like behaviour of the linewidth can be understood in terms of the Ruderman-Kittel-Kasuya-Yosida (RKKY) indirect interaction between the nuclear spins and the magnetic moments through the mobiles carriers[66-70]. By contrast, almost no thermal variation is

343

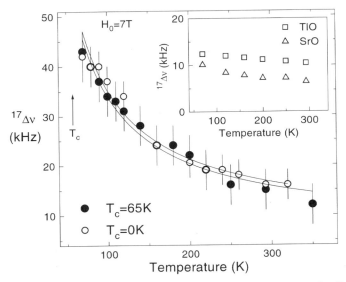

FIG. 10. T-dependence of the contribution to the linewidth due to localized magnetic moments for O in the CuO_2 plane in overdoped $(Tl_{0.5}Pb_{0.5})Sr_2CaCu_2O_7$ (T_c=65K) and $(Tl_{0.8}Pb_{0.2})Sr_2CaCu_2O_7$ ($T_c \approx 0K$). The solid curves are the best C/T fits. Inset: T-dependence of the linewidth of O in TlO and SrO planes. Reprinted from Physica C, Bellot *et al.* (ref.65).

detected for oxygen in apical (SrO) and reservoir (TlO) layers (fig. 10). These behaviours demonstrate that the magnetic moments are located in the conducting CuO_2 layers. It is worth to point out that, with macroscopic magnetic susceptibility measurements, the presence of localized magnetic moments was detected in other overdoped systems like $Tl_2Ba_2CuO_{6+x}$[72] and $La_{2-x}Sr_xCuO_4$[73]. In the latter cuprate, the presence of localized magnetic moments was also demonstrated by ^{63}Cu NMR linewidth measurements[46] and it was tentatively proposed that the magnetic moments were due to Sr dopant. Although the origin of the defects which produce magnetic moments is not yet clear, the fact that magnetic moments are induced shows that overdoped compounds are correlated systems. Besides, it raises a question about a possible connection between overdoping and the presence of defects.

These results demonstrate that strongly overdoped systems do not behave like conventional metals so that more studies are needed in overdoped systems before being able to describe their normal state.

CONCLUSION

Concerning the underdoped regime, most of the important results were obtained in YBCO and LaSrCuO systems before 1991. Since that time the studies on other cuprates have shown that the following characteristic behaviours are generic of all underdoped compounds i.e. the existence of a pseudogap which shows up at $T=T_0$ in the thermal behaviour of the Knight shift and at $T=T^*$ in the T-dependence of the copper spin-lattice relaxation with $T_c<T^*<T_0$. The physical origin of the pseudogap is not clearly understood. Furthermore its connection with the one observed by resistivity measurements or by ARPES is still an open question.

Because a lack of space we did not discuss the copper spin-spin relaxation data which yield complementary informations about the staggered spin susceptibility $\chi'(q=Q_{AF},\omega=0)$, we refer the reader to ref.74.

As it was shown, the nuclear spin Hamiltonian is expressed with only one relevant variable in the system, this variable being the electron spin S localised at the Cu site. One spin degree of freedom is enough to explain most of the data. The existence of a second degree of freedom due to an oxygen band has not yet been clearly evidenced.

Overdoped compounds were less studied. The fact that the T-dependence of the spin relaxation or the Knight shifts are not as "generic" as in underdoped compounds may be attributed or to the definition of overdoping i.e. to the determination of optimal doping or to the fact that overdoped samples have not the same high quality as YBCO systems.

In the overdoped regime, although the spin gap is suppressed, the AF fluctuations still persist. The fact that localized magnetic moments were detected in the CuO planes show that overdoped systems should not be too hastily considered as conventional metals. There may be also a relation between overdoping and defects which could give a contribution to the decrease of T_c in the overdoped regime due to pair breaking. The sample quality of overdoped compounds should be improved and more studies are needed before being able to describe their normal state.

ACKNOWLEDGEMENT

The author would like to thank P.-V. Bellot and H. Alloul for helpful discussions.

REFERENCES

1. F. C. Zhang and T. M. Rice, Effective Hamiltonian, *Phys. Rev. B* 37:3759 (1988).
2. C. H. Pennington, D. J. Durand, C. P. Slichter, J. P. Rice, E. D. Bukowski and D. M. Ginzberg, Static and dynamic Cu NMR tensors of $YBa_2Cu_3O_{7-\delta}$, *Phys. Rev. B* 39:2902 (1989).
3. M. Takigawa, P. C. Hammel, R. H. Heffner and Z. Fisk, Spin susceptibility in superconducting $YBa_2Cu_3O_7$ from ^{63}Cu Knight shift, *Phys. Rev. B* 39:7371(1989).
4. F. Mila and T. M. Rice, Analysis of magnetic resonance experiments in $YBa_2Cu_3O_7$, *Physica C* 157:561 (1989).
5. M. Takigawa, P. C. Hammel, R. H. Heffner, Z. Fisk, K. C. Ott and J. D. Thomson, ^{17}O NMR study of local susceptibility in aligned $YBa_2Cu_3O_7$ powder, *Phys. Rev. B* 43:247 (1989).
6. M. Horvatic, Y. Berthier, P. Butaud, Y. Kitaoka, P. Segransan, C. Berthier, H. Katayama-Yoshida, Y. Obabe and T. Takahashi, NMR Study of $YBa_2Cu_3O_{7-\delta}$ (T_c=92K) *Physica C* 159:689 (1989).
7. H. Alloul, T. Ohno and P. Mendels, ^{89}Y NMR Study of the electronic and magnetic properties of the 123 compounds, *J. Less. Common Metals* 164-165:1022 (1990).
8. A. Trokiner, L. Le Noc, A. Yakubovskii, K. N. Mykhalyov and S. V. Verkhovskii, ^{43}Ca NMR study of bismuth-based high-T_c superconductors, *Z. Naturforsch.* 49a:373 (1994)
9. H. Alloul, Experimental studies of magnetic properties of cuprates-from the antiferromagnetic to metallic State, *Scottish Universities Summer School in Physics. St Andrews*, edited by D. P. Tunstall and W. Badford (1991).
10. H. Alloul, T. Ohno and P. Mendels, ^{89}Y NMR evidence for a Fermi-liquid behaviour in $YBa_2Cu_3O_{6+x}$, *Phys. Rev. Lett.* 61:1700 (1989). Copyright(1989). The American Physical Society.
11. M. Takigawa, A. P. Reyes, P. C. Hammel, J. D. Thomson, R. H. Heffner, Z. Fisk and K. C. Ott, Cu and O NMR studies of the magnetic properties of $YBa_2Cu_3O_{6.63}$, *Phys. Rev. B* 43:247 (1991). Copyright(1991). The American Physical Society.
12. K. Ishida, Y. Kitaoka, G.-Q. Zheng and K. Asayama, 17O and 63Cu NMR investigation of high-Tc superconductor $La_{1.85}Sr_{0.15}CuO_4$, *J. Phys. Soc. Jap* 60:3516 (1991).

13. I. Mangelschots, M. Mali, J. Roos, D. Brinkman, S. Rusiecki, J. Karpinski and E. Kaldis, ^{17}O NMR study in aligned $YBa_2Cu_4O_8$ powder, *Physica C* 194:277 (1992).

14. Y. Yoshinari, Anisotropy of the Knight shift at planar oxygen sites in $YBa_2Cu_3O_x$:possible location of the doped holes, *Physica C* 276:147(1997).

15. B. S. Shastry, t-J Model and nuclear magnetic relaxation in high-Tc materials, *Phys. Rev. Lett.* 63:1288(1989).

16. T. Moryia, The effect of electron-electron interaction on the nuclear spin relaxation in metals, *J. Phys. Soc. Jpn.* 18:516 (1963).

17. C. P. Slichter, S. E. Barrett, J. A. Martindale, D. J. Durand, C. H. Pennington, C. A. Klug, K. E. O'Hara, S. M. DeSoto, T. Imai, J. Rice, T. A. Friedmann and D. M. Ginsberg, NMR studies of the superconducting state of copper oxide superconductors, *Appl. Magn. Reson.* 3:423 (1992).

18. M. Mehring (guest editor), *Appl. Magn. Res.* 3 (1992).

19. A. P. Kampf, Magnetic correlations in high temperature superconductivity, *Physics Reports* 249:219(1994).

20. C. Berthier, M. H. Julien, M. Horvatic and Y. Berthier, NMR Study of the normal state of high temperature superconductors,*J. Phys. I France* 6: 2205(1996).

21. I. Mangelschots, M. Mali, J. Roos, R. Stern, M. Bankay, A. Lombardi and D. Brinkman, Static and dynamic magnetic properties of Ba, Cu and O in $YBa_2Cu_4O_8$ and $Y_2Ba_4Cu_7O_{15}$, in *"Phase Separation in Superconductors"* K. A. Muller and G. Benedek Eds, World Scientific, 262 (1993).

22. T. Moryia, Y. Takahashi and K. Ueda, Antiferromagnetic spin fluctuations and superconductivity in two-dimensional metals- a possible model for high Tc oxides, *J. Phys. Soc. Jpn.* 599:2905 (1990).

23. A. J. Millis, H. Monien and D. Pines, Phenomenological model of nuclear relaxation in the normal state of $YBa_2Cu_3O_7$, *Phys. Rev. B* 42:167 (1990).

24. H. Monien, D. Pines and M. Takigawa, Application of the antiferromagnetic-Fermi-liquid theory to NMR experiments on $YBa_2Cu_3O_{6.63}$, *Phys. Rev. B* 43:258 (1991).

25. J. Rossat-Mignot, L. P. Regnault, C. Vettier, P. Bourges, P. Burlet, J. Bossy, J.Y. Henry and G. Lapertot, Spin dynamics in the high-Tc system $YBa_2Cu_3O_{6+x}$,*Physica B* 180-181:383 (1992).

26. J. Rossat-Mignot, L. P. Regnault, C. Vettier, P. Bourges, P. Burlet, J. Bossy, J.Y. Henry and G. Lapertot, Neutron scattering study of the $YBa_2Cu_3O_{6+x}$ system, *Physica C* 185-189:86 (1991).

27. W. W. Warren Jr., R. E. Walstedt, G. F. Brennert, R. J. Cava, R. Tycko, R. F. Bell and G. Dabbagh, Cu spin dynamics and superconducting precursor effects in planes above Tc in $YBa_2Cu_3O_{6.7}$,*Phys. Rev. Lett.* 62:1193 (1989).

28. M. Horvatic, P. Segransan, C. Berthier, Y. Berthier, P. Butaud, J. Y. Henri, M. Couach and J. P. Chaminade, NMR evidence for localized spins on Cu(2) sites from Cu NMR in $YBa_2Cu_3O_7$ and $YBa_2Cu_3O_{6.75}$ single crystals, Phys. Rev. B 39:7322 (1989).

29. R. E. Walstedt, R. F. Bell and D. B. Mitzi, Nuclear relaxation behavior of the superconducting cuprates: $Bi_2Sr_2CaCu_2O_{8+x}$, *Phys. Rev. B* 44:7760 (1991).

30. N. Winzek and M. Mehring, [205]Tl NMR in the cuprate superconductor $Tl_2Ba_2CaCu_2O_{8-\delta}$: evidence for a pseudo spin gap, Appl. Magn. Reson. 3:535(1992).

31. R. Stern, M. Mali, I. Mangelschots, J. Roos, D. Brinkmann, J. -Y. Genoud, T. Graf and J. Muller, Charge-carrier density and interplane coupling in $Y_2Ba_4Cu_7O_{15}$: a Cu NMR-NQR study, *Phys. Rev. B* 50:426 (1994).

32. K. Magishi, Y. Kitaoka, G.-q Zheng, K. Asayama, T. Kondo, Y. Shimakawa, T. Makano and Y. Kubo, Spin-gap behaviour in underdoped $TlSr_2(Lu_{0.7}Ca_{0.3})Cu_2O_y$:Cu and Tl NMR studies, *Phys. Rev. B*, 54:3070 (1996).

33. M.-H. Julien, P. Carretta, M. Horvatic, C. Berthier, Y. Berthier, P. Segransan, A. Carrington and D. Colson, Spin-gap in $HgBa_2Ca_2Cu_3O_{8+\delta}$ single crystals from [63]Cu NMR, *Phys. Rev. Lett.* 76:4238 (1996).

34. A. Trokiner, L. Le Noc, J. Schneck, A.- M. Pougnet, R. Mellet, J. Primot, H. savary, Y. M. Gao and S. Aubry, [17]O Nuclear magnetic resonance evidence for distinct carriers densities in the two types of CuO_2 planes of $(Bi,Pb)_2Sr_2Ca_2Cu_3O_y$, *Phys. Rev. B* 44:2426 (1991)

35. Z. P. Han, R. Dupree, A. P. Howes, R. S. Liu and P. P. Edwards, Charge distribution in $(Tl,Pb)Sr_2Ca_2Cu_3O_{9-\delta}$ (Tc=124K): an ^{17}O NMR study, *Physica C* 235-240:1709 (1994).

36. Yu. Piskunov, A. Geraschenko, K. Mikhalev, Yu. Zhdanov, S. Verkhovskii, A. Yakubovskii, A. Trokiner and P.-V. Bellot, Electric field gradient and hole concentration in inequivalent CuO_2 layers of $Tl_2Ba_2Ca_2Cu_3O_{10-\delta}$ with different Tc, *Physica C*, 282-287:1361(1997).

37. N. J. Curro, T. Imai, C. P. Slichter and B. Dabrowski, High temperature $^{63}Cu(2)$ Nuclear quadrupole and magnetic resonance measurements of $YBa_2Cu_4O_8$, *Phys. Rev. B* 56:877(1997). Copyright (1997). The American Physical Society.

38. H. Takagi, B. Batlogg, H. L. Kao, J. Kwo, R. J. Cava, J. J. Krajewski and W. F. Peck, Jr, Systematic evolution of temperature-dependence resistivity in $La_{2-x}Sr_xCuO_4$, *Phys. Rev. Lett.* 69:2975(1992); B. Battlogg, H. Y. Hwang, H. Takagi, R. J. Cava, H. L. Kao and J. Kwo, Normal state phase diagram of $La_{2-x}Sr_xCuO_4$ from charge and spin dynamics, *Physica C* 235-240:130(1994).

39. A. V. Chubukov, D. Pines and B. P. Stojkovic, Temperature crossovers in cuprates, *J. Phys.: Condens. Matter.* 8:10017(1996).

40. B. Wuyts, V. V. Moshchalkov and Y. Bruynseraede, Resistivity and Hall effect of metallic oxygen-deficient $YBa_2Cu_3O_x$ films in the normal state, *Phys. Rev. B* 53:9418(1996).

41. D. S. Marshall, D. S. Dessau, A. G. Loeser, C-H. Park, A. Y. Matsuura, J. N. Eckstein, I. Bozovic, P. Fournier, A. Kapitulnik, W. E. Spicer and Z.-X. Shen, Unconventional electronic structure evolution with hole doping in : angle-resolved photoemission results, *Phys. Rev. Lett.* 76:4841(1996).

42. N. Bulut, D. Hone, D. J. Scalapino and N. E. Bickers, Knight shifts and nuclear-spin-relaxation rates for two-dimensional models of CuO_2, *Phys. Rev. B* 41:1797(1990); N. Bulut, and D. J. Scalapino, Weak-coupling model of spin fluctuations in the superconducting state of the layered cuprates, *Phys. Rev. B* 45:2371(1992).

43. D. R. Grempel and M. Lavagna, Magnetic excitations in the t-J model: applications to neutron and NMR experiments in high T_c materials, *Solid St. Comm.* 83:595(1992); G. Stemmann, C. Pepin and M. Lavagna, Spin gap and magnetic excitations in the cuprate superconductors, *Phys. Rev. B* 50:4075(1994).

44. A. Sokol and D. Pines, Toward a unified magnetic phase diagram of the cuprates superconductors, *Phys. Rev. Lett.* 71:2813(1993); V. Barzykin and D. Pines, Magnetic scaling in superconductors, *Phys. Rev. B* 52:13585(1995).

45. T. Imai, Analysis of nuclear relaxation experiments in high-Tc oxides based on Mila-Rice Hamiltonian, *J. Phys. Soc. Jpn.* 59:2508 (1990).

46. Y. Itoh, M. Matsumura and H. Yamagata, Anomalous spin dynamics in $La_{2-x}Sr_xCuO_4$ (x=0,13 and 0,18) studied by Nuclear spin-lattice relaxation, *J. Phys.Soc. Jpn.* 65:3747 (1996).

47. S. Ohsugi, Y. Kitaoka, K. Ishida and K. Asayama, Cu NQR study of the spin dynamics in high-Tc superconductor $La_{2-x}Sr_xCuO_4$, *J. Phys. Soc. Jpn.* 60:2351 (1991).

48. S. Ohsugi, Y. Kitaoka and K. Asayama, Temperature dependence of spin susceptibility of $La_{2-x}Sr_xCuO_4$.-Cu Knight shift measurement- to appear in *Physica C*, special issue of M2S-HTSC conference, held Feb. 28-Mar. 4, 1997 in Beijing.

49. B. I. Altshuter, L. B. Ioffe and A. J. Millis, Theory of the spin gap in high temperature superconductors, *Phys. Rev. B* 53:415 (1996).

50. Y. Itoh, T. Machi, A. Fukuoka, K. Tanabe and H. Yasuoka, Pseudo spin-gap in single-layer high-Tc Cu oxide $HgBa_2CuO_{4+\delta}$, *J. Phys.Soc. Jpn.* 65:3751 (1996).

51. B. J. Suh, F. Borsa, J. Sok, D. R. Torgeson and M. Xu, ^{199}Hg and ^{63}Cu NMR in superconducting $HgBa_2CuO_{4+\delta}$ oriented powders, Phys. Rev. B 54:545(1996).

52. J. Bobroff, H. Alloul, P. Mendels, V. Viallet, J.-F. Marucco and D. Colson, ^{17}O NMR evidence for a pseudogap in the monolayer $HgBa_2CuO_{4+\delta}$, *Phys. Rev. Lett.* 78:3757 (1997).

53. S. Kambe, H. Yasuoka, A. Hayashi and Y. Ueda, NMR study of the spin dynamics in $Tl_2Ba_2CuO_{6+x}$ (Tc=85K), *Phys. Rev.B* 47:2825 (1993). Copyright(1993). The American Physical Society.

54. K. Ishida, Y. Kitaoka, K. Asayama, K. Kadowaki and T. Mochiku, Cu NMR study in single crystal $Bi_2Sr_2CaCu_2O_8$,- observation of gapless superconductivity-, *J. Phys. Soc. Jpn.* 63:1104 (1994).

55. A. Trokiner, K. Mikhalev, A. Yakubovskii, P.-V. Bellot, S. Verkhovskii, Y. Zhdanov, Y. Piskunov, A. Inyushkin and A. Taldenkov, ^{17}O NMR in high-Tc superconductors $Tl_2Ba_2CaCu_2O_{8+x}$, *Physica C* 255:204 (1995).

56. M. Horvatic, C. Berthier, Y. Berthier, P. Segransan, P. Butaud, W. G. Clark, J. A. Gillet and J. Y. Henry, Nuclear spin-lattice relaxation rate of planar oxygen in $YBa_2Cu_3O_{6.52}$ and $YBa_{1.92}Sr_{0.08}Cu_3O_7$ single crystals, *Phys. Rev. B* 48:13848 (1993).

57. Y. Kitaoka, K. Fujirawa, K. Ishida, K. Ayasama, Y. Shimakawa, T. Manako and Y. Kubo, Spin dynamics in heavily-doped high-Tc superconductors $Tl_2Ba_2CuO_{6+x}$, with a single CuO_2 layer studied by ^{63}Cu and ^{205}Tl NMR, *Physica C* 179:107(1991).

58. Y. -Q. Song, M. A. Kennard, K. R. Poeppelmeier and W. P. Halperin, Spin susceptibility in the $La_{2-x}Sr_xCuO_4$ system from undredoped to overdoped regime, *Phys. Rev. Lett.* 70:3131(1993).

59. L. Le Noc, A. Trokiner, J. Schneck, A. Pougnet, D. Morin, H. Savary, A. Yakubovskii, K. N. Mykhelyov, and S. Verkhovskii, Spin susceptibility behaviour in the $Bi_2Sr_2Ca_{n-1}Cu_nO_{2n+x}$ compounds: from underdoped to overdoped regime, *Physica C* 235-240:1703 (1994).

60. P. -V. Bellot, A. Trokiner, Y. Zhdanov, A. Yakubovskii, L. Shustov, S. Verkhovskii, S. Zagoulaev and P. Monod, Magnetic properties of $(Tl_{1-x}Pb_x)Sr_2CaCu_2O_7$ high-Tc oxide studied by ^{17}O NMR and SQUID: from superconducting to strongly overdoped metal, *Physica C* 282-287:1357(1997). Fig. 6 reprinted with kind permission of Elsevier Science-NL, Sara Burgerharstraat 75, 1055 KV Amsterdam.

61. P. -V. Bellot and A. Trokiner (private communication).

62. P. -V. Bellot, A. Trokiner, Y. Zhdanov and A. Yakubovskii, ^{43}Ca NMR in solid state, to appear in *J. Chim. Phys.*

63. M.-H. Julien, M. Horvatic, P. Carretta, C. Berthier, Y. Berthier, P. Segransan, S. M. Loureiro and J.-J. Capponi, ^{63}Cu and ^{199}Hg in overdoped $Hg_2Ba_2CaCu_2O_{6+x}$, *Physica C* 268:197(1996). Fig. 8 reprinted with kind permission of Elsevier Science-NL, Sara Burgerharstraat 75, 1055 KV Amsterdam.

64. A. Goto, H. Yasuoka, K. Otzschi and Y. Ueda, Carrier concentration dependence of the pseudo spin gap behaviour in $LaBa_2Cu_3O_y$, *J. Phys. Soc. Jpn.* 64:367 (1995).

65. P. -V. Bellot, A. Trokiner, A. Yakubovskii and L. Shustov, Intrinsic localized magnetic moments in overdoped high-Tc cuprate $(Tl_{1-x}Pb_x)Sr_2CaCu_2O_7$ probed by ^{17}O NMR, *Physica C* 282-287:1359(1997). Fig. 10 reprinted with kind permission of Elsevier Science-NL, Sara Burgerharstraat 75, 1055 KV Amsterdam.

66. H. Alloul, P. Mendels, H. Casalta, J.F. Marucco and J. Arabski, Correlations between magbetic and superconducting properties of Zn-substituted $YBa_2Cu_3O_{6+x}$, *Phys. Rev. Lett.* 67:3140(1991).

67. R. E. Walstedt, R. F. Bell, L. F. Schneemeyer, J. V. Waszczak, W. W. Warren, Jr., R. Dupree and A. Gencten, Absence of magnetic pair breaking in Zn-doped $YBa_2Cu_3O_7$, *Phys. Rev. B* 48:10646(1993).

68. A. V. Mahajan, H. Alloul, G. Collin and J. F. Marucco, ^{89}Y NMR probe of Zn induced local moments in $YBa_2(Cu_{1-y}Zn_y)_3O_{6+x}$, *Phys. Rev. Lett.* 72:3100(1994).

69. K. Ishida, Y. Kitaoka, K. Yamazoe and K. Asayama, Al NMR Probe of local moments induced by an Al impurity in high-T_c cuprates $La_{1.85}Sr_{0.15}CuO_4$, *Phys. Rev. Lett.* 76:531(1996).

70. G. V. M. Williams, J. L. Tallon, R. Dupree and R. Michalak, Transport and NMR studies of the effect of Ni substitution on superconductivity and the normal-state pseudogap in $YBa_2Cu_4O_8$, *Phys. Rev. B* 54:9532(1996).

71. J. Bobroff, H. Alloul, Y. Yoshinari, A. Keren, P. Mendels, N. Blanchard, G. Collin and J. F. Marucco, Using Ni substitution and ^{17}O NMR to probe the susceptibility $\chi'(q)$ in cuprates, *Phys. Rev. Lett.* (in press).

72. Y. Kubo, Y. Shimakawa, T. Manako and H. Igarashi, Transport and magnetic properties of $Tl_2Ba_2CuO_{6+\xi}$ showing a δ-depedent gradual transition from an 85K superconductor to a nonsuperconducting metal, *Phys. Rev. B* 43:7875(1991).

73. T. Nakano, M. Oda, C. Manabe, N, Momono, Y. Miura and M. Ido, Magnetic properties and electronic conduction of superconducting $La_{2-x}Sr_xCuO_4$, *Phys. Rev. B* 49:16000(1994).

74. C. H. Pennington and C. P. Slichter, Theory of nuclear spin-spin coupling in $YBa_2Cu_4O_{7-d}$, *Phys. Rev. Lett.* 66:381(1991).

FROM MAGNONS TO THE RESONANCE PEAK: SPIN DYNAMICS IN HIGH-T_C SUPERCONDUCTING CUPRATES BY INELASTIC NEUTRON SCATTERING

Philippe Bourges

Laboratoire Léon Brillouin,
CEA-CNRS, CE Saclay,
91191 Gif sur Yvette, France

1. INTRODUCTION

Over the last decade, inelastic neutron scattering (INS) experiments have provided a considerable insight in the understanding of the anomalous properties of high T_c-superconductors. Neutron measurements have shown the persistence of antiferromagnetic (AF) dynamical correlations over the whole metallic state of cuprates[1] which demonstrates the strong electronic correlations existing in metallic cuprates. Together with the nuclear magnetic resonance (NMR)[2, 3] and later with bulk magnetic susceptibility[4], INS has evidenced the "spin pseudogap" phenomenon in underdoped cuprates, a topic of intense current interest. Further, INS has shown that these spin excitations are very intimately linked to superconductivity as a sharp magnetic peak occurs when entering the superconducting state. This peak, referred to "resonance" since its first evidence by J. Rossat-Mignod et al[5], has spawned a considerable theoritical activity.

The structural peculiarity of high-T_c cuprates is that they are all built from stacking of CuO_2 planes separated by different kinds of layers, the "charge reservoirs", which are essential as they control the charge transfer mechanism. The CuO_2 plane is of central importance as it carries most of the anomalous physical properties in the normal state and, likely, contains the keypoint for the mechanism of high-T_c superconductivity as it has been proposed in many different approaches (See e.g. [6, 7, 8, 9, 10, 11]). The unusual properties of high-T_c cuprates are extensively reviewed in this book and will not be discussed here in details.

A generic phase diagram, sketched in Fig. 1, has been established on phenomenological grounds. Starting from an insulating and antiferromagnetically ordered state around zero doping with one electron per Cu $d_{x^2-y^2}$-orbital, the cuprates become metallic by introducing holes (or electrons) from the charge reservoirs to the CuO_2 plane.

The Gap Symmetry and Fluctuations in High - T_c Superconductors
Edited by Bok *et al.*, Plenum Press, New York, 1998

349

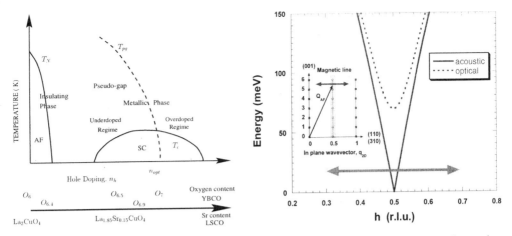

Figure 1: Schematic phase diagram of high-T_c cuprates. The dashed line corresponds to a crossover temperature, T_{pg}, below which most physical properties exhibit or infer a pseudogap behavior. Real systems are indicated versus hole doping in the bottom part. Note that T_N in YBCO disappears for $x \sim 0.4$ which actually corresponds to only about 2-4% of holes in each CuO_2 plane[1].

Figure 2: Spin-wave dispersion along the (110) direction in undoped YBCO. Inset displays the scattering plane in the reciprocal space commonly used in INS measurements with $Q = (h, h, q_l)$ or $Q = (3h, h, q_l)$. Squares represents nuclear Bragg peaks and circles AF Bragg peaks occuring in the Néel state. The two arrows indicate a typical Q-scan trajectory performed across the magnetic line.

A superconducting (SC) phase occurs at further doping defining an optimal doping, n_{opt} when the superconducting transition, T_c, is passing through a maximum. Doping rates, below and above n_{opt} define usually called underdoped and overdoped regimes, respectively. In the underdoped state, many physical properties exhibit anomalous behavior below a crossover temperature, T_{pg}: that is the case for the macroscopic spin susceptibility, NMR Knight shift, specific heat, transport properties[4]... In that doping range, it has been shown by NMR[2, 3] and INS[1, 5] experiments that the spin fluctuations are charaterized at low temperature by the opening of a spin pseudo-gap. Recently, photoemission experiments[12, 13] have evidenced that the single-particle excitation spectrum exhibits also a pseudo-gap in the normal state below T_{pg}. Similar observations have been done in optical conductivity[14] and Raman scattering[15] measurements. Concomitantly, these pseudo-gap observations disappear above the optimal doping. Few attempts have been made to describe the phase diagram which have raised questions such as: do the antiferromagnetic fluctuations alone explain the observed phases[6, 7, 8], or, do we need to consider a new critical point at optimal doping which would scale the physical properties[16] as proposed, for instance, in the "circulating current" phase[11] ? Conversely, a comprehensive microscopic description of all these gap observations is still missing at present.

Here, it is important to relate this generic phase diagram of cuprates to the phase diagram of the two systems on which INS experiments have been performed so far. In particular, in YBa$_2$Cu$_3$O$_{6+x}$ (YBCO) system, the relation between hole doping, n_h, and oxygen content is not obvious due to the charge transfer mechanism from the Cu-O chains to the CuO$_2$ planes[17]. Oxygen concentrations from $x \simeq 0$ to $x \simeq 1$ cover a major part of the phase diagram and are reported on Fig. 1 for typical doping regimes which will be discussed here: the optimal doping being realized for $x = 0.94$. In the

La$_{2-x}$Sr$_x$CuO$_4$ (LSCO) system, maximum T_c is reached for $x = 0.15$ and is generally assumed to match the optimal doping. A close inspection of the neutron results rather suggest that it could correspond to an underdoped regime. The superconductivity could be simply reduced at further doping because of the proximity of structural instabilities.

Furthermore, considering the unusual properties of these materials, one needs to know which kind of magnetism is observed in INS experiments: do we observe spin dynamics associated with the localized copper spins or rather related to the itinerant quasiparticles ? These different hypotheses have been widely addressed in the literature. As a limiting case, the slightly-overdoped YBCO$_7$ ($x \simeq 1$) could likely be described as a simple itinerant magnetism since, as expected for usual metals, antiferromagnetic spin fluctuations are practically not sizeable in the normal state[18, 19]. Unfortunately, this simple Fermi liquid approach fails at lower doping since dynamical AF correlations are unambiguously observed up to the optimal doping. One then needs to take into account the underlying antiferromagnetic background. Furthermore, the anomalous spectral lineshape detected in photoemission experiments[12] exhibits a behavior inconsistent with conventional band theory indicating that the single-particle spectra are necessarily renormalized due to electron-electron interactions. These "dressed" quasiparticles could even be strongly coupled to collective excitations centered at the AF momentum[20, 21], namely the spin fluctuations. The determination of the energy, momentum, doping and temperature dependences of the spin excitation spectrum is then of primarily importance to describe the anomalous properties of these materials.

Going from well-defined magnons (in the Néel undoped state) related to the localized copper spins to a Fermi liquid picture (at the highest doping available), INS experiments cover a whole range of situations where the spin and charge responses are intimately linked. More generally, INS observations provide direct information about the electronic interactions within CuO$_2$ planes, and even, between the two adjacent metallic CuO$_2$ layers in YBCO. Here, I shall review the doping evolution of the spin dynamics. After a short recall of the neutron technique (Sec. 2) and of the experimental difficulties (Sec. 3), magnons in the AF state are reported in Sec. 4. The momentum and energy dependences in the metallic state are presented in Sec. 5 and 6, respectively. The strong modifications of the spin dynamics induced by superconductivity are discussed in Sec. 7. The normal state spin susceptibility, characterized by a "spin pseudogap", is emphasized in Sec. 8. Most of the results described here concerns the YBCO system although comparisons with the LSCO system are occasionally made.

2. INELASTIC NEUTRON SCATTERING

The interaction of neutrons with the condensed matter is double[22]: nuclear interaction with the atomic nucleus and magnetic dipolar interactions between the neutron spin and magnetic moments in solids, spin of unpaired electrons for instance. Neutron scattering is then a very unique tool as it measures in the same time structural information (as can do X-ray scattering) but also magnetic properties. Furthermore, thermal neutrons which possess a wavelength of the order of the atomic distances, 0.5-10 Å, have in the same time an energy, 0.1 - 200 meV, which covers the large range of excitations in solids. Inelastic neutron scattering then gives invaluable information on both spatial and time-dependent of nuclear (like phonons) and magnetic correlations whose momentum and energy dependences are only accessible using INS. An extensive review of the possibilities of neutron scattering can be found in a recent course[23]. Here, it is worth emphasizing that single crystals are required to determine a complete momentum dependence of dynamical properties. Also, due to the weak interaction of neutrons with condensed matter, neutron scattering probes samples in bulk, but conversely requires large samples.

In high-T_c cuprates, INS using the triple-axis spectrometer technique[23] has been widely used to study phonons[24]. Here, we focus on magnetic scattering whose cross section is directly proportional to the scattering function which is identified to the Fourier transform in time and space of the spin-spin correlation function[22] as,

$$S^{\alpha\beta}(Q,\omega) = \frac{1}{2\pi\hbar} \int_{-\infty}^{+\infty} dt \exp(-i\omega t) < S_Q^\alpha S_{-Q}^\beta(t) > . \tag{1}$$

This scattering function is in turn related to the imaginary part of the dynamical generalized spin susceptibility, $Im\chi(Q,\omega)$, by the fluctuation-dissipation theorem. Here, we consider a single component of the dynamical generalized susceptibility tensor associated with Carthesian spin coordinates S^α and S^β with $\alpha, \beta = x, y, z$,

$$\chi^{\alpha\beta}(q,\omega) = -(g\mu_B)^2 \frac{i}{\hbar} \int_0^\infty dt \exp^{-i\omega t} < [S_Q^\alpha(t), S_{-Q}^\beta] > . \tag{2}$$

In isotropic magnetic systems, χ is simply identified to $Tr(\chi^{\alpha\beta})/3$. $Im\chi(Q,\omega)$ is a very useful quantity, especially when no theory can be used to describe the data. It basically contains all the physical interest and can be calculated in different microscopic models. In undoped Néel state, the spin susceptibility describes the excited states above the AF ground state, referred to as magnons[22], and neutron scattering cross-sections are well described using spin-wave theory in Heisenberg model. When adding a small amount of holes in CuO_2 planes ($n_h \leq 5\%$), the system remains in the insulating state, but very peculiar spin dynamics is observed: the low energy excitations are strongly enhanced at low temperature[25, 26, 27] likely due to electron-hole interactions. Increasing further the doping, spin fluctuations are still detected in the metallic state. Further, by calibration with phonon scattering, INS experiments provide absolute units for the dynamical susceptibility which are important for theoritical models. This absolute unit calibration is also necessary to compare the INS results in different systems as well as results obtained using different techniques (with NMR measurements for instance).

In undoped cuprates, the ordered AF phase is characterized by an in-plane propagation wave vector, $q_{AF} = (\frac{1}{2}, \frac{1}{2}, 0)$ and the magnetic neutron cross section is maximum at wave vectors like $Q = \tau + q_{AF}$, where τ denotes Bragg peaks of the nuclear structure (squares in the inset of Fig. 2). Roughly speaking, when going into the metallic state, the magnetic scattering remains always peaked around the in-plane component of the AF wave vector. This occurs for any values of the momentum transfer q_l perpendicular to the plane. The magnetic scattering is then concentrated around lines in the reciprocal lattice, like $(\frac{h}{2}, \frac{k}{2}, q_l)$ with h and k integers, denoted magnetic lines. Q-scans, performed within a Brillouin zone across these lines and at different energy transfers (sketched by the arrow in Fig. 2), exhibit a maximun around $q_{AF} \equiv (\pi, \pi)$ (Fig. 3) which is likely identified to the magnetic scattering.

3. EXPERIMENTAL EVIDENCE OF MAGNETIC EXCITATIONS

Unfortunately, INS experiments are not only measuring the magnetic scattering: other contributions occur either due to intrinsic nuclear scattering of the lattice, like phonons, or even due to spurious effects and impurities. Furthermore, owing to the

E= 10 meV **E= 17-26meV** **E= 35-40 meV**

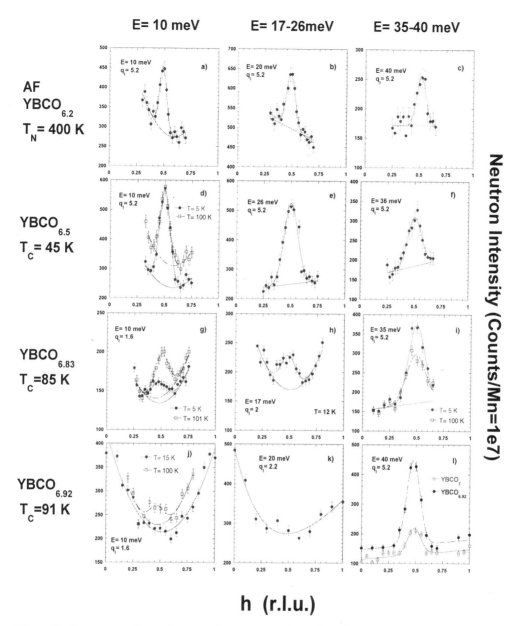

Neutron Intensity (Counts/Mn=1e7)

h (r.l.u.)

Figure 3: Q-scans performed across the magnetic line, $Q = (h, h, q_l)$, where h is scanning over two Brillouin zones. All neutron intensities are normalized to the same units, and are then directly comparable each others. The q_l value along the (001) direction was chosen to get the maximum of the magnetic structure factor[1]. In the AF state ($x = 0.2$), only a single peak is observed for the counterpropagating spin-waves; this is caused by resolution effect of the spectrometer. Note the strong reduction of the E= 40 meV peak intensity from $YBCO_{6.92}$ (nearly optimally doped regime) to $YBCO_7$ (overdoped regime) (Fig. 3.l).

relatively weak cross section of the magnetic scattering, its extraction from the total scattering is a major experimental problem encountered in INS experiments. In principle, polarized neutron beam experiments should easily separate these contributions, the magnetic scattering appearing in the spin-flip channel at the difference of nuclear scattering. However, due to the lack of statistics, polarized neutron results have yielded partly erroneous conclusions in high-T_c cuprates [28]. Therefore, unpolarized neutron experiments have been largely employed. Due to the experimental difficulty, new results always need to be crosschecked and confirmed by other measurements because a single isolated experiment can be unfortunately misled by spurious effects. It has led to extensive, sometimes contradictory, discussions over the last decade[1, 18, 19, 28, 29, 30, 31, 32]. *Only* the use of complementary methods as well as the accumulation of neutron data allow to overcome the experimental difficulties. Here, I shall briefly recall the main guidelines which have been used to estimate the magnetic signal.

Among these methods, let us emphasize the use of the momentum and temperature dependences because they are different for magnetism and phonons[22]. Especially, as a result of the magnetic form factor, magnetic scattering is known to decrease on increasing the amplitude of the wave vector over few Brillouin zones in contrast to the phonon scattering. Within a single Brillouin zone, one could also discriminate both signals by their q-dependences. Further, INS experiments have been also performed in different scattering planes to avoid some specific phonons[19].

Another powerful method developed for the YBCO system has been to conduct experiments in the different regimes of the phase diagram *on the same sample*[1, 30, 32]. Indeed, the $YBa_2Cu_3O_{6+x}$ system offers the great opportunity to cover the whole high-T_c cuprates phase diagram just by changing the oxygen content by thermogravimetry from the Néel state, $x \simeq 0$, to the overdoped metallic, $x \simeq 1$. Measurements scanning the wave vector along the (110) direction are shown in Fig. 3 at few different fixed energy transfers for few different states of YBCO. Further, these experiments have been performed on the same triple axis spectrometer (2T-Saclay) using the same experimental setup, i.e. the same spectrometer resolution. This gives the great advantage that impurity contributions and, in a less extent, the phonons are basically the same in all experiments: this facilitates the extraction of the magnetic scattering.

These scans exhibit at any doping well-defined maxima at $q = 0.5$, corresponding to the AF wave vector $\equiv (\pi, \pi)$, whose magnitude evolves with doping. At $E = 10$ meV, a correlated scattering signal is seen at lower doping whereas it is absent for optimally doped samples. In contrast, the correlated signal at $E = 40$ meV first increases with the oxygen content and then decreases for $x \simeq 1$. Measurements at nearly the same energy in the normal state exhibit, at most, a very weak magnetic signal (see Fig 11). These striking doping, energy and temperature dependences have given strong guidelines to analyse the data. In particular, two clear limits are well-defined: the undoped case where the theoritical spin-wave cross section is known and the case of the normal state in the overdoped regime where the very weak magnetic scattering can be neglected. A generic shape for the non-magnetic contributions is then deduced for any doping. The use of such empirical methods, as well as phonon calculations[33], give a self-consistent picture which has made possible to improve the data analysis all over the years[1, 5, 30, 32]. When using the triple-axis technique, a good confidence is now reached about the determination of the magnetic signal over a wide range of energy [34, 35]. However, it should be mentioned that this method excludes contributions which are weakly momentum-dependent. Unfortunately, no current neutron experiment is able to evidence such hypothetic magnetic contributions.

354

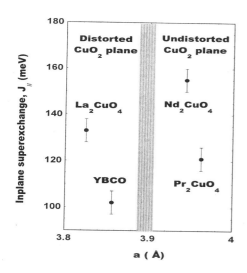

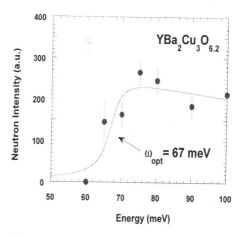

Figure 4: In-plane superexchange interaction determined by INS experiments versus Cu-Cu distance in different cuprates (from [38]).

Figure 5: Energy dependence of the neutron intensity at q_{AF} from spin-wave scattering at T=10 K (from [40]). The closed (open) symbols represent the intensity of the optical (acoustic) excitations. The full line is a fit by a step-like function.

4. ANTIFERROMAGNETIC STATE AND EXCHANGE PARAMETERS

Undoped parent compounds of high-T_c cuprates are Mott-Hubbard insulators which are usually described by a spin-$\frac{1}{2}$ antiferromagnetic Heisenberg model on a square lattice[36]. They exhibit an antiferromagnetic ordering below a Néel temperature ranging between 250 K and 420 K[1]. The most important parameter is the Cu-O-Cu nearest neighbor superexchange interaction, J, within the CuO_2 plane.

In ordered magnetic systems, INS experiments probe the spin-wave dispersion relations which relate the magnon energy, ω_q, to the scatterred wave vector (see Fig. 2). At sufficiently low energy (but notably above the small magnon gaps related to interlayer coupling and exchange anisotropies[1, 37]), the acoustic magnon dispersion relation starts linearly in AF systems as $\omega_q = cq$ (Fig. 2). J is then deduced from the measured spin-wave velocity c, as $c = 2S\sqrt{2}Z_c Ja$ (where a is the square lattice constant, S=$\frac{1}{2}$ and $Z_c \simeq 1.18$ represents quantum corrections of the AF ground state). Unfortunately, because of the large value of J and of resolution effect, counterpropagating spin-waves cannot be observed when scanning across the magnetic line (as sketched in Fig. 2): a single peak is usually measured (see Fig 3a-c). A special scattering geometry has thus been adapted to determine the spin-velocity with accuracy in YBa$_2$Cu$_3$O$_{6.1}$[37] and in three different monolayer undoped cuprates(La$_2$CuO$_4$, Nd$_2$CuO$_4$ and Pr$_2$CuO$_4$) [38]. The deduced in-plane antiferromagnetic superexchange coupling J is typically 130 meV (Fig. 4). However, J does not exhibit a monotonous behavior versus the bonding Cu-O-Cu length likely on account of detailed structure of each system. It underlines that the large enhancement of J is caused by other structural units like the Cu-O-O triangle[39].

Further, the unit cell of YBCO contains pairs of closely spaced CuO$_2$ layers, the bilayers. The intrabilayer coupling, referred as $J_\perp$, removes the degeneracy between even- and odd-parity electronic states. In the AF Néel state, these excitations correspond to optical (dashed line in Fig. 2) and acoustic (full line in Fig. 2) spin waves, respectively. These two modes display complementary dynamical structure factors along the momentum transfer perpendicular to the basal plane, q_l: $\sin^2(\pi z q_l)$ for the acoustic mode and $\cos^2(\pi z q_l)$ for the optical mode. (Here, $z = 0.29$ is the reduced distance between nearest-neighbor Cu spins within one bilayer). This allowed to distinguish these two excitations and to determine the optical gap[40, 41]. Fig. 5 show the energy dependence of the neutron intensity at q_{AF} obtained from our results[40] which gives an optical magnon gap at $\omega_{opt} = 67 \pm 5$ meV. Hayden et al[41] have reported a slightly larger value from less accurate data. Detailed spin-wave calculations[1, 37] reveal an optical magnon gap at[40], $\omega_{opt} = 2\sqrt{J_\perp J}$. Using the value of $J = 120$ meV[37, 41] for the in-plane superexchange, one deduces $J_\perp = 9.6$ meV. However, the above relation does not account from quantum corrections of the AF ground state. A more accurate treatment using Schwinger bosons representations[42], gives $J_\perp \sim 12$ meV which is in good agreement with band theory predictions[43], $J_\perp \sim 13$ meV. In classical superexchange magnetic theory, where J is proportional to the square of the overlap of the electronic wavefunctions, one can deduce the ratio between the intrabilayer and the inplane hopping matrix elements as $\frac{t_\perp}{t} = \sqrt{\frac{J_\perp}{J}} = 0.34$. This non-negligible ratio, which unlikely would vary with doping, shows that the electron transfer processes between direcly adjacent layers could play an important role in the high-T_c mechanism as it was suggested in the interlayer tunneling model[44].

Furthermore, by calibration with phonon cross section, one can determine the spectral weight of the spin susceptibility in absolute units. Surprisingly, the spin wave spectral weight is found smaller than expected from quantum corrections[38]. This reduction of about 30% is presumably due to covalent effects between copper d-orbitals and oxygen p-orbitals[45]. Reducing the absolute scale of the atomic form factor, such effects can also explain the reduction of the low temperature ordered magnetization value[1, 45].

5. WAVE VECTOR DEPENDENCES IN THE METALLIC STATE

We now turn to the results in the metallic state of cuprates where the magnetic scattering is only found inelastic, corresponding to dynamical fluctuations and peaked around the AF wave vector. The existence of these fluctuations is already surprising since in usual simple metals fluctuations arising from free electrons are too weak to be observed. This observation then signs the existence of strong electronic correlations in high-T_c cuprates.

Even and Odd excitations in YBCO: q_l-modulation in YBCO

In conventional band theory, interactions within a CuO_2-bilayer yield bonding and antibonding bands in the metallic state. Transitions between electronic states of the same type (bonding-to-bonding or antibonding-to-antibonding) and those of opposite types are characterized by even or odd symmetry, respectively, under exchange of two adjacent CuO_2 layers. As a result, odd and even excitations then exhibit a structure factor along q_l similar to the acoustic and optical spin-wave in the Néel state, respectively. As discussed in the previous section, this yields a $\sin^2$-type q_l-dependence of the lower energy excitations, the odd excitations. This structure factor has been effectively observed in any low energy (below ~ 45 meV) magnetic studies. It is, for instance, the case in Fig. 6 in the SC state of optimally doped YBCO. Similar results have been reported in the normal state as well[1, 46, 47]. In weakly

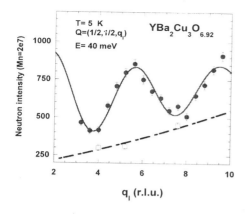

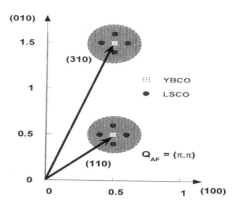

Figure 6: q_l-scan at $E = 40$ meV in YBCO$_{6.92}$ displaying a modulation typical of odd excitation (from [1]). The background, obtained from q-scans across the magnetic line (open squares), is represented by the dashed line. The full line correspond to a fit by $a + bF^2(Q)\sin^2(\pi z q_l)$ above the background and where $F(Q)$ is the Cu magnetic form factor.

Figure 7: In-plane Brillouin zone. The shaded area sketches the location of AF fluctuations. The closed circles sketch the four-peaks magnetic scattering observed at low energy in the LSCO system at $Q = (\pi(1 \pm \delta), \pi)$ and $Q = (\pi, \pi(1 \pm \delta))$, displaced from the AF momentum by an amount, $\delta = 0.245 = 0.28$ Å^{-1} for $x = 0.14$[50].

doped metallic state, $x \sim 0.5$, this modulation actually occurs because even excitations exhibit a gap around 53 meV[34], which is reminiscent of the optical magnon gap[40]. At higher doping, the even excitations gap is lowered further by doping (~ 35 meV in $x \simeq 0.7$)[35]. Surprisingly, the $\sin^2$-structure factor is still observed at energies above this even gap (Fig. 6). However, Fig. 6 suggests that this modulation is not complete as even excitations are sizeable at $q_l \simeq 3.5, 7$ but with a magnitude ~ 5 times smaller. Therefore, although the even excitations occur in the same energy range, they are surprisingly much weaker in amplitude than the odd excitations. Moreover, even excitations display unexpected temperature dependences[34, 35] as they are strongly reduced going fom 5 K to 200 K.

Commensurate fluctuations in YBCO

We then discuss the in-plane wave vector dependence observed at low energy results, i.e. below $E \sim 40$ meV. As shown in Fig. 3, the magnetic scattering is found in YBCO peaked at the AF momentum, (π, π), at any doping. However, it has been recently reported[48] that the magnetic fluctuations in an underdoped YBCO$_{6.6}$ ($T_c = 63$ K) sample would become incommensurate around $E \sim 25$ meV, but only below $T \sim 70$ K. This contradictory result needs to be confirmed as only a few measurements[49] have previously indicated, at most, a flat-topped shape of low-energy q-scans profiles. In any case, typical observed peaks are far from simple sharp Lorentzian-shape peaks. As shown for instance at $E = 17$ meV in YBCO$_{6.83}$ in Fig. 3h, two broad peaks could better described the observed profile. Actually, the extension in q-space of the magnetic scattering is quite important as its q-width, defined as the Full Width at Half Maximun (FWHM) Δ_q, is typically about one fifth of the Brillouin zone size. On increasing doping, the peak broadens (see Fig. 3), giving at most $\Delta_q \simeq 0.45$ Å^{-1}[30, 32] (after resolution deconvolution when fitting by a Gaussian shape) for highly doped samples,

$x \geq 0.9$ whereas $\Delta_q \sim 0.25$ Å^{-1} is found for weakly doped samples, $x \sim 0.5$. This q-width may be associated with a length in real space, $\xi = 2/\Delta_q$, which would correspond to an AF correlation length in a localized spins picture. Typically, this length is found very short[30], as $\xi/a \sim 1 - 2.5$ depending on doping (where a=3.85 Å). Furthermore, Δ_q, still obtained by a Gaussian fitting of the q-lineshape, displays no temperature dependence within error bars (see e.g. [32]).

Incommensurate fluctuations in LSCO

In LSCO, the low energy fluctuations clearly differ from what is observed in YBCO: a four-peaks structure is observed[50, 51] instead of a broad peak centered at (π, π) (Fig. 7). Each spin scattering peak occurs at an incommensurate wave vector which increases with doping. It exhibits a q-width much smaller than in YBCO such as the four peaks do not overlap. On increasing energy, the peaks broaden and concomitantly the incommensurability disappears: the spin scattering being maximum at (π, π) with a large q-width for energies above ~ 25 meV for $x = 0.15$[51].

The origin of the incommensurability in LSCO has been widely discussed in the framework of itinerant magnetism[52, 53, 54, 55]. Magnetic correlations are enhanced at the incommensurate wave vectors due to a near nesting property of the simple bidimensional Fermi surface deduced from simple tight-binding calculations. At the AF wave vector, low energy spin fluctuations are removed up to an energy which is twice the chemical potential[55]. This picture seems to nicely account for the observed features. However, the calculation of the chemical potential from the doping level cannot be deduced from conventional band structure calculations[54] and would be associated with new quasi-particles charge carriers[9]. This model can also explain why the LSCO and YBCO systems display different momentum dependences according to a different Fermi surface topology in the two systems[52, 53]. However, quite recently, another interpretation, based on real-space domains observed in insulating phases caused by charge ordering forming stripes[56], has been proposed to explain the discommensuration[57].

6. ENERGY DEPENDENCE OF THE SPIN SUSCEPTIBILITY

The most remarkable INS result in high-T_c cuprates is certainly the drastic change of the spin dynamics when entering the superconducting state. Basically, a gap opens at low energy, E_G, and a strong resonance peak appears in the odd spin susceptibility at a characteristic energy, E_r. These features are most likely related to coherence effects because of spin pairing of the superconducting electron pairs as it has been inferred in conventional BCS formalism[58]. However, the observed characteristics imply unconventional superconducting gap symmetry, i.e. anisotropic gap in k-space, most probably of d-wave symmetry. These features as well as their implications are described in the two next sections. The normal state properties will be discussed in a third section.

Doping dependence of the odd spin susceptibility and resonance peak

In order to emphasize these features, it is convenient to discuss the energy shape of the odd spin susceptibility. Fig. 8 depicts $Im\chi$ in the superconducting state for energies below 50 meV and at the AF wave vector: $Im\chi$ is displayed for 6 different oxygen contents, 4 in the underdoped regime and 2 in the overdoped regime. The

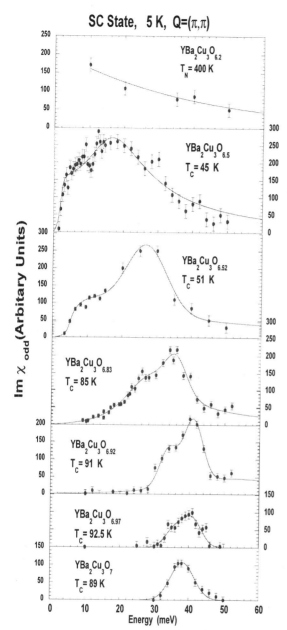

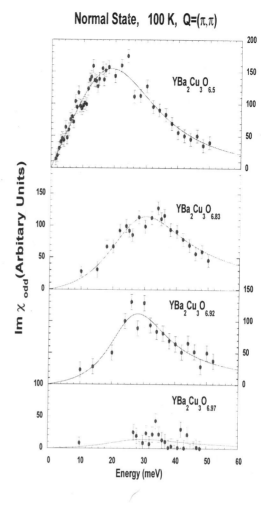

Figure 8: Imaginary part of the odd spin susceptibility at T= 5 K in the superconducting state for six oxygen contents in YBCO. At the top, the spin waves scattering of the undoped parent compound at T=8 K is displayed for comparison. (See Fig. 9 for other details).

Figure 9: Imaginary part of the odd spin susceptibility at T= 100 K in the normal state for four oxygen contents in YBCO. These curves are directly comparable each others from measurements performed on the same triple axis spectrometer (2T-Saclay). They have been normalized to the same units using standard phonon calibration[33]. Using further measurements[34], 100 counts in the vertical scale would roughly correspond to $\chi_{max} \sim 350 \, \mu_B^2/eV$ in absolute units. These curves are in the same units as those of Fig. 8. Lines are fit by Eq. 6.

normal state (T= 100 K) spin susceptibility is reported on Fig. 9 for four different samples equivalent to those reported in the SC state: 3 in the underdoped regime, 1 in the overdoped regime. All these measurements have been scaled to the same units from an analysis as discussed in section 3. The results have been obtained on the same sample, except for the $x = 0.97$ sample[18] which have been scaled using phonon calibration.

One first clearly notices that $Im\chi$ strongly evolves with doping. $Im\chi$ is characterized by a maximum which becomes sharper in energy for higher doping. In addition to that peak, magnetic correlations occur at other energies. For instance, the spin response in the nearly optimally case YBCO$_{6.92}$ displays a strong resolution-limited peak located at 41 meV and a plateau in the range 30-37 meV. Normal state AF fluctuations exist in a wider energy range with a broad maximum around 30 meV. The 41 meV enhancement has totally vanished at T= 100 K and has been, therefore, assigned to a "magnetic resonance peak", E_r[5].

Upon increasing doping, the resonance peak is slighly renormalized to lower energies whereas the plateau is substantially removed, so that, the SC spin susceptibility is basically well accounted for by a single sharp peak around E_r= 39-40 meV for fully oxidized YBCO$_7$ samples. A clear agreement is now established among the neutron community about this slightly overdoped regime [18, 19, 29, 32]. The most important result is that this resonance peak intensity disappears at T_c as shown in Fig. 10. Performing a q-scan at the resonance energy in both the SC state and in the normal state emphasizes the vanishing of the resonance peak above T_c: Fig. 11 displays a q-scan at $\hbar\omega = 39$ meV with very weak magnetic intensity at (π, π) and T= 125 K[47]. This striking temperature dependence demonstrates that this peak is intimately related to the establishement of superconductivity. Fig. 10 shows that the resonance intensity actually follows an order parameter-like behavior[5, 18, 29, 33] whereas the resonance energy is itself very weakly temperature dependent, if any[18, 28, 33]. The q-scan at 39 meV also underlines the strong reduction of the normal state AF correlations in the overdoped regime. This is in sharp contrast with the nearly optimally doped sample, YBCO$_{6.92}$[32]: on a more quantitative ground, Fig. 9 reveals that the AF correlations amplitude is reduced at T=100 K by a factor $\sim$ 2-4 (depending of the energy) going from $x = 0.92$ to $x = 1$.

On lowering doping, the spin susceptibility spreads out over a wide energy range. However, a close inspection of q- and temperature dependences of $Im\chi$ reveals that the resonance peak feature is still present in underdoped samples[30, 31, 32]. This can be demonstrated by making the difference between the neutron intensity measured at T= 5 K at Q=(π, π) and the same scan measured just above T_c[28, 59]: this difference exhibits a sharp peak at energy where $Im\chi$ at T=5 K is maximum. Here, the whole lineshape of $Im\chi$ is preferably reported as it can be directly compared with theoritical models. The resonance peak energy is shifted to lower energy in underdoped regime, actually following T_c[31, 59] (Fig. 12). The relative amplitude of the resonance peak as compared to the normal state intensity is decreasing with lowering oxygen content, so that, the resonance amplitude is about 50% of the maximum intensity for heavily doped samples ($x \geq 0.6$, $T_c > 60$ K , when the resonance energy is located around ~ 34 meV,)[28, 31, 35], and only about 15% for weakly-doped samples ($x \simeq 0.5$, $T_c \sim 50$ K and $E_r \simeq 25$ meV)[34]. In contrast, the intensity at the maximum at T= 5 K actually increases with decreasing doping. Conversely, the normal state excitations is larger in the weakly-doped range (see Fig. 9). Therefore, although the spectral weight of $Im\chi$ becomes larger, the resonance feature vanishes for smaller doping and T_c.

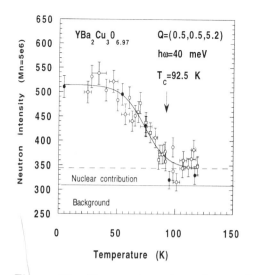

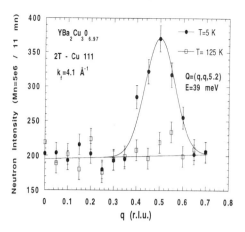

Figure 10: Temperature dependence of the resonance peak at E= 40 meV (from [18]). The reported "nuclear contribution" is due to a phonon peak whose maximum intensity is located at 42.5 meV[19].

Figure 11: Q-scans across the magnetic line at the resonance energy in the over-doped regime. These data are giving the most accurate upper limit of the normal state AF intensity which is about 8 times smaller than the resonance peak observed in the SC state.

Spin gap in the SC state

In the superconducting state, the spin susceptibility is limited at low energy by a gap below which no AF scattering is visible. This energy gap is then defined by the first inflexion point of $Im\chi$ curve and is referred to as "spin-gap", E_G. Below E_G, q-scans display no or very weak peak at (π, π), as displayed, at 10 meV in Fig. 3g and Fig. 3j. Going into the normal state (T=100 K), the spin-gap is suppressed because low energy excitations are sizeable at (π, π). Moreover, the peak intensity appears on heating at T_c[32] demonstrating that this spin gap is directly related to the superconducting gap. Further, E_G increases upon doping[30] but with a different dependence than the resonance energy.

In most cases, E_G is defined by a sharp resolution-limited step (Fig. 8). However, this low-energy step is found much broader for $x = 0.83$. Concomitantly, the superconducting transition of this sample was also broader than usual. Therefore, it is reasonable to attribute this behavior to the lack of sample homogeneity which could smear the sharp features of the spin response. Nevertheless, a resonance peak has been clearly observed in that sample[30] whose relative amplitude might be only reduced as compared with subsequent reports[28, 31, 59].

This comparison between spin-gap and T_C-width also gives some insight about two related issues. First, in the weakly-doped YBCO$_{6.5}$ regimes, this spin-gap has been widely reported by the french group[1, 5, 30, 32] and not by others[60] for sample having similar oxygen contents and T_c. As a matter of fact, a close inspection of the T_c-curves reveals that the samples exhibiting a spin-gap had clearly a sharper SC transition. Second, it may solve a controversy existing in LSCO system where a spin-gap has been reported only recently[51, 61]. Previous studies have not succeeded to detect it[50] likely because of sample inhomogeneities and/or impurity effects. More generally, the observation of the spin-gap appears to be very sensitive to sample defects

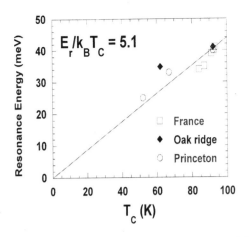
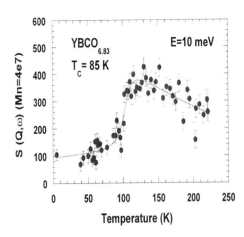

Figure 12: Resonance energy in YBCO versus T_c. French results are from [18, 30, 31, 32], Princeton results are from [19, 59], Oak Ridge results are from [28, 29].

Figure 13: Temperature dependence of the peak intensity at low energy in YBCO$_{6.83}$. Background has been subtracted using q-scans (see Fig. 3).

and impurities. As an example, the controlled substitution of 2% of zinc in overdoped YBCO (where T_c is reduced by ~ 20 K) induces low energy AF fluctuations below the spin-gap[26, 62].

Finally, it is very interesting to draw a parallel between the spin response in superconducting state of LSCO near maximum T_C ($x = 0.14$) and in weakly-doped underdoped YBCO$_{6.5}$. Both systems display a small spin gap in the SC state of about 2-5 meV. Both systems also exhibit a subtle enhancement of the spin susceptibility just above the spin gap: in LSCO, it has been underlined[63] that around 9 meV an additional enhancement occurs only in the SC state which is accompanied to a sharpening in momentum space. Similar observations have been previously reported around 7 meV in YBCO$_{6.52}$[64]. In light of the recent experiments in YBCO$_{6.5}$[59], it seems that this effect is different from the resonance peak feature occuring at higher energy. However, this issue is still under debate as no resonance peak has been detected so far in LSCO. Besides, this could be simply because the resonance intensity is expected to be only ~ 10 % of the total magnetic scattering for such low $T_c \leq 40$ K, still by comparison with YBCO$_{6.5}$[34].

7. COHERENCE EFFECTS IN THE SUPERCONDUCTING STATE

As the resonance peak and the spin-gap are both intimataly related to the superconductivity, a simple interpretation has then been proposed in itinerant magnetism approaches[42, 65, 66, 67, 68, 69, 70, 71, 72]. Beyond different hypotheses, these models basically show that the resonance peak occurs simply because of the BCS pairing. Starting from a Fermi liquid picture, the non-interacting electronic spin susceptibility (Lindhard Function) is written as,

$$Im\chi°(q,\omega) = (g\mu_B)^2 \lim_{\epsilon \to 0} \sum_k \frac{f_{q+k} - f_k}{\epsilon_{q+k} - \epsilon_k - \hbar\omega - i\epsilon} \qquad (3)$$

where ϵ_k is the electronic band dispersion and f_k is the associated Fermi function.

This spin susceptibility is described by the two-particle response function which usually gives featureless broad response due to sum over the reciprocal space. However, band structure singularities and nesting effects can induce rather complex lineshapes and q-dependences. Going into the SC state, one should account for the spin pairing of Cooper pairs[58], $Im\chi^\circ$ then becomes for T=0,

$$Im\chi^\circ(q,\omega) = (g\mu_B)^2 \lim_{\epsilon\to 0} \sum_k \left[1 - \frac{\Delta_k\Delta_{q+k} + \epsilon_{q+k}\epsilon_k}{E_{q+k}E_k}\right] \frac{1 - f_{q+k} - f_k}{E_{q+k} + E_k - \hbar\omega - i\epsilon} \qquad (4)$$

with $E_k = \sqrt{\Delta_k^2 + \epsilon_k^2}$ and Δ_k is the k-dependent SC energy gap. The term in brackets is a coherence factor related to the BCS pairing[58]. This function displays a soft edge behavior when the sum $(E_{q+k} + E_k)$ is mimimum[68, 72] but not a sharp peak. Further, the q-dependence of this function is not necessarily peaked around (π, π)[71]. Therefore, neutron data cannot be reproduced in a generic noninteracting electron model. To overcome these difficulties, a Stoner-like factor should be included which is related either to band structure singularities, to spin fluctuations, or even to interlayer pair tunneling effects[72]. In the case of a magnetic exchange, $J(q)$, the interacting spin susceptibility using Random Phase Approximation (RPA), is

$$\chi(q,\omega) = \frac{\chi^\circ(q,\omega)}{1 - J(q)\chi^\circ(q,\omega)} \qquad (5)$$

The RPA treatment have been held responsible for the sharpness of the peak in both energy and momentum. Using this expression, a sharp peak is obtained above a gap and both are proportional to the SC gap. In the normal state, featureless non interacting spin susceptibility for itinerant carriers is restored. Hence, this result in both the normal state and the SC state agrees with what is measured in the overdoped regime.

In this view, the resonance energy, E_r, closely reflects the amplitude of the SC gap as well as its doping dependence. Experimentally, this prediction is supported as the resonance energy is found to scale with T_c with $E_r/k_BT_c \simeq 5.1$ (Fig. 12). However, the true proportionality of the resonance energy with the maximum SC energy gap, Δ_{SC}^{max}, is not simply 2 as expected in simple BCS theory with an isotropic SC gap. Here, it crucially depends on band structure effect[9]. For the same reason, this relation is even more complex for the spin gap doping dependence[68]. Interestingly, this model requires $d_{x^2-y^2}$-wave symmetry of the SC gap function, Δ_k[65, 66, 67, 68, 69, 70], as a change of sign of $\Delta_k\Delta_{Q_{AF}+k}$ should occur in the coherence factor. However, other subtle scenario can occur[71]. Van-Hove singularities in band structure[67, 69] have been also invoked to play a role in the peak enhancement. As the peak position of the resonance exhibits very little temperature dependence[18, 28, 33], the SC gap should not change with temperature in this framework.

Finally, one can conclude that coherence effects related to the spin pairing seem to account for the marked modifications of the spin susceptibility in the SC state. Moreover, the additional enhancement in the SC state observed $\simeq 7 - 9$ meV[63, 64] can be also explained in this framework: a transfer of the spectral weight from the low energy (below the spin-gap, E_G) to energies just above E_G is indeed expected. In any case, the major limitation of this model is that in the underdoped regime (as well as for optimal doping) one observes another magnetic contribution existing in the SC state and remaining in the normal state. This more complex behavior of the spin susceptibility requires more sophisticated models than simple itinerant hole picture.

A correlated-electron model which is aiming to unify superconductivity and antifer-romagnetism[8], has been recently proposed to account for the resonance peak. A particle-particle collective excitation occurs in this model[73], whose matrix element coupling this excitation to the magnetic neutron cross section, vanishes in the normal state, but is nonzero in the superconducting state. In this model, because excitation energy is proportional to the hole doping[73], the resonance energy should increase with the oxygen content. Experimentally, this trend is found up to optimal doping. However, going from $YBCO_{6.9}$ to $YBCO_7$, E_r is rather reduced (Fig. 8) following actually T_c in contrast to the expectation.

8. SPIN PSEUDO-GAP AND QUASI-DISPERSION

As stressed in previous sections, the resonance feature is accompagnied in under-doped and optimally doped regimes by another broad contribution which is rapidly stronger in amplitude at lower doping. The odd spin susceptibility is actually max-imum at a characteristic energy, ~ 30 meV (see Fig. 9). This maximum of $Im\chi$, which directly corresponds to the pole of the spin susceptibility, then naturally defines a gap in the spin excitation spectrum. Obviously, this gap can be identified to the well-known "spin pseudo-gap"[2, 5, 32] although its definition is different from previous convention[5].

For the sake of clarity, it is useful to describe the normal state spin susceptibility by a damped Lorenztian function. This expression is usually applied to disordered short range localized spins. It is, for instance, the case of one dimensional magnetic system such as the so-called Haldane gap systems[74]. $Im\chi$ is then written,

$$Im\chi(q,\omega) = \frac{\chi_q \quad \omega \quad \Gamma}{(\omega - \omega_{pg})^2 + \Gamma^2} \qquad (6)$$

where ω_{pg} is the characteristic energy corresponding to the pole of the spin susceptibility. Eq. 6 is giving a good description of the normal state spin susceptibility (Fig 9) with $\omega_{pg} \simeq 28$ meV in heavily-doped samples ($x \geq 0.6$). In that doping range, the spin pseudo-gap, ω_{pg}, has no marked doping as well as temperature dependences. In contrast the spin gap, defined in Sec. 6, increases with doping. Therefore, these two gaps clearly differ. However, these two gaps "accidentally" occur in the same energy range yielding complex lineshapes in the SC state. Quite independently of the doping, the excitations display a strong damping with $\Gamma \simeq 12$ meV. As a result, sizeable excitations exist in the normal state down to low energy. This differs from the SC state where the low energy excitations are much more reduced by the spin-gap. Fig. 13 displays the temperature dependence of the magnetic scattering at (π, π) and at $\hbar\omega = 10$ meV for $x = 0.83$. In the SC state, small residual magnetic scattering occurs at this energy for that sample (Fig. 3). Upon heating, the AF fluctuations increase in the normal state passing through a maximum around a temperature $T^* \sim 120$ K larger than T_c. This temperature behavior reminds of that measured by copper NMR experiments in the underdoped regime[2, 3] where the spin-lattice relaxation rate $^{63}T_1$ is related to the spin susceptibility by $1/^{63}T_1T \propto \sum_q Im\chi(q,\omega)/\omega$. Similarly to Fig. 13, $1/^{63}T_1T$ displays a maximum at a temperature T^* in underdoped cuprates. This result has been widely interpreted by the opening of a spin pseudo-gap at the AF wave vector below T^*[2, 3]. More probably, this unusual temperature behavior as well as the value of T^* rather results from the interplay of the magnetic parameters, namely ω_{pg}, Γ and χ_q, which have different temperature dependences.

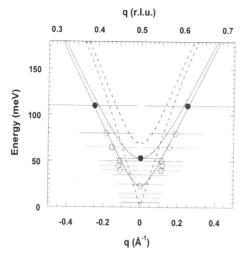

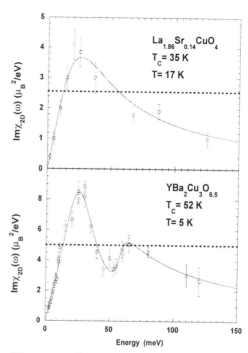

Figure 14: Spin excitation spectrum for odd - "acoustic" - (open circles) and even - "optical" - (closed circles) excitations at 5 K in YBCO$_{6.5}$ (from [34]). The lowest open circle represents the energy of the maximum of the odd susceptibility. The horizontal bars represent the intrinsic q-width (FWHM) after a Gaussian deconvolution from the spectrometer resolution. Full lines correspond to an heuristic quadratic fit like $\omega^2 = \omega_0^2 + c^2q^2$. Dashed lines represent the magnon dispersion curves of the undoped materials.

Figure 15: Comparison of the q-integrated spin susceptibility per CuO$_2$ plane in YBCO$_{6.5}$[34] and in La$_{1.86}$Sr$_{0.14}$CuO$_4$[75]. The dashed lines represent the 2D-integrated AF spin-wave contribution which is nearly constant in such an energy range (see[34, 38] for details).

In weakly-doped samples, YBCO$_{6.5}$, ω_{pg} is located $\sim$ 20 meV at T= 100 K (Fig. 9); ω_{pg} is also temperature dependent[32, 34] reaching 30 meV for T$\geq$ 200 K: this trend is caused by the fact that the low energy excitations (below $\simeq$ 25 meV) increase on decreasing temperature although the high energy part does not change[32]. This yields a critical-like behavior which can be associated with the proximity of the AF quantum critical point. Furthermore, similar energy and temperature dependences are found for samples in the insulating phase nearby $x \simeq 0.4$[26, 27].

Quasi-dispersion

This normal contribution actually extends to higher energies and a significant spectral weight at energies comparable to J is observed[34, 75]. Furthermore, it has been recently reported in YBCO$_{6.5}$ that the high energy spin excitations exhibit a quasi-dispersion behavior. Above $\hbar\omega \sim 50$ meV, a double peak structure emerges from constant energy q-scans[34]. This evidences a noticeable inplane propagating character of the spin excitations as, at each energy, one can define a characteristic wave vector which varies and matches a dispersion curve reproduced in Fig. 14. Similarly to magnons in an ordered magnetic system, one then observes propagating excitations in the metallic phase of cuprates.

Using classical spin-wave formalism for an Heisenberg model (which in principle cannot apply to such non-magnetic ground-state), one can describe the observed dispersion as a softening of the magnon dispersion of the undoped materials (Fig. 14). From a simple parabolic fit, one deduces a spin velocity, $c \simeq 420$ meV Å, 65% of the AF spin wave velocity of 650 meVÅ[37, 41]. This softening can be itself expressed in terms of an effective AF exchange as $J_{eff} \approx 0.65 J^{AF} \simeq 80$ meV at low temperature. A gap is found ~ 55 meV for the even excitations which slightly increases with temperature[34]. This even gap in the metallic state is found to be reduced from the magnon optical gap, $\omega_{opt}^{AF} = 67$ meV[40]. This effect can be readily accounted for by the same effective AF exchange using the classical spin-wave theory as, $\omega_{even} = 2\sqrt{J_{eff}J_\perp}$. However, as a limitation of this approach, counter-propagating excitations are not better resolved in the higher energy range[34]. Further, these observed peaks exhibit an intrinsic q-width which can be related to a real space correlation length of only about 9 Å. Therefore, instead of being well-defined propagating excitations, like magnons in the pure Néel state, one rather observes magnetic fluctuations which propagate only over small in-plane regions.

LSCO versus YBCO

Finally, it is instructive to compare the energy dependences of the spin response in LSCO near optimal doping ($x \sim 0.15$) and in weakly-doped underdoped YBCO$_{6.5}$. Above ~ 25 meV when the magnetic scattering becomes commensurate in LSCO[51], the neutron results show similar trends in the two systems. For instance, at about $E \sim 50$-60 meV, the magnetic spectrum is strongly reduced at (π, π)[5, 30, 32, 51]. This effect was previously attributed to an "high energy cut-off" but, more likely, this actually originates, at least in YBCO, due to the interplay of a few effects: first, above ~ 40 meV the spin susceptibility spectral weight at (π, π) effectively decreases with the energy. Moreover, $Im\chi$ displays a dip feature around 50 meV. Finally, the magnetic scattering strongly spreads out over the in-plane q-space for energies above 50 meV due to the quasi-dispersion behavior (Fig. 14). In LSCO, the quasi-dispersion has not been reported but, as a matter of fact, it can be inferred from high energy measurements for $x = 0.14$[75] as i) q-scan around $E \simeq 100$ meV exbibits significant broadening compatible with propagating excitations ii) a zone boundary peak is found as one would expect from a dispersion-like behavior. In any case, one clearly observes in the two systems a similar momentum broadening at high energy.

Actually, to emphasize the comparison of the two systems, it is more convenient to perform the q-integration of the spin susceptibility over the 2D in-plane wave vector q_{2D}, as $Im\chi_{2D}(\omega) = \int d\mathbf{q}_{2D} Im\chi(\mathbf{q}_{2D}, \omega)/ \int d\mathbf{q}_{2D}$. This also allows to overcome the difficulty that LSCO displays non overlapping incommensurate peaks below 25 meV. Further, the determination of $Im\chi$ in absolute units makes possible the direct comparison of LSCO ($x = 0.14$)[75] and YBCO$_{6.5}$[34]. The susceptibilities, calculated per single CuO$_2$-layer are reported in the same absolute units in Fig. 15. This q-integration has the effect to change the shape of the spin susceptibility as it enhances the high energy part due to the broadening in q-space of the AF fluctuations. For both systems, the spectral weight in the metallic state is very comparable to the spin wave spectral weight of the undoped materials (dashed lines in Fig. 15). The spin susceptibility exhibit a linear behavior at low energy, and then passes through a maximun around 25 meV, and extends to high energy. In YBCO, one finds in addition a dip feature around $E \simeq 50$ meV[34] (whose origin is not clear at present) which actually might also exist in LSCO. Finally, apart from incommensurate peaks at low energy, both systems exhibit very comparable momentum and energy dependences. As discussed above, modifications of $Im\chi$ in the SC state also show remarkable similarities. That strongly suggests that these two systems have same kind of hole doping per CuO_2 plane. In any case, this behavior is very different from the one observed in optimally doped YBCO. This is strongly suggesting that LSCO(x=0.15) does not correspond to an optimally doped cuprate in the generic phase diagram (Fig. 1).

Which models ?

What kind of models can explain these results in the normal state ? Some elements can be pointed out to describe at least part of the situation. First, these excitations can be associated with short range AF ordering of the localized copper spins. Indeed, the role played by localized copper spins in the metallic underdoped state is an important issue either for the nearly AF liquid model[7, 76] of in two dimensional quantum disordered approaches[68, 77, 78, 79]. Using RPA approximation, the magnetic exchange between copper spins $J(q)$ explains why the magnetic scattering is maximum at (π, π). Incidentally, the quasi-dispersion arises from $J(q)$ in this framework as heuristically discussed above. Further, an origin for the spin pseudogap is given in the t-J model[68, 78], which, similarly to what is found for the spin ladder compounds[79], occurs because of the formation of singlet RVB-states (spinon pairing). Second, these observations can also support itinerant magnetic models. As an example, it could remind the case of metallic Palladium where non-interacting electronic spin susceptibility is enhanced due to ferromagnetic interactions[80], leading to paramagnons behavior. In this approach, a dispersion-like behavior can simply occur due to the interplay of both band structure singularities and interaction renormalization. The observation of the even gap suggests the splitting of the Fermi surface caused by interlayer interaction. Interestingly, the even gap is shifted to low energies at higher doping[35]. Therefore, both dispersion curves in Fig. 14 tend to collapse at sufficiently high doping removing bands splitting. Such a doping dependence leads to an interesting issue regarding the photoemission experiments which have reported no "bilayer splitting" of the Fermi surface in $Bi_2Sr_2CaCu_2O_8$[13, 81]

10. CONCLUSION

The spin dynamics in metallic high-T_c cuprates as seen by Inelastic Neutron Scattering has been reviewed. Due the electronic interactions within a CuO_2-bilayer, one observes in the metallic state two different excitations: odd mode at lower energy (analogous of acoustic magnon) and even mode always weaker in amplitude (analogous of optical magnon). The odd susceptibility exhibits strong doping dependence which is characterized by two distinct contributions: a "magnetic resonance peak" which occurs only in the SC state, a normal contribution characterized by a spin pseudo-gap. Going into the overdoped regime in YBCO, the magnetic fluctuations in the normal state are strongly reduced[18]. Incidentally, it is very interesting to notice that the vanishing of magnetic correlations can be identified to the disappearance of the anomalous properties in overdoped cuprates (Fig. 1).

Using simple itinerant magnetism, it has been widely proposed that the resonance results from spin-flip charge carrier excitations across the SC energy gap. Such Fermi liquid approaches can also positively describe the normal state in the overdoped regime. It can even explain the occurence of incommensurate fluctuations in LSCO due to nesting effect of the Fermi surface. However, other aspects of the spin dynamics in underdoped and optimally doped samples, and in particular, the existence of the spin pseudo-gap cannot be accounted for. More generally, the observation of peaks around (π, π) implies the existence of strong electronic correlations in the metallic regime.

Other interesting properties, which have not been addressed here, are the effect of impurities. Zinc substitution is known to strongly reduce the SC temperature without changing the hole amount in the CuO_2 planes. Interestingly, zinc also strongly modifies the spin excitation spectrum[26, 62, 82, 83]: with only 2 % of zinc, the resonance peak intensity is strongly reduced and low energy AF excitations appear where nothing was measurable in zinc-free samples[26, 62]. Most likely, these results suggest that zinc enhances the AF correlations and induces a localization of the charge carriers.

Finally, Inelastic Neutron Scattering experiments have evidenced unusual spin dynamics in metallic cuprates which shed new light on the strong electronic correlations in these materials. Nevertheless, the way the copper spins are intrinsically coupled to the holes remains a question under discussion and yields a very interesting problem of localized-itinerant duality magnetism. For sure, quantitative comparison of the spin susceptibility measured by NMR and INS is needed to understand that issue.

Acknowledgments

The work reviewed here is the fruit of the collaboration of many people whose names appear all along the references. Here, I want particularly to acknowledge my close collaborators: L.P. Regnault (CENG-Grenoble), Y. Sidis (LLB-Saclay), B. Hennion (LLB-Saclay), H.F. Fong (Princeton University), B. Keimer (Princeton University), H. Casalta (ILL-Grenoble) and A.S. Ivanov (ILL-Grenoble).

REFERENCES

1. J. Rossat-Mignot, L.P. Regnault, P. Bourges, P. Burlet, C. Vettier, and J.Y. Henry in *Selected Topics in Superconductivity*, Frontiers in Solid State Sciences Vol. 1., Edited by L.C. Gupta and M.S. Multani, (World Scientific, Singapore, 1993), p 265. Early "neutron" studies are referenced in [30].
2. For a recent Nuclear Magnetic Resonance review in cuprates, see C. Berthier, M.H. Julien, M. Horvatic, and Y. Berthier, Journal de Physique I (France), **6**, 2205 (1997).
3. A. Trokiner, this book; M. Mehring this book.
4. B. Batlogg, H.Y. Hwang, H. Takagi, R.J. Cava, H.L. Kao, and J. Kuo, Physica C, **235-240**, 130, (1994); B. Batlogg and V.J. Emery, Nature, **382**, 20. (1996). B. Batlogg, this book.
5. J. Rossat-Mignod, L.P. Regnault, C. Vettier, P. Bourges, P. Burlet, J. Bossy, J.Y. Henry, and G. Lapertot, Physica C, **185-189**, 86 (1991).
6. N. Nagaosa, Science, **275**, 1078, (1997).
7. D. Pines, Z. Phys. B, **103**, 129 (1997); this book and references therein.
8. S.C. Zhang, Science, **275**, 1089, (1997).
9. F. Onufrieva, S. Petit, and Y. Sidis, Physica C, **266**, 101 (1996); Phys. Rev. B, **54**, (1996); J. of Low Temp. Phys., **105**, 597 (1996).
10. J. Labbé, and J. Bok, Europhys. Lett, **3**, 1225 (1987); J. Bouvier, and J. Bok, Physica C, **249**, 117 (1995); this book.
11. C.M. Varma, Phys. Rev. B, **55**, 14554, (1997).
12. H. Ding, T. Yokoya, J.C. Campuzano, T. Takahashi, M. Randeria,, M.R. Norman, T. Mochiku, K. Kadowaki, and J. Giapintzakis, Nature, **382**, 51, (1996).
13. J.C. Campuzano, M. Randeria, M.R. Norman, and H. Ding, this book.

14. T. Ito, K. Takenaka, and S. Uchida, Phys. Rev. Lett. **70**, 3995, (1993).

15. R. Nemetschek, M. Opel, C. Hoffmann, P.F. Müller, R. Hackl, H. Berger, L. Forroó, A. Erb, and E. Walker, Phys. Rev. Lett. **78**, 4837, (1997); R. Hackl, this book.

16. F. Onufrieva, P. Pfeuty, and M. Kisselev, (SNS 97) to appear in J. Chem. Phys. Solids, (1998).

17. see e.g. G. Uimin, and J. Rossat-Mignot, Physica C, **199**, 251, (1992).

18. P. Bourges, L.P. Regnault, Y. Sidis, and C. Vettier, Phys. Rev. B, **53**, 876, (1996).

19. H. F. Fong, B. Keimer, P.W. Anderson, D. Reznik, F. Dogan, and I.A. Aksay, Phys. Rev. Lett., **75**, 316 (1995).

20. Z.X. Shen, and J.R. Schrieffer, Phys. Rev. Lett. **78**, 1771, (1997).

21. M. Norman, H. Ding, J.C. Campuzanno, T. Takeuchi, M. Randeria, T. Yokoya, T. Takahashi, T. Mochiku, and K. Kadowaki, Phys. Rev. Lett. **79**, 3506, (1997) (cond-mat/9702144).

22. S.W. Lovesey, *Theory of Neutron Scattering from Condensed Matter*, Vol 1 & 2, (Clarendon, Oxford, 1984).

23. Neutron and Synchroton Radiation for Condensed Matter Studies, *Hercules, Ed. J. Baruchel, J.L. Hodeau, M.S. Lehmann, J.R. Regnard, and C. Schlenker*, (les Editions de Physique et Springer Verlag, 1993).

24. W. Reichardt, and L. Pintschovius, *Physical Properties of High Temperature Superconductors IV, ed by D.M. Ginsberg*, (World Scientific, 1993), p 295. W. Reichardt, J. of Low Temp. Phys., **105**, 807 (1996).

25. B. Keimer, N. Belk, R.J. Birgeneau, A. Cassanho, C.Y. Chen, M. Greven, M.A. Kastner, A. Aharony, Y. Endoh, R.W. Erwin, and G. Shirane, Phys. Rev. B, **46**, 14034 (1992).

26. Y. Sidis, thesis, University Paris-XI, Orsay (1995).

27. P. Bourges, Y. Sidis, B. Hennion, R. Villeneuve, G. Collin, and J. F. Marucco, Physica B, **213&214**, 54, (1995).

28. P. Dai, M. Yethiraj, H.A. Mook, T.B. Lindemer, and F. Dogan, Phys. Rev. Lett. **77**, 5425, (1996). P. Bourges, and L.P. Regnault, comment to Phys. Rev. Lett. to appear (1998).

29. H.A. Mook, M. Yehiraj, G. Aeppli, T.E. Mason, and T. Armstrong, Phys. Rev. Lett., **70**, 3490 (1993).

30. P. Bourges, L.P. Regnault, J.Y. Henry, C. Vettier, Y. Sidis, and P. Burlet, Physica B, **215**, 30, (1995).

31. P. Bourges, L.P. Regnault, Y. Sidis, J. Bossy, P. Burlet, C. Vettier, J.Y. Henry, and M. Couach, Europhysics Lett. **38**, 313 (1997).

32. L.P. Regnault, P. Bourges, P. Burlet, J.Y. Henry, J. Rossat-Mignod, Y. Sidis, and C. Vettier, Physica C, **235-240**, 59, (1994); Physica B, **213&214**, 48, (1995).

33. H.F. Fong, B. Keimer, D. Reznik, F. Dogan, and I. A. Aksay, Phys. Rev. B. **54**, 6708 (1996).

34. P. Bourges, H.F. Fong, L.P. Regnault, J. Bossy, C. Vettier, D.L. Milius, I.A. Aksay, and B. Keimer, to appear in Phys. Rev. B, **56**, R11439, (1997) (cond-mat/9704073).

35. H.F. Fong, P. Bourges, *et al* to be published.

36. E. Manousakis, Rev. Mod. Phys. **63**, 1 (1991), and references therein.

37. S. Shamoto, M. Sato, J.M. Tranquada, B. Sternlieb, and G. Shirane, Phys. Rev. B, **48**, 13817 (1993).

38. P. Bourges, H. Casalta, A.S. Ivanov, and D. Petitgrand, Phys. Rev. Lett. **79**, 4906, (1997) (cond-mat//9708060).

39. H. Eskes, and J.H. Jefferson, Phys. Rev. B, **48**, 9788 (1993).

40. D. Reznik, P. Bourges, H.F. Fong, L.P. Regnault, J. Bossy, C. Vettier, D.L. Milius, I.A. Aksay, and B. Keimer, Phys. Rev. B **53**, R14741 (1996).

41. S.M. Hayden, G. Aeppli, T.G. Perring, H.A. Mook, and F. Dogan, Phys. Rev. B **54**, R6905 (1996).

42. A.J. Millis, and H. Monien, Phys. Rev. B, **54**, 16172, (1996).

43. O.K. Andersen, *et al*, J. Phys. Chem. Solids, **56**, 1579, (1995).

44. S. Chakravarty, A. Sudbo, P.W. Anderson, and S. Strong, Science, **261**, 337,(1993).

45. T.A. Kaplan, S.D. Mahanti, and Hyunju Chang, Phys. Rev. B **45**, 2565 (1992).

46. J.M. Tranquada, P.M. Gehring, G. Shirane, S. Shamoto, and M. Sato, Phys. Rev. B, **46**, 5561, (1992).

47. P. Bourges, L.P. Regnault, Y. Sidis, J. Bossy, P. Burlet, C. Vettier, J.Y. Henry and M. Couach, J. of Low Temp. Phys., **105**, 377 (1996).

48. P. Dai, H.A. Mook, and F. Dogan, preprint, (1997) (cond-mat/9707112).

49. B. J. Sternlieb, J.M. Tranquada, G. Shirane, S. Shamoto, and M. Sato, Phys. Rev. B, **50**, 12915 (1994).

50. T.E. Mason, G. Aeppli, and H.A. Mook, Phys. Rev. Lett., **68**, 1414 (1992); T.R. Thurston, P.M. Gehring, G. Shirane, R.J. Birgeneau, M.A. Kastner, Y. Endoh, M. Matsuda, K. Yamada, H. Kojima, and I. Tanaka, Phys. Rev. B, **46**, 9128 (1992).

51. S. Petit, A.H. Moudden, B. Hennion, A. Vietkin, A. Revcolevschi, Physica B, **234-236**, 800, (1997); Physica C, **282-287**, 1375, (1997).

52. Q. Si, Y. Zha, and K. Levin, Phys. Rev. B., **47**, 9055 (1993); Y. Zha, K. Levin, and Q. Si, *ibid.*, **47**, 9124 (1993).

53. J.P. Lu et al Physica C, **179**, 191, (1991).

54. P.B. Littlewood *et al*, Phys. Rev. B., **48**, 487 (1993).

55. S.V. Maleyev, J. Phys. I. France, **2**, 181 (1992); S. Charfi-Kaddour, R-J. Tarento, and M. Héritier, J. Phys. I. France, **2**, 1853 (1992).

56. J.M. Tranquada, *et al*, Nature, **375**, 561, (1995).

57. V.J. Emery, and S.A. Kivelson, Physica C, **209**, 597, (1993).

58. J.R. Schrieffer, *Theory of Superconductivity*, (Frontiers in Physics (20), Addison Wesley), (1988).

59. H.F. Fong, B. Keimer, F. Dogan, and I.A. Aksay, Phys. Rev. Lett. **78**, 713 (1997).

60. J.M. Tranquada, W.J.L. Buyers, H. Chou, T.E. Mason, M. Sato, S. Shamoto, and G. Shirane, Phys. Rev. Lett. **64**, 800 (1990).

61. K. Yamada, *et al*, Phys. Rev. Lett. **75**, 1626, (1995).

62. Y. Sidis, P. Bourges, B. Hennion, L.P. Regnault, R. Villeneuve, G. Collin, and J.F. Marucco, Phys. Rev. B, **53**, 6811 (1996).

63. T.E. Mason, A. Schröder, G. Aeppli, H.A. Mook, and S.M. Hayden, Phys. Rev. Lett. **77**, 1604, (1996).

64. J. Rossat-Mignod, L.P. Regnault, P. Bourges, P. Burlet, C. Vettier, and J.Y. Henry, Physica B, **194&196**, 2131, (1994).

65. Y. Ohashi, and H. Shiba, J. Phys. Soc. Jpn., **62**, 2783 (1993); P. Monthoux, and D.J. Scalapino, Phys. Rev. Lett., **72**, 1874 (1994).

66. J. P. Lu, Phys. Rev. Lett., **68**, 125 (1992); N. Bulut, and D.J. Scalapino, Phys. Rev. B, **53**, 5149 (1996).

67. M. Lavagna, and G. Stemmann, Phys. Rev. B, **49**, 4235 (1994); G. Blumberg, B. P. Stojkovic, and M.V. Klein, Phys. Rev. B, **52**, R15741 (1995).

68. F. Onufrieva, and J. Rossat-Mignod, Phys. Rev. B, **52**, 7572 (1995). F. Onufrieva, Physica B, **215**, 41, (1995).

69. F. Onufrieva, Physica C, **251**, 348, (1995).

70. D.Z. Liu, Y. Zha, and K. Levin, Phys. Rev. Lett., **75**, 4130 (1995).

71. I.I. Mazin, and V. M. Yakovenko, Phys. Rev. Lett., **75**, 4134 (1995).

72. L. Yin, S. Chakravarty and P.W. Anderson, Phys. Rev. Lett., **78**, 3559 (1997).

73. E. Demler, and S.C. Zhang, Phys. Rev. Lett., **75**, 4126 (1995).

74. L.P. Regnault, I. Zaliznyak, J.P. Renard, and C. Vettier, Phys. Rev. B **50**, 9174 (1994).

75. S.M. Hayden, G. Aeppli, H.A. Mook, T.G. Perring, T.E. Mason, S-W. Cheong and Z. Fisk, Phys. Rev. Lett. **76**, 1344 (1996).

76. V. Barzykin, D. Pines, A. Sokol, and D. Thelen, Phys. Rev. B, **49**, 1544 (1994).

77. A.J. Millis and H. Monien, Phys. Rev. B **50**, 16606 (1994).

78. T. Tanamoto, H. Kohno, and H. Fukuyama, J. Phys. Soc. Jpn., **61**, 1886 (1992); G.

Stemmann, C. Pepin, and M. Lavagna, Phys. Rev. B, **50**, 4075, (1994).

79. T.M. Rice, S. Gopalan, and M. Sigrist, Europhys. Lett., **23**, 445, (1993).

80. R.M. White, Quantum Theory of Magnetism, *in* springer series to Solid-state Science, **32**, Springer Verlag Berlin (1983).

81. H. Ding, *et al*, Phys. Rev. Lett., **76**, 1533, (1996).

82. K. Kakurai, S. Shamoto, T. Kiyokura, M. Sato, J. M. Tranquada, and G. Shirane, Phys. Rev. B. **48**, 3485 (1993). H. Harashina, S. Shamoto, T. Kiyokura, M. Sato, K. Kakurai, and G. Shirane, J. Phys. Soc. Jpn **62**, 4009 (1993).

83. P. Bourges, Y. Sidis, B. Hennion, R. Villeneuve, G. Collin and J.F. Marucco, Czechoslovak Journal of Physics, **46**, 1155 (1996). P. Bourges, Y. Sidis, L.P. Regnault, B. Hennion, R. Villeneuve, G. Collin, C. Vettier, J.Y. Henry, and J.F. Marucco, J. Phys. Chem. Solids, **56**, 1937, (1995).

MAGNETIC PENETRATION DEPTHS IN CUPRATES: A SHORT REVIEW OF MEASUREMENT TECHNIQUES AND RESULTS

W. N. Hardy, S. Kamal, D. A. Bonn

Department of Physics and Astronomy
University of British Columbia
Vancouver, B.C., CANADA V6T-1Z1

INTRODUCTION

The London penetration depth, λ_L, is one of the most basic properties of a superconductor : it sets the scale for the exponential decay of magnetic fields within the superconductor . Furthermore, to within known constants, $1/\lambda_L^2$ is equal to n_s/m^* where n_s is the density of the carriers forming the superfluid condensate and m^* is the effective mass of the carriers. Therefore, a measurement of the temperature dependence of λ_L yields the temperature dependence of the superfluid density, something intimately related to the spectrum of excitations above the superconducting ground state. Such information has been particularly useful in the case of the HiT_c superconductors where many of the traditional techniques for characterizing the macroscopic properties have been difficult to apply. In many instances these difficulties are directly related to the large energy scale of the HiT_c superconducting state. An example is the electronic specific heat which, due to the high characteristic temperatures involved, is only a small fraction of the total specific heat. Therefore it has taken rather Olympian efforts to reliably extract the electronic specific heat. The penetration depth, something closely related to the specific heat, has intrinsic advantages of a) being relatively easy to measure, and b) having only electronic contributions. However, in spite of these advantages, the history of penetration depth measurements and their interpretation has been a rather bumpy one, with some fairly serious false starts and quite a bit of controversy, some of it extending up to the present.

As with just about any property of the HiT_c materials, there is an extensive literature that has accumulated over the past 10 years, although a substantial part of it has to be disregarded, for a variety of reasons. Of course there is not unanimous agreement on what should be believed and what should not, and the present review will certainly not be free of the prejudices of its writers, who have been involved with penetration depth measurements for some time. Given these circumstances we have chosen to spend rather a lot of time reviewing the methods of measurement* explaining, as best we can, the merits and weaknesses of each of the methods. This will serve to bring into focus the relevant issues, and in particular show when particular methods are appropriate and when they are not. It will also, hopefully, help the reader to make critical assessments of literature data.

ELELECTRODYNAMICS OF SUPERCONDUCTORS:
Theoretical models

At low frequencies the lossless part of the electromagnetic response of a superconductor is just the screening of a collisionless electron gas with superfluid number density $n_s(T)$ and effective mass m^*. The screening length $\lambda(T)$ is given by

$$\frac{1}{\lambda^2(T)} = \mu_0 \frac{e^2}{m^*} n_s(T)$$

A "normal" fluid density is thereby implied, $n_n = n - n_s$ which is a quantity related to the quasiparticle density of states $N(\omega)$ such that

$$\left(\frac{\lambda(0)}{\lambda(T)}\right)^2 = 1 - \frac{2}{kT} \int_0^\infty d\omega \frac{N(\omega)}{N_0} [f(\omega)(1 - f(\omega))]$$

where f is the fermi function, and where no account has been taken of Fermi liquid corrections and scattering. In general n_n is a tensor; references to more general formalisms can be found in Hirschfeld et al.[1].

Within a BCS type formalism, $N(\omega)$ is determined by the superconducting gap $\Delta(\vec{k})$ which in general, due to anisotropic fermi surfaces and $\vec{k}$ dependent pairing interactions, may have a substantial dependence on $\vec{k}$. What are the symmetry constraints on $\Delta(\vec{k})$? In the case of HiT_c superconductors where the superconducting ground state is a spin singlet, the spatial part of the wavefunction must be even. In a truely isotropic system (no lattice) only even values of the pair orbital angular momentum are allowed $l = 0, 2, 4 \cdots$ labelled s-wave, d-wave, g-wave etc. In a crystal lattice l is no longer a good quantum number so that one must use the group-theoretic labels. Table 1[2] shows the 4 singlet pairing states of a single plane with square symmetry, the approximate symmetry of most of the cuprates.

Figure 1 shows the shapes of representative gap functions[2]. We note that for S^+, any function that has the full symmetry of the lattice is allowed, and in particular the gap may change sign. The B_{1g} and B_{2g} states (informally called the "d-wave" states)

---

*Note that we are not going into the technical details of the measurement apparatuses. There are a number of reviews that can be referred to. Our purpose is to focus on the general physical principles involved in the measurement and to comment on the limitations.

Table 1. The four singlet pairing states of a single plane with square symmetry (Annett et al.[2])

Informal name	Group-theoretic notation	$\hat{R}_{\pi/2}$	$\hat{I}_{axis}$	Representative state
s^+	A_{1g}	+1	+1	const.
s^- ("g")	A_{2g}	+1	−1	$xy(x^2 - y^2)$
$d_{x^2-y^2}$	B_{1g}	−1	+1	$x^2 - y^2$
d_{xy}	B_{2g}	−1	−1	xy

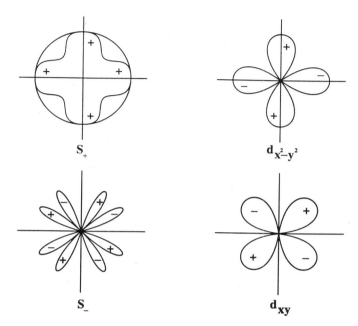

Figure 1. the four singlet irreducible representations possible in a single square CuO_2 plane (Annett et al. [2])

$d_{x^2-y^2}$ and d_{xy} break the symmetry of the lattice (as does the A_{2g} state) and are therefore formally distinct from the S^+ (s-wave) states. However, for orthorhombic crystals there is no longer any distiction and one will always have mixtures of, for example, s and d. However there is now considerable experimental evidence that the predominant

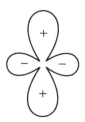

state is a "distorted" d-wave where the effect of orthorhombicity is mainly to make one set of opposing lobes somewhat larger than the other. This means that the nodes (positions where $\Delta(\vec{k}) \rightarrow 0$) may be shifted from the 45 degree positions; however at this point there is no evidence to show that this is a large effect. Therefore in general, when discussing the "d-wave" state in the HiT_c's we will have in mind a moderately distorted state such as that shown in the sketch.

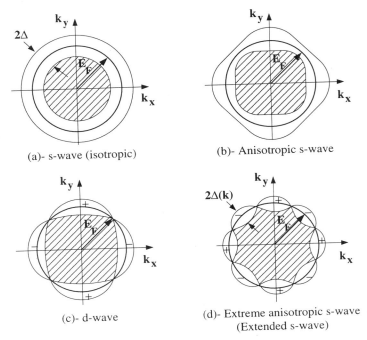

(a)- s-wave (isotropic) (b)- Anisotropic s-wave

(c)- d-wave   (d)- Extreme anisotropic s-wave
(Extended s-wave)

Figure 2. Some examples of gap symmetries for 2D square lattice.

Figure 2 a, b, c, d shows schematically the gap function for (a) isotropic s-wave, (b) moderately anisotropic s-wave, (c) d-wave and (e) an extreme case of anisotropic s-wave (sometimes called extended s-wave). Here we have used polar plots of energy, where the Fermi energy always gives a circle. Also we are assuming that the material is strongly 2 dimensional, with little dispersion in the z direction, so that the nodes in the d-wave and extended s-wave case are line nodes. In Fig. 3 we summarize the results for the densities of states and the expected low temperature dependencies for $\Delta\lambda(T) = \lambda(T) - \lambda(0)$. Note that extended s-wave and d-wave states would give identical results. We have also shown the effect of impurities on the d-wave density of states: a finite density of states is produced at the Fermi level which results in $\Delta\lambda(T) \propto T^2$ instead of T for the pure d-wave case [1].

Phenomenological Aspects

For the purposes of this review, it is useful to have in mind a simple picture of the temperature dependence of the electrodynamics of the systems we will be dealing with. Generally speaking, the cuprate superconductors can be treated in the local limit, where the current density $\vec{J}$ is related to the local electric field $\vec{E}$ by a (complex) conductivity tensor:

$$\vec{J} = \underline{\underline{\sigma}}\,\vec{E}$$

where $\underline{\underline{\sigma}} = \underline{\underline{\sigma}}_1 - i\underline{\underline{\sigma}}_2 \equiv \underline{\underline{\sigma}}(\omega, T)$ (for the convention $e^{i\omega t}$ for the time dependence of $\vec{E}$).

Non local effects occur when there are correlation lengths in the physical system that exceed the characteristic length for the penetration of the electromagnetic fields

376

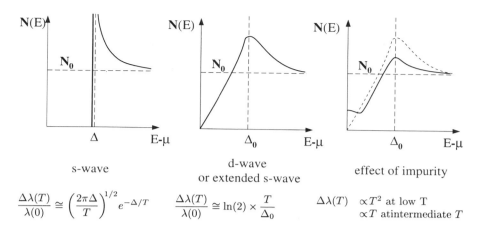

$$\frac{\Delta\lambda(T)}{\lambda(0)} \cong \left(\frac{2\pi\Delta}{T}\right)^{1/2} e^{-\Delta/T} \qquad \frac{\Delta\lambda(T)}{\lambda(0)} \cong \ln(2) \times \frac{T}{\Delta_0} \qquad \begin{aligned} \Delta\lambda(T) &\propto T^2 \text{ at low T} \\ &\propto T \text{ atintermediate } T \end{aligned}$$

Figure 3. Density of states for various gap symmetries and the associated low T behaviour.

into the sample. In the present case the latter length is the magnetic penetration depth, λ, which for the cuprates is typically 1000 to 2000 Å or more.

For the response of the superfluid, the appropriate correlation scale is the coherence length which is much smaller than λ for the cuprates and in most cases one is safely in the local limit. However Kosztin and Leggett[3] have shown that for applied H parallel to the c-axis of a clean d-wave superconductor, non-local effects in $\lambda_{ab}(T)$ may appear below a cross over temperature of about 1 K (due to the coherence length getting large at the nodes). Since up to now, measurements have been restricted to 1 K or above, we need not concern ourselves with these effects here. For the response of the "normal fluid", the situation is more complicated. For "clean" materials at low temperatures one can easily have in-plane quasiparticle mean-free-paths that are greater than the in-plane penetration depth , λ_{ab}. However, since the transport is strongly two dimensional, non-local effects are greatly suppressed for excitation fields applied parallel to the ab plane (i.e. the quasiparticles largely remain within the penetration depth between scattering events). Nevertheless, there are geometries where this condition does not hold; for example in the case of fields applied parallel to the ac face, the quasi-particles can exit the field penetration region at the surface <u>before</u> a scattering event[3]. In what follows we assume that local electrodynamics are valid.

In Fig. 4, we give a cartoon version of $\sigma_1(\omega, T)$ and $\sigma_2(\omega, T)$ for (a) $T > T_c$, (b) $T = 0$ and (c) $0 < T < T_c$. The shape in 4a is drawn as for a Drude metal, where

$$\sigma_1 = \frac{ne^2}{m^*} \frac{\tau}{1 + \omega^2\tau^2}$$

and

$$\sigma_2 = \frac{ne^2}{m^*} \frac{\omega\tau^2}{1 + \omega^2\tau^2}$$

although it is known that the high frequency dependence of $\sigma_1(\omega, T)$ is closer to $1/\omega$ than $1/\omega^2$. The measurement frequencies we will be discussing are generally much lower than the quasiparticle scattering rate $1/\tau$ in the normal state.

At $T = 0$ some fraction of the area under $\sigma_1(\omega, T > T_c)$ appears as a δ-function at $\omega = 0$. This is the response of the superfluid condensate (or just "superfluid ") which is therefore a very well defined quantity. The remaining area under $\sigma_1(\omega, T = 0)$ is relatively small for good quality undoped HiT_c samples (at least in the ab plane). This

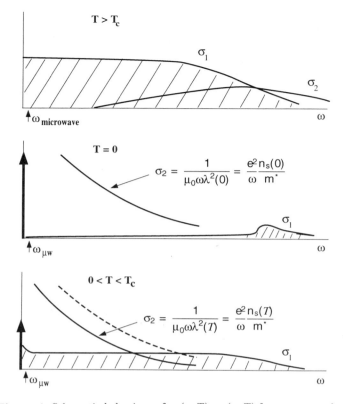

Figure 4. Schematic behaviour of $\sigma_1(\omega, T)$, $\sigma_2(\omega, T)$ for a superconductor.

is the so-called clean limit where the energy scale of the superconducting gap is larger than the quasiparticle (qp) scattering rate. Strictly speaking, the HiT_c 's are not really "clean" near T_c, but a rapid drop in $1/\tau$ as T is lowered quickly puts them into the very clean limit. We note that one of the reasons the superconducting energy gap is so difficult to observe by far infrared (Far IR) measurements of $\sigma_1(\omega, T < T_c)$ is that the materials are in the clean limit.

At finite temperatures, $\sigma_1(\omega > 0, T)$ increases over the $T = 0$ value, as quasiparticles become thermally excited. Correspondingly, the strength of the δ-function in σ_1 at $\omega = 0$ decreases by the increase in area under $\sigma_1(\omega > 0)$. Thus a measurement of the temperature dependence of the δ-function strength gives the increase in area under $\sigma_1(\omega > 0, T)$ which is determined by the spectrum of the quasiparticle excitations. In practice, one is usually measuring the contribution to $\sigma_2(\omega, T)$ from the δ-function in $\sigma_1(\omega, T)$. If we take the weight of the δ-function to be $(e^2/m^*)n_s(T)$, then the contribution to the imaginary part of σ (by Kramers Kronig) is

$$\sigma_2^{SF} = \frac{e^2}{m^*}n_s(T)\frac{1}{\omega}$$

We <u>define</u> the magnetic penetration depth λ by:

$$\sigma_2 = \frac{1}{\mu_0\omega\lambda^2(T)}$$

378

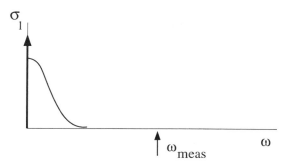

Figure 5. Situation where measurement frequency is too high for measurement of superfluid density.

such that the electromagnetic propagation constant $\kappa = [-i\mu_0\omega\sigma]^{1/2}$ is equal to $1/\lambda$ in the case that $\sigma = \sigma_1 - i\sigma_2$ is dominated by σ_2. Furthermore, if σ_2 at a particular frequency is mainly due to the superfluid response then

$$\sigma_2 = \frac{e^2}{m^*}n_s(T)\frac{1}{\omega} = \frac{1}{\mu_0\omega\lambda^2(T)}$$

and $\lambda(T)$ reduces to the usual London penetration depth $\lambda_L(T)$

At frequencies of a few GHz or below, σ_2 in the high T_c materials is completely dominated by the response of the superfluid condensate and contributions to σ_2 from the normal fluid are negligible except very close to T_c. However, depending on the qp scattering rate, the normal fluid may contribute at higher frequencies, even in the high microwave or low millimeter wave region. An extreme case is when $1/\tau$ is lower than the measurement frequency (see Fig. 5). Then $\sigma_2(\omega)$ approaches $\frac{e^2}{m^*}\frac{n_{total}}{\omega}$, where n_{total} includes both the superfluid and normal fluid (i.e. a narrow response of the normal fluid near $\omega = 0$, <u>looks</u> at some higher frequencies like a superfluid). Because of this one has to interpret Far IR (and even some mm wave) measurements that do not extend to $\omega \approx 0$ with care. If $1/\tau$ is strongly temperature dependent and falls below the minimum measurement frequency as one cools below T_c, one can easily get into the situation where the superfluid condensate <u>appears</u> to have very little temperature dependence at relatively high temperatures. Under these circumstances, the Far IR measurements may be useful for obtaining $\frac{n_s}{m^*}$ at $T = 0$ but not the temperature dependence of $\frac{n_s}{m^*}$ (they can also of course, in conjunction with independent "low" frequency measurements, give valuable information on the initial fall of $1/\tau$).

We note that there are wide variations in the temperature dependence of $1/\tau$: in materials doped deliberately with, for example, Zn or Ni or in most thin films, $1/\tau$ is extrinsic and has a much weaker temperature dependence. For some materials, however, such as high purity single crystals of YBCO, $1/\tau$ can drop into the microwave region (≈ 30 GHz $= 1$ cm^{-1}) by $T = 40$ K. This is well below the minimum Far IR measurement frequency usually encountered in single crystal work ($20 \rightarrow 50$ cm^{-1}), and even millimeter wave measurements have to be carefully scrutinized. Dähne et al.[4] for example were able to use the frequency dependence of λ in the mm wave region to extract τ. On the other hand, de Vaulchier et al.[5] saw no frequency dependence of λ in their films, up to 500 GHz. Further details will be given later in the MS.

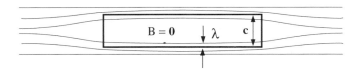

Figure 6. ‖ geometry where applied field at surface of sample is almost everywhere equal to the applied field.

METHODS FOR MEASURING $\lambda(T)$

It is convenient to separate the methods into those used for: A- single crystals, B-films and C- powders.

A. Single Crystals: Three Main Techniques
Excluded Volume Techniques

In this case the "signal" is the magnetic moment (either static or ac) produced by a sample in the Meissner state when it excludes an externally applied magnetic field. This method is used at frequencies all the way from DC (SQUID Magnetometry), to audio frequencies (AC susceptometers) to RF and microwaves (cavity perturbation). In the situation where there are no demagnetizing effects (i.e. the fields at the surface of the crystal are almost everywhere equal to the applied field), this magnetic moment is just $-H \times$ (geometrical volume of the crystal minus the volume penetrated). This is illustrated in the Fig. 6 where the excluded volume for a plate-like sample ($c \ll$ other two dimensions) is $(c - 2\lambda) \times$ area of plate. Since typical single crystals will have $d \geq 20\mu = 2 \; 10^5$ Å and $\lambda \approx 1000$ Å, in order to extract the absolute value of λ to within 10% from a measurement of the excluded volume, one would need to know the thickness of the crystal to better than $1/1000$. In addition there are demagnetizing corrections and calibration factors that would also have to be known to this accuracy. A direct attack in this direction is essentially impractical given the small and not perfectly regularly shaped crystals one is dealing with. Therefore one generally has to be content with a measurement of the temperature dependence of λ : $\Delta\lambda = \lambda(T) - \lambda(T_0)$ where T_0 is some reference temperature (usually the base temperature of the apparatus). To do this, one needs to be able to change the temperature of the sample, without affecting the calibration factors of the apparatus and at the same time avoiding (or correcting for) effects of sample holder materials that have temperature dependent magnetic properties. One arrangement that works extremely well is the use of a sapphire cold finger in vacuum, pioneered by Sridhar and Kennedy, and by Rubin et al.[6] and widely used. High purity sapphire has a very small magnetic susceptibility, is an excellent thermal conductor, and has very low dielectric loss even up to mm wave frequencies.

Thermal Expansion Effects. In many cases, it is imperative to correct for thermal expansion of the crystal itself. Consider first the geometry of figure 6 where we have in mind a typical HiT_c crystal with $c \ll a, b$, and where for concreteness we consider the shift in frequency δf of a microwave cavity when a (small) crystal is placed in a region where the microwave field is uniform. Then

$$\frac{\delta f}{f} = \frac{V_s}{2V_c}\left[1 - \frac{\tanh(c/2\lambda)}{c/2\lambda}\right] \simeq \frac{V_s}{2V_c}\left[1 - \frac{2\lambda}{c}\right] \tag{1}$$

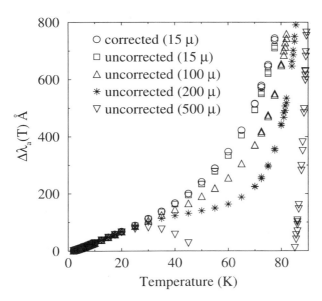

Figure 7. Thermal expansion effects on the apparent penetration depth of a YBCO single crystal for various thickness crystals.

for $c/2\lambda \gg 1$. Here V_s = sample volume and V_c = effective volume of cavity[8] from which

$$\frac{-V_c}{A}\frac{\Delta f}{f}\Big/\lambda(T) = \frac{\Delta\lambda(T)}{\lambda(T)} - \frac{c(T_0)}{2\lambda(T)}\frac{\Delta V}{V} - \frac{\Delta A}{A} \qquad (2)$$

where $\Delta\lambda(T) = \lambda(T) - \lambda(T_0)$, $\Delta f(T) = \delta f(T) - \delta f(T_0)$, $\Delta V = V(T) - V(T_0)$ etc, and A is the area of sample..

The second two terms in eqn. 2 are corrections to $\Delta\lambda/\lambda(T)$ due to thermal expansion. Since $\Delta V/V$ and $\Delta A/A$ are of similar magnitudes for YBCO, one sees that the second term dominates when $c/2\lambda \gg 1$, the usual situation. The effect of <u>not</u> making this correction is shown in Fig. 7 where the "apparent" $\Delta\lambda(T)$ is shown for crystals of thicknesses 15, 100, 200, and 500 μm. To produce this graph we have used real data for $\Delta\lambda_a(T)$ in YBa$_2$Cu$_3$O$_{6.95}$ and the thermal expansion data of Kraut et al.[7]. It is clear that the corrections are very important.

Now consider the geometry of Fig. 8, which corresponds to a typical situation when H_{rf} is applied $\parallel c$, in order to ensure that currents flow only in the ab plane. If we approximate the usual square or rectangular ab plane shape with a circle, then this geometry corresponds to one that has been extensively treated in the literature (see Brandt[9] and references therein). In general there is no analytical solution (so far), but Brandt [9] has given a numerical procedure involving the inversion of matrices whose size depend on the number of points in the coordinate grid. For large aspect ratios, $2a/c > 20$ or 30 , present computing techniques are inadequate when one wants accurate solutions for $\lambda \ll c \ll 2a$ [10]. This is in fact a highly relevant case, and the absence of a solution has led some researchers to <u>assume</u> that $\Delta\lambda = K\Delta f$ where K is a temperature independent calibration constant to be determined (by using another superconductor or the normal state properties of the HiT_c material). This is certainly <u>not</u> correct in detail, although at this point we do not know how serious the errors are. In the limit that $\lambda \to 0$ (perfect screening) then the frequency shift (for $c \ll 2a$) is[11]

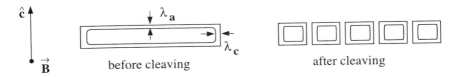

Figure 8. Sketch showing demagnetization effects in ⊥ geometry

$$\frac{\delta f}{f} = \frac{1}{2V_c}\frac{8}{3}a^3$$

For the purposes of discussing the effects of thermal expansion we make the approximation for finite λ that

$$\frac{\delta f}{f} = \frac{1}{2V_c}\frac{8}{3}(a - \lambda)^3$$

(which, as just discussed, is certainly <u>not</u> accurate). In this case $\Delta\lambda/\lambda(T)$ will have corrections of the order $\frac{a}{\lambda(T)}\frac{\Delta V}{V}$. Given that $a/\lambda(T)$ is typically a factor 10 larger than $c/2\lambda(T)$, one can see that thermal expansion effects are likely to <u>dominate</u> the measured frequency shifts over substantial regions of temperature. Given the absence of an EM solution for realistic sample dimensions, it is in fact not possible at this time to accurately correct data taken in this geometry. Therefore one should look very carefully at the sample dimensions and the details of the data analysis when using such data.

A number of groups (Gruner and coworkers[12], Maeda and coworkers [13], Sridhar and coworkers [14], Anlage and coworkers[15] and others) have extracted values of $\lambda(T \simeq 0)$ from cavity perturbation methods by making use of the fact that in the normal metal, the real and imaginary parts of the surface impedance are equal. To see how the method works, consider the geometry of Fig. 6 for which the total (complex) frequency shift due to the presence of the sample is

$$\frac{\delta f}{f} = \frac{1}{2}\frac{V_s}{V_c}\left[1 - \frac{\tanh(\alpha c/2)}{\alpha c/2}\right] \tag{3}$$

where $\alpha = (1 - i)/\delta$ for a metal (δ = skin depth) and $\alpha = 1/\lambda$ for a SC. Thus for $T > T_c$ and in the limit $c \gg \delta$:

$$2V_c\frac{\delta f(T)}{f} = V_s(T)\left[1 - \frac{\delta(T)}{c(T)} - i\frac{\delta(T)}{c(T)}\right] \tag{4}$$

and for $T \simeq 0$:

$$2V_c\frac{\delta f(0)}{f} = V_s(0)\left[1 - \frac{2\lambda(0)}{c(0)}\right]$$

where we have assumed that loss in the superconducting state is negligible. Assuming for the moment that $V_s(T) = V(0)$,

$$\frac{2V_c}{f}\ \mathrm{Re}\{\delta f(T) - \delta f(0)\} = -\frac{V_s}{c}\left[\delta(T) - 2\lambda(0)\right]$$

$$\frac{2V_c}{f}\ \mathrm{Im}\{\delta f(T)\} = -\frac{V_s}{c}\delta(T)$$

Since we have an experimental value of $\frac{V_s}{c}\delta(T)$ from $\mathrm{Im}\{\delta f(T)\}$, we can find $\lambda(0)$. A typical value of $\delta(T = 100\ K)$ at 10 GHz is 4.41 μ ($\rho \simeq 77\ 10^{-6}$ Ω-cm); this requires an absolute measurement of $\delta(1/Q)$ for the resonator with an accuracy of about 0.6 % , to extract a λ_0 of 1500 Å to within 10% . This requires care but is quite feasible. On the other hand, the thermal expansion corrections may be large. Assuming no error in $\delta(1/Q)$ we can form the experimental quantity

$$V_s(T) - 2V_c \frac{\delta f(0)}{f} = V_s(0) \left[\frac{2\lambda(0)}{c(0)} + \frac{\Delta V}{V(0)} \right]$$

The measurements of Kraut et al.[7] give $\frac{\Delta a}{a} = 0.208 \ 10^{-3}$, $\frac{\Delta b}{b} = 0.046 \ 10^{-3}$, $\frac{\Delta c}{c} = 0.492 \ 10^{-3}$ for the increase in lattice parameters from $T \simeq 0$ to $T = 100$ K. This gives $\Delta V(100 \text{ K})/V = 0.746 \ 10^{-3}$. For a crystal with $c = 20 \ \mu$ (which is about as thin a crystal as one could use with $\delta \approx 5 \ \mu$) and $\lambda(0) = 1500$ Å, $2\lambda(0)/c(0) = 1.5 \ 10^{-2}$, and therefore the thermal correction is about 5% . However, in the $H_{rf} \parallel c$ geometry, the thermal correction $\Delta V/V$ is to be compared to $\lambda(0)/a$ where a is the radius of the disk used to approximate the rectangular crystal. Here a is typically 0.5 mm so that $\lambda(0)/a \simeq 0.3 \ 10^{-3}$ and the thermal correction is substantially larger than the quantity we wish to extract.

We conclude that this method has to be applied with extreme care. In the $H \perp c$ geometry where the thermal corrections are relatively small, one has to contend with the contribution of λ_c or δ_c to the frequency shifts. In the $H \parallel c$ geometry where currents flow only in the ab plane, the thermal correction is generally large and the lack of a full solution to the EM problem makes it unlikely that this correction is under control. The correction also requires the availability of accurate thermal expansion data for the material under study.

Far Infrared Reflectivity: $|R|e^{i\theta}$

Here one measures $|R(\omega)| \equiv$ power reflectivity over a wide enough frequency range that one can perform a Kramers-Kronig on $|R(\omega)|$ to obtain $\theta(\omega)$. The Fresnel formula then gives σ_1 and σ_2. The superfluid contributes a δ-function in $\sigma_1(\omega, T)$ at $\omega = 0$ and a component to $\sigma_2 = (\mu_0 \omega \lambda_L^2)^{-1} = \frac{n_s e^2}{m^*} \frac{1}{\omega}$. Figure 9 shows the data of Basov et al.[16] for untwinned $YBa_2Cu_3O_{6.95}$ where the intercept of $(\mu_0 \omega \sigma_2)^{-1/2}$ plotted vs ω gives λ. The point is that if there is a wide region where the above quantity is constant, then it is reasonable to assume one is dealing with a $\sigma_1(\omega)$ concentrated near $\omega = 0$. As discussed previously, the value of λ obtained may not be the London penetration depth λ_L, if there are contributions from $\sigma_1(\omega)$ other than the SF δ-function. This is one of the drawbacks of the method. However, the method gives absolute values for λ and by changing the polarization and the faces one can measure λ_a, λ_b and λ_c independently.

An alternative method is to look for the plasmon associated with the superfluid wherein $1/|\epsilon(\omega)|$ peaks at the plasma frequency $\Omega_p = (Ne^2/\epsilon_0 m^*)^{1/2}$. Again one will include any narrow Drude component to $\sigma_1(\omega)$ in Ω_p. The method is particularly useful when, in grazing incidence, (with respect to the ab plane), one can pick up the c-axis plasmon[17] in highly anisotropic materials where other methods to measure λ_c may be problematic.

Measurement of Internal Field Distribution in Mixed State

In a type II superconductor for $H > H_{c_1}$ and for weak pinning, a regular lattice of vorticies with density B/Φ_0 is formed. For an isolated vortex the magnetic field falls to zero away from the vortex with scale length λ_L. For $H \gg H_{c_1}$ the density of vorticies

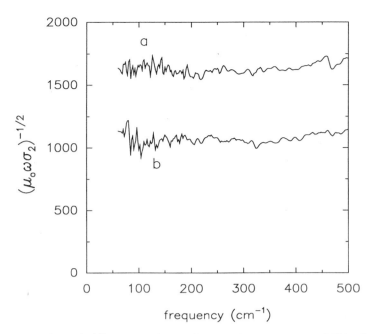

Figure 9. Quantity $(\mu_0\omega\sigma_2)^{-1/2}$ obtained from reflectivity data on untwinned $YBa_2Cu_3O_{6.95}$, plotted vs frequency. The extrapolated intercept gives $\lambda(0)$. (Basov et al.[16])

is high enough that the internal field B is relatively uniform, however $\overline{\Delta B^2}$ is set by $1/\lambda^2$ to within a constant.

The technique of Muon Spin Rotation (μSR) has been applied with great success to the cuprate superconductors beginning right after the discovery of HiT_c superconductivity, in many cases giving the first values of λ_L. The 100% spin polarized positive muons are implanted one at a time into the sample, where they quickly thermalize and take up a preferred interstitial position in the crystalline lattice. After an average lifetime of 2.2 μsec the muons decay with the emmision of an energetic positron, preferentially along the direction of the muon spin. With the use of a start counter and positional β^+ counters, a histogram is built up which contains information on the precession of the muon spin. In fact, something closely analogous to the "Free Induction Decay" in NMR is obtained, whose Fourier transform gives the distribution of magnetic fields within the sample. Figure 11 shows the magnetic field contours calculated in the so-called modified London model and the resulting field distribution[18]. Figure 11a shows time domain data for a twinned YBCO single crystal in a 0.5 T field ($T \simeq 10$ K) along with the FT, showing the characteristic assymetric lineshape[19]. The data in Fig. 11c was obtained after the external field was lowered by about 110 gauss; it shows that the flux lattice is strongly pinned at these temperatures.

The advantages of the method are that it is a bulk measurement (implant distance typically a few hundred microns), it gives absolute values for $\lambda(T)$ and most importantly, contains additional information. In particular the shape of the high field part of the distribution is sensitive to the details of the vortex core and is one of the very few methods that can measure the coherence length ξ directly (in principle low temperature STM can obtain similar information).

Some disadvantages are that rather large (~ 1 cm^2), thick (~ 0.5 mm) samples are still required at this time and it is difficult to measure λ_a, λ_b separately, or to measure λ_c at all. On the other hand the method works with ceramics or powders, although

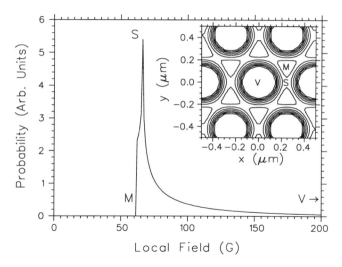

Figure 10. Insert showing the calculated contour lines from 65 to 100 G of the flux line lattice in an isotropic superconductor in a 100 gauss filed. The vortex cores are at V, and M and S mark the positions of the minimum field and saddle point respectively. Here $\kappa = \lambda/\xi$ is 60 and $\lambda = 0.1318~\mu$ (vortex spacing = 0.4889 μm at 100 G). (Riseman[18])

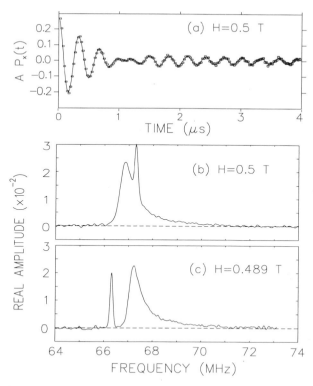

Figure 11. (a) Muon spin precession signal in $YBa_2Cu_3O_{6.95}$ single crystal cooled to 5 K with $H \parallel \hat{c}$. (b) Fourier transform of (a)(Gaussian apodization with 3 μsec τ). (c) Same as (b) except after field was lowered 0.0113 T. (Sonier et al.[19])

it is now clear that some early measurements gave misleading results. For example, in 1987 Harshman et al.[20] obtained a temperature dependence of $1/\lambda^2$ in a YBCO ceramic that was very flat at low temperatures and was interpreted as evidence for s-wave superconductivity . Contemporary results on single crystal YBCO now agree quite well with, for example, microwave methods. Finally, the μSR method can be used to measure the <u>field dependence</u> of quantities such as λ and ξ[21]. This capability will be of increasing importance as one tries to probe the superconducting state in more detail.

We note that in principle NMR can give the same information as μSR. However, the non-spin 1/2 species in the cuprates have NMR linewidths that are much too broad, and the spin-1/2 species, such as Y, give signals that are too weak. The technique of β-NMR where nuclear spin-polarized radioactive species are implanted has great promise[22] now that intense radioactive beams are becoming available. For one thing, smaller and thinner samples can be used.

B. Thin Films

Here the excluded volume techniques are very difficult to apply to a film whose lateral dimensions are of order of millimeters but whose thickness is of order 1000 Å. In these circumstances, demagnetizing effects are unavoidable, large, and not usually under control. For example when a field is applied parallel to the film, the slightest misorientation of the film, or inhomogeneities in the applied field will strongly distort the applied field.

Low Frequency Mutual Inductance Techniques

This is a powerful technique that works well for films that are thin enough. It is used by many groups[23] usually in the configuration shown, with the primary and secondary coils on either side of the films. It is easy to show that the film starts to screen the applied ac field when $t = \lambda \cdot \lambda/R$ where R = radius of coil and t is the thickness of the film. This means that substantial screening can occur for film thicknesses several orders of magnitude less than the penetration depth λ. Roughly, $H_{sec}/H_{prim} = (1 + Rt/\lambda^2)^{-1}$, so that for $R = 0.5$ cm, $t = \lambda \simeq 1500$ Å the field at the secondary is reduced by the factor 40,000.It is often difficult to reduce unwanted direct pickup and some groups use films with diameters as large as 4″ to solve this problem.[24]. Clearly one must carefully avoid physical defects in the films, and only in-plane currents can be probed. The method has the strong advantage of yielding absolute values of λ with quite good precision.

Microwave Techniques

The many microwave techniques have been reviewed by N. Klein [25]. Here we discuss 2 main groups.

Thin Film Resonator Techniques. In this type of measurement,the thin film is itself part of a resonant circuit. In the parallel plate method, one measures the TEM resonance(s) of a face to face pair of films separated by a thin dielectric. The fundamental mode occurs when the lateral dimension is equal to a half wavelength in the dielectric medium. The method is capable of very high resolution, but normally does not yield absolute values for λ; thermal expansion effects may also be important. Of course any sort of microstrip resonator can in principle be used to measure $\Delta\lambda(T)$. A variant developed by Andreone et al.[26] to avoid patterning of the film of interest (NCCO) uses a microstrip ring of YBCO with the NCCO as the ground plane. None of these methods give absolute values of λ.

Millimeter Wave Transmission. This technique is closely related to the previous one as far as measurement of λ is concerned, but can yield considerably more information such as R_s, the frequency dependence of λ, etc. Here again the transmitted signal can be greatly attenuated by the films: roughly, the fields are reduced by the factor $\lambda_{EM}t/2\pi\lambda^2$, where now the free space wavelength of the radiation takes the place of the coil dimension. For $\lambda_{EM} = 3$ mm, the reduction in the transmitted power can be very large at low temperatures, and leakage around the film can be a dominant problem. One must also take into account the effect of the substrate whose dielectric properties have to be measured separately. Phase coherent detection, which allows direct measurement of σ_1 and σ_2, has been used by Dähne et al.[4] and others. De Vaulchier[5] used light pipe type optics with the film electrically sealed to an aperture to obtain $\lambda(T)$ at low temperatures for fixed mm wave frequencies. Feenstra et al.[27] used the frequency dependent transmission of a focussed mm wave beam to take out substrate effects and also obtain $\lambda(T)$.

A relatively new technique, time domain terahertz spectroscopy, first applied by Nuss et al. to HiT_c films[28], expands the available frequency range to 1000 GHz or more, and also greatly increases the available power level for probing non-linear effects. The work of Orenstein et al.[29, 30] mainly on the electrodynamics of the vortex state, is a striking example of how powerful the method can be.

Far Infrared Reflection

This works well if one has films with good quality surfaces and relatively thick so that the second surface is not a problem. At grazing incidence the reflectivity is also sensitive to properties in direction $\perp$ substrate (usually but not always, the c-axis). Again, if one can observe the c-axis plasmon $(1/|\epsilon| \rightarrow \infty)$ one effectively measures n_s/m^*.

C. Powders

Although powders are not the preferred form for precision measurement of the electrodynamics properties of the cuprates, there are situations where the use of powders, especially grain-aligned, is particularly useful. One situation is where good quality single crystals or thin films are just not available; the second is when one wants to study the effect of impurities over a wide range of concentrations. In this latter case, controlling the impurity concentrations in single crystals is often problematic, whereas it is achieved with ease in powder samples.

A good example of the use of grain-aligned samples is the work of Panagopoulos et al.[31] on $HgBa_2Ca_2Cu_3O_{8+\delta}$. The grains were aligned by suspending them in epoxy cured in a 7 T magnetic field; the resulting angular distribution of the c-axis was $1.7°$

FWHM. Measurement of the penetration depth is achieved by measuring the temperature dependent magnetization in a weak $(1 - 10$ G$)$ probe field (DC SQUID or AC susceptometer). (This is an excluded volume technique where the thermal expansion effects are negligible for small grains). The magnetization is related to the penetration depth via the Shoenberg model [32], which requires knowledge of the grain size distribution and involves <u>assuming</u> the grains are spherical.

PENETRATION DEPTH RESULTS

A full review of the existing literature would take many times the length of the present article. The data included is considered to be <u>representative</u> of the best results available for the various members of the HiT_c family. Along the way we try to discuss the main issues of interest, but will have to pass over many of the interesting side issues. For more details, we refer the reader to the review by Bonn and Hardy[33].

$YBa_2Cu_3O_{7-\delta}$

This compound has received by far the most attention. It is the least anisotropic of the cuprates (which simplifies measurements of the electrodynamics), and electronically it is the most homogeneous. On the negative side, it is orthorhombic with highly conducting CuO chains. These chains are certainly incidental to the superconductivity, but interfere with the interpretation of the electrodynamics of the CuO_2 planes, the quantity of central interest. Ultimately, one is led to work with untwinned crystals in order to separate chain from non-chain effects.

In Fig. 12 we show the first data to exhibit what is believed to be close to the intrinsic behaviour for optimally doped material, that of Hardy et al.[8] on twinned $YBa_2Cu_3O_{6.95}(T_c = 93$ K$)$. A 1 GHz superconducting cavity perturbation method was

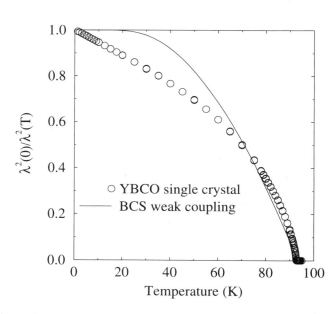

Figure 12. $\lambda_{ab}^2(0)/\lambda_{ab}^2(T)$ vs T for twinned $YBa_2Cu_3O_{6.95}$ using microwave values for $\Delta\lambda_{ab}(T)$ and $\lambda_{ab}(0) \approx 1400$ Å from μSR measurement on similar crystals. The solid line is the weak coupling BCS s-wave prediction. (Hardy et al.[8])

used to obtain $\Delta\lambda(T) = \lambda(T) - \lambda(1.3 \text{ K})$, and $\lambda_{ab} \simeq 1400$ Å was taken from μSR measurements on similar crystals. The data bears little resemblance to the weak-coupling BCS s-wave result (solid line) at both low and high temperatures, and is representative of most data obtained on the best materials (an exception is the recent work of Srikanth et al.[14] which will be discussed later). The linear low temperature part is universally seen in good single crystals and the very best thin films. There is less detailed agreement between measurements on the region near T_c where $1/\lambda^2$ appears to approach the temperature axis with infinite slope, indicating non-mean-field behaviour. This will be discussed separately in the section on fluctuations.

For reasons that are not fully understood, virtually all early measurements on all forms of YBCO (ceramics, thin films, single crystal) and by a variety of techniques, were interpreted in terms of a uniform single gap or two-gap BCS ground state. For instance measurements by Klein et al. of $\lambda(T)$ in YBCO films grown on NdGaO$_3$ and LaAlO$_3$ showed features that suggested the presence of two gaps, but also exhibited considerable sample dependence[34]. With the improvement in samples and measurement techniques, the picture changed and it is well established that $1 - \lambda^2(0)/\lambda^2(T)$ does not show activated behaviour at low T, being predominantly linear up to $T_c/3$, giving over to T^2 at lower temperatures, the cross-over temperature depending on the purity and perfection of the sample. For crystals grown in YSZ crucibles the crossover is in the 1 to 5 K range, but small concentrations (0.3 %) of Zn can raise this to 30 K. Mao et al.[15] show $\Delta\lambda_{ab}$ is essentially linear from 4.2 to 30 K. For thin films it is most common to see $1 - \lambda^2(0)/\lambda^2(T)$ varying as T^2 (which over a fairly wide range of T happens to be close to the BCS weak coupling result). The high resolution microstrip resonator data of Anlage and Wu fitted T^2 extremely well over the whole temperature range[35]. Later, Ma et al.[36] used the parallel plate resonator technique to study $\Delta\lambda(T)$ for a variety of films. As can be seen from Fig. 13, the higher the quality of the film, as measured by T_c, the closer $\lambda(T)$ matched the UBC single crystal data, always turning over, however, at low temperatures to a T^2 dependence. Lee et al.[37] using the low temperature mutual inductance technique on laser ablated films on $SrTiO_3$ ($T_c = 88$ K), had previously

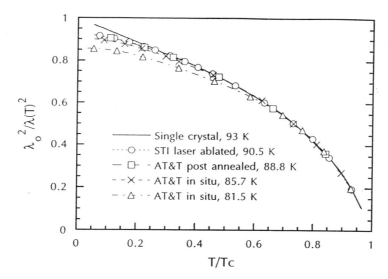

Figure 13. Penetration depth of YBa$_2$Cu$_3$O$_{7-\delta}$ grown by a variety of techniques on LaAlO$_3$ substrates. Data are normalized with a λ_0 that matches the curves to single crystal data at high T. (Z.X. Ma[36])

found good agreement between their films and single crystal results[8] from a few degrees below T_c to about 25 K. Gao et al.[38] reported a linear region between 6 and 30 K for a commercial film from Conductus (parallel plate resonator technique).

More recently, de Vaulchier et al.[5] used power transmission in the $120 - 500$ GHz range to measure $\lambda(T)$ absolutely. They showed that for films with a relatively large value of $\lambda(0)$ (3400 ± 200 Å; $T_c = 86$ K) $\Delta\lambda$ varied accurately as T^2. For a higher quality film ($\lambda_0 = 1570$ Å; $T_c = 92$ K) they observed $\Delta\lambda(T) \propto T$ from 4 to 40 K, with $\Delta\lambda/\Delta T$ essentially identical to that for twinned single crystals (Fig. 14). Thus it is now well established that the intrinsic low T behaviour of $\Delta\lambda$ for optimally doped YBCO is close to linear, and that the T^2 dependence observed in many films is due to some defect (unknown at this point).

The work of Dähne et al.[4] on very high quality films (high-pressure oxygen dc-sputtering on (110) $NdGaO_3$), illustrates the effect of very low elastic scattering rates on $\lambda(T)$. Using a quasi-optical Mach-Zehnder interferometer, they obtained $\lambda(T)$ at

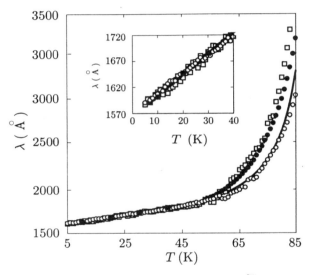

Figure 14. Temperature dependence of $\lambda(T)$ for a YBCO film at 3 mm wave frequencies, □ 120 GHz, • 330 GHz, ∘ 510 GHz (L.A. de Vaulchier) [5]. The filled squares come from the single crystal data of Hardy et al[8].

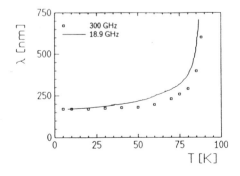

Figure 15. Data of Dähne et al.[4] which shows strong frequency dependence of $\lambda(T)$ for very clean YBCO films.

300 GHz and compared it to data at 18.9 GHz (Fig. 15). Below $0.6T_c$, λ is much less temperature dependent at 300 GHz than at 18.9 GHz. This is interpreted as the effect of a qp scattering rate that has dropped below 300 GHz, and they extract $1/2\pi\tau \approx 170$ GHz below 30 K. This is in contrast to the work of de Vaulchier, where no frequency dependence of $\lambda(T)$ was seen up to 500 GHz, presumably due to higher scattering rates. We note that Bonn et al. have inferred scattering rates in single crystals as low as 30 GHz [39]. The question remains: if the scattering rates in de Vaulchier et al.'s films were so large, why didn't they see $\Delta\lambda \propto T^2$ instead of T? This <u>may</u> be explained by the fact that two defects producing the same qp scattering rates need not be equally effective in changing T to T^2 (i.e. in producing states at the fermi energy).

Bonn et al.[39] and Achkir et al.[40] found that Zn impurities were much more effective in producing a change over to the T^2 dependence than was Ni, although these impurities were about equally effective as scattering centers (Bonn et al.[39, 41]). Fig. 16 shows the effect of $x = 0.15$ and 0.3% Zn on $\Delta\lambda(T)$ for $YBa_2(Cu_{1-x}Zn_x)_3O_{6.95}$. The addition of 0.7% Ni had less effect than 0.15% Zn. In contrast, Ulm et al.[42] studied the behaviuor of both $\lambda(0)$ and the temperature dependence of $1/\lambda^2(T)$ in films for 2 to 6% Ni and Zn impurities, and found Ni and Zn to have about equal effects. They found that $n_s(0)$ decreases by a factor of 2 for each % of dopant. This discrepancy is yet to be explained.

As an example of the quality of results that can now be obtained from μSR, we show in Fig. 17 field and temperature dependence of $1/\lambda_{ab}^2$ from the work of Sonier et al.[43] on single crystals. One clearly sees the linear temperature dependence (which agrees quite well with microwave measurements), but in addition a non-negligible field dependence. Although there are as yet no theories with which to compare, it is clear that μSR in the mixed state is a powerful probe of the non-linear response of the superfluid.

Penetration Depth Anisotropy in YBCO

Until relatively recently there has been little attention given to the ab plane anisotropy of the electrodynamics, although substantial anisotropies in the normal state

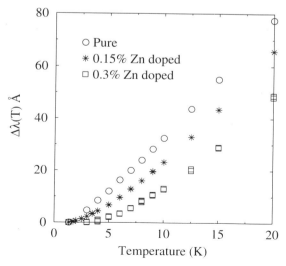

Figure 16. Low T penetration depth for $YBa_2(Cu_{1-x}Zn_x)_3O_{6.95}$ for $x = 0$ (nominally pure), 0.0015 and 0.003. (Bonn et al.[39])

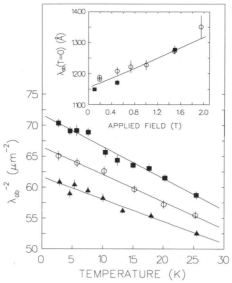

Figure 17. Temperature dependence of $1/\lambda_{ab}^2$ in $YBa_2Cu_3O_{6.95}$ for applied fields of 0.2 (solid square), 1.0 (circles) and 1.5 T (solid triangles). (Sonier et al.[19])

electrical conductivities and thermal conductivities(see Gagnon et al. and references therein[44]) had been observed. Zhang et al.[45] were the first to measure $\lambda_a(T)$ and $\lambda_b(T)$ separately using a combination of Far IR (see Fig. 9) and microwave techniques. The original microwave measurements were made at 1 GHz with $H \perp \hat{c}$ on a crystal thin enough that the effect of currents in the $\hat{c}$ direction could be ignored. IR measurements on the same crystal gave $\lambda_a(0) = 1600$ Å and $\lambda_b(0) = 1030$ Å, a rather large anisotropy. However this is not inconsistent with the n/m^* anisotropy observed in the normal state conductivities, nor the μSR results of Tallon et al.[46] for λ_a/λ_b inferred from the dependence of μSR relaxation on oxygen content. More recently the technique has been refined to the point where $\lambda_a(T)$, $\lambda_b(T)$ and $\lambda_c(T)$ can be determined absolutely from 1.3 K to T_c (see Hardy et al.[47], Bonn et al.[48]). This is accomplished by measuring $\Delta\lambda_a(T)$ and $\Delta\lambda_b(T)$ before and after the (approximately square) crystal is cleaved into 5 or more bars (figure 18). For $H \parallel$ to the long axis of the bars, the effect of λ_c is multiplied up by the number of pieces. A measurement perpendicular to the bars should not be affected to first order and serves as a control on the procedure.

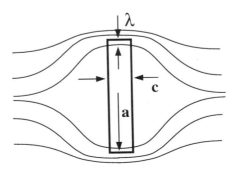

Figure 18. Arrangement to measure the c-axis penetration depth, $\Delta\lambda_c(T)$.

Table 2. $\lambda_i(0)$ for various oxygen contents of YBCO.

$\text{YBa}_2\text{Cu}_3\text{O}_{6+x}$ x	T_c (K)	$\lambda_a(T=0)$ (Å)	$\lambda_b(T=0)$ (Å)	$\lambda_c(T=0)$ (Å)
0.6 "underdoped"	59	2100	1600	65000
0.95 "optimally doped"	93.2	1600	1030	11000
0.99 "slightly overdoped"	89	1600[a]	800[a]	11000[b]

[a] $\lambda_a(0)$, $\lambda_b(0)$ not yet directly measured; we took $\lambda_a(0)$ to be same as for $x = 6.95$ and $\lambda_a(0)/\lambda_b(0)$ from Tallon et al[46].
[b] Obtained form microwave measurement.

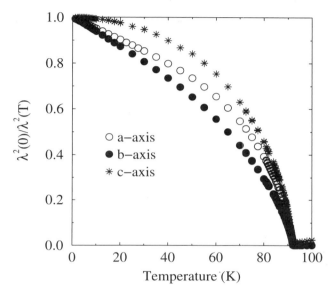

Figure 19. $\lambda_i^2(0)/\lambda_i^2(T)$ vs T for $i = a, b, c$ in $\text{YBa}_2\text{Cu}_3\text{O}_{6.95}$ obtained from microwave $\Delta\lambda_i(T)$ and Far IR derived values of $\lambda_i(0)$ (See Table 2). (Zhang et al.[45])

Figure 19 shows the final results, where the Far IR measurements of $\lambda_i(0)$ have been incorporated. One sees that $\lambda_a(T)$ and $\lambda_b(T)$ have rather similar temperature dependencies, which is very strong evidence that it is not the chains that are causing the linear low temperaure dependence. The c-axis behaviour is rather different, being much flatter, with perhaps a very weak linear term. This result contrasts sharply with the result of Mao et al[15] who found $\lambda_c^2(0)/\lambda_c^2(T)$ dropping much <u>faster</u> (factor 3 or more) than $\lambda_{ab}^2(0)/\lambda_{ab}^2(T)$. We believe the UBC result to be the correct one, given the robustness of the procedure used.We speculate that the use of data from geometries with $H \perp \hat{c}$ and $H \parallel \hat{c}$ without adequate control over thermal and demagnetizing effects produced the anomalous behaviour seen by the Maryland group.

Oxygen Doping Effects

As already noted, although $\text{YBa}_2\text{Cu}_3\text{O}_{7-\delta}$ is the "cleanest" material in many senses, it has strongly conducting chains and, except for $\delta = 0$, considerable disorder in the chains. Therefore it is extremely important to see how the electrodynamics evolve with δ to disentangle the effects of chains and planes. Surprisingly, although the c-axis transport is extremely sensitive to oxygen content, it is known that oxygen defects seem to have relatively little effect on the ab plane surface resistance (Bonn et al.[49], Fuchs et al.[24]) i.e. the O defects are weak electronic scatterers.

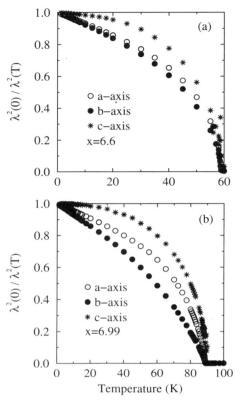

Figure 20. Same type of data as in Fig. 19 for a) underdoped $YBa_2Cu_3O_{6.6}$ ($T_c = 59$ K) and b) slightly overdoped $YBa_2Cu_3O_{6.99}$ ($T_c = 89$ K). (Hardy et al.[47], Bonn et al.[48])

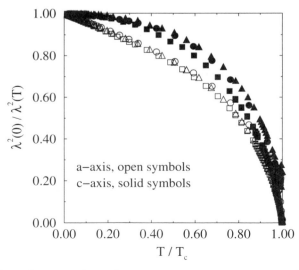

Figure 21. $\lambda_a^2(0)/\lambda_a^2(T)$ and $\lambda_c^2(0)/\lambda_c^2(T)$ vs T/T_c for $YBa_2Cu_3O_x$ for $x = 6.6$ (cicles), 6.95 (squares) and 6.99 (triangles), showing near universal behaviour versus doping.

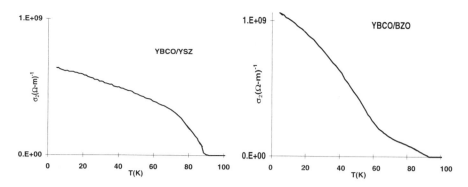

Figure 22. $1/\lambda_{ab}^2$ vs T for BZrO$_3$ grown YBCO crystals (right) and YSZ grown crystals (left). (Srikanth et al.[14])

Fig. 20a shows a complete set of results for $\lambda_i(T)$ for underdoped YBa$_2$Cu$_3$O$_{6.6}$ ($T_c = 59$ K) obtained in the same way as for the optimally doped case, and Fig. 20b shows the results for YBa$_2$Cu$_3$O$_{6.99}$ (slightly overdoped, $T_c = 89$ K). Table 2 gives the values of $\lambda_i(T = 0)$ used. What is striking is the overall similarity of the results, when plotted in this normalized fashion, especially given the large changes in $\lambda_c(0)$. This point is underlined in Fig. 21 where $\lambda_a^2(0)/\lambda_a^2(T)$ and $\lambda_c^2(0)/\lambda_c^2(T)$ are plotted for the 3 oxygen concentrations. The near universal behaviour is startling to say the least, particularly when one considers that the underdoped material is well into the region where a pseudogap is established well above T_c, and the overdoped material tends toward Fermi liquid behaviour.

We end this section on YBCO with a discussion of the very recent results of Srikanth et al.[14] on crystals grown in $BaZrO_3$ crucibles by the Geneva group (Fig. 22). Also shown is the data for crystals grown in YSZ crucibles (also by the Geneva group), which would be analogous to the data of Fig. 12 (Hardy et al.). At intermediate temperatures there is a very striking discrepancy between the two sets of data. There is no doubt that crystals grown in the BaZrO$_3$ crucibles have much higher chemical purity than previous crystals (by at least a factor of 10) and will come to represent the future "gold standard" for the YBCO system. However it is not yet clear that the new microwave results represent the intrinsic electrodynamics properties. Srikanth et al.[14] interpret their data in terms of two order parameters, which at first sight seems reasonable; for example, with fewer impurities, the chains might be susceptible to some new instability. However, to the best of our knowledge, no signature of such an occurence has been seen in specific heat measurements. In addition, near T_c the specific heat jump of the two types of crystals is nearly identical. The delayed condensation of part of the electronic system would surely <u>reduce</u> the specific heat jump at the higher T_c for the purer crystals, whereas the jump is slightly <u>larger</u> for the purer crystals[50]. We speculate that an oxygen inhomogeneity is involved, but there is no independent evidence of this. At the moment, the results present an intriguing puzzle.

$Bi_2Sr_2CaCu_2O_8$

The Stanford group[51] used cleaved $Bi2212$ crystals in the parallel plate method (here $H \perp \hat{c}$) and obtained a low temperature T^2 dependence for $1/\lambda^2$. However, they concluded this was likely contaminated by c-axis conduction due to the extreme anisotropy. This strongly 2D behaviour also complicates measurement of $\lambda(T)$ by μSR

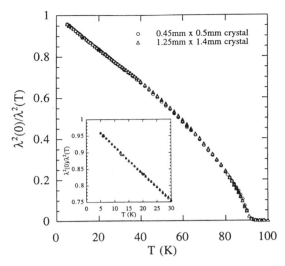

Figure 23. Temperature dependence of the superfluid fraction for $Bi_2Sr_2CaCu_2O_8$ single crystal, assuming $\lambda_{ab}(0) = 2100$ Å . (Lee et al.[54])

in the vortex state because the coherence of the vortices from layer to layer is strongly dependent on H and T[52]. The first study of λ_{ab} that showed a linear T dependence was that of Jacobs et al.[53] obtained by microwave cavity perturbation with $H \parallel \hat{c}$, for which $\lambda^2(0)/\lambda^2(T)$ vs T/T_c was quite similar to the a-axis data of Zhang et al.[45]. This was followed shortly after by the results of Lee et al.[54] which are shown in Fig. 23. The low temperature dependence is very linear, the overall dependence being rather similar to YBCO data, but falling closer to the weak coupling d-wave result.

$Tl_2Ba_2CaCu_2O_8$

Ma et al.[36] measured $\Delta\lambda_{ab}(T)$ for two pairs of commercial films (STI, Calif.) and fitted the data to $bT^2/(T+T^*)$ with $T^* = 25$ and 40 K respectively. For the first pair of films the overall dependence of $\lambda^2(0)/\lambda^2(T)$ vs T/T_c matched the YBCO single crystal data[8] rather well above T^*. The agreement for the second pair (considered inferior quality) was less good.

$Tl_2Ba_2CuO_{6+\delta}$

Very recently $\lambda_{ab}(T)$ has been measured for the tetragonal, single layer compound $Tl2201$ using a 35 GHz superconducting cavity technique with $H \parallel \hat{c}$ (Broun et al., preprint). The material is nearly optimally doped, with a T_c of 78 K. Below 20 K, $\Delta\lambda(T) \propto T$ with slope of 13 Å/K. This is larger than the corresponding slopes for YBCO and BSCCO (4.8 and 10.2 Å/K respectively). Figure 24 shows $\lambda^2(0)/\lambda^2(T)$ vs T with $\lambda(0) = 1650$ Å taken from μSR measurements [55]. It varies linearly with T over almost the whole temperature range (i.e. very similar to weak coupling d-wave), with a sudden downturn within a few degrees of T_c that may be due to fluctuations.

$La_{1-x}Sr_xCuO_4$

Shibauchi et al.[13] measured both $\Delta\lambda_{ab}(T)$ and $\Delta\lambda_c(T)$ of single crystals of $La_{1-x}Sr_xCuO_4$ for $0.09 < x < 0.19$ using platelets with $\hat{c} \perp$ and $\parallel$ to the face. The

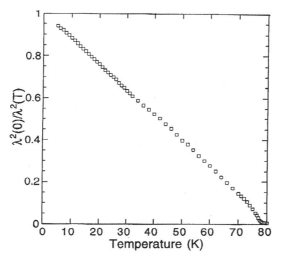

Figure 24. Temperature dependence of the superfluid fraction for $Tl_2Ba_2CuO_{6+\delta}$ using $\lambda_{ab}(0) = 1650$ Å from μSR measurements. (Broun et al., preprint)

large demagnetization factors were determined from Pb reference samples. They obtained $\lambda_{ab}(0) = 0.4$ μm and $\lambda_c(0) = 5$ μm (values consistent with optical and μSR measurements) by the method described earlier. The overall T dependence of $1/\lambda_{ab}^2(T)$ follows weak coupling s-wave BCS, but the resolution was insufficient to rule out a T^2 dependence at low T. $1/\lambda_c^2(T)$ is much flatter and the data was fit to a model involving Josephson coupled 2D superconducting layers (s-wave BCS).

$HgBa_2Ca_2Cu_3O_{8+\delta}$

Panagopoulos et al.[31] were able to measure both $\lambda_{ab}(T)$ and $\lambda_c(T)$ in grain aligned Hg1223 powders. They found $\lambda_{ab}(0)$ and $\lambda_c(0) = 2100$ and 61000 Å respectively, and a temperature dependence of $1/\lambda_{ab}^2(T)$ that fitted rather well to a d-wave model with $\Delta(0)/kT_c = 2.14$ (Fig. 25). On the other hand $1/\lambda_c^2(T)$ followed the behaviour of a Josephson coupled d-wave superconductor[56]. Overall the results again look remarkably similar to the YBCO data.

$Nd_{1.85}Ce_{0.15}CuO_4$ **Thin Films and Single Crystals**

This electron doped material has been studied by Wu et al. [57], Anlage et al.[58] and Andreone et al. [26]. Convincing fits of $\lambda_{ab}(T)$ to BCS s-wave have been made. There are no regions that follow T or T^2. Anlage et al., for example, were able to accurately fit their $\lambda_{ab}(T)$ data from 4 to 20.5 K and extracted $\Delta(0)/kT_c = 4.06\pm0.1$(Fig. 26). At this point, NCCO seems to be an anomaly. However one has to keep in mind that this material is very "dirty" electrodynamically, compared to most of the other established HiT_c materials[59].

Region near T_c: Fluctuations

Here we can only make the briefest survey of selected data and mention some of the issues. The first suggestion from electrodynamics of a wide critical region in the superconducting state came from the data of Kamal et al. on YBCO [60], Fig. 27, where it is seen (insert) that $n_s \propto 1/\lambda_{ab}^2$ does <u>not</u> approach zero linearly as $T \to T_c$,

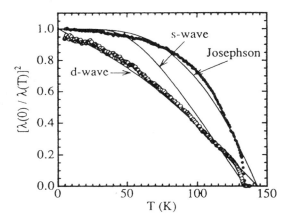

Figure 25. Plots of $\lambda_{ab}^2(0)/\lambda ab^2(T)$ (open circles) and $\lambda_c^2(0)/\lambda_c^2(T)$ (closed circles) vs T for Hg1223. The solid lines are theoretical predictions from weak-coupling BCS theory for s-wave, weak coupling BCS theory for d-wave, and a Josephson coupled d-wave superconductor. (Panagopoulos et al.[31])

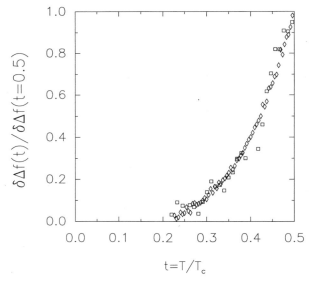

Figure 26. The normalized frequency shift (proportional to $\Delta\lambda$) of a 9.5 GHz resonator containing a $Nd_{1.85}Ce_{0.15}CuO_4$ film (squares) and a $Nd_{1.85}Ce_{0.15}CuO_4$ crystal (diamonds). (Anlage et al.[58])

as it would in low T_c superconductors. In the latter case, the number of Cooper pairs within a coherence volume is extremely large and mean field theory (i.e. BCS) works very well. In the HiT_c case the coherence lengths are exceedingly short and one could expect non-mean-field behaviour in an experimentally accessible region near T_c. Kamal et al. observed that $1/\lambda$ varied as $(T - T_c)^y$ with $y = 0.33 \pm 0.01$ over a region $8 - 10$ K wide. Anlage et al. saw essentially the same thing for $T_c - T < 5$ K [61]. In grain aligned Hg1223, Panagopoulos et al.[31] observed an exponent $y \simeq 0.33$ from $T/T_c = 0.8$ to ~ 0.97, but saw significant deviations nearer to T_c. In this material one expects a 2D→3D cross-over near $T/T_c \simeq 0.98$, so one seems to be observing $y = 0.33$ in the 2D regime.

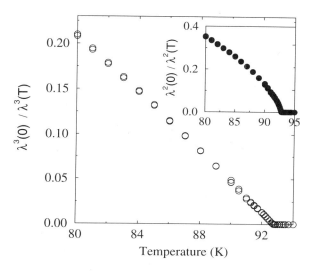

Figure 27. $[\lambda_{ab}(0)/\lambda_{ab}(T)]^3$ versus T for twinned $YBa_2Cu_3O_{6.95}$ single crystal, showing wide region of non-mean-field behaviour. (Kamal et al.[60])

Although the complete data for $\lambda_i(T)$ $i = a, b, c$ in $YBa_2Cu_3O_x$ with $x = 6.6$, 6.95, and 6.99 does not have the precision of that of Kamal et al. in an untwinned crystal, one can see that all components $\lambda_i(T)$ show non-mean-field behaviour (figs. 19, 20a and 20)b. On the other hand there are a number of studies on powders and thin films[62] that seem to show mean field behaviour. In general, the quality of the materials showing mean field behaviour (as indicated by their T_c) is certainly lower than that of the single crystals. However by the Harris criterion (negative specific heat exponent) small amounts of disorder should not affect the critical exponents. It is conceivable that the type (eg extended defects) or level of disorder invalidates the criterion.

A 3D (uncharged) superfluid with a single complex order parameter is in the 3D-XY model universality class and $1/\lambda$ should have a critical exponent of 0.33 (the regime where the charged nature of the SF is important is expected to be inaccesibly close to T_c). In spite of the fact that the data of Kamal et al. and Anlage et al. seem to show this behaviour rather cleanly, one would like to have corroborating evidence that it is not "accidental". In the case of lower quality materials where the nature of the defects is not known (especially the length scales associated with the inhomogeneities), no conclusions can be drawn.

DISCUSSION

We have chosen to present the penetration depth data with relatively little interpretation or fitting to models. Given the enormous evolution of the data (continuing up to today) this is still a wise policy. However we believe that the intrinsic character of the CuO_2 plane electrodynamics is finally becoming well established: when the sample quality is high enough, the results tend towards almost universal behaviour (the exceptions are LSCO and NCCO, both of which are electronically rather inhomogenous). In particular $n_s \propto 1/\lambda_{ab}^2(T)$ varies linearly at low temperatures, with non-mean-field behaviour near T_c. The simple $d_{x^2-y^2}$ pairing state (assuming tetragonal symmetry) gives $\Delta\lambda(T)/\lambda(0) \simeq \ln(2) \times (T/\Delta_0)^1$. For YBCO one can extract $\Delta_0/kT_c \simeq 2.4 - 3$, whereas for Tl2201 one obtains Δ_0/kT_c close to the weak coupling value of 2. Therefore, the observed behaviour of $\lambda^2(0)/\lambda^2(T)$ can be easily explained using very reasonable values of Δ_0/kT_c for the d-wave state. Of course on their own, penetration depth data cannot prove the existence of d-wave superconductivity, although the results are highly suggestive. However, if the d-wave state is the ground state, for which there is now almost overwhelming evidence from phase sensitive and gap anisotropy experiments, $\lambda(T)$ must vary linearly with T in the "clean samples", with about the expected $\Delta\lambda/\Delta T$ observed (where data for $\Delta(\vec{k})$ vs $\vec{k}$ is available from ARPES one should be able to check the correspondence rather directly. We have not done this). A number of alternative explanations have been proposed for the linear dependence of λ. Klemm and Liu[63] have invoked normal 2-D layers, and Pambianchi et al.[64] make a similar suggestion. Adrian et al.[65] and Kresin and Wolf[66] have considered models with 2 s-wave gaps (eg a large gap for the planes in YBCO and a smaller gap for the chains) and magnetic impurity scattering that induces gaplessness and power-law behaviour that depends on the degree of doping in the system. Roddick and Stroud[67], Coffey[68] and Emery and Kivelson[69] propose phase fluctuations in the order parameter, Combescot and Leyronas[70] have a model for YBCO involving anticrossing of the chain and plane bands, and Liu et al.[71] have explained the consequences of subbands with different phases. Gauzzi and Bok[72] also conclude that bilayer systems may show nodes in the gap function without invoking a d-wave state. Undoubtedly, some of these mechanisms contribute to the linear T dependence of $\Delta\lambda$. However the complexity of the materials is such that quantitative predictions are difficult. It is our opinion that the almost universal behaviour observed in 1 layer and 2 layer systems, with and without chains, etc, strongly mitigates against most of the above alternative explanations. A simple d-wave state for the CuO_2 planes explains the phase-sensitive experiments, the ARPES experiments and the penetration depth experiments, without the use of finely tuned parameters.

REFERENCES

1. P.J. Hirschfeld et al.,Phys. Rev. B **50**, 10250 (1994).
2. J.F. Annett et al., in Physical Properties of High Temperature Superconductors, Vol. V, edited by D.M. Ginsberg (World Scientific, Singapor, 1996), Chap. 6, p. 375.
3. Ioan Kosztin and A.J. Leggett, preprint
4. U. Dähne et al., J. Supercond. **8**, 129 (1995).
5. L.A. de Vaulchier et al., Europhys. Lett. **33**, 153 (1996).
6. S. Sridhar and W.L. Kennedy, Rev. Sci. Instrum. **59**, 531 (1988); D.L. Rubin et al., Phys. Rev. B **38**, 6538 (1988).
7. O. Kraut et al., Physica C **205**, 139 (1993). We use the actual data files kindly supplied to us by C. Meingast.

8. W.N. Hardy et al., Phys. Rev. Lett. **70**, 3999 (1993).
9. E.H. Brandt, Phys. Rev. B **54**, 4246 (1996).
10. E.H. Brandt, private communication.
11. E.H. Brandt, Phys. Rev. B **49**, 9024 (1994).
12. K. Holczer et al., Phys. Rev. Lett. **67**, 152 (1991).
13. T. Shibauchi et al., Phys. Rev. Lett. **72**, 2263 (1994).
14. H. Srikanth et al., Phys. Rev. B **55**, R14733 (1997).
15. J. Mao et al., Phys. Rev. B **51**, 3316 (1995).
16. D.N. Basov et al., Phys. Rev. Lett. **74**, 598 (1995) and private communication.
17. J.H. Kim et al., Physica C **247**, 297 (1995).
18. T. Riseman, Ph.D. Thesis, University of British Columbia, 1995.
19. J.E. Sonier et al., Phys. Rev. Lett. **72**, 744 (1994).
20. D.R. Harshman et al., Phys. Rev. B **36**, 2386 (1987).
21. J.E. Sonier et al., Phys. Rev. Lett. accepted Aug. 1997
22. H. Ackermann et al. in HF Ints. of Radioact. Nuclei ed. J. Christiansen, Topics in Current Physics 31 (Springer Verlag, 1983) P. 291.
23. J. Turneaure et al., J. Appl. Phys. **79**, 4221 (1996)
24. A. Fuchs et al., Phys. Rev. B **53** R14745 (1996).
25. N. Klein, in Synthesis and Characterization of HiT_c Superconductors, Mat. Sci. Forum **130-132**, 373 (1993)
26. A. Andreone, Phys. Rev. B **49**, 6392, (1994).
27. B.J. Feenstra et al., Physica C **278**, 213 (1997).
28. M. Nuss et al., Phys. Rev. B **66**, 3305 (1991). For more recent work see A. Frenkel et al., Phys. Rev. B **54** 1355 (1996).
29. J. Orenstein et al., to appear in Superconductors in a Magnetic Field, Carlos Sá de Melo, World Scientfic.
30. J. Orenstein et al., to appear in Physica C, Proc. of 5th Int. Conf. on MM HiT_c , Beijing, China, 1997.
31. C. Panagopoulos et al., Phys. Rev. B **53** R2999 (1996).
32. D. Schoenberg, Superconductivity, Cambridge UP 1954, p. 164
33. D.A. Bonn and W.N. Hardy, in Physical Properties of High Temperature Superconductors, Vol. V, ed. D.M. Ginsberg (World Scientific, Singapor, 1996), Chap. 2, p. 7.
34. N. Klein et al., Phys. Rev. Lett. **71**, 3355 (1993).
35. S.M. Anlage and D.-H. Wu, J. Supercond. **5**, 395 (1992).
36. Zhengxiang Ma, Ph.D. Thesis, Stanford University, 1995.
37. Ju Young Lee et al., Phys. Rev. B **50**, 3337 (1994).
38. F. Gao et al., Appl. Phys. Lett. **63**, 2274 (1993).
39. D.A. Bonn et al., Phys. Rev. B **50**, 4051 (1994).
40. D. Achkir et al., Phys. Rev. B **48**, 13184 (1993).
41. Kuan Zhang et al., Appl. Phys. Lett.**62**, 3019 (1993).
42. E.R. Ulm et al., Phys. Rev. B **51**, 9193 (1995).
43. J.E. Sonier et al., Phys. Rev. B **55**, 11789 (1997).
44. R. Gagnon et al., Phys. Rev. Lett. **78**, 1976 (1997).
45. Kuan Zhang et al., Phys. Rev. Lett. **73**, 2484 (1994).
46. J.L. Tallon et al., Phys. Rev. Lett. **74**, 1008 (1995).
47. W.N. Hardy et al., in Proc. Of the 10th anniversary HTS workshop, ed. B. Batlogg et al., (World Scientific, Singapor, 1996), p. 223
48. D.A. Bonn et al., Czechoslovak J. of Phys. **46**, 3195 (1996), Suppl. S6, Proc. of the 21st Int. Conf. on Low Temperature Physics. Prague, 1996.
49. D.A. Bonn et al., J. Supercond. **6**, 219 (1993).
50. A. Erb private communication.
51. Zhengxiang Ma et al., Phys. Rev. Lett. **71**, 781 (1993).
52. R. Cubitt et al., Nature **365**, 407 (1993)
53. T. Jacobs et al., Phys. Rev. Lett. **75**, 4516 (1995)
54. Shih-Fu Lee et al., Phys. Rev. Lett **77**, 735 (1996)
55. Y.J. Uemura et al., Nature **364**, 605 (1993).
56. W.E. Lawrence, S. Doniach, Proc 12th Int. Conf. on L.T. Physics, Academic, Kyoto 1971.
57. Dong-Ho Wu et al., Phys. Rev. Lett. **70**, 85 (1993).
58. Steven M. Anlage et al., Phys. Rev. B **50**, 523 (1994).
59. B. Maple, Private communication.

60. S. Kamal et al., Phys. Rev. Lett. **73**, 1845 (1994).
61. S.M. Anlage et a, Phys. Rev. B **53**, 2792 (1996).
62. see for example Z.H. Lin et al., Europhys. Lett. **32**, 573(1995).
63. R.A. Klemm et al., Phys. Rev. Lett. **74**, 2343 (1995).
64. M.S. Pambianchi et al., Phys. Rev. B **50**, 13659 (1994).
65. S.D. Adrian et al., Phys. Rev. B **51**, 6800 (1995).
66. V.Z. Kresin and S.A. Wolf, Phys. Rev. B **51**, 1229 (1995).
67. E. Roddick et al., Phys. Rev. Lett. **74**, 1430 (1995).
68. M.W. Coffey et al., Phys. Lett. A**200**, 195 (1995).
69. V.J. Emery et al., Phys. Rev. Lett. **74**, 3253 (1995).
70. R. Combescot and X. Leyronas, Phys. Rev. Lett. **75**, 3732 (1995); R. Combescot and X. Leyronas, Phys. Rev. B **54**, 4320 (1996).
71. D.Z. Liu et al., Phys. Rev. B **51**, 8680 (1995).
72. A. Gauzzi and J. Bok, J. Supercond. **9**, 523 (1996).

SPECIFIC HEAT EXPERIMENTS IN HIGH MAGNETIC FIELDS:
D-WAVE SYMMETRY, FLUCTUATIONS, VORTEX MELTING

Alain Junod, Marlyse Roulin, Bernard Revaz, Andreas Erb and Eric Walker

Département de physique de la matière condensée
Université de Genève
CH-1211 Genève 4

I. INTRODUCTION

Specific heat (C) is one of the few bulk thermodynamic probes of the superconducting state, and as such was studied extensively in both classic and high temperature superconductors. In classic superconductors, i.e. those known before 1986 with critical temperatures $T_c \leq 23$ K, C measurements established the existence of a gap in the electronic density-of-states (DOS) at the Fermi level, $\Delta(0) \cong 3.5 - 5k_B T_c$, showed an almost ideal example of a mean-field, second-order transition at T_c, with a specific heat jump $\Delta C(T_c) \cong 1.4 - 3\gamma T_c$ (γ =Sommerfeld constant), and generally allowed to correlate qualitatively the electron and phonon DOS and the critical temperature as predicted by the BCS theory,[1] taking into account suitable corrections for retardation effects.[2,3] Magnetic fields were mostly applied to extend normal-state data below T_c. For example, the largest field available in superconducting coils (≈ 20 T) can suppress completely superconductivity in a paramagnetically limited ≈ 11 K-superconductor. In Section II, we detail the behaviour of such a type-II superconductor in the mixed state to provide a comparison basis.

The discovery of high temperature superconductors (HTS) has led to a dramatic renewal of interest in type-II superconductivity. The origin of superconducting pairing is an unsettled question, and indirect informations on its mechanism, such as the symmetry of the order parameter, have been vividly debated.[4-7] The presence of line nodes in d-wave superconductors modifies the behaviour of C at $T \ll T_c$. This is the subject of Section III.2. Essential differences with respect to classic superconductors originate from the fact that coherence length ξ is small and anisotropic. Upper critical fields exceed 10^2 T. Laboratory fields only cause a modest shift of T_c, but nevertheless have a dramatic effect on the C anomaly at T_c. The smallness of the coherence volume and the quasi-2D structure enhance considerably fluctuations near T_c.[8] The transition of HTS is more closely related to the weak divergence of ^{4}He at the λ-point than to the usual mean-field step. This critical behaviour is discussed in Section IV.4. Finally it was recently discovered that the true phase transition of HTS in a magnetic field occurs well below the mean-field, upper critical field line, and is ideally of first order. The phenomenon of flux line lattice melting is discussed in Section IV.5. In this chapter, we review essentially experimental aspects.

The Gap Symmetry and Fluctuations in High - Tc Superconductors
Edited by Bok *et al.*, Plenum Press, New York, 1998

II. CONVENTIONAL TYPE-II SUPERCONDUCTORS

The transition metal alloy $Nb_{0.77}Zr_{0.23}$ is a type-II superconductor with $\kappa > 16$. Its specific heat for different magnetic fields (we do not make a distinction between H and B because the magnetisation is relatively small) is shown in the usual plot C/T versus T^2, Fig. 1a.[9] The specific heat difference $\Delta C/T \equiv C_s(T,B)/T - C_n(T)/T$, s = superconducting state, n = normal state, is an electronic quantity plotted in Fig. 1b. Such plots determine immediately the bulk critical temperature ($T_c = 10.8$ K), the full upper critical field curve and in particular its value at $T \to 0$ ($B_{c2}(0) = 7.9$ T $\cong 0.7 T_c |dB_{c2}/dT|_{Tc}$), the initial Debye temperature ($\Theta_D(0) \cong 230$ K), the specific heat jump at T_c ($\Delta C/T_c = 21$ mJ/K^2gat, 1 gat = 11.7 cm^3), and the Sommerfeld constant ($\gamma = 9.2$ mJ/K^2gat), which is proportional to the electron DOS at E_F renormalized by interactions. The integral of $\Delta C/T$ from T to T_c yields the entropy difference $\Delta S(T)$, with the result $\Delta S(T \to 0) = 0$ as required at equilibrium. The integral of $\Delta S(T)$ from T to T_c gives the thermodynamic critical field $B_c(T)$, in particular $B_c(T \to 0) = 0.26$ T. The ratios $\Delta C/\gamma T_c \cong 2.3$ and $T_c/\Theta_D \cong 0.04$ are indicators of the electron-phonon coupling strength.[10] The Ginzburg-Landau parameter $\kappa_1 = B_{c2}/2^{1/2} B_c$ is a function of T, which varies here from 16 near T_c to 22 at $T = 0$. Some of these parameters enter the Ginzburg-Landau free energy functional :

$$F\{\psi\} = F_{ns} + \int dV \{\frac{1}{8\pi}\mathbf{B}^2 + \alpha t |\psi|^2 + \frac{1}{2}b|\psi|^4 + \sum_j \frac{1}{4m_j^*}|(-i\hbar\nabla_j - \frac{2e}{c}\mathbf{A}_j)\psi|^2\} \qquad (1)$$

where $\mathbf{B} = rot\mathbf{A}$, F_{ns} is the free energy in the normal state, $t \equiv (T - T_c)/T_c$ and $m_j^* = (m_x^*, m_y^*, m_z^*)$ are the main values of the mass tensor m_{ik}^*. The relations are $\Delta C = \alpha^2/bT_c$, $(dH_c/d|t|)_{Tc} = (4\pi\alpha^2/b)^{1/2}$, and $(dH_{c2z}/d|t|)_{Tc} = 2c(m_x^* m_y^*)^{1/2}\alpha/e\hbar$. The information given by a set of specific heat curves in fields up to B_{c2} is rich. The situation for HTS is unfortunately much less favourable.

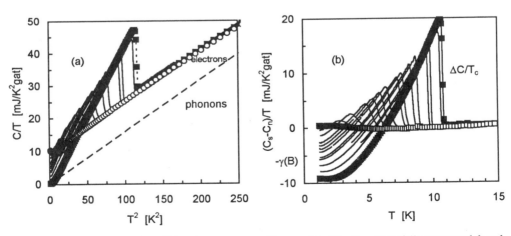

Figure 1. (a), total specific heat C/T versus temperature T squared for $Nb_{77}Zr_{23}$. $B=0$ (full squares and dotted line), 0.2, 1, 1.2, 2, 2.4, 3, 3.3, 4, 4.4, 4.8, 5.2, 6, 7.2 T (lines), 10 T (open circles). (b) electronic specific heat difference $\Delta C/T$ versus T, same fields except open squares : 7.2 T.

The shape of the specific heat in the mixed state was studied theoretically in the limit $T \to 0$.[11,12] The exponentially vanishing specific heat in zero field at low T is the signature of a gap. When the field increases, low energy states appear in vortex cores. The superficial vortex density is B / Φ_0. If each vortex core is counted with an effective area $2\pi\xi^2$, then the system in the mixed state may be considered as a mixture of a fraction $B / B_{c2}(0)$ in the normal state and a fraction $1 - B / B_{c2}(0)$ in the superconducting state, the latter with a normalized specific heat that depends only on the ratio $T / T_c(B)$:

$$\frac{C_s - C_n}{\gamma T} = \underbrace{\frac{B}{B_{c2}(0)}}_{\text{cores}} + \underbrace{\left(1 - \frac{B}{B_{c2}(0)}\right) f\!\left(\frac{T}{T_c(B)}\right)}_{\text{superconducting regions}} - \underbrace{1}_{C_n/\gamma T} = \left(1 - \frac{B}{B_{c2}(0)}\right)\left[f\!\left(\frac{T}{T_c(B)}\right) - 1 \right] \qquad (2)$$

To test this simple representation, we plot $(\Delta C / T)[1 - B / B_{c2}(0)]$ versus $T / T_c(B)$ in Fig. 2a. The curves collapse together except near T_c. The zero-field curve differs somewhat from the others : the transition into the Meissner state obeys Rutger's equation $\Delta C / T_c = (dB_c / dT)^2_{Tc} / 4\pi$, whereas the transition into the mixed state obeys the thermodynamic relation $\Delta C / T_c = (dB_{c2} / dT)^2 / 4\pi\beta(2\kappa_2^2 - 1)$ with $\beta = 1.16$ for a triangular lattice ; $\approx 16\%$ of the jump is lost. The specific heat of the regions outside of the vortex cores $\gamma T f(T)$ vanishes less rapidly at $T \to 0$ in large fields than for $B = 0$, suggesting that the gap gives way to a pseudo-gap in a strong field.[13]

The broadening of the transition is misleading. In clean HTS with a short coherence length, it would originate from thermal fluctuations. In the present case, it results from local variations of the mean free path, which cause a distribution of κ and B_{c2}.

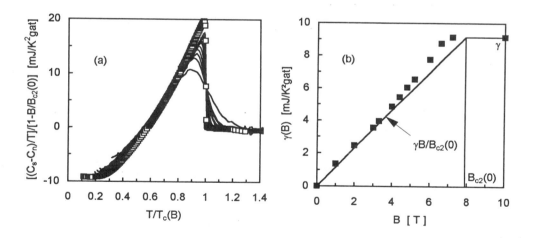

Figure 2. (a), zero-field (open squares) and mixed-state (lines) electronic specific heat difference $\Delta C/T$ for $Nb_{77}Zr_{23}$, normalized by the effective superconducting fraction $1-B/B_{c2}(0)$ versus reduced temperature. (b), coefficient of the mixed-state linear term $\gamma(B)$ at $T \to 0$ versus magnetic field.

The main points that should be remembered when the specific heat of classic superconductors in magnetic fields is compared with that of HTS are :
(1) the low temperature specific heat behaves as $C_s / T \propto \gamma B / B_{c2}(0)$, and in particular $\lim_{T \to 0} C_s(B = 0) / T \cong 0$ (Fig. 2b) ;
(2) for $B / B_{c2}(0) < 0.1$, the broadening of the specific heat jump is negligible ;

(3) the jump temperature is shifted by the field along the $B_{c2}(T)$ line;

(4) the lattice contribution near T_c is of the same order of magnitude as the electronic contribution, and obeys simple power laws (e.g. $C_\ell \propto T^3$).

None of these points remains true for HTS, and fields available in a laboratory are unable to suppress completely superconductivity to obtain reliable measurements of γ, $B_c(0)$, $B_{c2}(0)$, etc. All results concerning these parameters, estimations of the coupling strength, and even the definition of $B_{c2}(T)$ itself, result from artful but less well grounded methods than for the case just reviewed.

III. LOW-TEMPERATURE SPECIFIC HEAT OF HTS

By low temperature, we mean $T \leq \Theta_D / 50$ and $T \leq T_c / 10$. The phonon specific heat tends to its limiting form $C_\ell \propto T^3$, and the specific heat due to excitations across the gap $\gamma T f_s(T)$ is negligible. At first glance, we would expect that the electron specific heat arises only due to the vortex core term $\gamma T B / B_{c2}(0)$. But because the core radius is now small ($\xi \approx 10$ Å), the energy levels of core electrons are widely spaced, as expected on very general grounds ($\Delta p \Delta x \geq \hbar$). Theory[12] and experiment[14] agree that the first levels lie >5 meV above E_F. They cannot contribute to the specific heat below 10 K. So the electron specific heat of HTS at low T should be nearly zero. This is not the case.

III.1. Background Contributions

We refer to YBa$_2$Cu$_3$O$_7$ for definiteness. In addition to the lattice specific heat, it has become customary to identify a residual linear term $\gamma^* T$, magnetic terms due to spins, and a small extra contribution that is now considered to arise from d-wave symmetry, to say nothing of possible impurity contributions mainly due to BaCuO$_{2+x}$ or Y$_2$Cu$_2$O$_5$ in early samples. The separation of these contributions requires involved techniques. The assessment of their reliability is a difficult matter. Discussions may be found e.g. in Refs. [15,16].

The linear term is the main zero-field contribution. Its origin in unsettled. Two-level systems, normal-state regions, disordered magnetic structures and impurity scattering in d-wave superconductors[17] are suitable candidates. In all these cases the effect of a magnetic field should be isotropic and probably negligible. Experimentally, the better the sample, the lower γ^*; but it has never been possible up to now to reduce it (in units of J/K^{2}mol-Cu) below γ of pure copper in spite of the impressive progress in crystal growth. A purity in excess of 99.995% has now been achieved.[18]

The spin terms may be recognised from their characteristic dependence on T/B. Nuclear Cu spins show up at very low T ($\approx 10^{-1}$ K) as a predictable $C \propto T^{-2}$ upturn that moves up with B. Electron spins with $S = 1/2$ (and sometimes $S = 2$)[16] give rise to a well-defined Schottky anomaly in a field of a few teslas, but interactions complicate the behaviour in zero field. The anisotropy of these terms is small.[19]

Contributions related to the d-wave symmetry have attracted recently considerable attention. Specific heat measurements are insensitive to the phase of the order parameter. However they provide bulk information on the behaviour of the DOS near E_F, in particular on the topology of nodes if any. Alternatively, experiments that are sensitive to the phase tend to probe only the surface of the sample. The gap depends on the orientation in the reciprocal space in a d-wave superconductor. Its directional average yields in a V-shaped DOS at $E_F \equiv 0$, i.e. $N(E) \propto |E|$. The electron specific heat C / T is always proportional to $N(E)$ averaged over an interval $\approx k_B T$ around E_F. It follows that $C \propto T^2$. This is the characteristic quadratic d-wave term in zero field. Impurity scattering and magnetic field tend to «fill» the bottom of the «V» and to wash out the T^2 term. Attempts to demonstrate the existence of the $C \propto T^2$ term have been based on multiparameter fits of the total specific heat, including all above contributions plus lattice terms. The large number of degrees of freedom is a problem.[16,20] Positive results were obtained by differential measurements with respect to doping by impurities, which make use of more experimental information and thus yield less correlated parameters.[21] Other efforts to identify the characteristic signature of d-wave symmetry in the specific heat of HTS have been based on the peculiar response of vortex electrons in the vicinity of a node to magnetic fields, $C \propto TB^{1/2}$.[22] If the response of all other terms is known precisely, this (small) contribution can be included in a fit and estimated.[16,20] The robustness of the result depends on the assessment of errors and correlations.

III.2. D-wave Scaling

Progress in metallurgy,[18] experiment[19] and theory[23,24] has made possible a more straightforward demonstration of the consistency between bulk thermodynamic data and d-wave symmetry. As already mentioned, impurity scattering washes out some of the characteristic d-wave features, so that purity is essential. The present experiments were done using a 18 mg crystal with >99.995% purity grown in a non-reactive $BaZrO_3$ crucible.[18] The last significant residual «impurity» in such crystals is formed by clusters of oxygen vacancies. The latter are largely suppressed by annealing for weeks in 100 bars oxygen at 300°C. These crystals have allowed successful experiments such as imaging the vortex lattice by scanning tunnelling microscopy,[14] suppression of the magnetic «fishtail» effect,[25] observation of specific heat peaks at the flux line lattice melting transition in fields in excess of 16 T (Section IV.5), quasi-linear variation of the condensate density with T over a wide temperature range,[26] which is not the case at optimal doping, etc.

In the first experimental report on d-wave specific heat, published by Moler et al., the fitted d-wave vortex term was assumed to be of the form $C \propto TB^{1/2}$.[20] This is an asymptotic form in the low T and high B limit. In the opposite limit, $C \propto B$ is expected. General results were recently obtained in the form of scaling laws.[23,24] One focusses on the shape of the DOS near E_F in the presence of line nodes. Two effects are possible. Neglecting vortices (type-I superconductivity, or $B \perp c$), electron states undergo a Zeeman shift.[64] This shift closes the small gaps near to the nodes in fields $\mu_B B \approx \Delta_0(\mathbf{k})$, and gives rise to a non-zero DOS. This causes in turn a finite electronic C that scales with B/T, as it should for any spin or electron band. More precisely,[64]

$$\frac{C(T,B)}{C_n(T_c)} \propto (\frac{T}{T_c})^2 F(\frac{\mu_B B}{k_B T}) \tag{3}$$

$F(x)$ is a scaling function that depends on the DOS ; a numerical calculation is given in Ref. [15], Fig. III.18. Alternatively, in the presence of vortices, the shift of the electron states near E_F that determine the low-temperature specific heat is dominated by the Doppler effect due to the supercurrent flow $\mathbf{v}_s$: $E_s(\mathbf{k}) = m_e(\mathbf{v_k} + \mathbf{v}_s)^2/2 \cong E_0(\mathbf{k}) + m_e \mathbf{v}_F \cdot \mathbf{v}_s$. The shift $\propto v_s v_F \propto v_s \Delta_0 \xi \propto v_s T_c (\Phi_0/B_{c2})^{1/2}$ is now proportional to the square root of B because $\langle v_s \rangle \propto 1/R \propto (B/\Phi_0)^{1/2}$, where R is the intervortex distance. $\xi \ll R \ll \lambda$ is assumed, i.e., $H_{c1} \ll H \ll H_{c2}$. This leads to a scaling property:[23,24]

$$\frac{C(T,B)}{C_n(T_c)} \propto (\frac{T}{T_c})^2 F(\frac{T/T_c}{B^{1/2}/B_{c2}^{1/2}}) \tag{4}$$

The limits of the new scaling function are $F(x) = 1/x$, $x \ll 1$, and $F(x) = 1/x^2$, $x \gg 1$. A rough interpolation is given by $F(x) \cong 1/[x(x+1)]$.[24]

It is useful here to show how Eqs. (3) and (4) can be obtained intuitively. The specific heat of a fermion band is given by

$$C(T) = \int_{-\infty}^{\infty} EN(E) \frac{\partial f(E/k_BT)}{\partial T} dE \qquad (5)$$

$$f(x) = (1 + e^x)^{-1} \qquad \text{(Fermi statistics)}$$

This form is not quite general. It applies when the Fermi level does not move with T, which is true when $N(E)$ is symmetrical about $E_F(T = 0) \equiv 0$. We have also assumed that the DOS does not change with T, which applies at low T for a normal metal. However the change that results from the variation of $\Delta(T)$ in a superconductor can be included in Eq. (5) by replacing $N(E)$ by $N(E,T)$.[28] But we consider only the case $T \ll T_c$. Then

$$\frac{C}{T} = 2k_B^2 \int_0^{\infty} N(k_B Tx) Sch(x) dx \qquad (6)$$

$$Sch(x) = x^2 e^x (1 + e^x)^{-2} \qquad \text{(Schottky function)}$$

The electron specific heat C/T is given by a distribution of Schottky functions, which are well-known in magnetism. If $N(E) = const$, we recover the well-known result $C = \gamma T$ with $\gamma = (1/3)\pi^2 k_B^2 N(0)$. If $N(E) = \alpha|E|$, the V-shape due to line nodes, then

$$C = 2\alpha k_B^3 T^2 \int_0^{\infty} x Sch(x) dx = const \cdot T^2 \qquad (7)$$

Similarly, p-wave pairing with point nodes would give rise to a T^3 term, impossible to separate from phonon or antiferromagnetic spin wave contributions. Finally if the zero is shifted by a quantity $\delta E = \eta B^n$, one obtains the form of Eqs. (3) and (4) :

$$C = 2\alpha k_B^3 T^2 \int_{\eta B^n/k_B T}^{\infty} x Sch(x) dx \propto T^2 F(\eta B^n/k_B T) \qquad (8)$$

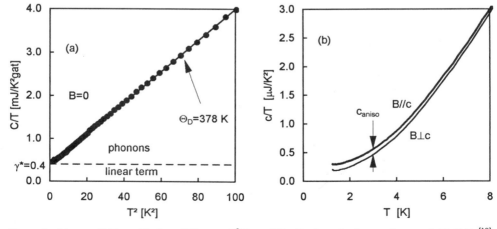

Figure 3. (a), zero-field specific heat C/T versus T^2 for an $YBa_2Cu_3O_{7.00}$ single crystal grown in $BaZrO_3$.[19] (b), heat capacity (including the sample holder) in $B = 8$ T versus T, two orientations.

In order to test relations (3) and (4), one has to remove first all contributions that have nothing to do with superconducting electrons. This is not trivial. Fig. 3a shows the total specific heat of Y-123 in zero field below 10 K : majority contributions arise from phonons and the linear term. This single crystal is remarkable because no low-temperature upturn is observed down to 0.2 K,[27] where the nuclear specific heat takes over, so that problems associated with Cu^{2+} magnetic contributions are greatly suppressed. In order to get rid of all unwanted contributions, one looks at the anisotropic component $C_{aniso} \equiv C(B_{//c}) - C(B_{\perp c})$, which can be measured directly in two runs. If the anisotropy factor was infinite, C_{aniso} would just be the specific heat of Eqs. (3) and (4). In the present case, the anisotropy ratio $\Gamma \equiv (m_c / m_{ab})^{1/2}$ is « only » 5, so that we lose a part of the c-axis specific heat. In counterpart, we are rewarded by the tremendous advantage that all isotropic contributions cancel out without having to model them. Fig. 3b shows the total heat capacity (including the sample holder) for $B = 8$ T in both directions. The difference, normalized to the sample mass, is reported in Fig. 4a for various fields. Within a numerical factor $f(\Gamma)$ of order 1, this is the specific heat of superconducting electrons. This set of curves can now be plotted in the scaling forms suggested by Eqs. (3) and (4), which differ only by the exponent of B in the abscissa. Figure 4b shows that the data collapse on a single curve in the scaling plot defined by Eq. (4). This first shows that the Zeeman effect is negligible, and secondly demonstrates the consistency between bulk thermodynamic data and d-wave symmetry. As a matter of fact, we have only probed the existence of line nodes. We can exclude point nodes and more generally nodes with a dimension $D \neq 1$, because scaling equation (4) takes more generally the form $C \propto T^2 B^{(1-D)/2} F(x)$.[24]

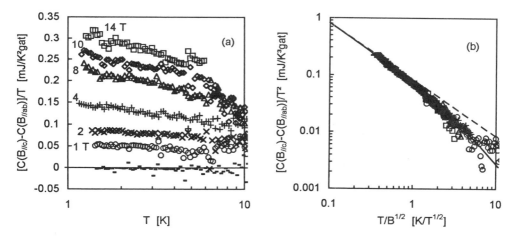

Figure 4. (a), difference between the specific heat for $B//c$ and $B//ab$ in fields 0, 1, 2, 4, 8, 10 and 14 T for YBa$_2$Cu$_3$O$_{7.00}$. (b), d-wave scaling plot of the same data. Dotted line : $F(x) \propto 1/x$; full line : $F(x) \propto 1/[x(x+1)]$.

The shape of the function $F(x)$ given by the scaling plot, Fig. 4b, indicates that the T, H domain that we have explored corresponds to a crossover regime where x is neither very small nor very large with respect to one. In other words, the asymptotic formula $C \propto TB^{1/2}$ overestimates the electron specific heat. This is more apparent in the sensitive scaling plot $C / TB^{1/2} = f(T / B^{1/2})$.[19] This remark is quite comforting; otherwise, the electronic entropy

$$S_{aniso}(T,B) = \int_0^T C_{aniso} d\ln T \qquad (9)$$

that is calculated using the low-T form $C \propto TB^{1/2}$ would exceed the entropy that is measured starting from high temperatures,

$$S_{aniso}(T,B) = -\int_T^{T_c} C_{aniso} d\ln T \qquad (10)$$

A fit of the interpolation scaling function $F(x) \propto 1/[x(x+1)]$ defines two parameters : $B_{c2}(0)$ and an effective normal-state γ corrected for the anisotropy factor $f(\Gamma)$. The Sommerfeld constant γ is consistent with previous estimations ($\approx$1-2 mJ/K^{2}gat) but $B_{c2}(0) \approx 320$ T is closer to $\Delta_0 / \mu_B \approx 350$ T than to independent estimations based on the reversible magnetization at higher temperatures (140-200 T).

IV. HIGH TEMPERATURE SPECIFIC HEAT OF HTS

IV.1. Normal-state electron specific heat

Typical examples of the specific heat of HTS over the range of temperatures from 15 to 250 K are shown in Fig. 5a. The electron specific heat is only a small fraction of the total.

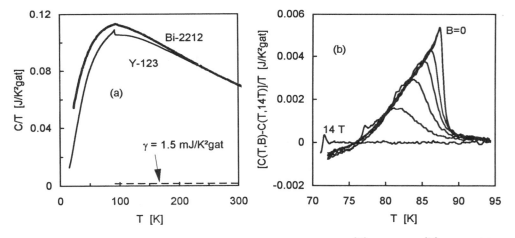

Figure 5. (a), specific heat C/T in $B=0$ of optimally doped crystals of Y-123[33] and Bi-2212[34]. (b), specific heat C/T (minus smoothed 14-T curve) of an overdoped Y-123 crystal near $T_c \cong 87.5$ K in $B=0$, 1, 2, 4, 8 and 14 T, $B//c$. The peak value, 70 mJ/K^{2}mol in $B=0$, and the area between the 0 and 14 T curves suggest an unusually large condensation energy. The small peaks on the left are due to vortex lattice melting.[35]

Because it is impossible to completely quench superconductivity in available fields, most attempts to separate the lattice and the electronic specific heat have been based on differential measurements with respect to doping.[29] Together with the normal-state susceptibility, the determination of the electron specific heat in the normal state over an extended temperature range can give key bulk information on the gaps that exist above T_c, either because superconductivity persists in a non-coherent state, or because a gap of different origin opens up in the normal-state DOS. High resolution differential specific heat measurements were discussed in this context by Loram *et al.*[30] The picture that emerges is that T_c reaches its maximum value with respect to doping when the value of the normal-state gap Δ_n nearly coincides with $k_B T_c$. Going to the overdoped side, the normal-state gap closes rapidly, the electronic specific heat C/T at $T \approx T_c$ and the specific heat « jump » tend to their maximum value, and both T_c and the superconducting gap Δ_0 decrease. On the underdoped side, Δ_n increases rapidly and tends to merge with Δ_0, which keeps increasing while T_c is suppressed. The electronic C/T at $T \approx T_c$ decreases, and so does the « jump ». In this picture, Δ_n and Δ_0 are independent entities. Other viewpoints have been expressed. In particular, scanning tunnelling spectroscopy (STS) has shown that the superconducting gap appears to persist above T_c (but the coherence peaks in the tunnelling DOS vanish) even on the overdoped side of Bi-2212.[31]

These studies of the normal-state electron specific heat are experimentally delicate because they depend on a subtle analysis of an essentially structureless curve above T_c (Fig. 5a). We shall rather review here phenomena which give rise to sharp features with an amplitude of a few % of the total specific heat : the anomaly at T_c and its dependence on $T - T_c$ and B on one hand, and the steps or peaks that occur when the vortex liquid freezes just below T_c. Although well-defined, these anomalies still require an experimental resolution of $\approx 0.01\%$, using small crystals (1-100 mg) and very high magnetic fields. Some experimental aspects are covered in Refs. [15] and [32].

IV.2. Background contributions

Crystal field splitting may give rise to large contributions in LnBa$_2$Cu$_3$O$_7$ compounds ; otherwise magnetic contributions due to localized moments are confined to low temperatures and may be neglected. The lattice specific heat can be modelled over a wide temperature range by a discrete distribution of Einstein δ-functions in the phonon DOS $g(\omega)$, with some analogy to Eq. (6) :

$$C(T) = k_B \int_0^\infty g(k_B T x) E(x) dx \tag{11}$$

$$E(x) = x^2 e^x (1 - e^x)^{-2} \qquad \text{(Einstein function)}$$

Typically 3 peaks constrained to contain $3N$ modes for N atoms (1 gat) are required to fit the specific heat to ≤0.2%, 50-250 K. This means 5 free parameters. Independent determinations of $g(\omega)$ (e.g. neutron scattering) are not precise enough to be used here. In other words, there is considerable freedom in the choice of a phonon baseline if one wants to study the small remaining electron specific heat. Based on the same data, one could show that superconducting fluctuations near T_c are either gaussian or critical, merely by shifting slightly T_c and the phonon baseline. In the following, we shall describe experiments that take advantage of the fact that the phonon specific heat is invariable with respect to the field, and study the measurable quantities $C(T,B) - C(T,0)$ or $\partial C / \partial B$, which are purely electronic. This requires accurate measurements of small differences. Note also that $\partial(C/T)/\partial T$ contains only a small lattice contribution for optimally doped Y-123 ; this is a particular case.

IV.3. The specific heat anomaly of HTS at T_c

We first illustrate the difference between the superconducting transition of HTS and that of classic superconductors, e.g. Nb$_{77}$Zr$_{23}$ (Fig. 1b). For the least anisotropic HTS, overdoped Y-123 with $T_c \cong 87.5$ K and $\Gamma \equiv (m_c / m_{ab})^{1/2} \cong 5$, a jump may be defined only in $B = 0$, with a peak at T_c and a tail above T_c that are due to fluctuations (Fig. 5b). The latter are relatively small and the gaussian approximation is suitable. A magnetic field shifts somewhat the bulk T_c, but the onset appears to be unchanged and the transition is smeared. Fluctuations are enhanced by the field, and the definition of $T_{c2}(B)$ becomes less clear.

Optimally doped Y-123 with $T_c \cong 93$ K and $\Gamma \cong 7$ is more puzzling (Fig. 6a). The loss of condensation energy below T_c with respect to overdoped Y-123 (Fig. 5b) is nearly 40%. It is not clear whether this is an effect of the opening of a normal-state gap, or the presence of submicroscopic regions in the immediate vicinity of oxygen vacancies which could remain non-superconducting down to ≈60 K. The strong curvature below T_c has no equivalent in classic superconductors. It becomes harder to separate the « jump » from fluctuations at T_c. The overall shape in $B = 0$ is closer to that of ^{4}He at the λ-point than to that of the BCS function. Therefore a model of critical fluctuations is a better starting point.

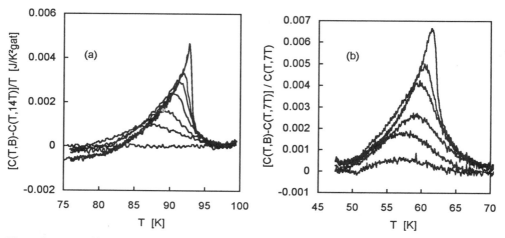

Figure 6. (a), specific heat C/T (minus 14-T baseline) of an optimally doped Y-123 crystal near $T_c \cong 93$ K in $B=0$, 0.5, 1, 2, 4, 8 and 14 T, $B//c$.[53] (b), *relative* specific heat change (with respect to the 7-T curve) of an underdoped Y-123 crystal with $T_c \cong 62$ K in $B=0$, 0.2, 0.5, 1, 2 and 3 T, $B//c$.[36] $C/T \approx 0.095$ J/K^{2}gat near T_c.

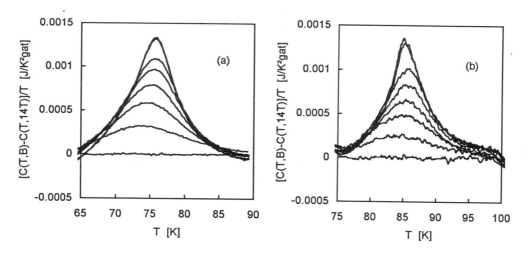

Figure 7. (a), specific heat C/T (minus 14-T baseline) of an overdoped Bi-2212 crystal near $T_c \cong 75$ K in $B=0$, 0.5, 1, 2, 4, 8 and 14 T, $B//c$.[37] (b), specific heat C/T (minus 14-T baseline) of a nearly optimally doped Bi-2212 crystal near $T_c \cong 85$ K in $B=0$, 0.5, 1, 2, 4, 8 and 14 T, $B//c$.[34]

Fig. 6b (note the different scale) shows the specific heat anomaly of underdoped Y-123,[36] the so-called 60-K phase with an estimated anisotropy $\Gamma \approx 15$. The amplitude is now quite small, 0.7% of the total C rather than 5% for overdoped Y-123, and tends to become more symmetrical.

We continue toward larger anisotropies. Fig. 7a shows the specific heat anomaly of overdoped Bi-2212 with $T_c \cong 75$ K and $\Gamma \cong 15-20$, and finally Fig. 7b that of nearly optimally doped Bi-2212 with $T_c \cong 85$ K and $\Gamma > 50$. The anomaly is almost completely symmetrical at any field. We can define neither a jump in $B = 0$, nor a critical temperature $T_{c2}(B)$ in a field; the center of gravity of the anomaly remains unchanged. A true phase transition seems to exist only in $B = 0$. Its symmetrical and quasi-logarithmic shape are features that have been predicted for ideal 2D systems.[38] The specific heat in a field is reminiscent of that of gases *above* the critical pressure. The trends described here are representative and are not due to variations of the homogeneity of the crystals.

IV.4. Fluctuations near T_c: scaling laws

Numerous papers have been devoted to the analysis of thermodynamic fluctuations at temperatures near to the superconducting transition temperature of HTS; see e.g. Ref. [39] and references therein. The probability to occupy a state with energy $\approx k_B T$ above the minimum equilibrium value becomes increasingly important when the size of the system is scaled down. This phenomenon is favoured by the high critical temperature, the small coherence volume $\xi_{//}^3 / \Gamma$ and the large anisotropy Γ of HTS. A magnetic field confines the electron motion and enhances these effects.

The Ginzburg criterion in zero field reads (note the uncertainty on $\xi_{//}^6$!)

$$t_G \equiv \left| \frac{T_G - T_c}{T_c} \right| = \frac{1}{2} \left[\frac{k_B T_c}{H_c^2(0) \xi_{//}^3 / \Gamma} \right]^2 \tag{12}$$

and separates the regions where fluctuations are weak (gaussian) or strong (critical). We focus on the latter. Attempts to fit the specific heat of optimally doped Y-123 with gaussian functions always yield residuals larger than those obtained by using critical functions, except possibly beyond ± 10 K from T_c where the fluctuation specific heat becomes too small to be significant.[15] On general grounds, it is expected that the appropriate scaling of the magnetic field in the intermediate critical regime is in terms of the number n of flux quanta per coherence area. Thus we expect the singular part of the free energy per unit volume to scale as[40-42]

$$-F = \frac{k_B T}{\xi_{//}^2 \xi_\perp} \varphi_1 (\frac{\xi_{//}^2}{n}) = \frac{\Gamma k_B T}{\xi_{//}^3} \varphi_1 (\frac{H_\perp \xi_{//}^2}{\Phi_0}) \tag{13}$$

where $\xi_{//}$ is the coherence length in the CuO_2 planes, $\xi_\perp = \xi_{//} / \Gamma$ the coherence length perpendicular to the planes, $H_\perp$ the field normal to the planes, and φ_1 some scaling function. Near to the critical point, $\xi(t) = \xi_0 |t|^{-\nu}$, $t \equiv T / T_c - 1$, so that $\varphi_1 (H_\perp \xi_{//}^2 / \Phi_0) = \varphi_2 (|t| \Phi_0^{1/2\nu} / H_\perp^{1/2\nu} \xi_{//0}^{1/\nu})$. Scaling relations follow for the entropy $S = -\partial F / \partial T$, the specific heat $C / T = -\partial^2 F / \partial T^2$, the magnetization $M = -\partial F / \partial H$, etc. We obtain e.g. for the latter

$$M = \frac{\Gamma k_B T H_\perp^{1/2}}{\Phi_0^{3/2}} \varphi_3(x), \quad x \equiv |t| / (H_\perp \xi_{//0}^2 / \Phi_0)^{1/2\nu} \tag{14}$$

413

as long as $\xi_\perp(t) > c$, where c is the distance between CuO$_2$ planes. Conversely, $\xi_\perp$ should be replaced by c in Eq. (13) when $\xi_\perp(t) < c$ and one would obtain the 2D limit :

$$M = \frac{k_B T}{\Phi_0 c}\varphi_3(x) \tag{15}$$

Eqs. (14) and (15) explain naturally the crossing points of the quantities $M/H_\perp^{1/2}$ for Y-123 and M for Bi-2212 and Bi-2223 at T_c ($x = 0$), where they become independent of the field.[43] An interesting quantity for the specific heat is the derivative $\partial(C/T)/\partial H$, which, as already mentioned, does not require any baseline subtraction :

$$\frac{\partial(C/T)}{\partial H} = \frac{\partial^2 M}{\partial T^2} = \frac{\Gamma k_B T H_\perp^{1/2}}{T_c^2 \Phi_0^{3/2}}\left(\frac{H_\perp \xi_{//0}^2}{\Phi_0}\right)^{-1/\nu}\frac{d^2\varphi_3}{dx^2}\left[1 + o\left\{\frac{H_\perp \xi_{//0}^2}{\Phi_0}\right\}^{1/2\nu}\right] \tag{16}$$

whereas for the 2D limit one would have

$$\frac{\partial(C/T)}{\partial H} = \frac{k_B T}{T_c^2 \Phi_0 c}\left(\frac{H_\perp \xi_{//0}^2}{\Phi_0}\right)^{-1/\nu}\frac{d^2\varphi_3}{dx^2}\left[1 + o\left\{\frac{H_\perp \xi_{//0}^2}{\Phi_0}\right\}^{1/2\nu}\right] \tag{17}$$

Note that the correction factor [...] is small because $H_\perp \xi_{//0}^2/\Phi_0 \ll H_\perp/H_{c2}(0) \ll 1$. We now proceed to compare these predictions with experiment. For definiteness, recall that the 3D-XY model gives $\nu = 0.669$ so that the scaling variable is $|t|/H^{0.747}$. The amplitude in Eq. (16) goes then with $H_\perp^{1/2-1/\nu} = H_\perp^{-0.995}$. This leads to the scaling plot shown in Fig. 8a, which is based on 16 fields from 0.5 to 14 T, using a massive Y-123 single crystal grown by the travelling solvent floating zone melting technique (TSFZM).[44,47] The surprizing result is that the « low-field » 3D-XY critical behaviour still holds in 14 T.

Alternatively, the high-field lowest landau level (LLL) approximation in 3D leads to a scaling equation of the form[15,45,46]

$$\frac{\partial(C/T)}{\partial H} \propto \frac{1}{(HT)^{2/3}}\left(\frac{dH_{c2}}{dT}\right)^{-1}\varphi_4\left(\frac{T - T_c(H)}{(HT)^{2/3}}\right) \tag{18}$$

In the derivation, only the leading terms that originate from the rapid variation of $T - T_c(H)$ in the transition region were considered. If it is assumed that $T_c(H)$ varies linearly with H, as it should in a Ginzburg-Landau model, then the scaling law predicted by Eq. (18) is poorly obeyed by data (Fig. 8b). However, $T_c(H)$ may be considered as a free set of parameters that must be chosen so that LLL scaling is verified. It turns out that this is possible with the same data if the curve $T_c(H)$ is taken from the 3D-XY model, i.e. $H_{c2}(T) \propto |t|^{2\nu}$ or $dH_{c2}/dT \propto H^{1-1/2\nu}$, so that the amplitude in Eq. (18) goes now with $H^{-0.919}$. The exponents that appear in the XY and LLL scaling plots are then too close to be distinguished, and both models can account for the measurements. The consideration of second-order terms does not modify this conclusion.[47] The LLL model requires an unusual dependence of $T_c(H)$, but is able to predict realistically the shape for the scaling functions. Otherwise, the low-field 3D-XY approximation describes well enough the specific heat of optimally doped Y-123 at all fields, even though it may merge smoothly into the 3D-LLL approximation at very high fields ; this is a theoretical rather than an experimental requirement. The conclusions of Overend *et al.*, based on a set of crystals which have a relatively small anomaly at T_c, are similar.[39]

414

The zero-field behaviour of the specific heat may also be derived from the very useful Eq. (13). A little algebra leads to

$$C(T, B = 0) = -T \frac{\partial^2 F}{\partial T^2} = \frac{\Gamma k_B}{\xi_{//0}^3} \frac{T^2}{T_c^2} |t|^{3\nu - 2} \varphi_5(\infty) \qquad (19)$$

where $\varphi_5(\infty) \equiv (R_\xi^+)^3 \cong 0.36$ and $3\nu - 2 \cong 0.007$ are determined in the 3D-XY model, which would further make a distinction between the amplitude (not the exponent !) above and below T_c.[48,49] The exponent of $|t|$ is small and practically equivalent to a logarithm ($A|t|^{-\varepsilon} \cong A - \varepsilon A \ln|t|$). It is possible to fit the zero-field specific heat of optimally doped Y-123 using Eq. (19) down to the experimental scatter level (0.015% standard deviation).[15] This determines a *critical* coherence volume $\xi_{//0}^3 / \Gamma \approx 100$ Å^3, where $\xi_{//0}$ is the prefactor in $\xi_{//}(t) = \xi_{//0} |t|^{-\nu}$. The *Ginzburg-Landau* coherence volume is larger because $\nu = 1/2$ is assumed from the beginning in this case, so that $\xi_{//0}$ is extrapolated differently.

Eq. (19) suggests that the specific heat anomaly should increase with anisotropy. This is not verified. For Y-123, Γ increases monotonously from ≈ 5 in the overdoped 87-K phase to ≈ 15 for the underdoped 60-K phase ; at the same time the specific heat anomaly at T_c *decreases* by a factor >10. Formally, an increase of $\xi_{//0} \propto v_F / T_c$ by a factor of ≈ 3 could account for it, but this does not go in the sense expected from a reduction of the carrier concentration. This discrepancy may be one of the manifestations of the opening of a normal-state gap.

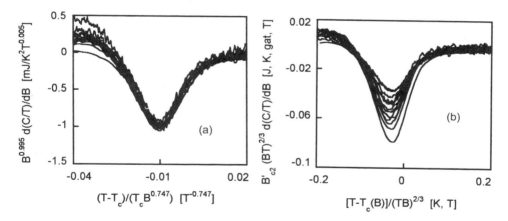

Figure 8. (a), 3D-XY scaling plot of $\partial(C/T)/\partial B$ for a TSFZM crystal of YBa$_2$Cu$_3$O$_{6.90}$. The fields $B//c$ are 1, 1.5, 2, 2.5, 3, 3.5, 4, 4.5, 5, 5.5, 6.5, 7.5, 9, 11 and 13 T.[44,47] (b), 3D-LLL scaling plot of the same data, assuming a linear variation of $T_c(B)$. If the variation of $T_c(B)$ is taken to be the same as for the 3D-XY model, then the modified 3D-LLL scaling plot becomes similar to that of Fig. 8a.[44,47]

We have discussed essentially Y-123 so far. Do we understand the transition of a very anisotropic superconductor such as Bi-2212 ? Specific heat and magnetization data are given e.g. in Refs. [15, 34, 43]. Eq. (19) in the 2D limit $\xi_{//0} / \Gamma < c$ is replaced by :

$$C(T, B = 0) = \frac{k_B}{\xi_{//0}^2 c} \frac{T^2}{T_c^2} |t|^{2\nu - 2} \varphi_6(\infty) \tag{20}$$

Note that the exponent of $|t|$ is always $D\nu - 2$ in D dimensions, as required by the hyperscaling relation. The 3D-XY model in this limit would imply $2\nu - 2 \cong -0.662$, i.e. a strong divergence at T_c. This is not verified experimentally. In contrast to this, both the magnetization and the specific heat of Bi-2212 can be empirically described by an exponent $\nu \approx 3/2$: the data collapse on a common curve in the corresponding scaling plots M and $H^{2/3} \partial(C/T)/\partial H$ versus $t/H^{1/3}$.[15] In zero field, Eq. (20) with $\nu \approx 3/2$ predicts a finite triangular peak at T_c, in qualitative agreement with idealized data.[15]

The preceding examples have shown that the nature of fluctuations depends considerably on the anisotropy of the compound. For Y-123 with a « small » anisotropy ($\Gamma \approx 5$), the main contribution comes from the mean-field jump, and corrections due to fluctuations are relatively small. Y-123 at optimal doping ($\Gamma \approx 7$) is probably the only good example of a 3D-XY transition in HTS, and appears to be a special case. The 3D-XY model does not describe very anisotropic (« quasi-2D ») superconductors. The basic assumption that fluctuations of the vector potential are negligible may be too crude. Another route was recently explored by Alexandrov et al., who assumed that electron pairs are preformed and condense at T_c.[61] The transition is then that of a weakly interacting charged Bose gas, with a characteristic triangular specific heat step in zero field at $T_c \cong 3.3 n_B^{2/3}/m$. An arbitrarily small field suppresses the Bose-Einstein condensation of an ideal charged Bose gas, but does not shift the characteristic temperature at which the chemical potential changes its slope, which is the cause of the specific heat anomaly. The difference $C(H) - C(0)$ calculated in this model matches closely experimental curves of the type shown in Figs. 7a and 7b, except that the $B = 0$ curve is a triangular peak ; it is natural to assume that experimental data are somewhat rounded by inhomogeneity in this limit. The suppression of the maximum value of $C(H) - C(0)$ should go as $\propto H^{1/2}$, in rough agreement with the observation $\propto H^{1/3}$.[34]

IV.5. Flux line lattice melting

Flux line lattice melting at a temperature $T_m(H)$ has been the subject of considerable attention recently. In a magnetic field, HTS do not exhibit a clear-cut transition on the $H_{c2}(T)$ line, which becomes, as already mentioned, a mere crossover ; the only true phase transition occurs at T_m. Vortex freezing manifests itself by a sudden drop of the resistivity to zero, a small jump ($\partial M/\partial T > 0$) in the magnetization, etc. In this section, we discuss the anomalies that are observed in the specific heat. Because of their smallness, they have been investigated only for Y-123. On very general grounds, the transition heat should be on the order of $k_B T_m$ per vortex, so the higher the field, the larger the anomaly. The transition disappears in fields on the order of 0.1 T or less in Bi-2212, to be compared with >18 T in Y-123.[36,50] A resolution of ≈ 1 ppm would be required to detect vortex lattice melting in Bi-2212, whereas 0.01% is sufficient for Y-123.

The problem of sample quality is crucial. As a rule of thumb, the smaller the crystal, the better - but the lower the absolute accuracy of the calorimetry. One experimental route consists in using mg-size single crystals and dynamic measuring methods allowing a high resolution ;[51] an other one consists in using conventional adiabatic calorimetric methods together with larger high purity crystals that can now be grown in BaZrO$_3$ crucibles.[35,50,52,53] Owing to these experimental difficulties, the crystals that have shown specific heat peaks on the freezing line can be counted on one hand's fingers. We review results reported up to now, keeping in mind that new data may still change our understanding.

Two kinds of transitions on the vortex freezing line have been predicted and reported. If the transition occurs between a crystalline solid and a liquid with different symmetries, the transition should be of first order.[54] If the solid is a glass, there is no symmetry change and the transition can be of second order or continuous.[55] The latter situation may be stabilized by pinning. By definition, a first-order transition gives rise to a discontinuity in the first derivatives of the free energy, e.g. the magnetization $M = -\partial F / \partial H$ and the entropy $S = -\partial F / \partial T$, or equivalently a δ-peak in the second derivatives $\chi_{differential} \equiv \partial M / \partial H$, $\partial M / \partial T = \partial S / \partial H$, $C / T = \partial S / \partial T$. Alternatively, a second-order transition gives rise to a change of slope (break) in the first derivatives, and a discontinuity (step) in the second derivatives. In practice, the situation may be less clear-cut, because ideal δ-peaks are broadened into finite peaks by residual inhomogeneity, and critical transitions give rise to a true divergence although they are of second order (e.g. ^{4}He at the λ point). When present (and not due to thermometer lag), hysteresis can make the difference. No hysteresis has been conclusively observed up to now. Experimental calorimetric results can be classified into two groups :

(1) specific heat steps have been observed on the freezing line with $\approx$50-300 mg Y-123 crystals produced by TSFZM at different oxygen concentrations in the overdoped range.[35,44,52,56] Small-angle neutron scattering experiments performed with similar crystals have shown a change of slope in the diffracted intensity at the freezing transition rather than a discontinuity.[62] TSFZM crystals contain numerous pinning centers.

(2) specific heat peaks have been observed for a 3 mg untwinned Y-123 crystal grown in a gold crucible,[51] for 18 and 40 mg twinned crystals grown in BaZrO$_3$ crucibles,[35,50,52,53] and for a $\approx$1 mg detwinned crystal also grown in BaZrO$_3$.[36] These flux-grown crystals are more reversible and contain very few pinning centers.

Therefore the present experimental situation seems to confirm that the transition is ideally of first order in the limit of perfect crystalline order in the vortex solid phase, whereas weak pinning leads to a second-order transition to a glassy phase. Strong pinning washes out any trace of the freezing transition. So does underdoping, possibly owing to the pinning properties of clusters of oxygen vacancies.[44,53]

Magnetization and specific heat are both static quantities, and it is possible to check their thermodynamic consistency. Measurements of M can be plagued by small-scale non-equilibrium effects that can escape attention in conventional SQUID magnetometers. Two groups could identify on the freezing line a magnetization jump $4\pi\Delta M \cong 0.3 \pm 10\%$ G superposed on a break of slope $4\pi\Delta(\partial M / \partial T) = 0.2 \pm 25\%$ G/K (these typical values correspond to a field of 4.2 T, $T_m \cong 84$ K).[57,58] Such data are not yet available for crystals grown in BaZrO$_3$ or prepared by TSFZM. The corresponding calorimetric discontinuities can be obtained from Clapeyron's equation :

$$\Delta S = \frac{L}{T} = -\Delta M \frac{dH_m}{dT} \qquad (21)$$

where L is the latent heat, and Clausius' equation :

$$\Delta C_H = \frac{dL}{dT} - \frac{L}{T} + \frac{L}{\Delta M} \Delta \left(\frac{\partial M}{\partial T} \right)_H \qquad (22)$$

About 85% of ΔC_H originates from the last term in Eq. (22) - which gives in fact one of Ehrenfest's relation in the limit $L \to 0$. The entropy jump required by Eq. (21), usually expressed in terms of $\Delta S \Phi_0 c / k_B B$, c=11.7 Å for Y-123, is in the range of 0.6-0.8 k_B per vortex per layer ;[57,58] the entropy integrated below the C / T peaks is indeed in the range of 0.45-0.6 k_B for the samples that exhibit peaks.[51,53] The specific heat jumps required by Eq.

(22) are also found to agree with experiment, both in samples that show specific heat peaks[51,53] and those which do not.[44,56] The jump represents the difference between the specific heat of the vortex liquid and that of the vortex solid.

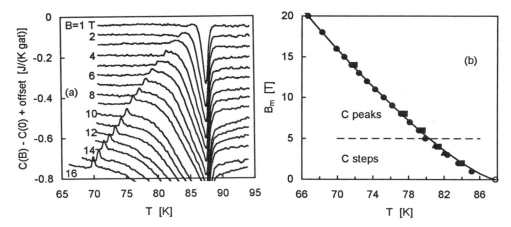

Figure 9. (a), difference between the specific heat C in a field $B//c$ = 1, 2, 3, ...16 T and the zero-field C as a function of T for a fully oxygenated $YBa_2Cu_3O_{7.00}$ crystal.[53] Curves are arbitrarily shifted for clarity. (b), phase diagram in the B,T plane, $B//c$. Closed symbols, top of C peaks or midpoint of C steps, 18 mg sample (circles[53]) and 40 mg sample (squares[35]). Triangles : M steps.[53] Line : fit $B_m \propto B_0|t|^{4/3}$.

Figure 9a shows the specific heat of a twinned Y-123 crystal (grown in $BaZrO_3$ and similar to that of Fig. 5b) in fields 0-16 T ;[53] the zero-field curve has been subtracted as a convenient baseline. The location of the peaks is reported in Fig. 9b and defines the freezing line. This line could be followed very recently up to 26.5 T in the Grenoble laboratories.[36] On the low field side, the « first-order » peaks give way to « second-order » steps below ≈5 T, even for detwinned crystals of the same batch. It is suspected that at low field, vortices fill preferentially residual pinning sites due e.g. to strains or twins, and cannot form a lattice until these sites are saturated. The positions of additional vortices in higher fields can then be determined by vortex-vortex interactions alone, allowing a lattice to form. The untwinned crystal of Ref. [51], grown in a gold crucible, shows peaks from 9 T down to 0.75 T. Its behaviour at >9 T is not yet known. Dissolved Au does not seem to build pinning centers.

The freezing line has the same shape for all samples, a shape that approaches very closely a power-law $H_m = H_0|t|^{4/3}$. Two scenarios are possible. In the first one, the transition is of second order, the C peak is a critical λ-peak and the exponent is given by 3D-XY model.[55] In the second one, the transition is of first order, the power law is coincidental and must be explained. The model of melting based on Lindemann's criterion predicts $H_m = H_0 t^2$,[59] but flexible corrections can bring it close to an effective power law $H_m = H_0|t|^n$, $n < 2$, at the expense of additional parameters.[60] The question cannot be answered based on specific heat data alone, even though these measurements provide presently the best way to map the freezing line over a wide range in high fields.

This ambiguity is also striking when one compares samples which show « first-order » specific heat peaks (pure flux-grown crystals) and those which show « second-order » steps (TSFZM crystals). The freezing line $H_m(T)$ is the same within experimental uncertainty. The TSFZM crystals facilitate some experiments, because samples are relatively large and allow an excellent absolute accuracy, and because they equilibrate fast and speed up the study of the effects of oxygen doping. Some results are reported in Figs. 10 and 11. Figure 10a shows the monotonous variation of the anisotropy and the maximum of the critical temperature in the vicinity of optimal doping. Both curves represent *bulk* data because they are derived from the specific heat anomaly at T_c. The anisotropy is determined by the ratio of two fields $B//c$ and $B \perp c$ which have the same effect on the specific heat anomaly. Typical C steps observed on the freezing line are shown in Fig. 10b. The conclusion of a recent study of these TSFZM crystals are :[44]

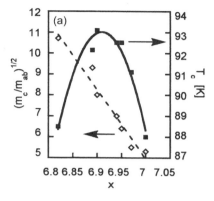 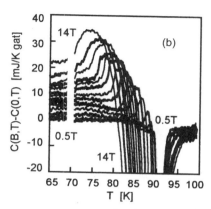

Figure 10. (a), anisotropy factor and critical temperature versus oxygen concentration for *one* YBa$_2$Cu$_3$O$_x$ crystal obtained by TSFZM. (b), difference between the specific heat C in a field $B//c$ = 0.5, 1, 1.5, 2.5, 3.5, 4.5, 5.5, 6.5, 7.5, 8.5, 9.5, 10.5, 12 and 14 T and the zero-field C as a function of T for a TSFZM crystal of YBa$_2$Cu$_3$O$_{6.97}$.[44,52,56]

(1) the melting line disappears in the underdoped regime ;
(2) the melting field H_m and the upper critical field H_{c2} have the same bulk anisotropy factor (Fig. 11b). This is also true in the case of first-order melting ;[51]
(3) H_m and H_{c2} move in opposite directions when the oxygen concentration (and the anisotropy) change (Fig. 11a). H_m is governed by λ, whereas H_{c2} is governed by ξ ; for a given condensation energy, λ and ξ are anticorrelated.
(4) H_m scales with anisotropy and temperature in the form $H_m = (m_c / m_{ab})^{-1/2} f(t)$;
(5) a critical point terminates the H_m line at a value of the field that increases with oxygen filling (Fig. 11a). This is consistent with the liquid and the solid (glass) having in this case the same symmetry ;
(6) the specific heat step at T_m obeys critical scaling relations, Eq. (13), corrected by the proportionality between the number of « particles » and B ;
(7) the step at T_m occurs always at the onset of strong fluctuations near T_c, i.e. for a fixed value of the critical scaling variable $t / H^{1/2\nu}$.

The image that emerges is that ideal first-order vortex freezing requires extremely clean samples, and is readily replaced by a critical glass transition in real samples with pinning. Further work is required to understand the interplay between critical fluctuations and vortex freezing.[63] The dependence between (multi)critical points on the $H_m(T)$ line at high and low fields and subtle crystal characteristics has still to be clarified. A better understanding is expected to follow from advances in the synthesis of ultra-pure crystals with controlled doping, and from high resolution measurements in ultra-high magnetic fields.

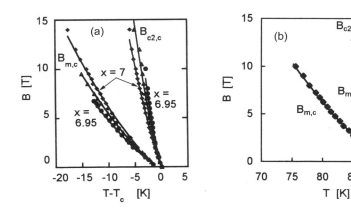

Figure 11. (a), phase diagram in the B,T plane showing the effect of doping on B_{c2} and B_m, $B//c$. (a), phase diagram showing the anisotropy of the upper critical field B_{c2} (open symbols) and of the freezing field B_m (closed symbols). Subscripts c and ab mean $B//c$ and $B//ab$, respectively. Sample : TSFZM crystal, Y-123.[44,52,56]

Acknowledgements

Collaboration of J.-Y. Genoud (Industrial Research Limited, Lower Hutt, New Zealand), A. Mirmelstein (Institute for Metal Physics, Ekaterinburg, Russia), discussions with C. Marcenat, R. Calemczuk, J. Muller and J. Sierro, and grants by the Fonds National Suisse de la Recherche Scientifique and NATO are gratefully acknowledged.

REFERENCES

1. J. Bardeen, L.N. Cooper and J.R. Schrieffer, *Phys. Rev.* **108**, 1175 (1957).
2. W.L. McMillan, *Phys. Rev.* **167**, 331 (1968).
3. P.B. Allen and R.C. Dynes, *Phys. Rev.* B **12**, 905 (1975).
4. R.C. Dynes, *Solid State Commun.* **92**, 53 (1994).
5. J.R. Schrieffer, *Solid State Commun.* **92**, 129 (1994).
6. C.C. Tsuei *et al.*, *Phys. Rev. Lett.* **73**, 593 (1994).
7. Ya.G. Ponomarev *et al.*, *Phys. Rev.* B **52**, 1352 (1995).
8. M.B. Salamon *et al.*, *Phys. Rev.* B **38**, 885 (1988).
9. A. Mirmelstein *et al.*, to be published in J. Supercond.
10. A. Junod, J.-L. Jorda and J. Muller, J. Low Temp. Phys. **62**, 301 (1986).
11. K. Maki, Physics 1, 21 (1964) ; 1, 127 (1964) ; Phys. Rev. **139**, A702 (1965).
12. C. Caroli, P.G. de Gennes and J. Matricon, Phys. Lett. **9**, 307 (1964) ; Phys. kond. Mater. **3**, 380 (1965).
13. S. Dukan and Z. Tesanovic, Phys. Rev. B **49**, 13017 (1994).
14. I. Maggio-Aprile C. Renner, A. Erb, E. Walker and Ø. Fischer, Phys. Rev. Lett. **75**, 2754 (1995).
15. A. Junod, Specific heat of high temperature superconductors in high magnetic fields, in : Studies of High Temperature Superconductors Vol. 19, A.V. Narlikar, ed., Nova Science, Commack N.Y. (1996) p. 1.
16. R.A. Fisher *et al.*, Physica C **252**, 237 (1995) ; J.P. Emerson *et al.*, J. Low Temp Phys. **105**, 891 (1996) ; D.A. Wright *et al.*, J. Low Temp. Phys. **105**, 897 (1996).
17. K. Maki and H. Won, Ann. Physik 5, 320 (1996).
18. A. Erb, E. Walker and R. Flükiger, Physica C **258**, 9 (1996).
19. B. Revaz, J.-Y. Genoud, A. Junod, A. Erb and E. Walker, submitted to Phys. Rev. Lett.
20. K.A. Moler *et al.*, Phys. Rev. Lett. **73**, 2744 (1994).
21. N. Momono and M. Ido, Physica C **264**, 311 (1996).
22. G.E. Volovik, Pis'ma ZhETF **58**, 457 (1993) ; JETP Lett. **58**, 469 (1993).

23. S.A. Simon and P.A. Lee, Phys. Rev. Lett. **78**, 1548 (1997).
24. G.E. Volovik, Pis'ma ZhETF **65**, 465 (1997) ; JETP Lett. **65**, 491 (1997) ; preprint Cond-mat/9702093.
25. A. Erb *et al.*, J. Low Temp. Phys. **105**, 1023 (1996).
26. S. Sridhar, Bull. Am. Phys. Soc. **42**, 698 (1997) ; H. Srikanth *et al.*, to be published in Phys. Rev. Lett.
27. K. Neumaier *et al.*, private communication.
28. J. Bok and J. Bouvier, Physica C **276**, 173 (1997).
29. J.W. Loram, K.A. Mirza, J.M. Wade, J.R. Cooper and W.Y. Liang, Physica C **235-240**, 134 (1994)
30. J.W. Loram, K.A. Mirza, J.R. Cooper and J.L. Tallon, in : M^2S-HTSC-V - Proceedings of the 5th Int. Conf. Materials and Mechanisms of Superconductivity High-Temperature Superconductors, Yu-Sheng He, ed., to appear in Physica C ; J.L. Tallon, G.V.M. Williams, N.E. Flower and E.M. Haines, id.
31. Ch. Renner, B. Revaz, J.-Y. Genoud, K. Kadowaki and Ø. Fischer, submitted to Phys. Rev. Lett.
32. A. Schilling and O. Jeandupeux, Phys. Rev. B **52**, 9714 (1995).
33. M. Roulin, Physica C **260**, 257 (1996).
34. A. Junod *et al.*, Physica C **229**, 209 (1994).
35. A. Junod *et al.*, in : M^2S-HTSC-V - Proceedings of the 5th Int. Conf. Materials and Mechanisms of Superconductivity High-Temperature Superconductors, Yu-Sheng He, ed., to appear in Physica C.
36. C. Marcenat and F. Bouquet, private communication.
37. B. Revaz, private communication.
38. L. Landau and E. Lifschitz, Physique statistique (Physique théorique Vol. V), Mir, Moscow (1967), §141.
39. N. Overend *et al.*, Phys. Rev. B **54**, 9499 (1996).
40. M.B. Salamon *et al.*, Phys. Rev. B **38**, 885 (1988) ; M.B. Salamon, J. Shi, N. Overend and M.A. Howson, Phys. Rev. B **47**, 5520 (1993).
41. D.S. Fisher, M.P.A. Fisher and D.A. Huse, Phys. Rev. B **43**, 130 (1991).
42. T. Schneider and H. Keller, Int. J. Mod. Phys. B **8**, 487 (1993) ; Physica C **207**, 366 (1993), and extensions of this work.
43. A. Junod, J.-Y. Genoud, G. Triscone and T. Schneider, submitted to Physica C.
44. M. Roulin, A. Junod and E. Walker, submitted to Physica C.
45. N.K. Wilkin and M.A. Moore, Phys. Rev. B **48**, 3464 (1993).
46. Z. Tesanovic and A.V. Andreev, Phys. Rev. B **49**, 4064 (1994).
47. M. Roulin, A. Junod and E. Walker, Physica C **260**, 257 (1996).
48. M.L. Kulic and H. Stenschke, Solid State Commun. **66**, 497 (1988).
49. V. Privman, P.C. Hohenberg and A. Aharony, in : Phase Transitions and Critical Phenomena, Vol. 14, C. Domb and J.J. Lebowitz, eds, Academic Press, London (1991) p. 1.
50. C. Marcenat, R. Calemczuk, A. Erb, E. Walker and A. Junod, in : M^2S-HTSC-V - Proceedings of the 5th Int. Conf. Materials and Mechanisms of Superconductivity High-Temperature Superconductors, Yu-Sheng He, ed., to appear in Physica C.
51. A. Schilling *et al.*, Nature **382**, 791 (1996) ; A. Schilling *et al.*, in : M^2S-HTSC-V - Proceedings of the 5th Int. Conf. Materials and Mechanisms of Superconductivity High-Temperature Superconductors, Yu-Sheng He, ed., to appear in Physica C ; A. Schilling *et al.*, to be published in Phys. Rev. Lett.
52. M. Roulin, A. Junod and E. Walker, J. Low Temp. Phys. **105**, 1099 (1996).
53. A. Junod *et al.*, Physica C **275**, 245 (1997).
54. E. Brezin, D.R. Nelson and A. Thiaville, Phys. Rev. B **31**, 7124 (1985).
55. M.P. Fisher, Phys. Rev. Lett. **62**, 1415 (1989) ; D.S. Fisher, M.P. Fisher and D.A. Huse, Phys. Rev. B **43**, 130 (1991).
56. M. Roulin, A. Junod and E. Walker, Science **273**, 1210 (1996).
57. U. Welp, J.A. Fendrich, W.K. Kwok, G.W. Crabtree and B.W. Veal, Phys. Rev. Lett. **76**, 4809 (1996).
58. R. Liang, D.A. Bonn and W.N. Hardy, Phys. Rev. Lett. **76**, 835 (1996).
59. G. Blatter, M.V. Feigel'man, V.B. Geshkenbein, A.I. Larkin and V.M. Vinokur, Rev. Mod. Phys. **66**, 1125 (1994).
60. G. Blatter and B.I. Ivlev, Phys. Rev. B **50**, 10272 (1994).
61. A.S. Alexandrov, W.H. Beere, V.V. Kabanov and W.Y. Liang, to be published in Phys. Rev. Lett.
62. M.T. Wylie *et al.*, Czech. J. Phys. **46**, 1569 (1996) ; T. Forgan and C.M. Aegerter, private communication.
63. S.W. Pierson, J. Bruan, B. Zhou, C.C. Huang and O.T. Valls, Phys. Rev. Lett. **74**, 1887 (1995) ; S.W. Pierson and O.T. Valls, submitted to Phys. Rev. Lett.
64. K. Yang and S.L. Sondhi, preprint Cond-mat/9706148.

THE SPECTRUM OF THERMODYNAMIC FLUCTUATIONS
IN SHORT COHERENCE LENGTH SUPERCONDUCTORS

Andrea Gauzzi

MASPEC – CNR Institute
Via Chiavari 18/A, I – 43100 Parma, Italy

1. Introduction

Our current understanding of phase transitions enables us to establish a *universal* correspondence between critical behaviour at the transition point and symmetry of the effective Hamiltonian $\mathcal{H}_{eff}$ of a given system, regardless of the specific nature of the microscopic interaction responsible for the transition. The universal validity of this correspondence requires that the critical behaviour is dominated by fluctuations of the order parameter having a wavelength much larger than the range of the above interaction. This is so for systems characterised by a long correlation (or coherence) length ξ of the order parameter. An example of these systems is given by superconducting metals and alloys with low transition temperature T_c, which are well described by the conventional $\mathcal{H}_{eff}$ of the BCS theory. In this paper we discuss how the above fluctuation spectrum is modified in the opposite case of short–ξ systems. This issue is relevant, since we can derive the $\mathcal{H}_{eff}$ of a system undergoing a second order phase transition from the fluctuation spectrum. This study might elucidate the still open question of generalising the conventional BCS theory to account for the unusual properties of cuprate superconductors. Fluctuation studies are suited for determining the $\mathcal{H}_{eff}$ of complex systems, such as the above materials, because only the relevant interaction driving the transition exhibits a critical behaviour in the fluctuation region, while this interaction is usually hidden among other interactions at temperatures far above T_c.

This paper is organised as follows. We first critically review selected general aspects on fluctuation phenomena near second order phase transitions in section 2 and the universal predictions of conventional theories of critical phenomena in section 3. The content of these sections can be found in any textbook (Landau and Lifshitz, 1980; Patashinskii and Pokrovskii, 1979; Ma, 1976; Stanley, 1971). However, we prefer presenting some key concepts, with particular emphasis on the limit of validity of the theoretical results, to facilitate our discussion. In section 4 we review the predictions of conventional fluctuation theories referring to the physical properties studied in the following sections. In section 5 we re–examine the application of standard universal predictions to selected experiments on conventional BCS superconductors with long ξ. In section 6 we present a similar analysis for cuprates and highlight the controversial interpretations reported for these superconductors with short ξ. In section 7 we re–analyse a selection of the experimental results which are a subject to controversy within the framework of a generalised mean–field approach. In particular, we examine the qualitative features of the fluctuation spectrum of cuprates as determined by fluctuation measurements within such generalised approach. Finally, in section 8 we compare these features with those of the conventional spectrum of the BCS $\mathcal{H}_{eff}$ in the long$-\xi$ limit and draw some conclusions. Several aspects discussed in this paper can be extended to other families of superconductors with short ξ, such as Bi–based oxides, Chevrel phases and alkali–doped C_{60}, and also to other second order phase transitions characterised by a short correlation length of the order parameter.

2. Relevance of thermodynamic fluctuations near second order phase transitions

According to general results of statistical mechanics, a thermally activated deviation (fluctuation) $\delta\psi$ of the order parameter ψ of a given second order phase transition with respect to the thermodynamic equilibrium value $\langle\psi\rangle$ at temperature T has probability

$$ w \propto \exp\left(- \delta\Phi/k_B T\right) \tag{1} $$

where $\delta\Phi$ is the extra–energy required to activate such deviation with respect to the value $<\Phi>$ of Φ at thermodynamic equilibrium. Φ is the difference of the appropriate thermodynamic potential associated with the phase transition

$$ \Phi \equiv \Phi(\psi) - \Phi(0) \tag{2} $$

and we assume for simplicity $\Phi(0) \equiv 0$. The quantity Φ is called *incomplete thermodynamic potential* and plays the role of the effective Hamiltonian $\mathcal{H}_{eff}$ of the system.

This paper is based on the concept of *fluctuation spectrum*, which is defined as the expectation value $\langle|\delta\psi_{k,\omega}|^2\rangle$ of the density of the fluctuation $\delta\psi$ with given wavevector $\mathbf{k}$ and with given frequency ω. We recall that $\langle|\delta\psi_{k,\omega}|\rangle = 0$ by definition of fluctuation. For simplicity, we consider the limit of slow fluctuations, corresponding to $\omega = 0$. The fluctuation spectrum is relevant to our discussion, since the type and the microscopic parameters of $\mathcal{H}_{eff}$ and the critical behaviour of all physical properties, i.e. their *critical exponents*, are derived from this spectrum. The fluctuation spectrum is expressed by the probability w as a function of $\delta\psi$ in eq. (1). To determine this probability, the exact functional dependence of Φ on ψ in eq. (2), i.e. the $\mathcal{H}_{eff}$ is required. We first consider the case of a uniform order parameter and note that the qualitative dependence of Φ on ψ above and below T_c must be like the one schematically represented in Fig. 1, according to the well–known arguments by Landau (1937). The equilibrium value $<\Phi>$ corresponds to a zero value of $\langle\psi\rangle$ above T_c and non–zero below T_c. Under reasonable physical assumptions, the functional $\Phi = f(\psi)$ is continuous and therefore admits an expansion in powers of ψ. Such an expansion is justified for sufficiently small values of ψ, which corresponds to temperatures near T_c. For simplicity, we shall consider only the high–symmetry state above the transition ($\langle\psi\rangle = 0$, $T > T_c$). The fluctuating potential can then be written as

$$ \delta\Phi = \frac{1}{2!}\left(\frac{\partial^2\Phi}{\partial\psi\partial\psi^*}\right)_{\psi=\langle\psi\rangle}|\delta\psi|^2 + \frac{1}{4!}\left(\frac{\partial^4\Phi}{\partial\psi^2\partial\psi^{*2}}\right)_{\psi=\langle\psi\rangle}|\delta\psi|^4 + ... $$

$$ = V\left(a|\delta\psi|^2 + \beta|\delta\psi|^4 + ...\right) \tag{3} $$

where V is the sample volume. In eq. (3), the linear term is zero, since $<\Phi>$ is a minimum; the cubic term is also zero if the critical points are not isolated in the phase diagram. From Fig. 1, one notes that the sign of the first coefficient a must change at the transition. It follows that,

424

at temperatures sufficiently near T_c, the temperature dependence of this coefficient can be approximated by the linear term $a = at$, where $t \equiv (T - T_c)/T_c$ is a reduced temperature. Usually, we can neglect the quadratic term $\propto t^2$ and the higher order terms in the expansion of a in powers of t if

$$t \ll 1 \tag{4}$$

The validity of the universal predictions for the critical exponents of the mean–field theory cited in section 4 and reported in Table 1 requires the validity of this linear approximation.

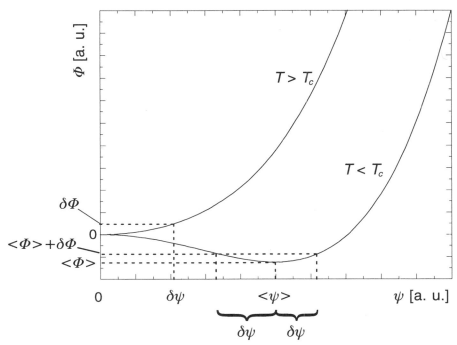

Figure 1. Schematic diagram of the dependence of the incomplete thermodynamic potential Φ of eq. (3) on the order parameter ψ in both regions above and below the transition (after Landau and Lifshitz (1980)). The fluctuation regions of $\delta\Phi$ and $\delta\psi$ are indicated by the broken lines.

2a. Weak fluctuations. Fluctuation spectrum

The power of the Landau formulation described above consists in its general validity: in principle, every physical system can be described by an appropriate choice of the coefficients in the expansion of eq. (3), provided that a sufficiently large number of terms is taken into account. This observation is valid for both weak and strong fluctuations. The former case can be treated by mean–field theories, since these are appropriate for weakly interacting systems. In this case, it is sufficient to retain the first quadratic term in eq. (3)

$$\delta\Phi \approx aVt|\delta\psi|^2 \tag{5}$$

Weak fluctuations are also called Gaussian, since they are independent of each other in first approximation and therefore their probability follows a Gaussian distribution function. This can also be seen mathematically: by inserting eq. (5) into eq. (1), the probability w becomes a Gaussian distribution function of the statistical variable $\delta\psi$. The Gaussian approximation is

425

found to be appropriate for weak–coupling superconductors and for certain ferroelectrics but, in general, not for magnetic transitions.

An important generalisation of the simple treatment described above consists in taking into account spatial variations of the order parameter. Such variations arise, for example, in the presence of a current flow in a superconductor. The functional in eq. (3) must then be replaced by an integral over the sample volume of a spatially varying functional density. Within the limit of slow spatial variations, it is sufficient to add a quadratic term in the first gradient of ψ to the above density. According to the original idea of Ornstein and Zernicke (Landau and Lifshitz, 1980), eq. (3) must be substituted by the following

$$\delta\Phi = \int_V \left(at\,|\delta\psi|^2 + \beta\,|\delta\psi|^4 + g_j^i \partial_i \delta\psi \partial^j \delta\psi^* \right) dV \tag{6}$$

where ∂_i are, in a superconductor, covariant gauge–invariant partial spatial derivatives

$$\partial_i \equiv \frac{\partial}{\partial x^i} - \frac{2ie}{\hbar} A_i \tag{7}$$

where $2e$ is the charge of a Cooper pair and $\mathbf{A}$ is the vector potential. It has been shown by Gor'kov (1959) that the functional of eq. (6) coincides with the BCS effective Hamiltonian in the vicinity of the transition by setting $\psi \propto \Delta$, where Δ is the superconducting energy gap. This equivalence establishes a correspondence between fluctuation spectrum and microscopic Hamiltonian of a given BCS superconductor. It would be important to extend this result to cuprates, since the microscopic Hamiltonian of these superconductors is still unknown. This method for determining the effective Hamiltonian is applicable to any second order phase transition, since eq. (6) formally describes the interaction of a charged scalar field with a vector potential. The same functional is then applicable not only to superconductors, but also to nematic–smectic–A transitions in liquid crystals (de Gennes, 1972) and to Higgs' mechanism of charged boson fields (Coleman and Weinberg, 1973).

To determine the probability of weak fluctuations in the general case of spatially–varying order parameter described by eq. (6), we again neglect the quartic term and expand the fluctuating field $\delta\psi$ into Fourier components with wavevector $\mathbf{k}$

$$\delta\psi(\mathbf{r}) = \sum_{|\mathbf{k}| \leq |\tilde{\mathbf{k}}|} e^{i\mathbf{k}\mathbf{r}} \delta\psi_{\mathbf{k}} \tag{8}$$

where we have emphasized that the sum is limited to wavevectors smaller than a cutoff value $\tilde{k}$. The existence of such cutoff is due to the fact that the order parameter ψ is defined in a sufficiently large volume $\tilde{V}$ containing a number of Cooper pairs much larger than unity. If this condition were not fulfilled, it would not be possible to define the quantity $|\psi|^2$ as the expectation value of the superfluid density. The value of $\tilde{k}$ is then determined by the condition that the spatial variations of ψ contain only Fourier components with a wavelength larger than the linear dimension of $\tilde{V}$. In sections 5 and 6 we discuss the fact that the above limitation is not relevant to conventional superconductors with long ξ, while it is fundamental for short–ξ superconductors. The substitution of eq. (8) in eq. (6) yields

$$\delta\Phi = V \sum_{|\mathbf{k}| \leq |\tilde{\mathbf{k}}|} \left(at + g_j^i k_i k^j \right) |\delta\psi_{\mathbf{k}}|^2 \tag{9}$$

This result enables us to explicitly evaluate the **k**–dependence of the probability w of the fluctuating field $\delta\psi(r)$. By replacing the above expression in eq. (1), we find that w becomes a product of Gaussian distribution functions w_k of each Fourier component $\delta\psi_k$

$$w \propto \prod_{k \le |\tilde{\mathbf{k}}|} \exp\left[-\frac{V}{k_B T}\left(at + g_j^i k_i k^j\right)|\delta\psi_k|^2 \right] \tag{10}$$

The expectation value of the component $\delta\psi_k$, i.e. the *fluctuation spectrum*, is found to be a *Lorentzian* function of **k** in a reference system where the tensor g is diagonal

$$\langle|\delta\psi_k|^2\rangle = \frac{k_B T}{2Vat} \frac{1}{\left(1 + \xi_x^2 k_x^2 + \xi_y^2 k_y^2 + \xi_z^2 k_z^2\right)} \tag{11}$$

where we have set

$$\xi_x \equiv \sqrt{\frac{g_x^x}{at}} = \xi_{x,0} t^{-\frac{1}{2}} \tag{12}$$

and similar expressions for the y– and z–directions. The above spectrum is called Ornstein–Zernicke spectrum and is represented in Fig. 2. The physical meaning of ξ is the (Ginzburg–Landau) correlation length of ψ, i.e. it is the characteristic length over which the correlation function of the order parameter decays significantly. The square–root divergence of ξ at the transition, which we note from eq. (12), arises from the linear approximation of the temperature dependence of the quadratic coefficient in eq. (6).

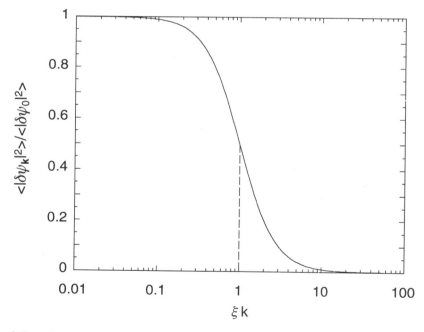

Figure 2. Dependence of the Ornstein–Zernicke fluctuation spectrum (see eq. (11)) as a function of wavevector. This spectrum is valid within the limit of weak (Gaussian) fluctuations and within the long wavelength limit $\xi\tilde{k} \gg 1$, where $\tilde{k}$ is the short–wavelength cutoff described in the text. The effect of this cutoff in the case of short coherence length ($\xi\tilde{k} \approx 1$) is represented schematically by the broken line.

We finally note that the **k**–dependence of the fluctuation spectrum can be expressed in the explicit Ornstein–Zernicke form of eq. (11) thanks to the statistical independence of each fluctuation with wavevector **k** within the weak–fluctuation approximation. Within this approximation, the fluctuating field is analogous to a perfect gas of quasiparticles with wavevector **k** and with density $\langle|\delta\psi_{\mathbf{k}}|^2\rangle$. The Ornstein–Zernicke spectrum is therefore the Green function of the free fluctuating field with no self–energy correction.

2b. Strong fluctuations

In this regime, interactions between fluctuations are important. Mathematically, this circumstance implies that the quartic term and higher–order terms in the Ginzburg–Landau expansion (3) are no longer negligible. The quartic term has the meaning of two–particle interaction, the sixth–order term that of three–particle interaction, etc. Strong fluctuations imply that the uniform fluctuation $\langle|\delta\psi_{\mathbf{k}=0}|^2\rangle$ in eq. (11) becomes comparable to the average value $\langle|\psi|^2\rangle = -\alpha t/2\beta$ of the order parameter in the coherence volume $\sim\xi^3$. This condition is known as Ginzburg condition and is expressed by the inequality (Levanyuk, 1959; Ginzburg, 1960)

$$t \lesssim t_G \equiv \left(\frac{\beta k_B T_c}{\alpha^2 \xi_{x,0}\xi_{y,0}\xi_{z,0}}\right)^2 = \frac{1}{32\pi^2}\left(\frac{k_B V}{\Delta C \xi_{x,0}\xi_{y,0}\xi_{z,0}}\right)^2 \tag{13}$$

where t_G is called Ginzburg temperature and ΔC is the specific heat jump at T_c. In the case of short ξ, one can have $t_G \gtrsim 1$, hence the Gaussian approximation is not valid at any temperature and the fluctuations are strong (or critical) at any temperature. The conditions leading to this case have been studied by Kapitulnik, Beasley, Di Castro and Castellani (1988) for cuprates. For the most studied compound $YBa_2Cu_3O_{6+x}$, with $T_c = 92$ K, we estimate $t_G \approx 3\times10^{-2} - 3\times10^{-3}$ (Gauzzi and Pavuna, 1995) by using experimental data of specific heat (Junod, 1990) and of ξ (Matsuda, Hirai and Komiyama, 1988; Lee, Klemm and Johnston, 1989; Lee and Ginsberg, 1992). We therefore expect that weak fluctuations dominate and that critical fluctuations become important below 0.3–3 K above T_c, but this question remains controversial.

The generalisation of the previous considerations to the case of strong fluctuations is straightforward. However, the partition function

$$Z = \sum_{\delta\psi} e^{-\frac{\delta\Phi(\delta\psi)}{k_B T}} \tag{14}$$

where the effective Hamiltonian $\delta\Phi$ has the Ginzburg–Landau form of eq. (6) cannot be calculated explicitly in presence of the quartic term. Numerical solutions have been obtained by renormalisation group methods. These methods were first applied by Wilson (1971) and are based on perturbation theories or on approximate schemes. Besides the Ginzburg–Landau Hamiltonian, alternative Hamiltonians can be used. Classic examples are the Ising, XY or Heisenberg Hamiltonians, corresponding to the case of respectively one, two and three scalar components of the order parameter. Contrary to the case of the Ginzburg–Landau model, these are discrete models defined on a lattice and are therefore suited for numerical simulations. The XY Hamiltonian describes phase transitions with one single complex order parameter where *phase* fluctuations dominate. This is opposite to the Ginzburg–Landau Hamiltonian, which accounts for predominant weak fluctuations of the *amplitude* of the order parameter.

3. Universality principle and scaling hypothesis for critical phenomena

On the basis of the foregoing considerations, we might ask whether the choice of a particular effective Hamiltonian is important or not to account for the behaviour of a given system in the fluctuation region. In other words, we shall ask under which conditions fluctuation phenomena are expected to be the signature of the particular microscopic interaction driving the phase transition or to follow a more general behaviour. Indeed, despite the variety of microscopic interactions responsible for second order phase transitions of magnetic, structural, electronic, etc. nature, it is often found that completely different systems exhibit the same type of behaviour at the transition. Specifically, the temperature dependence of measurable quantities, such as the specific heat, the magnetic susceptibility, the compressibility, etc. near the critical point follow the same power law whose exponent is called *critical exponent*. The heuristic explanation of such common behaviour is given by the following argument, known as *universality principle*; the ensemble of systems characterised by the same behaviour is therefore called *universality class*. As the critical point is approached, the coherence length ξ of the order parameter begins to diverge. Sufficiently close to the critical point, ξ becomes much larger than the range of relevant interactions responsible for the transition. In particular, if ξ becomes much larger than the cutoff wavelength of the fluctuations

$$\xi \gg 1/\tilde{k} \tag{15}$$

it is expected that the specific nature of the above interactions is irrelevant in determining the critical behaviour of the system. The critical behaviour must then depend only on the symmetry of $\mathcal{H}_{eff}$, which implies, in particular, a given dimensionality of space and a given number of components of the order parameter. The Ginzburg–Landau $\mathcal{H}_{eff}$ in the free–field approximation and the $3d$XY $\mathcal{H}_{eff}$ mentioned above are two classic examples of universality class with one complex order parameter. These two classes are suitable for describing the weak and strong fluctuation regimes respectively. As previously mentioned, most conventional BCS superconductors, which all have long coherence lengths, are well described by the first class. On the other hand, superfluid ^{4}He, which has a short coherence length, exhibits strong fluctuations in agreement with the Ginzburg criterion (eq. (13)), and its fluctuation behaviour is well described by the $3d$XY Hamiltonian as reported, for example, in Stanley's book (1971). Other superfluids and superconductors belong to different universality classes, since their order parameter has more than two real components. This is the case of superfluid ^{3}He and of some heavy fermion superconductors. In the former, for example, it was established that the order parameter has p–wave symmetry, corresponding to an orbital momentum $l = 1$ and to a total spin $s = 1$ of the Cooper pair. The number of real components of the order parameter is therefore 18. On the other hand, in cuprates, the available experimental data indicate that Cooper pairs form a singlet state ($s = 0$), as in conventional BCS superconductors and in superfluid ^{4}He. However, the question of the symmetry of the orbital part of the superconducting wave function is still controversial.

The predicted critical exponents of selected quantities are reported in Table 1 for the Ginzburg–Landau theory in the Gaussian approximation and for the $3d$XY model. In the next sections we present a comparison between these predictions and the experimental data on selected compounds of conventional and cuprate superconductors. We recall that the number of independent critical exponents is determined by applying scale invariance arguments to the length of the system (see, for example, Landau and Lifshitz, 1980). In the regime of weak fluctuations, there are two independent lengths, ξ and the dimension r_0 of the volume where the magnitude of the fluctuations is comparable to the magnitude of the equilibrium value of the order parameter. In this case, it can be shown that there are three independent exponents. In the regime of strong fluctuations, it is assumed that ξ is the only relevant length of the fluctuating system. This hypothesis is called *scaling hypothesis* and reduces to two the number of independent critical exponents.

The universality principle and the scaling hypothesis constitute the basis of the modern theory of critical phenomena, since they are expected to be generally valid, provided that the correlation length becomes sufficiently large at the transition, according to eq. (15). In real systems, however, it is not possible to study temperature regions closer than 0.1–1 mK to the critical point because of practical limitations. Therefore, in systems with sufficiently short ξ_0, the above inequality might be not fulfilled in the temperature region accessible to experiment. In strongly anisotropic systems, it could occur that the condition for universality is satisfied along a given direction, but not along another. We shall apply the above considerations to amorphous alloys and to cuprates.

Table 1. Critical exponents of selected physical quantities in the weak (or Gaussian) and strong (or critical) fluctuation regimes according to respectively the Ginzburg–Landau and the $3d$XY models for one complex order parameter.

Physical quantity		Critical exponent	Gaussian fluctuations Ginzburg–Landau theory	Critical fluctuations $3d$XY model
Specific heat	$\Delta C \sim t^{-\alpha}$	α	0	Logarithmic divergence
Order parameter	$\psi \sim t^{\beta}$	β	1/2	1/3
Coherence length	$\xi \sim t^{-\nu}$	ν	1/2	2/3
Magnetic pen. depth	$\lambda \sim t^{-\beta}$	β	1/2	1/3
Fluctuation of the magnetic susceptibility	$\delta\chi \sim t^{-\upsilon}$	υ	1/2	2/3
Fluctuation of the conductivity	$\delta\sigma \sim t^{-\eta}$	η	$2 - d/2$	2/3

4. Experimental aspects of fluctuation studies in superconductors

Generally speaking, the effects of thermodynamic fluctuations near a second order phase transition appear as a rounding of the discontinuities or as a cutoff of the divergencies of physical quantities, such as the specific heat, the thermal expansion coefficient, the conductivity, etc. In conventional superconductors, these effects are too weak to be measured, because of the low T_c and of the long ξ (see eq. (11) and eq.(13)). The situation is more favourable in amorphous alloys, since ξ is reduced by the amorphous structure down to 5–10 nm. In cuprates, fluctuation effects are particularly large, since T_c is large and ξ is extremely short.

An advantage of fluctuation studies is that fluctuating properties are determined by the condition of thermodynamic equilibrium even if the quantity to be measured is of kinetic origin, as in the case of the electrical conductivity. However, for some quantities, the signals to be measured are either small and/or the fluctuation contribution has to be subtracted from a large background. Moreover, the origin and temperature dependence of such background signal are unknown or difficult to determine. In superconductors, this is the case of the fluctuation contribution to the specific heat jump. A typical set of data reported by Roulin, Junod and Walker (1996) on the cuprate $YBa_2Cu_3O_{6+x}$ is shown in Fig. 3. In the case of other quantities, such as the magnetic susceptibility and the electrical conductivity, the fluctuation contribution is usually large and the background can be subtracted more easily, since it is well approximated by a constant or by a linear temperature dependence. An example of rounding of the superconducting transition of the resistivity in $YBa_2Cu_3O_{6+x}$ is shown in Fig. 4. Fluctuation measurements of the specific heat and of the conductivity have been reported by several groups on various compounds of cuprates and the results obtained exhibit good reproducibility. A comprehensive discussion on the results of specific heat measurements and on their interpretation was reported by Junod (1996), while a critical analysis of the experimental data on fluctuation conductivity was given by Gauzzi and Pavuna (1995).

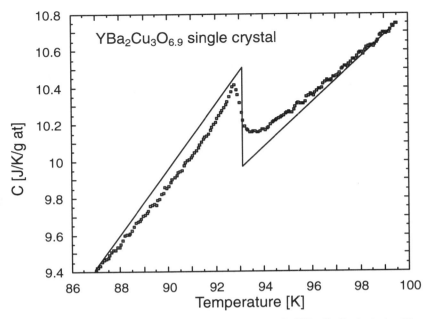

Figure 3. Experimental dependence of the specific heat of a single crystal of $YBa_2Cu_3O_{6.9}$ in the transition region. One notes the rounding of the jump at the transition attributed to the fluctuations of the order parameter. The solid line schematically represents the mean–field jump expected in the absence of fluctuations (after Roulin, Junod and Walker (1996)).

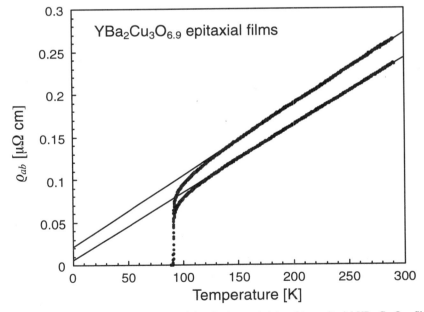

Figure 4. Experimental temperature dependence of the *ab*–plane resistivity of two epitaxial $YBa_2Cu_3O_{6.9}$ films. As in the preceding figure, the rounding of the jump at the transition is attributed to the fluctuations of the order parameter. In this case, the fluctuation contribution can be easily determined by extrapolating to low temperatures the linear temperature dependence at high temperatures where fluctuations are assumed to be absent. The fit of this linear dependence is represented by the solid lines. (Reproduced by permission of the American Physical Society from Gauzzi and Pavuna (1995)).

The analysis of experimental fluctuation data within the framework of the theories mentioned above is straightforward thanks to the simple universal predictions summarised in Table 1. The analytic derivation of these predictions can be found in any textbook. Here we just mention that the fluctuation of any physical property can be derived from the partition function of eq. (14) or from the incomplete thermodynamic potential of eq. (6). In the following, we give the result obtained in the regime of validity of the Gaussian approximation and in the region $T > T_c$ for 1) the fluctuation contribution $\delta \Delta C$ to the specific heat jump and 2) the fluctuation contribution $\delta \sigma$ to the electrical conductivity.

4.1. Specific heat jump

In first approximation, the fluctuating part $\delta \Phi$ of the incomplete thermodynamic potential is equal to the fluctuating part of the free–energy. Thus, by definition

$$\delta \Delta C = - \frac{1}{T_c} \frac{\partial^2 \delta \Phi}{\partial t^2} \tag{16}$$

By inserting eq. (9) in the above expression and approximating the series with an integral, the result is (Levanyuk, 1963)

$$\delta \Delta C = \frac{2V^3 a^2}{k_B T_c^2} \int_{|\mathbf{k}| \leq |\tilde{\mathbf{k}}|} \langle |\delta \psi_{\mathbf{k}}|^2 \rangle^2 \frac{d^3 k}{(2\pi)^3} \tag{17}$$

The same result was obtained by Aslamazov and Larkin (1968) from the microscopic BCS theory. In the case of sufficiently large ξ_0, the product $\tilde{\xi} k \gg 1$ at all temperatures in the Ginzburg–Landau region $t \ll 1$, hence the limit of integration can be extended toward infinity. Levanyuk's *universal* result is obtained

$$\delta \Delta C = \frac{k_B}{16\pi} \frac{V}{\tilde{\xi}_{x,0} \tilde{\xi}_{y,0} \tilde{\xi}_{z,0}} \frac{1}{\sqrt{t}} \tag{18}$$

Similar results are obtained in other dimensions and in the region $T < T_c$. We note that the critical exponent depends only on the dimension of space d, while the amplitude of the fluctuations depends on the microscopic parameters of the Ginzburg–Landau Hamiltonian, which are characteristic of the system.

4.2. Electrical conductivity. Paraconductivity

As in the case of the specific heat, also the electrical conductivity is usually enhanced by a term $\delta \sigma$ due to thermodynamic fluctuations near a superconducting transition. This phenomenon is called *paraconductivity*, in analogy with paramagnetism. This enhancement appears as a rounding of the resistivity jump at T_c, which is visible in Fig. 4. This phenomenon was experimentally observed for the first time by Glover (1967) in stabilised amorphous bismuth films over a temperature range of ≈ 3 mK. The experiment was suggested by Ferrel and Schmidt (1967) to verify their predictions on critical phenomena near the superconducting transition. These predictions were based on scaling theory (Widom, 1965; Kadanoff, 1966). Hence, they are valid in the region of critical fluctuations like, for example, in the region near the lambda point of ^{4}He. Ferrel and Schmidt (1967) proposed that $\delta \sigma$ diverges near T_c as the correlation length $\delta \sigma \sim \xi \sim t^{-2/3}$. Later, Lobb (1987) proposed the same dependence and a cross–over of the critical exponent to $-1/3$ in the immediate vicinity of the transition on the basis of the full dynamic theory by Hohenberg and Halperin (1977). Glover's results motivated Aslamazov and Larkin (1968) to formulate predictions on the effects of the fluctuations in the framework of the BCS theory. They calculated the fluctuation contribution to the electrical conductivity, to the specific heat and to the sound absorption coefficient in first approximation of perturbation theory. Their results can be obtained also from the Ginzburg–Landau time–dependent equations, as showed for the first time by Abrahams and Woo (1968); the same results have indeed been partly obtained by Levanyuk (1963) with this method.

In what follows, only the results by Aslamazov and Larkin are recalled. They found that the most singular diagram describing the response function operator $\delta \hat{Q}_\omega \equiv i\omega \delta \hat{\sigma}_\omega$ corresponds to the excess conductivity of fluctuating Cooper pairs. The contribution of this diagram to the dc conductivity ($\omega = 0$) is found to have the following expression

$$\delta\sigma \propto -\int_{|\mathbf{k}| \leq |\bar{\mathbf{k}}|} (\mathbf{k}\mathbf{u})\left(\mathbf{u}\frac{\partial}{\partial\mathbf{k}}\langle|\delta\psi_\mathbf{k}|^2\rangle^2\right)\frac{d^d k}{(2\pi)^d} \tag{19}$$

where $\mathbf{u}$ denotes the unit vector along the electric field vector. As in the previous case, a *universal* result is obtained in the long–wavelength limit $\xi\bar{k} \gg 1$. Within this approximation, by inserting eq. (9) into the above integral, Aslamazov and Larkin obtained the result reported in Table 1

$$\delta\sigma \sim t^{d/2-2} \tag{20}$$

The proportionality factor is

$$\begin{cases} \dfrac{e^2}{32\hbar\xi_0} & d = 3 \\[2mm] \dfrac{e^2}{16\hbar s} & d = 2 \\[2mm] \dfrac{\pi e^2 \xi_0}{16\hbar S} & d = 1 \end{cases} \tag{21}$$

where $s \ll \xi$ and $S \ll \xi^2$ are respectively the thickness of the film and the cross–sectional area of the whisker in the two– and one–dimensional cases respectively. As in the case of the specific heat, we note that the amplitude of the fluctuating term depends on the microscopic parameters of the system. Besides the diagram corresponding to eq. (19), the Aslamazov–Larkin theory predicts the existence of three other diagrams which are expected to give smaller contributions under ordinary conditions. Two of them give a negative contribution, which arises from the reduction of normal carrier density due to the appearance of fluctuating Cooper pairs. The third diagram gives an additional positive (paraconductivity) contribution, called Maki–Thompson or 'indirect' contribution, since it is associated with the interaction of fluctuating Cooper pairs with normal carriers. The reader can refer to Skocpol and Tinkham (1975) and to the original papers by Maki (1968; 1971) and by Thompson (1970) for further reading.

5. Application of standard fluctuation models to conventional superconductors

Fluctuation effects in clean three–dimensional superconductors of the conventional BCS type are found to be too small to be observed, in agreement with eq. (11) and eq. (13). These effects are enhanced in dirty samples and in samples with reduced dimensionality and were indeed detected in stabilised amorphous films of simple metals, such as Bi, Pb and Ga, and alloys, such as Bi–Sb. The most studied properties were the fluctuation of specific heat, electrical conductivity and magnetic susceptibility (fluctuation diamagnetism), probably because the effects observed are larger and more easily interpreted than in the case of other properties. In most reports, the experimental results were analysed within the framework of models based on the phenomenological Ginzburg–Landau theory or on the microscopic BCS theory. We recall that these two theories are equivalent in the transition region (Gor'kov 1959). The predictions of these theories, which are summarised above for the specific heat and for the electrical conductivity, are based on the simplest Gaussian approximation. Nevertheless, they are successful in accounting both qualitatively and quantitatively for the experimental data of most fluctuation properties, as reported in the review paper by Skocpol and Tinkham (1975). In particular, we mention the report by Glover (1969) showing the agreement between the universal Aslamazov–Larkin prediction in two–dimensions for the paraconductivity (see eq. (20) and eq. (21)) and the experimental behaviour observed in a series of amorphous films. More recently, it was reported by Xiang et al. (1993) that also Rb– and K–doped C_{60} follow the above predictions in three–dimensions, suggesting a rather conventional BCS behaviour in this new class of superconductors. On the other hand, experimental evidence for critical fluctuations of the conductivity according to the predictions by Ferrel and Schmidt (1967) was reported by Glover (1967) on amorphous Bi films in a temperature region a few mK above $T_c \approx 6$ K. Also in the case of the fluctuation specific heat, the universal prediction based on the Gaussian approximation (see eq. (18)) was found to account for the experimental data on conventional superconductors, such as Bi–Sb films (Zally and Mochel, 1971;1972). The same authors reported experimental evidence for a crossover to critical fluctuations in a temperature region close to the transition, as in the case of the fluctuation conductivity. On the other hand, a disagreement between predictions based on the Gaussian approximation and experimental data on conventional BCS superconductors was observed in the case of fluctuation diamagnetism. A universal prediction based on the Ginzburg–Landau theory was formulated by Prange (1970) after the first measurements reported by Gollub, Beasley, Newbower and Tinkham (1969) on several dirty metals. The departure of the experimental data from the above prediction was explained by taking into account a short wavelength cutoff in the fluctuation spectrum (Gollub, Beasley and Tinkham, 1970), thus suggesting the breakdown of the long–wavelength approximation. Similar conclusions were drawn by Johnson, Tsuei and Chaudhari (1978) to account for analogous deviations from the universal Aslamazov–Larkin predictions observed in the paraconductivity of amorphous alloys. We note that, in both cases, such cutoff effects were observed in samples where the correlation length was greatly reduced by the amorphous structure down to 5–10 nm.

6. Application of standard fluctuation models to cuprates. Controversial results

In the preceding section, we have mentioned that some universal predictions based on the Gaussian approximation fail to account for the experimental behaviour of fluctuations in amorphous samples of conventional BCS superconductors with $\xi_0 \approx 5$–10 nm. In this section, we discuss the same issue for cuprates. For these superconductors we expect an even larger departure of the experiments from the above universal predictions because of the extremely short correlation length. The short ξ follows directly from the high T_c through Heisenberg's uncertainty principle $\xi_0 \approx \hbar v_F / \Delta$, where v_F is the Fermi velocity and $\Delta \sim T_c$ is the superconducting energy gap. The expected small ξ is confirmed by susceptibility (Lee, Klemm and Johnston, 1989), specific heat (Lee and Ginsberg, 1992) and magnetoresistance (Matsuda, Hirai and Komiyama, 1988) measurements and is estimated to be ~1 nm in the ab plane of the CuO_2 planes and ~0.1 nm in the perpendicular (or c) direction. Under these conditions, *we expect that the universality principle and the scaling hypothesis are no longer valid*, at least in the c direction. In spite of this, with the exception of two studies (Freitas, Tsuei and Plaskett, 1988; Hopfengärtner, Hensel and Saemann–Ischenko, 1991), in previous reports, critical phenomena in cuprates have been extensively studied in several compounds by assuming the universal predictions of mean–field and scaling theories as valid. In these reports, the application of the models to the experimental data was aimed at studying the dimensionality d of the fluctuations by taking into account the dependence of the critical exponents on d (see Table 1) and at verifying whether the fluctuations are weak (Gaussian) or strong (critical). The motivation for these studies is that the electronic structure of cuprates in their normal state is quasi–two dimensional due to the presence of electronically active CuO_2 planes in their crystal structure. Therefore the role of the out–of–plane direction in determining the superconducting properties of these materials is considered as one of the fundamental open question related to high–temperature superconductivity in cuprates. Second, the short coherence length of cuprates combined with their small density of Cooper pairs has raised the question of the existence of critical fluctuations in the whole fluctuation region and, consequently, the breakdown of the mean–field regime. The experimental results of fluctuation measurements reported show a good reproducibility but the conclusions are highly controversial. The controversy concerns not only the dimensionality of the fluctuation spectrum, but also the number of components of the order parameter and the possible relevance of critical fluctuations. In what follows, we mention the most controversial points concerning 1) the specific heat in zero magnetic field and 2) in magnetic field; 3) the magnetic penetration depth and 4) the paraconductivity.

1) Specific heat in zero magnetic field. According to Table 1, in the case of Gaussian fluctuations, we would expect no divergence of the mean–field jump and a divergence of the fluctuation contribution with the critical exponent predicted by eq. (19). On the other hand, in the case of critical fluctuations described by the 3*d*XY model, a logarithmic divergence of the jump is expected. Most recent experimental data on $YBa_2Cu_3O_{6+x}$ single crystals reported by several groups are all in better agreement with the latter prediction than with the first one, as discussed in the review article by Junod (1996).

2) Specific heat in magnetic field. Both theories mentioned above predict a universal scaling of the specific heat jump as a function of magnetic field. Also in this case, controversial interpretations were reported: the papers by Salamon, Shi, Overend and Howson (1993) and by Overend, Howson and Lawrie (1994) supported a 3*d*XY scaling picture, while Roulin, Junod and Muller (1995) and Junod (1996) showed that neither the Ginzburg–Landau nor the 3*d*XY scaling theory can be ruled out, as in the case of zero–field.

3) Magnetic penetration depth. Near the transition, the magnetic penetration depth is expected to diverge with a critical exponent –1/2 or –1/3 according to the Ginzburg–Landau and 3*d*XY scaling theories respectively (see Table 1). Measurements consistent with the former were reported on $YBa_2Cu_3O_{6+x}$ by Fiory, Hebard, Mankiewich and Howard (1988) and by Lin *et al.* (1995), while Kamal *et al.* (1994) reported data supporting 3*d*XY behaviour.

4) Fluctuation conductivity. Paraconductivity. Several theoretical and experimental works on the paraconductivity in cuprates have been reported. Most experimental data were on $YBa_2Cu_3O_{6+x}$ but some data are available also on $La_{2-x}Ba_xCuO_4$ and on $Bi_2Sr_2CaCu_2O_{8+\delta}$. The experimental data show excellent reproducibility. The most recent data are those on high quality untwinned $YBa_2Cu_3O_{6+x}$ single crystals by Pomar et al. (1993), Holm, Eltsev and Rapp (1995) and Kim, Goldenfeld, Giapintzakis and Ginsberg (1997). A summary on this subject goes beyond the scope of this work and can be found elsewhere (Gauzzi and Pavuna, 1995). Here we mention the most important and controversial conclusions reported in the literature. As to the nature of the fluctuations, a few reports support the existence of critical fluctuations in zero magnetic field sufficiently close to T_c (Pomar et al., 1993; Kim, Goldenfeld, Giapintzakis and Ginsberg, 1997), while most reports are in favour of a predominant Aslamazov–Larkin contribution. In most reports the additional Maki–Thompson contribution was found to be negligible, in agreement with the original argument by Aslamazov and Larkin. Moreover, pair–breaking is expected to be important in all cuprates because of the large amount of impurities and disorder or of the large number of phonons at temperatures ≈ 100 K, which leads to a vanishing Maki–Thompson contribution. As to the dimensionality *d* of the fluctuation spectrum, all possible pictures have been supported in the case of $YBa_2Cu_3O_{6+x}$, according to the universal predictions by Aslamazov and Larkin: 1) 3*d*–fluctuations (Freitas, Tsuei and Plaskett, 1988; Xi *et al.*, 1989; Veira and Vidal, 1989; Sudhakar *et al.*, 1991); 2) 2*d*–fluctuations (Ausloos and Laurent, 1988; Hagen, Wang and Ong, 1988; Lang *et al.*, 1991); 1*d*–fluctuations (Ying and Kwok, 1990); 4) a cross–over from 2*d* to 3*d* associated with the layered structure of the cuprates (Oh *et al.*, 1988; Friedmann, Rice, Giapintzakis and Ginsberg, 1989; Gasparov, 1991; Baraduc *et al.*, 1992).

7. Evidence for the breakdown of the universality principle in short$-\xi$ systems

On the basis of the above summary of the state of the art on thermodynamic fluctuations in cuprates, we conclude that this field remains highly controversial. We argue that the controversy arises from three circumstances: 1) the different methods used to analyse the experimental data; 2) the extrinsic broadening of the transition due to local inhomogeneities and spatial distribution of T_c. In YBa$_2$Cu$_3$O$_{6+x}$, this can be easily produced by local deviations of the oxygen content x; 3) the assumption of validity of the universality principle to account for critical phenomena in cuprates, while this principle cannot be applied to systems with sufficiently short ξ. The first point is discussed in detail elsewhere (Gauzzi and Pavuna, 1995). The second point does not deserve any further comment. In this section, we draw our attention to the third point. In particular, we show that the controversial results obtained in cuprates can be explained by noting that both requirements $\tilde{\xi}k \gg 1$ and low anisotropy, which justify the universality principle and the scaling hypothesis, are not fulfilled in these materials. It follows that the critical behaviour of all fluctuating quantities generally depends on the cutoff of the fluctuation spectrum (Ivanchenko and Lisyanskii, 1992; Gauzzi, 1993).

We shall discuss only the case of the fluctuation of the conductivity, since this is the only quantity for which the analysis of the experimental data is relatively simple. A numerical calculation of the Aslamazov–Larkin expression (see eq. (20) and eq. (21)) with the inclusion of such a cutoff was reported by Freitas, Tsuei and Plaskett (1988) and by Hopfengärtner, Hensel and Saemann–Ischenko (1991) to account for the deviation of their experimental data on YBa$_2$Cu$_3$O$_{6+x}$ from the universal Aslamazov–Larkin prediction in the long wavelength approximation $\tilde{\xi}k \gg 1$. This generalised approach did indeed improve the agreement between theory and experiment at high temperatures but not at low temperatures, i.e. in the region close to the transition. In this region, the above authors assumed that the universal behaviour predicted by the long–wavelength approximation is recovered.

This discrepancy can be explained by an exact calculation of the Aslamazov–Larkin term (Gauzzi, 1993) which shows that, in the case of strongly anisotropic systems like cuprates, cutoff effects can be relevant not only at high temperatures, but also very close to the transition. In this case, a deviation from the universal predictions of eq. (20) and of eq. (21) is expected in the *whole* fluctuation region accessible to experiments. This possibility was indeed verified experimentally in YBa$_2$Cu$_3$O$_{6+x}$ (Gauzzi and Pavuna, 1995), as we can note from Fig. 5a. The same data are found to deviate also from the universal prediction of the scaling theory (see Table 1) in the whole fluctuation region down to 5 mK above T_c (see Fig. 5b). On the other hand, a quantitative account for the experimental data is given by the aforementioned exact calculation at both high and low temperatures down to ≈ 0.5 K above T_c (see Fig. 6a–b). We note that the results of the foregoing analysis are independent of the procedure of data analysis. In particular, the absence of a well–defined critical exponent is evident in the linear plot of the raw experimental data (see Fig. 5a–b). Such characteristic behaviour is insensitive to sample quality and has been reported also on best quality untwinned single crystal. This behaviour can be masked by the usual procedure of determining critical exponents from logarithmic plots, which can be misleading, since the slope of these plots is affected by the choice of T_c. Nevertheless, the use of logarithmic plots turns out to be useful *after* the data analysis to put into evidence the absence of a well–defined critical exponent (see Fig. 6b).

In order to compare the present analysis with apparently different analysis reported in the literature, we note that the analysis of the experimental data within the framework of the Lawrence–Doniach model of layered superconductors (Lawrence and Doniach, 1970) falls into the category of the present cut off approach, since the Lawrence–Doniach model is a particular cut off model where the short wavelength cut off of the fluctuation spectrum is $\tilde{\lambda} = s/2$, s being the spacing between adjacent superconducting layers (Gauzzi, 1993). More general Lawrence–Doniach models take into account the existence of different spacings within the unit cell of the cuprates (Klemm, 1990; Ramallo, Pomar and Vidal, 1996).

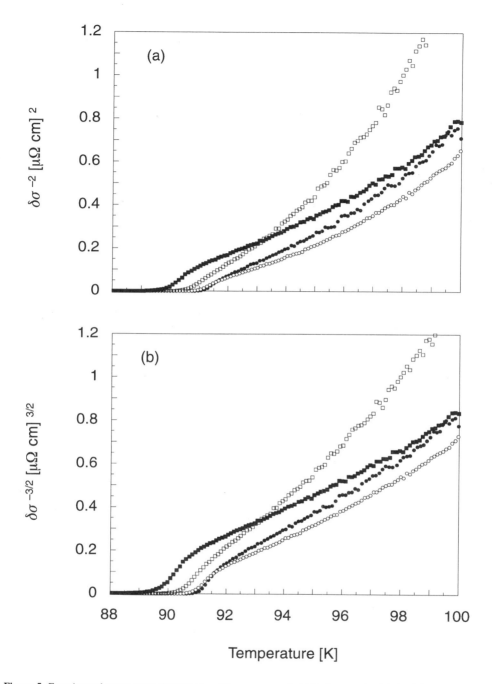

Figure 5. Experimental temperature dependence of the excess conductivity (paraconductivity) of four epitaxial $YBa_2Cu_3O_{6+x}$ films. **a)** The data are plotted with exponent 2 to verify a possible linear temperature dependence, which would support the universal prediction of mean–field theory in three dimensions (see Table 1, eq. (20) and eq. (21)). **b)** The same as in a) for the prediction of scaling theory $\delta\sigma \sim t^{-2/3}$ (see Table 1). Neither in a) nor in b) a linear dependence is observed, similarly to the case of the predictions of mean–field theory in one and two dimensions, which are not reported here. This analysis suggests the failure of the universality principle in $YBa_2Cu_3O_{6.9}$. (Reproduced by permission of the American Physical Society from Gauzzi and Pavuna (1995)).

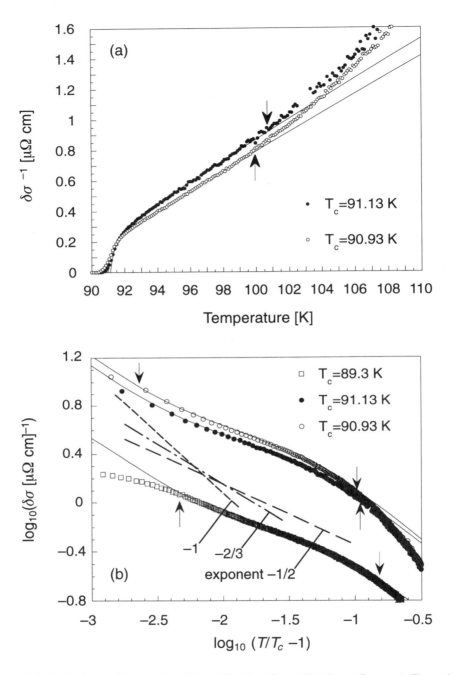

Figure 6. Analysis of some of the experimental data of the above figure within the cutoff approach. The result of the analysis is represented in both linear (a) and logarithmic (b) scales. In b), the deviations of the data from the universal predictions of mean–field and scaling theories are easily noted. These predictions are represented schematically by the broken lines with slopes −1/2, −2/3 and −1, corresponding to the critical exponents predicted by respectively mean–field theory in 3*d*, scaling theory in 3*d* and mean–field theory in 2*d* (see Table 1). The solid lines represent a fit of the data by using the exact calculation of the Aslamazov–Larkin term with cutoff included (Gauzzi 1993). The arrows indicate the range of temperatures where the data have been analysed. The values of the critical temperatures of the samples are indicated by the legend. (Reproduced by permission of the American Physical Society from Gauzzi and Pavuna (1995)).

The above results indicate that, in the short$-\xi$ superconductor $YBa_2Cu_3O_{6+x}$, the correlation length ξ of the fluctuations never becomes sufficiently large with respect to the interatomic distance to make the critical behaviour universal, i.e. independent of the microscopic interaction driving the transition. An experimental support for this picture could be provided by a study of the disorder–induced reduction of T_c in $YBa_2Cu_3O_{6+x}$ films (Gauzzi and Pavuna, 1997; Gauzzi, Jönsson, Clerc–Dubois and Pavuna, 1997). In these reports, we found that the minimum size r_c of structurally ordered domains necessary for obtaining the maximum value of T_c (≈ 91 K) is $r_c \approx 14$ nm (see Fig. 7). In more disordered samples, r_c is reduced and a rapid reduction of T_c is observed. The interpretation of these data is that the correlation length ξ of the superconducting order parameter in copper–oxide metals is limited by the value of the structural correlation length r_c. This is due to the strong directionality of the $3d$ orbitals of the copper in the CuO_2 planes forming the conduction band, while in conventional metals the conduction band is formed by the isotropic s–orbitals or by the less directional p–orbitals. As a result, the divergence of the superconducting correlation length ξ at the transition (see eq. (12)) is cut off by the value of the structural correlation length r_c, thus explaining the observed reduction of T_c in disordered samples. By assuming that this interpretation is valid, we conclude that r_c is a probe of the divergence of ξ at the transition. From the diagram of Fig. 7, we estimate that ξ does not become larger than ≈ 14 nm at the transition in well–ordered $YBa_2Cu_3O_{6+x}$ with $T_c \approx 91$ K. This is a surprisingly small value, which suggests the breakdown of the universality principle and of the scaling hypothesis in cuprates, in agreement with the experimental observations discussed above.

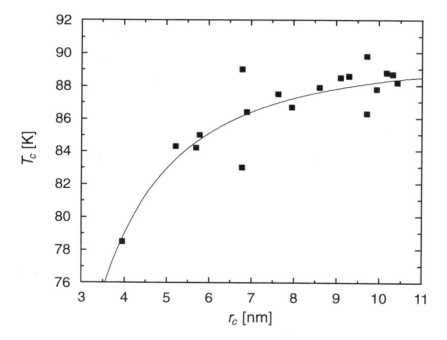

Figure 7. Experimental dependence of the disorder–induced reduction of critical temperature as a function of structural correlation length r_c in epitaxial $YBa_2Cu_3O_{6.9}$ films. The solid line represents the predicted behaviour by assuming that the square–root divergence of the Ginzburg–Landau superconducting coherence length ξ at the transition is limited by r_c. According to this interpretation, the above data indicates that, in $YBa_2Cu_3O_{6.9}$, ξ increases up to only ≈ 14 nm at the transition in well–ordered samples (Gauzzi, Jönsson, Clerc–Dubois and Pavuna (1997)).

8. Summary. Discussion and conclusions

In this paper we have presented experimental evidence for the deviation of the fluctuation behaviour of the cuprate superconductor $YBa_2Cu_3O_{6+x}$ from the universal predictions derived from both mean–field and scaling theories. Similar deviations were observed in other cuprates, such as $La_{2-x}Ba_xCuO_4$ and $Bi_2Sr_2CaCu_2O_{8+\delta}$ and also in amorphous alloys. In particular, no well–defined critical exponents of the fluctuation conductivity (paraconductivity) were found in the whole fluctuation region accessible to experiments by using a method of data analysis which is independent of any arbitrary choice of T_c. We have explained such a breakdown of universal predictions by applying an exact cutoff scheme to the mean–field fluctuation spectrum. The relevance of cutoff effects in cuprates is due to the extremely short correlation length ξ of the order parameter in these materials. This implies that ξ remains too small at the transition to allow the application of the universality principle to the critical behaviour at the transition. This circumstance is supported by an independent study carried out on the disorder–induced reduction of T_c in $YBa_2Cu_3O_{6+x}$ films. This study suggests that ξ increases up to only ≈ 14 nm at the transition in well–ordered $YBa_2Cu_3O_{6+x}$. This microscopic size is unusual for a coherent quantum state such as the superconducting state. Therefore, the expression 'divergence of the correlation length at the transition', which is commonly used in the field of second order phase transitions, does not seem to be appropriate for cuprates.

The loss of universality induced by the short ξ is not a negative statement as it might seem at first glance. Although the universality class of the effective Hamiltonian can no longer be recognised because of the non–universal behaviour at the transition, this behaviour is interesting, since it is distinctive of the type and of the parameters of the microscopic interaction driving the transition. For example, our study of the fluctuation spectrum within the cutoff approach leads to the *qualitative* conclusion that the fluctuations are rapidly attenuated with decreasing wavelength as compared to the case of conventional superconductors with long ξ (see Fig. 2). This conclusion is in agreement with previous strong–coupling calculations of the BCS–Eliashberg equations (Bulaevskii and Dolgov, 1988). Hence, the microscopic picture emerging from our fluctuation data is that of pair correlations between localised states, such as those predicted by the bipolaron theory of superconductivity of Alexandrov and Mott (1994).

We recall that our fluctuation measurements probe only the integral of the fluctuation spectrum over the entire momentum space (see eq. (17) and eq. (19)). It would therefore be essential to determine more precisely the momentum dependence of the spectrum of cuprates and compare it with the Lorentzian dependence of the conventional mean–field spectrum (see eq. (11)). This would enable us to determine the corrective terms of the conventional BCS Hamiltonian which are expected to give a better description of the superconducting state of cuprates. In this respect, a systematic study of other fluctuation properties, such as the fluctuation diamagnetism, would be important. Fluctuation studies on other short–ξ superconductors, such as Bi–based oxides, Chevrel phases and alkali–doped C_{60} would also be useful for exploring possible new aspects of the physics of phase transitions in short–ξ systems.

The author acknowledges A. Junod for providing the experimental data of Fig. 3, M. Bauer, V. A. Gasparov, M. Howson, D. Pavuna, A. Rigamonti and A. A. Varlamov for useful discussions and J. Berrocosa for his experimental help.

References

Abrahams, E. and Woo, 1968, J., Phys. Lett. **27A**, 117.
Alexandrov, A. S. and Mott, 1994, N., Rep. Prog. Phys. **57**, 1197.
Aslamazov, L. G. and Larkin, A.I., 1968, Fiz. Tverd. Tela **10**, 1104 [Sov. Phys. Solid State **10**, 875 (1968)].
Ausloos, M. and Laurent, Ch., 1988, Phys. Rev. B **37**, 611.
Baraduc, C., Pagnon, V., Buzdin, A., Henry, J. Y. and Ayache, C., 1992, Phys. Lett. A **166**, 267.
Bulaevskii, L. N. and Dolgov, O. V., 1988, Solid State Commun. **67**, 63.
Coleman, S. and Weinberg, E., 1973, Phys. Rev. D **7**, 1988.
de Gennes, P. G., 1972, Solid State Commun. **10**, 753.
Ferrel, R. A. and Schmidt, H., 1967, Phys. Lett. **25A**, 544.
Fiory, A. T., Hebard, A. F., Mankiewich, P. M. and Howard, R. E., 1988, Phys. Rev. Lett. **61**, 1419.
Freitas, P. P., Tsuei, C. C. and Plaskett, T. S., 1988, Phys. Rev. B **36**, 833.

Friedmann, T. A., Rice, J. P., Giapintzakis, J. and Ginsberg, D. M., 1989, Phys. Rev B **39**, 4258.

Gasparov, V. A., 1991, Physica C **178**, 445.

Gauzzi, A., 1993, Europhys. Lett. **21**, 207.

Gauzzi, A. and Pavuna, D., 1995, Phys. Rev. B **51**, 15420.

Gauzzi, A. and Pavuna, D., 1997, J. of Superconductivity, in press.

Gauzzi, A., Jönsson, B. J., Clerc–Dubois, A. and Pavuna, D., 1997, preprint.

Ginzburg, V. L., 1960, Fiz. Tverd. Tela **2**, 2031 [Sov. Phys. Solid State **2**, 1824 (1960)].

Glover, R. E., 1967, Phys. Lett. **25A**, 542.

Glover, R. E., 1969, *Proc. 11th Int. Conf. on Low Temperature Physics, St. Andrews, 1968,* Allen, J. F., ed.

Gollub, J. P., Beasley, M. R., Newbower, R. S. and Tinkham, M., 1969, Phys. Rev. Lett. **22**, 1288.

Gollub, J. P., Beasley, M. R. and Tinkham, M., 1970, Phys. Rev. Lett. **25**, 1646.

Gor'kov, L. P., 1959, Z. Eksp. Theor. Fiz. **36**, 1918 [Sov. Phys. JETP **36**, 1364 (1959)].

Hagen, S. J., Wang, Z. Z. and Ong, N. P., 1988, Phys. Rev. B **38**, 7137.

Hohenberg, P. C. and Halperin, B. I., 1977, Rev. Mod. Phys. **49**, 435.

Holm, W., Eltsev, Yu. and Rapp, Ö, Phys. Rev. B **51**, 11992 (1997).

Hopfengärtner, R., Hensel, B. and Saemann–Ischenko, G., 1991, Phys. Rev. B **44**, 741.

Ivanchenko, Yu. M. and Lisyanskii, A. A., 1992, Phys. Rev. A **45**, 8525.

Johnson, W. L., Tsuei, C. C. and Chaudhari, P., 1978, Phys. Rev. B **17**, 2884.

Junod, A., 1990, in *Physical Properties of High Temperature Superconductors II*, Ginsberg, D. M. ed.
(World Scientific, Singapore).

Junod, A., 1996, in *Studies of High Temperature Superconductors Vol. 19*, Narlikar, A. V., ed. (Nova
Science, Commack New York).

Kadanoff, L. P., 1966, Physics **2**, 263.

Kamal, S., Bonn, D. A., Goldenfeld, N., Hirschfeld, P. J., Liang, R. and Hardy, W. N., 1994, Phys. Rev. Lett.
73, 1845.

Kapitulnik, A., Beasley, M. R., Di Castro, C. and Castellani, C., 1988, Phys. Rev. B **37**, 537.

Kim, J.–T., Goldenfeld, N., Giapintzakis, J. and Ginsberg, D. M., Phys. Rev. B **56**, 118 (1997).

Klemm, R. A., Phys. Rev. B **41**, 2073 (1990).

Landau, L. D., 1937, Zh. Eksp. Theor. Fiz. **7**, 19 [Phys. Z. der Sowjet Union **11**, 26 (1937).

Landau, L. D. and Lifshitz, E. M., 1980, *Statistical Physics, part I* (Pergamon, Oxford).

Lang, W., Heine, G., Jodlbauer, H., Schlosser, V., Markowitsch W., Schwab, P., Wang, X.Z. and
Bäuerle, D., 1991, Physica C **185–189**, 1315.

Lawrence, S. and Doniach, W. E., *Proceedings 12th International Conference on Low Temperature Physics, Kyoto*
(1970), edited by Kanda, E. (Keigaku, Tokyo, 1971), p. 361.

Lee, W. C. and Ginsberg, D. M., 1992, Phys. Rev. B **45**, 7402.

Lee, W. C., Klemm, R. A. and Johnston, D. C., 1989, Phys. Rev. Lett. **63**, 1012.

Levanyuk, A. P., 1959, Zh. Eksp. Teor. Fiz. **36**, 810 [Sov. Phys. JETP **36**, 571 (1959)].

Levanyuk, A. P., 1963, Fiz. Tverd. Tela **5**, 1776 [Sov. Phys. Solid State **5**, 1294 (1964)].

Lin, Z.–H, Spalding, G. C., Goldman, A. M., Bayman, B. F. and Valls, O.T., 1995, Europhys. Lett. **32**, 573.

Lobb, C. J., 1987, Phys. Rev. B **36**, 3930.

Ma, S.–K., 1976, *Modern Theory of Critical Phenomena* (Benjamin, New York).

Maki, K., 1968, Prog. Theor. Phys. **40**, 193.

Maki, K., 1971, Prog. Theor. Phys. **45**, 1016.

Matsuda, Y., Hirai, T. and Komiyama, S., 1988, Solid State Commun. **68**, 103.

Oh, B., Char K., Kent, A. D., Naito, M., Beasley, M. R., Geballe, T. H., Hammond, R. H., Kapitulnik, A.
and Graybeal, J. M., 1988, Phys. Rev. B **37**, 7861.

Overend, N., Howson, M. A. and Lawrie, I. D., 1994, Phys. Rev. Lett. **72**, 3238.

Patashinskii, A. Z. and Pokrowskii, V. L., 1979, *Fluctuational Theory of Phase Transitions* (Pergamon,
Oxford).

Pomar, A., Díaz, A., Ramallo, M. V., Torrón, C., Veira, J. A. and Vidal, F., Physica C **218**, 257 (1993).

Prange, R. E., 1970, Phys. Rev. B **1**, 2349.

Ramallo, M. V., Pomar, A. and Vidal, F., Phys. Rev. B **54**, 4341 (1996).

Roulin, M., Junod, A. and Muller, J., 1995, Phys. Rev. Lett. **75**, 1869.

Roulin, M., Junod, A. and Walker, E., 1996, Physica C **260**, 257.

Salamon, M. B., Shi, J., Overend, N. and Howson, M. A., 1993, Phys. Rev. B **47**, 5520.

Skocpol, W. J. and Tinkham M., 1975, Rep. Prog. Phys. **38**, 1049.

Stanley, H. E., 1971, *Introduction to Phase Transitions and Critical Phenomena* (Oxford University Press,
Oxford).

Sudhakar, N., Pillai, M. K., Banerjee, A., Bahadur, D., Das A., Gupta, K. P., Sharma, S. V. and Majumdar,
A. K., 1991, Solid State Commun. **77**, 529.

Thompson, R. S., 1970, Phys. Rev. B **1**, 327.

Veira, J. A. and Vidal, F., 1989, Physica C **159**, 468.

Widom, B., 1965, J. Chem. Phys. **43**, 3892.

Wilson, K. G., 1971, Phys. Rev. B **4**, 3174.

Xi, X. X., Geerk, J., Linker, G., Li, Q. and Meyer, O., 1989, Appl. Phys. Lett. **54**, 2367.

Xiang, X. D., Hou, J. G., Crespi, V. H., Zettl, A. and Cohen, M., 1993, Nature **361**, 54.

Ying, Q. Y. and Kwok, H. S., 1990, Phys. Rev. B **42**, 2242.

Zally, G. D. and Mochel, J. M., 1971, Phys. Rev. Lett. **27**, 1710.

Zally, G. D. and Mochel, J. M., 1972, Phys. Rev. B **6**, 4142.

MULTILAYERING EFFECTS ON THE THERMAL FLUCTUATIONS OF COOPER PAIRS AROUND THE SUPERCONDUCTING TRANSITION IN CUPRATES.

Félix Vidal and Manuel V. Ramallo.

Laboratorio de Bajas Temperaturas y Superconductividad,
Departamento de Física de la Materia Condensada,
Universidad de Santiago de Compostela, E-15706.
Spain

1. INTRODUCTION.

The importance of thermal fluctuations in the physics of the so-called high temperature copper oxide superconductors (HTSC), already suggested in the pioneer work of Bednorz and Müller,[1] is now well established.[2-7] However, in spite of considerable theoretical and experimental efforts on that subject during the last ten years,[2-7] some of the central aspects of thermal fluctuations in HTSC are still open and controversial. One of these important open questions is how the thermal fluctuations are affected by the presence of various superconducting CuO_2 (ab) layers in the periodicity length, s, which is equal to the crystallographic unit cell length in the c direction (perpendicular to the layers) if the cell is primitive, or half if it is body centered. The important role that the multilayering may play in the understanding of thermal fluctuations in HTSC may be illustrated with a paragraph of the chapter devoted to these materials in the second edition of Michael Tinkham's book on superconductivity:[8] *"Quantitative comparisons between the Lawrence-Doniach (LD) theory and the real materials are problematic because the crystallographic unit cell typically contains two or more inequivalently CuO_2 planes which, in principle, would require a generalization of the LD model; thus, even the appropriate choice of the interplane space s in the model is not well defined..."*. In other words, even in the simple case where there are not magnetic vortices and mainly due to the fact that the strength of the Josephson coupling between the various CuO_2 layers per s may be not much different one from each other (see Section 2), in multilayered HTSC the effective layering periodicity length for the fluctuations of the Cooper pairs, s_e^c, could be very different of s, and it will depend on the reduced temperature, $\varepsilon \equiv |T-T_{co}|/T_{co}$, and on the Josephson coupling strengths between adjacent superconducting planes (here T_{co} is the mean field like critical temperature in absence of an applied magnetic field). The presence of different magnetic vortices (three dimensional Abrikosov vortices, two dimensional pancakes and Josephson vortices) introduces new (magnetic) couplings between the layers and, therefore, they may also modify the effective periodicity length for thermal fluctuations of vortices, s_e^v, which may differ from both s and s_e^c.

Here, we will summarize some recent results on the role that the multilayering may play on the thermal fluctuations of Cooper pairs around the superconducting transition in

The Gap Symmetry and Fluctuations in High - T_c Superconductors
Edited by Bok *et al.*, Plenum Press, New York, 1998

HTSC. In addition to their intrinsic interest, these results directly concern other important open aspects of the HTSC, as the presence or not of appreciable pair breaking and density of states (DOS) effects, also a controversial issue related in turn to their superconducting pairing state,[9-12] or the behaviour of the magnetic vortices around T_C, another important problem directly related to some of the most promising applications of these materials.[1-8]

2. QUALITATIVE CONSIDERATIONS: WHY MULTILAYERING MAY SO DEEPLY AFFECT THE THERMAL FLUCTUATIONS IN HTSC?

2.1. WHY THERMAL FLUCTUATIONS ARE SO IMPORTANT IN HTSC?

To answer at a qualitative level the two questions addressed in the main title of this section, let us first remember that the thermal fluctuations effects are so important in these materials mainly due to the smallness of their superconducting coherence length amplitude (at 0 K), $\xi(0)$, which in general is in all directions only a few times bigger (or even less, in the c direction) than the interatomic distances. Together with their layered nature (see next subsection), this property leads from one side to a very small superconducting coherent volume (in some cases quasi bi-dimensional), with a relatively small number of Cooper pairs inside such a small coherent volume. This also leads, from the other side, to very thin magnetic vortices, the diameter of their normal cores when the magnetic field, H, is applied perpendicularly to the ab planes being of the order of $\xi_{ab}(T)$, the in-plane supercoducting coherence length. In addition, T_C is indeed relatively high in HTSC and, therefore, even quite far away from T_C (a few degrees) the corresponding condensation energy of these vortices or of the Cooper pairs per coherent volume will be of the order of even less than the available (gratis!) agitation thermal energy (which is of the order of $k_B T$, where k_B is the Boltzmann constant). This makes possible, therefore, the creation and the annihilation of Cooper pairs or of vortices by this thermal agitation energy. There are, indeed, other possible thermal fluctuation effects, as for instance those that will appear when the available thermal agitation energy is of the order of the elastic energy of these thin vortices or of the pancakes. They will then fluctuate around their equilibrium positions, or even the vortex lattice may melt.[2,4,5] The relative importance of both types of vortex fluctuations will depend on the strength of the applied magnetic field relative to the upper critical field $H_{c2}(T)$.

2.2. WHY THE LAYERED NATURE OF THE HTSC MAY AFFECT THE THERMAL FLUCTUATIONS ?

Let us also remember here that the layered nature of a single-layered superconductor (with only one layer per periodicity length, as the one schematized on the right side of Fig.1) could play a crucial role in the thermal fluctuations of Cooper pairs around T_C if the superconducting coherence length in the c direction (perpendicular to the layers), $\xi_c(T)$, became of the order of s, the periodicity length of the layers. For $\xi_c(\varepsilon) \ll s$, the thermal fluctuations of the Cooper pairs will be essentially bi-dimensional (2D), whereas for $\xi_c(\varepsilon) \gg s$ the fluctuations will be three-dimensional (3D). The model well adapted to analyze the Gaussian thermal fluctuations of Cooper pairs in these single-layered superconductors is the Lawrence-Doniach (LD) approach,[13] which considers that the layers are coupled by Josephson tunnelling. The LD model predicts that the crossover between the 2D and the 3D fluctuations occurs when the "crossover LD parameter", $B_{LD} \equiv (2\xi_c(0)/s)^2$ is equal to ε, i.e., when $2\xi_c(\varepsilon) = s$, where $\varepsilon \equiv |T-T_{co}|/T_{co}$ is the reduced temperature.

Let us also remember that for H ∥ c these single layered superconductors exhibit also a crossover in the magnetic field dependence:[7,14,15] For $H < H_o \approx \phi_o/\mu_o(\gamma_a s)^2$, the vortices extend over many layers (like the conventional Abrikosov vortices) and their thermal fluctuations will be 3D in nature. For $H > H_o$, the vortices are concentrate in each individual layer (pancakes) and their fluctuations will be essentially 2D (in the above expression of H_o, $\gamma_a \equiv \xi_{ab}/\xi_c$ is the anisotropy factor and $\phi_o \equiv \pi\hbar/e$ is the flux quantum, where e is the electron charge and $\hbar$ is the reduced Planck constant). Here again, depending on the value of $H/H_{c2}(\varepsilon)$, both types of vortices may essentially fluctuate in number or in position. In the Ginzburg-Landau-like descriptions these two types of vortex fluctuations will correspond to fluctuations of the order parameter amplitude or, respectively, the phase. The pancake fluctuations have been calculated in the single layered scenario by Bulaevskii and coworkers[16,17] (in the low magnetic field regime, which correspond to fluctuations of the vortex positions) and by Tešanović and coworkers[18] (in the high magnetic field limit, which will correspond to fluctuations of the pancake number).

2.3. WHY THE "FINE" STRUCTURE OF THE MULTILAYERED HTSC MAY AFFECT THE THERMAL FLUCTUATIONS AROUND T_c?

Most of the HTSC are not single layered, with only one layer per periodicity length (N = 1). Instead, they have various superconducting CuO_2 layers per periodicity length,

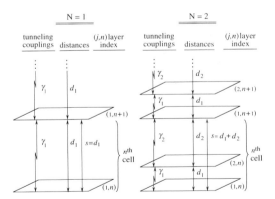

Figure 1. Schematic view of the superconducting layers in layered superconductors with one (N = 1) or two (N = 2) layers per unit cell length. The index (j,n) denotes the j = 1,..., N superconducting layer of the nth cell of length $s = d_1 +... d_N$. The tunnelling coupling constant between the (j+1,n) and the (j,n) layers is noted γ_j. Figure from Ref.11.

with possible different Josephson coupling strengths between layers. For instance, the YBa$_2$Cu$_3$O$_{7-\delta}$ (Y-123) or the Bi$_2$Sr$_2$CaCu$_2$O$_8$ (Bi-2212) compounds have two superconducting layers in s (N = 2), and their layering structure corresponds to the one schematized on the right side of Fig.1. *When such a multilayering structure will appreciably affect the intrinsic thermal fluctuations of Cooper pairs and of vortices?* In the next Section, we will summarize in more detail the available theoretical results on the multilayering effects on the Cooper pair fluctuations.[11,19-21] However, we may already obtain here a qualitative answer to that question by just considering two limiting cases of the bilayered scenario schematized in Fig.1. These two limits will depend on the relative Josephson coupling strength γ_1/γ_2 between adjacent superconducting layers. If $\gamma_1/\gamma_2 \to \infty$, the two more strongly coupled layers, with γ_1 as Josephson coupling strength, will behave as an unique single layer. This limit corresponds, therefore, to the conventional LD scenario with only a single layer per periodicity length. Such an approximation has been used, without other justification than its simplicity, by many authors[3] to analyze different in-plane effects of the thermal fluctuations of Cooper pairs in bilayered HTSC, such as the Y-123 and the Bi-2212 superconductors. The other limit is $\gamma_1/\gamma_2 \approx 1$. In this case the adequate approximation is again the conventional LD approach for single layered superconductors but with s/2 instead of s as characteristic length. We will see in the next Section that this apparently small modification has dramatic consequences on the physics of the thermal fluctuation effects in HTSC.

The results of Refs.11,21,24,25 and 27 that are going to be summarized in the next Section have also shown that such a LD-like scenario, but with $s_e^c = s/2$, is still an excellent approximation even for values of γ_1/γ_2 as high as 10^2. This conclusion clearly suggests that such a scenario not only is as simple as the conventional LD approach (with $s_e^c = s$), but also it is probably much better adapted to bilayered HTSC, as the Y-123 or the Bi-2212 compounds. The comparison with the experimental data summarized in the Section 4 fully confirms such a conclusion. So, we may conclude this paragraph by answering with a *practical rule* the question addressed above: In multilayering superconductors the *fine* layering structure is relevant for the in-plane thermal fluctuations around T$_c$ if in addition to $\xi_c(\varepsilon) \approx s$, also the relative Josephson coupling strengths between each superconducting CuO$_2$ layer with its two neighbour layers, γ_1 and γ_2, are not *very* different to each other.

In the case of the thermal fluctuation of vortices, the role played by the multilayering is still an open question. However, it was suggested[17,18] that these multilayering effects could be crudely accounted for by using some effective periodicity length s_e^v, instead of the crystallographic s, in the theoretical expressions calculated in the single layered scenario. The validity of this assumption, together with the value of s_e^v (or, equivalently, of the effective number, N_e^v, of independent fluctuating planes in s for the vortex fluctuation), has been analyzed recently through the crossing point of the magnetization in highly anisotropic multilayering HTSC.[22,23] Let us stress already here that the knowledge of s_e^v in multilayered HTSC (or, equivalently, of N_e^v) is important not only for the understanding of the thermal fluctuations of vortices but also for various other important aspects of the vortex behaviour in these materials, as the flux flow effects, through their elastic properties.[4]

3. GAUSSIAN-GINZBURG-LANDAU (GGL) SCENARIO FOR THE COOPER PAIR FLUCTUATIONS AROUND T$_c$ IN MULTILAYERED SUPER-CONDUCTORS.

The effects on the thermal fluctuations of Cooper pairs of the presence of various superconducting layers per periodicity length (multilayering effects) were first studied theoretically by Maki and Thompson.[19] These authors calculated, on the ground of a BCS-like approach, the *direct*, or Aslamazov-Larkin (AL), and the indirect anomalous Maki-Thompson (MT) contributions to the in-plane paraconductivity, $\Delta\sigma_{ab}$, in a bilayered superconductor, with two (N = 2) superconducting layers per periodicity length, s. A generalization for multilayered superconductors of the Ginzburg-Landau-like functional of Lawrence and Doniach (LD)[13] was first proposed by Klemm.[20] In this last work it was also provide some important but partial results (with not-explicit final expressions) on the in-plane paraconductivity and on the fluctuation-induced in-plane susceptibility, $\Delta\chi_{ab}$, and specific heat, $c_{fl}^{(+)}$, above T_C in bilayered (N = 2) superconductors. By using the Klemm's generalization of the LD functional, explicit expressions for $\Delta\chi_{ab}$ and c_{fl}, in this last case both above (+) and below (-) the superconducting transition, and for the direct paraconductivity contribution, $\Delta\sigma_{abAL}$, were first calculated by Ramallo and coworkers in bilayered superconductors.[11,21,24,25] These last authors also calculated for the first time the different contributions to the fluctuation induced in-plane magnetoconductivity, $\Delta\tilde{\sigma}_{ab}$, in bilayered superconductors,[11] by using a generalization of the procedure (based on a BCS-like approach) first proposed by Hikami and Larkin for single layered superconductors.[26] We have also calculated recently[27] the Levanyuk-Ginzburg (LG) crossover between the Gaussian mean-field-like and the full-critical behaviour of the thermal fluctuations in single layered and bilayered superconductors,[28] and also various relationships between $\Delta\sigma_{ab}$, $\Delta\tilde{\sigma}_{ab}$, $\Delta\chi_{ab}$ and $c_{fl}^{(\pm)}$ in the mean-field-like region (MFR).[24,25] Some of these results are summarized below.

3.1. EFFECTIVE NUMBER OF FLUCTUATING LAYERS: CROSSOVERS OF THE CRITICAL EXPONENT (DIMENSIONALITY) AND OF THE AMPLITUDE OF THE THERMAL FLUCTUATIONS OF THE COOPER PAIRS IN BILAYERED SUPERCONDUCTORS.

An useful new concept to take into account the multilayering effects on the thermal fluctuations of Cooper pairs is the so-called *effective number, N_e^c, of independent fluctuating layers per periodicity length,* s, first introduced by Ramallo and coworkers in Refs.11 and 21. In these works, it was shown that the GL free energy functional, $\langle\Delta F\rangle^{N=2}$, may be directly related, through N_e^c, to the one for a single layered superconductor, $\langle\Delta F\rangle^{N=1}$ as:

$$\langle\Delta F\rangle^{N=2} = N_e^c(\varepsilon,\gamma_1/\gamma_2) \langle\Delta F\rangle^{N=1} , \tag{1}$$

where N_e^c depends on both the reduced temperature and the ratio of the Josephson coupling strengths between adjacent planes (see Fig.1). In Eq.(1), it has been only included the terms relevant for thermal fluctuations.[24] The behaviour of N_e^c for a bilayered superconductor as a function of ε in units of the Lawrence-Doniach crossover parameter

$$B_{LD} \equiv (2\xi_c(0)/s)^2 , \tag{2}$$

and of γ_1/γ_2 is presented in Fig.2(a). In Fig.2(b), it is represented the corresponding behaviour of the critical exponent, x, of a bilayered superconductor, defined as

$$x(\varepsilon,\gamma_1/\gamma_2) \equiv \left(\frac{\partial \ln \Delta\sigma_{abAL}}{\partial \ln \varepsilon}\right)_{\gamma_1/\gamma_2} , \tag{3a}$$

where $\Delta\sigma_{abAL}$ is the so-called direct (or Aslamazov-Larkin[29]) contribution to the paraconductivity. Within the Gaussian-Ginzburg-Landau (GGL) approximation, the same critical exponent will arise in the mean-field-like region (MFR) in the fluctuation induced in-plane (H perpendicular to the layers) diamagnetism, $\Delta\chi_{ab}$, and specific heat above T_{co}, $c_{fl}^{(+)}$:

$$x(\varepsilon,\gamma_1/\gamma_2) \equiv \left(\frac{\partial \ln(\Delta\chi_{ab} / T)}{\partial \ln \varepsilon} \right)_{\gamma_1/\gamma_2} , \qquad (3b)$$

and

$$x(\varepsilon,\gamma_1/\gamma_2) \equiv \left(\frac{\partial \ln c_{fl}^{(+)}}{\partial \ln \varepsilon} \right)_{\gamma_1/\gamma_2} . \qquad (3c)$$

The results of Figs.2(a) and (b) clearly illustrate some of the main differences between the $N = 2$ and the single-layered (or $N = 1$) superconductors: Whereas the exponent crossover between the 3D limit ($x = -1/2$) and the 2D limit ($x = -1$) appears indeed in both cases (it is just related to $\xi_c(\varepsilon)/s$), the N_e^c crossover between $N_e^c = 1$ and $N_e^c = 2$ is present only in the bilayered ($N = 2$) case (it is related to the γ_1/γ_2 values). The combination of these two crossovers (in x and in N) leads to several different possible scenarios in bilayered superconductors as, for instance:

i) $x = -1$; $N_e^c = 2$. In this case the different superconducting layers per s are completely uncoupled, and the thermal fluctuations of the Cooper pairs in each plane are completely independent of the fluctuations in the other planes.

ii) $x = -1$; $N_e^c = 1$. In this case the two planes in s are strongly correlated to each other, and each bilayer may be considered, for the thermal fluctuations of the Cooper pairs, as an unique single 2D superconducting plane without internal structure.

iii) The case characterized by $x = -1$ and N_e^c crossing over the values 1 and 2. Here the system presents again uncorrelated superconducting bilayers, but now composed by two superconducting layers with some correlation between them (intermediate between the

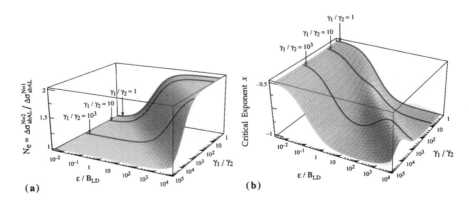

Figure 2. (a) Effective number, N_e^c, of independent fluctuating superconducting planes per unit cell length, s, and (b) mean-field critical exponent, x, in a bilayered superconductor as a function of the reduced temperature in units of the Lawrence-Doniach crossover parameter, B_{LD}, and of the relative tunnelling coupling strength, γ_1/γ_2, between adjacent superconducting layers. Figure from Ref.11.

448

former cases). In this last case, each bilayer presents a frustrated internal 3D structure, and in consequence the value of the exponent differs somewhat from the 2D value (x = -1), as it can be seen in Fig.2. To our knowledge, this mesoscopic effect in bilayered superconductors has not been tested experimentally up to now.

Finally, let us inspect here the influence of γ_1/γ_2 on N_e^c. In Ref.11 it has been shown that N_e^c may take the two limiting values (if $\xi_c(0) \neq 0$):

$$N_e^c = 1, \quad \text{for } \gamma_1/\gamma_2 \to \infty, \tag{4}$$

and,

$$N_e^c = 2 \left(\frac{1 + \dfrac{B_{LD}}{\varepsilon}}{1 + 4\dfrac{B_{LD}}{\varepsilon}} \right)^{-1/2}, \quad \text{for } \gamma_1/\gamma_2 = 1. \tag{5}$$

In the first limit ($\gamma_1 \gg \gamma_2$), the tunnelling interaction between the most coupled superconducting planes is so strong that these planes are, in all the mean-field-like region, equivalent to a single layer without internal structure.

In the opposite limit (i.e., for $\gamma_1 = \gamma_2$), the biperiodic system will become equivalent to a layered superconductor with a single periodicity, but now with s/2 as characteristic length, instead of the unit cell length s. In other words, in these γ_1/γ_2 limits, the $N_e^c = 2$ expressions for the *direct* fluctuation effects may be written as for $N_e^c = 1$, but substituting the unit cell length, s, by an effective interlayer periodicity, s_e^c, given by

$$s_e^c = \begin{cases} s, & \text{for } \gamma_1/\gamma_2 \to \infty, \\ s/2, & \text{for } \gamma_1 = \gamma_2. \end{cases} \tag{6}$$

In terms of the modulus of the superconducting wavefunctions, the different scenarios associated with x and N_e^c may be illustrated schematically as indicated in Figs.3(a) and (b). In the N = 1 case, the different scenarios are due only to the relative values of $\xi_c(\varepsilon)$ and s, and hence they depend only on the value of the critical exponent, x. For N = 2, there are new scenarios due to the possible different values of γ_1/γ_2. These different scenarios may be characterized by $N_e^c(\varepsilon, \gamma_1/\gamma_2)$ and $x(\varepsilon, \gamma_1/\gamma_2)$. In Fig.3, it is assumed $\gamma_1/\gamma_2 \geq 1$ (otherwise, the same graphics would be obtained but with a shift in the represented c-direction window).

The above results provide a first qualitative answer to the questions raised in the introduction about the multilayering influence on the thermal fluctuations of Cooper pairs. These results also show that this answer is much easier formulated in terms of N_e^c, an effective number of independent fluctuating layers, than in terms of s_e^c, an effective interlayer distance. In the case of a quasi-2D multilayered HTSC, the relationship between s_e^c and N_e^c is just, $s_e^c = s/N_e^c$, with $s_e^c \leq s$ because $1 \leq N_e^c \leq N$.

Let us also stress already here an important result that may be directly deduced from Figs. 2(a) and (b): even for γ_1/γ_2 as big as 10^2, the corresponding N_e^c and x take values much closer to those which correspond to $\gamma_1/\gamma_2 = 1$ than those for the $\gamma_1/\gamma_2 \to \infty$ limit. So, even in this case ($\gamma_1/\gamma_2 \lesssim 10^2$) the scenario with two layers per s is much more realist than the single layered scenario.

3.2. FLUCTUATION INDUCED IN-PLANE CONDUCTIVITY, MAGNETOCONDUCTIVITY, DIAMAGNETISM AND SPECIFIC HEAT IN BILAYERED SUPERCONDUCTORS: DIRECT CONTRIBUTIONS.

As noted before, these effects may be due to two types of thermal fluctuation contributions: Those called *direct*, because they are *directly* associated with the presence of the Cooper pairs created by the fluctuations.[29] And the so-called *indirect* effects, because they are due to the modifications that these fluctuating Cooper pairs may induced in the normal properties of the materials, as for instance the pair breaking (Maki-Thompson) effects or those associated with modifications of the electronic density of states (DOS effects). A thorough analysis of these indirect effects suggest that at least for the in-plane observables they are not relevant in HTSC.[10,11] So, we will summarize here how the multilayering affects the direct contributions in the so-called mean-field like (MFR) region, where the Ginzburg-Landau like approaches with Gaussian fluctuations are expected to be applicable (see below).

3.2.1. *Fluctuation-induced in-plane conductivity: paraconductivity and fluctuation-induced magnetoconductivity.*

One of the observables best adapted, from both the theoretical and the experimental point of view, to study the thermal fluctuations above T_c in HTSC is the so-called in-plane fluctuation-induced conductivity, which in presence of a magnetic field, H, applied perpen-

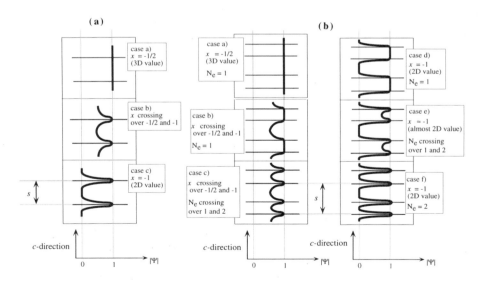

Figure 3. Schematic view of the qualitatively different possible order parameter c-direction correlations in a single-layered (a) and in a bilayered (b) superconductor. What we represent here is the typical decay of the modulus of the bulk superconducting wavefunction *if* the condition $\Psi = 1$ is imposed in the superconducting planes. Figure from Ref.24.

dicular to the ab (CuO_2) superconducting layers, may be defined as

$$\Delta\sigma_{ab}(\varepsilon,H) \equiv \sigma_{ab}(\varepsilon,H) - \sigma_{abB}(\varepsilon,H) , \tag{7}$$

where $\varepsilon \equiv |T-T_{CO}|/T_{CO}$ is the reduced temperature, T_{CO} is the mean-field superconducting transition temperature at zero magnetic field, $\sigma_{ab}(\varepsilon,H)$ is the measured in-plane electrical conductivity, and $\sigma_{abB}(\varepsilon,H)$ is the so-called in-plane background electrical conductivity (the normal conductivity above the superconducting transition if the thermal fluctuations were absent). The usefulness of $\Delta\sigma_{ab}(\varepsilon,H)$ is strongly enhanced, mainly in the case of HTSC, by the fact that it may be decomposed into two additive parts which may be calculated and measured separately: The zero-magnetic field fluctuation-induced conductivity, $\Delta\sigma_{ab}(\varepsilon,H=0)$, currently called paraconductivity and henceforth noted simply as $\Delta\sigma_{ab}(\varepsilon)$, and the so-called fluctuation-induced magnetoconductivity, $\Delta\tilde{\sigma}_{ab}(\varepsilon,H)$, which in terms of $\sigma_{ab}(\varepsilon,H)$ and of $\sigma_{abB}(\varepsilon,H)$ may be defined as

$$\Delta\tilde{\sigma}_{ab}(\varepsilon,H) \equiv [\sigma_{ab}(\varepsilon,H) - \sigma_{ab}(\varepsilon,0)] - [\sigma_{abB}(\varepsilon,H) - \sigma_{abB}(\varepsilon,0)]. \tag{8}$$

So, $\Delta\sigma_{ab}(\varepsilon,H)$ may be written as

$$\Delta\sigma_{ab}(\varepsilon,H) = \Delta\sigma_{ab}(\varepsilon) + \Delta\tilde{\sigma}_{ab}(\varepsilon,H) . \tag{9}$$

The central point here is that the available experimental results in HTSC clearly indicate that the background in-plane magnetoconductivity, $\sigma_{abB}(\varepsilon,H) - \sigma_{abB}(\varepsilon,0)$, measured in the normal region above T_C is always very small, orders of magnitude smaller than $\Delta\tilde{\sigma}_{ab}(\varepsilon,H)$ measured near the superconducting transition.[3,30-37] This result strongly suggests, therefore, that $\Delta\tilde{\sigma}_{ab}(\varepsilon,H)$ may be approximated, even through the transition, as

$$\Delta\tilde{\sigma}_{ab}(\varepsilon,H) \approx \Delta\tilde{\sigma}_{ab}(\varepsilon,H) \equiv \sigma_{ab}(\varepsilon,H) - \sigma_{ab}(\varepsilon,0) , \tag{10}$$

i.e., as the difference between two directly measurable observables, without any dependence on a never well settled background. However, for applied magnetic fields in the weak limit (see below) the fluctuation-induced magnetoconductivity measured in HTSC has been found to be also orders of magnitude smaller than the zero-field paraconductivity.[3,30-37] Therefore, the two contributions to the total fluctuation-induced conductivity, $\Delta\sigma_{ab}(\varepsilon)$ and $\Delta\tilde{\sigma}_{ab}(\varepsilon,H)$, are complementary from both the experimental and the theoretical point of view and their simultaneous study may provide useful information on the thermal fluctuation effects on HTSC.

In terms of $N_e^c(\varepsilon,\gamma_1/\gamma_2)$ the direct or Aslamazov-Larkin (AL) contribution to the in-plane paraconductivity in a bilayered superconductor has been calculated in Ref.18 and it may be written as

$$\Delta\sigma_{ab}^{N=2} = N_e^c \, \Delta\sigma_{ab}^{N=1} , \tag{11}$$

where the dependence of N_e^c on ε and γ_1/γ_2 has been summarized in Fig.2(a) (its explicit dependence may be seen in Ref.11), and $\Delta\sigma_{ab}^{N=1}$ is the in-plane paraconductivity in a single layered superconductor, given by

$$\Delta\sigma_{ab}^{N=1} = \frac{A_{AL}}{\varepsilon}\left(1 + \frac{B_{LD}}{\varepsilon}\right)^{-1/2} , \tag{12}$$

Here A_{AL} is the Aslamazov-Larkin paraconductivity amplitude given by

$$A_{AL} = \frac{e^2}{16\hbar s} \, . \tag{13}$$

The orbital Aslamazov-Larkin contribution, $\Delta\tilde{\sigma}_{abALO}^{N=2}(\varepsilon, H)$, to the fluctuation induced magnetoconductivity in bilayered superconductors, $\Delta\tilde{\sigma}_{ab}^{N=2}(\varepsilon, H)$, has been calculated in Ref.11 in the so-called weak magnetic field limit, defined through the condition,

$$\ell_H \equiv \left(\frac{\hbar}{2e\mu_o H} \right)^{1/2} \gg \xi_{ab}(\varepsilon) \, . \tag{14}$$

By taking into account that in the MFR $\xi_{ab}(\varepsilon) = \xi_{ab}(0)\varepsilon^{-1/2}$, Eq.(14) may be rewritten as

$$\varepsilon \gg h \equiv \frac{2\pi\mu_o H \xi_{ab}(0)}{\phi_o} = \frac{H}{H_{c2}^{\|c}(0)} \, , \tag{15}$$

where h is the so-called reduced magnetic field and $H_{c2}^{\|c}(0)$ is the amplitude (for $T = 0$) of the upper critical magnetic field parallel to the c-direction (perpendicular to the ab-planes). For any N, it was found[11]

$$\Delta\tilde{\sigma}_{abALO}(\varepsilon, H) = \frac{-h^2}{4} \frac{\partial^2 \Delta\sigma_{abAL}(\varepsilon)}{\partial \varepsilon^2}, \tag{16}$$

which allows us to obtain the explicit expression of $\Delta\tilde{\sigma}_{abALO}$ in bilayered superconductors from Eq.(11). This relationship shows already that $N_e^c(\varepsilon, \gamma_1/\gamma_2)$ will arise also in the direct contribution to the fluctuation induced in-plane magnetoconductivity. The resulting explicit expression may be seen in Ref.11. Here we just show in Fig.4 the relationship between $\Delta\tilde{\sigma}_{abALO}$ for bilayered (N =2) and single layered (N = 1) superconductors, as a function of

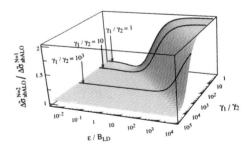

Figure 4. Relationship between the orbital Aslamazov-Larkin contribution to the fluctuation induced in-plane manetoconductivity for bilayered and for single layered superconductors, as a function of ε/B_{LD} and γ_1/γ_2.

both ε/B_{LD} and the relative Josephson coupling strength, γ_1/γ_2, between adjacent superconducting planes. The single layered limit will correspond again to $\gamma_1/\gamma_2 \gg 1$.

3.2.2. *Fluctuation-induced diamagnetism above T_C in the weak magnetic field limit.*

The presence above the superconducting transition of Cooper pairs with a finite lifetime, created by thermal fluctuations, decreases somewhat $\chi(T)$, the measured magnetic susceptibility.[38-40] The associated fluctuation induced diamagnetism above the superconducting transition, $\Delta\chi(T)$, may be defined (we choose here as sign criterion the one that makes $\Delta\chi(T)$ positive):

$$\Delta\chi(T) \equiv \chi_B(T) - \chi(T) \ , \tag{17}$$

where $\chi_B(T)$ is the so-called background susceptibility (the normal susceptibility above the transition if the thermal activated fluctuating Cooper pairs were absent). The first calculation of $\Delta\chi(T)$ in a single layered superconductor was done by Tsuzuki[41] and, independently, by Yamaji[42]. The first explicit calculation of $\Delta\chi(T)$ in bilayered superconductors was done by Ramallo, Torrón and Vidal (RTV).[21] Their resulting expression may be summarized as,

$$\Delta\chi_{ab}^{N=2}(\varepsilon,\gamma_1/\gamma_2) = N_e^c(\varepsilon,\gamma_1/\gamma_2) \ \Delta\chi_{ab}^{N=1}(\varepsilon) \ , \tag{18}$$

where

$$\frac{\Delta\chi_{ab}^{N=1}}{T} = \frac{A_S}{\varepsilon}\left(1 + \frac{B_{LD}}{\varepsilon}\right)^{-1/2} \ , \tag{19}$$

and

$$A_S \equiv \frac{\mu_0 \pi k_B \xi_{ab}^2(0)}{3\phi_o^2 s} \ , \tag{20}$$

is the Schmidt diamagnetism amplitude. The above expression corresponds to H perpendicular to the superconducting layers and in the weak magnetic field limit. For H parallel to the layers, RTV obtain[21]

$$\Delta\chi_c = \left(\frac{\xi_c(0)}{\xi_{ab}(0)}\right)^2 \Delta\chi_{ab} \ . \tag{21}$$

As in HTSC $\xi_c(0) \ll \xi_{ab}(0)$, $\Delta\chi_c$ is negligible in these compounds. Therefore, T_C may be directly estimated by just measuring the sharp discontinuity of $\chi_c(T)$ in the superconducting transition.[43]

3.2.3. *Specific heat induced by fluctuations above and below T_C in absence of an applied magnetic field.*

The effects of thermal fluctuations near T_{CO} on the specific heat in bulk (3D) superconductors were first calculated by Thouless.[44] These results were extended to superconductor films (2D) by Ferrel.[45] For a single layered superconductor, the first calculation of the contribution to the specific heat above the superconducting transition associated with fluctuations in zero applied magnetic field, $c_{fl}^{(+)}$, was done by Tsuzuki[46] and later by other authors.[46,47] For a bilayered superconductor, c_{fl} above (+) and below (-) T_{co} for $H = 0$ was first calculated by Ramallo and Vidal.[24,25,27] Here we will summarize these last results.

The measured specific heat per unit volume above (+) and below (-) T_c may be expressed for $H = 0$ as

$$c^{(\pm)} = c_B + c_o^{(\pm)} + c_{fl}^{(\pm)} , \qquad (22)$$

where c_B is the background (normal) contribution and $c_o^{(\pm)}$ the equilibrium superconducting contribution, which indeed is zero above T_{co} and

$$c_o^{(-)} = c_{jump}\left(\frac{T}{T_{co}}\right) , \qquad (23)$$

below T_{co}. The fluctuation contributions in bilayered superconductors may be expressed through the effective number, $N_e^{c(\pm)}$, of independent fluctuating superconducting layers above and below T_{co} as:[24,25,27]

$$c_{fl}^{(\pm),N=2} = N_e^{c(\pm),N=2}(\varepsilon,\gamma_1/\gamma_2)\frac{A_{TF}}{\varepsilon}\left(1+\frac{B_{LD}^{(\pm)}}{\varepsilon}\right)^{-1/2} , \qquad (24)$$

where A_{TF} is the Thouless-Ferrel *fluctuating specific heat amplitude* given by[44,45]

$$A_{TF} \equiv \frac{k_B}{4\pi\xi_{ab}^2(0)s} , \qquad (25)$$

and $B_{LD}^{(\pm)}$ are the so-called Lawrence-Doniach crossover parameters for $T \gtrless T_{co}$:

$$B_{LD}^{(+)} \equiv \left(\frac{2\xi_c(0)}{s}\right)^2 , \qquad \text{for } T > T_{co} , \qquad (26)$$

and

$$B_{LD}^{(-)} = \frac{1}{2}\left(\frac{2\xi_c(0)}{s}\right)^2 , \qquad \text{for } T < T_{co} . \qquad (27)$$

Both $N_e^{c(+)}(\varepsilon,\gamma_1/\gamma_2)$ and $N_e^{c(-)}(\varepsilon,\gamma_1/\gamma_2)$ are related as

$$N_e^{c(-)}(\varepsilon,\gamma_1/\gamma_2) = N_e^{c(+)}(2\varepsilon,\gamma_1/\gamma_2) , \qquad (28)$$

where $N_e^{c(+)}$ has been represented as a function of $\varepsilon/B_{LD}^{(+)}$ and of γ_1/γ_2 in Fig.2(a). Note that the specific heat contribution due to the fluctuations in a single layered superconductor ($N = 1$) may be directly obtained by just making $N_e^{c(\pm),N=2} = 1$ in Eq.(24). Note, finally, that $c_{fl}^{(+)}$ and $c_{fl}^{(-)}$ are related by[24,25]

$$c_{fl}^{(-)}(\varepsilon) = 2c_{fl}^{(+)}(2\varepsilon) , \qquad (29)$$

a relationship that holds for any value of N.

454

3.2.4. *Relationships between different fluctuation induced observables.*

By just combining the results summarized in the precedent subsections, we may obtain various very useful relationships, *which are independent* of N, between different fluctuation-induced observables (in MKSA units):[11,21,24,25,43]

$$- \frac{\Delta \chi_{ab}/T}{\Delta \sigma_{ab}} = \frac{16 \mu_o k_B}{3 \pi \hbar} \xi_{ab}(0) , \qquad (30)$$

$$\frac{c_{fl}^{(+)}}{\Delta \chi_{ab}/T} = \frac{3 \phi_o^2}{4 \pi^2 \mu_o} \xi_{ab}^{-4}(0) , \qquad (31)$$

$$\frac{c_{fl}^{(+)}}{\Delta \sigma_{ab}} = \frac{4 k_B \hbar}{\pi e^2} \xi_{ab}^{-2}(0) . \qquad (32)$$

In these equations, we have assumed the absence of appreciable *indirect* effects. By combining the above relationships, we may obtain an universal, parameter-free expression,[24,25]

$$\frac{c_{fl}^{(+)}}{\Delta \sigma_{ab}} \frac{\Delta \chi_{ab}/T}{\Delta \sigma_{ab}} = \frac{64 \mu_o k_B}{3 \pi^2 e^2} \approx 2 \text{ x } 10^{-14} \frac{\Omega^2 J}{mK^2} . \qquad (33)$$

3.3. THE WIDTH OF THE FULL-CRITICAL REGION FOR THERMAL FLUCTUATIONS OF COOPER PAIRS AROUND THE SUPERCONDUCTING TRANSITION IN LAYERED SUPERCONDUCTORS IN ZERO-APPLIED MAGNETIC FIELD.

An crucial and still open question about the thermal fluctuations of Cooper pairs around T_C in HTSC is the location and extent in temperature of the so-called mean-field-like region (where the fluctuations are relatively small and, therefore, they must be considered as Gaussian deviations from the mean-field contributions) and for the full-critical region (where the fluctuation contributions to each observable become of the order or even more important than its non-fluctuating part and, therefore, they must be treated through non-perturbative approaches).[2,5,7,27,28,49-55] An estimation of the crossover temperature between these two regions is provided by the so-called Levanyuk-Ginzburg (LG) temperature,[28] T_{LG}, which may be defined as the temperature at which the heat capacity due to fluctuations when T approaches T_{co} from below becomes equal to the heat capacity jump at the transition in absence of fluctuations.[56] The corresponding LG reduced temperature may be then defined as $\varepsilon_{LG} \equiv |T_{LG}-T_{co}|/T_{co}$. The original calculations of ε_{LG} by Levanyuk and Ginzburg were done for isotropic bulk superconductors.[28] Later, ε_{LG} has been calculated in anisotropic bulk superconductors and 2D superconducting films.[57] The first calculation of ε_{LG} in single layered and in bilayered superconductors has been done recently by Ramallo and Vidal.[24,27] Here, we will summarize some of these last results for a bilayered (N = 2) superconductor. The corresponding crossover temperature, $\varepsilon_{LG}^{N=2}$, may be written as:[27]

$$\varepsilon_{LG}^{N=2} = \frac{-c_1}{2} B_{LD} + \left\{ \left(\frac{c_1^2 - c_1}{4} \right) B_{LD}^2 + 2 \left(\varepsilon_{LG}^{film} \right)^2 \right. +$$

$$+ \left[\left(\frac{c_1}{4} B_{LD}^2 - 2\left(\varepsilon_{LG}^{film} \right)^2 \right)^2 + \left(c_1 B_{LD} \varepsilon_{LG}^{film} \right)^2 \right]^{1/2} \right\}, \qquad (34)$$

where $c_1 \equiv (\gamma_1/\gamma_2 + 1)^2/(2\gamma_1/\gamma_2)$, and

$$\varepsilon_{LG}^{film} = \frac{1}{4\pi} \frac{k_B}{s \xi_{ab}^2(0) c_{jump}}. \qquad (35)$$

The best way to see the influence of the multiperiodicity on ε_{LG} is compare the above result for N = 2 with ε_{LG} for a single layered superconductor,[27]

$$\varepsilon_{LG}^{N=1} = \left[\left(\frac{B_{LD}}{4} \right)^2 + \left(\varepsilon_{LG}^{film} \right)^2 \right]^{1/2}. \qquad (36)$$

To make easy this comparison we note that Eq.(34) may be rewritten as:

$$\varepsilon_{LG}^{N=2} = \frac{-B_{LD}}{4} + \left[\left(\frac{B_{LD}}{4} \right)^2 + \left(N_{LG} \varepsilon_{LG}^{film} \right)^2 \right]^{1/2}, \qquad (37)$$

where N_{LG} is just $N_e^c(\varepsilon_{LG})$ for $T < T_{co}$. The comparison of Eqs.(36) and (37) shows that $\varepsilon_{LG}^{N=2}$ has the same expression as the one for $\varepsilon_{LG}^{N=1}$, but with the substitution of the heat capacity jump by c_{jump}/N_{LG}. In consequence, the biperiodicity produces an increase of ε_{LG} equivalent to the reduction of c_{jump} by the factor N_{LG}. The general behaviour of ε_{LG} as a function of $\xi_c(0)$ and of γ_1/γ_2 is represented in Fig.5, where we have used values of s, $\xi_{ab}(0)$ and c_{jump} typical of HTSC compounds. The single-layered scenario corresponds to

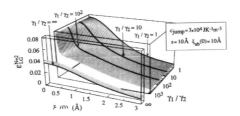

Figure 5. The Levanyuk-Ginzburg reduced temperature of a bilayered superconductor, $\varepsilon_{LG}^{N=2}$, as a function of the tunnelling coupling ratio between adjacent layers, γ_1/γ_2, and of the out-of-plane coherence length amplitude at T = 0 K, and by using values for s, $\xi_c(0)$ and c_{jump} typical of HTSC compounds. The case $\gamma_1/\gamma_2 = 1$ is equivalent to N = 1, but with s/2 as periodicity length, whereas, the $\gamma_1/\gamma_2 \to \infty$ limit is again equivalent to N = 1, but with s. Figure from Ref.27.

the limit $\gamma_1/\gamma_2 \to \infty$. As can be seen in Fig.4, $\varepsilon_{LG}^{N=2}$ is always larger than $\varepsilon_{LG}^{N=1}$ (twice in the 2D limit, associated with $\xi_c(\varepsilon) \ll s$). The physical explanation for this is that a bilayered superconductor has more fluctuating degrees of freedom per layer periodicity length than the single-layered one, the increase of those fluctuating degrees of freedom being proportional to $N_e^c(\varepsilon)$.

It is also interesting to compare the above results for $\varepsilon_{LG}^{N=1,2}$ with the well-known results valid for films and bulk superconductors. It is easy to check that in the limit $\xi_c(\varepsilon_{LG}) \gg s$, for which the fluctuations are expected to be 3D in nature at the Ginzburg temperature, both $\varepsilon_{LG}^{N=1}$ and $\varepsilon_{LG}^{N=2}$ recover the usual expression for ε_{LG} in an anisotropic bulk superconductor:[28,57]

$$\varepsilon_{LG}^{bulk} = \frac{1}{32\pi^2} \left(\frac{k_B}{\xi_c(0)\xi_{ab}^2(0)c_{jump}} \right)^2 . \tag{38}$$

Also, in the 2D limit ($\xi_c(\varepsilon_{LG}) \ll s$) the N = 1 expression for ε_{LG} reproduces the result valid for films of thickness s, i.e., ε_{LG}^{film} as given by Eq.(35). In this 2D limit, the N = 2 expression recovers, however, the LG reduced temperature of films of thickness s/2 (except if γ_1/γ_2 is strictly zero or infinity, in which case the 2D limit corresponds to the result for films of thickness s, see also below). The dimensional crossover of $\varepsilon_{LG}^{N=1}$ and $\varepsilon_{LG}^{N=2}$ is illustrated in Fig.5. It is also easy to conclude[27] that $\varepsilon_{LG}^{N=1}$ is never higher than ε_{LG}^{bulk} or ε_{LG}^{film} obtained by using the same values for the parameters entering in the corresponding expressions (Eqs.(34) to (38)). This is because the fluctuation effects on the heat capacity (per unit volume) are never higher in single-layered superconductors than in equivalent bulk or films of thickness s.

4. COMPARISON BETWEEN BILAYERED (GGL) SCENARIO AND SOME EXPERIMENTAL RESULTS IN $Y_1Ba_2Cu_3O_{7-\delta}$.

A thorough confrontation of the Gaussian-Ginzburg-Landau (GGL) results for bilayered superconductors summarized in the precedent subsections with some of the experimental results obtained in $Y_1Ba_2Cu_3O_{7-\delta}$ (Y-123) crystals has been done recently in Refs.11, 24 and 25. The data were those of Pomar et al.[36,52] for $\Delta\sigma_{ab}$, Torrón et al.[58] for $\Delta\chi_{ab}$ and Roulin et al.[59] for c_p. However, data on the same observables of other groups obtained in high-quality Y-123 crystals[60-63] do not change our conclusions, except very close to the transition ($\varepsilon < 10^{-2}$), at a temperature region where the extrinsic inhomogeneities may appreciably affect the measurements.[37,64-67]

Figure 6 shows the agreement obtained with the bilayered GGL expressions.[24,25] In performing this comparison, the first data points fitted by the theoretical expressions (Eq.(11)) were those corresponding to $\Delta\sigma_{ab}(\varepsilon)$. This leads to $\gamma_1/\gamma_2 \approx 1$ and $\xi_c(0) \approx 0.12$ nm. Then the experimental $\Delta\chi_{ab}/T$ was fitted to Eq.(18) with $\xi_{ab}(0)$ as free parameter. This leads to $\xi_{ab}(0) \approx 1.1$ nm. In both cases, the fits were done in the reduced temperature window above T_C given by $\varepsilon = 0.02$-0.1, which corresponds to the ε-region where the experimental uncertainties on $\Delta\sigma_{ab}$ and $\Delta\chi_{ab}/T$ (mainly due to non-fluctuating background substractions) do not excess 15%. As can be seen in Figs.6(b) and (c), the agreement between the theory and the experimental data is excellent. Then, the theoretical expressions for $c_{fl}^{(+)}$ (Eq.(24)) were fitted in the widest possible ε-range that includes the $T > T_C$ region $0.02 \leq \varepsilon \leq 0.1$. As the parameters entering in the fluctuation contribution to the heat capacity are already fixed from the $\Delta\sigma_{ab}$ and $\Delta\chi_{ab}/T$ fits, the heat capacity fit includes only as free parameter c_{jump} and the background contribution, which is supposed to have the functional form $a + bT + cT^2$. The resulting heat capacity jump is $c_{jump} \approx 3 \times 10^4$ JK^{-1}m^{-3}.

As it may be seen in Fig.6(a), the agreement is again excellent, in this case in all the region bounded by $0.015 \leq \varepsilon \leq 0.1$ above T_C and by $0.03 \leq \varepsilon \leq 0.1$ below T_C. The disagreement in the very close vicinity of T_C may corresponds to the penetration in a different fluctuation regime (maybe the GGL-to-3DXY crossover) or to effects of T_C inhomogeneities on the experimental data.[64,65,67] However, our $\Delta\sigma_{ab}$ and $\Delta\chi_{ab}$ results in high-quality Y-123 crystals[36,43,58] suggest the penetration in the 3DXY full-critical region for $\varepsilon < 10^{-2}$.

As shown in Fig.6(d) to (f), the GGL quotients among the three observables are also in good agreement with those deduced from these experimental data for Y-123 crystals, and that *without any free parameter* (by using for $\xi_{ab}(0)$ the value extracted from the $\Delta\chi_{ab}/T$ fit indicated before). Another important aspect of the comparison shown in Fig.6 is the resulting N_e^c values:

$$1.2 \leq N_e^{c(+)} \leq 1.6 , \qquad \text{for } 10^{-2} \leq \varepsilon \leq 10^{-1} , \qquad (39)$$

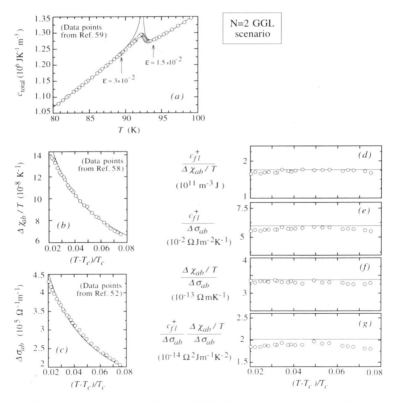

Figure 6. Comparison between the bilayered GGL theory and the experimental heat capacity, in-plane paraconductivity and fluctuation induced diamagnetism in $Y_1Ba_2Cu_3O_{7-\delta}$ crystals. The approximate rms errors are: (a) 1% -in the ε-region bounded by the arrows-, (b) 3%, (c) 3%, (d) 1%, (e) 3%, (f) 1%, (g) 7%. Figure from Ref. 25.

what *provides a direct indication of the relevance of the multilayering effects on the thermal fluctuations in Y-123 crystals*. Two important and complementary aspects of these last results on N_e^c must be stressed here: i) The arbitrarious use of the single layered scenario, by imposing $N_e^c = 1$, will not only change the numerical values of $\xi_{ab}(0)$ and $\xi_c(0)$ but, more important, it will have dramatic consequences on the physics of the thermal fluctuations. For instance, in the case of the paraconductivity the decrease of the direct AL contribution (which is directly proportional to N_e^c) will erroneously suggest the presence of *indirect* contributions (like the MT contributions). ii) The simultaneous analysis of various observables is crucial to discriminate between these very different scenarios. To further illustrate this point, in Fig.7 it is presented the relationship between $\Delta\sigma_{abAL}^{N=2}(\varepsilon)$, with $\gamma_1/\gamma_2 = 1$, and $\Delta\sigma_{abAL}^{N=1}(\varepsilon) + \Delta\sigma_{abMT}^{N=1}(\varepsilon)$, for different relative strengths of the anomalous indirect Maki-Thompson contribution, i.e., for different values of pair-breaking time, τ_ϕ, but always with $\xi_c(0) = 0.12$ nm and an electronic relaxation time $\tau = 10^{-15}$ s. As we may see in this figure, these two very different scenarios almost agree in all the MFR if in the N = 1 approach it is assumed the presence of an indirect MT contribution corresponding to $\tau_\phi \approx 10^{-15}$ s. So, as noted before, the discrimination between these two very different scenarios is only possible by analyzing simultaneously the thermal fluctuation effects on various observables.

Let us also stress here that the value found before for the lower limit of the MFR, $\varepsilon \approx 10^{-2}$, is in excellent agreement with the Levanyuk-Ginzburg reduced temperature for Y-123 in the bilayered scenario. By using $\xi_{ab}(0) = 1.1$ nm, $\xi_c(0) = 0.12$ nm, s = 1.17 nm, and $c_{jump} = 3 \times 10^4$ J K^{-1} m^{-3}, from Eq.(37), we get $\varepsilon_{LG}^{N=2} \approx 2.8 \times 10^{-2}$. Note, however, that in the case of the Y-123 crystals, the differences between the various values of ε_{LG} corresponding to the different scenarios are small. For instance, by using the above indicated values in Eq.(36), we obtain, $\varepsilon_{LG}^{N=1} \approx 1.7 \times 10^{-2}$. This is consistent with the fact that in this compound the thermal fluctuations are almost 3D, with $\xi_c(\varepsilon) \gtrsim$ s in a wide temperature region around the superconducting transition (in particular $\xi_c(\varepsilon_{LG}) \gtrsim$ s).

5. SOME COMMENTS ABOUT THE RELATIVE VALUES OF THE JOSEPHSON COUPLING STRENGTHS BETWEEN ADJACENT SUPERCONDUCTING LAYERS.

A central parameter in the description proposed in the preceding sections for the thermal fluctuation effects of Cooper pairs in bilayered HTSC is γ_1/γ_2, the relative

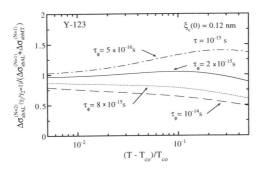

Figure 7. Relationship between $\Delta\sigma_{abAL}^{N=2}(\varepsilon)$, with $\gamma_1/\gamma_2 = 1$, and $\Delta\sigma_{abAL}^{N=1}(\varepsilon) + \Delta\sigma_{abMT}^{N=1}(\varepsilon)$, for different τ_ϕ values and $\tau = 10^{-15}$ s. We see that for $\tau_\phi \approx 2 \times 10^{-15}$ s, $\Delta\sigma_{abAL}^{N=2}(\varepsilon)$ almost coincides, in all the MFR, with $\Delta\sigma_{abAL}^{N=1}(\varepsilon) + \Delta\sigma_{abMT}^{N=1}(\varepsilon)$. Figure from Ref.11.

tunnelling coupling between adjacent superconducting layers. In the case of $Y_1Ba_2Cu_3O_{7-\delta}$ compounds, the analyses summarized in the precedent Section suggest that $1 \leq \gamma_1/\gamma_2 \lesssim 30$. These values indicate that it is in each CuO_2 layer where the superconducting condensate appears, and not in the CuO_2 bilayers as a whole. The implications of these results concern other aspects of these materials and not only the behaviour of their thermal fluctuations. Therefore, it is important to briefly summarize here an analysis of these γ_1/γ_2 values in connection with other experimental informations on these materials and also with some of the general aspects of the Josephson coupling strength between the superconducting CuO_2 layers in these materials. A more detailed account for such an analysis may be seen in Ref.24.

As far as the microscopic understanding of both the superconducting and the normal state of the HTSC have not been achieved until now, it is difficult to obtain a reliable theoretical dependence of γ_j on the interlayer characteristics of these materials. However, some simple arguments may shed some light here. First at all, the critical current of a single tunnelling junction is found (via various different approaches, including complete GL ones) to be proportional to its tunnelling constant,[71-73] $j_c \propto \gamma_j$. On the other hand, for most of the different superconducting tunnelling interactions (purely Josephson tunnelling, weak links, normal or insulating barriers .....) it is found $j_c \propto R_B^{-1}$, where R_B is the normal state resistance of the barrier per unit area.[73] So, we may assume $\gamma_j \propto 1/R_{Bj}$, where R_{Bj} is the electrical resistance in the c-direction in the normal state between the (j,n) and the (j+1,n) layers.

In the case of a bilayered superconductor, and if the resistivities in the c-direction were the same for the two junctions, the above relationship between γ_j and R_{Bj} leads to $\gamma_1/\gamma_2 = d_2/d_1$, were d_1 and d_2 are the two interlayer distances (see Fig.1(b)). This could perhaps be the case for some HTSC compounds, as $Bi_2Sr_2CaCu_2O_8$, in which the electrical properties of the different atomic structures separating the CuO_2 layers seem to be quite similar. In the case of Y-123 compounds, however, the existence of the CuO chains located between the more separated layers will probably favour a value for γ_1/γ_2 even closer to 1 than the d_2/d_1 value (which in this last case is $d_2/d_1 \approx 2.5$). Note also, the resemblance of the terms $\gamma_j|\Psi_{jn} - \Psi_{j+1,n}|^2$ in the free energy functional of the extended Lawrence-Doniach approach[11,20,21,24] to the discrete form of a z-derivative. So, one could also attribute to γ_j a dependence in the form $\gamma_j \propto d_j^{-2}$. In such a case, for bilayered superconductors we would have $\gamma_1/\gamma_2 = (d_2/d_1)^2$. For Y-123 this would lead to $\gamma_1/\gamma_2 \approx 6.25$. Let us stress, however, that both estimations are very crude.

A first experimental estimation of the γ_1/γ_2 values in Y-123 crystals may be obtained from the measurements of their band structure.[74] These data reveal a double periodic Fermi surface, with both components taking values comparable to each other. This leads to[24,74] $\gamma_1/\gamma_2 \lesssim 4$. Another very relevant kind of experiments to test the c-direction tunnelling interactions in HTSC are based on the direct measurements of the Josephson characteristics in this direction.[75-77] Although the first experiments of that kind do not reveal the double interlayer characteristic of the Y-123 compounds,[75,76] more precise measurements[77] have led to the observation of Josephson characteristics corresponding to two junctions of thickness 0.32 nm and 0.84 nm (the distances between the single CuO_2 layers), in excellent agreement with our conclusions from the analysis of the thermal fluctuations. A more detailed analysis of these aspects, including the case of the Bi-2212 crystals may be seen in Ref.24. Let us just stress here that for these last compounds $B_{LD}/\varepsilon \ll 1$ (i.e., $\xi_c(\varepsilon) \ll s$) and therefore, as can be easily deduced from Fig.2(a), the multilayering effects will be appreciable in these quasi 2D compounds even for relatively high values of γ_1/γ_2. In this case, these multilayering effects on the Cooper pair fluctuations will mainly affect their *amplitude*, i.e., N_e^c will be of the order of N, instead of $N_e^c = 1$, as for a single layer scenario.

6. CONCLUSIONS.

In conclusion, the results summarized in this paper strongly suggest that **if the multilayering effects are taken into account** the GGL approach is able to explain simultaneously and a quantitative level the available experimental data on the thermal fluctuation effects on c_p, σ_{ab}, $\tilde{\sigma}$, and χ_{ab} in Y-123 crystals, for H = 0 or in weak magnetic fields. In contrast, the 3DXY approach is ruled out,[24,25] at least for reduced temperatures up to $\varepsilon > 10^{-2}$, a value that is in agreement with the estimation of the Levanyuk-Ginzburg criterion in bilayered HTSC. Moreover, the results summarized here reinforce our earlier proposals[17] about the absence of appreciable indirect fluctuation effects (as, for instance, the so-called Maki-Thompson contribution), a result in turn consistent with a non-conventional (extended or non $^1s_0$), pair-breaking, pairing state in HTSC. These results also suggest that the experimental data obtained in good single crystals are *universal* and that the possible short wavelength fluctuation effects[68-70] or the so-called *dynamic* effects[7,20] are not appreciable in the ε-region bounded by $10^{-2} \leq \varepsilon \leq 10^{-1}$. Let us also note already here, however, that at present there are some measurements of other observables around T_c in HTSC which seem to favour a full critical behaviour instead of a mean field like one in this same ε-region.[54,55] Therefore, even for H = 0 new experimental and theoretical efforts are very desirable for a better understanding of the critical effects around T_c in HTSC.

7. ACKNOWLEDGEMENTS.

This work was supported by the Spanish CICYT, under the Grant No. MAT95-0279. Stimulating conversations with Julien Bok, Guy Deutscher, Arcadii Levanyuk and Davor Pavuna are gratefully acknowledge. Sections 3 and 5 of this paper are mainly based on the Ph.D. Thesis of Manuel V. Ramallo, whereas Section 4 is based on the Ph.D. Thesis of Alberto Pomat and Manuel V. Ramallo.

8. REFERENCES.

1. J.G. Bednorz and K.A. Müller, Z. Phys. B Condensed Matter **64**, 189 (1986).
2. See, e.g., M. Tinkham, *Introduction to Superconductivity*, (McGraw-Hill, New York, 1996), Chaps. 8 and 9, and references therein.
3. For earlier references of the effects of thermal fluctuations of Cooper pairs in HTSC, see, e.g., M. Akinaga, in *Studies of High Temperature Superconductors*, edited by A.V. Narlikar (Nova Science, Commack, NY, 1991), Vol. 8, p. 297. For more recent references, mainly on the fluctuation effects on the heat capacity, see, e.g., A. Junod, ibid. (1996), Vol. 19, p.1. See also, M. Ausloos, in *Physics and Material Science of High Temperature Superconductors* II, edited by R. Kossowsky, B. Raveau, D. Wohlleben and S.K. Patapis (Kluwer, Dordrecht, 1992), p. 775.
4. For a review and earlier references on the effects of thermal fluctuations of magnetic vortices in HTSC, see e.g., D. Feinberg, J. Phys. III France **4**, 169 (1994).
5. D.S. Fisher, M.P.A. Fisher and D.A. Huse, Phys. Rev. B **43**, 130 (1991).
6. For a recent review of the theoretical results on the non-direct (i.e., Maki-Thompson or density of states) effects associated with the Cooper pair fluctuations in single layered superconductors, see e.g., A.A. Varlamov, G. Balestrino, E. Milani and D. Livanov (to be published).

7. A discussion of the similarities between the thermal fluctuation effects in HTSC and in other layered superconductors may be seen in R.A. Klemm, in *Fluctuation Phenomena in High Temperature Ceramics*, edited by M. Ausloos and A.A. Varlamov (Kluwer, NATO Series, 1997), in press.

8. M. Tinkham in Ref.[2], p.325.

9. S.K. Yip, Phys. Rev. B **41**, 2612 (1990), and references therein.

10. J.A. Veira and F. Vidal, Phys. Rev. B **42**, R8748 (1990).

11. M.V. Ramallo, A. Pomar and F. Vidal, Phys. Rev. B **54**, 4341 (1996).

12. See, e.g., V.V. Dorin, R.A. Klemm, A.A. Varlamov, A.I. Buzdin and D.V. Livanov, Phys. Rev. B **48**, 12951 (1993).

13. W.E. Lawrence and S. Doniach, in *Proc. 12th Int. Conf. on Low-Temperature Physics*, Kyoto, Japan, 1970, edited by E. Kanda (Tokyo, Academic, 1971), p. 361.

14. R.E. Prange, Phys. Rev. B **1**, 2349 (1970).

15. R.A. Klemm, M.R. Beasley and A. Luter, Phys. Rev. B **8**, 5072 (1973).

16. L.N. Bulaevskii, M. Ledvij and V.G. Kogan, Phys. Rev. Lett. **68**, 3773 (1992).

17. V.G. Kogan, M. Ledvij, A. Yu. Simonov, J.H. Cho and D.C. Johnston, Phys. Rev. Lett. **70**, 1870 (1993).

18. Z. Tešanović, L. Xing, L.N. Bulaevskii, Q. Li and M. Suenaga, Phys. Rev. Lett. **69**, 3563 (1992).

19. K. Maki and R.S. Thompson, Phys. Rev. B **39**, 2769 (1989).

20. R.A. Klemm, Phys. Rev. B **41**, 2073 (1990).

21. M.V. Ramallo, C. Torrón and F. Vidal, Physica C **230**, 97 (1994).

22. J. Mosqueira, E.G. Miramontes, C, Torrón, J.A. Campá, I. Rasines and F. Vidal, Phys. Rev. B **53**, 15272 (1996).

23. J. Mosqueira, A. Maignan, Ch. Simon, C. Torrón, A. Wahl and F. Vidal, Physica C **282-287**, 1539 (1997); F. Vidal, J.A. Veira, C. Torrón, J. Mosqueira, A. Revcolevschi, I. Rasines, A. Maignan and J.A. Campá, to be published.

24. M.V. Ramallo, Ph. D. Thesis (Universidad de Santiago de Compostela, 1997), unpublished.

25. M.V. Ramallo and F. Vidal, to be published.

26. S. Hikami and A.I. Larkin, Mod. Phys. Lett. B **2**, 693 (1988).

27. M.V. Ramallo and F. Vidal, Europhys. Lett. **39**, 177 (1997).

28. A.P. Levanyuk, Sov. Phys. JETP **36**, 571 (1959); V.L. Ginzburg, Sov. Phys. Solid State **2**, 1824 (1960).

29. L.G. Aslamazov and A.I. Larkin, Phys. Lett. A **26**, 238 (1968).

30. Y. Matsuda, T. Hirai and S. Komiyama, Solid State Commun. **68**, 103 (1988); Y. Matsuda, T. Hirai, S. Komiyama, T. Terashima, Y. Bando, K. Ijima, K. Yamamoto and K. Hirata, Phys. Rev. B **40**, 5176 (1989).

31. M. Hikita and M. Suzuki, Phys. Rev. B **39**, 4756 (1989); **41**, 834 (1990).

32. G. Weigang and K. Winzer, Z. Phys. B Condensed Matter **77**, 11 (1989); G. Kumm and K. Winzer, Physica B **165-166**, 1361 (1990); K. Winzer and G. Kumm, Z. Phys. B Condensed Matter **82**, 317 (1991).

33. K. Semba, T. Ishii and A. Matsuda, Phys. Rev. Lett. **67**, 769 (1991).

34. N. Overend and M.A. Howson, J. Phys.: Condens. Matter **4**, 9615 (1992).

35. W. Holm, M. Andersson, Ö. Rapp, M.A. Kulikov and I.N. Makarenko, Phys. Rev. B **48**, 4227 (1993).

36. A. Pomar, M.V. Ramallo, J. Maza and F. Vidal, Physica C **225**, 287 (1994); A. Pomar, Ph. D. Thesis (Universidad de Santiago de Compostela, 1995), unpublished.

37. A. Pomar, M.V. Ramallo, J. Mosqueira, C. Torrón and F. Vidal, Phys. Rev. B **54**, 7470 (1996); J. Low Temp. Phys. **105**, 675 (1996).

38. See, e.g., W.J. Skocpol and M. Tinkham, Rep. Prog. Phys. **38**, 1094 (1975).

39. H. Schmidt, Z. Phys. **216**, 336 (1968).

40. A. Schmid, Phys. Rev. B **180**, 527 (1969).

41. T. Suzuki, Phys. Lett. A **37**, 154 (1971).

42. K. Yamaji, Phys. Lett. A **38**, 43 (1972).

43. F. Vidal, C. Torrón, J.A. Veira, F. Miguélez, and J. Maza, J. Phys.: Condens. Matter **3**, L5219 (1991); **3**, 9257 (1991); C. Torrón, O. Cabeza, J.A. Veira, J. Maza and F. Vidal, J. Phys.: Condens. Matter **4**, 4273 (1992).

44. D.J. Thouless, Ann. Phys., NY **10**, 553 (1960).

45. R.A. Ferrell, J. Low Temp. Phys. **1**, 241 (1969).

46. T. Suzuki, J. Low Temp. Phys. **9**, 525 (1972).

47. K.F. Quader and E. Abrahams, Phys. Rev. B **38**, 11977 (1988). Note that this paper contains a (probably typographical) error in the final result.

48. L.N. Bulaevskii, Int. J. Mod. Phys. B **4**, 1849 (1990).

49. See, e.g., C. Marcenat, R. Calemczuk and A. Carrington, in *Coherence in High Temperature Superconductors*, edited by G. Deutscher and A. Revcolevschi (World Scientific, Singapore 1995), p. 101 and references therein.

50. See, e.g., S.W. Pierson, Th.M. Katona, Z. Tšanović, and T.S. Valls, Phys. Rev. B **53**, 8638 (1996) and references therein.

51. N. Overend, M.A. Howson, I.D. Lawrie, S. Abell, P.J. Hirst, Ch. Changkang, Sh. Chowdhury, J.W. Hodby, S.E. Inderhess and M.B. Salamon, Phys. Rev. B **54**, 9499 (1996).

52. A. Pomar, A. Díaz, M.V. Ramallo, C. Torrón, J.A. Veira and F. Vidal, Physica C **218**, 257 (1993).

53. A. Andreone, C. Cantoni, A. Cassinese, A. Di Chiara and R. Vaglio, Phys. Rev. B **56**, 7874 (1997).

54. S. Kamal, D.A. Bonn, N. Goldenfeld, P.J. Hirschfeld, R. Liang and W.N. Hardy, Phys. Rev. Lett. **73**, 1845 (1994); J.C. Booth, D.H. Wu, S.B. Quadri, E.F. Skelton, M.S. Osofsky, A. Piqué and S.M. Anlage, Phys. Rev. Lett. **77**, 4438 (1996).

55. C. Meingast, A. Junod and E. Walker, Physica C **272**, 106 (1996).

56. As already noted in the works of Levanyuk (Ref. 40) and Ginzburg (Ref. 40), other definitions of T_{LG} are possible, as the ones based on the comparison of various order parameter averages (see also P.C. Hohenberg, in *Fluctuations in Superconductors*, edited by E.S. Gore and F. Chilton (Stanford Research Institute, Menlo Park, CA 1968) p.305). However, they lead to expressions of ε_{LG} similar to the one resulting from the heat capacity analysis, except for a numerical prefactor. In fact, such a prefactor ambiguity may be seen as a signature of the qualitativeness of any criterion for the crossover between both critical regions (see, e.g., Ref. 5). Let us stress, however, that the definition used here seems to be the more adequate one, as it as based on a physical observable rather than on more indirect considerations (see also V.L. Ginzburg, A. P. Levanyuk and A.A. Sobyanin, Ferroelectrics **73**, 171 (1987)). Let us also stress here, however, that the LG criterium is not directly related to the full critical region, which may appear well below ε_{LG}. Instead, it may be seen as an indication that for $\varepsilon \overset{>}{\sim} \varepsilon_{LG}$ the mean field like approaches plus Gaussian fluctuations could be a reasonable approximation.

57. L.N. Bulaevskii, V.L. Ginzburg and A.A. Sobyanin, Physica C 152, **378** (1988).

58. C. Torrón, A. Díaz, J. Jegoudez, A. Pomar, M.V. Ramallo, A. Revcolevschi, J.A. Veira and F. Vidal, Physica C **212**, 440 (1993); C. Torrón, A. Díaz, A. Pomar, J.A. Veira and F. Vidal, Phys. Rev. B **49**, 13143 (1994).

59. M. Roulin, A. Junod and E. Walker, Physica C **260**, 257 (1996).

60. W. Holm, Yu. Eltsev and O. Rapp, Phys. Rev. B **51**, 11992 (1995).

61. J.T. Kim, N. Goldenfeld, J. Giapintzakis and D.M. Ginsberg, Phys. Rev. B **56**, 118 (1997).

62. C.W. Lee, R.A. Klemm and D.C. Johnston, Phys. Rev. Lett. **63**, 1012 (1989).

63. S.E. Inderhess, M.B. Salamon, J.P. Rice and D.M. Ginsberg, Phys. Rev. Lett. **66**, 232 (1991).

64. An effective medium approach well adapted to the analysis of the influence on the electrical conductivity and magnetoconductivity of T_c inhomogeneities at long length scales and *uniformly* distributed was proposed by J. Maza and F. Vidal, Phys. Rev. B **43**, 10560 (1991). See also Ref.37.

65. The influence of T_c inhomogeneities at long length scales and *non-uniformly* distributed, including the so-called anomalous peaks, on the magnetoresistivity and on the thermopower around the average T_c in HTSC, has been studied by J. Mosqueira, A. Pomar, J.A. Veira and F. Vidal, Physica C **225**, 34 (1994); ibid **229**, 301 (1994); ibid **253**, 1 (1995); J. Appl. Phys. **76**, 1943 (1994).

66. The presence of T_C-inhomogeneities non-uniformly distributed in the sample surface may deeply affect the current density distributions, mainly around the average T_C. See, e.g., Th. Siebold, C. Carballeira, J. Mosqueira, M.V. Ramallo and F. Vidal, Physica C **282-287**, 1181 (1997). In turn, these temperature dependent current redistributios may affect the *measured* critical exponents of both the paraconductivity and the fluctuation induced magnetoconductivity (see also Ref.65).

67. F. Sharifi, J. Giapintzakis, D.M. Ginsberg and D.J. van Harlingen, Physica C **161**, 555 (1989).

68. P.P. Freitas, C.C. Tsuei and T.S. Plaskett, Phys. Rev. B **36**, 833 (1987); R. Hopfengärtner, B. Hensel and G. Saemann-Ischenko, Phys. Rev. B **44**, 741 (1991).

69. A. Gauzzi and D. Pavuna, Phys. Rev. B **51**, 15420 (1995).

70. M.R. Cimberle, C. Ferdeghini, E. Giannini, D. Demarré, M. Putti, A. Siri, F. Federici and A. Varlamov, Phys. Rev. B **55**, R14745 (1997).

71. V. Ambegaokar and A. Baratoff, Phys. Rev. Lett. **10**, 486 (1963); **11**, 104 (1963).

72. P.G. de Gennes, Rev. Mod. Phys. **36**, 225 (1964).

73. K.K. Likharev, Rev. Mod. Phys. **51**, 101 (1979). In this review numerous cases for tunnelling interactions are considered, all of them leading to the relationship $j_c \propto R_B^{-1}$.

74. L.C. Smedskjaer, J.Z. Liu, R. Benedek, D. Leguini, D.J. Laur, M.D. Stahulak, H. Claus, A. Bausil, Physica C **156**, 269 (1988).

75. R. Kleiner, F. Stenmeier, G. Kunkel and P. Müller, Phys. Rev. Lett. **68**, 2394 (1992).

76. R. Kleiner and P. Müller, Phys. Rev. B **49**, 1327 (1994).

77. D.C. Ling, G. Yong, J.T. Chen and L.E. Wenger, Phys. Rev. Lett. **75**, 2011 (1995).

C-AXIS CONDUCTIVITY AND THE ROLE OF D-WAVE SUPERCONDUCTIVITY AND FLUCTUATIONS ON ANISOTROPIC HIGH TEMPERATURE SUPERCONDUCTORS

Colin E Gough,

Superconductivity Research Group
The University of Birmingham,
Birmingham, B15 2TT, UK

1. INTRODUCTION

The cuprate HTc superconductors are the most anisotropic electrical conductors known, with ratios of normal state conductivity parallel and perpendicular to the CuO planes as high as 10^5, illustrated by the early compilation of measurements by Battlog[1]. A wide-ranging review of anisotropy and interlayer coupling in all the HTS cuprates has been given by Cooper and Gray[2]. In this paper we concentrate on the c-axis electrical properties of the most anisotropic HTc superconductors, largely confining our discussion to 2212-BSCCO.

2212-BSCCO ($Bi_2Sr_2CaCu_2O_{8+\delta}$) has been extensively investigated because (i) it is relatively easy to grow high quality, mm-sized, thin platelet, single crystals, (ii) it is easy to vary the coupling and anisotropy between the CuO planes by annealing in argon or oxygen, and (iii) the crystals can be cleaved between the BiO layers to expose chemically inert and atomically flat surfaces over very large areas ($\sim 100 \times 100$ micron^2) suitable for surface studies and for the patterning of high quality epitaxial mesa structures on the surface. 2212-BSCCO is the only known HTS having all the above properties, and has thus become the ideal material to study anisotropy, particularly for surface sensitive properties such as vacuum tunnelling and ARPES (Angular Resolved Photo-Emission Spectroscopy) - see the papers in this volume by Fischer and Campuzana.

Kleiner and Müller and co-workers[3] were the first to demonstrate that BSCCO crystals could be considered as a stack of near ideal, intrinsic Josephson junctions exhibiting both dc and ac Josephson effects across individual atomic layers within the crystal. A recent review of c-axis conductivity and associated Josephson effects in highly anisotropic superconductors has been given by Kleiner[4]. As we will describe in this paper, epitaxial structures containing only a few unit cells along the c-direction can be lithographically patterned on the cleaved surface of BSCCO and TBCCO ($Tl_2Ba_2Ca_2Cu_3O_{10+\delta}$) and other single crystals and whiskers[5-9]. We illustrate several of the important c-axis properties with mesa measurements by Yurgens and his collaborators at Chalmers and Birmingham. Using such structures it is possible to measure Josephson tunnelling characteristics between individual pairs of superconducting CuO_2 planes. Such structures provide the experimentalist with near perfect Josephson junctions within the crystal structure itself, which enable the electronic tunnelling density of states to be investigated in both the normal and superconducting states. In the normal state, we observe an energy gap in the low-lying density of states well above the superconducting transition, consistent with the presence of a pseudo- or spin-gap. This accounts for the increase in c-axis resistance observed in all underdoped HTc superconductors on decreasing temperature. The dynamic conductance also reveals a very sharp superconducting transition at T_c, with the appearance of further structure corresponding to a superconducting energy gap and a further redistribution of energy states from low to higher energies, and a supercurrent feature along the c-axis consistent with the rapid onset of three dimensional superconductivity. Well into the superconducting state, measurements can be made of the critical currents, VI characteristics and ac-Josephson effects   of the individual internal Josephson junctions formed between adjacent superconducting CuO bilayers.   Such measurements provide information on the correlation of the superconducting phase across the planes, which is of considerable interest in a magnetic field, as it involves the ordering and thermal fluctuations of pancakes in the individual planes.

In contrast to vacuum tunnelling measurements at the surface of BSCCO single crystals, which have hitherto provided the most reliable information on the tunnelling density of states[10], measurements on mesa structures are extremely reproducible and are sufficiently sensitive to reveal detailed information on the development of the energy dependent density of states in the immediate proximity of T_c. However, relatively large supercurrents flow between superconducting layers, which are only two atomic layers thick. This results in anomalous VI characteristics, which are almost certainly associated with strong quasi-particle injection effects, intrinsic to the c-axis conduction properties. Such anomalies are observed in both mesa structures and mm sized single crystals[11].

Measurements on mesa structures have a number of additional advantages over measurements on single crystals, as they can be made over a small number of well defined junctions, where the size is so small that the likelihood of crystal defects complicating measurement is considerably reduced. Very importantly, heating effects are significantly reduced, so that it is much easier to distinguish intrinsic quasi-particle injection effects from self-heating. Critical currents can be measured across individual intrinsic junctions, so that it becomes possible to investigate the very large changes in critical currents arising from the loss of phase coherence across adjacent pairs of CuO_2 planes resulting from the misalignment of vortex pancake lattice resulting from local pinning and thermal fluctuations.

Because BSCCO and other highly anisotropic superconductors retain a strongly two-dimensional character, the influence of thermal fluctuations is always large. Thermal fluctuations not only affect the intrinsic superconducting properties but also control the alignment of vortices across the planes and their pinning within the planes by lattice defects. As we will show, this has a marked affect on critical currents in the c-direction.

The aim of this paper is to introduce a number of fundamental aspects of superconductivity, particularly those relating to Josephson tunnelling, and then to use the properties of BSCCO single crystals and mesa structures as an illustration of their influence on the intrinsic c-axis properties of highly anisotropic HTc superconductors.

We start by considering general models for highly anisotropic superconductors, define the anisotropy factor and relate such models to the Lawrence-Doniach model for quasi-2D (two dimensional) layered superconductors[12] such as 2212-BSCCO.   We introduce the ideas of Josephson coupling, the properties of tunnel junctions, dc and ac Josephson effects, and dynamic models to describe the VI characteristics of Josephson junctions. We use these to develop an electrodynamic model for HTc superconductors in terms of a stack of intrinsic Josephson junctions. Using this model we introduce the Josephson penetration depth, the plasma frequency and relate these properties to the Josephson coupling energy and c-axis critical current.

We include a brief introduction to the magnetic properties of HTc superconductors, concentrating on those features that relate to the various regimes of 2D/3D, lattice, glass and liquid states, which characterise the degree of ordering and pinning of the pancake vortices, which in turn determine the c-axis properties in a magnetic field.

In addition to this pedagogic introduction, the main thesis of this paper is that BSCCO and other highly anisotropic superconductors provide us with naturally occurring, near perfect, intrinsic, junction structures. Measurements on such structures provide fundamental information about the electronic density of states in both the normal and superconducting states, which should help to distinguish between the large number of competing microscopic models proposed to describe high temperature superconductivity and associated properties.

466

2. ANISOTROPIC HTc SUPERCONDUCTORS

2.1 Anisotropic Normal and Superconducting Properties

In BSCCO the coupling between the double CuO layers is extremely weak, so that the normal state and superconducting properties approach those of an ideal 2-dimensional conductor. Despite the weakness of the interplanar-coupling, BSCCO, like all superconductors, is a bulk 3-dimensional superconductor, but having highly anisotropic characteristic superconducting lengths. To describe conventional anisotropic type-2 superconductors we use the anisotropic Ginsburg-Landau model with an anisotropy factor γ defined by

$$\gamma = \frac{\lambda_c}{\lambda_{ab}} = \frac{\xi_{ab}}{\xi_c} = \frac{H_{c2,ab}}{H_{c2,c}} = \frac{H_{c1,c}}{H_{c1,ab}} = \left(\frac{m_c}{m_{ab}}\right)^{1/2}, \tag{1}$$

where m_{ab} and m_c are the band masses in the ab-plane and c-directions, which are proportional to the resistivities in the two directions, if an isotropic relaxation time is assumed. λ_{ab} and λ_c and the similarly subscripted critical fields refer to the parameters involving current flow in the ab-plane or c-direction. The coherence lengths are measured in the subscripted directions. Further details on most of the physical properties and theories of conventional superconductors covered in this paper can conveniently be found in the textbook by Tinkham[13].

For highly anisotropic, quasi-2D, superconductors, a fully 3D band model involving free electron states along the c-axis may not be appropriate, as electrons may be localised within the CuO_2 planes. It is then more appropriate to use the Lawrence-Doniach model[12], in which a 2D form of the Landau-Ginsburg equations is used to describe superconductivity in the individual superconducting layers with weak Josephson coupling between them[13]. As we will show later, this model results in a critical current density $J_c = \varepsilon_J \, 2\pi \, / \Phi_o$, where ε_J is the inter-layer Josephson coupling parameter and Φ_o is the flux quantum h/2e. The Josephson penetration length λ_J is given by $\lambda_J^2 = \dfrac{\Phi_o}{2\pi \, J_c \mu_o s}$, where s is the average spacing of the CuO bilayers. It replaces λ_c in Equation 1 defining the anisotropic critical fields and superconducting parameters. λ_J can be relatively long (> 100 μm) in underdoped BSCCO single crystals. On approaching T_c, the critical current density becomes vanishingly small and the Josephson screening length becomes infinite. Therefore, even very weakly coupled 2D superconductors become bulk 3D superconductors at T_c. [Somewhat confusingly, a number of authors define a "Josephson length", $\lambda_J = \gamma s$, which is the interlayer separation scaled to mimic the properties of an isotropic superconductor].

For under-doped 2212-BSCC0, $\gamma \geq 300$, corresponding to a normal state resistance anisotropy $\sim 10^5$. This very large anisotropy is a consequence of the quasi 2D nature of conduction in these materials. The resistance in the c-direction can become so large that the electrons may well be localised within the individual pairs of CuO_2 planes, so that the concept of a 3D Fermi surface may not be valid. 3D Fermi surface calculations for BSCCO suggest that the double BiO layers may be weakly metallic rather than insulating, with a finite density of states at the Fermi surface[14,15]. Although there is no fundamental reason why the relaxation times for ab and c-axis conduction should be the same - they indeed have very different temperature dependences - the anisotropy in the normal and superconducting states is very similar, so that $\lambda_J / \lambda_{ab} \sim (\rho_c/\rho_{ab})^{1/2} \sim \gamma$. The anisotropy factor γ in the superconducting state can be determined from the highly anisotropic magnetic properties measured using torque magnetometry[16].

The temperature and field dependence of the resistivity of 2212-BSCCO in the ab-plane and c-directions[17] is illustrated in Fig. 1 by some recent mesa structure measurements for a slightly over-doped, oxygen annealed, crystal by Yurgens et al[18], for fields perpendicular to the CuO-planes. For optimally doped HTS, the ab-plane resistivity is metallic with a characteristic linear temperature dependence, which for many HTc superconductors extrapolates to near zero resistivity at zero temperature. In contrast, the c-axis resistivity of underdoped sample in the normal state (measured at small currents) increases on decreasing temperature. The c-axis resistivity passes through a maximum before decreasing to zero at a temperature close to the irreversibility line T_{irr} deduced from magnetic measurements. When superconductivity is suppressed by a large magnetic field, the increase in resistivity extends to lower temperatures, leading to a larger peak and a lower temperature for zero resistance[19]. The rate at which the normal state resistance increases on decreasing temperature and the peak height both increase with increasing anisotropy, which is achieved by annealing crystals in argon, helium or vacuum.

For many HTS superconductors, significant deviations from a linear temperature dependence of the ab-plane resistivity is observed on passing from the optimally doped to the underdoped regime, with an accompanying increase in magnitude and negative slope of the c-axis resistivity[20,21]. Both effects appear to be strongly correlated with the emergence of a spin- or pseudo-gap in the normal state electronic density of states deduced from inelastic neutron scattering[22], NMR Knight shifts[23-25], NMR T1 relaxation times[26], ARPES[27,28], and heat capacity[29] measurements.

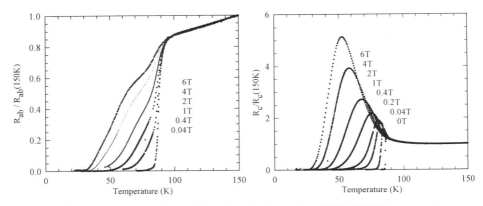

Figure 1. The ab-plane and c-axis resistivities of a slightly overdoped 2212-BSCCO mesa structure as a function of temperature and field perpendicular to the CuO -planes (Yurgens et al.[18])

2.2 Microscopic Models and Relevance to c-axis Properties

Soon after the discovery of the cuprate HTc superconductors, Anderson proposed that the anomalous normal and superconducting properties were intimately connected with tunnelling between the CuO-planes and predicted a critical temperature proportional to the Josephson coupling energy, $T_c \sim \varepsilon_J$ [30,31]. For CuO-bilayer HTc superconductors, like 123-YBCO and 2212-BSCCO, the relevant coupling is between the bilayers, so that any test of the Anderson mechanism needs to be made on single CuO layer cuprate compounds, such as Bi- and Tl-2201. In the Anderson a model, the incoherent interlayer tunnelling of electrons across the planes involves interactions with excitations in the planes, leading to a gap in the low lying tunnelling density of states. The c-axis conductivity is described in terms of interlayer hopping of electrons assisted by spinon-hole recombination[32]. The pairing of electrons leading to superconductivity leads to the possibility of coherent tunnelling, with the condensation energy gained by the delocalisation of electrons in the c-direction.

A model based on strong electron-electron interactions mediated by antiferromagnetic spin-fluctuations has been developed by Pines et al.[33] (see also the chapter in this volume). This accounts for many of the anomalous normal state properties and d-wave character of superconductivity. In this model, antiferromagnetic spin fluctuations result in a spin-gap in the density of states of electrons in the CuO planes, which is responsible for the observed increase in c-axis resistance[34].

Our tunnelling measurements on intrinsic junctions in 2212-BSCCO mesa structures provide experimental confirmation for a gap in the c-axis tunnelling current resulting in the observed increase in c-axis resistance, as originally proposed by Ong[35].

2.3 Importance of Fluctuations in the Layered Cuprates

A very different explanation for the rise in c-axis resistivity based on superconducting fluctuations in the normal state, has been proposed by Gray and Kim[36] and by Ioffe, Varlamov and co-workers[37-39]. The latter authors have recently written an extensive review developing these ideas[40]. Fluctuations are always important in the layered structures, especially in a magnetic field, which reduces the effective dimensionality by quantising the superconducting wave function into orbitals in the xy-plane, hence reducing the x- and y- degrees of freedom.

We can use the Ginsburg-Landau expression for the free energy to determine the spectrum of small amplitude fluctuating modes in the normal state along the c-direction. The spatial dependence of the fluctuating wave function can be described as a sum of Fourier components $\Sigma\, a_q e^{iqz}$. We insert such a wave function into the Ginsburg Landau expression for the free energy density and retain only the lowest order powers in the expansion. Equipartition of energy then determines the amplitudes as $a_q^2 = \dfrac{kT}{a\left[(t-1)+q^2\xi_0^2\right]}$, where $t = T/T$ and a is the condensation energy. The characteristic length scale for the superconducting fluctuations is therefore given by $q^{-1} \sim \xi_0/(t-1)^{1/2}$, which becomes infinitely long as the temperature approaches T_c from above (and below), however short the coherence length ξ_0. Hence, even extremely anisotropic HTc superconductors, where ξ_0 in the c-direction may be as short as 0.3Å, become 3-dimensional just above T_c (and at all lower temperatures). Later in this paper, we include measurements of the tunnelling characteristics for intrinsic junctions within a BSCCO crystal, which confirm the onset of a supercurrent along the c-direction (though thermally broadened) as soon as the crystal becomes superconducting, identified by the sudden appearance of additional gap structure in the low-lying tunnelling density of states.

2.4 Superconducting d-wave State

Despite the lack of an accepted microscopic model, it is now generally accepted that, at least for optimal doping, HTc superconductors are largely d-wave in character. However, the orthorhombic rather than tetragonal crystal symmetry requires some mixing of s- or d_z-wave pairing components, which is also implied by the observation of c-axis supercurrent tunnelling across HTc-conventional superconductor junctions[41,42]. The d-wave pairing of electrons leads to nodes in the energy gap along certain directions in k-space, consistent with ARPES measurements[43], the low temperature dependence of NMR relaxation times[44], and the linear low-temperature dependence of $\lambda(T)$ deduced from magnetic and microwave measurements in YBCO[45] and BSCCO[46] single crystals, as discussed by Hardy in this volume. However, the most persuasive evidence for d-wave comes from scanning SQUID microscope images of half-integral flux quanta on vortices trapped on specially prepared grain boundary junctions in YBCO and other HTc films[47] (see also the chapter by Kirtley in this volume).

Kleiner and colleagues[3] were the first to demonstrate that highly anisotropic HTc superconductors such as BSCCO could be considered as a linear array of intrinsic Josephson Junctions stacked on top of each other. Transport measurements along the c-axis therefore involves tunnelling across the insulating or weakly metallic BiO layers between adjacent superconducting CuO-bilayers. There are no symmetry arguments to preclude critical currents in the c-direction, though they are very much smaller than in the ab-plane, with typical values of 10^2 - 10^3 Acm^{-2}.

3. THEORETICAL MODELS

3.1 Lawrence- Doniach Model for a Layered Superconductor

We use the Lawrence and Doniach model[12] to describe the 2212-BSCCO crystal structure as a stacked array of Josephson junctions formed between adjacent CuO bilayers across the insulating (or weakly metallic) BiO charge-reservoir layers[3], illustrated schematically in Fig. 2. We also include the interlayer capacitance and a parallel resistance to represent the Giaever quasi-particle tunnelling current[48]. As indicated earlier, the dc properties of such a model reduce to the 3-dimensional anisotropic Ginsburg-Landau model for small anisotropies and for temperatures sufficiently close to T_c. However, it is important to recognise that in HTc superconductors the superconducting planes are extremely thin - only a single CuO layer in the 2201 compounds and only two CuO planes thick in the 2212 Bi and Tl compounds. Because of the extreme thinness of the superconducting layers, the Josephson and quasi-particle c-axis currents can strongly influence the superconducting properties of the individual planes, leading to the possibility of strong non-equilibrium, quasi-particle injection effects, as we will describe later.

The attractive Josephson coupling energy across individual pairs of superconducting CuO planes can be written in the form

$$\varepsilon_{J,n,n-1} = \varepsilon_J \cos \Delta\theta_n, \qquad (2)$$

where ε_J is the Josephson coupling energy per unit area and $\Delta\theta_n$ is the gauge-invariant, phase difference of the order parameter between the nth and $(n-1)$th superconducting planes,

$$\Delta\theta_n = \theta_n - \theta_{n-1} - \frac{2\pi}{\Phi_o} \int_{(n-1)s}^{ns} A_z \, dz, \qquad (3)$$

and s is the average spacing between the planes.

We show later that, in the absence of a magnetic field, the critical current density for such a model is just the same that one would derive for an individual junction with $J_c = \dfrac{\varepsilon_J 2\pi}{\Phi_o}$. This is derived by minimising the Free Energy, $F = -\varepsilon_J \cos\theta \ - I\dfrac{\Phi_o}{2\pi}\theta$, with respect to θ, where the second term is the free energy associated with the current source I.

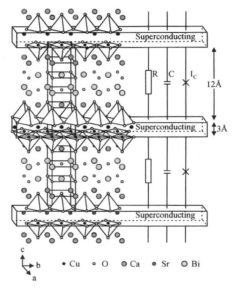

Figure 2. The 2212-BSCCO crystal structure and a simple electric circuit representation involving intrinsic Josephson junctions between the CuO bilayers, a parallel capacitance and a resistance to represent quasi-particle conduction.

3.2 Abrikosov and Josephson Vortices

For a magnetic field perpendicular to the superconducting planes, flux penetrates each layer in the form of *flux pancakes* or 2-dimensional quantum vortices, involving circulating currents within the ab-planes with a superconducting phase difference of 2π around each core[49]. A condensation core energy $\sim \pi\xi_{ab}^2 \, d \, \mu_o H_c^2 / 2$ can be associated with each pancake, where d is the superconducting CuO bilayer plane thickness (~ 3 Å). In an ideal material, the pancake vortices would interact magnetically to form a hexagonal lattice in the CuO planes, which was perfectly aligned with pancakes on adjacent layers by magnetic and Josephson interplane coupling. However, oxygen vacancies, local variations in chemical stoichiometry, thermal fluctuations, and microstructural defects will all tend to disrupt the lattice alignment, both within and between the planes. Both Josephson and magnetic coupling are important between the planes, though Josephson coupling will dominate the short-range alignment in the c-direction, as the magnetic coupling involves a much longer length scale $\sim \lambda_{ab}$. For fields $> H_{c1,ab}$ this will be much larger than the typical length over which the pancakes are misaligned, which has to be less than their spacing, $\sim (\Phi_0/B)^{1/2}$.

In a defect free crystal at low temperatures, the pancake vortices would be expected to form a regular hexagonal lattice with the pancakes aligned directly above each other, as indeed observed by Forgan and co-workers[50] in neutron diffraction measurements on BSCCO single crystals at low fields (<60 mT) and low temperatures (<40K). On increasing the field, the flux pancakes come closer together, so that the interlayer magnetic and Josephson coupling is reduced in strength. This results in the 3D to 2D decoupling transition, observed in both μSR and low-angle neutron diffraction measurements, with a loss of long-range order ($> \lambda_{ab}$) of the pancake arrays along the c-direction. The existence of such a transition can also be inferred from dc transport measurements and from the onset of the arrow-head pinning anomaly in hysteretic magnetisation measurements[51].

For fields less than the decoupling transition, the 3D flux line lattice persists up until $T_{irr}(B)$, the temperature at which the compound exhibits a finite dc resistivity and the magnetic properties become reversible. At T_{irr} the 3D flux lattice melts into what is believed to be a 2D pancake liquid with only small correlations between CuO planes.

For fields above the decoupling transition, the pancakes are believed to form an ordered 2D lattice within each plane, pinned be local impurities with little correlation between planes. The 2D lattice melts at $\sim 20K$ into what is believed to be a 2D pancake liquid. However, at these temperatures the motion of the pancakes remains fairly sluggish, so that correlations of pancakes both within and across the planes persist on relatively long time scales. Only when the correlation time becomes comparable to the characteristic measurement time, will a resistance develop at T_{irr}, so that the irreversibility line becomes quite a strong function of frequency (and voltage criterion) in addition to magnetic field. A 1μV voltage criterion for resistivity across a layer would correspond to a Josephson phase-slip frequency of ~500 MHz.

Even in the liquid state, our measurements on mesa structures demonstrate a remnant degree of ordering of the pancakes on the atomic interlayer scale. This results in a strongly depressed, but still finite, c-axis critical current. Above T_{irr} a finite thermally activated phase-slip resistance is evident in both the ab-plane and along the c-axis, as illustrated in Fig. 1. Measurements of the critical current as a function of magnetic field, enable the correlation of pancakes along the c-axis to be investigated on a truly microscopic scale, in contrast to μSR and neutron diffraction measurements, which can only probe such correlation on length scales $\geq \lambda_{ab}$.

A different situation arises if the magnetic field is directed parallel to the CuO planes. The quantised vortex state now involves currents flowing both parallel and perpendicular to the CuO planes. The associated energy is entirely associated with the circulating currents with no condensation core energy. Because the Josephson current density between the planes is so weak (relative to that in the ab-planes), the vortex is strongly elongated in the direction of the planes with the ratio of its major and minor axis $\dfrac{\lambda_J}{\lambda_{ab}} = \gamma$, which is typically ~ 300 for underdoped BSCCO. The resultant changes in the gauge-invariant phase across adjacent CuO bilayers, lead to a reduction of the interplanar critical current with a field dependence similar to the Fraunhofer pattern observed for single junctions[13]. However, extremely accurate alignment of the fields is necessary to observe such behaviour, to avoid nucleation of flux pancakes in the CuO planes by components of the field perpendicular to the planes, which have a far stronger effect on the c-axis critical current.

Even in the absence of a field, a c-axis current flowing through a junction or crystal larger in size than the Josephson penetration length can generate an internal Josephson vortex, which can traverse the sample giving rise to a self-generated phase-slip voltage $V = \dfrac{\Phi_o}{2\pi}\dfrac{d\theta}{dt}$. Kleiner et al.[52] have considered the electro-dynamics of such vortices, or soliton solutions, in a finite stack of junctions equivalent to a finite sized single crystal. Under certain conditions, phase locking can lead to the coherent motions of a number of such solitons leading to coherent Josephson radiation, as has indeed been observed experimentally[53].

3.3 Capacitatively Coupled RSJ Model

As indicated earlier, the electrodynamic properties of a BSCCO crystal for electric fields and currents perpendicular to the CuO planes can be understood in terms of the series array of capacitatively and resistively shunted Josephson junctions, illustrated schematically in Fig. 2. For wavelengths along the c-direction much larger than the spacing of the CuO planes, we can ignore the layered atomic structure and consider the electrodynamics in terms of bulk properties. This is described by the dielectric constant $\varepsilon(\omega)$, a dissipative conductivity associated with quasi-particle tunnelling between the CuO bi-layers (and any non-intrinsic current paths present; for example, from microscopic lattice defects), and a non-linear (phase dependent) "*inductivity*" associated with the Josephson tunnelling. These properties can be understood by first considering the properties of a simple Josephson junction, using the capacitatively shunted RSJ model (see Tinkham[13]).

The current across such a junction can be written as

$$i = \frac{V}{R} + C\frac{dV}{dt} + i_c \sin\theta \quad , \tag{4}$$

where θ is the phase difference across the junction, which is related to the voltage across the junction by the Josephson relationship $V = \frac{\Phi_o}{2\pi}\frac{d\theta}{dt}$. Equation 4 is equivalent to the equation of motion of a particle with mass $C\frac{2\pi}{\Phi_o}$ undergoing damped motion in a "washboard" potential $E_J\{-\cos\theta - i/i_c\}$, where $E_J = \frac{i_c\Phi_o}{2\pi}$. For currents less than the critical current i_c, we have steady state, zero-voltage, solutions of the form $i = i_c \sin\theta$, which determines the phase difference across the junction for a given applied current. Since $\sin\theta$ cannot exceed unity, for $i>i_c$ stable motion is no longer possible and the "*particle or phase*" starts to slide down the washboard potential at a rate determined by the current and parallel resistance R. This is referred to as phase-slip and results in a voltage across the junction, with an oscillating component at the Josephson frequency $f = 2eV/h$ (1 μV corresponds to 484 MHz). If the damping is large and the capacitance small (i.e. an *overdamped* junction), the *VI*-characteristics are reversible with $V = i_c R\sqrt{(i/i_c)^2 - 1}$, illustrated schematically in Fig 3a.

At a finite temperature, there is always a finite probability that the phase can be thermally excited out of a local minimum, giving rise to a thermally activated phase slip voltage. Ambegaokar and Halperin[54] showed that this led to an ohmic resistance at small currents given by $R_{th}/R = [Io(\gamma/2)]^{-2} \sim 3 \gamma e^{-\gamma}$ for small γ, where Io is the modified Bessel function and $\gamma = E_J/kT$ is the ratio of the Josephson coupling energy to the thermal energy (note that the same symbol γ is somewhat confusingly used for both this ratio and the ratio of anisotropic normal/superconducting properties).

Thermally activated phase slip therefore gives rise to a thermally activated ohmic resistance at small currents and a broadening of the resistive transition[54]. For intrinsic HTc junctions, significant corrections also need to be made for the finite capacitance[55], which increases the phase slip voltage and hence R_{th} by a factor $\sim (1+\beta_c)$, where β_c is the McCumber parameter[56] given by $\beta_c = \frac{I_c 2\pi}{\Phi_o} R^2 C$, which parameterises the influence of the capacitance on the phase-slip dynamics.

For HTc superconductors, thermally activated phase slip dominates the dc resistive properties above the irreversibility line, as we will discuss later in the context of the magnetic field dependence of c-axis properties.

For junctions with large capacitance and damping resistance, $\beta_c >1$. In this underdamped regime, once the critical current has first been exceeded, the *inertia* associated with the finite

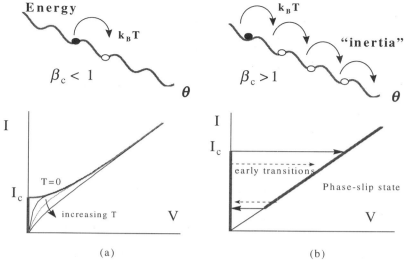

Figure 3. Schematic IV curves for a capacitatively shunted RSJ junction with and without thermal fluctuations in (a) the overndamped regime ($\beta_c < 1$) and (b) the underdamped regime ($\beta_c > 1$). Also illustrated are schematic representations of the thermally activated motions of the phase for the "washboard" potential in the overdamped and underdamped regimes.

capacitance enables the phase to move continuously across successive peaks of the Josephson potential, even when i is subsequently decreased below the critical current. The VI-characteristics therefore become hysteretic, as illustrated in Fig. 3b. On first exceeding i_c and then decreasing the current below i_c, phase-slip continues until the gain in energy on crossing successive maxima of the washboard potential is insufficient to compensate for the viscous losses. The phase then again becomes trapped in a local minimum and the junction reverts to the zero voltage state. The extent of the hysteresis can be used to estimate the McCumber parameter. In the hysteretic regime, thermal fluctuations can cause early transitions to the finite voltage phase-slip state and early returns to the zero voltage state on the return path, as illustrated. The critical current density measured at a finite temperature will therefore always be somewhat less than the intrinsic value $J_c = \frac{\varepsilon_J 2\pi}{\Phi_o}$.

3.4 The Plasma Resonance

When the phase is trapped in a local minimum, the phase can oscillate about its equilibrium position. Small excursions of the phase can then be described by linearising Equation 4 by replacing $\sin\theta$ by θ. The equation is then equivalent to that of a simple LCR oscillator with an equivalent inductance, $L_J = \frac{\Phi_o}{2\pi\, i_c}$ (the Josephson inductance). The phase can therefore oscillate about its equilibrium position with resonant frequency $\omega_P = \sqrt{\frac{i_c 2\pi}{\Phi_o C}} = \frac{2\pi}{\Phi_o}\sqrt{\frac{E_J}{C}}$, known as the *plasma frequency.*

The plasma resonance will be damped by the parallel resistance R and also by any thermally activated phase slip, which can be included as a phase slip resistance R_{th} in series with the Josephson inductance. When damping is dominated by thermally activated phase slip, the Q value of the resonance will be given by $\omega_p L\,/\,R_{th}$, while at low temperatures, the damping will be dominated by the parallel resistance, so that $Q = R/\omega_p L$. Measurement of the width of the plasma resonance at low temperatures should therefore provide information on the intrinsic conduction processes across the intrinsic junctions in BSCCO and other HTc superconductors.

3.5 An Electrodynamic Model for the c-axis Properties of Bulk Crystals

We can now write down a model for the bulk electrodynamic properties of a Josephson coupled layered superconductor in terms of a stack of ideal Josephson junctions. We initially restrict our discussion to low temperatures and small currents, so that we can ignore thermally activated phase slippage and can linearise the equations for the electromagnetic response. For E and j perpendicular to the CuO plane we may therefore write

$$j = \sigma E + \varepsilon \varepsilon_O \frac{\partial E}{\partial t} + \frac{1}{\Lambda_J} \int E \, dt \quad , \tag{5}$$

where we have defined a Josephson " bulk inductivity", $\Lambda_J = \Phi_o / 2\pi J_c s$. The first term represents the conduction from quasi-particle tunnelling between the superconducting planes; the second term is the Maxwell Displacement current, which turns out to be very important for HTc superconductors because of relatively small quasi-particle conductance and critical current density; while the third term corresponds to the changes in phase (hence voltage) associated with the Josephson coupling. Es is the voltage across each junction spaced on average a distance s apart (note that we use s, the half-unit cell length for 2212-BSCCO, so that the total voltage across a stack of junctions is the sum of the voltages across the individual stacks). The last term is often incorporated with the conductivity to give a complex, frequency dependent, conductivity $\sigma_1 - i\sigma_2$. We prefer to emphasise the physical origin of the separate terms, by introducing the concept of a Josephson inductivity.

To describe the low temperature, electromagnetic response for small currents parallel to the c-axis, the relevant Maxwell equations can be written as

$$curl \, E = -\mu_O \frac{\partial H}{\partial t} \tag{6}$$

and

$$curl \, H = \sigma E + \varepsilon \varepsilon_O \frac{\partial E}{\partial t} + \frac{1}{\Lambda_J} \int E \, dt \quad . \tag{7}$$

Note that these equations are exactly the same as for a conventional superconductor except that the London penetration length λ_L has been replaced by the Josephson penetration length λ_J. For E and J perpendicular to the planes, we can look for electromagnetic wave solutions propagating parallel to the planes of the form $\exp[i(k\,x-\omega t)]$. We then obtain

$$k^2 = \frac{\omega^2}{c^2} - \frac{1}{\lambda_J^2} + i\mu_O \omega \sigma \quad , \tag{8}$$

where $c = \frac{c_O}{\sqrt{\varepsilon}}$, $\lambda_J^2 = \frac{\Lambda_J}{\mu_O} = \frac{\Phi_O}{2\pi\mu_O} \frac{1}{J_c s}$, and $\delta = \sqrt{\frac{2}{\mu_O \omega \sigma}}$.

It is instructive to consider the above solutions for a number of limiting cases. At low frequencies, the solutions have the familiar Meissner screening current form with a spatial dependence $\varepsilon^{-x/\lambda_J}$, but involving the magnetic screening length λ_J rather than the London penetration depth λ_L. When the width of a crystal (or an individual junction) is much less than λ_J, it is referred to as being in the small junction limit, in which the magnetic fields and currents are essentially uniform over the width of the junction. In the opposite limit, the current flow across the junction area is non-uniform, and field is excluded from the bulk and only penetrates at the outer edges of the junctions or in the form of Abrikosov vortices between the superconducting layers. For layered cuprates the Abrikosov vortices have a finite size in the c-direction ($\sim \lambda_{ab}$) and therefore extend over many layers but are centered between a particular pair of superconducting CuO planes. The Josephson penetration depth λ_c for BSCCO ranges from $\sim$ 40-300μm as one passes from the over-doped to underdoped regimes. This is usually much less than the size of single crystals studied but is larger than the size of the mesa structures that we will describe later.

In the normal state, we have the familiar solution for penetration into a metal with an evanescent electromagnetic wave with a skin depth δ, which for BSCCO is typically >1 mm. This can often be significantly larger than the lateral dimensions of crystals available for experiments. Therefore, at some temperature on passing from the normal to fully superconducting state, the effective penetration length will match the lateral dimensions of the sample. This gives rise to a peak in the microwave losses, consistent with the interpretation of early microwave measurements on BSCCO single crystals[57].

At finite frequencies there are two important effects resulting from the time-varying magnetic and associated electric fields inside the superconductor. This leads to ac losses and a surface resistance $R_s = \frac{1}{2} \mu_o \lambda_J^3 \sigma \omega^2$. This is used to determine the "normal fluid", quasi-particle, conductivity σ in the superconducting state, which determines the losses and hence the performance of microwave devices. Secondly, it is important to recognise that the penetration depth of any superconductor (even conventional superconductors at sufficiently high frequencies) is *frequency dependent*, with

$$\lambda_{(\omega)} = \frac{\lambda_J}{\sqrt{1 - \left(\dfrac{\omega}{\omega_P}\right)^2}} \, , \tag{9}$$

where, for HTc superconductors for currents in the c-direction (λ_J), $\omega_P = \sqrt{\dfrac{J_c 2\pi s}{\Phi_o \varepsilon \varepsilon_o}} = c/\lambda_J$.

For strongly anisotropic superconductors, the plasma frequency in the c-direction is reduced by the anisotropy factor $\sim \gamma$ from its value for fields in the ab-direction, where $\omega_p = c/\lambda_{ab}$. For BSCCO, ω_p for electric fields in the E-direction (corresponding to ac magnetic fields parallel to the planes) is typically ~100 GHz. This frequency can be reduced into a more readily accessible experimental range, 10-50GHz, by decreasing the critical current by raising the temperature or applying a magnetic field.

As the measurement frequency approaches the plasma frequency, the penetration depth becomes very large and the field penetrates the whole sample. Above the plasma frequency, the wave vector acquires a real part so that a damped electromagnetic waves can propagate through the sample.

There been several reported measurements of the plasma resonance in the range 10-100GHz for HTc[58,59,60] and layered organic superconductors[61]: c-axis plasma resonances at infra-red frequencies have also been reported for a number of the more strongly coupled HTc superconductors[62]. Plasma resonances involving standing waves in the c-direction across the thickness of the crystal, rather like TEM wave-guide modes, have been observed in FIR transmission measurements[63].

Observation of the plasma frequency enables the Josephson coupling energy between layers, and hence the intrinsic critical current, to be determined without the need to attach electrical leads to a sample. Unlike conventional measurements of the critical current, the plasma frequency is not complicated by early transitions caused by thermal fluctuations, which simply broaden the resonance. Such measurements can therefore, in principle, provide a more reliable determination of the Josephson coupling energy. However, identification of the plasma resonance is not without its own problems. In order to achieve the sensitivity required to observe the plasma frequency, most experimentalists use a bolometric technique, The plasma resonance is then identified with a peak in the microwave losses, when the plasma resonance is tuned through the measurement frequency by varying either the temperature or field. However, a loss peak leading to sample heating is also expected, when the effective penetration depth matches the sample size[57], and it is not always clear from published data that such a possibility has been considered. Conclusive evidence for the plasma resonance ideally requires a measurement of both the reactive and resistive response of a crystal. Unlike the matching size peak, a plasma resonance would exhibit the characteristic dispersive and absorption features associated with any resonance.

We now return to the problem of the electrodynamic response at finite temperatures, especially above the irreversibility line in HTc superconductors, where thermally activated phase-slip results in a finite resistance R_{th} in series with the Josephson inductance. For frequencies $\omega << R_{th}/L$, the response will then be dominated by R_{th}, - as illustrated by the dc resistance measurements in Fig 1. A well-defined plasma resonance will only be observed when $\omega L >> R_{th}$.

3.6 Influence of Magnetic Fields on c-axis Tunnelling

The critical current across intrinsic HTc junctions is determined by the total Josephson coupling energy across the area of the junction, $-\varepsilon_J \int \cos\theta_{xy} dxdy$, where θ_{xy} is the phase difference across adjacent superconducting planes. Thermal fluctuations of the order parameter and any lack of alignment of pancake vortices along the c-direction result in spatial fluctuations of θ_{xy} and hence a reduction of the coupling energy and critical current.

If the pancake vortices were perfectly aligned along the c-direction, the phase changes around a vortex in any one plane would exactly match the phase changes around aligned vortices on each adjacent plane. There would then be no reduction in the coupling energy or critical current. However, if a pancake in one plane is displaced a distance u from its equilibrium position, there will be an area $\sim u^2$ between the centre of the displaced vortex and the vortices immediately above and below it where the phase is changed by $\sim \pi$, leading to decrease in coupling energy $\sim \varepsilon_J$ u^2 (see ref. 64). This provides the restoring force $\sim \varepsilon_J u$, to align the pancake lattice. We now consider $n = B/\Phi_0$ vortices per unit area, each displaced on average a small distance u, much less than their spacing $(\Phi_0/B)^{1/2}$, from perfect alignment. The critical current density will be then be decreased by a factor $[1- f(B/\Phi_0)u^2] \sim \exp[-fBu^2/\Phi_0]$, where $f = \pi/2$ [65]. At low fields, where the flux line spacing will be much larger than the average distance between the pinning centres - the large pinning density limit, an initial linear decrease in critical current with applied field is therefore expected, similar to that reported for 2212-BSCCO single crystals by Luo et al[11].

In the opposite limit, only a small fraction of the vortices will be trapped on pinning sites, and the vortices between them will be partially ordered. Josephson coupling between the planes will favour the alignment of ordered regions across adjacent planes, so that the exact structure will be a delicate balance between pinning, the Josephson coupling, and the degree of correlation of pancakes within the ab-planes. Correlations of the pancakes within the planes then leads to field dependences varying as $1/B^\mu$ as described by Bulaevskii et al[66]. Detailed theoretical analysis of such effects have been presented by several authors [65-68] (see also Bulaevskii et al[67] on the influence of correlations on the plasma frequency, which is directly related to the critical current density). The field dependence of the critical current across intrinsic junctions therefore provides a powerful probe to study correlations of the pancake vortices on the atomic interlayer scale in both the c-direction and the ab-planes.

In the liquid state above T_{irr}, the pancake vortices in each plane are assumed to be strongly disordered in the ab-planes, so that no correlation of the phases is expected in the c-direction. Nevertheless, Josephson coupling between vortices in adjacent layers leads to a thermally averaged correlation in their positions and a non-vanishing critical current.

To evaluate the critical current density in the liquid state, we determine the average coupling energy $\varepsilon_J < \cos\theta >$ between adjacent planes. If each pancake were to move completely at random in the spatially varying fluctuating potential of the randomly ordered pancakes in adjacent layers, the spatial average of $\cos\theta$ will therefore be zero. However, because of the Josephson coupling between layers, vortices will tend to spend more time in aligned positions than otherwise. The appropriate thermal average is therefore $< \cos\theta \exp[\varepsilon_J \cos\theta \, \Phi_0/BkT] >$, where Φ_0/B is the coupling area associated with each pancake. In the large temperature limit ($kT >> \varepsilon_J \Phi_0/B$), this reduces to $\varepsilon_J \Phi_0/2 \, kTB$, derived more rigorously by Koshelev[68]. This explains why the plasma resonance can still be observed in the liquid state well above the irreversibility line[59], and accounts for the finite critical currents observed in the mesa measurements to be presented later.

The above model also explains why little change in the critical current and plasma resonance frequency is observed on passing from the 2D flux lattice to the liquid state, since even in the flux lattice state the pancakes will be fairly mobile near T_c, which will allow a degree of correlation in the position of pancakes across the planes. Note that the larger the correlation lengths of pancakes within the planes the larger will be the correlation across the planes and hence the critical current.

Although the Josephson coupling energies and critical currents are not expected to change very dramatically on passing through the irreversibility line, the resistive losses arising from phase-slippage change dramatically. For a simple junction, the ratio L_J/R defines an effective time constant characterising the loss of phase coherence across the junction. In a magnetic field, we can similarly identify the length of time that pancakes spend in correlated positions across the planes with the ratio of the Josephson inductivity and dc c-axis resistivity, R_{th}.

476

4. C-AXIS MEASUREMENTS

4.1 Single Crystal Measurements

Several groups have published ab-plane and c-direction, field dependent, resistivities for 2212-BSCCO single crystals similar to Fig.1. Such measurements always tend to be made at very low currents to avoid sample heating and possible damage. A characteristic feature of the normal state c-axis measurements is the rapid increase in resistance on decreasing temperature, which we will argue later is associated with a gap in the spectrum of low lying tunnelling states. Superconducting fluctuations may also contribute to the increase in resistance[36,37]. Fluctuations are always important near T_c in highly anisotropic HTc superconductors, leading to a λ-transition in the heat capacity[69,70] in zero field and a very broad peak in finite fields with an onset close to $T_c(0)$ (see the paper by Junod in this volume). The excitation of superconducting pairs in the CuO-planes above T_c results in a reduction in the number of normal electrons available to tunnel across the planes. Although the superconducting fluctuations will enhance the conductivity in the ab-planes, they will not contribute significantly to the c-axis conductivity well above T_c, as the thermal fluctuations will occur on a sufficiently fast time scale to cause $< \cos(\Delta\theta) >$ to be averaged to zero. Nevertheless, on approaching T_c, the length scale and time scales over which the superconducting fluctuations occur increases, so that fluctuations on adjacent bilayers become increasingly correlated. The resulting c-axis supercurrent then leads to the observed decrease in resistivity towards zero as the irreversibility temperature T_{irr} is approached.

Magnetic fields suppress the superconducting transition, allowing the normal state resistance to achieve a higher peak value, as illustrated in Figs. 1a and 1b for both the single crystal and the mesa measurements to be described later. The mesa measurements suggest that, at least in a magnetic field, a contribution to the increase in resistance may be associated with a fluctuation term, which sets in strongly below $T_c(0)$ and is suppressed at sufficiently large fields. This results in an appreciable regime of negative magneto-resistance, illustrated in Fig. 1b, which has been investigated in some detail by Hashimoto et al.[71].

Below the peak in the c-axis resistance, the resistance drops continuously to zero with a similar thermally activated temperature dependence in both the ab-plane and c-directions[72]. Initially, the appearance of a relatively large, very broad resistive transition in the c-direction, for fields parallel to the current, was something of a puzzle, as this is the force-free configuration, with no Lorentz force acting on the flux lines to cause resistive flux flow. Tinkham[73] was the first to account for the thermally broadened resistive transition in this configuration by invoking the Ambegaokar-Halperin phase-slip model[54], assuming a thermally activated phase-slip resistance $R_{th} = R[I_0(E_J/2kT)]^{-2}$, with suitably chosen field and temperature dependent values for E_J appropriate for HTc superconductors.

However, below the irreversibility line $T_{irr}(B)$ in HTc superconductors, the dc resistance becomes vanishingly small. A number of authors have reported a thermally activated c-axis resistance varying more closely as $R \exp[-E^*/k(T-T_{irr})]$ [17, 74, 75] on approaching T_{irr} from above. All authors successfully fit their measurements to empirical models, but a rigorous theoretical model is lacking.

As remarked earlier, Kleiner et al[3] were the first to demonstrate VI characteristics of 2212-BSCCO single crystals showing multi-branched structures consistent with a stacked array of intrinsic junctions. It is tempting to identify each branch of the characteristics with the superconducting transition of an individual intrinsic junction - typically around 100 or so in a thin crystal (~1 micron thick). At each transition there is a jump of ~ 20 mV from one branch to the next, so that a very large voltage is developed across the crystal on passing from the superconducting to normal state. For typical critical currents of a few mA, the power dissipated can be many mW, which inevitably causes significant heating. There are also other problems in conventional measurements on single crystals, as the measured critical current will always be the weakest near the surface, which can easily be degrade by surface treatment associated with attachment of electrical bonds. Nevertheless, a number of groups have demonstrated the existence of a well defined critical current along the c-axis in single crystal measurements [11,76]. In these measurements, the c-axis characteristics were shown to be consistent with that expected from a series array of Josephson tunnel junctions, with thermally broadened, reversible, characteristics just below T_c and highly hysteretic characteristics at lower temperatures, when $\beta_c > 1$. Luo et al[11] also high-lighted the non-linear VI characteristics in the normal state, which is confirmed by measurements on small mesa structures.

More recently, a number of groups[5-7] have pioneered methods for fabricating very structures on the surface of BSCCO and other HTc superconductors involving a restricted number of junctions, so that problems of heating and of degradation from bonding to electrical contacts are largely overcome. For the remainder of this paper we focus on some recent measurements[18] made in Birmingham on mesa structures by August Yurgens and Birmingham colleagues, using mesa structures fabricated by Yurgens and his colleagues at Chalmers University.

5. MESA STRUCTURES

5.1 Fabrication

A typical mesa structure, which enables both ab-plane and c-axis measurements to be made, is shown in Fig.4. Gold is first sputtered onto the freshly cleaved surfaces of a BSCCO crystal to form the final electrical contacts on the top of each mesa. The epitaxial structure is then "sculpted" out of the single crystal by a combination of Ar-beam milling and chemical lithography using appropriate masking and photo-lithography[77]. The parent BSCCO crystal is typically about 1 x1 mm in area and ~100 μm thick. The central raised section is ~ 100x100 μm² micron square, on top of which four mesa structures of various areas (20x20 micron to 100x20 micron) are formed. The top level structures are patterned using ion beam milling at 77K, while simultane-ously measuring the superconducting characteristics. This enables mesas to be fabricated with an array containing a chosen number of junctions (half unit cell in height), typically in the range 5-15, though mesa structures with only a single CuO-BiO-CuO Josephson junction have been fabricated. The slot milled into the top of each mesa structure defines the uppermost atomic layers over which the VI-characteristics are made.

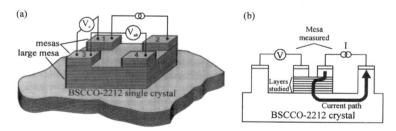

Figure. 4. A typical mesa structure fabricated by Yurgens et al.[77] and a schematic illustration of the 4-point resistance connections to the stack of intrinsic junctions within a particular mesa. The voltage contacts provide potential contacts to the uppermost and lowest members of the junction stack to be investigated.

4-point resistance measurements are made across an array of intrinsic junctions, the upper-most of which is just below the originally cleaved and coated surface, minimising problems form surface contamination or oxygen and metal diffusion. All the processing is performed at relatively low temperatures (<150°C), to prevent problems from inter-diffusion of the gold or oxygen into the array to be investigated. This might otherwise significantly decrease the superconducting properties of the uppermost layers, which is very difficult to avoid when bonding wires to the surface of larger sized crystal samples.

Current is supplied between the largest mesa and any one of the remaining three. Voltage contacts are made to one of the remaining two mesas and the mesa with the stack to be investi-gated, as illustrated schematically in Fig, 4b. Voltage contacts to the two mesas not connected to the current supply enable the ab-plane characteristics to be monitored, though such measurements can only provide relative values, as the geometry is ill-defined.

5.2 Normal State and Onset of Superconductivity

As emphasised earlier, almost all measurements of the c-axis resistance reported in the literature have been made in the small current and voltage, linear, ohmic regime, whereas in practice the characteristics become increasingly non-linear below room temperature[11]. The non-linearity of the normal state characteristics is illustrated in Fig. 5a by differential conductivity measurements across a stack of eight intrinsic junctions in a BSCCO mesa structure(determined from the number of resistive branches when superconducting).

A marked decrease in dynamic conductance develops at small voltages between room temperature and T_c. The voltage across each individual junction (a few mV) is one eighth of the voltage across the stack. This is less than thermal energies at T_c, so the characteristics will involve a considerable amount of thermal broadening. It is tempting to associate these features with the presence of a pseudo-gap and a redistribution of the density of states of electrons in the CuO planes from low to higher energies, though it could be a feature entirely confined to the c-axis conduction. The familiar increase in resistance along the c-axis *measured at small currents* is therefore a small voltage phenomenon only, and is a direct consequence of the reduction in the number of low lying carriers arising from the emergence of a gap feature in the c-axis tunnelling.

At T_c, new features occur corresponding to a redistribution of states consistent with a super-conducting energy gap. We assume that the c-axis tunnelling provides an average over all directions in the ab-plane, but as yet have no proper microscopic theory to interpret these measurements. The central, thermally broadened, feature arises from the c-axis critical current, which emerges immediately the additional superconducting features at higher energies are observed. The superconducting features appear to emerge with an already formed superconducting gap feature $(2\Delta/e)$ of a few mV per junction layer (one eighth of the voltage across the stack), unlike a BCS superconductor where the gap feature would increase from zero at T_c.

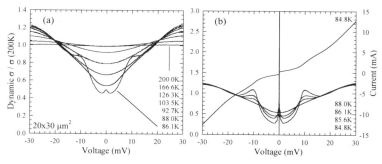

Figure 5. Measurements of the dynamic conductance across a mesa stack of eight intrinsic 2212-BSCCO junctions (a) from room temperature to just below T_c illustrating the emergence of the pseudo-gap features, and (b) in a 4K temperature range from just above to just below T_c, illustrating the emergence of superconductivity (Parker et al[82]).

It is interesting to compare these measurements with ultra-high vacuum tunnelling microscopy measurements on 2212-BSCCO crystals by Fischer's group[10] (see also the article in this volume). By carefully scanning over a large areas of a freshly single crystal, areas with reproducible tunnelling characteristics can be selected. When reproducible, the characteristics exhibit a similar dip in the normal state tunnelling conductance to that observed in our mesa experiments. Additional structure associated with the transition to the superconducting state is also observed at T_c. However, such measurements involve extremely small currents and are intrinsically rather noisy, whereas the noise on the intrinsic c-axis tunnelling mesa measurements is extremely small. We can therefore study the development of the superconducting gap features near T_c in considerably more detail than in the vacuum tunnelling measurements. However, vacuum tunnelling involves metal-insulator-superconductor (MIS) tunnelling and therefore has the advantage of being able to probe possible differences in the densities of state above and below the Fermi surface. The vacuum tunnelling measurements are indeed asymmetric with an apparent peak below the Fermi surface. Bok et al. associate this peak with a Kohn anomaly below the Fermi surface[78](see also the article by Bok in this volume), which the authors believe can also account for HTc superconductivity and the observed vacuum tunnelling within the framework of BCS theory. Unfortunately, tunnelling across the CuO-BiO-CuO (SIS) junctions is intrinsically symmetric with respect to the Fermi surface, which precludes the possibility of identifying any asymmetry in the density of states of BSCCO.

Nevertheless, the extremely high sensitivity and resolution of the mesa experiments and the intrinsic nature of the tunnelling involved suggest that considerable experimental and theoretical effort should be directed towards properly understanding the VI characteristics of the intrinsic junctions in BSCCO and other HTc superconductors. This will include a systematic study of the affect of doping across the underdoped to overdoped regimes. Such measurements should usefully complement vacuum tunnelling and ARPES determinations of the dependence of both the pseudo-gap and superconducting energy gap on doping.

5.3 Critical Currents and ac Josephson Effects

Immediately below T_c the VI characteristics are reversible, consistent with the thermally rounded characteristics expected for a series array of capacitatively shunted RSJ intrinsic junctions having eight, slightly different, critical currents. Such differences presumably arise from inhomogeneities in the bulk properties, slight changes in cross-section of the mesa structures, and a non-uniform flow of the supercurrent intrinsic to the geometry of the mesa structure.

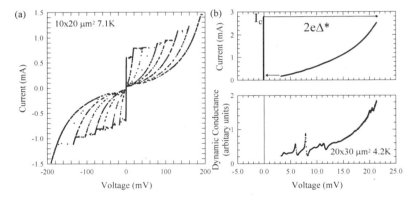

Figure 6. (a) Multi-branched hysteresis curves for a mesa including a stack of eight intrinsic junctions. (b) critical current, phase-slip voltage and resonant features for the first junction to enter the resistive regime (Parker et al.[82]).

5K or so below T_c', the McCumber parameter for the intrinsic junctions exceeds unity and the characteristics become hysteretic. The characteristics then become multi-branched with distinct critical current transitions with phase slip branches corresponding to the number of junctions in the mesa structure, as illustrated in Fig. 6a . By confining measurements to the first critical current and associated resistive branch, detailed VI measurements can be made across a single junction within the mesa structure, as illustrated in Fig. 6b. It is then possible to determine the dependence of the critical current on temperature and magnetic field, which can be applied parallel or perpendicular to the planes.

Earlier authors have associated the voltage discontinuity at I_c as $2e\Delta$, the superconducting energy gap. However, the observed values of typically 15–25 mV are only half the value expected from vacuum tunnelling measurements[10]. One possible explanation for the rather lower values deduced from intrinsic tunnelling is a reduction in the superconducting order parameter from strong quasi-particle injection effects. Yurgens et al.[5] have proposed an alternative explanation and argue that this is just what one would expect if the tunnelling between the layers is via intermediate metallic BiO layers.

In the phase-slip state at low temperatures, the McCumber parameter becomes very large, so that the junction persists in the phase-slip state until the current bias is decreased almost to zero. The VI characteristics in the phase slip regime is largely determined by the parallel conductance, which in turn is determined by the tunnelling of low lying excitations in the superconducting state. In this regime, the phase-slip current I is approximately as V^2, as evident from Fig.6b. This is qualitatively consistent with the expected linear density of low-lying states expected from the nodes in the energy gap of a d-state superconductor.

Superimposed on the phase-slip VI characteristics are highly reproducible features, common to all mesas on the same crystal, illustrated by the dynamic conductance measurements shown in the inset of Fig.6c. The "resonant structure" on the overall background characteristics occurs at different frequencies for different HTc superconductors [6,79]. These features strongly suggest resonances with those vibrational modes of the crystal that can be excited by Josephson radiation polarised with the E-field along the c-axis at a frequency $2eV/h$.[80]. Essentially, the McCumber parameter β_c, which controls the dynamics of phase slip, acquires a resonant structure through the dependence of the interlayer capacitance on the dielectric constant, $\varepsilon(\omega) = 1 + \sum \dfrac{A_n}{\omega_n - \omega + i\omega/Q_n}$,

where ω_n are the frequencies of the coupled lattice modes. The observed resonances extend to well above 5THz (1mV ~0.5 THz), demonstrating that that such structures could in principle be used for the voltage-tuneable generation and detection of radiation at THz frequencies.

5.3 Magnetic Field Dependence of the Critical Current

The critical currents derived from the intrinsic interlayer tunnelling characteristics exhibit a very strong dependence on the external field, particularly for fields along the c-direction, as shown in Fig. 7. The error bars indicate the variations in values of the measured critical current from typically 1000 measurements at each temperature and field and are intrinsic artifacts of the measurements rather than thermal or external noise. The black dots indicate the critical current determined from the main voltage transition to the first resistive branch, whereas the open circles represent the critical current determined using a 10μV voltage criterion across the whole stack. The latter are therefore essentially measurements of the thermally activated phase-slip conductance above the irreversibility line at T_{irr}, plotted in Fig.1. In zero field, the critical current decreases monotonically to zero at T_c on increasing temperatures, with a temperature dependence similar to the theoretical temperature dependence expected for a conventional BCS tunnel junction.

The fields used in these experiments were always larger than the fields required to destroy the 3D ordered pancake lattice at low temperatures, which are typically in the range 20-60 mT as deduced from both neutron diffraction, μSR studies of the flux lattice and observation of the onset of the arrowhead anomaly in magnetisation[50]. In the magnetic field, the pancakes therefore have no long range order along the c-axis. Nevertheless, they retain a degree of correlation on an atomic interlayer scale, as these measurements demonstrate.

Current interpretation of μSR and neutron diffraction measurements[81] suggests that, at high fields and low temperatures, the pancakes within the CuO planes are ordered, with little correlation between planes. On increasing temperature, an increasing number of dislocations in the ordered pancake lattice are excited ultimately leading to a 2D melting transition at around 20K, very close to the theoretically predicted temperature. The observed irreversibility line at high temperatures is then not the melting line, but rather the temperature when correlations in pancake alignment of the relatively viscous liquid matches the experimental time scale.

We can qualitatively account for the mesa measurements of the field dependence of the critical current in terms of such a model. At low temperatures ordered regions of pancakes will exist in the individual planes, but distorted randomly by local pinning. The spatial average of cos $\Delta\theta$ between planes and corresponding critical current will therefore be very small, with little thermal energy available for optimising correlations between the planes On increasing temperature, thermally excited dislocations within the planes will enable an increasing readjustment in position of the ordered pancakes within any one plane to take advantage of the Josephson coupling energy between the planes, leading to an increase in critical current with increasing temperature.

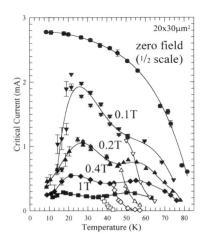

Figure 7. Temperature and field dependence of critical current in a 2212-BSCCO mesa (Yurgens et al. [18])

Close to the melting transition as ~20K, there will be sufficient mobility of dislocations to enable the pancakes to take full advantage of thermal averaging, as described earlier. This would account for the reduction in noise on the measurements and the absence of any marked change in the critical current on passing through the assumed 2D-lattice to liquid transition just above 20K. The critical current will then continue to decrease because of increased thermal fluctuations, as predicted by Koshelev[68]. However, even in the liquid state, there is little dissipation until one reaches T_{irr}, the temperature at which the crystal start to exhibit a significant dc-resistance. At around this temperature there appears to be a dip in the critical current determined from the characteristics and a measurable increase in c-axis resistivity. In the assumed liquid state above 25-30K, the field dependence is very close to the $1/B$ dependence predicted in the high tempera-ture limit[68] (see section 3.6).

These measurements provide direct evidence for a finite degree of correlation of pancakes at all temperatures and fields, even well into the assumed 2D liquid state. There is also little differ-ence in the Josephson coupling between planes, and hence critical current, as one passes from the glassy to liquid state. This is consistent with the lack of any significant change in the apparent plasma frequency derived from microwave measurements on passing into the liquid state.

A more complete description and detailed analysis of these mesa measurements will be published elsewhere[18,82].

6. SUMMARY

We have shown that thermal fluctuation play a very important role in determining the field dependent c-axis conduction properties of BSCCO single crystals both in zero and finite magnetic fields. Measurements on mesa structures fabricated from single crystals have been shown to provide highly reproducible measurements of the intrinsic c-axis properties, which are determined by atomic interlayer tunnelling across intrinsic Josephson junctions. Evidence for a gap in the c-axis tunnelling density of states is presented for both the normal and superconducting states, with extremely well defined additional superconducting features appearing at a sharp transition at T_c. A systematic experimental study is required to investigate the nature of these characteristics across the whole underdoped to over-doped regimes. More theoretical work is required to understand the detailed form of these characteristics in both the normal and superconducting states,. Measurements of intrinsic critical currents in magnetic field are shown to provide detailed information on the correlation of pancakes in the c-direction on an CuO_2 interlayer scale and demonstrate a finite, but much reduced, critical current extending well into the liquid state with no great change in value on passing from the glassy to liquid states. In the phase-slip state, the observed conductance varies approximately proportional to V^2, which is qualitatively consistent with a linear density of states for low-lying excitations expected from d-wave superconductivity. Resonant structure on such characteristics have been associated with induced resonances of structural modes of the crystal excited by intrinsic Josephson radiation with frequencies extending to ~5THz and above.

ACKNOWLEDGMENTS

I am particularly grateful to August Yurgens and my Birmingham colleagues, and Ian Parker in particular, for permission to quote as yet unpublished measurements and for many illuminating discussions. August Yurgens and his colleagues in the Department of Microelectronics and Nanoscience, Chalmers University, Sweden designed and fabricated the mesa structures on which the majority of the reported measurements were made. The crystals were grown by Guang Yang at the National Crystal Growth Facility for Oxide Superconductors, University of Birmingham. The research at Birmingham is supported by the EPSRC under a major programme on HTc superconductors. Finally, I should like to acknowledge NATO support for the Advanced Summer School on The Gap Symmetry and Fluctuations in High Tc Superconductors, which provided the incentive to write this paper.

REFERENCES

1. B. Battlog in Physics of High-Temperature Superconductors, Springer Series in Solid-State Sciences 106, (Springer-Verlag, Berlin Heidelberg 1992) pages 220-221
2. S.L. Cooper and K.E. Gray, Anisotropy and Interlayer Coupling in the High Tc Cuprates in Physical Properties of High Temperature Superconductors IV, ed. D.M. Ginsberg (World Scientific Publishing, Singapore, 1994)
3. R. Kleiner, F, Steimeyer, G. Kunkel and P. Muller, Phys. Rev. Lett. 68, 2394 (1992); R. Kleiner and P. Muller, Phys. Rev. B49, 2394 (1992).
4. R. Kleiner. Jnl. Low Temp. Phys. 106, 453 (1997)
5. A. Yurgens, D. Winkler, N.V. Zavaritsky and T. Claeson, Phys. Rev. B, R8887 (1996); A. Yurgens, D. Winkler, Y.M. Zhang, N. Zavaritsky, and T. Claeson, Physica C235-240, 3269 (1994)
6. K. Schlenga, G. Hechtfischer, R. Kleiner, W. Walkenhorst, P. Muller, H.L. Johnson, M Veith, W. Brodkorb, and E. Steinbeiss, Phys. Rev. Lett. 76, 4993 (1996)
7. Y.I. Latyshev, P. Monceau, and V.N. Pacvlenko, Physica C282-287, 387 (1997)
8. J. Takeya, S. Akita, S. Watauchi, J. Shimoyama, and K. Kishio, Physica C282-287, 2439 (1997)
9. K. Tanabe, Y. Hidaka, S. Karimoto, and M. Suzuki, Phys. Rev. B53, 9348 (1996)
10. Ø. Fischer, Ch. Renner, I. Maggio-Aprile, A. Erb, E. Walker, B. Revaz, and J.-Y. Genoud, Physica C282- 287, 315 (1997)
11. S. Luo, G. Yang and C.E. Gough, Phys. Rev. B51, 6655 (1995)
12. W.E. Lawrence and S. Doniach, in Proceedings of the 12th International Conference on Low Temperature Physics, edited by E. Kanda (Academic Press of Japan, Kyoto, 1971) p 361
13. M. Tinkham, Introduction to Superconductivity (McGraw Hill, Singapore, 1996)
14. H. Krakauer and W.E. Picket, Phys.Rev.Lett. 60, 1665 (1988)
15. B. Szpunar and V. Smith, Phys Rev. B45, 10616 (1992)
16. M. Tuominen, A.M. Goldman, Y.Z. Chang, and P.Z. Jiang, Phys. Rev. B42, 412 (1990)
17. G. Briceno, M.F. Crommie, and A. Zettl. Phys. Rev, Lett. 66, 2164 (1991) .
18. A. Yurgens, D. Winkler, T. Claeson, G. Yang, I.F.G. Parker, and C.E. Gough, to be published.
19. J.H. Cho, M.P. Maley, S. Fleshler, A, Lacerda, and L.N. Bulaevskii, Phys. Rev. B , 6493(1994).
20. K. Takanaka et al., Phys. Rev. B50, 6534 (1994)
21. S. Uchida, Physica C282-287, 12 (1997)
22. J. Rossat-Mignod et al., Physica C185-189, 86 (1991)
23. R.E. Walstedt and W.W. Warren, Appl. Mag, Res. 3, 469 (1992)
24. M. Takigawa et al., Phys. Rev. B43, 247 (1991)
25. G.V.M. Williams, J. Tallon, R. Michalak, and R. Dupree. Phys. Rev. B54, R6909 (1996)
26. H. Yasuoka, T. Imai, and T. Shimizu, in Strong Correlation and Superconductivity (Springer Verlag, 1989) ed. H. Fukuyama, S. Maekewa and A.P. Malezomoff, p.254.

27. D.S. Marshall, D.S. Dessau, AG. Loeser, C.H. Park, A.Y. Matsura, J. Eckstein, I. Bosovic, P. Fournier, A. Kapitulnik, W.E. Spicer, and Z,-X Shen, Phys. Rev. Lett. **76**, 4841, (1996)

28. A.G. Loeser, Z.-X. Shen, D.S. Dessau, D.S. Marshall, C.H. Park, P. Fournier, and A. Kapitulnik , Science **283**, 325 (1996)

29. J. Loram, K.A. Mirza, J.R. Cooper, and W.Y. Liang, Phys. Rev. Lett. **71**, 1740 (1993): J.W. Loram, K.A. Mirza, J.R. Cooper, W.Y. Liang and J.M. Wade, J. Supercond. **7**, 243 (1994)

30. P.W.Anderson, Phys. Rev. Lett. 59, 2497 (1987): P.W. Anderson, Physica C**185-189**, 11 (1991)

31. J.M. Wheatley, T.C. Hsu, and A.C. Anderson, Phys. Rev. B**37**, 5897 (1988)

32. N. Nagaosa, Phys. Rev. B**52**, 10561 (1995)

33. D. Pines, Physica C**235-240**, 113 (1994): Physica C**282-287**, 273 (1997)

34. A.J. Rojo and K. Levi, Phys. Rev. B**48**, 16861 (1997)

35. N.P. Ong, Science **273**, 32 (1996)

36. K.E. Gray and D.H. Kim, Phys. Rev. Lett. **70**, 1693 (1993)

37. L. Ioffe, A.I. Larkin, A.A. Varlamov, and L. Yu, Phys. Rev. B**47**, 8936 (1993)

38. G. Balestrino, M. Marinelli E, Milani, A.A. Varlomov and L. Yu, Phys. Rev. **47**, 6037 (1993): G. Balestrino, E. Milani, C. Aruta and A.A. Varlamov, Phys. Rev. B**54**, 3678 (1996)

39. V.V. Dorin, R.A. Klemm, A.A. Varlamov, A.I. Buzdin and D.V. Livanov, Phys. Rev. B**48**, 12,951 (1993)

40. A.A. Varlamov, Physica C**282-287**, 248 (1997); A.A. Varlamov, G. Balastrino, E. Milani and D.V. Livanov (review article- to be published).

41. A.G. Sun, D.A. Gajewsiki, M.B. Maple, and R.C. Dynes, Phys. Rev. Lett. **72**, 2267 (1994)

42. R. Kleiner, M. Mossle, W. Walkenhorst, K. Schlenga and P. Muller, Physica C**282**-287

43. Z.X. Shen et al., Phys. Rev. Lett. B**70**, 1553 (1993)

44. K. Ishida, Y. Kitaoka, N. Ogata, T. Kamino, K. Asayama, J.R. Cooper and N. Athanassopoulou, J. Phys. Soc. Jpn. **62**, 2803 (1993): K. Asayama et al . in Proc LT21, Prague, Czechoslavak Journal of Physics, 46-suppl, 3187 (1996)

45. W.N. Hardy, D.A. Bonn, D.C. Morgan, R. Liang, and K. Zhang, Phys. Rev. Lett. **70**, 3999 (1993)

46. S-F. Lee, D.C. Morgan, R.J. Ormeno, D.M. Broun, R.A. Doyle, J.R. Waldram and K. Kadawaki, Phys. Rev. Lett. **77**, 735 (1996)

47. J.R. Kirtley, C.C. Tsuei, J.Z. Sun, C.C. Chi, L.S. Yu-Hannes, A. Gupta, M. Rupp and M.B. Ketchen, Nature **373**, 225 (1995)

48. I. Giaever, Phys. Rev. Lett. **5**, 147 (1960).

49. J.R. Clem, Phys. Rev. B**43**, 7837 (1991)

50. R. Cubitt, E.M. Forgan, G. Yang, S.L. Lee, D. McK. Paul, H.A. Mook, M. Yehthiraj, P.H. Kes, T.W. Li, A.A. Menocsky, Z. Tarnawski, and K. Mortensen, Nature **365**, 407 (1993)

51. G. Yang, P. Shang, S.D. Sutton, I.P. Jones, J.S. Abell and C.E. Gough, Phys. Rev. B**48**, 4054 (1993)

52. R. Kleiner, P. Muller, H. Kohlstedt, N.F. Pedersen and S. Sakai, Phys. Rev. B**50**, 3942 (1994)

53. R. Kleiner, B. Aigner, B. Avenhaus, C. Kreuzer, P. Pospischil, F. Steinmeyer, P. Muller, and K. Andres, Physica B**194-6**, 1753 (1994)

54. V. Ambegaokar and B.I. Halperin, Phys. Rev. Lett. **22**, 1364 (1969)

55. P.A. Lee, J. Appl. Phys. **42**, 325 (1971)

56. D.E. McCumber, J.Appl.Phys. **39**, 2503 (1968); ibid 3113 (1968)

57. C.E. Gough and N.J. Exon, Phys. Rev. B**50**, 488 (1994)

58. O.K.C. Tsui, N.P. Ong, Y. Matsuda, Y.F. Yan, and J.B. Peterson, Phys. Rev. Lett. **73**, 724 (1994)

59. Y. Matsuda, M.B. Gaifullin, K. Kumagai, K. Kadawaki, and T. Mochiku, Phys. Rev. Lett. **75**, 4512 (1995)

60. S. Sakamoto, T. Maeda, T. Hanaguri, Y. Kotaka, J. Shimoyama, K. Kishio, Y. Matsuchita, M. Hasegawa, H. Takei, H. Ikeda, and R. Yoshizaki., Phys. Rev. B**53**, R14749 (1996)

61. T. Shibauchi, M. Sato, A. Mashio, T. Tamegai, H. Mori, S. Tajami, and S. Tanaka, Phys. Rev. B**55**, 11971 (1997)

62. H. Shibata and T. Yamada, Phys. Rev. B**54**, 7500 (1996).

63. S. Uchida - data in M. Tachiki, Physica C**282-287**, 383 (1997)

64. L.N. Bulaevskii, M. Levij, and V. Kogan, Phys. Rev. B**46**, 366 (1992)

65. L.L. Daemen, L.N. Bulaevskii, M.P. Maley, and J.Y. Coulter. Phys Rev. B**47**, 11291 (1993)

66. L.N. Bulaevskii, V.L. Pokrofsky, and M.P. Maley, Phys. Rev. Lett. **76**, 1719 (1996)

67. L.N. Bulaevskii, M.P. Maley, and M. Tachiki, Phys. Rev. B**74**, 801 (1995)

68. A.E. Koshalev, Phys.Rev.Lett. **77**, 3901 (1997)

69. W. Schnelle, P. Ernst, and P. Wohlleben, Ann. Physik **2**, 109 (1993)

70. S.E. Inderhees et al. , Phys. Rev. B**47**, 1053 (1993): N. Overend, M.A. Howson, I.D. Lawries, S. Abell, P.J. Hirst, C.K. Chen, S. Chowdhury, J.W. Hodby, S.E. Inderhees, M.B. , Phys. Rev. B**54**, 9499 (1996)

71. K. Hashimoto, K. Nakao, H. Kado, and N. Noshizuka, Phys. Rev. B**53**, 892 (1996)
72. V.R. Busch, G. Ries, H. Werthner, G. Kreiselmeyer, and G. Saemann-Ischnko, Phys. Rev. Lett. **60**,. 2194 (1988)
73. M. Tinkham, Phys. Rev. Lett. **61**, 1658 (1988)
74. K. Kadowaki, S.L. Yuan, K. Kishio, T. Kimura, and K. Kitazawa, Phys. Rev. B**50**, 7230 (1994)
75. M. Giura, S. Sarti, E. Silva, R. Fatampa, F. Murtas, R. Marcon, H. Adrian, and P. Wagner, Phys. Rev. B**50**, 12920 (1994)
76. J.H. Cho, M.P. Maley, S. Fleshler, A. Lacerda, and L.N. Bulaevskii, Phys. Rev. B**50**, 6493 (1994).
77. A. Yurgens, D. Winkler, and T. Claeson, Appl. Phys. Lett. **70**, 1760 (1997)
78. J. Bok and J. Bouvier. Physica C**274**, 1(1997): Physica C**282-287**, 299 (1997)
79. A. Yurgens, D, Winkler, N. Zavaritsky, and T. Claeson, SPIE Proceedings, Conference on Superconductor Physics and Nano-Engineering II, 2697 (1996)
80. F. Forsthofer, Ch. Helm, Ch. Preis, J. Keller, K. Schlenga, R. Kleiner, and J. Keller, Physica C**282**, 287, 2431 (1997).
81. E.M. Forgan (private communication)
82. I.P. Parker, A. Yurgens, C.E. Gough, M. Endres (to be published)

SCANNING TUNNELING SPECTROSCOPY ON HIGH TEMPERATURE SUPERCONDUCTORS

Øystein Fischer, Christophe Renner and Ivan Maggio-Aprile

Département de physique de la matière condensée
Université de Genève
24, Quai E.-Ansermet, 1211 Genève 4
Switzerland

INTRODUCTION

The discovery of the high temperature superconductors in 1986[1] radically changed the situation in the field of superconductivity. Not only did the critical temperature increase drastically, but it becomes more and more clear that the nature of the superconducting state in the high temperature superconductors is radically different from the one found in the well known low temperature superconductors. In spite of a considerable world wide effort to study these new superconductors we still do not know the origin and the nature of the pairing interaction. It remains one of the central challenges of solid state physics today to elucidate this mechanism.

One essential ingredient in the quest for an understanding of the superconducting state is the quasiparticle density of states or in other words the density of single particle excitations above the superconducting ground state. In 1960 Ivar Giaever [2] demonstrated that tunneling between a superconductor and a normal metal allowed a direct measurement of the quasiparticle density of states in the superconductor. After this path breaking work, tunneling became a main tool for investigating the various low temperature superconducting materials known at the time. In particular the inversion scheme developed by Rowell and McMillan[3] to extract the electron phonon coupling from the tunneling spectra made a quantitative verification of the BCS model for these low temperature materials possible. Since the discovery of the high temperature superconductors, numerous investigations have been carried out with the hope that such tunneling studies will finally allow to unravel the mechanisms leading to high temperature superconductivity. However, the cuprates showing high temperature superconductivity are complex materials and it has proven difficult to control the tunneling barriers to the degree necessary for the various tunneling methods to give reliable results. One possible way out of this situation is to use the advantages offered by a scanning tunneling microscope (STM). This instrument uses vacuum as a tunneling barrier and offers the possibility to modify the barrier and thus the junction in-situ. This in turn allows to separate extrinsic from intrinsic effects in the tunneling spectra. In this paper we shall present the method of scanning tunneling spectroscopy (STS) and review recent results we have obtained in our laboratory on high temperature superconductors. In particular we shall focus on the temperature and doping dependence of $Bi_2Sr_2CaCu_2O_{8+\delta}$ and the observation of vortices and vortex core spectroscopy on $YBa_2Cu_3O_{7-\delta}$ and $Bi_2Sr_2CaCu_2O_{8+\delta}$.

The early work of Giaever and coworkers used planar junctions. Usually these would be made of a thin film whose surface would be oxidized to form the tunnel barrier and subsequently one would deposit a second thin film as the counter electrode. The central property of such a tunnel junction is that if the normal metal has a constant density of states around the Fermi level then the tunnel conductivity ($\sim dI/dV$) at low temperatures is proportional to the quasiparticle density of states $N_s(E)$.

The scanning tunneling spectroscopy we shall consider here corresponds to a situation where the surface of the superconductor is being scanned with a normal tip and where the tip does not touch the surface so that the barrier is ideally the vacuum separating the two. This geometry avoids in principle the difficulty encountered with other tunneling methods where the barrier is an insulating material which is difficult to control microscopically. One may of course also encounter problems with STS, for instance the tip may touch the surface or the surface may have some unwanted particles or impurities and as a consequence the measured spectra do not reflect the superconducting density of states. However, the great advantage of the STM is that the barrier can be changed in-situ and thus such non-ideal situations can be detected and avoided.

In the idealized case where the tip can be considered as a point electrode and $T << T_c$, the STM measures the local density of states $N_s(E,r)$:

$$\frac{dI}{dV}(V,r) = G_{nn} \int_{-\infty}^{\infty} N_s(E,r) \left[-\frac{\partial f(E + eV)}{\partial(eV)} \right] dE \qquad (1)$$
$$\approx G_{nn} N_s(eV,r))$$

Here G_{nn} is the tunneling conductance when both electrodes are completely in the normal state. In the case of a uniform superconductor $N_s(E,r)$ is just proportional to the density of states itself. However, if the surface is non uniform either because of material inhomogeneity or because of intrinsic inhomogeneity, as is the case if vortices are present, the STM will detect these and this offers an extra advantage of the STS. Note that in the STM geometry we measure an average over all k-directions in the plane being investigated.

1. INVESTIGATION OF THE TUNNELING SPECTRA ON $Bi_2Sr_2CaCu_2O_{8+\delta}$

Typical Spectra.

Several groups have published tunnel spectra on $Bi_2Sr_2CaCu_2O_{8+\delta}$ obtained with an STM[4-8]. Basically these spectra agree on the main features. A typical spectrum obtained on an optimally doped $Bi_2Sr_2CaCu_2O_{8+\delta}$ sample is shown in Fig. 1. The dominating feature seen here is two well developed peaks, situated approximately at ± 40 meV. Contrary to a BCS superconductor with such well developed peaks, we find a large conductivity inside the gap. It has been shown that the low bias behavior is consistent with the nodes corresponding to a d-wave if we assume that the small but finite zero bias conductivity reflect impurity scattering[6]. However the weight of the peaks is much more important than expected from a simple d-wave superconductor. If we determine the maximum gap from the peak positions ($\pm\Delta_p$) we find $2\Delta_p/k_BT_c = 10 \pm 1.5$, a value much higher than the BCS value 3.5. A particular feature is the dip seen at approximately $-2\Delta_p$, usually not seen at positive sample bias[6,7]. Sometimes a weaker dip can also be seen around $+2\Delta_p$[8], but there is always an asymmetry in the spectra at these energies.

Testing intrinsic behavior.

An obvious question that comes to mind when non standard results like these are found, is whether or not they reflect the intrinsic properties of the superconductor. How do we, for instance, know that this spectrum is not perturbed by some extrinsic effects as for instance resulting from the tip touching the surface? We would therefore like here to illustrate how the STM can be used to test that the obtained spectra are reflecting intrinsic

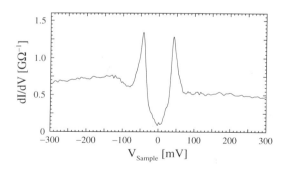

Figure 1. Typical tunneling spectrum for an optimally doped sample measured at 4.2K

effects. There are two tests that can verify this. The first one is to let the tip scan the surface at a constant height and take spectra at regular intervals along the surface. If the tip touches the surface it will usually not be possible to obtain good spectra at all. Furthermore, if the tip did not touch but some particle or impurity influenced the data, scanning along a sufficiently long line should reveal this. In Fig. 2a we show a case where the reproducibility of the data are demonstrated along a line of 34nm[6]. In this experiment spectra were taken every 0.7nm along this line. As can be seen, the spectra do not vary appreciably from one spot to the next along the whole line. If some perturbation would be present this would easily be seen by the loss of the structure reflecting the superconducting state. Fig. 2b show the spectra directly superposed on top of each other. All the main features discussed above are also clearly visible in this plot. The inset show the average spectrum over the 34nm line. The second test that can be carried out is to vary the tip sample distance. In the case the tunneling geometry is not simply superconductor/vacuum/metal-tip the shape of the spectra will be distance dependent and in that case we have the problem to determine at which distance the obtained spectra reflect the superconducting density of states. In Fig. 3 we show an example of tunneling on a $Bi_2Sr_2CaCu_2O_{8+\delta}$ surface where it is demonstrated that the spectra are independent of the tip to sample distance [6]. Fig. 3a shows the spectra taken at different distances

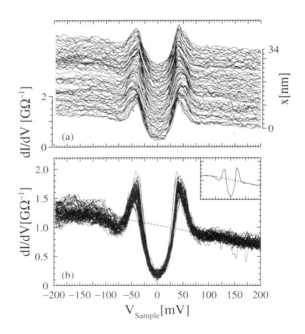

Figure 2. (a) Spectra taken along a 34 nm long line at 4.2K ; (b) superposition of the 50 spectra shown in (a). The inset show the average over all these spectra (From Ref. 6)

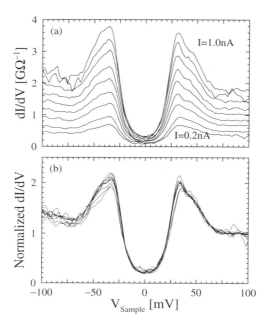

Figure 3. (a) Tunneling spectra measured as a function of tip/sample distance at 4.2K. The tunneling current is changed from 1.0 nA to 0.2 nA with 0.1 nA increments at constant voltage. (b) Same curves plotted in a normalized conductance scale. (From Ref. 6)

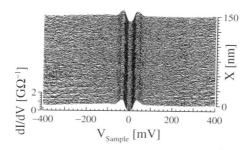

Figure 4. Tunneling spectra taken on an overdoped sample along a 150 nm long line at 4.2K illustrating the homogeneity both of the superconducting part and the ohmic high bias part.

and in Fig.3b these have been normalized and superposed. The latter clearly shows that the shape of the spectra are distance independent. The position and distance independence shown in Figs. 2 and 3 are not always verified on a given surface and in that case the spectra usually deviate from the characteristic shape, indicating that some inhomogeneity or other perturbation influences the experimental spectrum. The spectra discussed in this article were taken on surfaces where these criteria have been verified. As an example of the homogeneity obtained in our more recent samples we show in Fig. 4 a scan over 150 nm on an overdoped sample. Note the ohmic behavior at bias voltages above the gap.

Doping dependence

As described in other contributions of this volume, high temperature superconductors have a very characteristic temperature/carrier-concentration phase diagram. At low carrier concentration the compounds are antiferromagnetic insulators but at a certain doping level one observes an insulator- metal transition. Beyond this doping level the compounds become superconducting and the critical temperature T_c increases upon doping until T_c goes through a maximum and decreases upon further

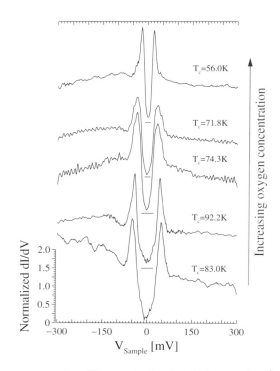

Figure 5. Tunneling spectra for different doping levels (increasing from bottom to top) at 4.2K.

doping. We refer here to the carrier concentration giving the maximum T_c as optimum doping and to the lower and higher doping levels as underdoped and overdoped respectively. Because all high temperature superconductors seem to have such a phase diagram its understanding is one of the central questions to answer.

It is generally expected that in the underdoped region the superconductor is dominated by strong electron-electron correlations and phase fluctuations, whereas in the overdoped region the superconductor may possibly become more BCS-like. Since such an evolution of the superconducting state may result in major changes in the quasi particle density of states, it is of interest to investigate how the tunneling spectra evolve as the doping is changed. This has been reported in references 6, 9,10. In Fig. 5 we show a series of spectra obtained for one underdoped sample (T_c = 83K), one optimally doped sample (T_c = 92K) and three overdoped samples, characterized by T_c = 74.3K, T_c = 71.3K and T_c = 56K. Contrary to T_c, the gap is reduced monotonically upon doping. Thus the ratio $2\Delta_p/k_B T_c$ decreases from 12 in the underdoped sample to 9 in the strongly overdoped sample. However, the shape of the spectra are basically the same. The presence of two well developed peaks as well as the asymmetric dip at -$2\Delta_p$, is present in all spectra although the dip feature seems to move away from the gap as the doping is increased. In particular we find that the normalized conductivities at low bias are mainly the same. This is illustrated in Fig. 6 where we have plotted the spectra for the underdoped and the strongly overdoped sample in normalized units. As can be seen the two shapes are really very similar. This tell us that the node
structure, which determines the low bias spectra, is very similar in the two. This is a strong sign that the basic pairing state is not changing with doping in this material.

Temperature dependence

A number of experiments have provided evidence of an unusual behavior of the electronic states in the normal state of underdoped

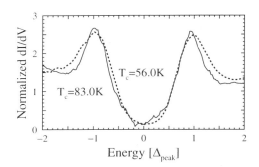

Figure 6. Tunnel spectra for the underdoped (T_c = 83K, full line) and the strongly overdoped (T_c = 56K, dotted line) samples in Fig. 5 plotted in reduced units.

compounds. Analysis of NMR[11], infrared conductivity[12], neutron scattering[13], transport properties[14], specific heat[15], thermoelectric power[16], spin susceptibility[17], and Raman spectroscopy[18], have shown evidence for the presence of a gap structure in the electronic excitation spectra below a characteristic temperature T^* ($> T_c$). More recently angular resolved photoemission (ARPES) studies [19,20] have given more direct evidence for the presence of a pseudogap in the density of states above T_c in underdoped samples.

An important question now is whether or not there is a direct relation between the pseudogap and the superconducting gap. Are the two of different origin or is the pseudogap somehow a manifestation of the presence of pairs above T_c? Tunneling spectroscopy is a direct way to obtain information on the pseudogap and thus on this question. In Fig.7 we show the temperature dependence of the tunneling density of states for an underdoped sample (T_c = 83K)[7]. The striking result is that the peak position does not vary at all with temperature up to T_c. (The highlighted line is a spectrum taken 1K above T_c). At T_c the peak at negative bias has disappeared but the peak at positive bias is still visible and moves to somewhat higher energy upon crossing T_c. However the most remarkable result is that even above T_c we observe a clear gap, which we identify with the pseudogap, with the same energy scale as the gap in the superconducting state. This structure is gradually reduced in amplitude as the temperature is increased, but there is no sign that the energy scale, i. e. the pseudogap, is reduced. Thus the superconducting density of states shows a temperature independent gap which evolves gradually into the pseudogap structure.

So far the pseudogap was only observed in underdoped compounds. However, in Ref.8 it was shown that the pseudogap is also observed in optimally doped and overdoped samples. In Fig. 8 we show the spectra obtained for an overdoped sample at various temperatures. The striking result here is qualitatively the same as in the underdoped sample. The peak position of the spectra does not change even for the spectrum taken at 69K, just 5K below T_c. As in the underdoped sample the peak at negative bias disappears at T_c and there remains a peak at positive bias moving to a slightly higher value as T_c is crossed. Again there remains a gap-like feature above T_c which seems to fill in faster as the temperature is raised than we observed in the underdoped sample. An important fact here is that the gap in the overdoped sample is considerably smaller than in the underdoped one and that the same is true for the pseudogap. This means that the pseudogap scales with the superconducting gap and thus shows that the two are intimately related.

In order to quantify the observation that the superconducting gap is temperature independent, we have calculated the thermal smearing of the 4.2K spectrum assuming that the gap is temperature independent. The result is shown for T = 69K. As can be seen, the peak position is well reproduced even for the 69K sample. If the gap would have a BCS-like temperature dependence the peak to peak distance would have been reduced by more than a factor two. The dashed line show the calculated spectrum for the case where the gap at 69K is reduced by 20%. Clearly this also does not fit the data. We arrive thus at the conclusion that the superconducting gap is temperature independent up to T_c where, without closing, it evolves into the pseudogap structure.

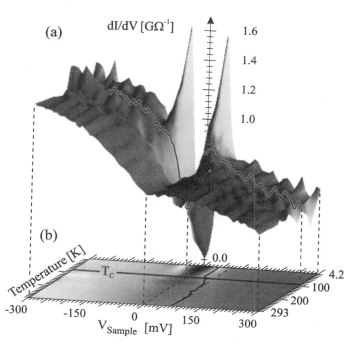

(a) dI/dV [GΩ⁻¹]

1.6

1.4

1.2

1.0

0.0

(b)

Temperature [K]

T_c

4.2

100

200

293

-300   -150   0   150   300

V_{Sample} [mV]

Figure 7. (a) Three dimensional view of the tunneling conductance versus voltage and temperature. The highlighted curve is the spectrum measured at T_c. (b) projection onto the energy-temperature plane. The line at positive bias indicates the position of the positive bias conductance peak which clearly shifts to higher energies above T_c. (From Ref. 7).

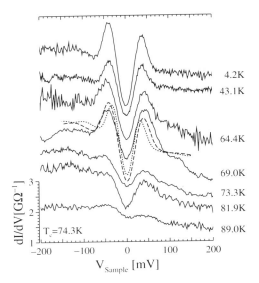

Figure 8. Tunneling spectra measured as a function of temperature on overdoped Bi-2212. The conductance scale corresponds to the 89K spectrum, the other spectra are offset for clarity. The dashed spectrum corresponds to the 4.2 spectrum thermally smeared to 69K ($\Delta_{69K} = \Delta_{4.2K}$). The dotted spectrum is obtained in the same way, but assuming a reduced gap at 69K ($\Delta_{69K} = 0.8\Delta_{4.2K}$) (From Ref. 7)

2. VORTEX CORE SPECTROSCOPY

The local quasiparticle density of states becomes position dependent if vortices are present in the superconductor. If we consider a situation where the magnetic field is perpendicular to the surface under consideration the tunneling spectra will depend on the position of the tip with respect to the vortex cores. When the tip is placed between vortices we expect to see a spectrum close to the zero field one. However, when the tip is situated at the

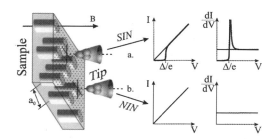

Figure 9. Vortex lattice imaging by STM is based on locally changing tunneling characteristics as illustrated here for a dirty BCS superconductor : (a) the tip sits between vortices and we have a local SIN junction, (b) the tip is positioned over a vortex core and we have a local NIN junction.

center of the vortex core we expect to see a completely different spectrum. If we naively consider the vortex core as a normal metal we would expect to see an ohmic behavior, i. e. a constant density of states. as illustrated in Fig. 9. In reality this is only seen in the so-called dirty limit, whereas in the clean limit the vortex core spectra contain important information about the superconducting state.

The first observation of vortices by this technique was reported by Hess et al.[21]. These experiments were carried out on the low temperature superconductor NbSe$_2$. A central result of this work was the observation of a zero bias conductance peak in the vortex cores. Following this work Renner et al.[22] showed in the alloy series Nb$_{1-x}$Ta$_x$Se$_2$, that this zero bias peak disappears as x is increased and the compound enters the dirty limit where the mean free path l becomes smaller than the coherence length ξ.

The vortex lattice of YBa$_2$Cu$_3$O$_{7-\delta}$ was observed by STM by Maggio-Aprile et al.[23]. Fig. 10a shows a view of the vortex lattice in this compound. It was found that this lattice is oblique and not hexagonal due to the a-b plane anisotropy and possibly also due to the d-wave component in the pairing state. Note also the elliptic shape of the vortex cores, reflecting the a-b plane anisotropy of the coherence length. That this deformation of the vortex lattice is related to the a-b plane anisotropy can be seen in Fig.10b which shows an image where a twin boundary is present[24]. As can easily be seen, the anisotropy switches by 90° when we cross the twin boundary. The latter is seen as a dark line in the figure due to the high density of trapped vortices. In the following we shall focus on the vortex core spectroscopy in the high temperature superconductors. For a more complete review on the observation of vortices by STM we refer to the review by Renner at al.[25].

The spectra obtained inside the vortex cores and outside the vortex cores are shown in Fig. 11 for NbSe$_2$ and YBa$_2$Cu$_3$O$_{7-\delta}$. The striking difference is that whereas in NbSe$_2$ we find a zero bias anomaly, in YBa$_2$Cu$_3$O$_{7-\delta}$ we find

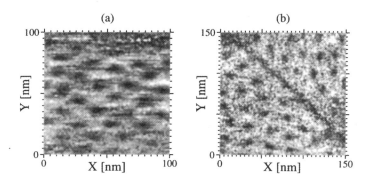

Figure 10. Vortex lattice images on an YBa$_2$Cu$_3$O$_{7-\delta}$ crystal at 4.2K. (a) Distorted lattice with elongated vortex cores at 6 Tesla (From ref. 23), (b) two domains of the vortex lattice near a twin boundary at 3 Tesla. The twin boundary appears as a dark line as a result of a large density of vortices trapped (From ref. 24).

a split peak. The origin of these structures is to be found in the internal electronic structure of a vortex. This was investigated already in 1964 by Caroli et al.[26] who found that the quasiparticle states are localized inside the vortex cores and that they are quantized with an equal spacing of about Δ^2/E_F. When the STM tip is situated in the vortex center only the first two states will show up in the tunnel characteristics. In NbSe$_2$ this spacing is of the order of μV and cannot be resolved. One therefore only observe a zero bias peak. Hess et al.[27] showed that this peak splits as the tip moves away from the vortex center demonstrating thereby the presence of a quasi continuos spectrum of states in the core. In YBa$_2$Cu$_3$O$_{7-\delta}$ we find two

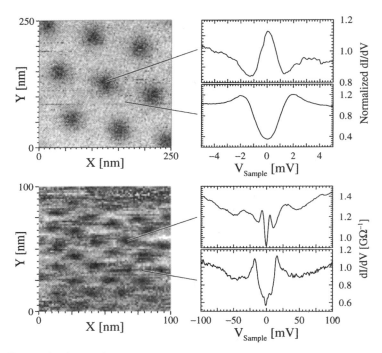

Figure 11. Vortex lattice and vortex core spectroscopy : Top : NbSe$_2$ at 1.3 K and 0.3 Tesla ; bottom : YBa$_2$Cu$_3$O$_{7-\delta}$ at 4.2 K and 6 Tesla.

peaks and these two do not shift when the tip moves away from the vortex core[23]. This demonstrates that the vortex core only contains two states which we identify with the first two states $\pm \Delta^2/2E_F$. The fact that we can observe these states is of course directly related to the fact that the gap is much larger in YBa$_2$Cu$_3$O$_{7-\delta}$ than in NbSe$_2$. This result tell us that we have an extreme quantum situation in that only two states are present and this in turn tell us that the gap is becoming of the order of magnitude of the Fermi energy E_F. Using the result of Caroli et al.[26] we estimate that $E_F \approx 2\Delta_p$. The exact position of the two states will of course depend on the exact pairing state. It has been found that a clean d-wave superconductor may only show a zero bias anomaly instead of a split peak[28]. However, if an s-wave component is present around the vortex core a split peak will appear[29].

The imaging of the vortex lattice on $Bi_2Sr_2CaCu_2O_{8+\delta}$ has turned out to be very difficult in spite of the very good and well developed quasiparticle tunneling spectra. Recently it has appeared that this is because the difference between the quasiparticle spectra inside and outside the vortex core is unexpectedly small. When proper criteria for mapping the vortex lattice are used it is indeed possible to observe the vortices[30]. In Fig. 12 we show a series of spectra taken along a line from the vortex center to a point outside the vortex core where the full superconducting signal is recovered.

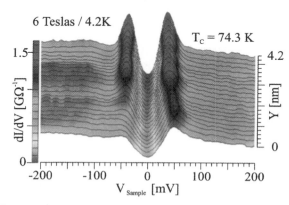

Figure 12. Tunneling spectra taken along a 4.2 nm long line from the center of the vortex (y = 0) to a point outside the vortex core (y = 4.2 nm)

The striking result here is that the low bias part of the spectra does not change appreciably when going from outside to inside the vortex core. There is no sign of either a zero bias anomaly as in $NbSe_2$ nor a split peak as in $YBa_2Cu_3O_{7-\delta}$. Furthermore, the peak at negative bias is suppressed as we enter the vortex core. However, the peak at positive bias is only weakened and shifts to somewhat higher energy. This is precisely the characteristics of the pseudogap structure observed in zero field above T_c. It was shown in ref. 30 that the spectra seen in the vortex core of $Bi_2Sr_2CaCu_2O_{8+\delta}$ have indeed the low temperature pseudogap structure. Another fact that can be seen in Fig. 12 is that the size of the vortices, i. e. the coherence length ξ, is very small in $Bi_2Sr_2CaCu_2O_{8+\delta}$, typically a factor of two smaller than in $YBa_2Cu_3O_{7-\delta}$. This is consistent with the fact that the gap in optimally doped $Bi_2Sr_2CaCu_2O_{8+\delta}$ is about a factor of two larger than the main gap structure in optimally doped $YBa_2Cu_3O_{7-\delta}$. The absence of a signature of localized quasiparticle states in the vortex core is probably due to a combination of the d-wave nature of the superconducting state and the very large value of the gap.

3. DISCUSSION

The results presented in part 1 and 2 have important implications for our understanding of the nature of the superconducting state in high temperature superconductors. Considering first $Bi_2Sr_2CaCu_2O_{8+\delta}$ we note that $k_BT_c << \Delta_p$. This strongly suggests that at $T = T_c$ the thermal energy is not enough to completely destroy the Cooper pairs. The fact that the gap is not reduced up to T_c and seems to extend beyond T_c in the form of the pseudogap is consistent with this statement. We have furthermore shown above that the pseudogap scales with T_c as the doping is varied. This is even more clearly seen in the pseudogap observed in the vortex cores[30]. It appears therefore that the pseudogap as seen in tunneling is some manifestation of the presence of pairs above T_c.

How does this fit with the vortex core spectra ? The interpretation of the double peak in $YBa_2Cu_3O_{7-\delta}$ as the two first localized states implies that the size of a pair, the coherence length ξ, is only a factor of 2 or 3 larger than the average distance between carriers. On the other hand, the larger gap in $Bi_2Sr_2CaCu_2O_{8+\delta}$ is consistent with the absence of localized states. This in turn implies that the coherence length must be smaller in $Bi_2Sr_2CaCu_2O_{8+\delta}$ than in $YBa_2Cu_3O_{7-\delta}$ and this is precisely what we observe. We arrive therefore at the conclusion that the size of the pairs in $Bi_2Sr_2CaCu_2O_{8+\delta}$ must be of the order of the average distance between the pairs. For low temperature superconductors the BCS theory tells us that the size of a pair is much larger than the average distance between particles. There is therefore a strong correlation between the pairs and as a result the pairs do not exist separately but are destroyed when the coherent state is destroyed. On the other hand, in an hypothetical case where the size of a pair would be much smaller than the average particle distance, one would expect some kind of Bose-Einstein condensation to take place. The case of $Bi_2Sr_2CaCu_2O_{8+\delta}$ therefore seems to represent an intermediate case where possibly, in one way or another, the pairs can survive the destruction of the coherent state. In this picture the pseudogap structure observed above T_c reflects the breaking up of individual pairs, existing without the simultaneous presence of long range coherence. The fact that we find the same structure in the vortex cores at low temperature suggests that the normal core is composed of pairs which are non-coherent with the superconducting state rather than the presence of localized quasiparticles at the Fermi level. In the case of $YBa_2Cu_3O_{7-\delta}$, both the smaller gap, the larger size of the vortices, as well as the presence of one localized state in the vortex core suggest that this compound is more BCS-like in that the size of a pair is larger than the average particle distance.

Summarizing, we have presented tunneling spectra for high temperature superconductors which clearly demonstrate that the properties of these materials can only be described by a theory that take into account

500

that the size of the pairs are of the order of the distance between the particles and that such pairs may possibly exist as independent entities above T_c and in the vortex cores.

ACKNOWLEDGEMENTS

It is a pleasure to thank B. Revaz, J.-Y. Genoud, K. Kadowaki, A. Erb, E. Walker and R. Flükiger for supplying the samples on which most of the results reported here were obtained. We also thank G. Bosch et A. Stettler for technical assistance. This work was supported by the Swiss National Science Foundation.

REFERENCES

1. J. G. Bednorz and K. A. Müller, *Z. Phys.* B 64 :189 (1986)
2. I. Giaever, Phys. Rev. Lett. 5 :147 (1960) ; *ibid*, 5 :464 (1960)
3. See for instance W. L. McMillan and J. M. Rowell, in *Superconductivity*, R. D. Parks ed., Marcel Dekker Inc. New York (1969)
4. Ch. Renner, Ø. Fischer, A.D. Kent, D. B. Mitzi, and A. Kapitulnik, *Physica* B194-196 :1689 (1994)
5. E. L. Wolf, A. Chang, Z. Y. Rong, Yu. M. Ivanchenko, and Farun Lu, J. *Supercond.* 7 :355 (1994)
6. Ch. Renner and Ø. Fischer, *Phys. Rev.* B51 :9208 (1995)
7. Ch. Renner, B. Revaz, J.-Y. Genoud, K. Kadowaki and Ø. Fischer, *Phys. Rev. Lett.* 80 :149 (1998)
8. Y. DeWilde, N. Miyakawa, P. Guptasarma, M. Iavarone, L. Ozyuzer, J. F. Zasadzikski, P. Romano, D. G. Hinks, C. kendziora, G. W. Crabtree amd K. E. Gray, *Phys. Rev. Lett.* 80 :153 (1998)
9. Ch. Renner, B. Revaz, J.-Y. Genoud and Ø. Fischer, *J. Low. Temp. Phys.* 105 :1083 (1996)
10. M. Oda, K. Hoya, R. Kubota, C. Manabe, N. Momono, T. Nakano amd M. Ito, *Physica* C 282-287 :1499 (1997)
11. H. Alloul, T. Ohno and P. Mendels, *Phys. Rev. Lett.* 63 :1700 (1989) ; W. W. Warren et al. *Phys. Rev. Lett.* 78 :721 (1989)
12. S. L. Cooper et al. *Phys. Rev.* B40 :11358 (1989)
13. J. Rossat Mignod et al. Physica B169 :58 (1991)
14. H. Y. Huang et al. *Phys. Rev. Lett.* 72 :2636 (1994) ; B. Batlogg et al. Physica C 235 :130 (1994)
15. J. W. Loram et al. *Phys. Rev. Lett.* 71 :1740 (1993)
16. J. L. Tallon et al. *Phys. Rev. Lett.* 75 :4114 (1995)
17. W. Y. Liang et al. *Physica* C263 :277 (1996)
18. R. Nemetschek et al. *Phys. Rev. Lett.* 78 :4837 (1997)
19. A. G. Loeser et al. Science 273 :325 (1996)
20. H. Ding et al. *Nature* 382 :51 (1996)
21. H. F. Hess, R. B. Robinson, R. C. Dynes, J. M. Valles, Jr., and J. V. Waszczak, *Phys. Rev. Lett.* 62 :214 (1989).

22. Ch. Renner, A. D. Kent, Ph. Niedermann, Ø. Fischer and F. Lévy, *Phys. Rev. Lett.* 67 :1650 (1991)

23. I. Maggio-Aprile, Ch. Renner, A. Erb, E. Walker and Ø. Fischer, *Phys. Rev. Lett.* 75 :2754 (1995)

24. I. Maggio-Aprile, Ch. Renner, A. Erb, E. Walker and Ø. Fischer, *Nature* 390 :487 (1997)

25. Ch. Renner, I. Maggio-Aprile and Ø. Fischer, To appear in *Superconductors in high magnetic fields*, Carlos A. R. Sa de Melo, editor

26. C. Caroli, P. G. de Gennes, and J. Matricon, *Phys. Lett.* 9 :307 (1964)

27. H. F. Hess, R. B. Robinson, R. C. Dynes, J. M. Valles, Jr., and J. V. Waszczak, *J. Vac. Sci. Technol.* B A6 :450 (1991)

28. Y. Wang and A. H. MacDonald, *Phys. Rev.* B50 :13883 (1995)

29. A. Himeda, M. Ogata, Y. Tanaka, and S. Kashiwaya, *Physica* C 282-287 :1521 (1997)

30. Ch. Renner, B. Revaz, K. Kadowaki, I. Maggio-Aprile and Ø. Fischer, submitted for publication.

FROM THE ANDREEV REFLECTION TO THE SHARVIN CONTACT CONDUCTANCE

G.Deutscher(1) and R.Maynard(2)

(1) School of Physics and Astronomy
University of Tel-Aviv
Tel-Aviv, Israel
(2) Physique et Modelisation des Milieux Condenses
CNRS/Universite Joseph Fourier
BP 166
38042 Grenoble Cedex
France

1.   INTRODUCTION

This Chapter deals essentially with weak contact experiments that are sensitive to the structure of the gap, including phase sensitive experiments somewhat different from those described by Kirtley in his Chapter. The experiments that we describe and analyze, are *one junction* experiments, involving either Sharvin contacts or tunneling junctions. They are based upon the idea that a surface affects differently the superconducting order parameter, depending upon its symmetry, and that the conductance of a weak contact to the surface is sensitive to this effect. Our discussions include the case of Superconductor/Normal metal contacts, and that of Josephson junctions.

2.   THEORETICAL

1 - *Andreev reflection*

The complete description of the excitations of a metal in the superconducting state has been given by Bogoliubov and de Gennes[1] in terms of coupled equations for the wave functions of the particle (electron) and the corresponding hole:

$$i\hbar\frac{\partial f}{\partial t} = \left[-\frac{\hbar^2}{2m}\nabla^2 - \mu\left(x\right) + V\left(x\right)\right] f\left(x,t\right) + \Delta\left(x\right) g\left(x,t\right)$$

$$i\hbar\frac{\partial g}{\partial t} = -\left[-\frac{\hbar^2}{2m}\nabla^2 - \mu\left(x\right) + V\left(x\right)\right] g\left(x,t\right) + \Delta^*\left(x\right) f\left(x,t\right)$$

where $\mu\left(x\right)$ is the local chemical potential, $V\left(x\right)$ the electrical potential and $\Delta\left(x\right)$ the pairing potential. In the normal state $\Delta\left(x\right) = 0$ and $f\left(x,t\right)$ describes the wave function of an electron, while $g\left(x,t\right)$ describes the hole wave function. In addition to the hole-particle symmetry taken into account in these equations, one can explicit the time reversal symmetry present in situations where there is no applied magnetic field. For stationary plane waves in the normal state, 4 waves are involved in the framework of the Bogoliubov-de Gennes equations: electron and hole of given wave vectors k^+ and k^- corresponding to a given energy E above the Fermi level, as well as $-k^+$ and $-k^-$ obtained by time reversal symmetry.

In the superconducting state, $\Delta(x)$ is not 0 and the excitations waves are strongly modified around the Fermi level for $k \cong k_F + \kappa$ ($\kappa \ll k_F$) :

1 -the dispersion relation is :

$$E(\kappa) \cong \sqrt{|\Delta| + 4\kappa^2 k_F^2}$$

2 - the composite wave function is for the electron-like excitation :

$$\begin{bmatrix} u_0 \\ v_0 \end{bmatrix} e^{\pm i k^+ x}$$

and for the hole-like excitation:

$$\begin{bmatrix} v_0 \\ u_0 \end{bmatrix} e^{\pm i k^- x}$$

where:

$$1 - v_0^2 = u_0^2 = \frac{1}{2} \left(1 + \frac{\sqrt{E^2 - \Delta^2}}{E} \right)$$

For $E < \Delta$, u_0 and v_0 are complex numbers that describe damped waves in the superconductors.

The pure Andreev reflection case arises for a clean contact - $V(x) = 0$ - between a normal metal and a superconductor, having the same $\mu(x)$ .In this situation, the reflected wave is dominantly of the hole type for an incident electronic wave, with a reflection coefficient equal to 1 for $E < \Delta$.This property is analogous to the phenomenon of backscattering from a phase conjugate mirror in non linear. optics(3] When the normal side of the contact is in the form of a slab of finite thickness d_n,Saint James (?] has shown that the Andreev reflections at the S/N interface, combined with the normal reflections at the outer surface of the N slab, give rise, for energies smaller than Δ,to bound states of finite energy. The normal state density of states in N is then replaced by a series of sawteeth, peaked at energies roughly quantized in units of (v_F / d_n).

The more realistic case of a non vanishing potential barrier at the interface between the normal metal and the superconductor has been analyzed in details in the article of Blonder,Tinkham and Klapwijk (?] .The main result of this theory, where the barrier is treated as a $\delta(x) - function$, is a strong attenuation inside the gap of the Andreev reflection while the normal reflection is increased. This barrier would have the effect of attenuating the amplitude of the Saint James density of states oscillations, without changing their position.

2 - The interface between a normal metal and a superconductor

It is well known that clean metallic contacts between two normal metals - without a dielectric barrier -have a finite resistance. This is the case of the Sharvin contact, for which the conductance is limited by the quantization effect of the finite lateral dimension of the contact. For a constant transmission coefficient Υ of electrons at the Fermi surface, the Sharvin conductance is given by:

$$G_N = \frac{e^2}{\hbar} (k_F w)^2 \Upsilon$$

where $(k_F w)^2$ counts the number of transverse modes determined by the cross-section of the contact. For clean contacts, the transmission coefficient Υ is in general less than 1, because of the mismatch of the group velocities of the two metals at the Fermi surface. Actually, on can show(?] that :

$$\Upsilon = \frac{4 v_1 v_2}{(v_1 + v_2)}$$

where v_1 and v_2 are respectively the unrenormalized Bloch group velocity on each side of the interface, for standard electron-phonon or electron-electron interactions (5] .

The current-voltage I(V) characteristic of an interface between a normal metal and a superconductor with an isotropic order parameter and an isotropic Fermi Surface, but with a finite potential barrier at the interface, has been treated in detail by Blonder, Tinkham and Klapwijk (?) . The most spectacular result of their treatment comes from the Andreev reflected holes which count as charge carriers: hence the transmission coefficient is augmented by the Andreev reflection coefficient and can be twice as large as that in the normal state, in the regime where the applied bias is smaller than the gap!

The problem of the anisotropy of the order parameter at an interface between a normal metal and a d-wave superconductor, has been discussed recently by Hu(6) and by Tanaka and Kashiwaya(7) . As shown by Hu, when the order parameter has an odd mirror symmetry at the interface plane (interface perpendicular to the [110] direction). bound states are obtained in the normal slab, near the Fermi surface (in the middle of the gap). This result is to be contrasted with the finite energy bound states found by Saint James for a s-wave order parameter. These zero (or near zero) energy states persist even when the thickness of the normal slab vanishes, and can be considered as true surface bound states. These surface states do not exist when the order parameter has an even symmetry with respect to the perpendicular to the interface ([100] direction).

The implication of this sensitivity of the matching condition to the symmetry of the order parameter is very important for the transport coefficient, particularly for the contact conductance between a normal metal and the superconductor. In the Tanaka and Kashiwaya paper, the normalized conductance is calculated for different situations:
- variable potential barrier
- variable orientation of the order parameter relative to the interface plane.

For a clean interface, the normalized conductance decreases as a function of the applied bias from the value of 2 at zero bias to 1, with a slope which reflects the presence of nodes in the order parameter on the Fermi surface. The value of this slope is insensitive to the orientation of the d-wave order parameter relative to the interface. It may be interesting to note that the zero bias cusp in the conductance versus the applied bias, looks like the cusp in the backscattering reflection versus the angle of laser light from a random suspension. For significant barriers, the relative tunneling conductance is strongly dependent upon the orientation of the order parameter: it takes the form of a dip in the case of an even mirror symmetry, or of a sharp peak in the case of an odd mirror symmetry. These contrasted results originate from the matching condition, where the wave parameter is sensitive or not , following the symmetric or antisymmetric combination, to Hu's surface bound states. The sharp conductance peak for the odd mirror symmetry is a direct result of the existence of surface bound states for this orientation.

3 - the bi-crystal interface

The standard formula for the Sharvin conductance G was established for a spherical Fermi surface. It is necessary to generalize this expression to a Fermi surface presenting singularities, in order to apply it to the case of High Tc superconductors, for which the Fermi surface is expected to be close to a square perimeter in the basic Brillouin zone. The first step is to start from the Transmission coefficient Υ (k) depending on the k vector on the Fermi surface:

$$\Upsilon(\mathbf{k}) = \frac{4 v_{k_z} v_{k_z'}}{\left(v_{k_z} + v_{k_z'}\right)^2}$$

where k_z and k_z' are Fermi wave vectors having equal components parallel to the plane of the interface. At the singular points of the Fermi surface, the group velocities vanish, and hence Υ (k) also does. This remark explains the sensitivity of the conductance to the Van Hove singularities of the Fermi surface.

As a second step, we consider the case of a Sharvin contact between two unconventional superconductors, having a d-wave order parameter (as well as possibly singular Fermi surfaces), and wish to calculate the value of the Josephson critical current through the contact. Starting from a classical expression of the supercurrent J_S through a potential barrier due to Ambegaokar and Baratoff(8) :

$$J_S \approx G \Delta_1 (T) \Delta_2 (T)$$

near Tc, where Δ_1 and Δ_2 are the pair potentials in the two superconductors, an anzatz has been proposed (?) for a weak Sharvin contact:

$$J_S \approx \sum_{\mathbf{k}} G(\mathbf{k}) \Delta_1 (\mathbf{k}) \Delta_2 (\mathbf{k})$$

where $G(\mathbf{k}) = (e^2/\hbar) \Upsilon(\mathbf{k})$ is the Sharvin contact conductance for the mode $\mathbf{k}$. The standard expression is restored when $\Delta(\mathbf{k}) = \Delta$. It must be noticed that a cancellation of J_S is possible when $\Delta_1(\mathbf{k})$ or $\Delta_2(\mathbf{k})$ have alternate signs on the Fermi surface.

For thick barriers (tunneling contacts), only the states k almost perpendicular to the junction are important. For symmetric bicrystals, whith rotation angles of the crystals $\pm \varphi$ as measured from the interface plane, two strong differences have been predicted between Sharvin and tunneling contacts:

- cusps in the angular variation of $J_S(\varphi)$, originating from the Van Hove singularities of the Fermi surface through the transmission coefficient $\Upsilon(\mathbf{k})$,

- a zero value of $J_S(\varphi)$ in the intermediate range of φ, instead of a zero at the node angle $\varphi = \pi/4$. This result comes from the integration of contributions to J_s over the entire Fermi surfaces, with alternating signs of the order parameters.

3. EXPERIMENTAL

3.1 Sharvin contacts

Sharvin contacts are realized as small metallic contacts between a sharp needle electrode and the surface of the sample to be studied. The needle-shaped electrode is for example a fine gold wire cut with a razor edge. It is brought into contact with the sample by a mechanical device, the size of the contact being in the sub-micron range. The normal state conductance of the contact is ideally limited by its size, as explained in the theory section.

An important practical concern is to avoid heating effects at the contact. This is necessary, because when the bias reaches values of the order of the gap, the current density is quite large. In fact, it is of the order of the depairing critical current. The way to avoid heating effects at the contact is to have its size smaller than the inelastic mean free path in the electrodes. In good metals, this mean free path is typically larger than 1micron at cryogenic temperatures. In the cuprates, it is also quite large at low temperatures (several thousand Angstrom), due to the long inelastic scattering times achieved at T<<Tc (see the Chapter by Hardy in this Volume). Therefore, a contact size of a few tens of nanometers is adequate for the study of Andreev reflections.

These reflections offer, in fact, a simple and pratical way to determine the actual (electrical) size of the contact. When the small bias conductance is increased by a factor close to 2, compared to the normal state (high bias) conductance, one knows that the metallic contact is almost perfect, namely that Eq.() holds with a transmission coeffient close to unity, and it is possible to calculate the size of the contact with this relation. The typical Sharvin contacts that have been studied have resistances of the order of 10Ω, this corresponds to sizes of the order of 10 nm, well below the mean free path.

3.1.1 Sharvin YBCO contacts

Andreev reflections on single crystal quality cuprate samples were first reported by Hass et al.(10) , on bulk textured YBCO. With a gold tip oriented in the ab plane, along one of the principal axis, they observed that the zero-bias conductance was about 50% larger than the normal state value (measured at high bias). The conductance was essentially flat up to a bias of about 20 meV, where it showed a sharp decrease, which the authors interpreted as the gap edge (Fig.1). From the normalized zero bias conductance value, and assuming that there was no signifiant dielectric barrier at the interface, they calculated the Fermi velocities mismatch and obtained for YBCO a value of $7.10^7 cm/s$.

The very observation of a large fraction of Andreev reflections below the gap, means that the structure of the superconducting state in the cuprate is not essentially different from that in a conventional superconductor, namely that the Bogoliubov - de Gennes description of the condensed state applies. This point is by no means trivial. As mentioned in the introductory Chapter to this Volume, it is not clear that this statement is completely correct, since the excitation (tunneling) and coherence (Andreev) gaps may well be distinct in the cuprates. Unconventional theories of the condensed state in the cuprates have not yet been sufficiently developed to make predictions concerning the Andreev gap. Yey, we do expect the symmetry aspects to be correctly described by the de Gennes-Bogoliubov equations, as exposed in the theory section, and we shall proceed with our discussion on that basis.

The fact that contacts with transmission coefficients close to unity could be made with the cuprates may look at first surprising. It implies that the Fermi velocity ot YBCO and that of the normal tip (typically gold), are close to each other (see Eq.()). The renormalized Fermi velocities of the cuprates are known, from photo-emission measurements, to be of the order of a few $10^7 cm/s$; this is smaller by a factor of about 5, compared to the Fermi velocity of gold. But, as already mentioned in the theory Section, the Fermi velocity to be used in the boundary condition is not renormalized by mass enhancing effects [5], and thus can indeed approach that of a usual metal.

Another point that was not understood at the time were the Sharvin contact results on YBCO became available, was the detailed shape of the conductance characteristic. At first glance, it resembles that for an s-wave superconductor, because of the sharp gap edge. But no significant conductance peak is observed at the gap value, contrary to the BTK predictions for an s-wave superconductor. In fact, it is now clear that the experimental result is in much better agreement with the calculations of Tanaka [7] for a d-wave order parameter, with the tip oriented along the [100] direction. For the calculated Fermi velocities mismatch, the conductance is predicted to be almost constant up to the gap, as observed experimentally. The small zero-bias anomaly, often observed in these characteristics, is also understood in terms of a d-wave order parameter, for an imperfect [100] alignement of the tip, which is normally the case.

Thus, the Sharvin contact results on YBCO are now basically well understood. They are in agreement with a d-wave symmetry, and give an Andreev gap of about 20 meV.

3.1.2   Sharvin LSCO contacts

Using the same technique as Hass et al. [10], Achsaf et al. [11] and Deutscher et al. (?] reported Andreev measurements on an LaSrCuO single crystal near optimum doping, the tip being oriented along the [110] direction. A zero-bias conductance increase by a factor close to 2 was observed when the surface preparation was adequate [11], but the shape of the conductance characteristic could not be interpreted in details in terms of a d-wave order parameter. Its general shape is that of an inverted V, as predicted by Tanaka [7] for this symmetry when the conductance ratio is close to 2 (independent of the tip's orientation, as underlined in the theory section); but the behavior at small bias rather resembles that predicted by BTK for an s-wave order parameter, with an initial *increase* of the conductance, rather than a decrease as predited for d-wave (Fig.2).

A satisfactory fit to the LSCO data was reported by Deutscher et al. (?], assuming an extended s-wave order parameter without nodes; alternatively, a fit to a (s+d) order parameter is also possible, with a dominant s fraction.

These results on LSCO are rather disturbing, in view of the fact that, so far, all hole-doped cuprates have been reported to have a dominant d-wave order parameter, at least near and below optimum doping. It may be that the dominant s-wave character seen in LSCO is only a surface property [13]. But the d-wave character of YBCO appears, from similar Sharvin contact experiments, to persist up to the surface. The intepretation of multi-junction experiments on YBCO and other hole doped cuprates, also implies this persistence.

So LSCO appears to be different from most hole doped cuprates. At the very least, the Sharvin contact results imply that it has a sub-dominant, but rather strong, s-wave order parameter. Detailed conductance calculations for that case - a dominant d-wave order parameter in the bulk, but s-wave at the surface - are not yet available, so that this hypothesis cannot at the moment be checked quantitatively.

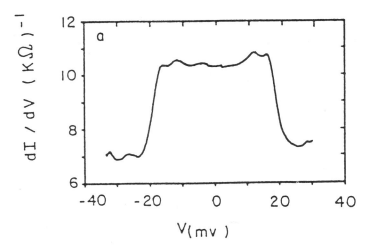

Fig.1 Conductance of a Sharvin contact to a textured YBCO bulk sample, the Au tip being oriented parallel to the CuO planes. The characteristic is consistent with the tip being in the [100] direction, and a d-wave gap having a magnitude of 20 meV.

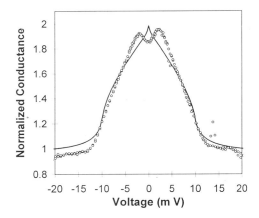

Fig.2 Conductance of a Sharvin contact to a LSCO single crystal, slightly underdoped. The Au tip is oriented along the [110] direction. The circles are measured conductance values, and the continuous line an attempt to fit the data to a d-wave gap. The fit fails at low bias, because ot the initial rise of the conductance, which indicates the absence of nodes.

3.2 Tunneling contacts

The normal state conductance of tunneling contacts is limited by the barrier between the electrodes. In conventional tunneling, this barrier consists of a thin dielectric -typically Aluminium oxide - having a thickness of the order of 20A, giving resistances of the order of 1.10^2 to 1.10^4 Ω, for a contact area of the order $1mm^2$. Because the barrier thickness is much larger than the decay length of the electronic wave function, the coefficient of transmission is sharply peaked for a wave vector perpendicular to the interface. Contrary to Sharvin contacts, conventional tunneling contacts will only probe a small fraction of the Fermi surface. This difference is important when the contact is probing a superconductor having an unconventional order parameter symmetry. Conventional, highly directional, tunnel barriers are appropriate for phase sensitive experiments that use multiple contacts, each contact probing the phase of the order parameter in a certain direction, as described by Kirtley in his Chapter. They are not appropriate for the one-junction phase sensitive experiments that we discuss here.

Another kind of contact is that used in Scanning Tunneling Spectroscopy. Here, tunneling occurs between a very sharp tip, of atomic size, and the surface of the sample (for details, see the Chapter by Fischer in this Volume). The tip is then very close to the sample - otherwise the resistance of the contact would be too high, and the tunneling current immeasurably small. Typically STM contacts have a resistance in the $G\Omega$ range. Because it is very close to the surface, the STM contact is not as directional as the conventional tunneling contact. It is therefore more appropriate for one-junction phase sensitive experiments than conventional tunnel contacts. Alternatively, point contacts (as used for Sharvin contacts), but with a significant barrier, are also adequate for these experiments. Nb tips, which develop naturally an oxide barrier, and can be used for this purpose.

3.2.1 YBCO

Zero Bias Conductance Peaks (ZBCP) have been widely reported in the literature for tunneling contacts with YBCO, and diversely interpreted. Early interpretations focused on the presence of magnetic impurities in the barrier, which can produce such peaks, as predicted by Appelbaum (14) and Anderson (15) . More recent interpretations have emphasised the d-wave symmetry as the origin of these peaks.

Alff et al. (16) performed STM experiments on YBCO films, having either the [100] of the [110] orientation, with contact resistances of the order of $0.1G\Omega$. As predicted for a d-wave symmetry, they did not observe a ZBCP for [100] oriented films (or only sometimes a small one, which they attribute to surface roughness), but they did consistently observe a large one for [110] films. This ZBCP was found to be spatially continous, which also argues against the magnetic interpretation. The zero bias conductance was found to be about 3 times larger than the conductance measured at the gap edge. When a ZBCP is seen for orientations where it should in principle be absent (such as the [100] or the [001] orientations), it is much smaller, typically 10% of the conductance at the gap edge, which makes an interpretation in terms of surface roughness very reasonable.

These observations of Alff et al. on YBCO are fully consistent with the Sharvin contact measurements summarized in the preceding section, indicating a d-wave gap having an amplitude of about 20 meV. Alff et al. note, however that they could not fit in all its details the [110] conductance characteristic to the calculations of Tanaka (7) . The selected set of parameters that fits well the ZBCP structure (which, by the way, includes a broad sensing angle of the order of $\pi/10$), does not give a good fit at biases above the gap value of 20 meV, where a broad peak extends up to twice that value. It may be that the Bogoliubov-de Gennes equations do not give a complete description of the gap structure in the cuprates, as we have already mentioned above in conjunction with a possible distinction between a coherence gap (which would give here the ZBCP structure), and a higher value excitation gap.

3.2.2 NCCO

Alff et al. (16] have conducted on NCCO the same experiments on [100] and [110] films, as performed on YBCO. The difference between the two cuprates is spectacular. Whatever the film orientation, no ZBCP is seen on NCCO films. In fact, the conductance characteristics are essentially identical for both orientations studied, giving a gap edge of 5 meV.

This is, as far as we know, the only phase sensitive experiment that has been performed on NCCO. In fact, the multi-grain boundary junction experiments described by Kirtley cannot be used to investigate NCCO, because this cuprate does not form "natural" grain boundary junctions. This may well have something to do with the fact that this cuprate is not a d-wave superconductor.

We consider the s-wave character of the gap in NCCO to be well established by the experiments of Alff et al..They confirm other -not phase sensitive - experiments reviewed by Fournier in his Chapter, such as the low temperature measurements of the London penetration depth, which fit the exponential BCS behavior, with the same gap value as reported by Alff et al.

3.2.3 LSCO

Achsaf et al. (11) have used Nb tips to make point contacts to the same LSCO single crystal used for the Sharvin experiments summarized above, and along the same [110] direction. Because Nb oxidizes rapidly in air, the resistance of these contacts is increased by an interface dielectric barrier. For that case, Tanaka et al. (7) predict a large ZBCP, if the order parameter is d-wave. But no ZBCP was found in these experiments, an observation consistent with the conclusions drawn from the study of high transparency Sharvin contacts reported above.

It may of course well be that in these experiments, the tip orientation was not exactly along the [110] direction, because of surface roughness. However, the tip was certainly not *exactly* along the [100] or the [001] orientations, the only ones for which no ZBCP is predicted. Therefore, we conclude that tunneling results on LSCO are not compatible with a d-wave symmetry for concentrations near optimum doping. On the other hand, Froehlich et al. (17) have reported a ZBCP for LSCO grain boundary tunnel junctions. But we note that these films had a lower Tc than optimally doped single crystals; also, grain boundaries tend to be naturally underdoped. It may be that there is a change in symmetry as a function of doping in LSCO.

[1] P.G.de Gennes, "Superconductivity of Metals and Alloys", W.A.Benjamin, New York (1966).
[2] D.Saint James, J. de Physique, vol 25, 899(1964)
[3] D.M.Pepper, "Applications of optical phase conjugation", SCI.AM., January 1986
[4] G.E.Blonder, M.Tinkham and T.M.Klapwidjk, Phys.Rev.B25,4515(1982).
[5] G.Deutscher and P.Nozieres, Phys.Rev.B50,13557(1994).
[6] C.R.Hu, Phys.Rev.Lett.72,1526(1994).
[7] Y.Tanaka and S.Kashiwaya, Phys.Rev.Lett.74,3451(1995);see also Y.Tanaka in "Coherence in High Temperature Superconductors", Eds.G.Deutscher and A.Revcolevschi, World Scientific 1996, p.393.
[8] V.Ambegaokar and A.Baratoff, Phys.Rev.Lett.10,486(1963).
[9] G.Deutscher and R.Maynard, Europhys,Lett., 30,(6), 361 (1995)
[10] N.Hass, D.Ilzycer, G.Deutscher, G.Desgardin, I.Monot and M.Weger, J.Supercon.5,191(1992).
[11] N.Achsaf, G.Deutscher, A.Revcolevschi and M.Okuya, in "Coherence in High Temperature Superconductors", Eds.G.Deutscher and A.Revcolevschi, World Scientific 1996, p.428.
[12] G.Deutscher, N.Achsaf, D.Goldschmidt, A.Revcolevschi and A.Vietkine, Physica C283-287,140(1997).
[13] M.Fogelstrom, D.Rainer, J.A.Sauls, Phys.Rev.Lett.79,281(1997).
[14] J.Appelbaum, Phys.Rev.Lett.17,91(1966).
[15] P.W.Anderson,Phys.Rev.Lett.17,95(1966)
[16] L.Alff, H.Takashima, S.Kashiwaya, N.Terada, T.Ito, K.Oka, M.Koyanagi and Y.Tanaka, in "Advances in Superconductivity IX", Springer-Verlag, Tokyo 1997.
[17] O.M.Froehlich, P.Richter, A.Beck, R.Gross and G.Koren, J.Low Temp.Phys.106,243(1997)

TUNNELING IN HIGH Tc SUPERCONDUCTING CUPRATES

J. Lesueur[1], B. Leridon[2], M. Aprili[1,3], X. Grison[1]

[1]C.S.N.S.M. CNRS-IN2P3, Bât 108 91405 Orsay (France)
[2]E.S.P.C.I. 10 rue Vauquelin, 75005 Paris (France) &
UMR Thomson CSF-CNRS, Domaine de Corbeville, 91401 Orsay (France)
[3]U.I.U.C. Dpt of Physics, 1110 W. Green St, Urbana IL 61801(USA)

INTRODUCTION

In the sixties [1, 2] , tunneling experiments on superconductors (SC) provided key proofs that the BCS theory (and its extensions) were correct. Probing the Density of States (DOS) of these materials at the Fermi energy with a very high resolution (in the μeV range at low temperature), they evidenced the opening of a gap, allowed its study in temperature, magnetic field ..., displayed specific features related to the coupling excitations (phonons) involved in SC, played a key role in the understanding of the proximity effect between normal-metals and SC, confirmed the existence of bound states in special geometrical configurations ... Provided the elastic tunneling is the only conducting channel through an insulating barrier, a tunnel junction acts as a spectrometer : therefore, quantitative comparisons have been made between theories and experiments with a high degree of accuracy. Moreover, the pioneer work of Anderson and Rowell [3] clearly showed the existence of the Josephson effect in Superconductor-Insulator-Superconductor (SIS) tunnel junctions, opening the doors to fantastic experiments and devices using the quantum mechanical coherence in SC.

This is why numerous tunneling experiments have been carried out since the discovery of the High Temperature Superconducting Cuprates (HTSC) [4, 5] to probe the DOS of these fascinating materials on energy scales relevant for a superconducting gap (from a few μeV to a few 10 meV), to look for specific pairing excitations, and more recently, to test the symmetry of the order parameter through Josephson experiments.

After a short survey of the basis of tunneling in Low Tc Superconductors (LTSC) and the well established results on the Quasi-Particle (QP) spectroscopy and the Josephson effect, we will focus on tunneling in HTSC. We will give a brief overview of some recent theoretical aspects which are important to discuss the tunneling results, and then present the main experimental results on *planar* tunnel junctions based on HTSC. The reasons for this choice are threefold : STM results are presented by O Fisher in another chapter of this book ; *planar* tunnel junctions have been the corner-stones of the LTSC tunneling experiments presented in the first part ; and more important, their geometry is fully controlable, temperature independent and stable in time. Even if their macroscopic size may induce some problems in short coherence length materials, their stability is essential to use them as performing spectrometers and to get reliable and reproducible results.

TUNNELING IN LOW-TEMPERATURE SUPERCONDUCTORS

A) Basic aspects of tunneling

Tunneling occurs between two conductors separated by an insulating barrier provided that it is sufficiently thin and that the barrier potential is not too high. Due to the quantum mechanical nature of the electrons, the tails of the wave functions on both sides of the barrier overlap and there is a non zero probability per unit time for the electrons to cross the potential barrier. This process requires global conservation of the energy. This means that for the one-electron process the transfer is made horizontally from one energy level on one side of the barrier to another energy level on the other side. When two electrons are involved in the process, the energy lost by one electron must be equal to the energy gained by the other.

One can intuitively derive that the tunneling probability is proportional to the DOS of the corresponding energy levels on both sides, to the statistical occupation of the states (the electron must tunnel from an occupied state towards an empty state) and to the barrier transmittivity. In the analysis of the tunneling effects we will refer to the transfer hamiltonian formalism suggested by Cohen, Falicov and Phillips[6] and used by Ambegaokar and Baratoff[7]:

$$H_T = \sum_{kq\sigma} \left[T_{kq} c_{k\sigma}^+ d_{q\sigma} + T_{kq}^* d_{q\sigma}^+ c_{k\sigma} \right]$$

where

$$|T_{kq}|^2 \propto k_z q_z \exp\left(-\frac{1}{\hbar} \sqrt{2mU} t \right) \delta_{k_y q_y} \delta_{k_x q_x}$$

and T_{kq} is the matrix element from the state k to the state q, U the barrier potential, t the barrier thickness, m the mass of the particle. This expression implies k-space selectivity, and is derived under the approximation of energy independent tunneling probability. Actually, tunneling through a potential barrier favors essentially transitions with k perpendicular to the surface and within an angle of 2-5° from the normal to the surface.

B) Tunneling spectroscopy

Tunneling through a superconductor-insulator-normal metal junction gives direct information about the superconducting DOS singularities, as the normal metal DOS is constant. The conductance as a function of the applied voltage is directly proportional to the superconducting DOS normalized to the normal DOS. Such an experiment allowed Giaever[1, 2] and Nicol, Shapiro and Smith[8] to measure the superconducting gap of lead with a very high resolution (3.5 $k_B T$ is a few μeV at low temperature).

If the Fermi surface, or the electron-phonon matrix elements, or the phonon spectrum are anisotropic, the superconducting gap can be anisotropic itself. By using single crystals and a serie of junctions formed on different faces of the crystal it is possible to measure the gap as a function of the angle. The angular resolution will then be about 5-10° as given by the tunneling effect selectivity.

This allowed the observation of gap anisotropy and multiple gaps in Pb which correspond to different sheets of the Fermi surface [9-11], and remarkably large gap anisotropy in Sn [12-14]. Scattering effects in the vicinity of the surface can reduce the measured gap even if the bulk superconductor residual resistance ratio measurements show a clean limit behavior. Tunneling measurements are essentially surface-sensitive and the measured value of the gap is valid only close to the junction (within a superconducting coherence length roughly). They played also an important role in confirming the predictions of gapless superconductivity when doping with magnetic impurities[15, 16].

In the case of strong coupling superconductivity, the Eliashberg function $\alpha^2F(\omega)$ related to the phonon DOS was deduced from the tunneling data[17]. The Eliashberg function, the Coulomb pseudopotential μ^*, the pair potential or gap function $\Delta(\omega)$ and the renormalization function $Z(\omega)$ can be determined from the phonon range conductance normalized to the BCS conductance σ/σ_{BCS} - 1 and the gap edge value $\Delta(0)=\Delta_0$. The phonon spectrum was derived for Pb and gave a remarkable agreement with neutron scattering measurements which constituted a stringent test for the validity not only of the inversion procedure but also of the Eliashberg theory itself [17].

C) Josephson effect in low Tc superconductors

If two superconducting electrodes are separated by a distance sufficiently small (typically a few 10 Å), they are able to exchange Cooper pairs. The resulting current could be simply considered as a natural extension of the bulk supercurrent, but it contains really peculiar features that Josephson predicted in 1962[18] before the first observation[3] of the effect that bears his name. Originally, Josephson had in mind the SIS case, where the exchange is made by tunneling through a dielectric barrier but it applies more generally to weak links i.e. when the order parameter is locally depressed as in Dayem bridges, point contacts, proximity effect junctions, etc...

The Josephson current J is a function of the phase difference ϕ between the two superconducting electrodes:

$$J = J_c \sin(\phi)$$

If the external magnetic field is zero, the phase difference depends only on the voltage applied across the barrier, through the relation:

$$\frac{\partial \varphi}{\partial t} = \frac{2eV}{\hbar}$$

When the voltage across the barrier is zero, the phase difference is constant for a given bias current. When the bias current exceeds the critical value J_c, the junction jumps to a finite voltage, the dc tunneling current is then a quasiparticle current associated with pair breaking . The Josephson current is given by :

$$J = J_c \sin\left(\frac{2eV}{\hbar}t + \varphi_0\right)$$

Therefore a constant non zero voltage bias produces an ac pair current, with a frequency-voltage relation given by f/V =483.6 MHz/μV.

1) Field dependence of the phase

Analogously to the Meissner-Ochsenfeld effect in a type I superconductor, the Josephson supercurrents in an SIS junction are confined to the edges of the junction over a distance equal to the Josephson penetration length :

$$\lambda_J = \left(\hbar c^2/8\pi e dJ_c\right)^{1/2}$$

where d is the junction effective thickness ($d=\lambda_1+\lambda_2+t$). λ_1 and λ_2 are the London penetration depths of the two superconducting electrodes, t is the barrier thickness.

In low Tc superconductors, the London penetration depth is of the order of hundreds of angströms and the Josephson length of hundreds of microns. This distance represents consequently the distance over which the phase is able to bend. If the junction length is smaller than λ_J, the phase will be able to rotate and the current amplitude will be modulated over the whole junction length. This case is usually referred to as "small" junction, as opposed to "large" junction, with L larger than λ_J,. In the "small" junction case, the variation of the critical Josephson current as a function of the magnetic field looks like the Fraunhofer diffraction pattern of an aperture of the same shape as the junction surface, which constitutes a stringent test for Josephson coupling. The function $Ic(B_y)$ is proportional to the Fourier transform of the current distribution $I(x)$. In the large junction case, however, the penetration of Josephson fluxons strongly modifies this sketch and the Fraunhofer pattern is no longer recovered [19-21].

2) Shapiro steps

We explained above that a constant voltage bias through the junction leads to ac Josephson currents. This is a manifestation of the high non-linearity of the SIS junction. On the contrary, a high-frequency current applied through the junction (rf radiation) will induce a non-zero bias of the junction (Shapiro steps)[22, 23]. The amplitude of these current steps at voltages $V_n = n\hbar\omega/2e$ reflects the rf current amplitude, hence giving access to the frequency and power dependence and to the singularities of this rf current (Riedel peak). These steps are also an important signature of the Josephson effect. The total current in the time-domain for a voltage biased junction in the case of magnetic field and microwave irradiation has been calculated by Werthamer[24] and by Larkin and Ovchinnikov[25] .

3) Fiske steps

The same system consisting of the two superconducting electrodes separated by a thin insulating layer can also be considered as a transmission line, with the electric field confined within the insulator region and the magnetic field penetrating the electrodes over the distance λ. In the presence of an alternating Josephson current, the high frequency radiation will couple to the cavity formed by this transmission line of finite length, leading to self resonant modes appearing as current steps in the I-V characteristic. These effects are known as Fiske steps[26].

4) Temperature dependence of the Josephson current and RnIc product

The temperature dependence of the Josephson current between two superconductors was derived by Ambegaokar and Baratoff[8, 27] using the tunneling Hamiltonian formalism described above and in the framework of the BCS theory.

$$J_c = \frac{\pi\Delta(T)}{2eR_n}\tanh\frac{\Delta(T)}{2kT}$$

An important parameter is the RnIc product which gives some information about the strength of the coupling. An analogous calculation has been made in the case of proximity effect junction and leads to a different variation. This measurement constitutes then an indication of the nature of the transport through the interlayer, provided that it is the same over the whole temperature range.

In low Tc superconductors, tunneling provides a spectroscopic analysis of the material in the vicinity of the surface, through the study of the quasiparticle branch as well as the Josephson currents. It gives information about the DOS structure anomalies, the gap anisotropy, the phonon structure and the junction surface geometry. Quantitative agreement with the BCS theory is so good that the measurement of the above-described specific superconducting features is the most stringent test to establish that a given interlayer is a tunnel barrier.

The small coherence length in high Tc materials and the complexity of the order parameter modifies the outputs from tunneling techniques as we will see in the next section.

TUNNELING IN HIGH-TEMPERATURE SUPERCONDUCTORS

A) Brief overview of some recent theoretical developments

1) The superconducting gap and the order parameter.

If HTSC were conventionnal BCS superconductors, a gap should open in the QP spectrum below Tc. The expected value for YBCO should be 13.4 meV for instance in the weak coupling limit, up to 18-20 meV in the strong coupling one. Therefore, the dynamical

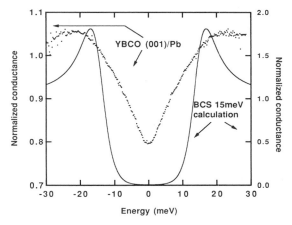

Fig. 1 Measured tunneling conductance in a YBCO/Pb junctions (a) compared to a standard BCS calculation for the DOS.

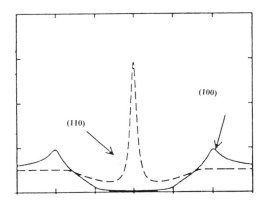

Fig. 2 Schematic drawing of an engineered ramp junction YBCO/PBCO/YBCO. ref 78

conductance of a SIN tunnel junctions should drop to zero exponentially below this energy in the zero-temperature limit (Fig. 1). In most cases, the result is very different (Fig 1.). Among others, this reason was a starting point for theorists to promote new ideas, new concepts to explain High-Temperature Superconductivity. In the most elaborate scenarios, the DOS of the QP is sketched, and the conductance of SIN tunnel junctions computed in various situations. For obvious reasons, we cannot present all of them here, but we will focus on the main ones, with a special emphasis on those which are consistent with most of the tunneling data.

The major ingredient one has to put in when describing HTSC is the anisotropy : properties in the basal CuO_2 plane are distinct to those in the c-direction [28]. The Fermi surface is quasi-cylindrical, and a directional probe as tunneling should see different behaviors according to the used geometrical configuration. Focusing on in-plane properties where SC develops, one expects Van-Hove singularities to occur in the normal state DOS.

Bok and Labbe [29] proposed that the Fermi level being in the vicinity of this singularity, the usual electron phonon coupling mechanism can give Tc's up to 100K. In this model, a gap fully develops in the SC state [30], anisotropic in the plane due to band structure effects, with special features in the DOS reminiscent of the V-H singularity. This gap does not change sign along the Fermi surface. It is usually called "extended s-wave" [31], by reference to the regular "s-wave" of the BCS one.

This model is essentially based on itinerant electrons. However, behaviors have been observed in the SC cuprates, which do not follow the standard properties of metals (NMR [32] , high-temperature resistivity [33], IR conductivity [34], tunneling [35] (as we shall see) ...). Such results suggest that electron correlations may be important in the metallic phase of the cuprates, leading to non Fermi-liquid features ; furthermore, they could participate to the pairing mechanism itself, leading eventually to non-conventional superconductivity.

Originaly proposed by Bickers, Scalapino and White [36], the electron coupling by anti-ferromagnetic excitations scenario has been developed in several directions, using NMR results as a starting point to know the excitation spectrum of the normal state electrons [37], or more microscopic calculations. More recently, Emery and Kivelson [38] proposed in this context that spin excitations and Cooper-pairs may separate spatially in a quasi-1d structure named "stripes", reminiscent of the charge-spin separation in 1d correlated systems (Luttinger liquids). Disregarding the details of those scenarios, the most important common points of interest for tunneling experiments are the following ones : (i) the DOS in the normal state may be different than the free electrons one; (ii) the SC order parameter (OP) has a "d-wave" component, and changes sign along the FS.

This last point will have major consequences. In the case of a pure d_{x2-y2} OP parameter for instance, they are nodes along the (110) directions ; therefore, low energy excitations may take place and the the QP DOS will increase slowly with the energy below the gap at low temperature, and not exponentially as in the BCS case. Moreover, the observed gap will have different values according to the tunneling direction. In a "d-wave" SC case, the Anderson theorem does not apply [39], and disorder strongly affects the SC state. In addition to a reduction of the amplitude of the order parameter, it mixes the wave vectors on the FS, producing available states within the "gapless" region around nodes. This will give a non-zero tunneling conductance at zero bias for example.

Another consequence of a such a symmetry for the OP, is the specific role of surfaces on SC, especially the (110) one. It has been pointed out by Hu [40] and Tanaka et al [41-43] that electrons specularly reflected on a (110) surface will experience a change in sign of the OP (see Figure. 11), and therefore an Andreev reflexion [44] occurs naturaly on such a surface together with a strong depairing. Quasi-Particles are trapped in the vicinity of the surface, and a bound state is formed. It is equivalent to those occuring in N/S bilayers with Low Tc compounds [45], but the "d-wave" nature of the pairing leads to a single peak in the DOS at zero energy, instead of multiple peaks at finite energies in an "s-wave" model. This effect is expected to be stronger in the (110) direction, and absent in the (100), (010) and (001) ones. This should be easily probed by tunneling experiments along different crystallographic directions. It is worth noticing [46] that in case of diffuse scattering at a surface, the randomization of the wave vectors will lead to the same behavior than the one observed in disordered "d-wave" superconductors (cf above) : a depressed order parameter and "in-gap" states in the tunneling characteristics.

Up to now, we have chosen the d_{x2-y2} symmetry for the OP to illustrate the consequences of a non-conventional SC on the DOS as it can be probed by tunneling. In fact, the situation can be different. Firstly, other d-wave symmetry like dxy may be relevant, depending on the details of the pairing. Secondly, mixed symmetries are also possible (like d+is, s+id ...) : in these cases the two components are dephased by $\pi/2$ to minimize the Ginzburg-Landau free energy [47]. Moreover, the weight of the components may spatially vary from bulk to surface [47, 48], leading to time-reversal symmetry breaking behaviors. Finally, the OP symmetry has to match the underlying crystallographic one : all the above considerations rely on a square symmetry in the basal plane. But most of the HTSc materials have an orthorhombic distortion (like the widely used YBCO) : in this case a pure d-wave OP is prohibited and s+d mixtures have to be considered. It has been shown however that the orthorhombicity can be important to achieve high Tc's in certain cases [49, 50], leading to specific signatures in the tunneling DOS.

To end this section, it is important to notice that the symmetry of the order parameter is a crucial question nowadays. Instead of checking one by one all the pairing mechanisms one can think about, it is more efficient to test the main features of a model through the symmetry of the OP it generates. A detailed revue of the possibilities has been made by Annett, Goldenfeld and Leggett recently [51]. Focusing on the DOS of QP as it can be probed by tunneling, we would like to stress the following points :
- in the pure "s-wave" case, a complete single value gap is expected at low temperature.
- in the "anisotropic s-wave" case, different gaps may be observed according to the tunneling direction, and even nodes leading to non-exponential decrease of the tunneling conductance below the gaps.
- if a "d-wave" component is present, a linear [52] increase of the tunneling conductance vs energy is expected below the gap : a peak at zero-bias should be seen when tunneling along the (110) direction due to the bound state, and disorder should strongly affect the tunneling conductance.

This last point is the only criteria to distinguish between s and d-wave symmetry by spectroscopy of the QP DOS. However, Josephson tunneling experiments will provide another crucial test.

2) The Josephson effect

As it has been presented in the first part of this paper, the Josephson effect relies on the coupling between the order parameters of two superconductors, i.e. the basic quantities which describe the coherence of the superfluid condensate. Therefore, one can study the fondamental properties of the OP in a SC (like its symmetry for instance) by measuring its coupling to another SC through the Josephson effect. In a conventionnal SC (and by extension in any SC having an anisotropic "s-wave" OP), the OP can be described by an amplitude, which may vary on the FS. In a "d-wave" SC, the phase of the OP is internally modulated along the FS. Therefore, one can realize suitable geometrical configurations of Josephson junctions to evidence this internal phase shift.

Three main configurations have been proposed and experimentally tested : (i) the coupling between two or more pieces of HTSc materials, (ii) the coupling between two parts of *the same HTSC sample* through a piece of LTSC material, and (iii) the coupling between a HTSC and a LTSC. To the best of our knowledge, tunnel junctions have been used only in the last type of experiments, providing informations on the QP DOS and the Josephson effect on the same samples. The other experiments used SNS type junctions, where the Josephson branch only is measurable. Excellent revue articles have been published on this subject [31, 51], and the paper of J. Kirtley in this book also deals with it. In the present article, we would like to enlighten the main points of these experiments and focus on the Josephson tunneling data.

The experiments (i) [53-56] are based on an idea proposed by Sigrist and Rice [57] : a ring configuration of HTSC materials is prepared, with JJ coupling regions where the (100) directions are misaligned with an angle of 30° or so. If the order parameter is "d-wave", when the number of JJ is odd, a frustration occurs because of the phase shift in the OP induced by the misalignement ; a spontaneous circulating supercurrent (and therefore magnetic flux) is generated to compensate this phase shift. This does not apply if an even number of JJ is used. C. C. Tsuei et al [54] claim that they did observe this spontaneous magnetization on tricrystal thin films (patterned in a ring-shape or not) of different HTSC materials, and concluded that a d_{x2-y2} OP is relevant in these superconducting compounds.

The experiments of type (ii) have been first made at UIUC by Wollman et al [58]. A single crystal of HTSC is connected to a piece of LTSC through two JJ. The LTSC material is supposed to be "s-wave", and therefore to keep a constant phase for the OP. If the OP of the HTSC is "d-wave", when the JJ are made on the (100) and (010) surfaces respectively, a frustration occurs again, and a π shift appears when one calculates the circulation along the ring formed by the HTSC and LTSC materials. This shift will modify the Fraunhofer pattern in magnetic field (cf the first section) ; this will be the signature of an unconventional OP. If both JJ are made on the same surface, such a phenomenum is not expected to occur since the same sign of the OP is connected to the LTSC material. Different experiments addressed this question [58-61] : they all conclude to a "d-wave" scenario also, at least in YBCO.

Type (iii) experiments will test the coupling properties of the HTSC OP with a conventional "s-wave" OP. If a JJ is made between a LTSC and a HTSC in the basal plane, the coupling between the two OP will follow the amplitude of the HTSC one, and no phase effect is expected to occur. However, when tunneling into the c-direction of a material, one couples to different parts of the FS. If the OP changes sign, the coupling will be reduced, or even null in the case of a pure d_{x2-y2} symmetry. Three groups have tried such experiments up to know, in UCSD [62-65], in Japan [66] and in France [67-69]J. Lesueur, 1997 #58]. Their conclusions are different.

It is worthy noticing that type (i) and (ii) experiments rely on magnetic measurements, which would be strongly affected if vortices are present in the system. However, type (iii) is based on the IcRn value in the JJ as a measure of the strength of the coupling, which may be affected by surface effects. All of them suppose that there is a given orientation of the OP on a macroscopic scale, i.e. the size of the device (at least few tens of microns) ; this requirement is certainly not fulfilled in thin films or twinned crystals.

Although it is beyond the scope of this paper, we would like to mention that Josephson Effect can occur internally in layered cuprates [68-70]. The CuO2 superconducting basal planes are coupled through conducting or insulating layers ; the c-axis coherence length is shorter than the lattice spacing, and the Josephson coupling between the planes can be seen experimentally on I-V characteristics of small mesas isolated on single crystals or thin films.

3) High energy effects

Tunneling may also evidence specific contributions to the DOS in addition to the gap (see first part). We cannot review here all the models proposed in the litterature, but we list below some of them. Phonons structures may appear of course if the electron-phonon coupling is strong enough, but other excitations involved in the pairing mechanism and strongly coupled to the electrons will also give specific features on the tunneling DOS. In the V-H scenario [71], a peak is expected on one side of the tunneling conductance. In the Marginal Fermi Liquid picture, a peak at 3Δ has been predicted, symmetric with respect to the origin [72]. Modeling the HTSC cuprates as an assembly of superconducting layers, Tachiki et al [73], calculated the tunneling conductance : small modulations are expected as a function of the energy above the main gap.

The background conductance itself is a matter of interest. In LTSC junctions, the non-ohmic behavior comes from the shape of the junction. It appears that most of the tunnel junctions made on HTSC (and almost all the planar junctions) display a steeper background than predicted by standard models (cf Fig. 16). Many interpretations have been proposed, some of them on new theoretical basis. For instance the Marginal Fermi Liquid (MFL) model [72, 74] suggests that the unusual excitation spectrum in the cuprates directly generates a

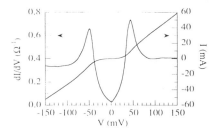

Fig. 3 I-V and dI/dV vs voltage characteristics of a YBCO/PBCO/HoBCO junction at T=4.2 K). ref 82

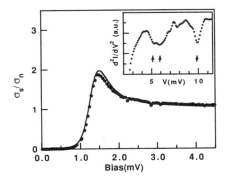

Fig. 4 Tunneling conductance as a function of energy in a YBCO/Pb junction at T=1.2 K. the circles are the data and the solid line, a BCS fit with Δ=1.38 meV. The inset shows the phonon dips in the second derivative curve.

linear one-particle DOS, and therefore a linear tunneling conductance as a function of the energy. An alternative proposition has been made by Kirtley and Scalapino [75] concerning the c-axis tunneling experiments. Since the FS is quasi-cylindrical, electrons coming along the c-direction suffer a strong scattering when entering the HTSC material. It is proposed that AF fluctuations with a broad excitation spectrum provide an efficient inelastic channel for this mechanism. The calculated DOS is linear in energy, so is the tunneling conductance. Experiments on LTSC junctions with AF impurities (Cr oxide) have been made to test this general statement [76]. In the first years of the HTSC story, the RVB model of P. W. Anderson et al [77] also provided a "natural" explanation of the linear background conductance in terms of spinon-holon DOS.

B) How to make HTSC junctions ?

1) Engineered tunnel junctions.

Because HTSC materials are oxides, it is quite difficult to make artificial planar tunnel junctions, and most attempts failed. In fact, the crucial point is to prepare a HTSC sample with a surface flat at an atomic scale, before the deposition of a thin insulator (usually another oxide) on top of it. The high temperature required to grow oxides, together with the chemical and structural complexity of these phases make the task fairly hard. Different oxides have been used, and we present here a few examples of junctions displaying QP-like I-V characteristics. However tunneling is not always clearly established.

T. Becherer et al [78] used CeO_2 layers between two YBCO layers in a ramp-type junction [79]. They mask half of the $SrTiO_3$ substrate during a first evaporation of c-axis oriented YBCO. Then they deposit a CeO_2 insulating layer, followed by a 4000 Å thick insulating $LaAlO_3$ layer. After removing the mask, a layer of CeO_2 is deposited (100 to 200 Å thick), followed by a final YBCO layer. In the region corresponding to the edge of the mask, there is a junction between the two YBCO layers through the CeO_2 one along the basal plane (Fig. 2) . The junction area itself is then defined by standard lithography to $0.5*10 \ \mu m^2$. Its resistance increases exponentially with the CeO_2 layer thickness. The tunneling conductance displays structures which look like a superconducting gap, on top of a steep linear background.

Although the conducting mechanism in PrBCO is still unclear [80], it has been used as an insulating barrier [81]. G. A. Alvarez et al [82] deposited sequentially c-axis NdBCO and PrBCO layers onto a $SrTiO_3$ substrate, and then used standard lithography method to define small junctions ($8*14 \ \mu m^2$). They observed non-linear I-V characteristics and Shapiro steps in these structures, and they conclude to an observation of the superconducting gap. However, the interpretation is difficult because several junctions have been grown in series first, and second, the "normal state" (which happens to take place at 32K) tunnel conductance is also strongly non-linear, as we do not expect in a regular tunnel junction. A. M. Cucolo et al [83] defined the junction area with shadow masks ($0.5*0.5 \ \mu m^2$) during the deposition of YBCO/PrBCO/HoBCO trilayers structures. A well defined dip in the tunneling conductance is observed, which is attributed to a gap of 21.5 meV (Fig. 3). The normal state conductance has a strong temperature dependence (a factor of 10 between 4K and 100 K), which means that other processes than tunneling occur simultaneously. Finally, the reproducibility of this result is poor (2 among 12 junctions). This is common to almost all the engineered planar tunnel junctions on HTSC. This is why most of the results we will discuss have been obtained on "natural barrier" junctions, for which the reproducibility is much higher.

2) Natural barrier junctions.

Surfaces of most of the HTSC materials appear to have an insulating behavior. Whether an oxygen deficiency occurs (like in YBCO), or whether there is an interaction with the surrounding atmosphere, the result is always a less conducting surface on depths compatible with tunneling (a few nm). Therefore, it has been shown back in the early days of HTSC history [84] that very good tunnel junctions can be reproducibly made by depositing a metallic layer on top of a HTSC crystal or thin films : the surface layer of HTSC is therefore a "natural" tunnel barrier . Using a low Tc superconducting counter-electrode (CE), one can test the junction quality by measuring its specific DOS features (gap, temperature dependence, phonon related strucures for a strong coupling material) and comparing with the calculated BCS ones. Although many CE have been used (Ag, Au, Sn, Al ...), the most popular one is Pb, whose Tc (7.23K) is rather high, and has well-defined phonon peaks. Therefore, hundreds of YBCO/Pb junctions have been used by several groups [35, 65, 67, 85-87]. Fig. 4 displays the tunneling conductance of typical YBCO/Pb junctions at low energy and temperature where the Pb gap opens : the BCS fit to the data proves that 99.9% of the total current is a tunnel one, and the phonon peaks are at the correct energy with the right amplitude. These requirements being fulfilled gives us confidence that these tunnel junctions are good spectrometers, and therefore that the differential conductance actually measures the DOS of the HTSC material.

C) Main experimental results on the Quasi-Particle branch.

1) The "gap-like feature"

In Fig. 5 typical tunneling curves of YBCO/Pb planar junction are presented (T<Tc of the Pb CE). Besides the strong feature due to Pb gap near zero voltage, a clear dip in the conductance occurs (often referred to as a "gap-like feature" GLF), roughly symmetric in bias, is observed around 20 meV.

This has been observed by numerous authors [35, 84, 86, 87] using different CE, both on films and crystals based junctions, and whatever their crystallographic orientation is. As we have already noticed, this does not resemble the standard BCS gap : the edge peaks are weaker, and the zero-bias conductance does not vanish even at T/Tc<0.1. The temperature dependence displayed in Fig. 6 indicates that this structure disappear above Tc ; it is therefore related to superconductivity. This gap-like feature has been observed more clearly in the BSCCO system, by STM [88], but also on BSCCO/Pb planar junctions [85] as we can see in Fig. 7. In this case the gap is computed to be 26 meV, and the zero-bias conductance is

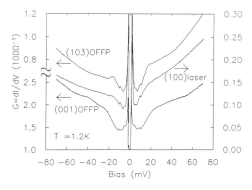

Fig. 5 Tunneling conductance vs energy for YBCO/Pb junctions made on different crystallographic orientations (T=1.2K).

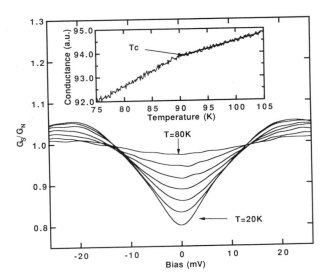

Fig. 6 Temperature dependence of the tunneling conductance as a function of the voltage on YBCO. The GLF disappear at Tc. The inset shows the conductance at zero bias vs temperature.

smaller, although the shape of the structure does not follow the BCS one. Let us make the following remarks :

- there is a definite depression of the DOS of HTSC below their Tc, which corresponds to typical energies of 20-25 meV for 90K materials, i.e. $2\Delta/k_BTc \approx 5$, a strong coupling value. In most experiments, the missing states are pushed at the gap edge as it is expected for a DOS feature, leading to a more or less pronouced peak. There is a gap-like structure in the excitations spectrum of QP in the HTSC materials, corresponding to a strong, but reasonable coupling strength.

- the low bias conductance is always finite at low temperature. Besides very low energy effects we will discuss later, the DOS remains finite at low energy. It is lower in the BSCCO system than in the YBCO one : this is probably related to their respective surface quality. This means, either there is an additionnal process which modifies the tunnel DOS (bound

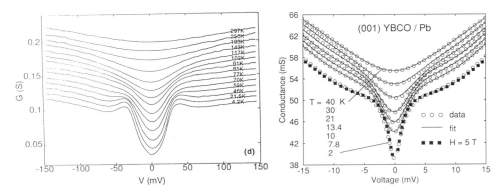

Fig. 7 Temperature dependence of BSCCO/Pb planar tunnel junctions. ref 84

Fig. 8 Tunneling conductance in a c-axis YBCO/Pb junction at low energy showing a small dip below 5 meV, which is magnetic field independant. This feature disappears at around 40K, as it is expected taking into account thermal broadening effects (solid line).

states, proximity effect, depairing), either there is an anisotropic gap with nodes on the FS, i.e. low energy excitations can occur in certain directions and contribute to the tunneling conductance.

- the shape of the feature never matches the usual BCS one, unlike LTSC compounds. It has been very often proposed to introduce an "ad-hoc" lifetime for the QP [89], which broadens the gap feature, according to the so called "Dynes formula" [90] :

$$N_{sf}(E) = \text{Re}\left[(E+i\Gamma)^2 \Big/ \left((E+i\Gamma)^2 - \Delta^2\right)\right]^{1/2}$$

Under certain circumstances, a reasonable agreement is obtained between the computed tunneling conductance and the data [89]. However, this parameter does not follow a clear dependence on whatever may produce this depairing (impurities, disorder ...), and its temperature dependence is not standard. Even if we cannot completly remove this possibility, another natural explanation is again an anisotropic gap. In a pure $d_{x^2-y^2}$ case, a linear dependance is expected [52] at low temperature, but other symmetries lead to different gapless-shape for the DOS (see for instance Monod-Maki [50]). Moreover, disorder can mix the wave vectors and "widens" the node region on the FS.

- when the two electrodes are in the superconducting state (T<7.2K for YBCO/Pb junctions for instance) there is no structure at $\Delta_{YBCO}+\Delta_{Pb}$ and $\Delta_{YBCO}-\Delta_{Pb}$, as expected if both gaps were complete. The Pb gap structure is observed around zero-bias (cf Fig. 5) which also means that the YBCO DOS is finite at low energy.

- the temperature dependence of the gap is somewhat controversial for two reasons : firstly, in STM measurements, the gap is very often rather well defined, but thermal drift prevent the measuring of the very same junction at different temperatures and secondly, a functionnal form of the DOS has to be known to extract the gap from the data at each temperature.

- finally, the gap dependence with Tc is not yet clear. The spread in the measured gap values for each compound is rather large among the available data [4], and it is difficult to compare between HTSC materials. Concerning planar junctions on YBCO, Pr doped YBCO samples have been used by A. M. Cucolo et al [91], and by M. Covington et al [92] : in the first set of experiments, the GLF vanishes when Tc decreases ; in the other one it scales with Tc. Surprisingly, underdoped YBCO films display a weak GLF at energies higher than the 90K ones [93]. Once again, the absence of functionnal form for the gap makes its absolute value determination difficult.

Tunneling being a somewhat directional measurement (see first section), different crystallographic orientations have been used to make junctions. YBCO/Pb planar junctions have been made on (001), (100), (110) and (103) surfaces (thin films and crystals). In every configuration, the GLF is present (in a very reproducible fashion at around 20 meV) with the above listed characteristics (Fig. 5). This is not surprinsing if the gap were isotropic of course ; in the anisotropic case, GLF is expected to be seen at the same energy for (001) and (100) orientations since the maximum gap amplitude is probed. It appears that (110) and (103) films have a surface roughness higher than the others : therefore part of the tunneling electrons can enter along different directions in the basal plane.

In summary, tunneling data on HTSC are compatible with the opening of an anisotropic gap on the FS below Tc, presenting nodes along certain directions, whose maximum amplitude correspond roughly to a strong coupling value $2\Delta/k_B Tc \approx 5$. Let us now move onto *low* energy features.

2) Low energy features.

Once again, let's talk about YBCO/Pb planar junctions mainly. When tunneling in the c-direction only, a small dip develops at low temperature below 5 meV (Fig. 8). It has been shown [35] that this feature is insensitive to magnetic field, and that usual thermal smearing of the DOS accounts for its temperature dependence. At first sight it could be a smaller gap in the DOS. This has been proposed by different authors essentially on the basis of anisotropic FS [30, 94], internal proximity effect [95] or mixed d+s pairing state [50]. In these cases, it should be observed also when tunneling into the ab-plane, which is never the case. Tachiki et al [96] developped an interesting model of layered superconductors where a dip is expected at low energy in the less conducting direction, namely the c-axis. However, this is still an open question.

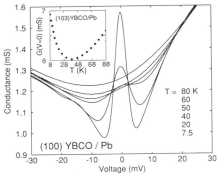

Fig. 9 Tunneling conductance in an a-axis YBCO/Pb junction at low energy showing a pronounced peak below 5 meV. The inset shows the temperature dependence of the zero-bias conductance.

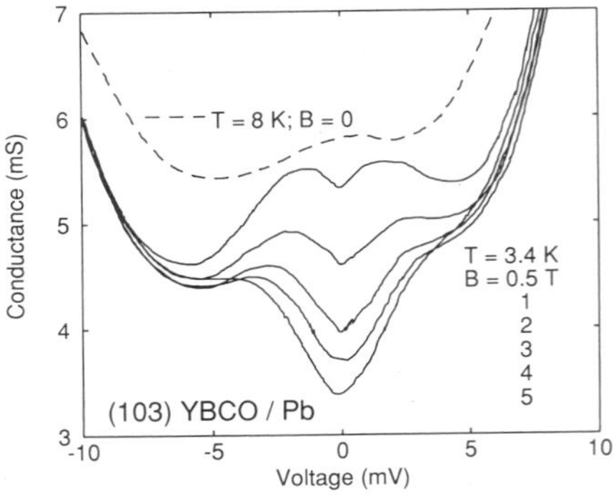

Fig. 10 The Zero Bias Conductance Peak (ZBCP) reverses when a magnetic field is applied.

Let us turn now to the tunneling conductance in the ab-plane. We have shown [35] that a well defined conductance peak develops below 5 meV at low temperature in this case (Fig. 9). This has been also reported in the literature [35, 84, 97-102]. Moreover, a strong magnetic field dependence was observed [35] (cf Fig. 10). A tentative explanation was done in terms of magnetic impurities in the barrier [103-105], but a quantitative agreement could not be found. In fact, this is mostly visible when tunneling into (110) and (103) oriented films. Recently, an alternative explanation has been proposed in terms of a bound state due to strong Andreev reflexions in the (110) direction in a d-wave superconductor.

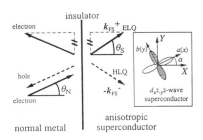

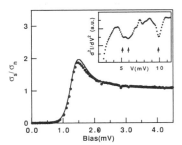

Fig. 11 Schematic of the normal and Andreev reflections at the surface of an anisotropic superconductor. These processes lead to a bound state located at the interface. ref 42

Fig. 12 DOS in the (110) and (100) directions in the case of a two components OP with a phase shift $\pi/2$. A strong peak (Andreev bound state) is seen at low energy. ref 48

3) Andreev bound state.

C. R. Hu [40] showed that a QP bound state forms at the Fermi energy (E_F) when the node of a d-wave OP is normal to a specularly reflecting surface. QP reflecting from the surface experience a change in the sign of the order parameter and therefore undergo strong Andreev reflection (see Fig. 11). Constructive interferences between incident and reflected QP lead to a bound state at E_F confined to the surface. Following this idea, calculations of the tunneling DOS have been performed [41, 48, 106, 107] in various situations. For example, in the case of a d_{x2-y2} symmetry, an Andreev bound state will be present for every specular surface misoriented from (100), and a peak in the DOS at zero bias will appear (Zero-Bias Conductance Peak ZBCP) (see Fig. 12). It is worth noticing that this behavior is not expected to occur when tunneling in the c-direction.

New experiments have been made recently [41, 100, 106-109], which agree very well with this hypothesis. It is confirmed that the ZBCP is always present provided the tunneling direction is not (001) or (100). It is related to the presence of the superconducting gap (as seen by the GLF), as expected for an Andreev bound state. Its temperature dependence follows the expected one [100]. These observations strongly suggest that a d-wave component is present in YBCO.

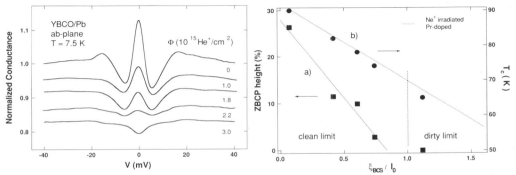

Fig. 13 Reduction of the ZBCP in (110)YBCO/Pb junctions as a function of the irradiation dose (He+ 1 MeV). From top to bottom, the mean free path l_0 is computed to be 285, 48, 33, 27 Å respectively. Curves are shifted for clarity. ref 109

Fig. 14 Decrease of Tc as a function of the QP scattering rate and the Impurity scattering rate (see text). $1/\tau_S$ is the estimated surface scattering rate. ref 109

Further on, Aprili et al [109] studied the sensitivity of the ZPCP to disorder. Since the bound state comes from constructive interferences between partial electronic waves, the reduction of the mean free path l_0 will strongly affect this coherent process, and the ZBCP should be quenched for $l_0 \approx \xi_{ab}$ (the in-plane superconducting coherence length). (103) and (110) YBCO/Pb junctions have been ion irradiated (He 1 MeV) to create a weak atomic disorder, mainly in the film. Fig. 13 displays the evolution of the ZBCP in weakly disordered (103)YBCO films. It vanishes out for a fluence as low as $2.2 \cdot 10^{15}$ He/cm^2, which corresponds to $l_0 \approx 27$ Å as computed from resistivity measurements, close to the in-plane coherence length (20-25Å). Finally, the broadening of the ZBCP with increasing disorder has been studied. The width of the ZBCP is a direct measure of the QP scattering rate. For a d-wave pairing state, normal impurities scattering lowers Tc as magnetic impurities scattering does in a BCS superconductor [110]. Using the same functionnal form (Abrikosov-Gorkov pair-breaking), the impurity scattering rate has been deduced from the Tc reduction as a function of disorder (see Fig. 14) [109], and correspond to the QP one. Therefore, the QP scattering rate is a direct measure of the pair breaking strength. All the above results reinforced the idea that the ZBCP is indeed an Andreev bound state at a surface of an unconventional superconductor.

Moreover, Palumbo et al [47, 48] proposed that if there are two components of different symmetries in the OP (in a quadratic material), surfaces will induce Broken Time-Reversal Symmetry (BTRS) for the following reasons. If two components are present (for example d_{x2-y2} and s as calculated in their paper), it is shown that the dominant component in the bulk (d for instance) will be suppressed at surfaces (mainly in (110) direction in this case), and the subdominant one reinforced. In this context, the minimization of the Ginsburg-Landau free energy implies [51] that there is a $\pi/2$ phase shift in between, i.e. d+is or s+id. As a consequence, a supercurrent is spontaneously generated at surfaces and the ZBCP is moved towards finite energies. An external magnetic field will increase this splitting [48].

Covington et al [100] have proposed that this accounts for the observed magnetic field dependence of the ZBCP. Fig. 15 displays the peak splitting as a function of an applied magnetic field. Data on HTSC are in good agreement with the BTRS model, whereas those on regular metals follow the magnetic impurity scattering scenario.

To sum up this section, the observation of the ZBCP in tunnel junctions is very well interpreted as an Andreev bound state at surfaces of a d-wave superconductor. This is probably the clearer evidence of the presence of a "d-wave" component in HTSC coming from QP tunneling.

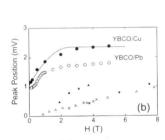

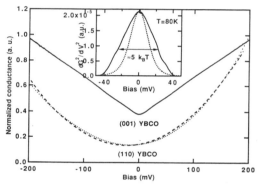

Fig. 15 (a) Splitting of the ZBCP at low temperature and under magnetic field. (b) The peak position as a function of field (solid line) agrees with the calculation of Fogelström et al (ref 48). Data in the lower part of the plot correspond to magnetic impurities in the barrier from the literature. ref 100

Fig. 16 Tunneling conductance as a function of voltage in (110) (parabolic) and (001) (linear) directions. Fit with the BDR model gives reasonable numbers for the barrier properties (Φ=300 meV, $\Delta\Phi$=400 meV, w=33 Å). The inset shows the second derivative of the conductance for the (001) junction. Its width is >5k_BT instead of the usual 3.5 k_BT for elastic tunneling.

4) High energy features

Many tunneling data (mainly STM) have been reported [111, 112] showing wiggles in the conductance for energies ranging from 10 to 100 meV. They have been interpreted as phonon structures and sometimes the $\alpha^2F(\Omega)$ function (related to the phonon density of states) computed from the experiments and compared with neutrons data. But the phonon spectra of HTSC materials are so complex that such a comparison is very hazardous. Moreover, many barrier effects may occur in this energy range (see for instance Wolf's book [113]), and one has to worry about the actual DOS origin of those features. Finaly, no common agreement appears on the peaks positions and height based on reproducible data.

Therefore, we do not consider that tunneling experiments have yet probed the electron-phonon coupling in the HTSC.

More interesting is the study of the background conductance. In many tunneling data, the background conductance (up to 200-300 meV) is not flat (as expected from naive models). Steep linear backgrounds together with parabolic ones are routinely observed in STM or planar junctions experiments [4, 86, 87, 114]. The question arises whether these are DOS related features (like those proposed by Varma et al [72, 74] or Anderson et al [77]), or barrier effects. In YBCO/Pb planar junctions, we found a symmetric linear conductance in c-axis oriented junctions, and an asymmetric parabolic one in ab-plane junctions ((100),(103),(110) oriented films) (see Fig. 16). If YBCO was isotropic, such a difference could not be explained in terms of DOS.

In a quasi-2D material, the tunneling probability along the c-axis should be very low, and therefore DOS features are expected to be seen mostly in the ab-plane tunneling conductance. However, the above quoted theories account for a linear shape, and not for the parabolic one as observed in this case. Moreover, we can fit the parabola using the Brinkman-Dynes-Rowell (BDR) for a trapezoidal barrier (see first section) with reasonable parameters (see Fig. 16), like a barrier height of 300 meV. So, if the ab-plane background conductance can be explained by standard barrier shape models, how can we understand the linear c-axis tunneling conductance ? Following Kirtley and Scalapino (see above), we propose that inelastic events needed to scatter the incoming electrons into the available states of the FS can account for this behavior. Provided a flat distribution of inelastic scattering states is available, tunneling conductance will increase linearly with the energy. It has been proposed [75] that AF fluctuations can lead to such a process. Notice that the width of the thermal smearing in the c-axis tunneling conductance is bigger than $3.5k_BT$, as expected for inelastic events (inset Fig. 16). As a conclusion, the unusual shape of the high energy tunneling conductance seems to be mainly due to inelastic events.

Recent experiments by Dagan et al on YBCO/Al junctions [115] give a new insight in this problem. By cycling such junctions in temperature, they obtained a serie of tunneling conductance curves which display a specific relationship between the linear coefficient of the background conductance and the zero-bias conductance value. Analysing their data in the framework of the MFL theory [72, 74], they compute a coupling parameter for the interaction and a characteristic energy of the model which are reasonable. This picture is consistent with a quasi-linear background originated from an intrinsic feature of the HTSc : a non Fermi-Liquid behavior. However, one has to check carefully that the *tunneling conductance in the c-direction* is indeed measured in this experiment, to validate strong conclusions on the basis of a somewhat sophisticated analysis.

Renner et al [88] studied BSCCO crystals by STM along the c-axis exclusively. They come to the conclusion that linear and parabolic shapes originate from non-vacuum tunneling processes, and that a flat background is a necessary condition to obtain reliable data and gap structure. However their steep background conductances are related to extremely low work functions (10 meV), and therefore to a non-tunneling process. In our case, sharp Pb gap structures and phonons are always present : this insures that no other conductance channel is open, at least at low energy.

D) Main experimental results on the Josephson effect.

Although Josephson effect has been observed in numerous SNS situations, we will focus here on planar *tunnel* junctions where both DOS features and Josephson supercurrent are present. To the best of our knowledge, there is no report of Josephson tunneling in SIS' junctions with artificial barriers, neither with point-contact or STM junctions. The only available data come from HTSC/I/LTSC planar junctions on YBCO.

1) c-axis coupling

The first report of Josephson tunneling in SIS' structures has been made by Kwo et al [116] back in 1990. At this time, the s versus d wave controversy was not launched and the result was not widely publicised. Moreover, the poor magnetic field dependence of the critical current Ic made short-circuits as a possible explanation for the Josephson behavior. In 1994, Sun et al [63] published very clear data on c-axis YBCO/Pb planar junctions on crystals : an almost ideal Fraunhofer pattern observed together with a rather high value of the IcRn product in some cases (more than 20% of the AB value) that there were serious arguments in favor of a true SIS' tunnel junction behavior. Therefore, it was argued that a substantial s component was present in the OP of YBCO. Experiments on films were also carried out [65] and gave roughly the same results, although the IcRn product was two orders of magnitude smaller than in best crystals. Iguchi et al reported similar experiments [66] and they concluded in favor of a d-wave scenario, based on an unusual Ic(B) pattern. However, their explanation relies on unproved critical current trajectories. Moreover, they find an increasing Ic as a function of temperature, which is most difficult to understand.

In the above quoted experiments, two conditions were required to get reliable YBCO/Pb JJ, which make people wonder about the actual tunneling geometry, even if serious studies have been made on this subject [99]. Firstly, a Br-ethanol etch was needed to clean the surface prior to the junction fabrication (which induce pits on the surface and expose ab-planes). Secondly, a thin (1-4nm) Au or Ag layer was evaporated in most cases between YBCO and Pb. Therefore, we decided to make all *in-situ* YBCO/Pb tunnel junctions [67].

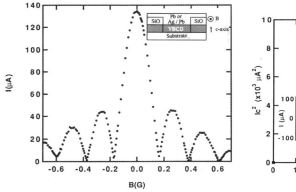

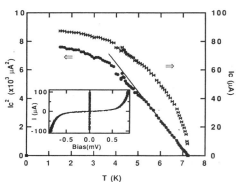

Fig. 17 Ic(B) curve for a c-YBCO/Ag/Pb junction (100Å of silver). The pattern is highly symmetric in field and current, and the periodicity corresponds to the junction geometry (inset).

Fig 18. Ic(T) for a c-YBCO/Ag/Pb junction (10Å of silver). The inset shows the I(V) characteristics with both the QP branch and the Josephson one.

As reported earlier, high quality tunnel junctions (see Fig. 4) can be made by an *in-situ* process. A c-axis oriented film is deposited on MgO or SrTiO3 by a co-evaporation technique [117] through a shadow mask in the shape of a rectangle. After cooling it down to room-temperature, other masks are used to deposit contact pads, an insulating SiO window and a counter-electrode (Pb or Ag/Pb) in a cross geometry [113] without any exposure of the surface to air (see inset Fig 17). Let us emphasize that *all* junctions display a very clean tunnel behavior (as we can see on the QP branch) together with a Josephson critical current (inset Fig. 18). Fig. 17 displays the Ic(B) pattern for a typical junction : it is symmetric in field and current, the maximum is observed for B=0, and Ic goes to zero for each flux quantum as expected from the geometry of the window (0.5*0.9 mm^2), and the thickness of the layer (d=80 nm) which limits the extension of the screening currents (d<λ_{ab}=140 nm). Finally Ic(T) exhibits a conventionnal behavior (Fig. 18) and vanishes for bulk Pb Tc (7.2K). No specific role of silver has been found for thicknesses ranging from t=0 to t=10 nm, except that the normal state resistance of the junctions decreases as t increases [118]. The IcRn product is roughly constant in the whole set of studied junctions and found to be 5 μeV, a very small value. Therefore, a coupling does exist between the "s-wave" superconductor Pb and YBCO along the c-axis : YBCO is not a "pure d-wave" superconductor, as it is in fact expected from symmetry arguments [51] . By aligning the magnetic field along a and b directions of detwinned YBCO single crystals, Sun et al [64] measured the anisotropy ratio (λ_a/λ_b)^2≥2 in this compound, compatible with a significative contribution of the chains to the pairing state (see Fig. 19). Although there was an important scattering within the experimental λ_a values, these data suggest that orthorhombicity plays an important role in YBCO.

The question thus arises whether YBCO is a pure "s-wave" SC, which couples naturally to Pb, or a mixed OP superconductor (with a d- and an s- component for instance). In the latter case, the Josephson coupling would originate from the "s-wave" part, and the IcRn product a "measure" of its relative weight according to naive arguments (see discussions in [50, 119]). Although twins may play a significant role in this problem (see discussion below), we think that our reproducible very low IcRn product on films whose surfaces are very smooth [117] (we are in a real situation of c-axis tunneling) is in favor of a mixed "s+d" OP, with a small s component.

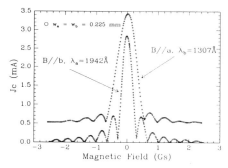

Fig. 19 Ic(B) for c-YBCO/Pb junctions for untwinned crystals. The anisotropy of the penetration depth is computed from the periodicity in field. ref 64

2) ab-plane coupling

Sun et al [99] repeated their measurements on the edge surfaces of twinned YBCO crystals. ab-plane tunneling was evidenced by the ZBCP constantly observed in the QP branch [35] and JJ were succesfully made in this geometry. Although most of them displayed behaviors not as ideal as the c-axis ones (rough surfaces together with easier flux trapping were serious drawbacks in this configuration), a clear Josephson coupling was established, with an IcRn product comparable to the previous ones. The coupling strength seems to be isotropic in their experiments, i.e. with a dominant (or alone) s-component [99]. However authors remain cautious about this statement, because of the spread in IcRn products and the question of twins.

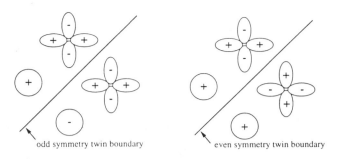

odd symmetry twin boundary   even symmetry twin boundary

Fig. 20 Ic(B) on bicrystal based c-YBCO/Pb junctions. The field is rotated (Φ) with respect to the boundary line. ref 128

3) Twins and Josephson coupling

Most of Josephson tunneling data concern YBCO. Albeit small (2%), its crystallographic orthorhombicity may induce rather anisotropic electronic properties within the basal plane, as seen for instance by penetration depth measurements [64, 120]. Therefore, the pairing potential can be anisotropic too, and one has to worry about twin boundaries effects in the above quoted experiments. This point has been the subject of numerous papers [51, 99, 121-125] and the reader will find more precise arguments than those briefly exposed here. For the sake of simplicity, we will first suppose that there is no "internal" mixing of different components (like s+id for instance), but only a mixing coming from the low symmetry of an orthorhombic phase (d+s for example). If there is a "d-wave" component, two distinct situations can occur at a twin boundary (see Fig. 21) (referred to as gyroscopic or not in [51], or odd-even in [122]). There is no reason a-priori to select one or the other. However, since the result on corner-SQUIDs experiments has been obtained on highly twinned single crystals [59], there is a rather large agreement that the gyroscopic situation is favored. One has to keep in mind that this hypothesis is supported by experimental facts only. However, the presence of twins has strong consequences : a "twin-averaged OP" has to be defined [51], and the experiments mostly test the twin-boundary symmetry [123] rather than the intrinsic OP one. For instance, if the gyroscopic case applies, c-axis Josephson current in YBCO/Pb junctions should average to zero in the thermodynamic limit (infinite number of twins in the junction area), when it should be equal to that of untwinned samples in the other case.

If we go on with the gyroscopic situation, then the relative weights between s and d can be roughly probed. In the untwinned case, c-axis tunneling measures the s component only, and strong anisotropy in the ab-plane is expected (d-s vs d+s in the a and b directions) provided one component is not very small. The situation is more complicated in twinned samples (cf table II in [99]). However, if s>d, the s-component is measured in all the directions, and the IcRn product looks isotropic. If d>s, IcRn should be zero in c-axis tunneling, and proportionnal to the d-component in the ab plane. Let us now analyze the available data.

According to the UCSD group [99], the IcRn product does not scale with the number of twins N according to statistical laws ($1/\sqrt{N}$), and high IcRn values in c-axis tunneling are obtained in heavily twinned samples. There are three possibilities : (1) there is no d-component, and the low IcRn products are explained by extrinsic reasons (disorder, degraded surface [126]...); (2) the gyroscopic hypothesis is wrong, and the coupling is not twin-averaged to zero, and (3) there is a very large imbalance between "+" and "-" domains, and a blind statistical average does not apply. Hypothesis (1) is unlikely, based on many experimental results [51] and on the QP spectroscopy described above. To remove number (2), one has to find another explanation for Wollman's results [59] : trapped flux can always been invoked but needs to be proved seriously. There is one possible support for number (3): strains in samples could favor one type of domain. Thermal gradients may induce anisotropic strain fields in crystals (high IcRn value), when thin films are relaxed (low IcRn values); but this has to be verified. Nothing can be said in this situation about the relative strength of the two components. However, since IcRn is roughly the same in all directions, there should be a substantial amount of s-component in YBCO OP.

As already mentionned, it has been proposed that two distinct components coexist in the OP, and that the subdominant one is present at surfaces, inducing BTRS in the vicinity of the junctions, and therefore spontaneous currents. On the other hand, regular bias currents can induce non d-wave components in a d-wave superconductor [127]. Specific phase-shifts will occur in the Josephson coupling equations, which should strongly modify the tunneling supercurrent. This has not been observed yet ; however, Josephson junctions on (110) or (103) orientations need to be done to further test these ideas.

We have not discussed here the surface roughness problem. One may think for instance that pits (which expose the ab plane to tunneling in the c-axis experiments) can account for the high and isotropic IcRn product observed in crystals, when films display low IcRn due to their smooth surface. This argument has been dismissed by Sun et al [99] on a basis of carreful studies of crystal surfaces. One can also say that defects in films modify the tunneling geometry at an atomic scale. As always when it comes to sample quality in HTSC material, one has to remain open-minded and cautious.

To sum up, let us say that the Josephson tunneling experiments in YBCO/Pb junctions are compatible with a mixed order parameter (d+s) with imbalance number of "+" and "-" domains. The latest experiment reported in a preprint by Kouznetsov et al [128] in Berkeley confirms this idea.

4) c-axis tunneling across a single twin-boundary.

It happens that detwinned samples exhibit only two domains, i.e. a single twin-boundary. The Berkeley-UCSD-UIUC collaboration made tunnel junctions across the twin-boundary of YBCO single crystals using standard techniques for c-axis YBCO/Pb junctions. According to the gyroscopic hypothesis, and provided a mixed symmetry (s+d) with a dominant d-wave part, the s component changes sign across the boundary : therefore the

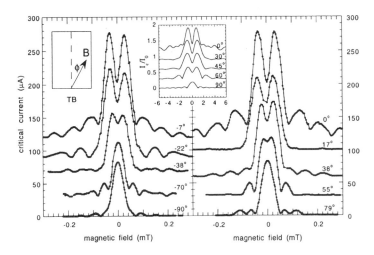

Fig. 21 Gyroscopic (or odd) and non-gyroscopic (or even) configuration at a twin boundary. ref 122

Josephson current can be expressed as follow for a magnetic field B applied *parallel* to the boundary :

$$I_c^2(\Phi, \gamma) = (I_0\Phi_0/\pi\Phi)^2 \left[1 + \cos^2(\pi\Phi/\Phi_0) - \cos(2\gamma\,\pi\Phi/\Phi_0) - \cos[2(\gamma-1)\,\pi\Phi/\Phi_0]\right]$$

where γ is the ratio of the surfaces on both sides of the boundary line, Φ the flux in the junction, and I_0 the maximum critical current in the homogeneous case. In this case screening supercurrents test both sides of the junctions and probe the sign change. When the field is perpendicular to the boundary, regular pattern is expected. Fig. 20 clearly shows that the Ic(B) pattern is strongly modified when B is rotated, going from the standard one (Ic(0) maximum), to an unconventional one (local minimum for Ic(0)). Junctions made on one side of the crystal display a regular Fraunhofer pattern whatever the orientation of the field is. This is quite a convincing result that a d+s pairing state (predominantely d) exists in YBCO. However discrepancies appear between the optically measured γ and the computed one from the Ic(B) patterns, and one has to reconcile this result with the high IcRn products reported previously.

CONCLUSION

Tunneling spectroscopy together with Josephson experiments give an interesting insight within the physics of High Temperature Superconductors. An incomplete gap is observed in the QP excitation spectrum, as if they were either states in the gap or nodes in some regions of the FS, like in an anisotropic superconductor. A zero bias conductance peak is seen in the in-plane tunneling geometry, which can be explained as a surface induced Andreev bound state in a "$d_{x^2-y^2}$" superconductor. The Josephson coupling between conventional superconductors and HTSC materials is also compatible with this symmetry for the order parameter, provided the orthorhombicity is taken into account (small s component in YBCO).

REFERENCES

[1] Giaever, I., Phys. Rev. Lett., **5**, 147 (1960).

[2] Giaever, I., Phys. Rev. Lett., **5**, 464 (1960).

[3] Anderson, P.W. &J.M. Rowell, Phys. Rev. Let., **10**, 230 (1963).

[4] T. Hasegawa, H. Ikuta, K. Kitazawa, *revue,* in *Properties of High Temperature Superconductors,* D. M. Ginsberg Editor. (1992), World Scientific: Singapore.

[5] J. R. Kirtley, Int. Jour. Mod. Physics B, **4**, 201 (1990).

[6] M. H. Cohen, L. M. Falicov, J. C. Phillips, Phys. Rev. Lett. **8**,316 (1962)

[7] Ambegaokar, V. &A. Baratoff, Phys. Rev. Lett., **10**, 486 (1963) & **11**, 104 (1963).

[8] J. Nicol, S. Shapiro, P. H. Smith, Phys. Rev. Lett. **5**, 461 (1960).

[9] B. L. Blackford, R. H. March Phys. Rev. **186**, 397 (1969).

[10] G. I. Lykken, A. L. Geiger, K. S. Dy, E. N. Mitchell Phys. Rev. B **4**, 1523 (1971).

[11] J. P. Carbotte in *Anisotropy effects in Superconductors,* H. W. Weber Editor. (1977) Plenum : New York p 183.

[12] N. V. Zavaritskii Zh exp. theor. fiz **45**, 1839 (1964) & JETP **18**, 1260.

[13] N. V. Zavaritskii Zh exp. theor. fiz **48**, 837 (1965) & JETP **21**, 557.

[14] B. L. Blackford, K. Hill J. Low Temp. Phys. **43**, 25 (1981).

[15] E. Guyon, A. Martinet, J. Matricon, P. Pinkus, Phys. Rev **138**, A746 (1965).

[16] Gennes, P. G. de, Rev. Mod. Phys., **36**, 225 (1964)

[17] W.L. McMillan, & J. M. Rowell, *Tunneling and Strong-Coupling,* in *Superconductivity,* R. D. Parks Editor. (1969), Marcel Dekker: New York. p. 561

[18] Josephson, B. D., Physics Letters, **1**, 251 (1962).

[19] Rowell, J. M., Phys. Rev. Lett., **11**, 200 (1963)

[20] Owen, C. S. &D. J. Scalapino, Phys. Rev., **164**, 538 (1967).

[21] Balsamo, E. P., *et al.*, Phys. Status Solidi (a), **35**, 173 (1976).

[22] Shapiro, S., Phys. Rev. Lett., **11**, 80 (1963).

[23] Shapiro, S., A.R. Janus & S. Holly, Rev. Mod. Phys., **36**, 223 (1964).

[24] Werthamer, N. R., Phys. Rev., **147**, 255 (1966).

[25] Larkin, A. I. &Yu. N. Ovchinnikov, Zh. Eksp. Teor. Fiz., **51**, 1535 (1966).

[26] Fiske, M. D., Rev. Mod. Phys., **36**, 221 (1964).

[27] Ambegaokar, V. &A. Baratoff, Phys. Rev. Lett., **11**, 104 (1963).

[28] C. P. Poole Jr., T. Datta, H. A. Farach, *Copper Oxides Superconductors.* (1988), New-York: Wiley.

[29] J. Labbé, J. Bok, Europhys. Lett., **3**, (1987).

[30] J. Bouvier, J. Bok, Physica C, **249**, 117 (1995).

[31] D. J. Van Harlingen, Rev. of Mod. Phys., **67**, 515 (1995).

[32] Y. Kitaoka et al, J. Phys. Chem. Solids, **54**, 1385 (1993).

[33] A. T. Fiory, S. Martin, R. M. Flemming, L. F. Schneemeyer, J. V. Waszczak, A. F. Hebard, S. A. Sunshine, Physica C, **162-164**, 1195 (1989).

[34] N. Bontemps, D. Davidov, P. Monod, R. Even, Phys. Rev. B, **43**, 11512 (1991).

[35] J. Lesueur, Greene L.H., Feldmann W.L. & Inam A., Physica C, **191**, 325 (1992).

[36] N.E. Bickers, D. J. Scalapino & S. R. White, Physical Review Letters, **62**, 961 (1989).

[37] P. Monthoux, A. Balatsky & D. Pines, Physical Review B, **46**, 14803 (1992).

[38] V. J Emery, S. A. Kivelson, Phys. Rev. Lett., **74**, 3253 (1995).

[39] P. W. Anderson, J. Phys. Chem. Solids, **11**, 26 (1959).

[40] Chia-Ren Hu, Physical Rev iew Letters, **72**, 1526 (1994).

[41] Y. Tanaka, S. Kashiwaya, Phys. Rev. Lett., **74**, 3451 (1995).

[42] S. Kashiwaya, Y. Tanaka, M. Koyanagi, K. Kajimura, Phys. Rev. B, **53**, 2667 (1996).

[43] Y. Tanaka, *Theory of Tunneling Effect and Proximity Effect in high Tc Superconductors,* in *Coherence in High Temperature Superconductors,* Alex Revcolevschi Guy Deutscher Editor. (1995), World Scientific, Singapore: Tel Aviv. p. 393

[44] A. F. Andreev, Sov. Phys. JETP, **19**, 1228 (1964).

[45] J. M. Rowell, Phys. Rev. Lett., **30**, 167 (1973).

[46] G. Deutscher, *Aspects of coherence in High Temperature Superconductors,* in *Coherence in High Temperature Superconductors,* G. Deutscher and A. Revcolevschi Editor. (1995), World Scientific: Singapore. p. 3

[47] M. Palumbo, L.J. Buchholtz, D. Rainer and J.A. Sauls, preprint, (1996).

[48] M. Folgelström, D. Rainer, J. A. Sauls, Phys. Rev. Lett., **79**, 281 (1997).

[49] K. Maki, M. T. Béal-Monod, Phys. Lett. A, **208**, 365 (1995).

[50] M. T. Béal-Monod, K. Maki, Phys. Rev. B, **53**, 5775 (1996).

[51] J. F. Annett, N. Goldenfeld, A. Leggett, *Experimental Constraints on the Pairing State of the Cuprates superconductors : an emerging consensus.,* in *Physical Properties of High Temperature Superconductors V,* D. M. Ginsberg Editor. (1995), World Scientific: Singapore-New Jerseu-London-Hong Kong. p. 375

[52] H. Won, and K. Maki, Physical Review B, **49**, 1397 (1994).

[53] C.C. Tsuei, J.R. Kirtley, C.C. Chi, L.S. Yu-Jahnes, A. Gupta, T. Shaw, J.Z. Sun & M.B. Ketchen, Physical Review Letters, **73**, 593 (1994).

[54] C.C. Tsuei, J.R. Kirtley, Z.F. Ren, J.H. Wang, Letters to Nature, **387**, 481 (1997).

[55] P. Chaudari, & Shawn-Yu Lin, Physical Review Letters, **72**, 1084 (1994).

[56] J.H. Miller, Jr., Q.Y. Ying, Z.G. Zou, N.Q. Fan, J.H. Xu, M.F. Davis, and J.C. Wolfe, Physical Review Letters, **74**, 2347 (1995).

[57] M. Sigrist, & T. M. Rice, Journal of the Physical Society of Japan, **61**, 4283 (1992).

[58] D.A. Wollman, D. J. Van Harlingen, W. C. Lee, D. M. Ginsberg & A. J. Leggett, Physical Review Letters, **71**, 2134 (1993).

[59] D.A. Wollman, D. J. Van Harlingen, J. Giapintzakis and D. M. Ginsberg, Physical Review Letters, **74**, 797 (1995).

[60] D.A. Brawner, and H.R. Ott, Physical Review B, **50**, 6530 (1994).

[61] A. Mathai, Y. Gim, R.C. Black, A. Amar, and F.C. Wellstood, Physical Review Letters, **74**, 4523 (1995).

[62] A.G. Sun, L.M. Paulius, D.A. Gajewski, M.B. Maple & R.C. Dynes, Physical Review B, **50**, 3266 (1994).

[63] A.G. Sun, D.A. Gajewski, B.M. Maple & R.C. Dynes, Physical Review Letters, **72**, 2267 (1994).

[64] A.G. Sun, S.H. Han, A.S. Katz, D.A. Gajeweski, M.B. Maple and R.C. Dynes, Physical Review B, **52**, (1995).

[65] A.S. Katz, A.G. Sun, and R.C. Dynes, K. Char, Appl Phys Lett, **66**, 105 (1995).

[66] I. Iguchi, Z. Wen, Physical Review B, **49**, 12388 (1994).

[67] J. Lesueur, M. Aprili, A. Goulon, T. J. Horton & L. Dumoulin, Physical Review B, **55**, R3398 (1997).

[68] F. X. Régi, J. Schneck, B. Leridon, M. Drouet, F. R. Ladan. *Series array of intrinsic Josephson junctions in mesas patterned on BSCCO single crystals*. in *EUCAS*. (1995). Edimbourg:

[69] R. Kleiner, P. Müller, Phys. Rev. B, **49**, 1334 (1994).

[70] R. Kleiner, F. Steinmeyer, G. Kunkel, P. Müller, Phys. Rev. Lett., **68**, 2394 (1992).

[71] J. Bok, J. Bouvier. *Gap anisotropy and Van-Hove singularities in high Tc superconductors*. in *Spectroscopic Studies of Superconductors*. (1996). San José (CA): SPIE. vol 2696, p 122

[72] C.M. Varma, P.B. Littlewood and S. Schmitt-Rink, E. Abrahams and A.E. Ruckenstein, Physical Review Letters, **63**, 1996 (1989).

[73] M. Tachiki, S. Takahashi, F. Steglich, H. Adrian, Z. Phys. B, **80**, 161 (1990).

[74] P. B. Littlewood, C. M. Varma, Phys. Rev. B, **45**, 12636 (1992).

[75] J.R. Kirtley, and D.J. Scalapino, Physical Review Letters, **65**, 798 (1990).

[76] J.R. Kirtley, S. Washburn and D.J. Scalapino, Physical Review B, **45**, 336 (1992).

[77] P. W. Anderson, Z. Zou, Phys. Rev. Lett., **60**, 132 (1988).

[78] Th. Becherer, M. Kunz, C. Stölzel, and H. Adrian, Proc MS-HTSC IV, (1994).

[79] Th. Becherer, C. Stözel, G. Adrian and H. Adrian, Physical Review B, **47**, 14 650 (1993).

[80] Kawabasa, Phys. Rev. Lett., (1993).

[81] E. Polturak, G. Koren, D. Cohen, E. Aharoni, G. Deutscher, Phys. Rev. Lett., **67**, 3038 (1991).

[82] G.A. Alvarez, T. Utagawa and Y. Enomoto, Appl Phys Lett, **69**, 2743 (1996).

[83] A.M. Cucolo, R. Di Leo, A. Nigro, P. Romano, and F. Bobba, E. Bacca and P. Prieto, Physical Review Letters, **76**, 1920 (1996).

[84] J. Geerk, X. X. Xi, G. Linker, Z. Phys. B, **73**, 329 (1988).

[85] H. J. Tao, A. Chang, F. Lu, E. L. Wolf, Phys. Rev. B, **45**, 10622 (1992).

[86] M. Lee, M. Naito, A. Kapitulnik, M. Beasley, Solid State Com., **70**, 449 (1989).

[87] M. Gurvitch, J. M. Valles, A. M. Cucolo, R. C. Dynes, J. P. Garno, L. F. Schneemeyer & J. V. Waszczak, Physical Review Letters, **63**, 1008 (1989).

[88] C. Renner, O. Fischer, Phys. Rev. B, **51**, 9208 (1995).

[89] E. W. Wolf, H. J. Tao, B. Susla, Solid State Com., **77**, 519 (1991).

[90] R. C. Dynes, V. Narayanamurti, J. P. Garno, Phys. Rev. Lett., **41**, 1509 (1978).

[91] A. M. Cucolo, J. M. Valles, R. C. Dynes, M. Gurvitch, J. M. Phillips, J. P. Garno, Physica C, **161**, 351 (1989).

[92] M. Covington, and L.H. Greene. *Tunneling investigation of the gap-like feature in superconducting YPBCO thin films.* in *SPIE - Spectroscopic Studies of Superconductors.* (1996). San Jose, CA, USA: vol 2696, p 8.

[93] D. Racah, G. Deutscher, Physica C, **263**, 218 (1996).

[94] V. Z. Kresin, S. A. Wolf, Physica C, **169**, 476 (1990).

[95] G. Deutscher, V. Z. Kresin, S. A. Wolf, Bull. APS, **36**, 375 (1991).

[96] M. Tachiki, S. Takahashi, Physica B, **169**, 121 (1991).

[97] S. Kashiwaya, Y. Tanaka, M. Koyanagi, H. Takashima, K. Kajimura, Phys. Rev. B, **51**, 2155 (1995).

[98] D. Mandrus, L. Forro, D. Koller, L. Mihaly, Nature, **351**, 460 (1991).

[99] A. G. Sun, A. Truscott, A. S. Katz, R. C. Dynes, B. M. Veal, C. Gu, Phys. Rev. B, **54**, 6734 (1996).

[100] M. Covington, M. Aprili, E. Paraoanu, and L.H. Greene, F. Xu, J. Zhu and C.A. Mirkin, Physical Review Letters, **79**, 277 (1997).

[101] J.S. Tsai, I. Takeuchi, J. Fujita, S. Miura, T. Terashima, Y. Bando, K. Iijima and K. Yamamoto, Physica C, **157**, 537 (1989).

[102] J.S. Tsai, I. Takeuchi, J. Fujita, T. Yoshitake, S. Miura, S. Tanaka, T. Terashima, Y. Bando, K. Iijima and K. Yamamoto, Physica C, 1385 (1988).

[103] P. W. Anderson, Phys. Rev. Lett., **17**, 95 (1966).

[104] J. A. Applebaum, Phys. Rev. Lett., **17**, 91 (1966).

[105] L. Y. L. Shen, J. M. Rowell, Phys. Rev, **165**, 566 (1968).

[106] J. Yang, C. R. Hu, Phys. Rev. B, **50**, 16766 (1994).

[107] L. J. Buchholtz, M. Palumbo, D. Rainer, J. A. Sauls, J. Low Temp. Phys., **101**, 1099 (1995).

[108] L. Alff, H. Takashima, S. Kashiwaya, N. Terada, and H. Ihara, Y. Tanaka, M. Koyanagi and K. Kajimura, Physical Review B, **55**, R14 757 (1997).

[109] M. Aprili, M. Covington, E. Paraoanu, B. Niedermeier and L.H. Greene, preprint, (1997).

[110] S. K. Tolpygo et al, Phys. Rev. B, **53**, 12454 (1996).

[111] N. Miyakawa, Y. Shiina, T. Kaneko and N. Tsuda, Journal of the Physical Society of Japan, **62**, 2445 (1993).

[112] M. Ohuchi, D. Shimada and N. Tsuda, Jpn J Appl Phys, **32** (**Part 2, N°2B**), L 251 (1993).

[113] E. L. Wolf, *Principles of Electron Tunneling Spectroscopy.* (1985), New-York: Oxford University Press.

[114] A.M. Cucolo, R. Di Leo, A. Nigro, P. Romano and M. Carotenuto, Physical Review B, **49**, 1308 (1994).

[115] Y. Dagan, A. Kohen, G. Deutscher, C. M. Varma, preprint, (1997).

[116] J. Kwo, T.A. Fulton, M. Hong, and P.L. Gammel, Appl Phys Lett, **56**, 788 (1990).

[117] J. Lesueur, Aprili M., Horton T.J., Lalu F., Guilloux-Viry M., Perrin M. & Dumoulin L., Alloys and Compounds, **251**, 1 (1997).

[118] X. Grison, J. Lesueur, M. Aprili, A. Goulon, to be published,

[119] J.H. Xu, J.L. Shen, J.H. Miller, Jr. & C.S. Ting, Physical Review Letters, **73**, 2492 (1994).

[120] D. N. Basov, R. Liang, D. A. Bonn, W. N. Hardy, B. Dabrowski, M. Quijada, D. B. Tanner, J. P. Rice, D. M. Ginsberg, T. Timusk, Phys. Rev. Lett., **74**, 598 (1995).

[121] M.B. Walker, & J. Luettmer-Strathmann. *Macroscopic Symmetry Group Describes Josephson Tunneling in Twinned Crystals.* in *MOS 96.* (1996). Karlsruhe:

[122] M.B. Walker, & J. Luettmer-Strathmann, Physical Review B, **54**, 588 (1996).

[123] M.B. Walker, Physical Review B, **53**, 5835 (1996).

[124] A. G. Sun. *Tunneling Studies of the High Temperature Superconductor YBCO*. (1996). University of California.

[125] C. O'Donovan, M.D. Lumsden, B.D. Gaulin and J.P. Carbotte, Physical Review B, **55**, 9088 (1997).

[126] M. Ledvij, & R. A. Klemm, Physical Review B, **52**, 12552 (1995).

[127] M. Zapotocky, D. L. Maslov and P.M. Goldbart, Physical Review B, **55**, 6599 (1997).

[128] K. A. Kouznetsov, A. G. Sun, B. Chen, A. S. Katz, S. R. Bahcall, J. Clarke, R. C. Dynes, D. A. Gajewski, S. H. Han, M. B. Mapple, J. Giapintzakis, J. T. Kim, D. M. Ginsberg, Phys. Rev. Lett., **79**, 16 (1997).

FLUX QUANTIZATION EXPERIMENTS IN CUPRATE SUPERCONDUCTORS

J.R. Kirtley,[1] C.C. Tsuei,[1] and K.A. Moler[2]

[1]IBM T.J. Watson Research Center
P.O. Box 218
Yorktown Heights, NY 10598
[2]Department of Physics
Princeton University
Princeton, NJ 08544

INTRODUCTION

Magnetic flux penetrates conventional superconductors as fluxoids quantized in units of $\Phi_0 = hc/2e$. This quantization results from the requirement that the order parameter must be single valued, and applies not only to fluxoids in type II superconductors, but also to magnetic flux in rings, Superconducting Quantum Interference Devices (SQUIDs), vortices in Josephson junctions, and interplanar vortices in layered superconductors. The possibility of vortices with non-integer multiples of the conventional fluxoid was first suggested for rotational vortices in He[3],[1] which has an order parameter with nontrivial phases. Such nonintegral vortices could also appear in superconductors which contain Josephson weak links, either because of unconventional properties of the weak link[2] or because of an unconventional order parameter[3, 4] Unconventional order parameters can change sign (a phase change of π) upon rotation. In the right geometry, this phase change results in a spontaneously generated half-integer flux quantum $\Phi_0/2$, both theoretically[3, 4] and experimentally.[5, 6, 7, 8, 9, 10] In addition, "fractional" vortices, vortices with total flux different from integer (N) or half-integer(N+1/2) multiples of the conventional superconducting flux quantum hc/2e, are possible. The existence of fractional vortices would imply phase changes in the order parameter different from integer multiples of π, and could result from a complex order parameter, breaking time-reversal symmetry.[11, 12]

Here we summarize work we have done on flux quantization in unconventional superconductors, as directly imaged using a scanning SQUID microscope.[13] We first review measurements on sparsely twinned single crystals of YBa$_2$Cu$_3$O$_{6.95}$(YBCO). Sigrist et al, in an effort to explain the observation of non-zero Josephson pair current in c-axis YBCO-Pb junctions[14], has suggested that a complex order parameter may develop locally on the twin boundaries of YBCO.[15] Our measurements detected only vortices with the conventional flux quantization $\Phi = hc/2e$.[16] However, fractional vor-

The Gap Symmetry and Fluctuations in High - T$_c$ Superconductors
Edited by Bok *et al.*, Plenum Press, New York, 1998

537

tices apparently are seen in biepitaxial grain boundaries, [17] and in bicrystal asymmetric 45 degree grain boundaries.[18] We believe that these results can be understood in terms of faceting of the grain boundary, combined with d-wave symmetry.

The presence or absence of vortices quantized with half-integer multiples of the superconducting flux quantum $(N+1/2)hc/2e$ is a powerful tool for testing the symmetry of the order parameter in the cuprate superconductors. We have performed a series of experiments that use the scanning SQUID microscope to directly measure the flux state of multiple-grain superconducting films with various specially designed geometries.[7, 19, 20, 21, 22] These experiments have the advantages that: 1) They are non-invasive. The flux state of the rings is imaged with a scanning SQUID microscope that has very small mutual inductance coupling between the sample and the sensor. This means that the rings are measured in their ground state, and no corrections for the effects of an externally applied current need to be made. 2) The imaging of magnetic field is extremely sensitive. Any flux trapped in the sample, down to a level of about $10^{-3}\Phi_0$, is imaged. 3) The geometry can be varied to test the effects of symmetry on the flux quantization observed. 4) This class of experiments can be applied to many unconventional superconductors. 5) The interface areas in our experiments are 1000 times smaller than in the bulk geometries, and the vortices have much smaller distances to travel to escape, so that flux trapping is extremely unlikely to occur.

We have performed these experiments with 4 different substrate geometries, in thin films patterned into rings and disks as well as unpatterned films, and for four different cuprate superconductors. We conclude by describing a set of experiments in a geometry which depends only on symmetry arguments for its conclusions, and which provides strong evidence for pure d-wave symmetry in the Tl2201 system.

SCANNING SQUID MICROSCOPE

Images were obtained using a Scanning Superconducting Quantum Interference Device (SQUID) Microscope (SSM) at 4.2K. The SQUID sensor was mounted on a flexible cantilever and mechanically scanned relative to the sample. The pickup loop, which was integrated into the SQUID design, was located a few microns from an etched edge of the sensor chip, which was pressed against the sample. The plane of the pickup loop was oriented at a shallow angle ($\sim$ 20 degrees) with respect to the plane of the sample. The spatial resolution is limited by the size of the pickup loop (4 to 10 microns) and by the height of the pickup loop above the sample (typically a few microns). Since the signal from the SQUID is proportional to the flux through the pickup loop, the SSM images show the perpendicular component of the magnetic field a few microns above the surface of the sample integrated over the area of the pickup loop. Our SSM has been described in detail elsewhere.[13]

Although our SQUID's can have a noise as low as $2 \times 10^{-6}\Phi_0/\sqrt{\text{Hz}}$, under normal scanning conditions the noise of the SSM is dominated by other factors. The digitization noise is 2×10^{-4} full scale, where the full scale is usually several Φ_0 or less. Also, surface roughness inductively couples to the pickup loop. These two factors result in an effective noise, defined as the standard deviation of the background signal in regions with no clear magnetic features, of $10^{-3}\Phi_0$ under normal operating conditions. The surface roughness can also lead to potential systematic errors by changing the orientation of the SQUID pickup loop with respect to the surface.

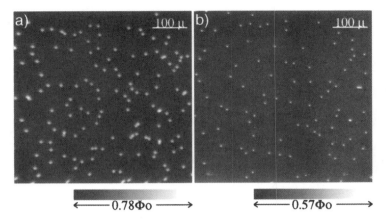

1. Magnetic images of a sparsely twinned $YBa_2Cu_3O_{6.95}$ single crystal. a) Imaged with an 8.2 micron square pickup loop. b) Separate cooldown, imaged with a 4 micron octagonal pickup loop.

TWIN BOUNDARIES

$YBa_2Cu_3O_{7-\delta}$ single crystals were grown by a flux growth technique in $BaZrO_3$-coated yttria stablized zirconia crucibles and annealed on YBCO ceramic in a tube furnace under high purity oxygen flow.[23] This study required twin boundaries which were separated on a length scale of the SQUID imaging. Two crystals were chosen which had regions of isolated twin boundaries which were more than $10\mu m$ from the nearest neighboring boundaries. The twin boundaries were oriented at 45 degrees to the edges of the crystal. Both crystals were studied at both optimum doping ($7 - \delta = 6.95$, $T_c=93.1K$) and underdoping ($7 - \delta = 6.60$, $T_c=60K$).

Figure 1 shows two images of a 512 micron $\times$ 512 micron area of one of the crystals, taken in separate cooldowns with an 8.2 micron square pickup loop and a 4 micron octagonal pickup loop, with one data point per micron. Previous work on similar crystals has shown that $H_{c1} = 1100$ Gauss at 4 Kelvin for fields applied parallel to the c-axis.[24] Despite the fact that the applied field is much less than either H_{c1} or the geometry-dependent first penetration field, there are many vortices trapped inside the crystal. At 4 Kelvin, the number of vortices in the interior does not change with changing magnetic field for the small fields (< 1Gauss) used in this study. Presumably, the fact that vortices are observed in the sample at fields so far below H_{c1} is an indication of the high pinning potential in these materials.

The density of trapped vortices closely follows the field which is applied during cooling divided by Φ_0. This occurs in all of the cuprates studied, for thin films as well as single crystals, and appears to be independent of the shape of the sample. These observations indicate that the vortices each have $hc/2e$ flux trapped in them, in agreement with previous work.[25]

There are several magnetic dipoles appearing in the Figure 1 images. Similar dipoles have also appeared in images of other superconductors, including films of various high-T_c materials and niobium films. These dipoles are correlated with topographical features on the surface, probably $BaCuO_x$ flux spots, which are visible under optical inspection. The apparent magnetic dipoles may result from the SQUID sensor "rocking" as it passes over the features.

The main qualitative results are the uniformity of the vortices in both spatial extent and total flux, and the lack of any clear feature associated with the twin boundaries. Closeups of vortices chosen at random (Figure 2) show that the vortices have a uniform

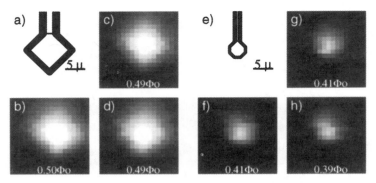

2. a) Sketch of an 8.2 micron square pickup loop. b-d) Closeups of individual vortices from Figure 1a, imaged with the 8.2 micron square pickup loop. e) Sketch of a 4 micron octagonal pickup loop. f-h) Closeups of individual vortices from Figure 1b, imaged with the 4 micron pickup loop. Labels indicate the peak flux through the pickup loop for each image.

appearance within the limits of the imaging technique, and that the apparent size and shape of the vortices is determined by the size and shape of the pickup loop.

The structure of the twin boundaries are known to be quite sensitive to oxygen content. With the assumption that the detailed behavior of the order parameter at the twin boundaries would also be sensitive to oxygen content, the measurements were repeated on underdoped crystals. Here, we concentrate on data obtained for one optimally doped crystal. Quantitatively similar results were obtained on a second optimally doped crystal, and on both crystals after reoxygenation to $7 - \delta = 6.60$.

The in-plane penetration depth in similar crystals has been shown to be $\lambda_L = 0.15\mu m$,[26] much less than the size of the SQUID pickup loop. It is thus expected that the images in Figure 2 will be consistent with vortices which are essentially point sources, unless there are two or more neighboring fractional vortices contributing to each flux maximum. On length scales long compared to the London penetration depth but small compared to the size of the crystal, a vortex can be modeled as a monopole source of magnetic field,[27] so that the z-component of the field is given by $B_z = \Phi_0 z/2\pi r^3$. This field can be integrated over the SQUID's pickup loop to give the expected magnetic flux signal. Figure 3a shows the result of this calculation, neglecting the tilt of the pickup loop with respect to the sample, for an 8.2 micron square pickup loop at a height of $2.5\mu m$ above the surface. The calculation is compared with cross-sections of the magnetic flux data from the vortices in Figures 2b-d. Figure 3b shows the calculated flux for the 4 micron octagonal loop at a height of $1.5\mu m$, compared with cross-sections of the magnetic flux data from the vortices in Figures 2f-h. The heights of the pickup loops have been treated as the only free parameter, and the resulting heights are consistent with the geometry of the SQUID sensors.

There do not appear to be any fractional vortices in these images. It is possible that what appears to be a single integral vortex may actually be two or more fractional vortices, whose flux sums to one flux quantum, which are too close to be resolved by the SQUID pickup loop. In order to indicate time-reversal symmetry breaking, such vortices must be isolated, that is, they must be far apart compared to a London penetration depth. As an example of the difficulty of resolving such vortices, Figure 3c shows a calculation of the flux through the 4 micron octagonal SQUID pickup loop for two half-integral vortices separated by 0, 2, 4, or 6 microns. Conservatively, the images in Figures 1 and 2 are not sufficient to resolve vortices which are separated by less than the size of the pickup loop. However, if different sets of fractional vortices

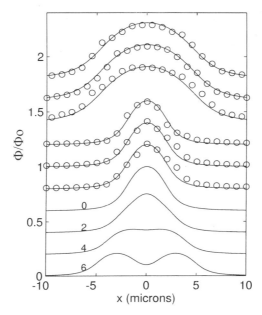

3. Top three curves: calculated flux through an 8.2 micron square pickup loop scanning over a single hc/2e vortex at a height of 2.5 microns, compared with experimental cross-sections of the three vortices in Fig. 2b-d. Middle three curves: calculated flux through a 4 micron octagonal pickup loop scanning over a single hc/2e vortex at a height of 1.5 microns, compared with cross-sections of the three vortices in Fig. 2f-h. Bottom four curves: calculated flux through a four micron pickup loop scanning at a height of 1.5 microns above two half-integral vortices separated by 0, 2, 4, and 6 microns. Successive curves have been offset for clarity by 0.2 Φ_0.

were separated by different amounts, then there would be a corresponding spread in the maxima in Figure 4.

A histogram of the flux maxima (Figure 4) thus serves to quantify the uniformity of the vortices, and the possibility that separated fractional vortices appear as one integral vortex. For this purpose, a maximum is defined as a data point that is at least three standard deviations above the local background and is larger than its eight nearest neighbors. Note that the x-axis of Figure 4, flux through the SQUID pickup loop, is not the same as the total flux carried by the vortex. The peaks at $0.5\Phi_0$ in Figure 4a and $0.4\Phi_0$ in Figure 4b represent integral vortices, as has been confirmed by cooling in various applied magnetic fields and counting the number of trapped vortices. A few maxima in Figure 4 result from one of the dipoles, mentioned above, which we believe result from bumps on the surface. A maximum resulting from a single isolated pixel, whose neighboring pixels are within the background, cannot be taken to represent an actual magnetic feature, which should have a spatial extent of at least the SQUID pickup loop diameter. None of the maxima which are below $0.1\Phi_0$ meet this criteria. The peaks in Figure 4 thus represent integral vortices with a spread of a few percent, which is consistent with the sampling density of 1 micron per data point. This result suggests that any pairs or triplets of fractional vortices are closer than about a micron.

In images taken in higher magnetic fields than the ones shown here, the vortices appear to sit preferentially in more heavily twinned regions or near twin boundaries. Otherwise, there is no clear magnetic feature associated with the twin boundaries. In particular, there is no evidence for currents flowing near the twin boundaries, which

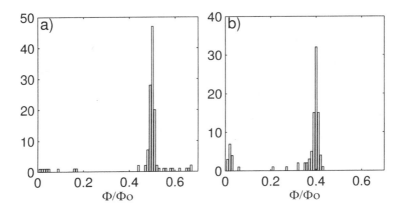

4. Distribution of local maxima in data sets taken with a) an 8.2 micron square pickup loop and b) a 4 micron octagonal pickup loop.

would be expected to exist over a length scale of the London penetration depth. Since the London penetration depth is quite small compared to the size of the pickup loop, and the currents on opposite sides of the twin boundary would flow in opposite directions, it is probable that the theoretical magnetic field from the currents would average to a flux which is below the sensor resolution. As an extremely crude model for the magnetic field from such currents, we take one-dimensional infinite wires located a distance $\lambda_L = 0.15\mu m$ to each side of the twin boundary, carrying currents flowing in opposite directions. While currents associated with the twin boundaries would be expected to be strips as thick as the crystal rather than wires at the surface, the magnetic field from currents which are much more than a penetration depth would be shielded by the bulk of the crystal. Modeling the twin boundaries as two current-carrying wires is only intended to indicate the scale of the possible magnetic fields. Guessing a current of $I = j_c\lambda_L^2 \approx 20\mu A$, where $j_c \approx 10^5 Amps/cm^2$ is the critical current density, and integrating over an 8 micron square SQUID loop scanning at a height of $2.5\mu m$, results in a flux profile which would be just barely detectable. Although no such signal is seen in these images, more realistic models would probably result in an even smaller signal, and cannot be ruled out with the present data.

BIEPITAXIAL GRAIN BOUNDARIES

The fabrication of biepitaxial grain boundary samples has been described elsewhere.[28] Briefly, an epitaxial film of MgO is grown onto a LaAlO$_3$ single crystal substrate, and patterned into the desired shapes. An epitaxial film of CeO$_2$ is followed by a 250 nm film of YBa$_2$Cu$_3$O$_{7-\delta}$. The YBa$_2$Cu$_3$O$_{7-\delta}$ film on the MgO is rotated by 45° relative to the film off the MgO. The critical current density of the grain boundaries, measured by transport techniques, is about 2×10^3 Amps/cm^2 at 4.2 K, suggesting that the Josephson penetration depth is about $10\mu m$.

Figure 5a shows a scanning SQUID image of an array of triangular shaped biepitaxial grain boundaries. The full grey-scale color variation in Figure 5a is $0.25\Phi_0$ through the pickup loop. The edges of the triangles are decorated with flux concentrations of both signs. Also apparent in this image, outside but not inside the triangles, is a mottled background pattern. This structure is quite repeatable from scan to scan in a particular cooldown, and is of uncertain origin, although we suspect that it results from inductive coupling of topographical features into the SQUID output signal.

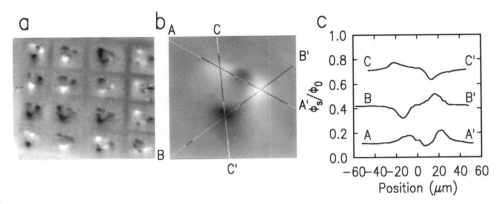

5. a)Scanning SQUID microscope image of an array of YBCO triangular biepitaxial grain boundaries, cooled and imaged in nominally zero field, taken with a 10 micron diameter octagonal pickup loop. The YBCO was removed between the triangles by laser ablation to limit supercurrents over large length scales. b) An expanded view of the upper left hand triangle. c) Cross-sections through the data along the edges of the triangles.

Despite extensive efforts to cool slowly in zero field, we always observe some flux trapped in our grain boundaries, often with both signs. The positions of the trapped flux varied from cooldown to cooldown and from triangle to triangle. These effects are not particular to the triangle geometry. Similar patterns are seen in photolithographically patterned hexagons.[17]

Figure 5b shows an expanded view of the upper left hand triangle in Figure 5a. Figure 5c shows cross-sections taken along the grain boundaries in this triangle in the directions indicated. The spatial resolution is not sufficient to determine whether the flux which is concentrated at the corners for many triangles should be considered to be isolated. A model has previously been presented[17] in which the cross-sections in Figure 5 have been fit using nine delta-function sources of magnetic flux for each triangle. The individual sources had fluxes much smaller that Φ_0, but the sum of the flux sources in any particular triangle was consistent with either 0 or 1 Φ_0.

This data is thus consistent with isolated fractional vortices. It is also consistent with a complicated pattern of distributed flux. To make an unambiguous distinction between these two models would require measurements with probes which are small compared to the smallest possible Josephson penetration depth and compared to the separation between point sources in Figure 5, or probes of about a micron.

BICRYSTAL GRAIN BOUNDARIES

Although many high-T_c grain boundary junctions have superconducting properties that are well described by conventional models of strong coupled Josephson junctions, an exception is given by asymmetric 45° [001]-tilt boundaries in YBCO, which show anomalous dependence of their critical current on applied field. This anomalous field dependence can be attributed to d-symmetry of the superconducting gap, combined with the presence of grain-boundary faceting.[29, 30] We tested this model by studying macroscopically straight grain boundaries produced by epitaxial growth on bicrystalline SrTiO$_3$ substrates. The samples used were [001]$\pm$1 deg oriented films of YBa$_2$Cu$_3$O$_{7-\delta}$, grown by standard pulsed-laser deposition on various bicrystalline SrTiO$_3$ substrates to a thickness of 20-150 nm. The critical temperatures T_c were 88.5-90 K and the critical current densities J_c (77 K) were $\sim 2 - 8 \times 10^6$A/cm^2. Figure 6 is a scanning

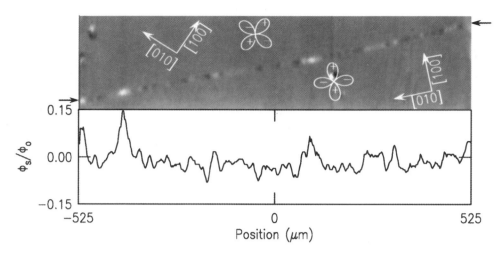

6. Scanning SQUID microscope image of asymmetric 45 degree epitaxial bicrystal grain boundary, cooled and imaged in nominally zero field. The white overlays on the image show the orientations of the underlying SrTiO$_3$ substrate crystalline axes, and polar plots of assumed d-wave order parameters aligned with the principle axes of the films. The bottom portion of the figure shows a cross-section through the data along the grain boundary, as indicated by the arrows on the image.

SQUID microscope image of a YBCO asymmetric 45° [001] tilt boundary, cooled in a field less than 1mG, and imaged at 4.2K with a 4 micron diameter octagonal pickup loop.[18] Below the image is a cross-section of the flux taken along the grain boundary line. This flux is randomly distributed and also changes its sign randomly, with fine scale variations limited only by the 4 micron spatial resolution of the instrument. The average flux is nearly zero. Experiments with 18.4°/18.4°, 22.5°/22.5°, and 45°/12° grain boundary samples do not show this effect. In addition, as can be seen from Figures 9 and 10, 0°/30° grain boundaries also do not show this effect, so that it has no influence on the tricrystal symmetry test experiments described in this paper.

The appearance of self-generated flux has several significant implications for the understanding of grain-boundary properties and for their use.[30] For the purposes of this paper, however, the most important question is whether the flux is evidence of a locally complex order parameter, or in other words, whether the data indicate the presence of isolated fractional vortices. The Figure 6 image does not indicate that the flux is isolated. It is possible that the flux is isolated on a length scale which is small compared to the size of the SQUID, but this is unlikely considering that the smallest Josephson penetration depth, estimated from critical current measurements as $J_c \leq 10^5 \mathrm{A/cm^2}$ on commonly used grain boundaries, should be greater than a micron. Relative pair phases of other than 0 or π, which would indicate complex order parameters, are not needed to explain this data. They are also not ruled out, however.

The observed flux distribution can be qualitatively reproduced using the same reasoning as for the anomalous magnetic field dependence of the critical current: considering the grain boundary faceting together with the d-wave component of the order parameter.[30, 18] In an asymmetric 45 degree grain boundary, a lobe of the order parameter is normal to the boundary on one side, while a node is normal on the other. Depending on the local microstructure, the pair transfer integral in different regions of the grain boundary can have a relative pair phase of either 0 or π. The magnitude of the coupling may also change. If the length scale for the disorder in the grain boundary were long compared to the Josephson penetration depth, a half-integral flux quantum

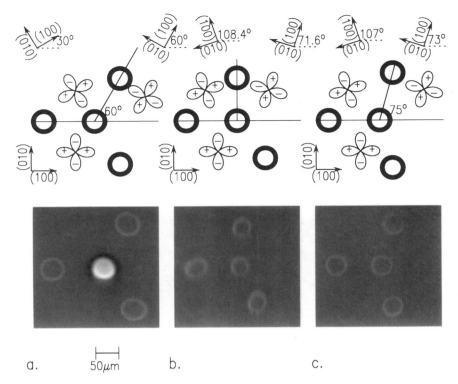

7. Scanning SQUID microscope images, taken with the sample cooled and imaged in nominally zero field, with a 10 micron diameter octagonal pickup loop, of epitaxial thin film ring YBCO samples with three different SrTiO₃ tricrystal geometries.

would form spontaneously each time the relative pair tunneling phase changed by π. Since the length scale for disorder is small compared to the Josephson penetration depth, the spontaneously generated flux shows a distribution of signs and amplitudes, with a characteristic amplitude much smaller than Φ_0.

This is also a likely explanation for experiments in which apparently spontaneous magnetization was observed in biepitaxial 45° grain boundaries as described in the previous section,[17] and may explain why evidence for d-symmetry was not seen in samples using such grain boundaries.[31]

SYMMETRY TESTS USING HALF-INTEGER FLUX QUANTIZATION

The symmetry of a pair wavefunction can best be probed at the junction interface as the Cooper pairs tunnel across a Josephson junction or weak link.[2] The sign of the Josephson current of a junction between two unconventional superconductors depends on the relative orientation of their order parameters with respect to the junction interface. As shown recently by Sigrist and Rice for d-wave superconductors,[4] the supercurrent I_s^{ij} can be expressed by:

$$I_s^{ij} = A^{ij} \cos 2\theta_i \cos 2\theta_j \sin \Delta\phi_{ij} = I_c^{ij} \sin \Delta\phi_{ij}, \qquad (1)$$

where A^{ij} is a constant characteristic of junction ij, and θ_i and θ_j are angles of the crystallographic axes (or equivalently wave vectors k_x and k_y) with respect to a junction interface between superconductors i and j. In our geometry the junctions are formed from grain boundaries.

In the case of s-wave symmetry, the sign of I_s^{ij} is independent of θ_i, θ_j, but its magnitude can vary due to gap anisotropy. Sigrist and Rice showed that the lowest energy state of a ring with a single junction with one sign change to the normal component of the order parameter has a spontaneous magnetization if the critical current is sufficiently large. The magnetic flux threading through such a π-ring is exactly half of the flux quantum ($\Phi_0/2 = h/4e = 1.035x10^{-7}G - cm^2$) when the external field H_{ext} = 0 and the condition $L \mid I_c \mid >> \Phi_0$, where L is the self-inductance of the ring, is satisfied. In the case of a multiple-junction ring, a ring with an odd number of sign changes will also exhibit $\Phi_0/2$ spontaneous magnetization.[4, 7]

The Sigrist-Rice formula, Eq. 1, is based on the implicit assumption that the junction is perfectly smooth and without any disorder. In reality, the electron wavevector orthogonal to the junction face can be significantly distorted by interface roughness, impurities, strain, oxygen deficiency, etc.. To model the disorder, one can consider the consequence of an angular deviation $\Delta\theta$ from the perfect interface. A straightforward calculation leads to a maximum-disorder formula for the circulating current:

$$I_s^{ij} = A^{ij} cos2(\theta_i + \theta j)sin\Delta\phi_{ij}, \tag{2}$$

Given Eq. 1 and Eq. 2, one can design a d-wave pairing symmetry test for cuprate superconductors that is valid for all cases. Our original experimental geometry consists of a ring containing three grain boundary junctions of high-T_c superconductor. A scanning SQUID microscope[13] is then used to image the magnetic flux threading through the superconducting tri-grain ring to search for the $\Phi_0/2$ spontaneous magnetization.

The three-junction rings used in this work are fabricated from epitaxial films of cuprate superconductors deposited on a (100) tri-crystal $SrTiO_3$ substrate using standard pulsed laser deposition. In the initial experiment, four rings (inner diameter = 48 μm, width = 10 μm) are patterned using a standard photolithographic process.

For sufficiently large I_cL the product of the signs of the critical currents of the junctions making up the ring determines whether it will be a 0-ring (the product of signs is positive) with integer flux quantization, or a π-ring (negative sign) with $1/2$ integer flux quantization.[7] Three sets of ring samples were made as outlined above. From the I_c^{ij} value measured from test bridges and the estimated self-inductance of the rings (L = 100 pH) one finds that the LI_c^{ij} product is about 100 Φ_0, easily satisfying the condition $LI_c^{ij} >> \Phi_0$. Therefore, a spontaneous magnetization of $\Phi_0/2$ at $\Phi_{ext} = 0$ should be observable in our 3-junction π-ring if allowed by symmetry. The only difference between the samples was the tricrystal substrate geometry. The 0-junction rings and 2-junction rings should show integer flux quantization. Only the 3-junction π-ring should show the $1/2$ integer quantum effect if it is due to nodes in the superconducting order parameter, but all 3-junction rings should show the effect if it is due to a symmetry-independent mechanism. In each case, the design was chosen to show the desired effect in either the clean (Eq. 1) or dirty (Eq. 2) limits, and therefore presumably in all cases. The ratio of the mutual inductance between loop and ring to the self-inductance of the ring is about 0.02, so that the effect of the SQUID flux coupling back into the ring should be small.

Figure 7 compares scanning SQUID microscope images from a YBCO tricrystal ring sample with the π-ring geometry (a) and two 0-ring geometries (b,c), cooled to the measuring temperature of 4.2K in fields less than 2 mGauss. Our interpretation of

546

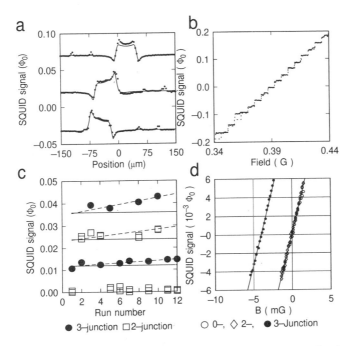

8. Summary of four techniques used for calibrating the total flux in superconducting rings using the scanning SQUID microscope: a) Direct calculation. b) Measurement of mutual inductance between ring and pickup loop by fluxoid entry into the ring. c) Measurement of SQUID signal amplitudes for the four rings after succesive cooldowns. d) Measurement of SQUID signal for the four rings as a function of externally applied field.

these images is that the 3-junction ring shows spontaneous magnetization of a half-flux quantum (hc/4e) in the first geometry a), but in all other cases the rings generate no spontaneous flux. The outer control rings are visible through mutual inductance coupling between the rings and the SQUID loop. Images taken with these samples cooled in different fields showed that the 3-junction π-ring always had $(N+1/2)\ \Phi_0$ (N an integer) flux in it, while the 3-junction 0-rings always had $N\Phi_0$. These results are highly reproducible, and as discussed below, there can be no serious question that we are observing half-integer flux quantization in these samples. However, several authors have suggested symmetry independent mechanisms to cause π-phase shifts at the junctions,[32, 33] resulting in the 1/2 integer flux quantum effect in our 3-junction rings. In principle, since disorder in the grain boundaries is angle dependent, it is possible that the magnetic scatterers could also be angle dependent. However, it seems unlikely that this could be an explanation for our results, since two of the grain boundary misorientation angles chosen for the control geometries are the same (and should therefore cancel out for this mechanism) while the third is within a few degrees of the 30 degrees chosen for the π-ring geometry. The fact that the results for 3 samples with slightly different crystalline and grain boundary angles are as predicted for a d-wave superconductor, confirm that this effect is symmetry dependent.

We have used four techniques for calibrating the amount of flux trapped in our superconducting rings (Figure 8):

a) Direct calculation: A given flux Φ threading a superconducting ring with self-inductance L induces a circulating current $I_r = \Phi/L$ around the ring, which in turn induces a flux $\Phi_s(\vec{\rho}) = M(\vec{\rho})\Phi/L$ in the pickup (sensor) loop. We calculate the inductance of our rings to be 99 ± 5 pH. The mutual inductance between a ring and the pickup loop $M(\vec{\rho})$ is calculated using the Biot-Savart law.[7] The solid lines in Figure 8a are model calculations for three cross-sections through the image of Figure 7a, assuming $\Phi = \Phi_0/2 = h/4e$ in the 3-junction ring. The experimental cross-sections (dots) are taken through the center of the 3-junction ring, at an angle parallel to the horizontal grain boundary (top), rotated 60 degrees clockwise (middle) and 60 degrees counterclockwise (bottom). The asymmetry in the images results from the tilt of the pickup loop, as well as the asymmetric pickup area from the unshielded section of the leads. Clearly using $\Phi_0/2$ for the flux in the 3-junction ring results in much better agreement than would be obtained using Φ_0.

b) Our value for M(0) was checked by positioning the pickup loop in the centers of the rings and measuring the SQUID output vs field characteristic. Representative results for the 3-junction ring are shown in Figure 8b. In this Figure a linear background, measured by placing the loop over the center of the 0-junction control ring, has been subtracted. At low fields stepwise admission of flux into the ring leads to a staircase pattern, with progressively smaller heights and widths to the steps, until over a small intermediate field range, shown for increasing field in Figure 8b, single flux quanta are admitted. We interpret the "noise" in this data as flux motion in the grain boundaries and the other rings. At larger fields the steps disappear and the SQUID flux vs. field characteristic slowly oscillates about a mean line. The heights of the single flux-quantum steps in the intermediate field region, derived by fitting the data to a linear staircase (dashed line) are $\Delta\Phi_s = 0.0237\Phi_0$. This is in good agreement with our calculated value of $\Delta\Phi_s = M(0)\Phi_0/L = 0.024 \pm 0.003\Phi_0$. Twelve repetitions of this measurement, including measurements of both the 2-junction and 3-junction rings, gave values of $M(0)\Phi_0/L = 0.028 \pm 0.005\Phi_0$.

c) Figure 8c summarizes the results from 12 cooldowns of the sample. We plot the absolute value of the difference between the SQUID loop flux in the centers of the 2-junction or 3-junction rings, and the 0-junction control ring. The solid lines are the expected values for the flux difference, calculated as described above. In all of our measurements $\Delta\Phi$ always fell close to $(N+1/2)$ (h/2e) for the 3-junction ring, and close to Nh/2e for the 2-junction rings (N an integer). However, there is some drift to the data, which we associate with tip wear. The wear of the tip was clearly visible upon optical inspection after this series of experiments, supporting this hypothesis. A fit to the eight $\Phi_0/2$ points in Figure 8c, assuming exactly h/4e flux threads the 3-junction rings, implies that the mutual inductance M(0) = 2.4 pH for the as fabricated tip, and increases to 2.9 pH at the end of the series. For comparison, our calculations give 2.4 pH for the center of the loop 10 μm from the tip end, and 2.7 pH for the tip end just at the edge of the pickup loop. The dashed lines, including this correction to the mutual inductance, agree with the data.

d) Figure 8d shows the SQUID sensor signal at the center of the ring, relative to the signal outside the ring, for all of the rings of the YBCO sample of Figure 7a, as a function of field applied by a coil surrounding the microscope. The coil was calibrated by replacing the sample with a large area pickup loop SQUID magnetometer. For this experiment the sample was cooled in sufficiently low field that all of the rings had no flux, except for the 3-junction ring, which contained $\Phi_0/2$ spontaneously generated flux. As an external field is applied, the rings screen the field with a circulating supercurrent until a critical field, typically about 50 mG, is exceeded, at which point flux enters the rings through the grain boundaries.[7] The small ratio of mutual inductance between

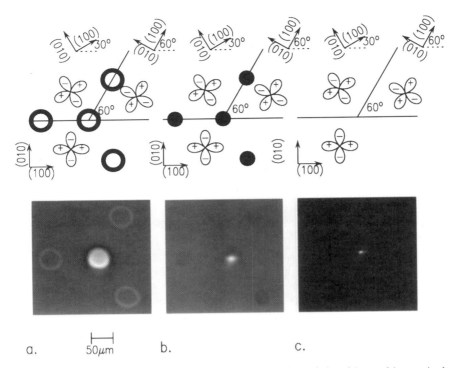

9. Scanning SQUID microscope images, taken with the sample cooled and imaged in nominal zero field, of three tricrystal YBCO samples in a substrate geometry that should show the half-integer flux quantum effect at the tricrystal point for a d-wave superconductor, for a) a ring sample, imaged with a 10 micron octagonal pickup loop b) a disk sample, imaged with an 10 micron octagonal pickup loop, and c) a sample with no photolithographic patterning, imaged with a 4 micron octagonal pickup loop.

the SQUID pickup loop ($\sim$ 2.4 pH) and the self inductance of the ring ($\sim$ 100pH) means that the ring currents are only very weakly perturbed by the measurement. In the absence of flux penetration, the SQUID difference signal goes to zero when there is as much field inside the ring as outside the ring. In the limit $2\pi L I_c / \Phi_0 \gg 1$ the difference in flux through the rings is just the difference in applied field required to make the signal go to zero times the effective area of the ring.[4] Estimating the effective area of the rings to be $\sim \pi((r_{in} + r_{out})/2)^2 = 2642 \pm 80\mu m^2$, where $r_{in} = 24\mu m$ and $r_{out} = 34\mu m$ are the inner and outer radii of the rings respectively, the 3-junction ring in Figure 3a has $0.505 \pm 0.02\Phi_0$ more flux threading through it than the 0-junction, or 2-junction rings. Further, the difference in flux between any of the other rings is $| \Delta\Phi | < 0.01\Phi_0$. We estimate that any e.g. s-wave component to a presumed s+id superconducting order parameter would alter the flux quantization condition away from exactly $\Phi_0/2$ by roughly the fractional portion that is s-wave. We therefore conclude that the superconducting order parameter in YBCO has lobes and nodes consistent with d-wave symmetry and put experimental limits of about 4% on any out of phase s-wave component.

If the half integer flux quantum effect is intrinsic to YBCO grown on a tricrystal substrate in the π-ring geometry, then it should not depend on the macroscopic geometry that the YBCO takes. In fact, as illustrated in Figure 9, we have observed

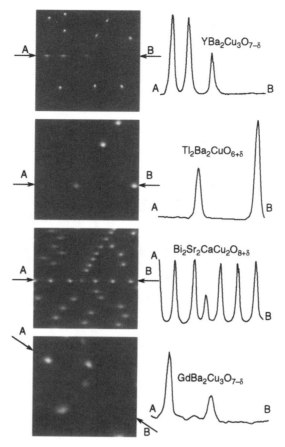

10. Scanning SQUID microscope images, taken with the sample cooled and imaged in a field of a few mG, of tricrystal samples in a substrate geometry that should show the half-integer flux quantum effect at the tricrystal point for a d-wave superconductor, for epitaxial thin films of a) $YBa_2Cu_3O_{7-\delta}$, b) $Tl_2Ba_2CuO_{6+\delta}$, c) $Bi_2Sr_2CaCu_2O_{8+\delta}$, and d) $GaBa_2Cu_3O_{7+\delta}$. In each case a hc/4e Josephson vortex is spontaneously generated at the tricrystal point.

half-integer flux quantization in rings, disk, and blanket coverages of YBCO epitaxially grown on π-ring tricrystal substrates. Experiments on blanket (unpatterned) samples provide direct measurements of the Josephson penetration depth in the samples, as well as providing a means for making symmetry tests without photolithographic definition of the samples. This makes it much easier to do symmetry tests on unconventional superconductors.

Figure 10 shows comparison images of blanket coverages of 4 different cuprate superconductors grown epitaxially on tricrystal substrates with the π-ring geometry of Figure 7a. Each image is of an area 300 microns by 300 microns. The YBCO image was taken with an octagonal pickup loop 4 microns in diameter, the Tl2201 and BSCCO images with an 8.2 micron square pickup loop, and the GdBCO sample with a 12 micron square pickup loop. The samples were cooled in fields from 1-10 mG, and imaged at 4.2K. In each case h/2e flux quanta are trapped in the sample away from the tricrystal point, and a h/4e half-flux quantum is trapped at the tricrystal point. When the samples were cooled in nominally zero field, only the half-flux quantum at the tricrystal point was present. Modelling as outlined below confirms that there is in fact close to h/4e flux trapped at the tricrystal point in these samples.[20]

It is experimentally well-established that the crystal structure of YBCO is orthorhombic, has two CuO_2 planes per unit cell, has CuO chains in between the planes, and can be twinned. Each of these complicating factors have been used to explain the observation of the half-integer flux quantum effect in YBCO.[34, 35, 36, 37] These questions can best be resolved by a d-wave tricrystal experiment with a cuprate system such as $Tl_2Ba_2CuO_{6+\delta}$ (Tl2201), which has a single plane per unit cell, no chains, no twins, and has tetragonal symmetry. We have repeated our ring geometry experiments using Tl2201 films.[21] These films have relatively low critical current grain boundaries, so that the LI_c product of the rings is not as large as in YBCO. This means that the flux trapped in the rings is less that $(N+1/2)\Phi_o$. Nevertheless the measured phase shift is still very close to π.[21] In addition, as shown in Figure 10, the half-integer flux quantum effect is observed in blanket samples in the Tl2201 system, indicating that its pair wavefunction has nodes and lobes consistent with d-symmetry.[21]

Photoemission experiments indicate a gap in BSCCO which is highly anisotropic, consistent with d-wave symmetry[38, 39] Unfortunately, these experiments are not sensitive to the sign of the gap function. Our experiments indicate that the gap function in BSCCO does in fact change sign consistent with d-symmetry. When combined with the photoemission results, these experiments conclusively show that BSCCO has d-symmetry, and rule out, for example, extended s-symmetry.

MODEL INDEPENDENT SYMMETRY TEST

As outlined above, several recent tests of the order parameter symmetry using the half-integer flux quantum effect [5, 6, 7, 19, 9, 10, 21] have yielded convincing evidence for d-wave pairing in the high-T_c cuprate superconductors. However, none of these experiments were able to unambiguously rule out the possibility of an admixture of s-wave with d-wave (i.e. a mixed s+d pair state). This is because, with the exception of the tricrystal experiment in tetragonal $Tl_2Ba_2CuO_{6+\delta}$ (Tl2201), all these phase-sensitive experiments were done in orthorhombic cuprates such as $YBa_2Cu_3O_{7-\delta}$ (YBCO). The effect of orthorhombicity in YBCO (i.e. the inequivalence of the a and b directions in the CuO_2 planes) on certain normal-state and superconducting properties are well established in the literature. Based on a group theoretic symmetry argument and Josephson junction measurements on untwinned YBCO crystals, a recent study[37] has concluded that s+d pairing mixing is an unavoidable consequence of orthorhombicity. The consistent observation of c-axis pair tunneling in YBCO argues strongly in favor of such a mixed pair state.[14] The angular position of the node line in an orthorhombic superconductor depends on the relative proportions in the s+d admixture. Since the Pb-YBCO corner SQUID (or single Josephson junction) interference experiments rely on a sign change between the a and b faces of YBCO, they can not distinguish pure d-wave from mixed s+d pair states as long as d/s $\geq$ 1. The tricrystal experiments with YBCO and Tl2201 can, in principle, locate the nodes on the Fermi surface. However this requires a systematic series of tricrystal experiments and a detailed model describing pair tunneling across a grain boundary weak-link, including the effects of disorder at the junction interface. The success of such experiments depends on whether one can obtain a reliable interpolation between the Sigrist-Rice clean limit[4] (Eq. 1) and the maximum disorder (Eq. 2) formula.[7]

In the face of these difficulties in determining the degree of s and d mixing in pairing, it is important to demonstrate unambiguously the existence of a pure $d_{x^2-y^2}$-wave high-T_c cuprate superconductor. Such an experiment is significant in view of the

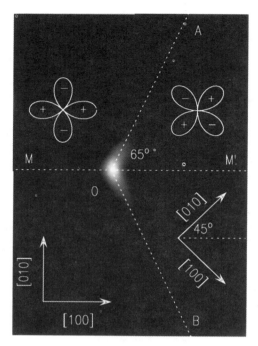

11. Scanning SQUID microscope image, taken with the sample cooled and imaged in nominal zero field, with an 8.2 micron square pickup loop, of an unpatterned epitaxial thin film of Tl_2Ba_2 on a tetracrystal of $SrTiO_3$, with a geometry chosen to produce the half-flux quantum effect at the tetracrystal point.

substantial amount of theoretical work[40, 41, 42, 43] on the effect of s+d wave pairing on the properties of cuprate superconductors, including the origin of high temperature superconductivity. The demonstration of a pure d-wave superconductor can help to understand whether the coexistence of d and s singlet pairing channels is essential or just accidental to high-T_c superconductivity. It will also serve as a well-defined starting point for understanding the more complex (d+s)-wave superconductors.

We have performed a phase-sensitive test of pairing symmetry, suggested by Walker and Luettmer-Strathmann,[37] which depends only on symmetry considerations for its interpretation. Our results provide model-independent pair tunneling evidence for pure $d_{x^2-y^2}$ symmetry in tetragonal single layer $Tl_2Ba_2CuO_{6+\delta}$ superconductors.

Tetragonal cuprate superconductors have C_{4h} symmetry. Following the notation of Annett et al.[40], the singlet pair states in the CuO_2 plane can transform under the allowed symmetry operations as the identity(s^+), as $xy(x^2-y^2)$ (s^-), x^2-y^2 $(d_{x^2-y^2})$, or xy (d_{xy}). The original experiment proposed by Walker and Luettmer-Strathmann consists of a ring containing two c-axis oriented tetragonal cuprate grains rotated about the c-axis by an angle of $\pi/4$ with respect to each other. The symmetrical placement of the two grain boundary junctions in the ring assure that for a pair state with $d_{x^2-y^2}$ or d_{xy} symmetry the currents across the two junctions are equal in magnitude but opposite in sign. As a result of this π phase shift, the ring will exhibit a spontaneously generated half-flux quantum ($\Phi_0/2 = h/4e = 1.035 \times 10^{-7} G-cm^2$). However, for s^+ or s^- symmetry, there is no sign reversal, and the ring should show the standard integer flux quantization.

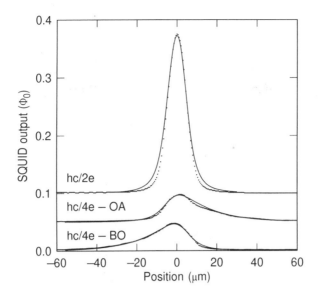

12. Modelling of the SSM data in Fig. 11 for cross-sections along the directions OA, OB, and also through the peak of an isolated bulk hc/2e flux quantum in the same Tl2201 film. The solid lines are fits to the data. Successive curves are offset by 0.05 units for clarity.

In the present work, instead of a ring geometry, we use a blanket film of single-layer tetragonal Tl2201 deposited on a (100) SrTiO$_3$ (STO) substrate with the bicrystal geometry proposed by Walker and Luettmer-Strathmann (Fig.11). It is extremely difficult to make a bicrystal STO substrate with the required $\pi/4$-rotated wedge as well as two identical grain boundaries that are smooth and free of microscopic voids. To overcome these technical difficulties, we achieved the desired bicrystal wedge configuration by effectively fusing two bicrystals along the dividing line MM′ in Fig. 11. The grain boundary MM′ does not involve any grain misorientation, and is irrelevant to this experiment. For a given total misorientation angle $\theta = \theta_1 + \theta_2$, the pair tunneling current is maximized[37] for a symmetric grain boundary $\theta_1 = \theta_2 = \pi/8$. We chose the angle $\alpha = \theta_1$ between the vertical ([010] in grain 1) and the grain boundaries (OA and OB) as 25°. Also shown in Fig. 11 are the polar plots of the assumed $d_{x^2-y^2}$ gap functions (i.e. pair wavefunctions Ψ_1 and Ψ_2) aligned with the crystallographic axes of the STO substrate.

In the design of our bicrystal STO substrate, there is a built-in reflection symmetry with respect to the equal-partition line (MOM′ in Fig. 11). The mirror reflection symmetry operation is even in grain 1 and odd in grain 2. For $d_{x^2-y^2}$ or d_{xy} pairing states a reflection symmetry operation m on I_s results in a pair tunneling across the grain boundary OA that is related to its counterpart along OB by:

$$(I_s)_{OB} = m(I_s)_{OA} = -(I_s)_{OA} \tag{3}$$

(with the understanding that the tunneling Hamiltonian is invariant under the reflection operator m about the line MOM′). Due to this sign change between the supercurrents across junctions interfaces OA and OB, any superconducting loop enclosing the wedge tip O will result in a net π phase shift which manifests itself as a spontaneously gener-ated half flux quantum $(\Phi_0/2)$ as the ground state of the loop.

The tetragonality and the absence of twinning of our Tl2201 epitaxial films has been demonstrated by the combined use of Raman spectroscopy, x-ray diffraction,[44] bright field imaging transmission electron microscopy, selected area electron diffraction, and convergent beam electron diffraction.[45]

Figure 11 shows an SSM image of the Tl2201 blanket film deposited on a (100) $SrTiO_3$ substrate with the geometry described above. The oxygen optimized Tl2201 film has a superconducting transition temperature of 83K. The sample was cooled in a field of less than one mG and imaged at 4.2K. There is only one magnetic vortex trapped in the $170 \times 230 \mu m^2$ field of view, at the bicrystal meeting point. Detailed modelling, presented below, shows that this vortex has $\Phi_0/2 = h/4e$ total flux included in it. The fact that this is the only flux trapped in a relatively large area suggests that this is spontaneously generated flux, as expected for the thermodynamic ground state of a superconducting loop with an odd number of sign changes in the pair tunneling current in the loop. The fact that the magnetic flux trapped at the wedge tip is evenly and symmetrically distributed, and that there is no extra flux spreading along the OM or OM' directions provide a strong testamony that the grain boundaries OM and OM' are strongly coupled: The sample is functionally a bicrystal as intended in the design.

Fig. 12 shows cross-sections through the wedge point of the SSM image of Fig. 11 along the directions OA (curve (a)) and OB (curve (b)). A cross-section through the peak of an isolated bulk single flux quantum in the same Tl2201 film is also shown in Fig. 12. The flux per unit length at the sample surface of a Josephson vortex can be written as[20]:

$$\frac{d\Phi}{dx} = \frac{\Phi_0}{2\pi} \frac{-4a}{\lambda_j} \frac{e^{-|x|/\lambda_j}}{1 + a^2 e^{-2|x|/\lambda_j}}. \tag{4}$$

For a half-flux quantum with two symmetric branches $a = \tan(\pi/8) = 0.414$. The fields above the surface are determined using Fourier transform propagation techniques, and then integrated over the known pickup loop geometry to obtain the flux vs position curves shown as solid lines in Fig. 12. The height of the pickup loop above the surface was determined by fitting the data from a single bulk flux quantum. In this case the fit height was $5.15\mu m$, in reasonable agreement with that expected from the spacing between the pickup loop and etched step and the angle between the sample and SQUID substrate. If we assume that the total flux is h/4e, the best fit to the data (lower two solid curves in Fig. 12) was obtained for a Josephson penetration depth of $12.5\mu m$. If we allow both the total flux and the Josephson penetration depth to float, we find $\Phi/\Phi_0 = 0.54 \pm 0.04$ and $\lambda_j = 13.8 \pm 1.5\mu m$, using a criterion of a doubling of the best-fit χ^2 value.

The observation of the half-integer flux quantum effect in this geometry represents a model-independent pair tunneling evidence for pure $d_{x^2-y^2}$ pairing symmetry. Al-though the present experiment cannot distinguish between the $d_{x^2-y^2}$ and d_{xy} states, d_{xy} symmetry has been ruled out by photoemission.[38, 39] Our results can also unam-biguously rule-out symmetry independent mechanisms, such as spin-flip tunneling at the grain boundary, for the half-flux quantum effect. Such a mechanism would not have a sign reversal of I_s for the symmetry operation about the line MOM', and the half-integer flux quantum effect should not be observed. The present symmetrical ge-ometry can not rule out the mixed two-parameter s+d state in a tetragonal system. It should be pointed out that such a state would be expected to produce two bulk su-perconducting phase transitions, which have not been observed.[40] However, by varying the angle between the grain boundary OA and/or OB and the vertical line (Fig. 11) in the design of the wedge, one could determine the degree of s+d admixture in an orthorhombic cuprate such as YBCO by using the presence or absence of the half-integer flux quantum effect.

In conclusion, flux quantization studies using the scanning SQUID microscope have proven to be invaluable for fundamental phase sensitive studies of the order parameter in unconventional superconductors. Several years of intense effort has conclusively demonstrated that half-integer flux quantization exists in the cuprate superconductors in special geometries, and that this effect is the result of predominantly d-wave symmetry of the order parameter. "Fractional" flux quantization, while apparently observed in several systems, has not been demonstrated to result from time reversal symmetry breaking, and may result instead from complex interface geometries combined with sign changes in the order parameter.

We would like to acknowledge the many contributions co-workers made to the research reviewed in this paper: The SQUIDs used in this study were designed by M.B. Ketchen and were made by IBM Research's Silicon facility and by M. Bhushan. Our SQUID microscopes were designed and built with the assistance of S.H. Blanton and A. Ellis. The sparsely twinned YBCO crystals were provided by Ruixing Liang, D.A. Bonn, and W.N. Hardy. The biepitaxial samples were provided by Conductus, and the work was done in collaboration with P. Chaudhari, S.Y. Lin, N. Khare, and T. Shaw. The bicrystal work was done in collaboration with J. Mannhart, H. Hilgenkamp, B. Mayer, and Ch. Gerber. The tricrystal experiments were done in collaboration with J.Z. Sun, A. Gupta, C.C. Chi, S. Yu-Jahnes, and T. Shaw. The tricrystal and tetracrystal $SrTiO_3$ substrates were provided by Shinkasa. The Tl2201 thin films were grown and characterized by C.A. Wang, Z.F. Ren, and J.H. Wang. The thin Bi2212 films were grown and characterized by H. Raffy and Z.Z. Li.

REFERENCES

1. G.E. Volovik and V.P. Mineev, *JETP Letters* 24:561(1976).
2. V.B. Geshkenbein and A.I. Larkin, *JETP Letters* 43:395(1986).
3. V.B. Geshkenbein, A.I. Larkin, and A. Barone, *Phys. Rev.* B 36:235(1987).
4. M. Sigrist and T.M. Rice, *Journ. Phys. Soc. Japan* 61:4283(1992).
5. D.A. Wollman, D.J. van Harlingen, W.C. Lee, D.M. Ginsberg, and A.J. Leggett *Phys. Rev. Lett.* 71:2134(1993).
6. D.A. Brawner and H.R. Ott, *Phys. Rev.* B 50:6530(1994).
7. C.C. Tsuei, J.R. Kirtley, C.C. Chi, S. Yu-Jahnes, A. Gupta, T.Shaw, J.Z. Sun, and M.B. Ketchen, *Phys. Rev. Lett.* 73:593(1994).
8. D.A. Wollman, D.J. van Harlingen, J. Giapintzakis, and D.M. Ginsberg, *Phys. Rev. Lett.* 74:797(1995).
9. J.H. Miller, Jr., Q.Y. Ying, Z.G. Zou, N.Q. Fan, J.H. Xu, M.F. Davis, and J.C. Wolfe, *Phys. Rev. Lett.* 74:2347(1995).
10. A. Mathai, Y. Gim, R.C. Black, A. Amar, and F.C. Wellstood, *Phys. Rev. Lett.* 74:4523(1995).
11. M.R. Beasley, D. Lew, and R.B. Laughlin, *Phys. Rev.* B 49:12330(1994).
12. M. Sigrist, D.B. Bailey, and R.B. Laughlin, *Phys. Rev. Lett.* 74:3249(1995).
13. J.R. Kirtley *et al. Appl. Phys. Lett* 66:1138(1995).
14. A.G. Sun, D.A. Gajewski, M.B. Maple, and R.C. Dynes, *Phys. Rev. Lett.* 72:2267(1994).
15. M.Sigrist, K. Kuboki, P.A. Lee, A.J. Millis, and T.M. Rice, *Phys. Rev.* B 53:5835(1996).
16. K.A. Moler, J.R. Kirtley, Ruixing Liang, D.A. Bonn, and W.N. Hardy, *Phys. Rev.* B 55:12753(1997).
17. J.R. Kirtley, P. Chaudhari, M.B. Ketchen, N. Khare, Shawn-Yu Lin, and T. Shaw, *Phys. Rev.* B 51:12057(1995).
18. J. Mannhart, H. Hilgenkamp, B. Mayer, Ch. Gerber, J.R. Kirtley, K.A. Moler, and M. Sigrist, *Phys. Rev. Lett* 77:2782(1996).
19. J.R. Kirtley, C.C. Tsuei, J.Z. Sun, C.C. Chi, Lock See Yu-Jahnes, A. Gupta, M. Rupp, and M.B. Ketchen, *Nature* 373:225(1995).

20. J.R. Kirtley, C.C. Tsuei, Martin Rupp, J.Z. Sun, Lock See Yu-Jahnes, A. Gupta, M.B. Ketchen, K.A. Moler, and M. Bhushan, *Phys. Rev. Lett.* 76:1336(1996).

21. C.C. Tsuei, J.R. Kirtley, M. Rupp, J.Z. Sun, A. Gupta, C.A. Wang, Z.F. Ren, J.H. Wang and M. Bhushan, *Science* 272:329(1996).

22. C.C. Tsuei, J.R. Kirtley, Z.F. Ren, J.H. Wang, H. Raffy, and Z.Z. Li, *Nature* 387:481(1997).

23. Ruixing Liang, P. Dosanjh, D.A. Bonn, D.J. Baar, J.F. Carolan, and W.N. Hardy, *Physica C* 195:51(1992).

24. Ruixing Liang, P. Dosanjh, D.A. Bonn, W.N. Hardy, and A.J. Berlinsky, *Phys. Rev. B* 50:4212(1994).

25. P.L. Gammel, D.J. Bishop, G.J. Dolan, J.R. Kwo, C.A. Murray, L.F. Schneemeyer, and J.V. Waszczak, *Phys. Rev. Lett.* 59:2952(1987).

26. T.M. Riseman, J.H. Brewer, K.H. Chow, W.N. Hardy, R.F. Kiefl, S.R. Kreitzman, R. Liang, W.A. McFarlane, P. Mendels, and G. Morris, *Phys. Rev. B* 52:10569(1995).

27. A.M. Chang, H.D. Hallen, H.F. Hess, H.L. Kao, J. Kwo, A. Sudbo, and T.Y. Chang, *Europhys. Lett.* 20:645(1992).

28. K. Char, M.S. Colclough, S.M. Garrison, N. Newman, and G. Zaharchuk, *Appl. Phys. Lett.* 59:733(1991).

29. C.A. Copetti et al., *Physica C* 253:63(1995).

30. H. Hilgenkamp, J. Mannhart, and B. Mayer, *Phys. Rev. B* 53:14586(1996).

31. P. Chaudhari and S.Y. Lin, *Phys. Rev. Lett.* 72:1084(1994).

32. L.N. Bulaevski, V.V. Kuzii, and A.A. Sobyanin, *JETP Lett.* 25:290 (1977).

33. B.I. Spivak and S. Kivelson, *Phys. Rev. B* 43:3740 (1991).

34. D. Z. Liu, K. Levin, and J. Malay, *J. Supercond* 8:663(1995), *Phys. Rev. B* 51:8680 (1995).

35. A. I. Liechtenstein, I. I. Mazin, and O. K. Andersen, *Phys. Rev. Lett* 74:2303 (1995); A. A. Golubov, and I. I. Mazin, *Physica C* 243:153 (1995).

36. R. Combescot and X. Leyronas, *Phys.Rev. Lett.* 75:3732(1995); Kazuhiro Kuboi and Patrick A. Lee, *J. Phys. Soc. Jpn.* 64:3179(1995).

37. M.B. Walker and J. Luettmer-Strathmann, *Phys. Rev. B* 54:598(1996).

38. Z.-X. Shen, D.S. Dessau, B.O. Wells, D.M. King, W.E. Spicer, A.J. Arko, D. Marshall, L.W. Lombardo, A. Kapitulnik, P. Dickinson, S. Doniach, J. Dicarlo, A.G. Loeser, and C.H. Park, *Phys. Rev. Lett.* 70:1553(1993).

39. H. Ding, M.R. Norman, J.C. Camupzano, M. Randeria, M. Bellman, A.F. Yokoya, T. Takahashi, T. Mochiku, and K. Kadawaki, *Phys. Rev. B* 54:9678(1996).

40. A recent review of this area is in J.F. Annett, N. Goldenfeld, N., and A.J. Leggett, *J. of Low Temp. Phys.* 105:473(1996), and references therein.

41. D.J. Scalapino, *Phys. Rep.* 250:329(1995).

42. Q.P.Li, B.E.C. Koltenbach, and R. Joynt, *Phys. Rev. B* 48:437(1993).

43. M.T. Monod and K. Maki, *Europhys. Lett.* 33:309(1996).

44. C.A. Wang, Z.F. Ren, J.H. Wang, D.K. Petrov, M.J. Naughton, W.Y. Yu, and A. Petrou, *Physica C* 262:98(1996).

45. Z.F. Ren, J.H. Wang, and D.J. Miller, *Appl. Phys. Lett.* 69:1798(1996).

CONCLUDING REMARKS

Our Advanced Study Institute, on "The Gap Symmetry and Fluctuations in High-Tc Superconductors" has provided a stimulating forum for discussing several important results very relevant to the overall research in this field. Here we briefly summarize some of the most relevant of the new results that have been discussed in our Institute.

Great progress has been made worldwide in preparation and in-depth characterisation of samples of various cuprate families, including the detailed studies of electron doped oxides. The reproducible, high resolution experimental results by all leading groups now provide a remarkable body of knowledge concerning (1) thermodynamic properties : specific heat, magnetic susceptibility, variation of penetration depth with temperature, (2) spectroscopy of quantum states and excitations : optical spectroscopy, angular resolved photoemission, neutron scattering, NMR, etc, (3) transport properties. All these results are difficult to explain with one all inclusive microscopic theory.

In the normal state there is now well established evidence, shown by all experiments, for the existence of two transition lines in the electronic phase diagram of HTSC cuprates, even though the nature of the pseudogap and the exact nature of the insulator to metal (superconductor) transition remain disputed.

In the superconducting state, a majority of experiments indicate a dominant d-wave symmetry in the underdoped to optimally doped samples. There are, however, some notable exceptions: there is no evidence for d-wave component in the electron doped cuprates (results from the Maryland Center), Sharvin experiments on LSCO give a finite minimum gap (G. Deutscher et al) and electronic Raman experiment on Hg-2201 compound (A. Sacuto et al) is not compatible with d-wave but rather with extended s-wave (with nodes).

Only few experiments were performed systematically in the overdoped regime. Recent photoemission experiments in overdoped BSCCO, performed by Onellion et al, indicate a finite minimum gap that contradict pure d-wave symmetry scenario. Moreover, most recent measurements by same authors on this system, by using polarized synchrotron light, reveal that the symmetry of the states changes from odd in optimally doped samples to even in highly overdoped system. This unexpected result is broadly in agreement with generally accepted view that the over-doped cuprates exhibit more Fermi-liquid and BCS alike behavior in the normal state and in the superconducting state, respectively.

More advanced experiments will be performed to verify these latest results and to clarify the remaining puzzles, especially close to the insulator transition and in the extreme overdoped regime. And, it is still not clear why different spectroscopies give different value of the gap : the $2\Delta/kTc$ varies from about 3-5 up to 7 and even 12 for different techniques; this may well be due to the different characteristic 'depth' of the sample that each technique measures. Also, the results obtained by neutron scattering are difficult to reconcile with the NMR data, and the STM results give a much larger gap that the Sharvin contact studies. Perhaps two gaps, one excitation gap of large value and one smaller coherent gap exist in these materials (see the article by G. Deutscher).

The Gap Symmetry and Fluctuations in High - T_c Superconductors
Edited by Bok *et al.*, Plenum Press, New York, 1998

557

There is at present still no concensus on the exact mechanism of superconductivity although main conceptual options have been clearly stated. More theoretical progress is also needed to fully account for the very rich physics of electronic phase diagram of these remarkable solids, especially close to the non-trivial metal-insulator transition, a difficult topic that remains one of the most challenging topics in physics of correlated electron systems.

A detailed predictive theory is important for knowing which parameter is most effective for high Tc and finding new superconductor materials with higher Tc and also for developing applications.

Julien Bok, director of the ASI

INDEX